D1709829

Signal Processing for Active Control

Signal Processing and its Applications

Books in the series

P. M. Clarkson and H. Stark, *Signal Processing Methods for Audio, Images and Telecommunications* (1995)

R. J. Clarke, *Digital Compression of Still Images and Video* (1995)

S-K. Chang and E. Jungert, *Symbolic Projection for Image Information Retrieval and Spatial Reasoning* (1996)

V. Cantoni, S. Levialdi and V. Roberto (eds.), *Artificial Vision* (1997)

R. de Mori, *Spoken Dialogue with Computers* (1998)

D. Bull, N. Canagarajah and A. Nix (eds.), *Insights into Mobile Multimedia Communications* (1999)

D. S. G. Pollock, *A Handbook of Time-Series Analysis, Signal Processing and Dynamics* (1999)

N. Fourikis, *Advanced Array Systems, Applications and RF Technologies* (2000)

Signal Processing for Active Control

S. J. ELLIOTT
Institute of Sound and Vibration Research
University of Southampton, UK

ACADEMIC PRESS

A Harcourt Science and Technology Company

San Diego San Francisco New York
Boston London Sydney Tokyo

This book is printed on acid-free paper

Academic Press
A Harcourt Science and Technology Company
Harcourt Place, 32 Jamestown Road, London NW1 7BY, UK
http://www.academicpress.com

Academic Press
A Harcourt Science and Technology Company
525 B Street, Suite 1900, San Diego, California 92101-4495, USA
http://www.academicpress.com

ISBN 0-12-237085-6

A catalogue for this book is available from the British Library

Typeset by Phoenix Photosetting, Chatham, Kent
Printed in Great Britain by Redwood Books, Trowbridge, Wiltshire

01 02 03 04 05 06 RB 9 8 7 6 5 4 3 2 1

Contents

Series Preface

Signal processing applications are now widespread. Relatively cheap consumer products through to the more expensive military and industrial systems extensively exploit this technology. This spread was initiated in the 1960s by the introduction of cheap digital technology to implement signal processing algorithms in real-time for some applications. Since that time semiconductor technology has developed rapidly to support the spread. In parallel, an ever increasing body of mathematical theory is being used to develop signal processing algorithms. The basic mathematical foundations, however, have been known and well understood for some time.

Signal Processing and its Applications addresses the entire breadth and depth of the subject with texts that cover the theory, technology and applications of signal processing in its widest sense. This is reflected in the composition of the Editorial Board, who have interests in:

(i) Theory—The physics of the application and the mathematics to model the system;
(ii) Implementation—VLSI/ASIC design, computer architecture, numerical methods, systems design methodology, and CAE;
(iii) Applications—Speech, sonar, radar, seismic, medical, communications (both audio and video), guidance, navigation, remote sensing, imaging, survey, archiving, non-destructive and non-intrusive testing, and personal entertainment.

Signal Processing and its Applications will typically be of most interest to postgraduate students, academics, and practising engineers who work in the field and develop signal processing applications. Some texts may also be of interest to final year undergraduates.

Richard C. Green
The Engineering Practice,
Farnborough, UK

Dedication

'All brontosauruses are thin at one end, much much thicker in the middle, and then thin again at the far end. That is my theory.' (A. Elk, 1972).

This book is dedicated to all those whose theories become glaringly obvious in retrospect.

Preface

Active control is a method for attenuating unwanted disturbances by the introduction of controllable 'secondary' sources, whose outputs are arranged to interfere destructively with the disturbance from the original 'primary' source. This technique has found its widest application in the control of sound and vibration, and is referred to as 'active' to distinguish it from passive methods of control, in which the sound or vibration is absorbed or reflected by elements that are not themselves capable of generating power. The general principle of controlling a disturbance by the superposition of an equal and opposite disturbance is of course common in many fields of science and engineering. The active control of sound and vibration presents a rather specific set of problems, however, largely because of the wave nature of these disturbances, so that control at one point in space does not guarantee control at other points.

Interest in active control has grown significantly over the past decade, fuelled by the rapid developments in microelectronics, which have enabled fast, multichannel controllers to be implemented at a reasonable cost, and by an increasing need for lightweight solutions to low-frequency sound and vibration problems. Current industrial applications include the active control of sound in headsets, in air-conditioning ducts and in propeller aircraft, as well as the active isolation of vibration in industrial machines and helicopters.

The successful practical application of an active control system requires a detailed knowledge of both the physical principles and the limitations of active control, and also of the electronic control techniques which must be used to implement such a system. The physical principles involved in the active control of sound and the active control of vibration are now reasonably well understood (as described, for example, in Nelson and Elliott, 1992; Fuller, Elliott and Nelson, 1996; Preumont, 1997; Clark, Saunders and Gibbs, 1998). The objective of this book is to explain the common principles of electronic control that run through both of these applications, and also to put this work more clearly into the context of conventional signal processing and conventional automatic control.

The book has been written from a signal processing perspective, to be accessible to senior undergraduate and postgraduate students, and to practising engineers. Historically, the control techniques used in practical active control systems probably owe more to the signal processing literature than to the automatic control literature. Although the two disciplines address similar problems, differences in perspective, assumptions and even in notation can cause considerable difficulty in translating from one body of literature to another. The signal processing community has tended to concentrate on real-time implementation at fast sample rates, particularly using adaptive methods, and so has emphasised relatively simple algorithms. Considerable effort has been expended by the control community in developing off-line methods for the optimal design of controllers with provable stability properties. These have tended to be operated at relatively slow sample rates, so that their computational expense is not too much of an issue. Recent developments in control theory, however, particularly in adaptive controllers and those that are robust to system uncertainties, are very relevant to the issues faced in active control, and some of these developments will be discussed here within the context of the earlier signal processing work.

The duality between discrete time-domain approaches and discrete frequency-domain methods is particularly emphasised in this book, both for the calculation of the optimum performance, and as a framework for adaptive control. Both approaches can be used to calculate the performance from directly measurable parameters in a practical control system, in particular from correlation and spectral density measurements for random data. The use of state variables is deliberately avoided in most cases, partly because the approach is so extensively discussed elsewhere, and partly to avoid the difficulties encountered in the practical problem of fitting a state-space description to a measured system response. It is particularly difficult to identify reliable state-space models for many acoustic systems, which can be of very high order and contain significant delays.

The development of a typical active control system can very broadly be divided into the following stages (which are similar to the levels of performance described by Morgan, 1991), although it is a fortunate designer who never has to apply these stages recursively:

(1) Analysis of the physical system using analytical models of simplified arrangements, in order to determine the fundamental physical limitations of active control in the given application.
(2) Calculation of the optimum performance using different control strategies under ideal control conditions, by using off-line calculations based on measurements taken from the physical system that is to be controlled.
(3) Simulation of different control strategies using data from the physical system that is to be controlled, under a variety of operating conditions.
(4) Implementation of a real-time controller and testing of the system under all conditions to ensure that its behaviour is as predicted.

It is thus very important to have a clear idea of the physical limitations of active control, as well as those imposed by the particular electronic control strategy employed. Before venturing into the control aspects of active systems, the underlying physical basis of the active control of both sound and vibration is first reviewed. Many attempts at applying active control have foundered on unrealistic assumptions about the limits of its physical performance and so this first chapter is an important precursor to the remainder of the book. Chapter 1 provides a brief guide to the simple analytical models which can be used to determine the feasibility of active control in a variety of applications. These models are based on continuous-time representations of the system under control. The disturbance can also be assumed to be tonal for these calculations, which simplifies the formulation of the model and allows direct physical insight into the processes of active control. Although the detailed development of these models may be unfamiliar to readers with a signal processing or control background, it is hoped that the graphical representation of the results from a number of model problems will allow the physical limitations of a variety of active control problems to be assessed.

After a review of optimal and adaptive electrical filters in Chapter 2, the design of feedforward controllers takes up the following three chapters, starting with the single-channel control of tonal, or narrowband, disturbances and leading up to the multichannel control of random, or broadband, disturbances. Exact least-square solutions to these control problems are emphasised, since one stage of control system development is generally the calculation of the optimum performance of a particular control strategy, using data measured from the real application. This optimal performance can also provide a benchmark against which the measured performance of the implemented system can be judged. The practical importance is also emphasised of accounting for the control effort, which can regularise the control problem and ensure that the adaptive algorithms are robust to changes in disturbance or plant response. Initially the control effort is included as a term in the quadratic cost function being minimised, but practical methods are also discussed of constraining the system so that the control effort is below a prescribed maximum value. Various adaptive control algorithms are discussed, and a particular attempt is made to clarify the connections between the filtered reference and the filtered error approaches.

Feedback control systems are covered in Chapters 6 and 7, with an emphasis on a particular structure, or parametrisation, of the controller that reduces the feedback control problem to one which is very close to the feedforward problem considered previously. A central issue for feedback controllers is their stability in the face of uncertainties in the response of the system under control, and it is shown how this requirement for robust stability can be conveniently incorporated as a constraint within the framework already developed. Whereas the design of fixed feedback controllers, as discussed in Chapter 6, is well established, the theoretical issues involved in making such a feedback controller adaptive are somewhat less well developed. Nevertheless, Chapter 7 reviews the architectures that could be used for such an adaptive feedback controller, particularly those that are adaptive to non-stationary disturbances. The maintenance of robust stability in the controller during adaptation is particularly important in active control applications. The complementary properties of analogue and digital controllers are also emphasised, which leads us to a combined arrangement, with a fixed analogue inner feedback loop and an adaptive digital outer feedback loop.

Chapter 8 is concerned with the control of nonlinear systems. In active control applications, the system under control is often weakly nonlinear, particularly if structural actuators are used. Methods of compensating for weak nonlinearities are discussed, and it is shown that such nonlinearities do not necessarily degrade the performance of an active control system. The more challenging case of controlling a strongly nonlinear system, which can display chaotic behaviour, is also briefly introduced in this chapter, although rather different strategies of control have to be used in this case.

Another important practical aspect of active control is covered in Chapter 9, which is the optimisation of the positions of the actuators and sensors. This optimisation problem is significantly different from that encountered in most signal processing problems in that the transducer placement problem often reduces to a large combinatorial search. Guided random search procedures, rather than gradient descent methods, must often be used to find good transducer locations, and the use of both genetic algorithms and simulated annealing is discussed in this chapter. The need to choose transducer positions so that the performance of the control system is robust to changes in the plant response and disturbance is particularly emphasised.

Chapter 10 then describes some of the hardware issues involved in the real-time implementation of active control systems. This includes a discussion of processors, converters and anti-aliasing filters, as well as a discussion of the finite-precision and other numerical effects that can occur in practical active control systems. A brief review of linear algebra is finally included in the Appendix, which also describes the use of matrices in the description of multichannel control problems.

Although I must take full responsibility for any shortcomings of this book, I would like to thank all those who read drafts of individual chapters and offered much useful feedback, including Dr. D. Anthony, Dr. K-H. Baek, Professor. S. Billings, Dr. J. Cook, Dr. E. Friot, Dr. P. Gardonio, Professor C. H. Hansen, Dr. L. Heck, Dr. M. Jolly, Dr. P. Joseph, Professor A. J. Keane, Dr. A. Langley, Dr. V. Martin, Professor M. Morari, Dr. A. Omoto, Dr. B. Petitjean, Dr. S. Popovich, Dr. B. Rafaely, Dr. D. Rosetti, Dr. K-H. Shin, Dr. R. W. Stewart, Dr. T. J. Sutton and Dr. M. C. M. Wright. I would particularly like to thank Professor R. L. Clark and Professor P. A. Nelson, who looked through the final draft, and Dr. D. R. Morgan, whose careful reading of the whole manuscript saved me from many notational and grammatical errors.

My thanks are also due to Mrs. J. Shotter for wordprocessing the manuscript and calmly coping with numerous changes of mind and changes of notation; Mrs. M. Hicks and Mr. T. Sors as well as Mrs. J. Shotter for the preparation of many of the figures; and finally to my wife and family for their patience and understanding, without which this book would not have been possible.

Stephen Elliott
June 2000

Glossary

Signal Conventions

$x(t)$	Scalar continuous-time signal
T	Sampling time, so $t = nT$ where n is integer
f_s	Sample rate, equal to $1/T$
$x(n)$	Scalar sampled-time signal as a function of the discrete time variable n
$\boldsymbol{x}(n)$	$\left[x_1(n)\ x_2(n)\ \cdots\ x_K(n)\right]^{\mathrm{T}}$ vector of a set of sampled signals, or $\left[x(n)\ x(n-1)\ \cdots\ x(n-I+1)\right]^{\mathrm{T}}$ vector of past samples of a single signal
$X(z)$	z-transform of the signal $x(n)$
$\mathbf{x}(z)$	Vector of z-transforms of the signals in $\boldsymbol{x}(n) = \left[x_1(n)\ x_2(n)\ \cdots\ x_K(n)\right]^{\mathrm{T}}$
$X\left(e^{j\omega T}\right)$	Fourier transform of $x(n)$
$X(k)$	Discrete Fourier transform (DFT) of $x(n)$
ω	Angular frequency, i.e. 2π times the actual frequency in hertz
ωT	Normalised angular frequency, which is dimensionless
$R_{xx}(m)$	$E\left[x(n)\, x(n+m)\right]$, autocorrelation sequence
$R_{xy}(m)$	$E\left[x(n)\, y(n+m)\right]$, cross-correlation sequence
$S_{xx}(z)$	z-transform of $R_{xx}(m)$
$S_{xy}(z)$	z-transform of $R_{xy}(m)$
$S_{xx}\left(e^{j\omega T}\right)$	Power spectral density, also $E\left[\left\lvert X\left(e^{j\omega T}\right)\right\rvert^2\right]$ subject to conditions described in the Appendix
$S_{xy}\left(e^{j\omega T}\right)$	Cross spectral density, also $E\left[X^*\left(e^{j\omega T}\right) Y\left(e^{j\omega T}\right)\right]$ subject to conditions described in the Appendix
$\boldsymbol{R}_{xx}(m)$	$E\left[\mathbf{x}(n+m)\, \mathbf{x}^{\mathrm{T}}(n)\right]$, autocorrelation matrix
$\boldsymbol{R}_{xy}(m)$	$E\left[\mathbf{y}(n+m)\, \mathbf{x}^{\mathrm{T}}(n)\right]$, cross-correlation matrix
$\mathbf{S}_{xx}\left(e^{j\omega T}\right)$	Fourier transform of $\boldsymbol{R}_{xx}(m)$, also $E\left[\mathbf{x}\left(e^{j\omega T}\right) \mathbf{x}^H\left(e^{j\omega T}\right)\right]$ subject to conditions described in the Appendix
$\mathbf{S}_{xy}\left(e^{j\omega T}\right)$	Fourier transform of $\boldsymbol{R}_{xy}(m)$, also $E\left[\mathbf{y}\left(e^{j\omega T}\right) \mathbf{x}^H\left(e^{j\omega T}\right)\right]$ subject to conditions described in the Appendix
$\mathbf{S}_{xx}(z)$	z-transform of $\boldsymbol{R}_{xx}(m)$
$\mathbf{S}_{xy}(z)$	z-transform of $\boldsymbol{R}_{xy}(m)$

Symbols

a	Mode amplitude, feedback coefficients in IIR filter
b	Feedforward coefficients in IIR filter linear
c	Constant coefficient in quadratic form
d	Desired or disturbance signal
e	2.718 . . .
e	Error signal
f	Force, frequency, filtered error signal
g	Coefficients of plant impulse response
h	Impulse response coefficients
i	Index
j	$\sqrt{1}$, index
k	Wavenumber, discrete frequency variable, index for reference signals
l	Length, index for error signals
m	Mass per unit area, index for control signals
n	Time index, number of poles
p	Pressure
q	Volume velocity
r	Filtered reference signal, separation distance
s	Laplace transform variable, sensed or observed reference signal
t	Continuous-time, filtered output signal
u	Plant input, or control, signal
v	Velocity, noise signal, transformed control signal
w	FIR filter and controller coefficients
x	Coordinate, reference signal
y	Coordinate, output signal
z	Coordinate, z-transform variable
A	IIR filter denominator, wave amplitude, quadratic coefficient
B	IIR filter numerator, wave amplitude, bandwidth, bound on uncertainty
C	Noise filter numerator, wave amplitude, transformed controller
D	Directivity, bending stiffness
E	Young's modulus, energy, expectation operator
F	Spectral factor
G	Plant response
H	Overall controller transfer function, Hankel function
I	Number of controller coefficients, moment of inertia, directivities
J	Number of plant impulse response coefficients, cost function
K	Number of reference signals, stiffness
L	Number of error signals, dimension
M	Number of control signals, mobility, modal overlap, misadjustment
N	Block size
P	Primary plant response, control effort
Q	Matrix of eigenvectors or right singular vectors
R	Real part of mobility or impedance, correlation function, matrix of eigenvectors or left singular vectors
S	Area, spectral density, sensitivity function

T	Transfer function, sampling time, complementary sensitivity function
U	Complex control signal
V	Volume
W	Controller transfer function
Z	Acoustic impedance
α	Convergence coefficient
β	Filter coefficient weighting
γ	Leakage factor
Δ	Uncertainty
δ	Kronecker delta function
ε	Error
ζ	Damping ratio
θ	Angle
κ	Discrete frequency variable
λ	Eigenvalue, forgetting factor, wavelength
μ	Convergence factor
ν	Transformed and normalised coefficients
ξ	White noise signal
Π	Power
π	3.1415 ···
ρ	Density, control effort weighting
σ	Singular value
τ	Delay
ϕ	Angle
ω	Angular frequency
$\mathcal{L}$	Laplace transform
$\mathcal{Z}$	z-transform

Matrix Conventions (with reference to equation numbers in the Appendix)

$\mathbf{a}$	Lower-case bold variables are column vectors (A1.1)
$\mathbf{a}^{\mathrm{T}}$	The transpose of a column vector is a row vector (A1.2)
$\mathbf{a}^{\mathrm{H}}$	The Hermitian or conjugate transpose, which is a row vector (A1.3)
$\mathbf{a}^{\mathrm{H}}\mathbf{a}$	The inner product of **a**, which is a real scalar (A1.4)
$\mathbf{a}\,\mathbf{a}^{\mathrm{H}}$	The outer product of **a**, whose trace is equal to the inner product (A5.1)
$\mathbf{A}$	Upper-case bold variables are matrices (A2.1)
$\mathbf{I}$	The identity matrix (A2.2)
$\mathbf{A}^{\mathrm{H}}$	The Hermitian or conjugate transpose (A2.12)
det ($\mathbf{A}$)	The determinant of **A** (A3.5)
$\mathbf{A}^{-1}$	The inverse of **A** (A3.9)
$\mathbf{A}^{-\mathrm{H}}$	The inverse of the Hermitian of **A** (A3.13)
trace ($\mathbf{A}$)	The trace of **A** (A4.1)
$\partial J/\partial \mathbf{A}$	The matrix, whose size is equal to that of the real matrix **A**, with elements that are the derivatives of the real function J with respect to the elements of **A** (A4.8)
$\lambda_i(\mathbf{A})$	The i-th eigenvalue of **A** (A7.10)

$\rho(\mathbf{A})$	The spectral factor of **A** (A7.12)
$\sigma_i(\mathbf{A})$	The i-th singular value of **A** (A8.4)
$\overline{\sigma}(\mathbf{A})$	The largest singular value of **A** (A8.7)
$\underline{\sigma}(\mathbf{A})$	The smallest singular value of **A** (A8.7)
$\mathbf{A}^{\dagger}$	The generalised inverse of **A** (A8.13)
$\kappa(\mathbf{A})$	The condition number of **A** (A8.15)
$\lVert\mathbf{a}\rVert_p$	The p-norm of the vector **a** (A9.1)
$\lVert\mathbf{a}\rVert_2$	The 2-norm of **a**, which equals the square root of its inner product (A9.2)
$\lVert\mathbf{A}\rVert_F$	The Frobenius norm of the matrix **A** (A9.4)
$\lVert\mathbf{A}\rVert_{ip}$	The induced p-norm of the matrix **A** (A9.6)
$\lVert y(t)\rVert_p$	The temporal p-norm of a signal $y(t)$ (A9.15)
$\lVert\mathbf{H}(s)\rVert_2$	The H_2 norm of the system with transfer function matrix $\mathbf{H}(s)$ (A9.22)
$\lVert\mathbf{H}(s)\rVert_\infty$	The H_∞ norm of the system with transfer function matrix $\mathbf{H}(s)$ (A9.25)

Abbreviations

ADC	Analogue-to-digital converter
BPF	Blade passage frequency
DAC	Digital-to-analogue converter
FFT	Fast Fourier transform
FIR	Finite impulse response
IFFT	Inverse fast Fourier transform
IIR	Infinite impulse response
IMC	Internal modal control
LMS	Least mean square
LSB	Least-significant bit
MIMO	Multiple-input multiple-output
MSB	Most-significant bit
RLS	Recursive least square
SISO	Single-input single-output
SNR	Signal-to-noise ratio
SVD	Singular value decomposition

1

The Physical Basis for Active Control

1.1. INTRODUCTION

In this chapter the physical basis for the active control of both sound and vibration is described. Although examples of the active control of specific acoustical and structural systems will be presented, an attempt has been made to use a common description for both of these applications, to emphasise their underlying physics. It is important at the early stages of an active control project to develop simple analytical models of the physical system under control so that the fundamental physical limitations of the proposed control strategy can be assessed. Several of the most common types of physical model are presented in this chapter. The objectives of this simple model are:

(1) to investigate the effects of different physical strategies of control, e.g. cancellation of pressure or minimisation of power output;
(2) to determine the parameters that affect the physical limits of performance in a given control arrangement, e.g. the distance from the primary to the secondary source;
(3) to allow the effect of different kinds of control actuator to be assessed;
(4) to determine the number and type of sensors that are required to measure the physical quantity of interest and thus maintain control.

In most cases these physical limitations can be made clear by assuming that the disturbance is tonal, which considerably simplifies the analysis. It should be emphasised, however, that active control can be applied to systems that are excited by disturbances with considerably more complicated waveforms than pure tones, but in order to determine the limitations of performance due to the physical aspects, rather than the electronic control aspects, of the problem it is convenient to initially assume a single-frequency excitation. In this chapter it will also be assumed that all signals are continuous functions of the time variable, t.

1.1.1. Chapter Outline

In the remainder of this section the complex form of the acoustic wave equation for tonal signals is introduced, before considering the active control of plane sound waves in a one dimensional duct in Section 1.2. Section 1.2 goes on to describe the physical consequences of using a single secondary source to reflect the sound wave or of using a pair of secondary sources to absorb the sound wave.

Section 1.3 considers the limitations of an active control system that uses multiple secondary sources to minimise the power radiated by a compact primary source in an infinite medium. The two-dimensional case is discussed first, by considering flexural waves of vibration propagating in a plate, and this is extended to three dimensions by considering acoustic waves in free space. It is shown that in order to achieve significant reductions in radiated power, a single secondary source must be placed a distance from the primary source that is less than one-tenth of a wavelength and that if an array of secondary sources is used, each must be separated by a distance of less than half a wavelength.

A closed duct provides a simple example of a finite acoustic system that can resonate at certain frequencies, and in Section 1.4 this system is used to illustrate the variety of active control strategies that can be employed in a finite system. These include reflection and absorption of the incident wave, the maximisation of the power absorbed by the secondary source, and the minimisation of the total power generated by both the primary and secondary source. It turns out that the minimisation of total power is a very useful strategy in finite systems, and in Section 1.5 it is shown that such a strategy is almost equivalent to minimising the energy stored in the system. Section 1.5 goes on to present simulations of the minimisation of vibrational kinetic energy in a finite plate and of acoustic potential energy in an enclosure. The physical performance limitations of an active control system are shown to be imposed by the number of modes that significantly contribute to the response at any one frequency in both cases.

Although in these sections the active control of sound and the active control of vibration have been considered individually, vibration control is sometimes implemented with the objective of minimising the sound radiation from the vibrating structure, as discussed in Section 1.6. The interaction between the sound radiation of the structural modes of a system is described in this section, together with a set of velocity distributions that radiate sound independently: the so-called radiation modes. Active control of sound radiation by cancelling the lowest-order radiation mode is almost equivalent to controlling the volume velocity of the structure, and the strengths and limitations of such an approach are outlined in this section.

Finally, in Section 1.7 the local control of sound or vibration is described, where a secondary source is used to cancel the disturbance at one point in space. If the distance from the cancellation point to the secondary source is small compared with a wavelength then, in the acoustic case, a shell of quiet is generated all around the secondary source. If the cancellation point is farther away from the secondary source, then only a spherically shaped zone of quiet is generated, which is centred on the point of cancellation and has an average diameter of one-tenth of a wavelength.

1.1.2. The Wave Equation

What distinguishes most active sound and vibration problems from many more conventional control problems is that the disturbances propagate as waves from one part of the physical system to another. Sound waves propagating in only the x direction, for example, obey the one-dimensional *wave equation* (see for example Morse, 1948; Kinsler et al., 1982; Nelson and Elliott, 1992)

$$\frac{\partial^2 p(x,t)}{\partial x^2} - \frac{1}{c_0^2}\frac{\partial^2 p(x,t)}{\partial t^2} = 0 , \tag{1.1.1}$$

where $p(x,t)$ denotes the instantaneous acoustic pressure at position x and time t, and c_0 is the speed of sound, which is about 340 m s^{-1} for air at normal temperature and pressure. The acoustic pressure is assumed to be small compared with the atmospheric pressure to ensure the linearity of equation (1.1.1), and this assumption is very closely satisfied for the noise levels normally encountered in practice, which may have an acoustic pressure of the order of 1 Pa, compared with a typical atmospheric pressure of about 10^5 Pa. The one-dimensional wave equation is satisfied by an acoustic pressure wave of arbitrary waveform, but whose dependence on space and time takes the form

$$p(x,t) = p(t \pm x/c_0) , \tag{1.1.2}$$

where a $(t + x/c_0)$ dependence represents a wave travelling in the negative x direction and a $(t - x/c_0)$ dependence represents a wave travelling in the positive x direction. Such a wave propagates without change of amplitude or waveform in an infinite homogeneous medium which is assumed to have no acoustic sources and no *dissipation*. All frequency components of $p(x,t)$ that obey equation (1.1.1) thus propagate with the same speed and are therefore not subject to *dispersion*. One example of a positive-going acoustic wave that obeys the wave equation is a tonal pressure signal, which can be represented in terms of the complex quantity A as

$$p(x,t) = \mathrm{Re}\left[A\,\mathrm{e}^{j\omega(t-x/c_0)}\right] = \mathrm{Re}\left[A\,\mathrm{e}^{j(\omega t-kx)}\right], \tag{1.1.3 a,b}$$

where Re[ ] denotes the real part of the quantity in brackets, ω is the angular frequency and $k = \omega/c_0$ is the acoustic *wavenumber*. Equation (1.1.1) would also be satisfied by the imaginary part of $A\,\mathrm{e}^{j(\omega t-kx)}$, corresponding to a sinusoidal rather than a cosinusoidal wave, and indeed by the whole complex number $A\,\mathrm{e}^{j(\omega t-kx)}$. In the analysis of distributed acoustic systems it is convenient to define the *complex pressure* to be

$$p(x) = A\,\mathrm{e}^{-jkx}, \tag{1.1.4}$$

so that equation (1.1.1) is satisfied by

$$p(x,t) = A\,\mathrm{e}^{j(\omega t-kx)} = p(x)\,\mathrm{e}^{j\omega t}. \tag{1.1.5}$$

Substituting equation (1.1.5) into equation (1.1.1) allows the wave equation to be written as

$$\left[\frac{\mathrm{d}^2 p(x)}{\mathrm{d}x^2} + \frac{\omega^2}{c_0^2}\,p(x)\right]\mathrm{e}^{j\omega t} = 0, \tag{1.1.6}$$

which is only satisfied at all times providing

$$\frac{\mathrm{d}^2 p(x)}{\mathrm{d}x^2} + k^2 p(x) = 0. \tag{1.1.7}$$

Equation (1.1.7) is known as the one-dimensional *Helmholtz equation*. As well as being satisfied by the complex pressure for a positive-going acoustic wave, the Helmholtz equation is also satisfied by the complex pressure corresponding to a negative-going acoustic wave of amplitude B,

$$p(x) = B\,\mathrm{e}^{+jkx}. \tag{1.1.8}$$

For problems where the excitation is tonal, the pressure field can thus be conveniently expressed in a compact complex form.

A consequence of the *linearity* of the wave equation given by equation (1.1.7), is that if the two complex pressures $p_1(x)$ and $p_2(x)$ are both solutions of this equation, then $p_1(x) + p_2(x)$ must also be a solution. This is a statement of the principle of *superposition* and the vast majority of active control systems work using this simple principle. To take a trivial example, if $p_1(x) = A\,\mathrm{e}^{-jkx}$, i.e. a positive-going wave, and $p_2(x) = -A\,\mathrm{e}^{-jkx}$, another positive-going wave of equal amplitude but 180° out of phase, then the net response of a system to both of these disturbances acting simultaneously would be *zero* acoustic pressure at all times and at all points in space.

One of the earliest suggestions for active control was for the suppression of exactly such a one-dimensional acoustic wave, and was contained in a patent granted to Paul Lueg in 1936. The illustrations page from this patent is shown in Fig. 1.1 and the active control of one-dimensional sound waves propagating in a duct is shown as 'Fig. 1' of these illustrations. The waveform of the original primary wave is shown as a solid line here, S_1, whose waveform is detected by the microphone, *M*, and used to drive a loudspeaker acting as a secondary source, *L*, after being manipulated with an electronic controller *V*. The secondary source generates another acoustic wave, whose waveform is shown as the dotted line, S_2, which is arranged to have an amplitude equal to that of the primary wave but to have the opposite phase. These two acoustic waves interfere destructively, which significantly attenuates the sound wave propagating down the duct. The physical consequences of such wave cancellation will be considered in Section 1.2. Lueg also considered the cancellation of more complicated waveforms, 'Fig. 3', and the possibility of controlling sources radiating into a three dimensional space, 'Fig. 4', which will also be briefly described in Section 1.3.

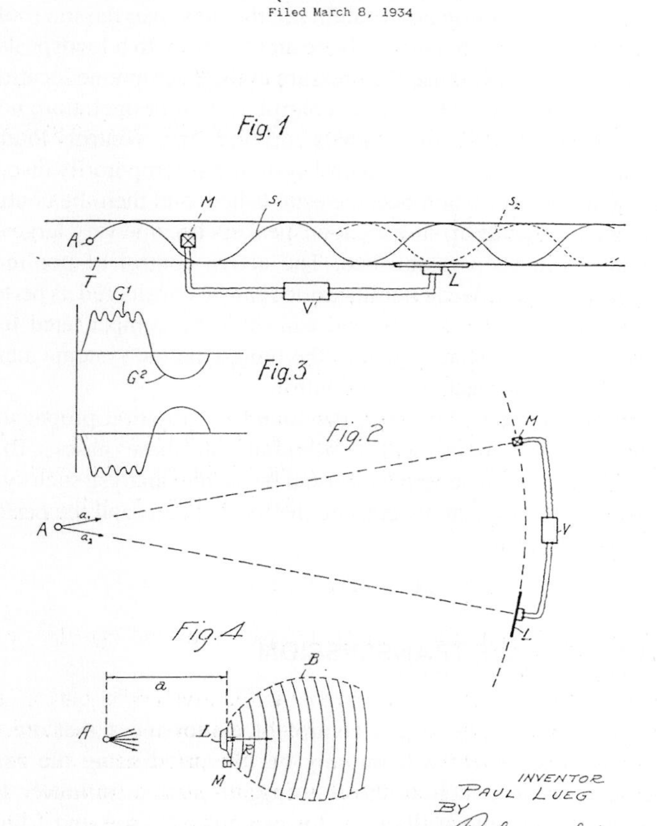

Figure 1.1 Illustrations page from an early active noise control patent by Lueg (1936).

1.1.3. Source Control

The equations governing the pressure variations in a system are not, however, always linear. When sound is generated by flow, for example, the full nonlinear equations of fluid dynamics must generally be used to describe sound generation, in which case active control by simple superposition may no longer be successful. It is sometimes possible, however, to make rather small changes to the variables that control the sound generation, and thus *suppress* the sound by direct manipulation of the sound generation mechanism. This strategy has been used to suppress the flow-induced pressure oscillations in a cavity, as described by Ffowcs-Williams (1984) for example, who pointed out that only rather small control signals are needed to prevent such an oscillation from starting up, although the control signals would have to be much larger if control were attempted once the oscillations had become established. Billout et al. (1991) have also shown that an adaptive controller can suppress flow instabilities, even under time-varying conditions.

These observations were confirmed in a practical implementation of such a control system, reported by Stothers et al. (1993), which was used to control the low frequency pressure oscillations in a car due to an instability in the flow over the sun roof. The pressure inside the car was measured with a microphone and fed back to a loudspeaker via a digital filter, which was adapted to minimise the pressure at the microphone location. If the speed of the car was increased from rest with the control system in operation, no flow-induced oscillation was observed and the signals driving the control loudspeaker were correspondingly small. If, however, the control system was temporarily disconnected while driving along, so that the oscillation became established, and then the control system was again connected up, the signals driving the loudspeakers became very large until control of the oscillation had again been established. The active control of nonlinear systems is discussed briefly in Chapter 8. Weak nonlinearities can be considered as perturbations from the strictly linear response of a system and can often be compensated for accordingly, whereas strongly nonlinear systems, such as the fluid dynamic systems mentioned above, require a rather different approach to their control.

In the majority of situations, however, the sound or vibration propagates very nearly linearly and most of this book will concentrate on these cases. To a good first approximation, the principle of superposition can be used to analyse such systems, and this can provide considerable physical insight into the mechanisms and the performance limits fundamental to active control.

1.2. CONTROL OF WAVE TRANSMISSION

In this section, we consider the active control of disturbances that are transmitted as propagating waves. This subject will initially be illustrated using the example of one-dimensional acoustic waves, such as those propagating in an infinitely long duct with uniform cross-section and rigid walls (see, for example, Nelson and Elliott, 1992). For sufficiently low frequency excitation only *plane waves* of sound can propagate in such a duct. In that case the waves have a uniform pressure distribution across any section of the duct and obey the one-dimensional wave equation, equation (1.1.1). The fundamental approaches to active wave control can be illustrated using this simple example, but some of the complications involved in controlling other types of propagating wave are also briefly discussed at the end of the section.

1.2.1. Single Secondary Actuator

To begin with, it will be assumed that an incident tonal wave of sound, travelling in the positive x direction along a duct, is controlled by a single acoustic secondary source such as a loudspeaker mounted in the wall of the duct, as illustrated in Fig. 1.2. The duct is assumed to be infinite in length, to have rigid walls and to have a constant cross section.

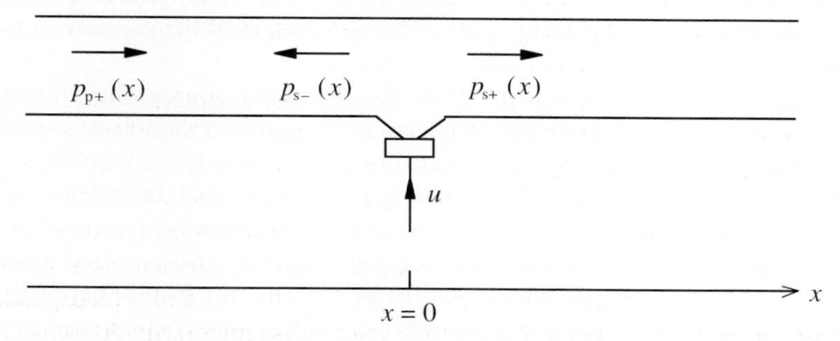

Figure 1.2 Active control of plane sound waves in an infinite duct using a single secondary source.

The complex pressure due to the incident primary wave is

$$p_{p+}(x) = A\,e^{-jkx}\ , \quad \text{for all } x\ , \tag{1.2.1}$$

where the subscript p+ denotes the primary wave travelling in a positive x direction or *downstream*. An acoustic source such as a loudspeaker driven at the same frequency as that of the incident wave will produce acoustic waves travelling both in the downstream direction and in the *upstream* or negative-x direction, whose complex pressures can be written as

$$p_{s+}(x) = B\,e^{-jkx}, \quad \text{for } x > 0, \quad p_{s-}(x) = B\,e^{+jkx}, \quad \text{for } x < 0\ , \tag{1.2.2 a,b}$$

where the secondary source has been assumed to be at the position corresponding to $x = 0$, and B is a complex amplitude, which is linearly dependent on the electrical input to the secondary source, u, in Fig. 1.2. If this electrical input is adjusted in amplitude and phase so that $B = -A$, then the total downstream pressure will be

$$p_{p+}(x) + p_{s+}(x) = 0, \quad \text{for } x > 0\ , \tag{1.2.3}$$

i.e. the pressure will be perfectly cancelled at all points downstream of the secondary source. This suggests that a practical way in which the control input could be adapted is by monitoring the tonal pressure at any point downstream of the secondary source and adjusting the amplitude and phase of the control input until this pressure is zero. We are mainly interested here in the physical consequences of such a control strategy, however, and so we calculate the total pressure to the left, i.e. on the upstream side, of the secondary source, which in general will be

$$p_{p+}(x) + p_{s-}(x) = A\,e^{-jkx} + B\,e^{+jkx}, \qquad x < 0 \ . \tag{1.2.4}$$

If the secondary source is adjusted to cancel the pressure on the downstream side, then $B = -A$, and the pressure on the upstream side becomes

$$p_{p+}(x) + p_{s-}(x) = -2jA \sin kx, \qquad x < 0 \ , \tag{1.2.5}$$

since $e^{jkl} - e^{-jkl} = 2j \sin kl$. Thus a perfect acoustic *standing wave* is generated by interference between the positive-going primary wave and the negative-going wave generated by the secondary source. Notice that this standing wave has nodes of pressure at $x = 0$, i.e. at the position of the secondary source, and $x = \lambda/2$, $x = -\lambda$ etc., where λ is the *acoustic wavelength*, and that when $x = -\lambda/4$, $x = -3\lambda/4$ etc., its amplitude is exactly twice the amplitude of the incident primary wave. The distribution of the pressure amplitude in the duct under these circumstances is shown in Fig. 1.3.

In cancelling the pressure downstream of the secondary source, the pressure at the secondary source location has been driven to zero by the effect of the active control system. The secondary source thus acts to create a pressure-release boundary condition as far as the incident wave is concerned, and effectively *reflects* this wave back up the duct with equal amplitude and inverted phase, which gives rise to the standing wave observed in Fig. 1.3. The acoustic power generated by a loudspeaker is equal to the time-averaged product of its volume velocity and the acoustic pressure in front of it. The fact that the acoustic pressure at the secondary source location is zero means that the secondary source can generate no acoustic power when operating to cancel the incident wave, and it acts as a purely reactive element.

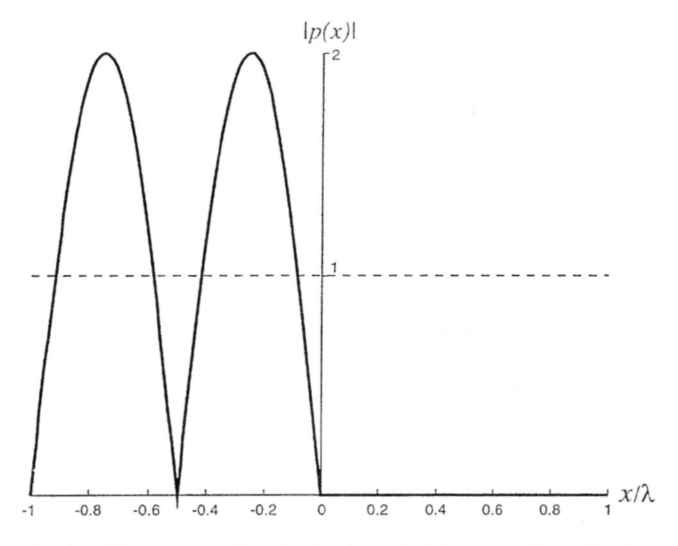

Figure 1.3 Amplitude of the pressure distribution in an infinite duct after a single secondary source, at $x = 0$, has been adjusted to cancel an incident plane wave of unit amplitude travelling in the positive x direction.

1.2.2. Two Secondary Actuators

It is also possible to use an active control system to *absorb* the whole of the incident primary wave, instead of reflecting it back upstream, but such a strategy requires a pair of secondary sources. This is illustrated in Fig. 1.4, where two loudspeakers act as secondary sources, at positions corresponding to $x = 0$ and $x = l$, which can individually generate complex pressures on their upstream and downstream sides which are given by

$$p_{s1+}(x) = B\mathrm{e}^{-jkx}, \quad \text{for } x > 0, \quad p_{s1-} = B\mathrm{e}^{+jkx}, \quad \text{for } x < 0 , \tag{1.2.6 a,b}$$

and

$$p_{s2+}(x) = C\mathrm{e}^{-jk(x-l)}, \quad \text{for } x > l, \quad p_{s2-} = C\mathrm{e}^{+jk(x-l)}, \quad \text{for } x < l , \tag{1.2.7 a,b}$$

where the complex amplitudes B and C are proportional to the complex inputs to the two secondary sources u_1 and u_2 in Fig. 1.4.

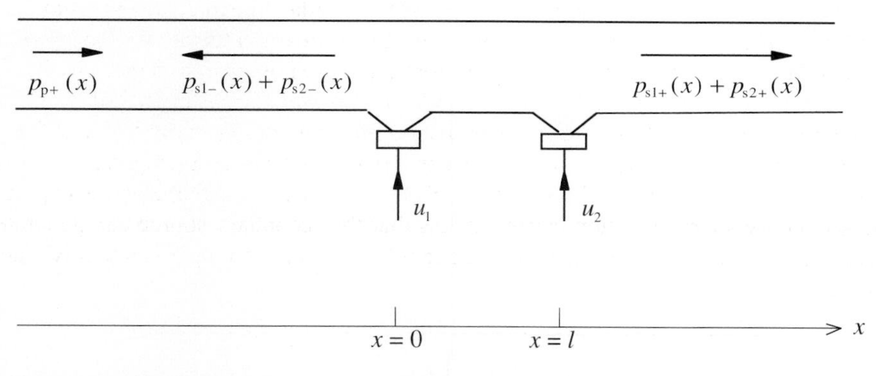

Figure 1.4 Active control of plane waves in an infinite duct using a pair of secondary sources.

The two secondary sources can be driven so that they only affect the downstream wave, if the net upstream pressure they produce in the duct is zero, so that,

$$p_{s1-}(x) + p_{s2-}(x) = 0, \quad x < 0 . \tag{1.2.8}$$

This requires that the control input u_1 is adjusted relative to u_2 so that

$$B = -C\mathrm{e}^{-jkl} , \tag{1.2.9}$$

i.e. the loudspeaker at $x = 0$ produces a delayed and inverted version of that produced by the loudspeaker at $x = l$ (Swinbanks, 1973), and the two loudspeakers act as a secondary source array.

The total pressure produced downstream of the two secondary sources is then

$$p_{p+}(x) + p_{s1+}(x) + p_{s2+}(x) = \left(A + C(\mathrm{e}^{jkl} - \mathrm{e}^{-jkl})\right)\mathrm{e}^{jkx}, \quad x > l . \tag{1.2.10}$$

The loudspeaker array can thus be arranged to cancel the primary incident wave provided u_2 is adjusted to ensure that

$$C = \frac{-A}{2j \sin kl} . \tag{1.2.11}$$

It should be noted that at some frequencies C will have to be very large compared with A. Both secondary sources will have to drive very hard if $\sin kl \approx 0$, which occurs when $l \ll \lambda$, $l \approx \lambda/2$ etc. The frequency range over which such a secondary source array can be operated is thus fundamentally limited (Swinbanks, 1973), although this does not cause problems in many practical implementations that only operate over a limited bandwidth (Winkler and Elliott, 1995).

The pressure distribution between the two secondary sources can be calculated using equations (1.2.6) and (1.2.7) with the conditions given by (1.2.9) and (1.2.11), and turns out to be part of another standing wave. The total pressure distribution after the absorption of an incident primary wave by a pair of secondary sources is shown in Fig. 1.5 for the case in which $l \approx \lambda/10$. The pressure amplitude upstream of the secondary source pair is now unaffected by the cancellation of the downstream incident wave. In this case the secondary source farthest downstream, driven by signal u_2, still generates no acoustic power, because it has zero acoustic pressure in front of it, but the other secondary source, driven by signal u_1, must absorb the power carried by the incident primary wave.

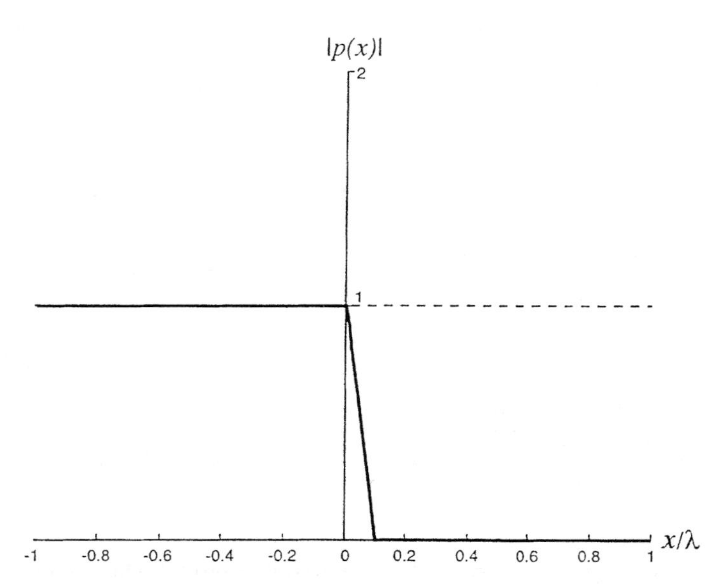

Figure 1.5 The amplitude of the pressure distribution in an infinite duct after a pair of secondary sources, at $x = 0$ and $x = \lambda/10$, has been adjusted to cancel an incident plane wave of unit amplitude travelling in the positive x direction, and also to suppress the wave generated by the secondary sources travelling in the negative x direction.

1.2.3. Control of Multiple Modes

As the excitation frequency is increased and the acoustic wavelength becomes comparable with the dimensions of the duct cross section, then it becomes possible for more than just plane acoustic waves to propagate in the duct. The other types of acoustic wave that can propagate in the duct have pressure distributions that are not uniform across the cross section of the duct and are referred to as *higher order modes*. For a rectangular duct with a height, L_y, which is greater than its depth L_z, then the first higher-order mode can propagate in the duct when it is excited above its first *cut-on frequency* given by

$$f_1 = \frac{c_0}{2L_y} . \tag{1.2.12}$$

The pressure distribution of the first higher-order mode in the y direction, i.e. measured across the height of a rectangular duct, is illustrated in Fig. 1.6, together with the pressure distribution of the plane wave or zeroth order mode. If two identical secondary sources are used, one on either side of the duct, which are driven at the same amplitude, then a plane wave will be produced when they are driven in phase, and when they are driven out of phase they will only excite the higher-order mode. These two secondary sources can thus be used to excite any combination of amplitudes of these two modes and so could be used to actively control both modes. In general it will require N secondary sources to control N modes, provided the combination of secondary sources is able to independently excite each of these modes.

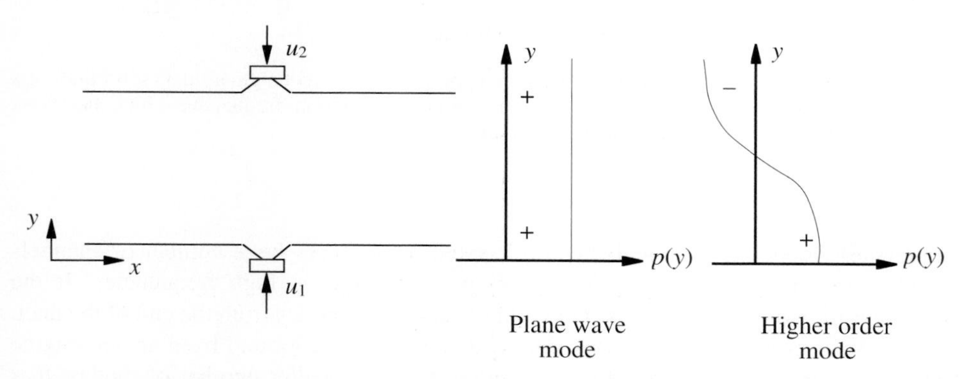

Figure 1.6 Cross section of the part of a duct driven by two loudspeakers mounted in the same plane, which can be driven in phase to produce a plane wave mode and out of phase to produce a higher order mode. The pressure distribution of these two modes across the duct are also shown.

Although it is, in principle, possible to actively control higher-order modes using multiple secondary sources, there are several additional problems which are not encountered in the active control of plane acoustic waves, as discussed in more detail, for example, by Fedorynk (1975), Tichy (1988), Eriksson et al. (1989), Silcox and Elliott (1990), Stell and Bernhard (1991) and Zander and Hansen (1992). There is an increasing interest in the active control of higher-order acoustic modes in short ducts because of the potential application in controlling the fan tones radiated from the inlets of aircraft engines, particularly as an aircraft is coming in to land (see for example Burdisso et al., 1993; Joseph et al., 1999). It should be noted, however, that the number of higher-order modes which are able to propagate in a duct increases significantly with the excitation frequency, as illustrated in Fig. 1.7 for both a rectangular and a circular duct. The average number of propagating modes is proportional to the square of the excitation frequency when it is well above the lowest cut-on frequency. If a significant amount of energy is being transmitted in

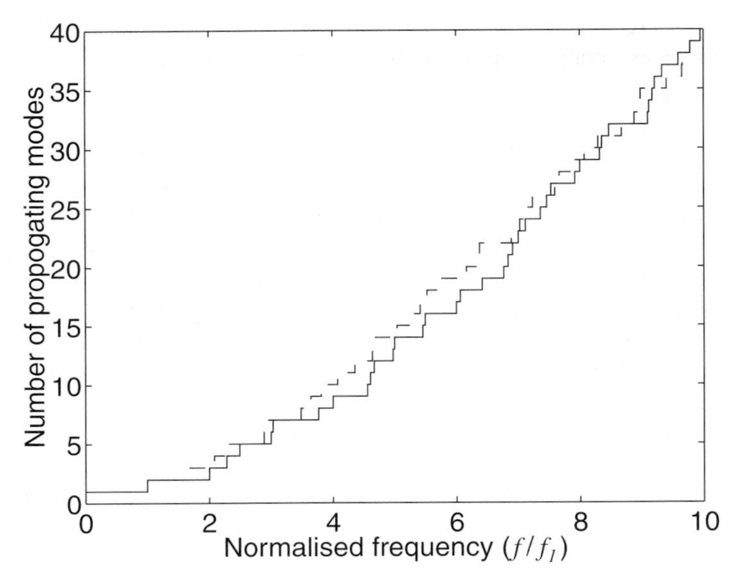

Figure 1.7 The number of acoustic modes that can propagate in a rectangular duct (solid line) or a circular duct (dashed line) as a function of excitation frequency, which has been normalised by the lowest cut-on frequency of the duct, f_1.

each of these modes then an active control system with a very large number of channels would be required to attenuate the overall pressure level at high frequencies. If the objective is only to reduce the sound radiating at particular angles from the end of the duct, however, such as those which causes significant sound on the ground from an aeroengine inlet duct, this could be achieved by controlling a much smaller number of modes. It is possible to detect the amplitude of these modes using an axial array of sensors placed inside the duct (Joseph et al., 1996).

Multiple types of wave also can propagate in one-dimensional structural systems such as beams, pipes and struts. These waves appear in their clearest form in a structure like a solid beam, within which three distinct types of wave can be transmitted at low frequencies (Cremer and Heckl, 1988; Fuller et al., 1996):

(1) Longitudinal, with motion in the same direction as wave propagation, as for an acoustic wave, but with wave speeds which may be many time higher than that of sound in air.
(2) Flexural, which involves motion both at right angles to the direction of wave propagation and rotation in a plane parallel to the direction of propagation. Such waves can have both evanescent and propagating components at all excitation frequencies. Two independent flexural waves can be generated on a uniform beam with motion in each of the two directions at right angles to the direction of propagation. Since the flexural waves in each of these two planes can have both propagating and evanescent components, a total of four flexural components of motion can transmit vibration from one part of the structure to another.

(3) Torsional, whose physical mechanism is similar to that of longitudinal waves except that it involves rotational motion in a plane at right angles to the direction of propagation.

With the four components of the flexural waves and one each of longitudinal and torsional waves, there are thus six wave components that can transmit vibration along such a structure. In general, it is important to control all of these components if complete isolation of the vibration is required (Pan and Hansen, 1991), but in practice several of the components may be naturally isolated, and so fewer control channels are required to control the remaining components. With only a small amount of rotary compliance it is often possible to passively attenuate torsional waves, for example, and provided the structure is at least as long as the flexural wavelength, then the evanescent parts of the flexural waves will also decay away naturally along the length of the structure, leaving only three components that need to be controlled. This natural filtering of some components of vibration was used to reduce the number of actuators required to actively control the vibration being transmitted along a helicopter strut (Brennan et al., 1994; Sutton et al., 1997) and reduce the total kinetic energy of the structure to which the strut was attached in a laboratory experiment by an average of 40 dB at excitation frequencies from about 300 Hz to 1200 Hz.

The wave types that can propagate in a shell, cylinder or pipe can be considerably more complicated than those that can propagate in a solid body (see, for example, Fuller et al., 1996). The complexity of such systems is further increased in a pipe that contains fluid. In order to actively control vibration transmission under the most general circumstances it would be necessary to control all of the waves propagating in this fluid-structural system. Once again, however, it is often possible in practice to use passive methods to attenuate some of the wave types and thus considerably simplify the active control system required for good overall attenuation (Brennan et al., 1996).

1.3. CONTROL OF POWER IN INFINITE SYSTEMS

In the previous section, we saw that waves travelling in one direction could be either reflected using a single secondary source, or absorbed using a pair of secondary sources. In this section, we consider the active control of disturbances propagating as waves in two or three dimensions. The physical interpretation is clearest if we initially restrict ourselves to the control of waves propagating only away from the sources, i.e. in infinite systems with no reflections. The waves propagating in such two- and three-dimensional systems cannot be perfectly cancelled unless the secondary source is physically collocated with the primary source. Significant reductions in wave amplitude can still be achieved, however, if the separation between primary and secondary sources is not too great compared with a wavelength. The total power generated by the combination of sources provides a convenient way of quantifying the space-average or *global* effect of various control strategies. In this section we will introduce the idea of minimising such a global measure of performance by adjusting the secondary source strengths, rather than arranging for the secondary sources to perfectly cancel the primary field. An example in which waves propagate only outwards in a two-dimensional plane is given by flexural disturbances

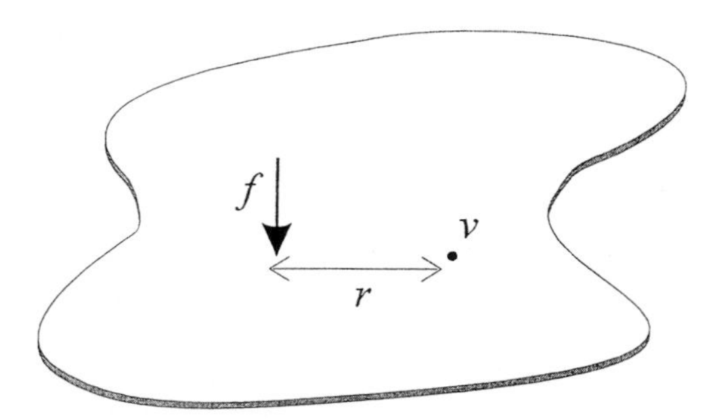

Figure 1.8 The out-of-plane velocity v due to a point force f, a distance r away, on an infinite panel.

travelling in an infinite flat plate or panel. This problem will be considered first, before the full three-dimensional problem is considered, taking the propagation of sound in an infinite medium as an example.

1.3.1. Point Forces on an Infinite Panel

Consider the velocity field generated on an infinite thin panel by a point force, as shown in Fig. 1.8. The point force could, for example, be an approximation to the low frequency excitation of a reciprocating engine mounted on the panel. The complex out-of-plane velocity at any position on the panel will only depend on its distance r from the point force, and for excitation at a single frequency this can be written as

$$v\,(r) = M(r)\,f\,, \tag{1.3.1}$$

where f is the complex amplitude of the point force and $M(r)$ is the transfer response between these two points, which for this physical example is termed the *mechanical transfer mobility*. An expression for $M(r)$ on an infinite thin panel is given by Cremer and Heckl (1988) as

$$M(r) = \frac{1}{8\sqrt{Dm}}\left[H_0^{(2)}(k_\mathrm{f}r) - H_0^{(2)}(-jk_\mathrm{f}r)\right]\,, \tag{1.3.2}$$

where ω is the angular excitation frequency, D is the bending stiffness of the panel $\left(D = Eh^3\big/\left[12\left(1-\nu^2\right)\right]\right.$ where E and ν are the Young's modulus and Poisson's ratio for the plate material and h its thickness) and m is the mass per unit area ($m = \rho h$ where ρ is the material density), $H_0^{(2)}$ denotes the Hankel function of the second kind and k_f is the wavenumber of the flexural wave which is equal to

$$k_\mathrm{f} \;=\; \left(\frac{\omega^2 m}{D}\right)^{1/4}\,. \tag{1.3.3}$$

The *power* that is generated by this point force acting alone on the infinite panel is given by the time average product of the force and associated velocity, which can be expressed in terms of the complex force and velocity as

$$\Pi = \frac{1}{2}\mathrm{Re}\left(v^*(0)f\right) , \tag{1.3.4}$$

where the superscript * denotes conjugation and $v(0)$ is the out-of-plane velocity at the point of application of the force, f. Using equation (1.3.1), the input power can be written as

$$\Pi = \frac{1}{2}|f|^2 \mathrm{Re}(M(0)) , \tag{1.3.5}$$

and using the properties of the Hankel functions we find that

$$M(0) = \frac{1}{8\sqrt{Dm}} , \tag{1.3.6}$$

which is already entirely real in this case, and is also independent of frequency.

1.3.2. Minimisation of Power Input

We are now in a position to calculate the power that would be supplied to the panel by two forces. If one of these forces, f_p, generates the primary disturbance and the other acts as the secondary source, f_s, we can calculate the best possible reduction in the total power supplied to the panel by the adjustment of the amplitude and phase of the secondary source. The *primary source* may be the engine described above and the secondary force could be supplied by an electromagnetic shaker driven at the same frequency as the engine rotation, for example, whose amplitude and phase could be adjusted relative to that of the engine to minimise the total mechanical power supplied to the panel. Because we are interested in the action of multiple primary sources and multiple secondary sources, it is convenient to formulate the problem in matrix form. Let

$$\mathbf{f} = \left[f_1, f_2 \;\cdots\; f_N\right]^{\mathrm{T}} , \tag{1.3.7}$$

be the vector of complex primary and secondary forces acting on the panel and

$$\mathbf{v} = \left[v_1, v_2 \;\cdots\; v_N\right]^{\mathrm{T}} , \tag{1.3.8}$$

be the vector of complex velocities at the points of application of each of these forces. The total power supplied to the panel by all of these forces can now be written as

$$\Pi = \frac{1}{2}\mathrm{Re}\left(\mathbf{v}^{\mathrm{H}}\,\mathbf{f}\right) , \tag{1.3.9}$$

where the superscript H denotes the Hermitian, i.e. the complex conjugate, transpose of the vector, as described in the Appendix.

The velocity at every point depends on each of the forces and the total velocity can be calculated by the superposition of a number of terms of the form of equation (1.3.1). The vector of velocities can thus be represented in matrix form as

$$\mathbf{v} = \mathbf{M}\,\mathbf{f} , \tag{1.3.10}$$

where $\mathbf{M}$ is the matrix of input and transfer mobilities between each of the forces. The matrix $\mathbf{M}$ is symmetric because of reciprocity (Cremer and Heckl, 1998), i.e. $\mathbf{M}^{\mathrm{T}} = \mathbf{M}$.

The total power supplied to the panel, equation (1.3.9), can now be written as

$$\Pi = \frac{1}{2}\mathrm{Re}\left(\mathbf{f}^{\mathrm{H}}\,\mathbf{M}^{\mathrm{H}}\,\mathbf{f}\right) = \frac{1}{2}\mathbf{f}^{\mathrm{H}}\,\mathbf{R}\mathbf{f} \,, \tag{1.3.11a,b}$$

where $\mathbf{R} = \mathrm{Re}(\mathbf{M})$.

The vectors of velocities and forces are now partitioned into primary and secondary components (Bardou et al., 1997) so that

$$\mathbf{v} = \begin{bmatrix} \mathbf{v}_{\mathrm{p}} \\ \mathbf{v}_{\mathrm{s}} \end{bmatrix} = \mathbf{M}\mathbf{f} = \begin{bmatrix} \mathbf{M}_{\mathrm{pp}} & \mathbf{M}_{\mathrm{ps}} \\ \mathbf{M}_{\mathrm{sp}} & \mathbf{M}_{\mathrm{ss}} \end{bmatrix} \begin{bmatrix} \mathbf{f}_{\mathrm{p}} \\ \mathbf{f}_{\mathrm{s}} \end{bmatrix} , \tag{1.3.12}$$

and we define the partitioned matrix $\mathbf{R}$ to be

$$\mathbf{R} = \begin{bmatrix} \mathbf{R}_{\mathrm{pp}} & \mathbf{R}_{\mathrm{ps}} \\ \mathbf{R}_{\mathrm{sp}} & \mathbf{R}_{\mathrm{ss}} \end{bmatrix} = \mathrm{Re}\begin{bmatrix} \mathbf{M}_{\mathrm{pp}} & \mathbf{M}_{\mathrm{ps}} \\ \mathbf{M}_{\mathrm{sp}} & \mathbf{M}_{\mathrm{ss}} \end{bmatrix} . \tag{1.3.13}$$

The total power supplied to the panel by both the primary and secondary forces, equation (1.3.11b), is thus given by

$$\Pi = \frac{1}{2}\left(\mathbf{f}_{\mathrm{s}}^{\mathrm{H}}\,\mathbf{R}_{\mathrm{ss}}\,\mathbf{f}_{\mathrm{s}} + \mathbf{f}_{\mathrm{s}}^{\mathrm{H}}\,\mathbf{R}_{\mathrm{sp}}\,\mathbf{f}_{\mathrm{p}} + \mathbf{f}_{\mathrm{p}}^{\mathrm{H}}\,\mathbf{R}_{\mathrm{sp}}^{\mathrm{T}}\,\mathbf{f}_{\mathrm{s}} + \mathbf{f}_{\mathrm{p}}^{\mathrm{H}}\,\mathbf{R}_{\mathrm{pp}}\,\mathbf{f}_{\mathrm{p}}\right) , \tag{1.3.14}$$

where $\mathbf{R}_{\mathrm{ps}} = \mathbf{R}_{\mathrm{sp}}^{\mathrm{T}}$ by reciprocity. Equation (1.3.14) is of Hermitian quadratic form, as described in the Appendix, so that the power is a quadratic function of the real and imaginary parts of each element of the vector $\mathbf{f}_{\mathrm{s}}$. This quadratic function must always have a minimum rather than a maximum associated with it, since otherwise, for very large secondary forces, the total power supplied to the panel would become negative, which would correspond to the impossible situation of an array of forces absorbing power from the otherwise passive panel. It is shown in the Appendix that the minimum possible value of power is given by a unique set of secondary forces provided the matrix $\mathbf{R}_{\mathrm{ss}}$ is positive definite, and that this optimum set of secondary forces is given by

$$\mathbf{f}_{\mathrm{s,opt}} = -\mathbf{R}_{\mathrm{ss}}^{-1}\,\mathbf{R}_{\mathrm{sp}}\,\mathbf{f}_{\mathrm{p}} \,. \tag{1.3.15}$$

The positive definiteness of $\mathbf{R}_{\mathrm{ss}}$ is guaranteed on physical grounds in this case, since the power supplied by the secondary forces acting alone is equal to $\mathbf{f}_{\mathrm{s}}^{\mathrm{H}}\mathbf{R}_{\mathrm{ss}}\mathbf{f}_{\mathrm{s}}$ which must be positive for all $\mathbf{f}_{\mathrm{s}}$ provided they are not collocated. The minimum value of the total power that results from this optimum set of secondary forces is given by

$$\Pi_{\mathrm{min}} = \frac{1}{2}\mathbf{f}_{\mathrm{p}}\left[\mathbf{R}_{\mathrm{pp}} - \mathbf{R}_{\mathrm{sp}}^{\mathrm{T}}\,\mathbf{R}_{\mathrm{ss}}^{-1}\,\mathbf{R}_{\mathrm{sp}}\right]\mathbf{f}_{\mathrm{p}} \,. \tag{1.3.16}$$

Each of the elements in the matrix $\mathbf{M}$ in equation (1.3.10) can be calculated by using equation (1.3.2), using the geometric arrangement of the primary and secondary forces to compute the distance between each of them. The maximum possible attenuation of the input power for this geometric arrangement of primary and secondary sources can thus be calculated by taking the ratio of the power before control, given by equation (1.3.14) with $\mathbf{f}_{\mathrm{s}}$ set to zero, and after control, equation (1.3.16).

The maximum attenuation in input power has been calculated for the arrangements of primary and secondary forces shown in Fig. 1.9, in which a single primary source is controlled by one, two, four or eight secondary forces each uniformly spaced at a distance d from the primary source. The results of such a calculation are shown in Fig. 1.10, in which the attenuation for an optimally adjusted set of secondary forces is plotted against the separation distance normalised by the flexural wavelength (Bardou et al., 1997). It must be emphasised that these results have been obtained one frequency at a time by performing a series of calculations assuming tonal excitation at an appropriate frequency. No claims are being made about broadband attenuation at this stage and the objective of the exercise is to determine the limits of performance due to the physical limitations of this geometrical arrangement only.

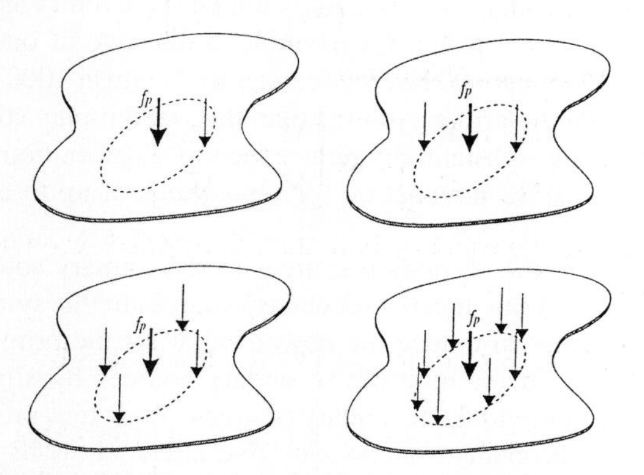

Figure 1.9 Arrangement of primary (bold arrows) and secondary forces (light arrows) on an infinite panel used in the calculation of the attenuation of total power output for the cases in which one, two, four or eight secondary sources are acting. Each of the secondary forces is positioned a distance of d from the primary source.

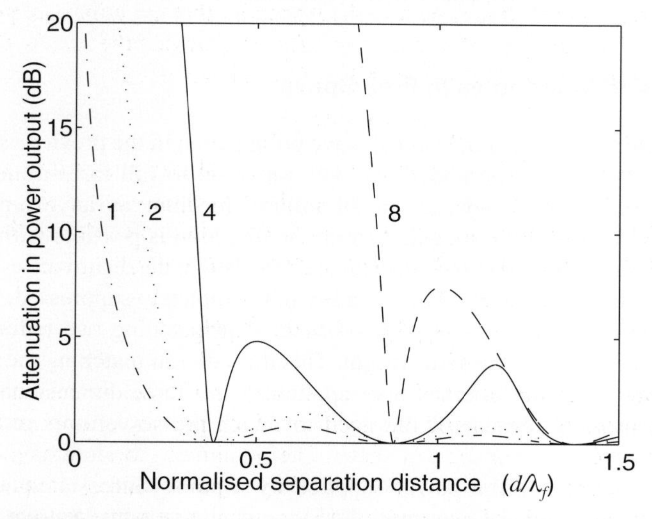

Figure 1.10 Attenuation in the total power supplied to an infinite panel by a point primary force when optimally controlled by one (-·-·-·-·), two (······), four (——) or eight (------) secondary sources, as a function of the distance from the primary to the secondary sources, d, normalised by the flexural wavelength.

It can be seen that if the secondary source is closer to the primary source than about one-tenth of a flexural wavelength, then attenuations in total power of up to 10 dB can be achieved with only a single source. The flexural wavelength on a thin panel is equal to $2\pi/k_{\mathrm{f}}$, where k_{f} is given by equation (1.3.3), so that

$$\lambda_{\mathrm{f}} = 2\pi \left(\frac{D}{\omega^2 m} \right)^{1/4} , \tag{1.3.17}$$

where D is the bending stiffness, which is equal to the product of the modulus of elasticity and the second moment of inertia for the panel, and m is the mass per unit area, which is equal to the product of the panels density and its thickness (Cremer and Heckl, 1988).

Thus for a 2-mm thick steel panel, for example, a distance of one-tenth of a flexural wavelength is equal to 32 mm at 100 Hz, but reduces to 10 mm at 1000 Hz. If the secondary force is farther away from the primary source than this, then it may still be possible to use multiple secondary forces to substantially reduce the input power. Four secondary sources, for example, can give a 10 dB attenuation for separations of up to about 0.34 times the flexural wavelength.

When the distance from the secondary sources to the primary source takes on certain values, however, e.g. $d \approx 0.38\lambda_{\mathrm{f}}$, then the secondary sources in this symmetric array are not capable of working together to reduce the power output of the primary source. This ill-conditioning can be avoided for multiple secondary sources by slightly perturbing the relative positions of the individual secondary sources. With this proviso, the secondary source array with eight secondary sources can give attenuations in power of 10 dB for separation distances up to about 0.84 times the flexural wavelength. The overall trend is that the number of secondary sources required to achieve significant reductions in the power supplied to the panel is *linearly* proportional to their normalised distance from the primary, d/λ_{f}. The maximum separation between each of the multiple secondary sources required to give at least 10 dB attenuation is about half of a flexural wavelength.

1.3.3. Acoustic Monopoles in Free Space

Having progressed from one-dimensional wave propagation in the previous section to two-dimensional wave propagation above, we now move on to full three-dimensional wave propagation, using acoustic sources in an infinite medium as an example. Whereas complete suppression of wave propagation in one direction is possible with only a single source, we have seen that, for two-dimensional propagation, the disturbances caused by the primary source can be attenuated, but generally not completely suppressed, by an array of secondary sources. This is because of the difficulty of matching two radially expanding wavefronts that do not have the same origin. This difficulty in matching the wavefronts of the primary and secondary sources also applies to the three-dimensional case, but is somewhat more severe because of the need to match the wavefronts in the additional dimension.

In a mechanical system, the power supplied by a point source is equal to the time-averaged product of a single force multiplied by a single velocity, but we must be more careful with acoustic sources since the force is a distributed quantity, determined by the acoustic pressure, and an acoustic source can also have a distribution of velocity. If we assume, however, that the source is vibrating with equal amplitude in all directions, such as a pulsating sphere, and that this sphere is small compared with the acoustic wavelength, so that the pressure on its surface is reasonably uniform, then its power output can be simply calculated by taking the time-average product of the acoustic pressure at the surface and the *volume velocity* of the source. The volume velocity, q, is equal to the radial surface velocity multiplied by the surface area. Such an acoustic source is known as a *monopole*, and is used to set up a similar model problem for three-dimensional acoustic wave propagation as was discussed for two-dimensional structural wave propagation above. A practical example of such an acoustic source may be the end of an engine's exhaust pipe radiating sound at low frequencies, which can be acting as the primary source in an active control system.

The relationship between the complex acoustic pressure at a distance r from an acoustic monopole source operating at a single frequency may be expressed as

$$p(r) = Z(r)\, q \,, \tag{1.3.18}$$

where q is the complex volume velocity of the monopole source and $Z(r)$ is the complex *acoustic transfer impedance* which plays the same role here as the mechanical transfer mobility defined in equation (1.3.1). In an infinite medium with no acoustic reflections, i.e. a free field, the acoustic transfer impedance is given by (Morse, 1948, see also Nelson and Elliott, 1992)

$$Z(r) = \frac{\omega^2 \rho_0}{4\pi\, c_0}\left(\frac{je^{-jkr}}{kr}\right) , \tag{1.3.19}$$

where ρ_0 and c_0 are the density and the speed of sound in the medium and k is the acoustic wavenumber which is equal to ω/c_0, or $2\pi/\lambda$, where λ is the acoustic wavelength.

The acoustic power generated by a monopole source is thus equal to

$$\Pi = \frac{1}{2}\operatorname{Re}\left(p^*(0)q\right) , \tag{1.3.20}$$

and using equation (1.3.18), this can be written as

$$\Pi = \frac{1}{2}\,|q|^2 \operatorname{Re}(Z(0)) , \tag{1.3.21}$$

where $\operatorname{Re}(Z(0))$ is the real part of the acoustic input impedance, which in a free field is given from equation (1.3.19) by

$$\operatorname{Re}(Z(0)) = \frac{\omega^2 \rho_0}{4\pi\, c_0} . \tag{1.3.22}$$

By direct analogy with the structural case, equation (1.3.12), the vectors of pressures at the positions of the primary and secondary sources can now be written as

$$\begin{bmatrix} \mathbf{p}_p \\ \mathbf{p}_s \end{bmatrix} = \begin{bmatrix} \mathbf{Z}_{pp} & \mathbf{Z}_{ps} \\ \mathbf{Z}_{sp} & \mathbf{Z}_{ss} \end{bmatrix} \begin{bmatrix} \mathbf{q}_p \\ \mathbf{q}_s \end{bmatrix} , \tag{1.3.23}$$

where $\mathbf{q}_p$ and $\mathbf{q}_s$ are the vectors of primary and secondary source volume velocities.

An exactly similar analysis to that used above can now be employed to calculate the minimum power output from a monopole primary acoustic source when controlled by an array of secondary acoustic sources (Nelson et al., 1987; Nelson and Elliott, 1992). In this case, however, the sources must be arranged in three-dimensional space around the primary source, as shown in Fig. 1.11 in which the solid sphere denotes the primary source and the open spheres denote the secondary sources. A practical arrangement corresponding to this model problem may be the use of an array of loudspeakers used as secondary sources placed around the end of an engine's exhaust pipe in the open air.

The result of optimising the real and imaginary parts of the volume velocities of the secondary sources to minimise the total acoustic power output is shown in Fig. 1.12. When the separation distance is small compared with the acoustic wavelength, the attenuations achieved in power output for a single secondary source are similar to those for the two-dimensional case on the panel shown in Fig. 1.10. The fall-off of attenuation as the normalised separation distance increases for multiple secondary sources, however, is considerably more rapid for the three-dimensional case than for the two-dimensional case. In the three-dimensional case the general trend is that the number of secondary sources

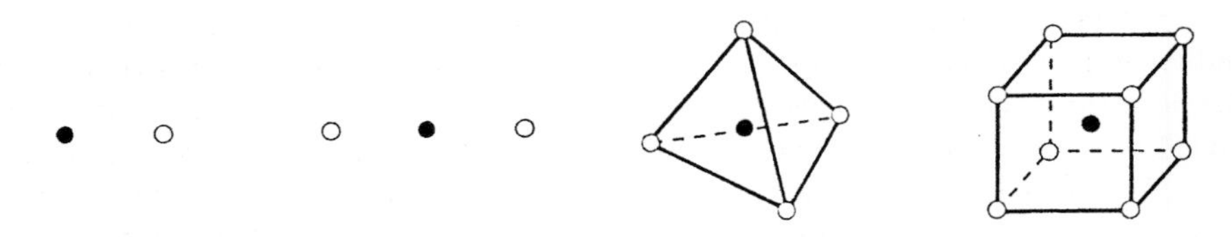

Figure 1.11 Arrangement of primary (solid sphere) and secondary acoustic monopole sources (open spheres) in free space, used in the calculation of the attenuation of total power output. Each of the secondary sources is positioned a distance of d from the primary source.

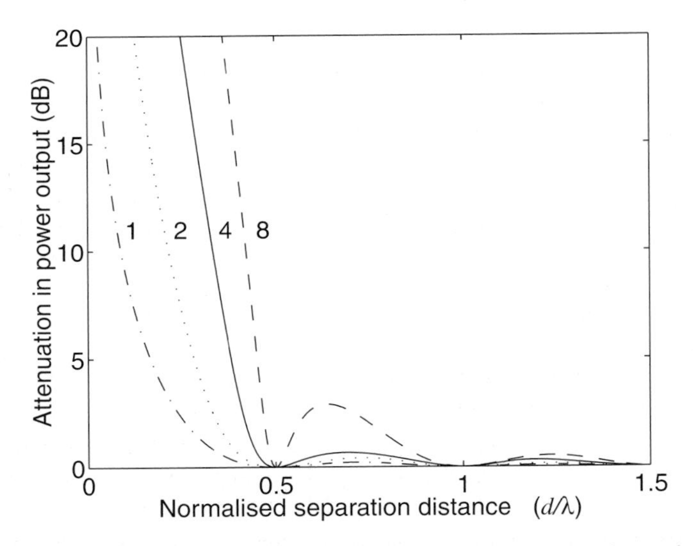

Figure 1.12 Attenuation in the total acoustic power radiated into free space by a primary monopole source when optimally controlled by one (-·-·-·-·), two (······), four (——) or eight (------) secondary monopole sources, as a function of the distance from the primary to the secondary sources, d, normalised by the acoustic wavelength.

required to achieve a given attenuation increases as the *square* of the normalised separation distance d/λ . The maximum separation distance required between each of the multiple secondary sources to achieve 10 dB attenuation in total power output for this three-dimensional problem is about half a wavelength, which is the same figure as was calculated for the two-dimensional problem above.

With a single layer of point forces, or acoustic monopoles, all at equal distances from the primary source, the power output of the primary source can be reduced by reflecting the outward-going wave back towards the primary source. It can be shown that the power output of all the secondary sources is identically zero when controlling a single primary source (Elliott et al., 1991), emphasising their entirely reactive role, and this case is thus directly analogous to the one-dimensional case whose pressure distribution is shown in Fig. 1.3. It is also possible to use a double layer of forces or acoustic monopoles, at unequal distances from the primary source, to *absorb* the waves radiated by the primary source, by direct analogy with the one-dimensional acoustic case, whose pressure distribution is shown in Fig. 1.5. When operating in this manner, the pairs of monopole sources can be shown to be synthesising a single point monopole/dipole source (Nelson and Elliott, 1992), which is sometimes called a tripole (Jessel, 1968; Mangiante, 1994). The two- and three-dimensional equivalent problems to those considered above for the absorption of sound have been studied by Zavadskaya et al. (1976) and Konaev et al. (1977), who showed that the number of such sources required for a given level of control was again proportional to d/λ for the two-dimensional case and $(d/\lambda)^2$ for the three-dimensional case, where d is the distance from the primary source to the secondary source.

1.4. STRATEGIES OF CONTROL IN FINITE SYSTEMS

A source driving an infinite system will only generate waves moving away from the source position. In practice most systems have boundaries, which reflect the outgoing waves from a source within them, and the interference between the incident and reflected waves creates *resonances* at certain excitation frequencies. In the following section, the response of finite two- and three-dimensional systems will be described entirely in terms of the *modes* that are associated with these resonances. For one-dimensional systems, however, it is sometimes clearer to retain a model of the system in terms of the constituent waves, since this provides a more direct way of calculating its response.

There are many different active control strategies that can be implemented in a finite system. In this section, the physical effects of several of these strategies are illustrated using the simple one-dimensional acoustic system shown in Fig. 1.13. This consists of a long duct with rigid walls, in which only plane acoustic waves can propagate, with a primary source of volume velocity q_p at the left hand end, $x = 0$, and a secondary source of volume velocity q_s at the right hand end, $x = L$. This simple system was originally used by Curtis et al. (1987) to illustrate the effect of different control strategies on the energy in the duct. The modification presented here, based on input power, was originally presented by Elliott (1994), who also considered the equivalent mechanical system of a thin beam excited by moments at the two ends. A more comprehensive treatment of the structural case is provided by Brennan et al. (1995).

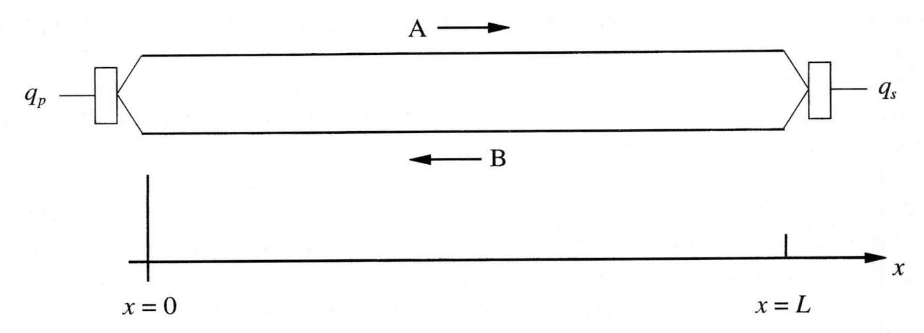

Figure 1.13 A long duct, driven at one end by a primary source of volume velocity q_p, and at the other end by a secondary source of volume velocity q_s, which is used to illustrate the effect of different control strategies on a finite system.

1.4.1. Acoustic Impedances in a Finite Duct

In this section, the acoustic input and transfer impedances in the duct are derived, and the acoustic power supplied by the primary source is defined. The complex pressure at position x along such a duct, $p(x)$, can be written as the sum of two travelling waves, one of amplitude A travelling in the positive x direction and one of amplitude B travelling in the negative direction, so that

$$p(x) = A\,\mathrm{e}^{-jkx} + B\,\mathrm{e}^{+jkx} \; . \tag{1.4.1}$$

The complex particle velocity associated with a plane acoustic wave is $+1/\rho_0 c_0$ times the pressure for waves travelling in the positive x direction and $-1/\rho_0 c_0$ times the pressure for wave travelling in the negative x direction, where ρ_0 and c_0 are the density and speed of sound in the medium (Kinsler et al., 1982; Nelson and Elliott, 1992). The complex particle velocity in the duct is thus

$$u(x) = \frac{1}{\rho_0 c_0}\left[A\,\mathrm{e}^{-jkx} - B\,\mathrm{e}^{+jkx}\right] . \tag{1.4.2}$$

Before active control is applied, the volume velocity of the secondary source, q_s, is zero. The particle velocity at $x = L$ in the duct is thus also zero, and can be written as

$$u(L) = \frac{1}{\rho_0 c_0}\left[A\,\mathrm{e}^{-jkL} - B\,\mathrm{e}^{+jkL}\right] = 0 \; , \tag{1.4.3}$$

so that the amplitude of the reflected wave is related to the amplitude of the incident wave by the expression

$$B = A\,\mathrm{e}^{-j2kL} \; . \tag{1.4.4}$$

The particle velocity at the other end of the duct is imposed by the primary source and is equal to q_p/S where q_p is the volume velocity of the primary source and S is the cross-sectional area of the duct, and so, using equation (1.4.2) and (1.4.4),

$$u(0) = A\left[1 - \mathrm{e}^{j2kL}\right] / \rho_0 c_0 = q_p / S \; , \tag{1.4.5}$$

which can be used to calculate the amplitude of the downstream wave, as a function of the primary volume velocity,

$$A = \frac{Z_c q_p}{1 - e^{-j2kL}} , \tag{1.4.6}$$

where Z_c is the characteristic acoustic impedance of the duct, which is equal to $\rho_0 c_0 / S$ The complex pressure at the two ends of the duct can be written using equations (1.4.1) and (1.4.4) as

$$p(0) = A\left(1 + e^{-j2kL}\right) , \tag{1.4.7}$$

and

$$p(L) = 2A\,e^{-jkL} . \tag{1.4.8}$$

Using equation (1.4.6), expressions for the acoustic input and transfer impedance of the duct can now be derived. These are given by

$$Z(0) = \frac{p(0)}{q_p} = Z_c \frac{1 + e^{-j2kL}}{1 - e^{-j2kL}} = -jZ_c \cot kL , \tag{1.4.9}$$

and

$$Z(L) = \frac{p(L)}{q_p} = Z_c \frac{2e^{-jkL}}{1 - e^{-j2kL}} = \frac{-jZ_c}{\sin kL} . \tag{1.4.10}$$

With these definitions of the input and transfer impedances, and using the symmetry of the duct, expressions for the total pressure at the two ends of the duct with both primary and secondary source operating can be readily deduced as

$$p_p = Z(0)\,q_p + Z(L)\,q_s , \tag{1.4.11}$$

and

$$p_s = Z(L)\,q_p + Z(0)\,q_s , \tag{1.4.12}$$

where in this case $p_p = p(0)$ and $p_s = p(L)$. Note that $Z(L)$ is equal to the transfer impedance from the secondary source to the primary source position, as well as that from the primary source to the secondary source position. These expressions can now be used to calculate the effect of various active control strategies. The strategies are compared by calculating their effect on the acoustic power output of the primary source, which, from equation (1.3.20) above, is found to be

$$\Pi_p = \frac{1}{2}\mathrm{Re}\left(p_p^* q_p\right) . \tag{1.4.13}$$

With the primary source operating alone in the duct, this quantity is plotted as the solid line in Fig. 1.14, and in the subsequent figures, normalised with respect to the power output of the primary source in an infinite duct which is equal to

$$\Pi_{p,\mathrm{infinite}} = \frac{1}{2}\left|q_p\right|^2 Z_c . \tag{1.4.14}$$

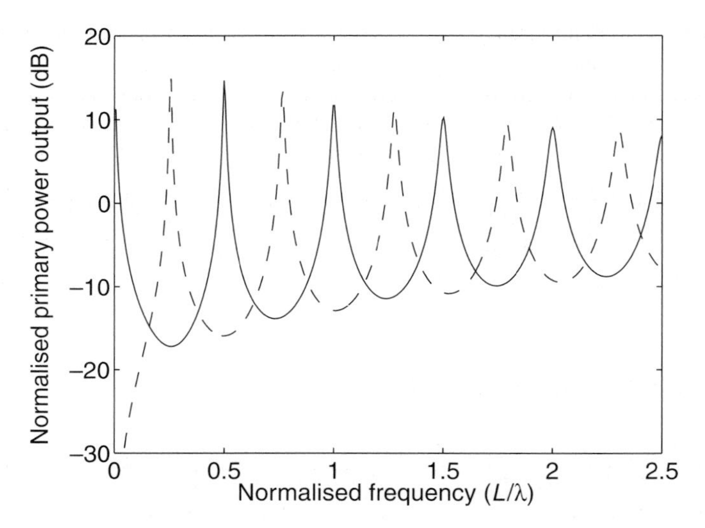

Figure 1.14 Power output of the primary source in the finite duct before control (solid line) and after the secondary source has been adjusted at each frequency to cancel the pressure in front of the secondary source (dashed line).

The normalised power output is plotted as a function of the normalised frequency, which is equal to $L/\lambda = kL/2\pi = \omega L/2\pi c_0$ where k is the wavenumber and L the length of the duct. It is important to include some dissipation or loss in the duct, since otherwise the input impedance, equation (1.4.9), is entirely reactive and no power is ever supplied by the primary or secondary sources. This dissipation is provided here by assuming a complex wavenumber

$$k = \frac{\omega}{c_0} - j\alpha , \tag{1.4.15}$$

where α is a positive number that represents the attenuation of a wave propagating in the duct, and is assumed to be small compared with ω/c_0 . The value of α was chosen in these simulations to give an effective damping ratio of about 1%.

1.4.2. Cancellation of Pressure

The first control strategy considered is that of cancellation of the pressure at the secondary source position, at the right-hand end of the duct. This pressure is equal to p_s in equation (1.4.12), which can be set to zero if the secondary source is driven so that

$$q_{s1} = -\frac{Z(L)}{Z(0)} q_p . \tag{1.4.16}$$

Using equations (1.4.9) and (1.4.10), this expression reduces to

$$q_{s1} = -\frac{q_p}{\cos kL} . \tag{1.4.17}$$

The effect of this control strategy on the power output of the primary source is shown in Fig. 1.14. It can be seen that the power supplied by the primary source has been reduced at the previous resonant excitation frequencies, which occurred when $L/\lambda = 0, \frac{1}{2}, 1$, etc, but is now a maximum when $L/\lambda = \frac{1}{4}, \frac{3}{4}$, etc. This is because the secondary source is acting to create a pressure release boundary at the right-hand end of the duct, which now has the resonance frequencies associated with a closed/open duct, rather than a closed/closed duct. The fact that the duct can still be resonant after the pressure has been cancelled at the secondary source position means that the power supplied to the duct by the primary source is significantly increased by the active control action at the new resonant frequencies.

1.4.3. Absorption of Incident Wave

The second strategy we will consider is that in which the secondary source is driven to suppress the wave reflected from the right-hand end of the duct, which corresponds to absorption of the incident wave. The required condition for the secondary source in this case can be obtained by calculating the amplitude of the reflected wave, B, as a function of the primary and secondary source strengths, which gives

$$B = \frac{Z_c\left(q_p\,\mathrm{e}^{-jkL} + q_s\right)}{\left(\mathrm{e}^{+jkL} - \mathrm{e}^{-jkL}\right)} . \tag{1.4.18}$$

The reflected wave amplitude is thus zero if the secondary source volume velocity is equal to

$$q_{s2} = -q_p\,\mathrm{e}^{-jkL} , \tag{1.4.19}$$

see for example section 5.15 of Nelson and Elliott (1992), where this is referred to as the absorbing termination, and the earlier reference of Beatty (1964). Practical realisations of such absorbing terminations have also been developed by Guicking and Karcher (1984), Orduña-Bustamante and Nelson (1991) and Darlington and Nicholson (1992). The effect of such a termination on the power output of the primary source is shown in Fig. 1.15, from which it can be seen that after control the power output is the same as that produced by the primary source in an infinite duct, equation (1.4.14), as expected.

The control strategies of cancelling the pressure at the secondary source position and of absorbing the plane wave incident upon the secondary sources can be considered as two special cases of the general approach of local pressure feedback. In the general case the secondary source volume velocity, q_s, is arranged to be a real gain factor, $-g$, times the pressure in front of it, p_s, so that

$$q_s = -g\,p_s . \tag{1.4.20}$$

The acoustic impedance presented by the secondary source to the duct using local pressure feedback is thus equal to

$$Z_s = \frac{p_s}{-q_s} = \frac{1}{g} . \tag{1.4.21}$$

If the gain factor is zero, the volume velocity of the secondary source is also zero, and Z_s then corresponds to a hard boundary, as assumed for the duct with no control. If the gain

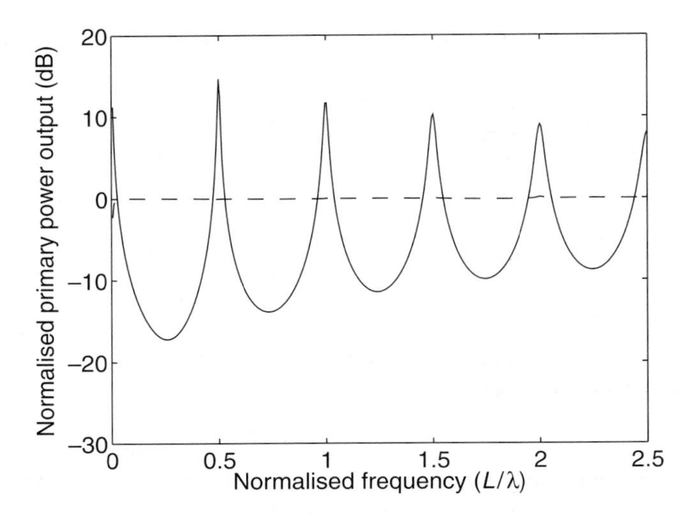

Figure 1.15 Power output of the primary source in the finite duct before control (solid line) and after the secondary source has been adjusted at each frequency to cancel the sound wave reflected by the secondary source (dashed line).

factor is very high, then the pressure in front of the secondary source is almost cancelled by the effect of the feedback and Z_s tends to zero, with the result shown in Fig. 1.14. If the gain factor is set equal to the reciprocal of the characteristic impedance, $g = 1/Z_c$, then the impedance presented to the duct by the secondary source, Z_s, is equal to the characteristic impedance of the duct, which creates the absorbing termination whose effects are shown in Fig. 1.15. The dashed line in Fig. 1.15 can thus be considered as being mid-way between the two extremes of a continuous set of responses which can be obtained by varying the gain in a local pressure feedback system, and the limits of these responses are shown by the solid and dashed lines in Fig. 1.14.

In principle, local pressure feedback is stable for any gain factor g, because the acoustic response of the system under control equation (1.4.9) is *minimum phase*, as explained in more detail in Chapter 6. In practice, however, the volume velocity of a loudspeaker will not be a frequency-independent function of its electrical input, because of the loudspeaker's dynamics. Thus the response of the system under control, or plant as it is called in the later chapters, is not as simple as the acoustic input impedance given by equation (1.4.9) and some care must be taken to maintain the stability of such a feedback system, as described by Guicking and Karcher (1984) and Darlington and Nicholson (1992), for example.

1.4.4. Maximisation of Secondary Power Absorption

The final two control strategies that will be considered in this section are based on the sound power output of the sources. In the first of these we naively attempt to do an even better job than just absorbing the incident wave, as in Fig. 1.15, by *maximising* the power absorbed by the secondary source (Elliott et al., 1991). The power output of the secondary source can be written as

$$\Pi_s = \frac{1}{2}\mathrm{Re}\left(p_s^* q_s\right) = \frac{1}{4}\left(p_s^* q_s + q_s^* p_s\right) , \tag{1.4.22}$$

and using equation (1.4.12) this can be expressed in the Hermitian quadratic form

$$\Pi_s = \frac{1}{4}\left(2 q_s^* \mathrm{Re}(Z(0)) q_s + q_s^* Z(L) q_p + q_p^* Z^*(L) q_s\right) . \tag{1.4.23}$$

The power output of the secondary source is minimised, so that the power absorbed by the secondary source is maximised, by the secondary volume velocity corresponding to the global minimum of this Hermitian quadratic form, which is

$$q_{s3} = -Z(L) q_p / 2\,\mathrm{Re}(Z(0)) . \tag{1.4.24}$$

The results of implementing a secondary source strength corresponding to equation (1.4.24) in the duct are shown as the dashed line in Fig. 1.16, from which it can be seen that, at most excitation frequencies, the power output of the primary source has been *increased* by this control strategy. In general the maximisation of the power absorbed by the secondary source is a rather dangerous control strategy, particularly for narrowband disturbances in resonant systems, because of the ability of the secondary source to increase the power output of the primary source. The secondary source achieves this increase by altering the impedance seen by the primary source and making the physical system appear to be resonant over a much broader range of frequencies than would naturally be the case. A more detailed analysis of the power balance within the duct reveals that only a fraction of the power that is induced out of the primary source is actually absorbed by the secondary source, and the remaining power is dissipated in the duct, which consequently has a

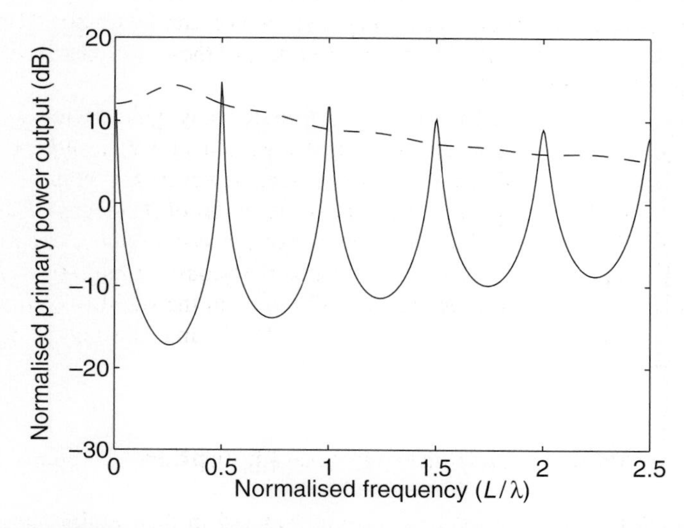

Figure 1.16 Power output of the primary source in the finite duct before control (solid line) and after the secondary source has been adjusted at each frequency to maximise the power absorption by the secondary source (dashed line).

significantly increased level of energy stored within it. It should be pointed out, however, that the effectiveness of a maximum power absorbing strategy will also depend on the nature of the primary excitation and, if this has a random waveform, on the bandwidth of the excitation (Nelson, 1996).

1.4.5. Minimisation of Total Power Input

The final control strategy we consider in this section is that of minimising the total power input to the duct, which is equal to the sum of the power outputs of the primary and secondary sources. The total power input to the duct can thus be written, by analogy with equation (1.3.14) as

$$\Pi_{\mathrm{T}} = \frac{1}{2}\left(q_{\mathrm{s}}^{*}\, R_0\, q_{\mathrm{s}} + q_{\mathrm{s}}^{*}\, R_L\, q_{\mathrm{p}} + q_{\mathrm{p}}^{*}\, R_L\, q_{\mathrm{s}} + q_{\mathrm{p}}^{*}\, R_0\, q_{\mathrm{p}}\right) , \tag{1.4.25}$$

where $R_0 = \mathrm{Re}(Z(0))$ and $R_L = \mathrm{Re}(Z(L))$. This Hermitian quadratic form is minimised by the secondary source volume velocity

$$q_{\mathrm{s}4} \;=\; -\,R_L\, q_{\mathrm{p}} / R_0 \ . \tag{1.4.26}$$

The resulting power output of the primary source is shown in Fig. 1.17, which shows that the power output of the primary source is always reduced by this control strategy and that the residual level of output power at each frequency is then similar to that produced under anti-resonant conditions. The action of the secondary source in this case is to alter the impedance seen by the primary source so that the physical system appears to be anti-resonant over a much broader range of frequencies than would naturally be the case. In order to implement this control strategy in practice, the power output of the two acoustic sources could be deduced from measurements of their source strength and the pressure in front of

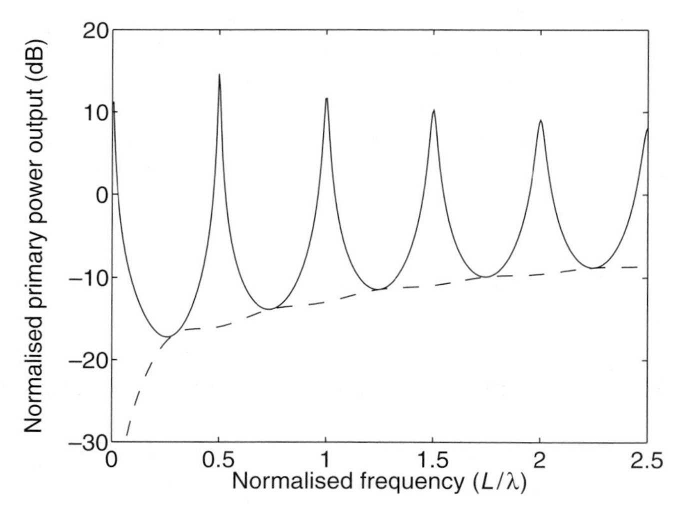

Figure 1.17 Power output of the primary source in the finite duct before control (solid line) and after the secondary source has been adjusted at each frequency to minimise the total power output of both the primary and secondary sources (dashed line).

them, or the amplitudes of the waves propagating in the two directions in the duct. We will see in the next section, however, that minimising the total power supplied to a system is very nearly the same as minimising the energy stored in the system, and that this energy can be readily estimated using an array of sensors distributed within the system under control.

1.5. CONTROL OF ENERGY IN FINITE SYSTEMS

One measure of the overall or *global* response of a finite system is the energy stored within it. In this section we discuss the active minimisation of the vibrational energy in a panel and the acoustic energy in an enclosure. Before we consider these specific examples, it is useful to establish the relationship between the energy stored in a system and the power supplied to it. These two quantities are related by the dissipation or damping in the system, since under steady state conditions the power supplied to a system must increase its stored energy until the power lost through dissipation is the same as the power supplied by the sources. We would thus expect that if an active control system were adjusted to minimise the total power supplied to a system, then the energy stored by the system would also be minimised. This turns out to be true in the majority of cases of practical interest, and although the minimisation of total input power and of total stored energy does give slightly different analytical results, the differences are generally of academic rather than practical importance.

1.5.1. Power Input and Total Energy

The connection between input power and stored energy can be illustrated by considering the response of a finite panel to a single point force. This analysis is also used to introduce the modal model for finite systems, and is exactly analogous in the acoustic case. The response of a finite distributed system can always be represented by a summation in terms of its modes (Meirovitch, 1990). The steady state complex velocity distribution on a panel in the x,y plane when excited at a frequency ω can thus be written, using the notation of Fuller et al. (1996), as

$$v(x,y,\omega) = \sum_{n=0}^{\infty} a_n(\omega)\psi_n(x,y) \;, \tag{1.5.1}$$

where $a_n(\omega)$ is amplitude of the n-th mode of vibration, which has a mode shape given by $\psi_n(x,y)$. The mode shapes are orthogonal and will be assumed here to be entirely real and normalised so that

$$\frac{1}{S}\int_S \psi_n(x,y)\psi_m(x,y)\,\mathrm{d}x\,\mathrm{d}y = \begin{cases} 1 \text{ if } n = m \\ 0 \text{ if } n \neq m \end{cases}, \tag{1.5.2}$$

where S is the surface area of the panel. The structural mode shapes for a uniform panel of dimensions L_x by L_y, for example, which is constrained not to have any linear motion at its edges but whose edges can have angular motion, i.e. it is *simply supported*, are given by

$$\psi_n(x,y) = 4\sin\left(\frac{n_1\pi x}{L_x}\right)\sin\left(\frac{n_2\pi y}{L_y}\right), \tag{1.5.3}$$

where n_1 and n_2 are the two modal integers denoted by the single index, n, in equation (1.5.1).

If the panel is excited by a point force, f, applied at position (x_0,y_0) then the mode amplitude can be written as

$$a_n(\omega) = \frac{A_n(\omega)\psi_n(x_0,y_0)}{M}f, \tag{1.5.4}$$

where M is the total mass of the panel. The term $A_n(\omega)$ in equation (1.5.4) denotes the modal resonance term which can be written as

$$A_n(\omega) = \frac{\omega}{B_n\omega + j\left(\omega^2 - \omega_n^{\ 2}\right)}, \tag{1.5.5}$$

where B_n is the modal bandwidth, and ω_n is the natural frequency of the n-th mode. If viscous damping is assumed then $B_n = 2\omega_n\zeta_n$, where ζ_n is the modal damping ratio.

The power supplied to the panel by the point force, f, at (x_0, y_0) can be written, as in equation (1.3.4), as

$$\Pi(\omega) = \frac{1}{2}\operatorname{Re}\left(v^*(x_0,y_0)f\right), \tag{1.5.6}$$

and using the modal expansion for $v(x,y)$ the input power can be expressed as

$$\Pi(\omega) = \frac{|f|^2}{2M}\sum_{n=0}^{\infty}\operatorname{Re}\left(A_n(\omega)\right)\psi_n^2(x_0,y_0). \tag{1.5.7}$$

The *total kinetic energy* stored in the panel is equal to the surface integral of the local panel mass multiplied by the mean-square velocity, which for a uniform panel is equal to

$$E_k(\omega) = \frac{M}{4S}\int_s |v(x,y,\omega)|^2\,dx\,dy. \tag{1.5.8}$$

If the velocity distribution is expressed in terms of the modal expansion, equation (1.5.1), and using the orthonormal properties of the modes, equation (1.5.2), then the total kinetic energy can be written as

$$E_k(\omega) = \frac{M}{4}\sum_{n=0}^{\infty}|a_n(\omega)|^2, \tag{1.5.9}$$

and so is proportional to the sum of the squared mode amplitudes. Using equation (1.5.4) for $a_n(\omega)$, and the fact that the mode resonance term, equation (1.5.5), has the interesting property that

$$|A_n(\omega)|^2 = \frac{\operatorname{Re}\left(A_n(\omega)\right)}{B_n}, \tag{1.5.10}$$

then the total kinetic energy can be written as

$$E_k(\omega) = \frac{|f|^2}{2M} \sum_{n=0}^{\infty} \mathrm{Re}\big(A_n(\omega)\big)\psi_n^2(x_0, y_0)/2B_n \ . \tag{1.5.11}$$

Comparing the terms in equation (1.5.11) for the total kinetic energy with those in equation (1.5.7) for the power supplied to the panel, we can see that they differ by a factor of twice the modal bandwidth. If the panel is lightly damped and excited close to the resonance frequency of the m-th mode, ω_m, then only the m-th term in the modal summations for the energy and power will be significant and in this case we can express the total kinetic energy as

$$E_k(\omega_m) \approx \Pi(\omega_m)/2B_m \ , \tag{1.5.12}$$

so that the stored energy is equal to the supplied power multiplied by a time constant, which depends on the damping in the system. An entirely equivalent result can also be obtained for acoustic systems (Elliott et al., 1991). If the modal bandwidth is similar for a number of modes adjacent to the m-th mode then equation (1.5.12) will continue to be approximately true over a frequency band that includes several modes. It is thus clear that an active control strategy which minimises stored energy will have a similar effect to one which minimises input power.

1.5.2. Control of Vibrational Energy on a Finite Panel

We now consider the effect of minimising the total kinetic energy of a finite sized panel when it is driven at a single frequency. When the panel is excited by a primary force distribution and M secondary point forces, $f_{s1},\ldots,f_{sM}$, then the complex amplitude of the n-th mode at a single frequency can be written as

$$a_n = a_{np} + \sum_{m=1}^{M} B_{nm}\, f_{sm} \ , \tag{1.5.13}$$

where B_{nm} now denotes the coupling coefficient between the m-th actuator and n-th mode, which is proportional to $\psi_n(x,y)$ at the point of application of the m-th secondary force, and a_{np} is the amplitude of the n-th mode due to the primary force.

We now assume that the response of the panel can be accurately represented by the sum over a finite number of modes (N) in equation (1.5.1). Equation (1.5.13) can then be written in vector form as

$$\mathbf{a} = \mathbf{a}_p + \mathbf{B}\mathbf{f}_s \ , \tag{1.5.14}$$

where $\mathbf{a}$ is the vector of complex mode amplitudes of the N modes, $\mathbf{B}$ is the matrix of coupling coefficients, and $\mathbf{a}_p$ is the vector of primary mode amplitudes.

The total structural kinetic energy can now be written, from equation (1.5.9), as

$$E_k = \frac{M}{4}\mathbf{a}^H\mathbf{a} = \frac{M}{4}\left[\mathbf{f}_s^H\mathbf{B}^H\mathbf{B}\mathbf{f}_s + \mathbf{f}_s^H\mathbf{B}^H\mathbf{a}_p + \mathbf{a}_p^H\mathbf{B}\mathbf{f}_s + \mathbf{a}_p^H\mathbf{a}_p\right] , \tag{1.5.15}$$

where superscript H denotes the Hermitian transpose. The kinetic energy E_k is thus a Hermitian quadratic function of the real and imaginary parts of each of the secondary forces, which is guaranteed to have a global minimum, since the matrix $\mathbf{B}^{\mathrm{H}}\mathbf{B}$ must be positive definite. The minimum kinetic energy in the panel is achieved for a secondary force vector given by

$$\mathbf{f}_{\mathrm{s,opt}} = -\left[\mathbf{B}^{\mathrm{H}}\,\mathbf{B}\right]^{-1}\mathbf{B}^{\mathrm{H}}\,\mathbf{a}_{\mathrm{p}} \;, \tag{1.5.16}$$

which results in a minimum value of kinetic energy given by

$$E_{\mathrm{k,min}} = \frac{M}{2}\,\mathbf{a}_{\mathrm{p}}^{\mathrm{H}}\left[\mathbf{I} - \mathbf{B}\left[\mathbf{B}^{\mathrm{H}}\,\mathbf{B}\right]^{-1}\mathbf{B}^{\mathrm{H}}\right]\mathbf{a}_{\mathrm{p}} \;, \tag{1.5.17}$$

where both expressions can be deduced from the general case of the Hermitian quadratic form discussed in the Appendix. Thus, given the primary force distribution, the properties and boundary conditions of the panel, from which the mode shapes and natural frequencies can be calculated, and the positions of the secondary sources, the minimum possible total kinetic energy can be calculated after active control for a set of discrete excitation frequencies.

Figure 1.18 shows the physical arrangement assumed for a computer simulation of active minimisation of total kinetic energy on a panel. The steel panel was assumed to have dimensions 380 mm × 300 mm × 1 mm thick, with a damping ratio of 1% in all modes, and to be simply supported at the edges. The total kinetic energy of the panel due only to the primary point force, f_{p}, positioned at $(x,y) = (342,270)$ mm, is plotted for a range of discrete excitation frequencies as the solid line in Fig. 1.19. The geometry of this example is more complicated than that of the examples previously considered, with several characteristic distances, and so it is not so easy to plot these results in non-dimensional form. The sizes and frequency range have thus been chosen to give an intuitive feel for the range of results. The dotted line in Fig. 1.19 shows the residual level of total kinetic energy after it has been minimised using a single secondary force, f_{s1} at position (38,30) mm in Fig. 1.18, and the dashed–dotted line shows the residual level after minimisation with this and two additional secondary forces, f_{s2} and f_{s3}, at positions (342,30) mm and (38,270) mm in Fig. 1.18. The

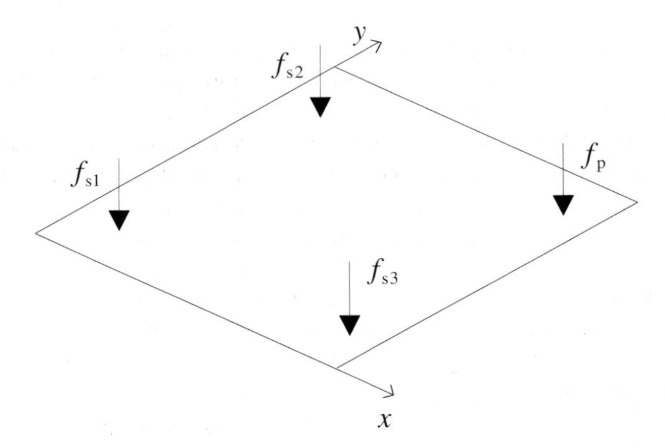

Figure 1.18 Physical arrangement for active control of vibration on a finite plate with a primary point force, f_{p}, near one corner and either a single secondary point force, f_{s1}, near the opposite corner, or three secondary point forces, f_{s1}, f_{s2} and f_{s3} in the other corners.

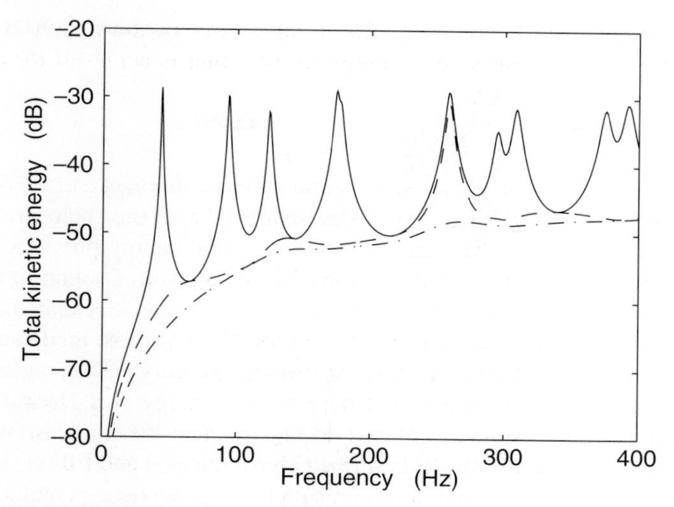

Figure 1.19 Total kinetic energy of vibration on a plate when driven by the primary point force, f_p, alone at discrete frequencies (solid line) and after the total kinetic energy has been minimised by a single secondary force, f_{s1}, optimally adjusted at each excitation frequency (dashed line), or three secondary forces, f_{s1}, f_{s2} and f_{s3}, optimally adjusted at each excitation frequency (dash–dot line).

single secondary force is clearly able to suppress the resonant response of individual modes of the panel below about 200 Hz. Above this frequency, several modes with similar natural frequencies can be simultaneously excited by the primary source, which cannot all be controlled independently by this single primary source, as is the case at 260 Hz for example. The use of three secondary forces solves this problem, since the combination of secondary forces can independently couple into the two modes contributing to the primary response at 260 Hz and thus achieve control. Notice, however, that between the frequencies at which the resonant responses occur, only relatively small reductions in energy can be achieved because of the large number of modes that contribute to the response at these frequencies.

In a practical control system, the total kinetic energy of a structure can be estimated from the outputs of a number of discrete vibration sensors. Although the outputs from these sensors could be manipulated to estimate the amplitudes of a number of structural modes, it is only the sum of the squares of these amplitudes that is important in estimating the total kinetic energy. If a large number of these sensors are reasonably well distributed, the sum of their mean-square output will give a good estimate of the surface integral of the mean-square velocity, which is the more direct measurement of total kinetic energy, as in equation (1.5.8), and this quantity could thus be used as a practical performance measure in such a control system.

1.5.3. Control of Acoustic Energy in an Enclosure

The complex acoustic pressure in an enclosure due to an acoustic source distribution operating at a single frequency can be represented by a modal summation exactly analogous to equation (1.5.1) (Nelson and Elliott, 1992). The *total acoustic potential energy* in such an enclosure is proportional to the space average mean-square pressure and can be written as

$$E_{\mathrm{p}}(\omega) = \frac{1}{4\rho_0 c_0^2} \int_V |p(x, y, z, \omega)|^2 \mathrm{d}V \ , \tag{1.5.18}$$

where ρ_0 and c_0 are the density and speed of sound in the medium, $p(x,y,z,\omega)$ is the complex acoustic pressure at the point (x,y,z) and at the frequency ω in the enclosure, and V is the total volume of the enclosure. The total acoustic potential energy provides a convenient cost function for evaluating the effect of global active control of sound in an enclosure. Because of the assumed orthonormality of the acoustic modes, E_{p} can again be shown to be proportional to the sum of the squared mode amplitudes, and these mode amplitudes can again be expressed in terms of the contribution from the primary and secondary sources, as in equation (1.5.14). Thus the total acoustic potential energy is a Hermitian quadratic function of the complex strengths of the secondary acoustic sources, which can be minimised in exactly the same way as described above (Nelson and Elliott, 1992).

A simulation has been carried out of minimising the total acoustic potential energy in an enclosure of dimensions $1.9 \times 1.1 \times 1.0$ m, as illustrated in Fig. 1.20, in which the acoustic modes have an assumed damping ratio of 10%, which is fairly typical for a reasonably well-damped acoustic enclosure such as a car interior at low frequencies. The acoustic mode shapes in a rigid-walled rectangular enclosure are proportional to the product of three cosine functions in the three dimensions. The lowest order mode, for the zeroth order cosine functions, has a uniform mode amplitude throughout the enclosure and corresponds to a uniform compression of the air at all points. The mode with the next highest natural frequency corresponds to fitting a half wavelength into the longest dimension of the enclosure, and this first longitudinal mode has a natural frequency of about 90 Hz for the enclosure shown in Fig. 1.20, whose size is similar to the interior of a small car.

Figure 1.21 shows the total acoustic potential energy in the enclosure when driven only by the primary source placed in one corner of the enclosure and when the total acoustic potential energy is minimised by a single secondary acoustic source in the opposite corner (dashed line) or by seven secondary acoustic sources positioned at each of the corners of the enclosure not occupied by the primary source (dash–dot line). Significant attenuations in the total acoustic potential energy are achieved with a single secondary source at excitation

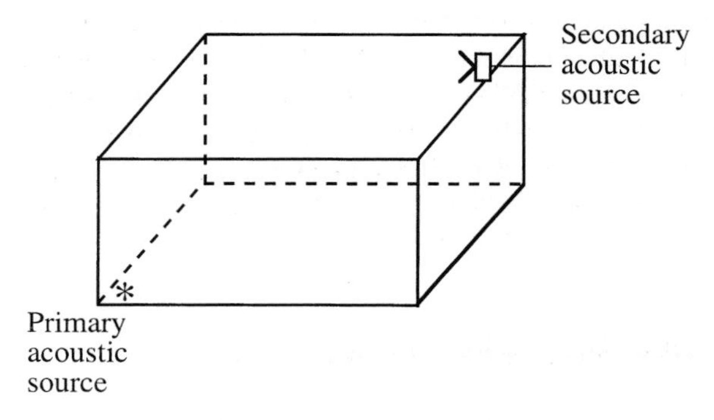

Figure 1.20 Physical arrangement for the simulation of active control of tonal sound in a rectangular enclosure, which is about the size of a car interior, excited by a primary acoustic source in one corner and a secondary acoustic source in the opposite corner.

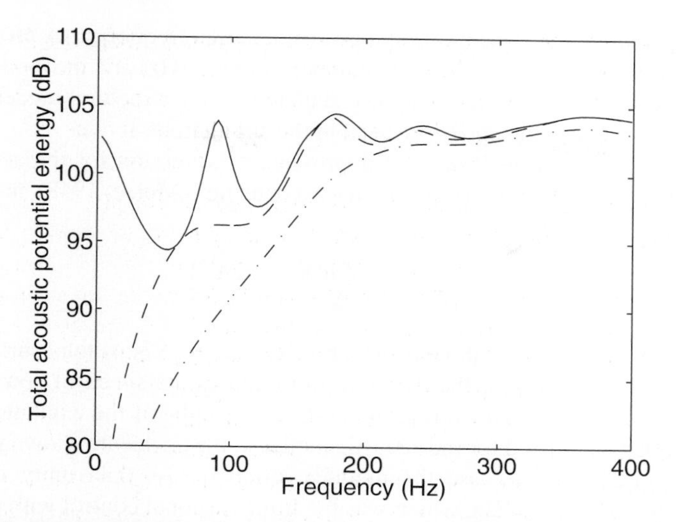

Figure 1.21 Total acoustic potential energy in the enclosure when driven by the primary acoustic source alone at discrete frequencies (solid line) and when the total potential energy has been minimised using either the single secondary source shown in Fig. 1.20, optimally adjusted at each excitation frequency (dashed line), or seven secondary acoustic sources placed in all the corners of the enclosure not occupied by the primary source (dash–dot line).

frequencies below about 20 Hz, where only the zeroth order acoustic mode is significantly excited, and for excitation frequencies close to the natural frequency of the first longitudinal mode at about 90 Hz.

The response of the system does not, however, show clear modal behaviour for excitation frequencies above about 150 Hz, and very little attenuation can be achieved with a single secondary source above this frequency. This is because the spacing of the natural frequencies of the acoustic modes in a three-dimensional enclosure becomes closer together with higher mode order. Even introducing seven secondary sources into the enclosure does not allow global control to be maintained at frequencies above about 250 Hz in this case.

The amplitude of an individual mode of a physical system could be completely controlled using a single secondary source, provided it was not placed on a nodal line. In controlling this mode, however, the secondary source will tend to increase the excitation of other modes of the system. The minimisation of total energy generally involves a balance between cancelling the dominant modes and not overly exciting the other, residual, modes of the system. This balance is automatically maintained when the total energy in the system is minimised. The attenuation which can be obtained in the energy at any one excitation frequency will generally depend on the number of modes that significantly contribute to the response.

1.5.4. The Effect of the Modal Overlap

The number of modes that are significantly excited in a system at any one excitation frequency can be quantified by a dimensionless parameter known as the *modal overlap*, $M(\omega)$. This is defined to be the average number of modes whose natural frequencies fall within the bandwidth of any one mode at a given excitation frequency, ω. $M(\omega)$ is equal to the product of the modal density (average number of modes/Hz) and the modal bandwidth (in Hz), and both of these quantities can be calculated for the structural modes in the panel and the acoustic modes in the enclosure used in the simulations above.

For a three-dimensional enclosure, an approximate expression for the acoustic modal overlap can be calculated from that for the modal densities (Morse, 1948), and is given by (Nelson and Elliott, 1992)

$$M(\omega) = \frac{2\zeta\omega L}{\pi c_0} + \frac{\zeta\omega^2 S}{\pi c_0^2} + \frac{\zeta\omega^3 V}{\pi^2 c_0^3}, \qquad (1.5.19)$$

where L is the sum of the linear dimensions of the enclosure, S is its total surface area, V is the volume of the enclosure, ζ the damping ratio and c_0 the speed of sound. At higher frequencies the acoustic modal overlap increases as the cube of the excitation frequency. The modal overlap calculated for the acoustic modes in the enclosure shown in Fig. 1.20 is plotted in Fig. 1.22. In this case the modal overlap is less than unity for excitation frequencies below about 150 Hz, which was the limit of global control with one source in Fig. 1.21, and is under seven for excitation frequencies below about 250 Hz, which was the limit of global control with seven sources in Fig. 1.21.

For the panel, the structural modal density is independent of frequency but the modal bandwidth increases proportionally to frequency if a constant damping ratio, ζ, is assumed. The modal overlap in this case is approximately given by

$$M(\omega) = S\zeta\left(\frac{m}{D}\right)^{1/2}\omega, \qquad (1.5.20)$$

where S is the area of the panel, ζ the modal damping ratio, D its bending stiffness, and m its mass per unit area. The modal overlap for the panel described above is plotted as a

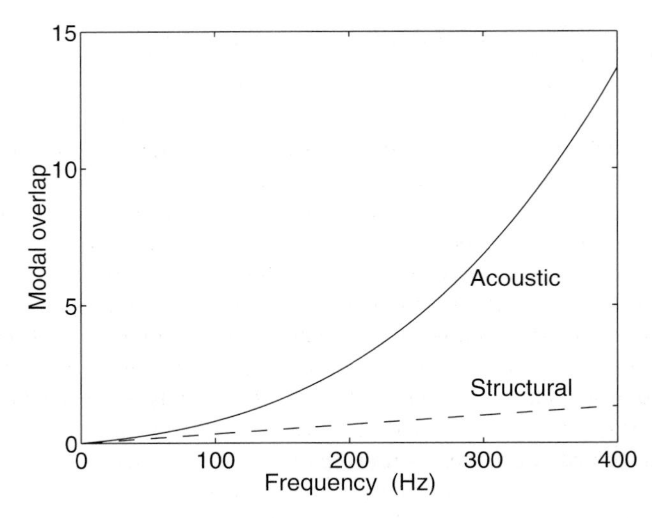

Figure 1.22 The modal overlap, $M(\omega)$, for the structural modes on the plate used for the simulations shown in Fig. 1.19 (dashed line), and for the acoustic modes in the enclosure used for the simulations shown in Fig. 1.21 (solid line).

dashed line in Fig. 1.22 as a function of frequency. The modal overlap in this case increases linearly with frequency and becomes greater than one at about 250 Hz, which is the frequency at which control of the energy in the panel cannot be obtained with a single secondary force because more than one mode is significantly excited.

The modal overlap can thus be seen to be a useful method of characterising the complexity of the modal structure in a system at a given excitation frequency. It can also be used as a very approximate guide to the number of secondary sources required to achieve a given level of global control, i.e. a given reduction in energy level in a system. The difference between the behaviour of the modal overlap with frequency in a typical system undergoing structural vibration, and in an enclosure excited acoustically, explains the very significant difference in the physical limitations of active control in these two cases.

The total acoustic potential energy in an enclosure is proportional to the volume integral of the mean-square pressure, equation (1.5.18). In a practical active control system, the total potential energy can be estimated using the sum of the squared outputs of a number of pressure microphones. The number of microphones required to obtain an accurate estimate of the total acoustic potential energy is proportional to the number of significantly excited acoustic modes within the enclosure and thus increases sharply at higher excitation frequencies. Microphones can be placed at a smaller number of locations if they simultaneously measure pressure and pressure gradient in the three directions (Sommerfeldt and Nasif, 1994). The active control of sound in the passenger cabin of a propeller aircraft is used as an example of the application of a multichannel system for tonal noise in Chapter 4, and in Chapter 5 the active control of road noise inside a car is used as an example of the application of a multichannel system for the control of random noise. If control over only a part of the enclosure is required, then the microphones can be concentrated in this region. The zone of quiet generated in the limiting case where only one microphone is used is discussed in Section 1.7.

1.6. CONTROL OF SOUND RADIATION FROM STRUCTURES

So far in this chapter we have considered the active control of sound radiation using acoustic sources and the active control of vibration using structural actuators. In a number of practical applications the main objective of using structural actuators is not to control the vibration of a structure but to control the sound that it radiates. If the structural actuators could completely cancel the velocity of the structure then the radiated sound would also be zero, and the objectives of active sound control and active vibration control would be identical. In practice, however, the vibration can often only be reduced by the action of the secondary actuators, and minimising the total kinetic energy of a structure does not generally result in the minimisation of radiated sound. In fact a reduction of vibration may be accompanied by an increase in sound radiation, as originally pointed out by Knyazev and Tartakovskii (1967). The use of secondary vibration actuators on a structure to reduce the sound radiation has been extensively studied by Fuller and his co-workers (1985, 1988) and has been termed *active structural acoustic control* (ASAC). The approach can be applied to both interior sound fields, e.g. inside an enclosure driven by a vibration panel, or exterior sound fields, e.g. the sound radiated into an infinite space by a vibrating panel. It is the latter problem that we will concentrate on here to demonstrate the principles of this approach.

1.6.1. Sound Radiation from a Vibrating Panel

In order to illustrate the difference between controlling vibration and sound power radiation, we return to the modal description of the vibration of a panel, equation (1.5.1), and the expression for the total kinetic energy of the panel (equation 1.5.15) in terms of the vector of amplitudes of the N significant structural modes,

$$E_{\mathrm{k}} = \frac{M}{4} \mathbf{a}^{\mathrm{H}} \mathbf{a} , \qquad (1.6.1)$$

where M is the mass of the panel. We have seen in Section 1.5 that if the panel is excited by a primary excitation which generates the vector of mode amplitudes $\mathbf{a}_{\mathrm{p}}$ and these modes are also driven by a set of secondary forces $\mathbf{f}_{\mathrm{s}}$ via modal coupling matrix $\mathbf{B}$, so that

$$\mathbf{a} = \mathbf{a}_{\mathrm{p}} + \mathbf{B}\mathbf{f}_{\mathrm{s}} , \qquad (1.6.2)$$

then the total kinetic energy is minimised by the set of secondary forces given by

$$\mathbf{f}_{\mathrm{s,opt}:E_{\mathrm{k}}} = -\left[\mathbf{B}^{\mathrm{H}}\mathbf{B}\right]^{-1}\mathbf{B}^{\mathrm{H}}\mathbf{a}_{\mathrm{p}} . \qquad (1.6.3)$$

The sound power radiated by the panel is a more complicated function of the structural mode amplitudes than the total kinetic energy, however. If one mode is vibrating on its own, the ratio of the radiated sound power to the mean-square velocity of the vibration on the panel, which is proportional to the *self radiation efficiency* of the mode (Wallace, 1972; Fuller et al., 1996), is a function of both the mode shape and the excitation frequency. When the velocity of the panel is due to the excitation of more than a single mode, however, the sound fields due to each of the modes will generally interact. The radiated sound power is then related to the amplitudes of the structural modes by the expression

$$\Pi_{\mathrm{R}} = \sum_{i=1}^{N}\sum_{j=1}^{N} M_{ij}\, a_i^{*}\, a_j , \qquad (1.6.4)$$

where M_{ij} are entirely real frequency-dependent numbers (Fuller et al., 1996), and so the radiated power can be written in matrix form as

$$\Pi_{\mathrm{R}} = \mathbf{a}^{\mathrm{H}}\mathbf{M}\mathbf{a} , \qquad (1.6.5)$$

where $\mathbf{M}$ should not be confused with the mobility matrix used in Section 1.3. The diagonal terms in the matrix $\mathbf{M}$ are proportional to the *self radiation efficiencies* of the structural modes, and the off-diagonal terms are proportional to the *mutual radiation efficiencies*, or combined mode radiation efficiency, which account for the interaction of the sound fields generated by different structural modes. The radiation efficiency is a dimensionless quantity which is equal to the ratio of radiated sound power of the panel to the sound power radiated by an equal area of infinite rigid panel having the same mean-square velocity.

If we consider the modes of a simply supported panel, for example, the mode shapes have the particularly simple form

$$\psi_n(x,y) = 4\sin\left(n_1\pi x/L_x\right)\sin\left(n_2\pi y/L_y\right) , \qquad (1.6.6)$$

where the factor of 4 is chosen so that the mode satisfies the orthonormality condition, equation (1.5.2), L_x and L_y are the dimensions of the panel in the x and y directions and n_1 and n_2 are the modal integers. The structural mode with modal integers $n_1 = 2$ and $n_2 = 1$, for example, is commonly referred to as the (2,1) mode. The shapes of some of these modes on a simply supported panel are shown in Fig. 1.23.

The self and mutual radiation efficiencies of the (1,1), (2,1) and (3,1) structural modes, for a panel with $L_y/L_x \approx 0.79$, mounted in an infinite baffle, are shown in Fig. 1.24 as a function of normalised excitation frequency, L_x/λ. The self radiation efficiency of the (1,1) structural mode, for example, is denoted $\sigma_{11,11}$ and the mutual radiation efficiency of the (1,1) structural mode and the (3,1) structural mode is denoted $\sigma_{11,31}$. It can be seen that all the modes considered radiate with similar efficiency if $L_x \geq \lambda$, and there is a decreasing amount of interaction between the modes as the excitation frequency is increased. In the low frequency region, however, when the size of the panel is small compared with the acoustic wavelength, the (1,1) mode and the (3,1) mode radiate considerably more efficiently than the (2,1) mode, but also there is considerable interaction between the sound radiation due to the (1,1) and the (3,1) modes. The important difference between the (1,1) and (3,1) modes and the (2,1) mode is that the latter does not have any net volumetric component. When the panel is vibrating in the (2,1) mode at low frequencies, most of the air in front of the panel is transferred from one side of the panel to the other as it vibrates, generating very little sound. The modes with an (odd, odd) mode number, however, have a space-average volumetric component and so displace fluid into the medium as they vibrate. It is this

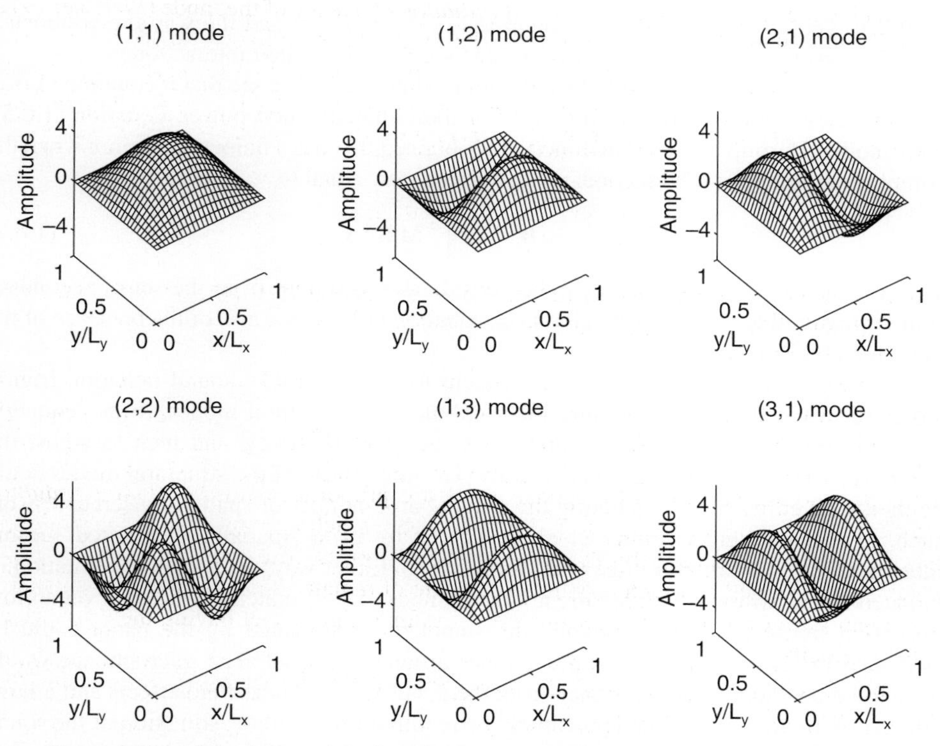

Figure 1.23 Mode shapes of the first six structural modes of a simply supported panel.

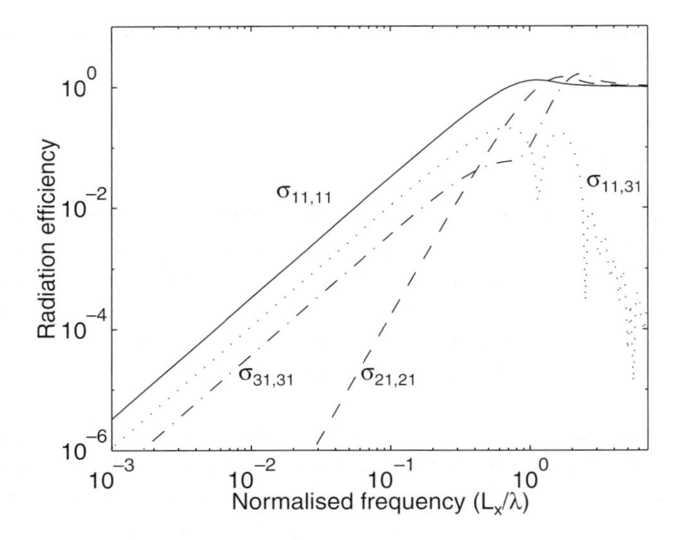

Figure 1.24 Sound radiation efficiencies of the modes of a simply supported panel with $L_y/L_x \approx 0.79$ when radiating into free space. $\sigma_{11,11}$, $\sigma_{21,21}$ and $\sigma_{31,31}$ are the self radiation efficiencies of the modes vibrating alone and $\sigma_{11,31}$ is the mutual radiation efficiency of these two modes vibrating together.

volumetric component that creates sound at low frequencies, and the way the volumetric contributions from the (odd, odd) modes add together causes their interaction.

To return to the active control of such sound radiation, we can see that if equation (1.6.2) is substituted into the matrix expression for the radiated sound power, equation (1.6.5), then another Hermitian quadratic function is obtained that has a unique minimum when the complex amplitudes of the secondary sources are now equal to

$$\mathbf{f}_{\mathrm{s,opt},\Pi_\mathrm{R}} = -\left[\mathbf{B}^\mathrm{H}\mathbf{M}\mathbf{B}\right]^{-1}\mathbf{M}\mathbf{B}^\mathrm{H}\mathbf{a}_\mathrm{p} \; . \tag{1.6.7}$$

This vector of secondary forces can be considerably different from the set of secondary sources required to minimise the vibration, equation (1.6.3), because of the presence of the radiation efficiency matrix, **M**.

One way of implementing an active system for the control of sound radiation from a structure would thus be to measure the amplitudes of structural modes, to use equation (1.6.5) to estimate the radiated sound power at each frequency, and then to adjust the secondary sources to minimise this quantity. The amplitudes of the structural modes could be measured either by manipulating the outputs of a number of spatially discrete sensors such as accelerometers (Fuller et al., 1996) or by using spatially distributed sensors integrated into the structure (Lee and Moon, 1990). Either way, the accurate estimation of radiated sound power would require a large number of well-matched sensors, even at low excitation frequencies. Alternatively, the sound power radiated by the panel could be estimated by using an array of microphones around the panel. The microphones would have to be positioned some distance from the panel to avoid near-field effects and a large number of microphones would potentially be required for accurate estimation of the sound power. A more efficient implementation would involve controlling the output of a number of structural sensors that measure only the velocity distributions on the panel that are efficient at radiating sound.

1.6.2. Radiation Modes

It is possible to transform the problem of actively controlling sound radiation into a form that gives a clearer physical insight into the mechanisms of control, and also suggests a particularly efficient implementation for an active control system, particularly at low frequencies. This transformation involves an eigenvalue/eigenvector decomposition of the matrix of radiation efficiencies, which can be written as

$$\mathbf{M} = \mathbf{P}\boldsymbol{\Omega}\mathbf{P}^{\mathrm{T}} , \tag{1.6.8}$$

where $\mathbf{P}$ is an orthogonal matrix of real eigenvectors, since $\mathbf{M}$ has real elements and is symmetric, and $\boldsymbol{\Omega}$ is a diagonal matrix of eigenvalues, which are all real and positive, since $\mathbf{M}$ is positive definite. The radiated sound power, equation (1.6.5) can thus be written as

$$\Pi_{\mathrm{R}} = \mathbf{a}^{\mathrm{H}}\mathbf{M}\mathbf{a} = \mathbf{a}^{\mathrm{H}}\mathbf{P}\boldsymbol{\Omega}\mathbf{P}^{\mathrm{T}}\mathbf{a} , \tag{1.6.9}$$

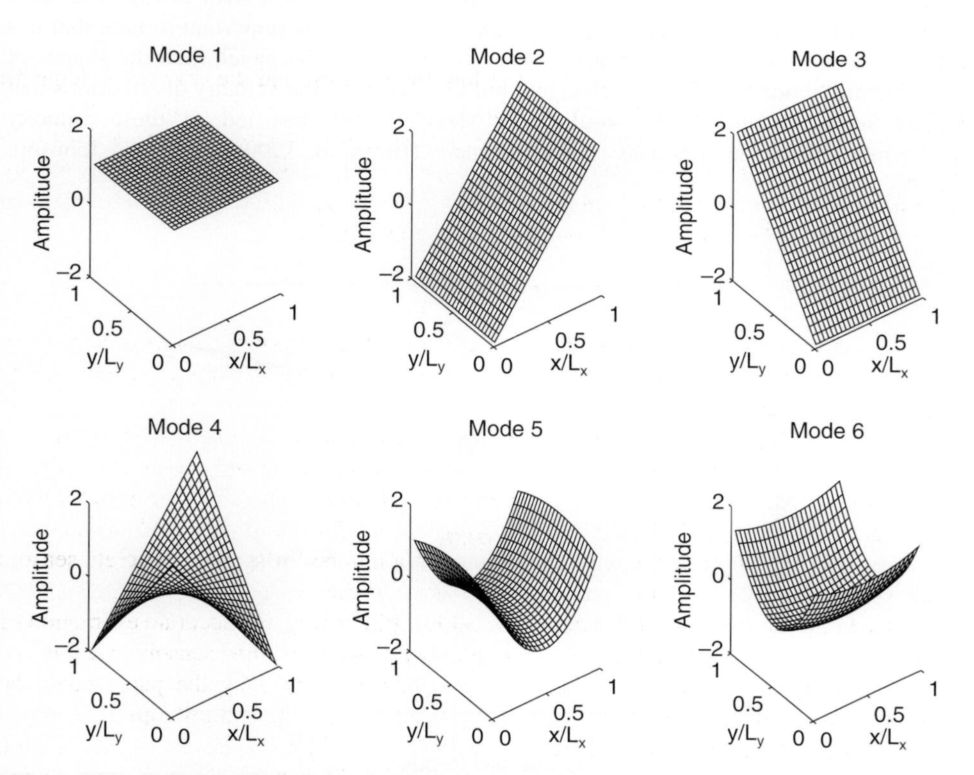

Figure 1.25 Velocity distributions that radiate sound independently and thus correspond to the shapes of the first six radiation modes.

and if we define a vector of transformed mode amplitudes to be $\mathbf{b} = \mathbf{P}^{\mathrm{T}}\mathbf{a}$, the radiated sound power can be written as

$$\Pi_{\mathrm{R}} = \mathbf{b}^{\mathrm{H}}\boldsymbol{\Omega}\,\mathbf{b} = \sum_{n-1}^{N}\Omega_n\left|b_n\right|^2 . \tag{1.6.10}$$

These transformed modes thus have no mutual radiation terms and hence radiate sound *independently* of each other (Borgiotti, 1990; Photiadis, 1990: Cunefare, 1991; Baumann et al., 1991; Elliott and Johnson, 1993).

The velocity distributions that correspond to this transformed set of modes are shown in Fig. 1.25 for a panel of the size used to calculate the structural modes shown in Fig. 1.23 (Elliott and Johnson, 1993). The radiation efficiencies of these velocity distributions are shown in Fig. 1.26. The velocity distributions shown in Fig. 1.25 were calculated for an excitation frequency corresponding to $L_x \approx \lambda/5$, but these velocity distributions vary to only a small extent with frequency provided $L_x \leq \lambda$. The radiation efficiencies for all of these velocity distributions are of similar magnitudes if $L_x \geq \lambda$, but not at low frequencies, when the size of the panel is small compared with the acoustic wavelength; then the first velocity distribution radiates far more efficiently than any of the others. All of the radiation efficiencies shown in Fig. 1.25 are self radiation efficiencies since, by definition, the mutual radiation efficiency terms in this case are zero.

Although it is possible to express the velocity distributions shown in Fig. 1.25 as a function of the mode shapes of a simply-supported panel, it is important to note that if a different set of boundary conditions had been assumed for the panel, then the shapes of the structural modes would have changed, but the shapes of the velocity distributions that radiate sound independently would not. This property has led to these velocity distributions being described as *radiation modes* (Borgiotti, 1990; Elliott and Johnson, 1993).

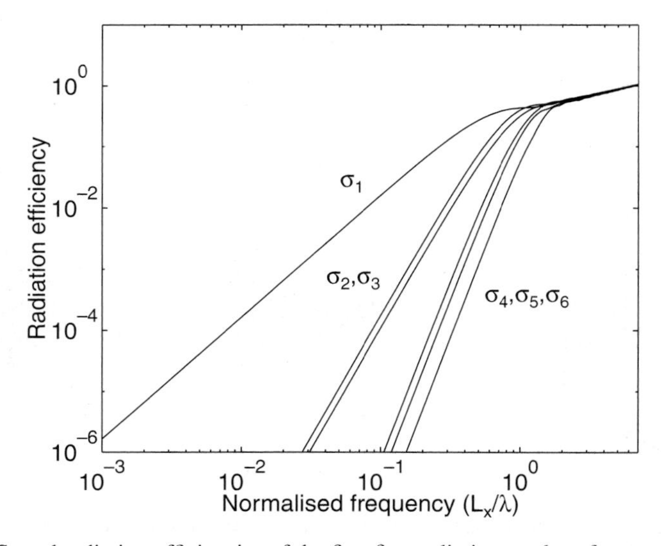

Figure 1.26 Sound radiation efficiencies of the first five radiation modes of a rectangular surface.

1.6.3. Cancellation of Volume Velocity

It is particularly interesting to note the simple form of the velocity distribution corresponding to the first radiation mode in Fig. 1.25. The amplitude of this radiation mode at low excitation frequencies is equal to the surface integral of the complex velocity over the whole of the radiating surface, i.e. the net *volume velocity* of the surface. The fact that this is such an efficiently radiating mode at low frequencies suggests that significant reductions in the sound power radiated from a surface can be achieved by the active control of this single quantity. The net volume velocity of a clamped panel can be directly measured using a single distributed sensor (Rex and Elliott, 1992; Guigou and Berry, 1993; Johnson and Elliott, 1993, 1995a), so that it is not necessary to measure the amplitudes of all the individual structural modes.

As an example of the effect of controlling the net volume velocity on the sound radiation from a panel, Johnson and Elliott (1995b) have presented simulation results for the arrangement shown in Fig. 1.27, in which an incident plane acoustic wave excites the vibration of a baffled panel, which then radiates sound on the other side. The panel was made of aluminium, had dimensions 380 mm × 300 mm × 1 mm with a damping ratio of 0.2%, and was excited by a plane acoustic wave incident at an angle of $\theta = 45°$, $\phi = 45°$. Figure 1.28 shows the ratio of the sound power radiated by the panel to the incident sound power, which is equal to the *power transmission ratio*, and is the inverse of the transmission loss. The power transmission ratio with no control is shown as the solid line in Fig. 1.28. This is large when the panel is excited near the natural frequencies of its (odd, odd) modes, i.e. about 40 Hz, 180 Hz, 280 Hz, etc., since the panel then vibrates strongly and these modes radiate sound efficiently. The volume velocity of the panel is then cancelled at each frequency by the action of a piezoceramic actuator mounted on the panel as a secondary source, and the power transmission ratio is again calculated as shown by the dashed–dotted

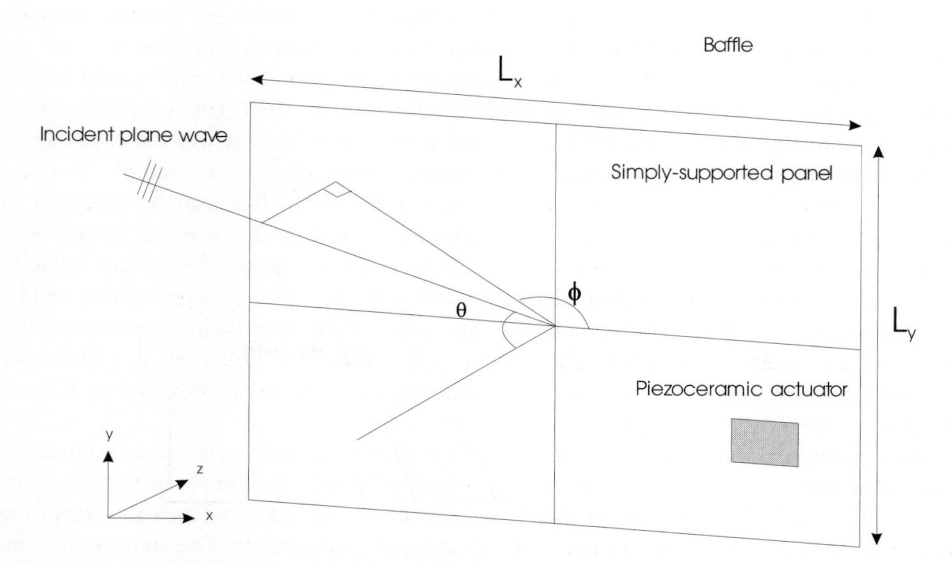

Figure 1.27 Arrangement used for the calculation of the effect of volume velocity cancellation on the sound transmitted through a rectangular panel mounted in a rigid wall.

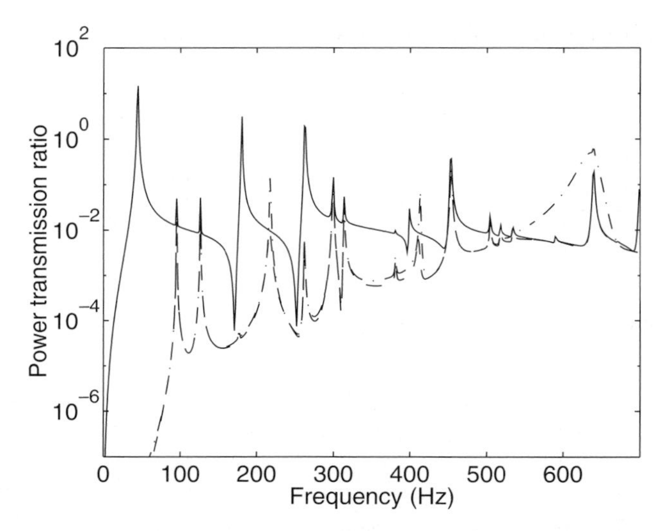

Figure 1.28 The ratio of the transmitted sound power to that incident on a panel when excited by an incident plane wave as a function of excitation frequency without control (solid line), after minimisation of the radiated sound power (dashed line), and after cancellation of the volume velocity (dash–dot line).

line in Fig. 1.28. The sound power radiated by the panel has been significantly reduced by the cancellation of its volume velocity at excitation frequencies below about 400 Hz, for which the largest dimension of the panel is less than half an acoustic wavelength. The residual levels of sound power that are radiated after control are due to the weakly radiating structural modes, which do not contribute to the volume velocity. At some higher frequencies, however, the sound radiation is slightly increased by volume velocity cancellation because the amplitude of these weakly radiating modes is increased.

The effect of using the piezoceramic actuator to minimise the total sound power radiated by the panel is also plotted as the dashed line in Fig. 1.28. This would be a very difficult control problem to implement in practice since it would involve the use of large numbers of microphones to estimate the radiated sound power, for example. It does, however, provide a benchmark for the best possible results that can be obtained in controlling sound transmission with this actuator. Apart from a few isolated frequencies, it can be seen that below about 400 Hz the reduction of sound power by volume velocity cancellation is almost as great as the best possible reduction of sound power that can be obtained using this piezoceramic actuator. The mechanism by which the piezoceramic actuators can generate vibration on a panel is described by Fuller et al. (1996) for example, and a more detailed review of both actuators and sensors is provided by Hansen and Snyder (1997).

The control of sound radiation using structural actuators can thus be seen to require a rather different strategy from the control of vibration. Relatively simple control systems can be effective in reducing sound radiation from structures at low frequencies when the size of the structure is small compared with the acoustic wavelength. The control problem becomes progressively harder as the frequency increases, again setting an upper limit on the frequency range of a practical control system.

1.7. LOCAL CONTROL OF SOUND AND VIBRATION

Apart from minimising a global cost function, such as radiated power or total energy, an active control system can also be designed to minimise the local response of a system. The physical effect of such a local control system will be illustrated in this section by considering the cancellation of the vibration at a point on a plate and the cancellation of the pressure at a point in a room.

1.7.1. Cancellation of Vibration on a Large Panel

The sinusoidal velocity at one point on a panel can be cancelled with a single secondary force, using the arrangement shown in Fig. 1.8 for example, where $r = L$ at the point of cancellation. The shaded area in Fig. 1.29 shows the region within which an average of 10 dB vibration reduction is achieved on a large panel, when a single secondary force, at the origin, is adjusted to cancel the velocity at a point denoted by the cross (Garcia-Bonito and Elliott, 1996). The primary disturbance is assumed to be *diffuse*, i.e. to be made up of waves with random amplitudes and phases arriving from all directions on the panel. The cancellation of velocity produces a disc-shaped *zone of quiet* round the cancellation point. The diameter of the disc within which the velocity has been reduced by at least 10 dB is about $\lambda_f/10$, where λ_f is the flexural wavelength on the panel. The secondary force is thus within this disc if its separation from the cancellation point is less than $\lambda_f/10$, as in Fig. 1.29(a), and the secondary force is outside the disc if its separation from the cancellation point is greater than $\lambda_f/10$, as in Fig. 1.29(b). If the primary disturbance were perfectly diffuse, the contours of equal attenuation would be symmetrical about the y axis,

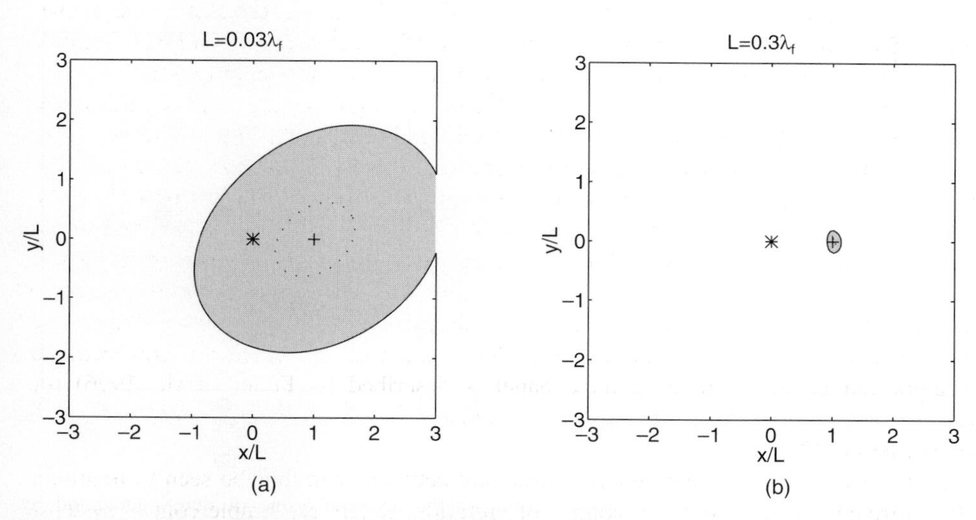

Figure 1.29 The spatial extent of the vibrational 'zone of quiet' generated by cancelling the out-of-plane velocity at $x = L$ on an infinite plane with a point force at the origin of the coordinate system for (a) $L = 0.03\lambda_f$ and (b) $L = 0.3\lambda_f$, where λ_f is the flexural wavelength. The shaded area within the solid line corresponds to a 10 dB attenuation in the diffuse primary field and the dotted line to a 20 dB attenuation.

and the slight asymmetry in Fig. 1.29 is due to the finite number of waves used to approximate a diffuse field in this simulation.

1.7.2. Cancellation of Pressure in a Large Room

Cross sections through the corresponding acoustic zones of quiet, generated by cancelling the pressure at a point in a large three-dimensional space, are shown in Fig. 1.30. The soundfield in the room is again assumed to be diffuse, which occurs when the excitation frequency is such that the modal overlap in equation (1.5.19) is well above one. In this case, when the distance from the secondary source to the cancellation point is very small compared with the acoustic wavelength, λ, then the zone of quiet forms a *shell* around the secondary source, as indicated by the shaded area in the two-dimensional cross section shown in Fig. 1.30(a). The difference between the acoustic and structural case at low frequencies, Figs. 1.29(a) and 1.30(a), is due to the fact that an acoustic monopole source has a very intense near-field, within which the pressure is proportional to the reciprocal of the distance from the source and thus rises dramatically at small distances from the source. The effect of a point force on an infinite panel, however, is to produce a 'dimple' within which the velocity is a smoothly-varying function of position, with no discontinuities. This contrast is illustrated in Fig. 1.31 (Garcia-Bonito and Elliott, 1996). If the velocity is cancelled at a point in the near-field of a point force on a panel at low frequencies, both the primary and secondary fields will be uniform enough for there to be significant attenuation over a whole disc around the secondary force. If the pressure is cancelled in the near-field of an acoustic monopole source, however, the secondary field will be equal and opposite to the primary field only at distances from the secondary source that are about the same as the

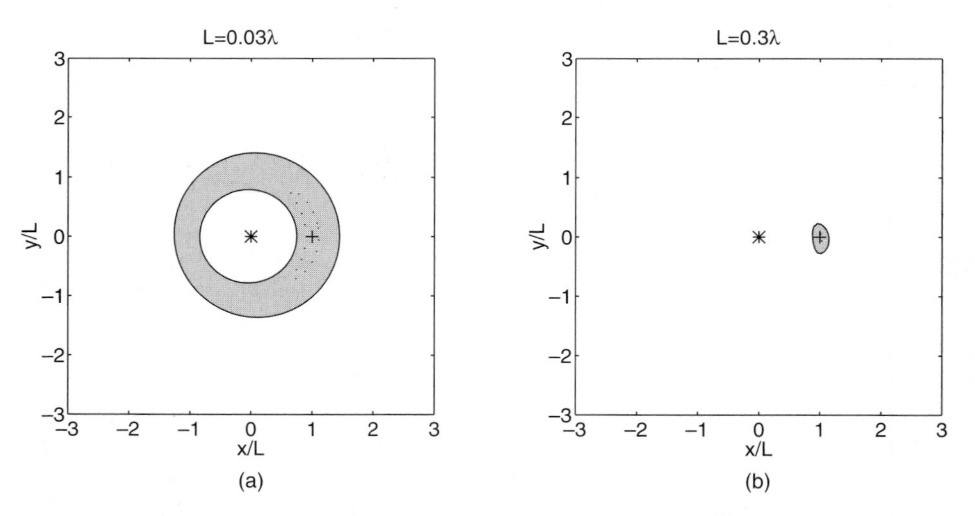

Figure 1.30 Cross section of the spatial extent of the acoustic 'zone of quiet' generated by cancelling the pressure at $x = L$ in a three-dimensional free field using a point monopole acoustic secondary source at the origin of the coordinate system for (a) $L = 0.03\lambda$ and (b) $L = 0.3\lambda$, where λ is the acoustic wavelength. The shaded area within the solid line corresponds to a 10 dB attenuation in the diffuse primary field, and the dotted line to a 20 dB attenuation.

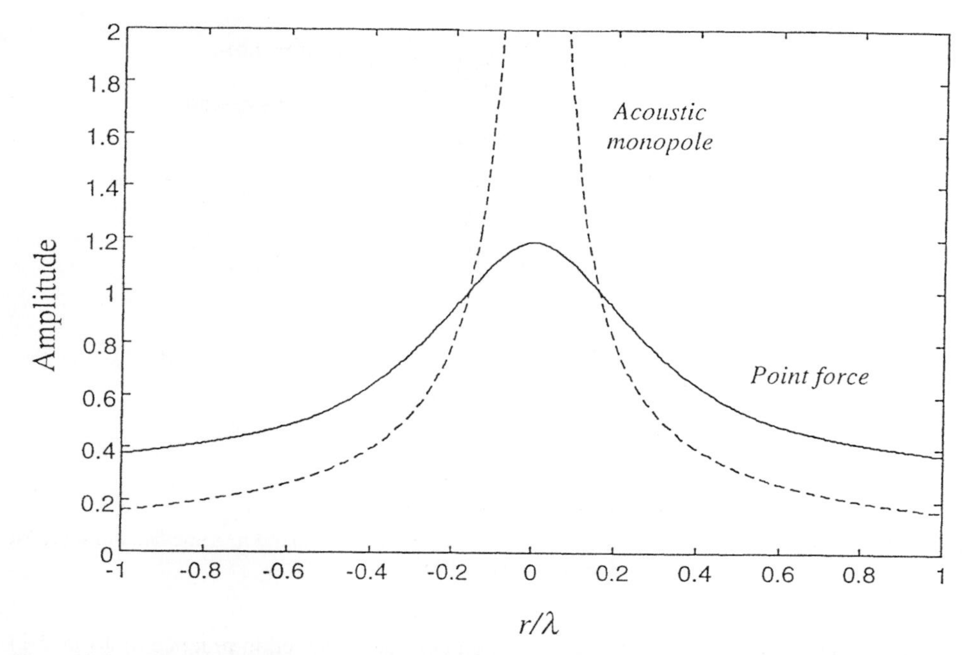

Figure 1.31 The near-field velocity due to a point force acting on an infinite plate (solid line) and the near-field pressure due to an acoustic monopole source in free space (dashed line).

distance to the cancellation point, thus generating a shell of cancellation. At higher frequencies in the acoustic case, when the distance from the secondary source to the cancellation point is not small compared with the wavelength, then the zone of quiet does not form a complete shell round the secondary source but is now concentrated in a sphere centred on the cancellation point, whose diameter is about $\lambda/10$ (Elliott et al., 1988), as shown in Fig. 1.30(b).

The advantage of a local control system is that the secondary source does not have to drive very hard to achieve control, because it is very well coupled to the response at the cancellation point. Thus local zones of quiet can often be generated without significantly affecting the overall energy in the system. Local active control systems also have the advantage that, because the secondary actuator and the error sensor are close together, there is relatively little delay between them, which can improve the performance of both feedforward and feedback control systems. One of the earliest designs for a local active control system was put forward by Olsen and May (1953), who suggested that a loudspeaker on the back of a seat could be used to generate a zone of quiet round the head of a passenger in a car or an aircraft, as shown in Fig. 1.32. Olson and May had the idea of using a feedback control system to practically implement such an arrangement, although in the 1950s most of the design effort had to be spent in reducing the phase lag in the audio amplifier to ensure stability of the feedback loop. More recent investigations (Rafaely et al., 1999) have demonstrated the trade-off in such a system between good acoustic performance and robust stability of the feedback loop and this will be discussed in more detail in Chapter 6. The upper frequency of control in such a system will be fundamentally determined by the acoustic considerations outlined above, and the extent of the listener's

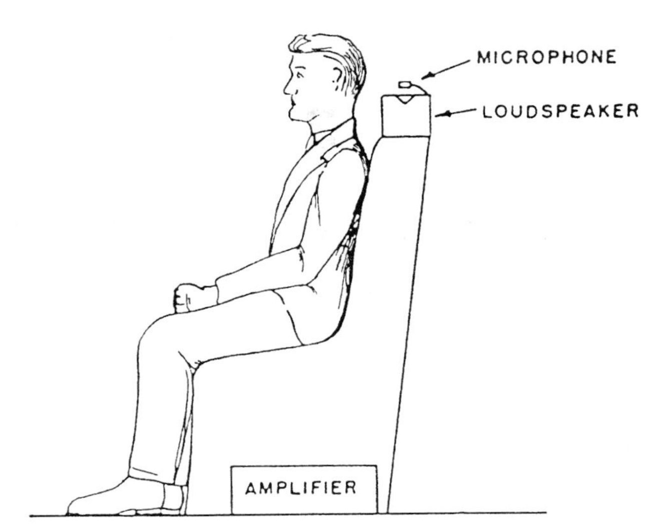

Figure 1.32 Illustration of a local system for the active control of sound near the head of a seated passenger, taken from Olsen and May, 1953.

head movements. Practically sized zones of quiet can be achieved in practice up to several hundred hertz. In an active headset, loudspeakers are fitted inside the headset and are thus kept close to the ears as the wearer's head moves. The performance of active headsets will be discussed in Chapter 7, but it is clear that the acoustic limitations of such a system will not occur until a higher frequency than those of the system shown in Fig. 1.32, and do not in practice occur until about 1 kHz.

In this chapter, the physical principles of active control have been briefly introduced in order to illustrate the fundamental limitations to its use. Examples have been taken of the active control of sound in ducts and in enclosures together with the local control of sound, which could be used in headrests and headsets, all of which will be returned to as practical applications in later chapters. Before concentrating on the main chapters of the book, however, which are concerned with algorithms for the practical realisation of active control, Chapter 2 reviews some background material on digital filters, which introduces most of the important analysis techniques and notation that will be used in subsequent chapters.

2

Optimal and Adaptive Digital Filters

2.1. INTRODUCTION

In this chapter some of the important properties of optimal and adaptive digital filters are introduced. Digital filters operate on *sampled* and *quantised data*. A discussion of the effects of quantisation is left until Chapter 10. In contrast to the previous chapter, in which the variables were functions of a continuous time variable, t, the time variable in this chapter will be assumed discrete and to represent the sampled version of the physical signals discussed in Chapter 1. The *sampling time* is assumed to be equal to T, so that the sampling frequency is $1/T$. The continuous time signals are sampled at $t = nT$ where the variable n can only take integer values. Although, strictly speaking, the sampled version of the continuous time signal $x(t)$ should thus be denoted as $x(nT)$ we will use the simplified form $x(n)$ for notational convenience. We will also assume that sequences of time histories are entirely real.

The process of controlling a continuous-time physical system with a sampled data controller has to be approached with some care. This point is described in some detail by, for example, Kuo (1980), Åström and Wittenmark (1997) and Franklin et al. (1990), but we will postpone a detailed discussion of the physical significance of the sampled signals until later chapters and concentrate on the manipulation of such signals by digital filters. We will concentrate on digital filters that behave linearly when they are not being adapted, although the adaptation process itself will generally introduce some nonlinear behaviour. Digital filters that are designed to have a nonlinear steady-state response are described in Chapter 8. Adaptive digital filters are often used to form an internal model of the response of the physical system under control in many control systems, and the direct application of the adaptive filters described in this chapter to active control is mainly associated with the identification of this response. This chapter also reviews a good deal of background material which is important when adaptive controllers are considered in Chapter 3.

2.1.1. Chapter Outline

After a review of the z transform in the remainder of this section, the structure and notation associated with digital filters is introduced in Section 2.2. The optimal filter response, in the sense of minimising a mean-square error criterion, is derived separately using a time-domain formulation in Section 2.3 and a transform-domain formulation in Section 2.4. These formulations are extended to the rather less familiar case of multichannel digital filters in Section 2.5 as preparation for the discussion of the multichannel control problems encountered in later chapters.

Various methods are then discussed for adapting the coefficients of a digital filter to automatically minimise a mean-square error criterion, starting with the well-known LMS algorithm in Section 2.6. The LMS algorithm is a form of steepest descent algorithm which has some very attractive properties in practice, but may not converge as fast as an exact least-squares algorithm, such as the RLS algorithm described in Section 2.7. Frequency-domain implementations of these two adaptive approaches are then discussed in Section 2.8. Finally, the more difficult problem of adapting recursive digital filters is outlined in Section 2.9.

2.1.2. The *z* transform

The two-sided, or bilateral, z transform of a sequence $x(n)$ is defined to be

$$X(z) = \sum_{n=-\infty}^{\infty} x(n)\, z^{-n} \ , \tag{2.1.1}$$

where z is a complex variable. The z transform of a delayed or shifted sequence takes a particularly simple form. If the z transform of $x(n)$ is defined by equation (2.1.1), then the z transform of this sequence delayed by one sample, $x(n-1)$, is

$$\sum_{n=-\infty}^{\infty} x(n-1)z^{-n} \ = \ \sum_{m=-\infty}^{\infty} x(m)z^{-(m+1)} \ = \ z^{-1}X(z) \ , \tag{2.1.2}$$

where $m = n - 1$.

The z transform of a sequence can often be represented as the ratio of two polynomials in z,

$$X(z) \ = \ \frac{N(z)}{D(z)} \ . \tag{2.1.3}$$

The *poles* of $X(z)$ are the roots of the denominator, i.e. the values of z that will ensure $D(z) = 0$ and so $X(z)$ is infinite. The *zeros* of $X(z)$ are the roots of the numerator, i.e. the values of z that will ensure $N(z) = 0$ and so $X(z)$ is zero.

The *region of convergence* of the z transform is the range of values of z over which the series defined by equation (2.1.1) will converge, as discussed, for example, by Oppenheim and Schafer (1975). For a right-sided sequence, i.e. $x(n) = 0$ for $n < n_1$, $X(z)$ converges for all z exterior to the circle whose radius is equal to that of the pole of $X(z)$ that is farthest from the origin. If the sequence is left-sided, i.e. $x(n) = 0$ for $n > n_2$, then $X(z)$ converges for all z interior to the circle whose radius is equal to the pole of $X(z)$ that is closest to the origin.

The inverse z transform enables the sequence $x(n)$ to be calculated from the polynomial $X(z)$. It can be calculated either using contour integration or partial fraction expansion (Oppenheim and Schafer, 1975). Some care needs to be taken with either process to ensure that a physically meaningful sequence is produced, since different sequences are generated depending on the assumed region of convergence. If, for example, we assume that the region of convergence of the z transform $\left(1 - a z^{-1}\right)^{-1}$ is for $|z| > |a|$, i.e. it is a right-sided sequence, then $|a/z| < 1$ and the z transform may be expanded as

$$\frac{1}{1 - a z^{-1}} \ = \ 1 + a z^{-1} + a^2 z^{-2} + \cdots \ . \tag{2.1.4}$$

Since z^{-1} corresponds to a delay of one sample, the inverse z transform of equation (2.1.4) is equal to $x(0) = 1$, $x(1) = a$, $x(2) = a^2$, etc., which is *causal*, i.e. $x(n) = 0$ if $n < 0$, and also decays to zero provided $a < 1$, in which case it is *stable*. If the same region of convergence is assumed but $a > 1$, the sequence is still causal but diverges with positive time and is said to be *unstable*. Notice that, for stability of the causal sequence, the pole of equation (2.1.4) must lie inside the *unit circle*, defined by $|z| = 1$. A sequence is said to be *stable* if it is absolutely summable, i.e.

$$\sum_{n=\infty}^{\infty} |x(n)| < \infty \ . \tag{2.1.5}$$

If, on the other hand, the region of convergence of equation (2.1.4) is assumed to be for $|z| < |a|$, i.e. it corresponds to a left-sided sequence, then $|z/a| < 1$ and it may now be expanded as

$$\frac{1}{1 - a\,z^{-1}} = -a^{-1}\,z - a^{-2}\,z^{2} - \cdots \ . \tag{2.1.6}$$

This corresponds to the sequence $x(0) = 0$, $x(-1) = -a^{-1}$, $x(-2) = -a^{-2}$, etc., which is *entirely noncausal*, and decays to zero as $n \to -\infty$ provided $a > 1$, i.e. the pole of equation (2.1.6) lies outside the unit circle. If the same region of convergence is assumed but $a < 1$, so that the pole of $X(z)$ lies inside the unit circle, the noncausal sequence diverges and can again be said to be unstable, but now in negative time.

It is generally not possible to determine both the stability and causality of a sequence from its z transform, and so a z transform cannot be associated with a single unique sequence. If it is known that a sequence is stable, however, i.e. it decays to zero in both positive and negative time, it can be uniquely reconstructed from its z transform with the causal components associated with the poles that lie inside the unit circle and the noncausal components with the poles outside the unit circle. This interpretation is important in the discussion of optimal filters in Section 2.4.

The initial value of a causal sequence is given by considering each of the terms in equation (2.1.1) as z tends to ∞, which is inside the region of convergence for such a sequence, in which case it can be seen that

$$\lim_{z \to \infty} X(z) = x(0) \ , \tag{2.1.7}$$

which is known as the *initial value theorem* (see, for example, Tohyama and Koike, 1998). For a causal sequence to have a finite initial value, and thus be stable according to equation (2.1.5), it must have at least as many poles as zeros.

The Fourier transform of the sequence $x(n)$ can be considered to be the special case of the z transform when $z = \mathrm{e}^{j\omega T}$, where ωT is the non-dimensional *normalised angular frequency*. The true frequency, in hertz, is equal to $\omega/2\pi$. The Fourier transform of $x(n)$ is thus given by

$$X(\mathrm{e}^{j\omega T}) = \sum_{n=-\infty}^{\infty} x(n)\mathrm{e}^{-j\omega n T} \ . \tag{2.1.8}$$

The inverse Fourier transform for sampled signals is then given by

$$x(n) = \frac{1}{2\pi}\int_0^{2\pi} X(\mathrm{e}^{j\omega T})\,\mathrm{e}^{j\omega n T} d\omega T \ . \tag{2.1.9}$$

2.2. STRUCTURE OF DIGITAL FILTERS

The general block diagram of a digital or sampled-data system is shown in Fig. 2.1. If the system is causal, each sample of the output sequence, $y(n)$, is only affected by the current and past values of the input sequence, $x(n)$, $x(n-1)$, etc. which can be written

$$y(n) = H\left[x(n), x(n-1), \ldots\right] , \tag{2.2.1}$$

where H is a function which, in general, may be nonlinear. If the digital system is linear, it must obey the principle of *superposition*, and so the function H must take the form of a linear summation, and so in the general case of a causal linear system, where the output signal is affected by all the previous input signals, we can express the output as

$$y(n) = \sum_{i=0}^{\infty} h_i x(n-i) , \tag{2.2.2}$$

which describes the discrete time *convolution* of $x(n)$ with the sequence h_i.

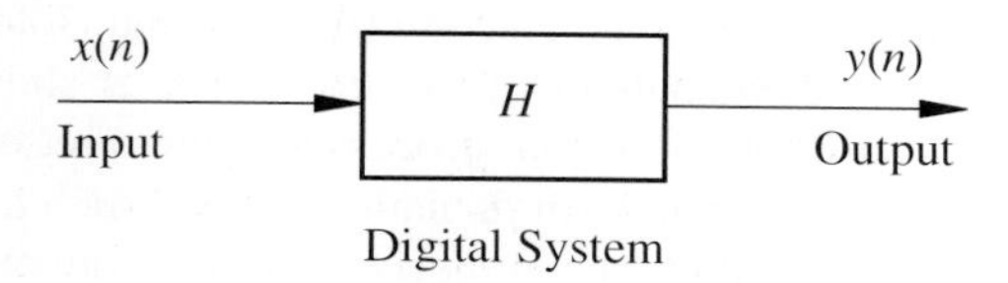

Figure 2.1 General block diagram of a digital system.

The parameters h_i form the samples of the *impulse response* of the system, i.e. if the input sequence is equal to the *Kronecker delta function*, $x(n) = \delta(n)$, which is equal to 1 if $n = 0$ and is otherwise zero, then

$$y(n) = \sum_{i=0}^{\infty} h_i \delta(n-i) = h_n , \quad n=0 \text{ to } \infty . \tag{2.2.3}$$

The definition of a *stable system* used here is one for which a bounded input produces a bounded output. It is shown by Rabiner and Gold (1975), for example, that a necessary and sufficient condition for such stability is that the impulse response sequence is absolutely summable, i.e. that equation (2.1.5) is satisfied when $x(n) = h_n$.

2.2.1. FIR Filters

To be able to implement a digital filter in practice, the calculation of each output sample can only take a finite time. One way of practically realising a digital filter would be to truncate the infinite summation in equation (2.2.2) to give

$$y(n) = \sum_{i=0}^{I-1} w_i x(n-i) , \tag{2.2.4}$$

where w_i are now the *coefficients*, or *weights*, of the digital filter, and it is assumed that there are I such coefficients. Note that the current output, $y(n)$, is assumed to depend on the current input, $x(n)$. This pre-supposes that the digital filter can act *instantaneously* in calculating the output, which is unrealistic in a real-time system. It is thus common to assume a one-sample delay in a real-time digital filter, to allow for the computation time, and to write its output as

$$y(n) = \sum_{i=1}^{I} w_i x(n-i) \ . \tag{2.2.5}$$

The sampled response of a physical system will also typically have a delay of at least one sample, particularly if anti-aliasing and reconstruction filters are used, which is sometimes implicit in the formulation of its sampled response (see for example Åström and Wittenmark, 1997). In order to be consistent with the signal processing literature, however, the more general form of equation (2.2.4) will be used below.

The response of the digital filter described by equation (2.2.4) to a Kronecker impulse excitation, $\delta(n)$, will be a sequence of finite length, i.e. $y(n) = w_n$ for $n = 0$ to $I - 1$ only. Such a filter thus has a *finite impulse response* and is said to be an *FIR* filter. Such filters are also described as 'moving average' (MA), 'transversal' or 'non-recursive', although it is sometimes possible to implement them recursively (Rabiner and Gold, 1975).

The *unit delay operator* will be denoted z^{-1} in this book, as shown in Fig. 2.2. This convention is very widely used in the signal processing literature, even though z is also used as the complex variable of a transform domain equation, such as in equation (2.1.2). In some parts of the control literature, the unit delay operator is denoted by the symbol q^{-1} (Åström and Wittenmark, 1997; Goodwin and Sin, 1984; Johnson, 1988; Grimble and Johnson, 1988) and the operator notation is also widely used in equation form, so that, for example, one could write

$$q^{-1} x(n) = x(n-1) \ . \tag{2.2.6}$$

Using this operator notation, equation (2.2.4) for the FIR filter would be written as

$$y(n) = W\left(q^{-1}\right) x(n) \ , \tag{2.2.7}$$

where

$$W\left(q^{-1}\right) = w_0 + w_1 q^{-1} + w_2 q^{-2} + \cdots + w_{I-1} q^{-I+1} \ . \tag{2.2.8}$$

Note that it is conventional in this notation to write the operator that corresponds to the FIR filter as $W\left(q^{-1}\right)$ rather than $W(q)$ (Åström and Wittenmark, 1997; Goodwin and Sin, 1984; Johnson, 1988; Grimble and Johnson, 1988) since it is a polynomial in q^{-1}, although other conventions are also used (Ljung, 1999). This is a rational but unfortunate choice since it leads to the transfer function of the FIR filter being written as $W\left(z^{-1}\right)$. The frequency response would thus be written as $W\left(\mathrm{e}^{-j\omega T}\right)$, which is completely inconsistent with the

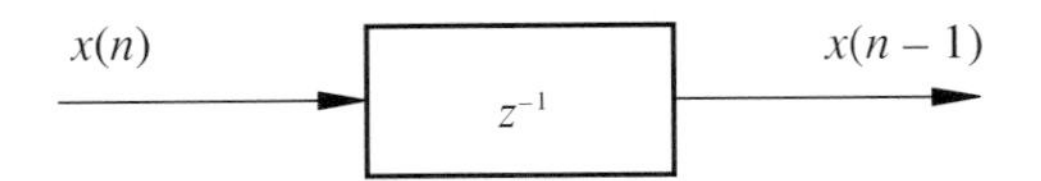

Figure 2.2 The use of z^{-1} as a unit delay operator.

signal processing literature. Although the use of the operator notation is very attractive as a shorthand notation for the convolution operation given by equation (2.2.5) and can be particularly elegant when it comes to describing time-varying systems, it is not used any further here in order to avoid the notational inconsistency that has grown up between the two bodies of literature.

It can be seen from equation (2.2.4) that the output of an FIR filter is the weighted sum of a finite number of previous input values. The FIR filter could thus be implemented using the architecture shown in Fig. 2.3. The *transfer function*, which relates the z transform of the output sequence to the z transform of the input sequence, will be written here for an FIR filter as

$$W(z) = w_0 + w_1 z^{-1} + w_2 z^{-2} + \cdots + w_{I-1} z^{-I+1} , \tag{2.2.9}$$

so that an algebraic equation for the z transform of $y(n)$ can be written as

$$Y(z) = W(z)\, X(z) , \tag{2.2.10}$$

where $Y(z)$ and $X(z)$ are the z transforms of the sequences $y(n)$ and $x(n)$. Equation (2.2.9) is a polynomial in z^{-1}, which can be represented as a polynomial in z, $w_0 z^{I-1} + w_1 z^{I-2} + \cdots + w_{I-1}$, divided by z^{I-1}. The $I-1$ *zeros* of the transfer function are given by the roots of this polynomial in z, and the *poles* of the transfer function are given by the roots of $z^{I-1} = 0$ in this case. A causal FIR filter with I coefficients thus has $I - 1$ poles at $z = 0$, and the commonly-used description of such filters as being 'all-zero' is rather misleading. FIR filters have the important properties that (1) they are always stable provided the coefficients are bounded, and (2) small changes in the coefficients will give rise to small changes in the filter's response.

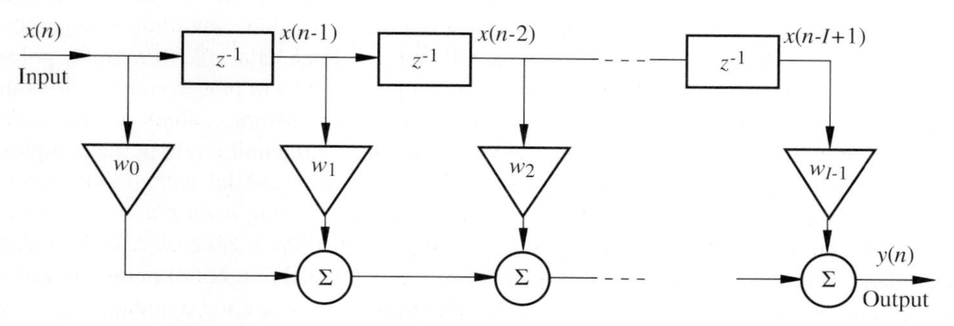

Figure 2.3 Implementation of an FIR filter by the direct calculation of a weighted sum of past inputs.

2.2.2. IIR Filters

It would be necessary to use a very large number of coefficients in equation (2.2.4), to describe the response of a lightly damped resonant system. A more efficient representation of such a response can be obtained by transforming the differential equation that describes the continuous-time system into a discrete-time difference equation, which then has the general linear form

$$y(n) = \sum_{j=1}^{J} a_j \, y(n-j) + \sum_{i=0}^{I-1} b_i \, x(n-i) \ , \tag{2.2.11}$$

where there are J 'feedback' coefficients, a_j, and I 'feedforward' coefficients, b_i.

The response of such a system to an impulse would take an infinite time to decay and so such a system is said to have an *infinite impulse response*, or an *IIR* form. They are also sometimes described as 'pole-zero', 'recursive' or 'autoregressive, moving average' (ARMA) systems, although in the time series analysis literature ARMA filters implicitly have white-noise excitation (Haykin, 1996).

The z transform of equation (2.2.11) can be written as

$$A(z)Y(z) = B(z)\,X(z) \ , \tag{2.2.12}$$

where

$$A(z) \ = \ 1 - a_1 \, z^{-1} \cdots - a_J \, z^{-J} \ , \tag{2.2.13}$$

$$B(z) \ = \ b_0 + b_1 \, z^{-1} \cdots + b_{I-1} \, z^{-I+1} \ , \tag{2.2.14}$$

Equation (2.2.12) can be rearranged in the form

$$H(z) \ = \ \frac{Y(z)}{X(z)} \ = \ \frac{B(z)}{A(z)} \ , \tag{2.2.15}$$

which is the transfer function of the system defined by the difference equation (2.2.11). Once again the *zeros* of the system are the roots of the numerator, i.e. the solutions of $B(z) = 0$, and the *poles* of the system are the roots of the denominator, i.e. the solution of $A(z) = 0$. In order for a causal system to have a finite initial value, according to equation (2.1.7), and so satisfy equation (2.1.5) so that it is stable, it must have at least as many poles as zeros. In contrast to an FIR filter, an IIR filter can have poles in non-trivial positions in the z-plane. This opens up the possibility of the filter being *unstable*, which occurs when the position of any of the poles of a causal system is outside the unit circle in the complex z plane, defined by $|z| = 1$. If a causal system, $H(z)$, has no overall delay and all of its zeros, as well as its poles, are inside the unit circle, it is said to be *minimum phase* as well as stable. A minimum phase system has the important property that its inverse, $1/H(z)$, is also stable and causal. A minimum phase system must have an equal number of poles and zeros if the initial value of its impulse response and that of its inverse are to be finite.

The number of coefficients required to accurately represent the sampled response of a physical system using either an FIR or an IIR filter depends very much upon the nature of the physical system being modelled. A system with a small number of lightly damped modes will have a frequency response with the same number of well-defined resonances, and can be efficiently represented by an IIR filter. A system that has a large number of heavily damped modes will not tend to display clear peaks in its frequency response, but will tend to exhibit sharp dips caused by destructive interference between the modal contributions. The impulse response of such a system will also generally be well contained. Such systems can be most efficiently modelled using an FIR filter. These features will be illustrated in Section 3.1 where the plant responses of typical structures and acoustic systems are discussed.

2.3. OPTIMAL FILTERS IN THE TIME DOMAIN

Optimal filters are those that give the best possible performance under a given set of circumstances. The best performance is generally defined in terms of minimising a *mean-square*, or H_2, *cost function*, since this has a physical interpretation in terms of minimising the power of an error signal, and also leads to a linear set of equations to be solved for the optimal FIR filter. It is also possible to design a filter that minimises the maximum value of the error at any frequency, which is called minimax optimisation in the signal processing literature (Rabiner and Gold 1975) and H_∞ minimisation in the control literature (Skogestad and Postlethwaite, 1996). The symbols H_2 and H_∞ denote different signal norms, as described in the Appendix. Alternatively, the optimum parameter values could be calculated for each example of a random set of data, so the parameters are themselves random variables, and maximum a posteriori (MAP) or maximum likelihood (ML) estimates can be defined (Clarkson, 1993).

2.3.1. Electrical Noise Cancellation

In this section we will consider the optimal FIR filter that minimises the mean-square error in the prototypical electrical noise cancellation problem shown in Fig. 2.4 (see for example Kailath, 1974; Widrow and Stearns, 1985; Haykin, 1996). In this figure the *error signal*, $e(n)$, is given by the difference between the *desired signal*, $d(n)$, and the *reference signal*, $x(n)$, filtered by an FIR filter with coefficients w_i, so that

$$e(n) = d(n) - \sum_{i=0}^{I-1} w_i\, x(n-i) \quad . \tag{2.3.1}$$

The reference signal is referred to as the input signal by some authors, although this can be confusing because there is generally more than one 'input' signal.

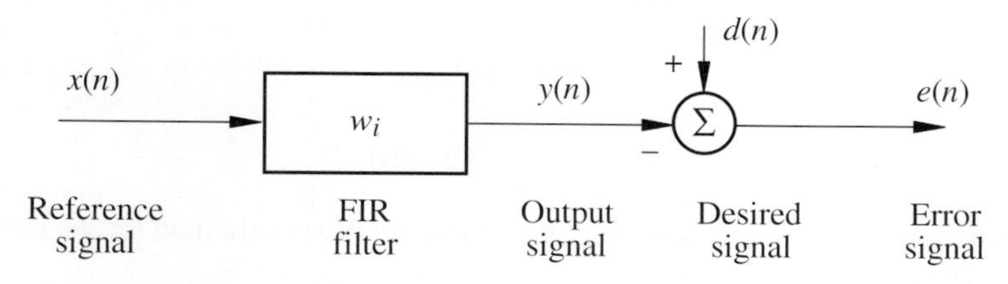

Figure 2.4 Model problem for the derivation of the optimal FIR filter.

The summation over $w_i x(n-i)$ in equation (2.3.1) can be conveniently represented as a vector inner product, such that

$$e(n) = d(n) - \boldsymbol{w}^{\mathrm{T}} \boldsymbol{x}(n) = d(n) - \boldsymbol{x}^{\mathrm{T}}(n)\, \boldsymbol{w} \ , \tag{2.3.2a,b}$$

where

$$\boldsymbol{w} = \begin{bmatrix} w_0 & w_1 & \cdots & w_{I-1} \end{bmatrix}^{\mathrm{T}} \ , \tag{2.3.3}$$

$$\boldsymbol{x}(n) = \begin{bmatrix} x(n) & x(n-1) & \cdots & x(n-I+1) \end{bmatrix}^{\mathrm{T}} \ , \tag{2.3.4}$$

and the superscript T denotes the transpose of the vectors, which are assumed to be column vectors, as described in the Appendix.

The objective is to find the values of each of the filter coefficients $w_0 \cdots w_{I-1}$ that minimise the quadratic *cost function* given by the mean square error,

$$J = E\left[e^2(n)\right] , \tag{2.3.5}$$

where E denotes the *expectation operator* (see, for example, Papoulis, 1977). If $x(n)$ and $d(n)$ were not stationary, then J, and the optimal filter coefficients, would be functions of time. We assume here that all the signals are stationary and ergodic, so that the expectation is time invariant, and can be calculated by averaging over time. The cost function given by equation (2.3.5) is thus equal to the average mean-square value of the error signal. This very restricted use of the expectation operator will be assumed throughout this chapter.

Using Equations (2.3.2a and b), the cost function can be written as

$$J = \boldsymbol{w}^{\mathrm{T}} \boldsymbol{A} \boldsymbol{w} - 2\, \boldsymbol{w}^{\mathrm{T}} \boldsymbol{b} + c , \tag{2.3.6}$$

where

$$\boldsymbol{A} = E\left[\boldsymbol{x}(n)\boldsymbol{x}^{\mathrm{T}}(n)\right] , \tag{2.3.7}$$

$$\boldsymbol{b} = E\left[\boldsymbol{x}(n)d(n)\right] , \tag{2.3.8}$$

$$c = E\left[d^2(n)\right] . \tag{2.3.9}$$

In a quadratic equation having the general form of equation (2.3.6), the matrix $\boldsymbol{A}$ is known as the *Hessian matrix*, and in this case its elements are equal to the values of the autocorrelation function of the reference signal,

$$\boldsymbol{A} = \begin{bmatrix} R_{xx}(0) & R_{xx}(1) & \cdots & R_{xx}(I-1) \\ R_{xx}(1) & R_{xx}(0) & & \\ \vdots & & \ddots & \\ \vdots & & & \ddots \\ R_{xx}(I-1) & & & R_{xx}(0) \end{bmatrix} , \tag{2.3.10}$$

where $R_{xx}(m)$ is the symmetric *autocorrelation function* of $x(n)$, defined for the entirely real time sequences assumed here to be

$$R_{xx}(m) = E\left[x(n)x(n+m)\right] = R_{xx}(-m) . \tag{2.3.11}$$

The Hessian matrix is not necessarily equal to this matrix of autocorrelation functions, and in order to be consistent with the discussion below the Hessian matrix is written here as $\boldsymbol{A}$ rather than the more specific form $\boldsymbol{R}$, which is widely used for FIR filters in the signal processing literature. An example of this more general form for $\boldsymbol{A}$ is when the cost function to be minimised includes a term proportional to the sum of the squared filter weights, $\boldsymbol{w}^{\mathrm{T}}\boldsymbol{w}$, so that

$$J = E\left[e^2(n)\right] + \beta\, \boldsymbol{w}^{\mathrm{T}} \boldsymbol{w} , \tag{2.3.12}$$

where β is a positive, real *coefficient-weighting parameter*. This cost function can also be written as a quadratic equation of the form of equation (2.3.6), but now the Hessian matrix has the form

$$\boldsymbol{A} = \boldsymbol{R} + \beta \mathbf{I} , \tag{2.3.13}$$

where $\boldsymbol{R}$ is the autocorrelation matrix given by the right-hand side of equation (2.3.10), and $\mathbf{I}$ is the identity matrix.

Minimisation of the cost function given by equation (2.3.12) can prevent excessively large values of the filter coefficients which do little to reduce the mean-square error for a reference signal with the assumed autocorrelation structure, but may significantly increase the mean-square error if the statistics of the reference signal changes. The incorporation of the coefficient weighting parameter thus makes the optimum solution more robust to variations in the statistics of the reference signals, and for many problems this can be achieved without a significant increase in the mean-square error under nominal conditions. Small values of the coefficient weighting parameter β in equation (2.3.13) can also make it easier to invert the matrix $\boldsymbol{A}$, and β is said to *regularise* the least-square solution.

Returning to Fig. 2.4 it is interesting to note that if a white noise signal, $v(n)$, were added to the original reference signal, $x(n)$, then the autocorrelation function of the modified reference signal would be of exactly the form of equation (2.3.13), in which case β would be equal to the mean-square value of $v(n)$ (Widrow and Stearns, 1985). If regularisation is only required in certain frequency ranges, this equivalent noise signal could also be coloured, as is discussed in more detail for the design of feedback controllers in Chapter 6.

The vector $\boldsymbol{b}$ in equation (2.3.8) has elements that are equal to the values of the *cross-correlation function* between the reference signal and the desired signals so that

$$\boldsymbol{b} = \left[R_{xd}(0) \; R_{xd}(1) \; \cdots \; R_{xd}(I-1)\right]^{\mathrm{T}} , \tag{2.3.14}$$

where for the entirely real and stationary time sequences assumed here

$$R_{xd}(m) = E\left[x(n)d(n+m)\right] = E\left[x(n-m)d(n)\right] . \tag{2.3.15}$$

Finally, c is a scalar constant equal to the mean-square value of the desired signal.

When written in the form of equation (2.3.6), it is clear that the mean-square error is a quadratic function of each of the FIR filter coefficients. This quadratic function always has a minimum rather than a maximum, since J will become very large if any one filter coefficient takes large positive or negative values. This minimum is only unique, however, if the matrix $\boldsymbol{A}$ in equation (2.3.6) is *positive definite*. If $\boldsymbol{A}$ is given by equation (2.3.7), it can either be positive definite (in which case it is also non-singular) or positive semi-definite (in which case it is singular), depending on the spectral properties of the reference signal and the number of coefficients in the FIR filter. If there are at least half as many spectral components as filter coefficients, the reference signal is said to be *persistently exciting* or 'spectrally rich', which ensures that the autocorrelation matrix given by equation (2.3.7) is positive definite and so equation (2.3.6) has a unique minimum (see for example Treichler et al., 1987; Johnson, 1988). The cost function when plotted against any two coefficients will thus define a parabolic bowl-shaped performance surface or *error surface*, as shown in Fig. 2.5. If a finite value of β is used in equation (2.3.12), then $\boldsymbol{A}$ is guaranteed to be positive definite, since the equivalent noise on the reference signal ensures that it is persistently exciting.

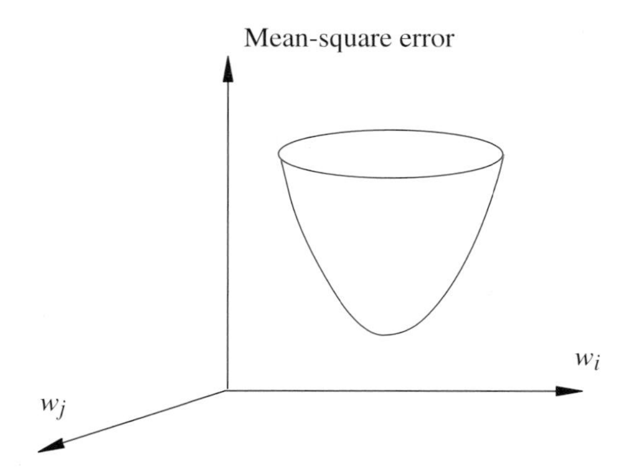

Figure 2.5 The quadratic error surface formed by plotting the mean-square error in Fig. 2.4 against two coefficients of the FIR filter.

2.3.2. The Wiener Filter

The value of the coefficients of the FIR filter that reduces the mean-square error to a minimum can be found by differentiating the cost function with respect to each coefficient and setting all of the resulting derivatives to zero. It is convenient to express this differentiation using vector notation, and so we define the vector of derivatives to be

$$\frac{\partial J}{\partial \boldsymbol{w}} = \left[\frac{\partial J}{\partial w_0} \;\; \frac{\partial J}{\partial w_1} \;\; \cdots \;\; \frac{\partial J}{\partial w_{I-1}} \right]^{\mathrm{T}} . \qquad (2.3.16)$$

Using the definition of J in equation (2.3.6) and the properties of vectors of derivatives outlined in the Appendix, equation (2.3.16) can be expressed as

$$\frac{\partial J}{\partial \boldsymbol{w}} = 2\left[\boldsymbol{A}\,\boldsymbol{w} - \boldsymbol{b}\right] . \qquad (2.3.17)$$

Assuming that $x(n)$ is persistently exciting, so that $\mathbf{A}$ is not singular, the vector of optimal filter coefficients can be obtained by setting each element of (2.3.17) to zero, to give

$$\boldsymbol{w}_{\mathrm{opt}} = \boldsymbol{A}^{-1}\,\boldsymbol{b} . \qquad (2.3.18)$$

The filter that has these optimal coefficients is often called the *Wiener filter* after the pioneering work of N. Wiener in the 1940s (Wiener, 1949), although the explicit solution to the discrete-time form of the problem, equation (2.3.18), is due to Levinson (1947). Levinson also put forward an efficient method for solving equation (2.3.18) which exploits the *Toeplitz* nature of the matrix $\boldsymbol{A}$, i.e. all of its elements are equal along each diagonal, and this method is widely used in speech coding (see, for example, Markel and Gray, 1976). Levinson modestly prefaces his paper by stating that ‘A few months after Wiener's work appeared, the author, in order to facilitate computational procedure, worked out an approximate, and one might say, mathematically trivial procedure’. In fact the discrete form of Wiener's formulation leads to an elegant matrix formulation which is very well suited to efficient numerical solution.

Using the definition of $\boldsymbol{A}$ and $\boldsymbol{b}$ in terms of the auto- and cross-correlation functions, equation (2.3.17) for the optimal filter can also be written as a summation of the form

$$\sum_{i=0}^{I-1} w_{i,\mathrm{opt}} R_{xx}(k-i) - R_{xd}(k) = 0 \quad \text{for} \quad 0 \le k \le I-1 \; . \tag{2.3.19}$$

This set of equations is known as the *normal equations* and represents a discrete form of the *Wiener–Hopf equation* (Kailath, 1981).

The vector of cross-correlations between the I past values of the reference signal and the error signal can be written, using equation (2.3.2), as

$$E[\boldsymbol{x}(n)e(n)] = E\left[\boldsymbol{x}(n)\left((d(n) - \boldsymbol{x}^{\mathrm{T}}(n)\boldsymbol{w}\right)\right] = \boldsymbol{b} - \boldsymbol{A}\boldsymbol{w} \; . \tag{2.3.20}$$

It is clear that all the elements of this vector are zero when the FIR filter is adjusted to minimise the mean-square error, as in equation (2.3.18). In minimising the mean-square error, the Wiener filter thus also sets to zero the cross-correlations between the reference signal and the residual error signal over the length of the filter, so that the residual error signal contains no component that is correlated with the current and previous $I - 1$ values of the reference signal. This is a statement of the *principle of orthogonality* (Kailath, 1981; Haykin, 1996).

The values of the auto- and cross-correlation functions required to define the elements of $\boldsymbol{A}$ and $\boldsymbol{b}$ in Equations (2.3.7) and (2.3.8) can often be most efficiently estimated from measured data by first calculating the power spectral density of $x(n)$ and the cross spectral density between $x(n)$ and $d(n)$. The correlation functions are then obtained using the Fourier transform (Rabiner and Gold, 1975). From these average properties of the reference and desired signal, the coefficients of the Wiener filter can be calculated using equations (2.3.10, 14 and 18). The minimum value of the mean-square error can also be obtained directly by substituting equation (2.3.18) into equation (2.3.6) to give

$$J_{\min} = c - \boldsymbol{b}^{\mathrm{T}} \boldsymbol{A}^{-1} \boldsymbol{b} \; . \tag{2.3.21}$$

The residual mean-square error can thus be calculated directly from the statistical properties of the reference and desired signals. This can be very useful in the early stages of a design, to help understand the trade-off between performance and filter length, for example.

2.3.3. Linear Prediction

A particularly interesting form of the model problem shown in Fig. 2.4, which has important implications for feedback control, is when the reference signal is a delayed version of the desired signal. The objective of the Wiener filter is then to minimise the mean-square error by making the output of the FIR filter as close as possible to the current desired signal, using only past values of this signal, i.e. to act as a *linear predictor* for the desired signal. The block diagram for this arrangement is shown in Fig. 2.6, in which a delay of Δ samples between the desired and reference signals has been assumed, and $\hat{d}(n)$ is now the estimate of $d(n)$ obtained by linear prediction.

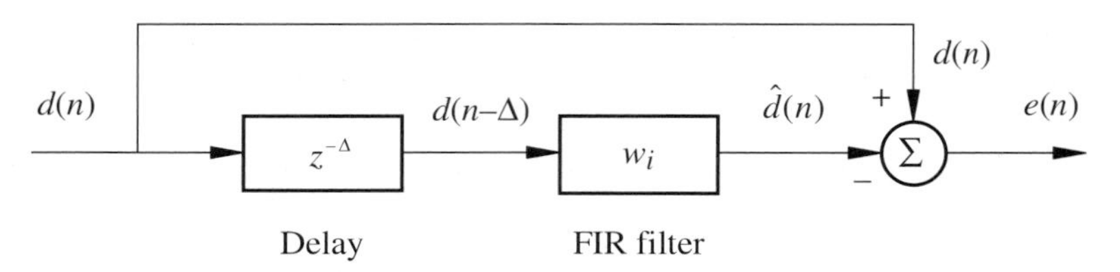

Figure 2.6 Block diagram for the derivation of the optimal prediction filter.

The elements of both the matrix $\boldsymbol{A}$, equation (2.3.10), and the vector $\boldsymbol{b}$, equation (2.3.14), will now be equal to various values of the autocorrelation function of the desired signal, $d(n)$, only. The Wiener filter, equation (2.3.18), and the residual mean-square error, equation (2.3.21), are thus only dependent on the autocorrelation structure of the desired signal. The more correlated the desired signal is, over the length of the filter, the lower the residual mean-square value will be, as demonstrated for example in the simulations presented by Nelson (1996). Thus if $d(n)$ is white noise, the FIR filter can make no prediction of its future value, so that $\hat{d}(n) = 0$, and no attenuation in the error signal is obtained for any non-zero Δ. In contrast, if $d(n)$ were a pure tone, this would be perfectly predictable, and so we could make $\hat{d}(n) = d(n)$ in this case and large attenuations of the error signal would be possible. In fact the arrangement shown in Fig. 2.6 is widely used to enhance the tonal components of signals, which give rise to spectral lines, and this arrangement is thus commonly known as a *line enhancer*.

The residual mean-square error of a linear predictor for a given desired signal can be calculated directly from its autocorrelation function using the formulation outlined above. As an example of such a calculation, Fig. 2.7 shows the level of the residual mean-square error, relative to the mean-square value of the desired signal, for the optimal linear prediction of two different signals using a 128-coefficient FIR filter with various prediction

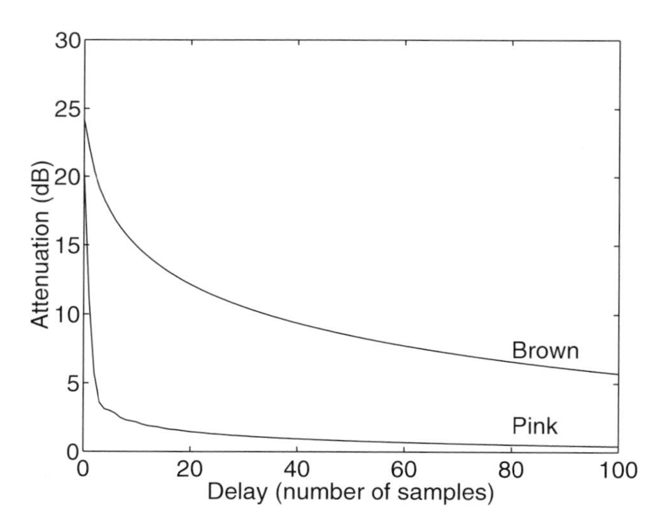

Figure 2.7 Attenuation of the error signal against prediction delay Δ, for the optimal prediction of 'pink noise', whose power spectral density falls off at 3 dB/octave, and 'brown noise', whose power spectral density falls off at 6 dB/octave.

delays Δ. The first signal analysed is 'pink noise', which has an equal energy in each octave band and thus a power spectral density which falls off at 3 dB/octave with frequency. This is also referred to as $1/f$ noise. An interesting discussion of the problems of defining the zero-frequency spectrum of such signals and of the prevalence of $1/f$ noise in nature is provided by Buckingham (1983). The second signal is obtained by passing white noise through an integrator, so that the power spectral density falls off at 6 dB/octave, which is sometimes called 'brown noise', or more properly Brownian noise (Schroeder, 1990). To perform the simulations required to produce these curves the autocorrelation functions were calculated from the power spectral densities, which were set to nought in the zero frequency bin to avoid the singularity which otherwise occurs. It can be seen that the brown noise, with a power spectral density proportional to $|\text{frequency}|^{-2}$, is more predictable than pink noise, with a power spectral density proportional to $|\text{frequency}|^{-1}$, since it is 'more coloured', i.e. further from being purely white. The linear predictor thus has more success in estimating the current output from the previous ones and the attenuation in the error in Fig. 2.7 is considerably higher for brown noise than it is for pink noise.

2.4. OPTIMAL FILTERS IN THE TRANSFORM DOMAIN

2.4.1. Unconstrained Wiener Filter

Returning to the expression for the normal equations, equation (2.3.19), we find that we can obtain a particularly simple expression for the frequency response of the optimal filter if we relax the assumptions that the filter must be causal and of finite length. If the filter is non-causal and of infinite duration, the normal equations can be written as

$$\sum_{i=-\infty}^{\infty} w_{i,\text{opt}}\, R_{xx}(k-i) \;=\; R_{xd}(k) \quad \text{for} \quad -\infty < k < \infty \;. \tag{2.4.1}$$

Taking the Fourier transform of equation (2.4.1) allows us to express the Wiener filter in this completely unconstrained case, which is sometimes referred to as the *unconstrained* or *two-sided Wiener filter*, as

$$W_{\text{opt}}(\mathrm{e}^{j\omega T}) \;=\; \frac{S_{xd}(\mathrm{e}^{j\omega T})}{S_{xx}(\mathrm{e}^{j\omega T})} \;, \tag{2.4.2}$$

where $S_{xx}\left(\mathrm{e}^{j\omega T}\right)$ is the *power spectral density* of the reference signal and $S_{xd}\left(\mathrm{e}^{j\omega T}\right)$ is the *cross spectral density* between the reference and desired signals. The power spectral density is defined to be the Fourier transform of the autocorrelation function $R_{xx}(n)$, but can also be expressed as

$$S_{xx}(\mathrm{e}^{j\omega T}) \;=\; \lim_{N\to\infty} \frac{1}{N}\; E\left[X_m^*(\mathrm{e}^{j\omega T})X_m(\mathrm{e}^{j\omega T})\right] \;, \tag{2.4.3}$$

where $X_m\left(\mathrm{e}^{j\omega T}\right)$ is the Fourier transform of the m-th record of the sequence $x_m(n)$, which has N samples and is assumed to be stationary, and the expectation operator refers to the average over the ensemble of records (Bendat and Piersol, 1986). In practice, an estimate of the power spectral density can be derived by averaging over a series of successive data lengths of finite duration provided the signal is ergodic. It is notationally convenient to abbreviate equation (2.4.3) by simply writing

$$S_{xx}(\mathrm{e}^{j\omega T}) = E\left[X^*(\mathrm{e}^{j\omega T})X(\mathrm{e}^{j\omega T})\right] , \tag{2.4.4}$$

where the expectation operator now implies ensemble averaging over records whose length tends to infinity. This use of the expectation operator in the frequency domain will be used from now on. Clearly the power spectral density is an entirely real function of ωT.

The cross-spectral density is defined to be the Fourier transform of the cross-correlation function $R_{xd}(m)$, but can also be written, using the notation defined above, as

$$S_{xd}(\mathrm{e}^{j\omega T}) = E\left[X^*(\mathrm{e}^{j\omega T})D(\mathrm{e}^{j\omega T})\right] , \tag{2.4.5}$$

where $D\left(\mathrm{e}^{j\omega T}\right)$ is the Fourier transform of a segment of $d(n)$ whose length tends to infinity. The cross spectral density is in general complex but has the property that $S^*_{xd}(\mathrm{e}^{j\omega T}) = S_{dx}(\mathrm{e}^{j\omega T})$. For completeness we also define the z transform of $R_{xx}(m)$ to be $S_{xx}(z)$, which can also be written as

$$S_{xx}(z) = E\left[X(z^{-1})X(z)\right] , \tag{2.4.6}$$

and the z transform of $R_{xd}(m)$ to be $S_{xd}(z)$, which can also be written as

$$S_{xd}(z) = E\left[X(z^{-1})\,D(z)\right] , \tag{2.4.7}$$

where again the expectation implies that the z transform of finite sections of data have been taken, whose length tends to infinity, as explained in detail by Grimble and Johnson (1988).

2.4.2. Causally Constrained Wiener Filter

It is easy to compute the frequency response of the optimal filter given by equation (2.4.2), but its use could be misleading because it is not constrained to be causal, and so might be impossible to realise in a real-time system. In this section, we consider the calculation of the frequency response of the optimum filter when it is constrained to be causal but not constrained to be of finite duration. In practice, this gives an equivalent result to solving the normal equations for an FIR filter with very many coefficients and then taking the Fourier transform of the resulting impulse response. Although the original formulation for such a filter was obtained by Wiener (1949), the physical interpretation of the derivation presented here was originally due to Bode and Shannon (1950) and Zadek and Ragazzini (1950).

The normal equations for the optimal filter when it is constrained to be causal, but not constrained to be of finite length, can be written from equation (2.3.19) as

$$\sum_{i=0}^{\infty} w_{i,\mathrm{opt}}\, R_{xx}(k-i) = R_{xd}(k) \quad \text{for} \quad 0 \le k < \infty . \tag{2.4.8}$$

We now assume the very special condition that the reference signal, $x(n)$ is given by a sequence $v(n)$, that is completely uncorrelated from sample to sample, i.e. it is white noise, and also has zero mean and unit variance, so that $R_{xx}(m) = R_{vv}(m) = \delta(m)$, which equals 1 if $m = 0$ and which otherwise equals zero. Under these conditions equation (2.4.2) has the form

$$\sum_{i=0}^{\infty} w_{i,\text{opt}:v}\, \delta(k-i) = R_{vd}(k) \quad \text{for} \quad 0 \le k < \infty , \tag{2.4.9}$$

and because of the sifting property of $\delta(m)$, equation (2.4.9) is equivalent to

$$w_{k,\text{opt}:v} = R_{vd}(k) \quad \text{for} \quad 0 \le k < \infty . \tag{2.4.10}$$

Thus, if the reference signal is white noise, the impulse response of the optimal causal filter is equal to the causal part of the cross-correlation function between the reference and desired signal. The z transform of the optimal filter under these conditions can be written as

$$W_{\text{opt}:v}(z) = \{S_{vd}(z)\}_+ , \tag{2.4.11}$$

where $S_{vd}(z)$ is the z transform of the cross-correlation function above and $\{\ \}_+$ denotes the fact that the causal part of the inverse z transform of the quantity inside the brackets has been taken, so that

$$\{S_{vd}(z)\}_+ = \mathcal{Z}[h(m)\, R_{vd}(m)] , \tag{2.4.12}$$

where $\mathcal{Z}$ denotes the z transform and $h(m)$ is the discrete step function which is equal to 1 for $m \ge 0$ and is equal to zero for $m < 0$.

2.4.3. Spectral Factors

In the more general case, the reference signal is not a white noise sequence. The problem can still be reduced to that of a white noise reference, however, by considering a *pre-whitening* of the reference signal. Specifically, we assume that the reference signal was originally generated by passing a zero mean white noise signal, $v(n)$, with unit variance (so that its power spectral density is also unity) through a minimum phase *shaping filter*, with transfer function $F(z)$. The two-sided z transform of the autocorrelation function $R_{xx}(m)$, can be written as $S_{xx}(z)$, and because $R_{xx}(m)$ is symmetrical in time, then $S_{xx}(z) = S_{xx}(z^{-1})$. Also, if $S_{xx}(z)$ is *rational*, i.e. it has a number of well defined poles and zeros, then it can be expressed in the form

$$S_{xx}(z) = A\frac{(1-az^{-1})(1-az)(1-bz^{-1})(1-bz)\cdots\cdots}{(1-\alpha z^{-1})(1-\alpha z)(1-\beta z^{-1})(1-\beta z)\cdots\cdots} , \tag{2.4.13}$$

where the parameters $a,b, \ldots \alpha,\beta, \ldots$ may be assumed to be less than one (Widrow and Walach, 1996). The pole-zero diagram of $S_{xx}(z)$ thus has zeros inside the unit circle at $z = a,b, \ldots$ and outside the unit circle at $1/a$, $1/b, \ldots$ and poles inside the unit circle at $z=\alpha,\beta, \ldots$ and outside the unit circle at $1/\alpha$, $1/\beta, \ldots$. We can now factorise $S_{xx}(z)$ into two parts, $F^{+}(z)$, which has all its poles and zeros inside the unit circle, and is thus causal, stable and minimum phase, and $F^{-}(z)$, which has all its poles and zeros outside the unit circle, so that

$$S_{xx}(z) = F^{+}(z)\, F^{-}(z) , \tag{2.4.14}$$

where

$$F^+(z) = \sqrt{A}\,\frac{\left(1-az^{-1}\right)\left(1-bz^{-1}\right)\cdots\cdots}{\left(1-\alpha z^{-1}\right)\left(1-\beta z^{-1}\right)\cdots\cdots} , \tag{2.4.15}$$

and

$$F^-(z) = \sqrt{A}\,\frac{(1-az)(1-bz)\cdots\cdots}{(1-\alpha z)(1-\beta z)\cdots\cdots} . \tag{2.4.16}$$

which are called the *spectral factors* of $S_{xx}(z)$. It is important to note that

$$F^-(z) = F^+(z^{-1}) , \tag{2.4.17}$$

so that $F^-(z)$ can be interpreted as having the same impulse response as $F^+(z)$, but reversed in time.

If $x(n)$ is generated by passing $v(n)$ through a filter with the transfer function $F(z)$, then

$$X(z) = F(z)\,V(z) , \tag{2.4.18}$$

and so using the definition of the z transform of the autocorrelation function in equation (2.4.6),

$$S_{xx}(z) = F(z)\,F(z^{-1})\,S_{vv}(z) . \tag{2.4.19}$$

But $S_{vv}(z)$ is equal to unity, since $v(n)$ is white noise of unit variance, and so this equation reduces to the form of (2.4.14), confirming that $F(z) = F^+(z)$ is indeed the minimum phase shaping filter we are looking for. An example of the pole-zero plot of $S_{xx}(z)$ for one particular reference signal is shown in Fig. 2.8, together with the pole zero plots for the spectral factors, $F(z)$ and $F(z^{-1})$.

Clearly we could transform the reference signal back into a white noise sequence by passing it through a *whitening filter* given by the inverse of the shaping filter, which is stable since $F(z)$ is minimum phase, to give the signal $v(n)$ with the z transform

$$V(z) = \frac{X(z)}{F(z)} . \tag{2.4.20}$$

This signal can be regarded as the result of a 'distillation' of the reference signal to produce only the new information contained in each sample, and it is thus sometimes called the innovations process.

The transfer function of the optimum causal filter which would operate on this white noise reference signal is given by equation (2.4.11). The transfer function of the complete optimum filter, operating on the reference signal $x(n)$, is thus equal to

$$W_{\text{opt}}(z) = \frac{1}{F(z)}\left\{S_{vd}(z)\right\}_+ , \tag{2.4.21}$$

as illustrated in Fig. 2.9. Using equation (2.4.20), the z transform of the cross-correlation function between $v(n)$ and $d(n)$ can be written as

$$S_{vd}(z) = \frac{S_{xd}(z)}{F\left(z^{-1}\right)} . \tag{2.4.22}$$

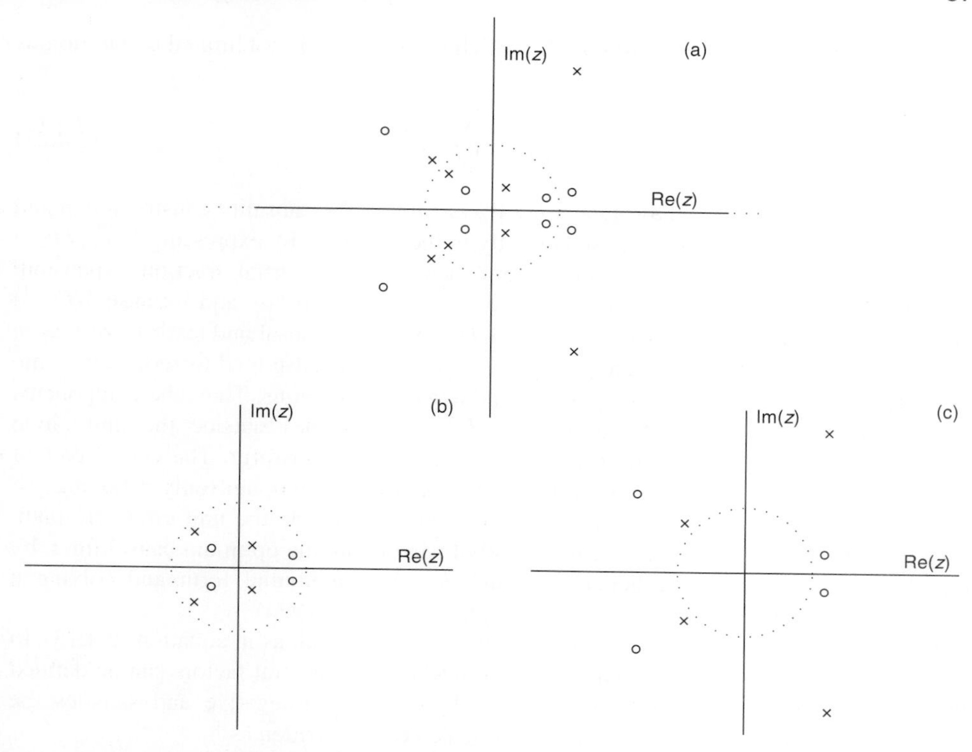

Figure 2.8 Pole-zero plot of the power spectral density of a reference signal, $S_{xx}(z)$ (a) which is the double-sided z-transform of the reference signal's autocorrelation function, and the pole-zero plots of its spectral factors, $F(z)$, with an entirely causal impulse response (b) and $F(z^{-1})$, with an entirely noncausal impulse response (c), where the dotted line corresponds to the unit circle for which $|z| = 1$.

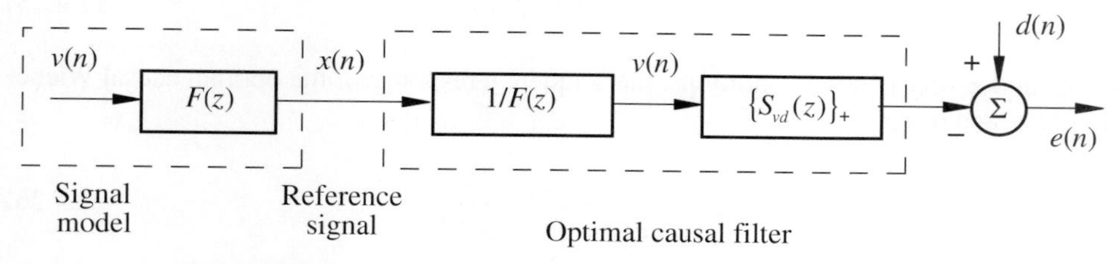

Figure 2.9 Signal model for the reference signal, $x(n)$, in which it assumed to be generated by passing a white noise signal, $v(n)$, through a shaping filter $F(z)$, and the optimal causal filter, which then takes the form of a pre-whitening filter, $1/F(z)$, and the Fourier transform of the causal part of the cross-correlation function between $v(n)$ and $d(n)$, $\{S_{vd}(z)\}_+$.

The transfer function of the optimum filter which is causal, but is not limited in duration, is thus given by

$$W_{\text{opt}}(z) = \frac{1}{F(z)} \left\{ \frac{S_{xd}(z)}{F(z^{-1})} \right\}_+ , \tag{2.4.23}$$

which is sometimes called the *single-sided Wiener filter*. The causality constraint denoted by the operator $\{\ \}_+$ can be imposed directly in the z domain by expressing $S_{xd}(z)/F(z^{-1})$ as the ratio of two polynomials in z^{-1} and then taking a partial fraction expansion. Because $R_{xd}(m)$ must tend to zero as m tends to both $+\infty$ and $-\infty$, and because $1/F(z^{-1})$ corresponds to the time reversed version of $1/F(z)$, which is causal and tends to zero as m tends to ∞, then the inverse z transform of $S_{xd}(z)/F(z^{-1})$ must also tend to zero as the time index tends to infinity in both the positive and negative directions. Thus, the components of the partial fraction expansion of $S_{xd}(z)/F(z^{-1})$ with poles outside the unit circle correspond to the entirely noncausal part of its inverse z transform. The causal part of $S_{xd}(z)/F(z^{-1})$ can be calculated by discarding these components and only retaining the components of the partial fraction expansion with poles inside the unit circle (Kailath, 1981; Therrien, 1992). An equivalent method of calculating the optimum causal filter, by expressing the filters that generate $x(n)$ and $d(n)$ in polynomial form and solving a Diophantine equation, is described, for example, by Kučera (1993).

It is not, however, necessary to assume that $S_{xx}(z)$ is rational, as in equation (2.4.13), to define the spectral factors. Papoulis (1977) shows that the spectral factors can be defined provided the power spectral density, $S_{xx}(e^{j\omega T})$, is real, non-negative and satisfies the discrete form of the Paley–Wiener condition, which can be written as

$$\int_0^{2\pi} \left|\ln S_{xx}(e^{j\omega T})\right| d\omega T < \infty , \tag{2.4.24}$$

where ln denotes the natural logarithm. Thus the spectral factorisation can be generally applied to the power spectral density, regardless of whether its pole-zero structure is known. One way in which the frequency response of the shaping filter $F(e^{j\omega T})$ can be computed from the power spectral density in practice is to use the *Hilbert transform* to calculate the minimum-phase component of a system whose modulus is equal to $\sqrt{S_{xx}(e^{j\omega T})}$ (Kailath, 1981). The spectral factorisation can then be defined in the frequency domain as being

$$S_{xx}\left(e^{j\omega T}\right) = F\left(e^{j\omega T}\right) F^*\left(e^{j\omega T}\right) , \tag{2.4.25}$$

so that under very general conditions, the frequency response of the optimal causal Wiener filter can be written as

$$W_{\text{opt}}\left(e^{j\omega T}\right) = \frac{1}{F\left(e^{j\omega T}\right)} \left\{ \frac{S_{xd}\left(e^{j\omega T}\right)}{F^*\left(e^{j\omega T}\right)} \right\}_+ . \tag{2.4.26}$$

If the constraint of causality is removed, and using the definition of the shaping filter in equation (2.4.25), then the single-sided, causal, Wiener filter in equation (2.4.26) reduces to the expression derived above for the unconstrained, potentially noncausal, Wiener filter, equation (2.4.2).

The frequency-domain formulation can be used to directly calculate the optimum causal filter in practice if the normalised frequency, ωT, is divided into N equal increments or *bins*, as in the discrete Fourier transform (DFT), and the discrete frequency variable is $k = 0, 1, \ldots, N-1$. The spectral factorisation can now be performed directly in the discrete frequency domain by using the cepstral method to calculate the discrete Hilbert transform (Oppenheim and Shafer, 1975). The minimum phase part of $F(k)$ can thus be calculated from its magnitude, which is equal to the square root of $S_{xx}(k)$, so that

$$F(k) \;=\; \exp\left(\mathrm{FFT}\left[c(n)\,\mathrm{IFFT}\,\ln\left(S_{xx}(k)\right)\right]\right) , \tag{2.4.27}$$

where the causality constraint is given by $c(n) = 0$ for $n < 0$, $c(n) = 1$ for $n > 0$ and $c(0) = \frac{1}{2}$ and the fast Fourier transform (FFT) is used to calculate the DFT. The causality constraint can also be implemented directly in the discrete-frequency domain by transforming $S_{xd}(k)/F^*(k)$ into the time domain, setting the noncausal part to zero and transforming back to the discrete-frequency domain, so that

$$\left\{\frac{S_{xd}(k)}{F^*(k)}\right\}_+ \;=\; \mathrm{FFT}\left[h(n)\,\mathrm{IFFT}\left(\frac{S_{xd}(k)}{F^*(k)}\right)\right] , \tag{2.4.28}$$

where $h(n)$ is the unit step function, $h(n) = 1$ for $n \geq 0$ and otherwise $h(n) = 0$.

The discrete frequency version of the optimum causal filter can now be written as

$$W_{\mathrm{opt}}(k) \;=\; \frac{1}{F(k)}\left\{\frac{S_{xd}(k)}{F^*(k)}\right\}_+ . \tag{2.4.29}$$

Equation (2.4.29) will be equal to the continuous frequency version of the optimum filter at the frequencies corresponding to each bin provided $\left\{S_{xd}(k)/F^*(k)\right\}_+$ is an accurate estimate of $\left\{S_{xd}(\mathrm{e}^{j\omega T})/F^*(\mathrm{e}^{j\omega T})\right\}_+$ at the frequencies corresponding to each bin. This will be guaranteed provided the DFT size is sufficiently large, so that the causal part of the inverse Fourier transform of $S_{xd}\left(\mathrm{e}^{j\omega T}\right)/F^*\left(\mathrm{e}^{j\omega T}\right)$ has a duration of less than $N/2$ samples, as illustrated in Fig. 2.10 (Elliott and Rafaely, 2000).

The numerical conditioning of the solution is improved if the natural logarithm of $S_{xx}(k) + \beta$ is taken instead of $S_{xx}(k)$ in equation (2.4.27), since large negative values of the log are then avoided if $S_{xx}(k)$ becomes very small in any frequency bin. This form of regularisation is equivalent to adding a low level white noise signal to the reference signal, as discussed in Section 2.3, which minimises a cost function of the form of equation (2.3.12). The causal filter which minimises this modified cost function is of exactly the same form as equation (2.4.23), but the spectral factors which must be used are now given by

$$S_{xx}(z) + \beta \;=\; F(z)\,F(z^{-1}) . \tag{2.4.30}$$

2.5. MULTICHANNEL OPTIMAL FILTERS

In some electrical cancellation problems, multiple reference signals are available which are operated on by a matrix of filters to give the best approximation to a number of

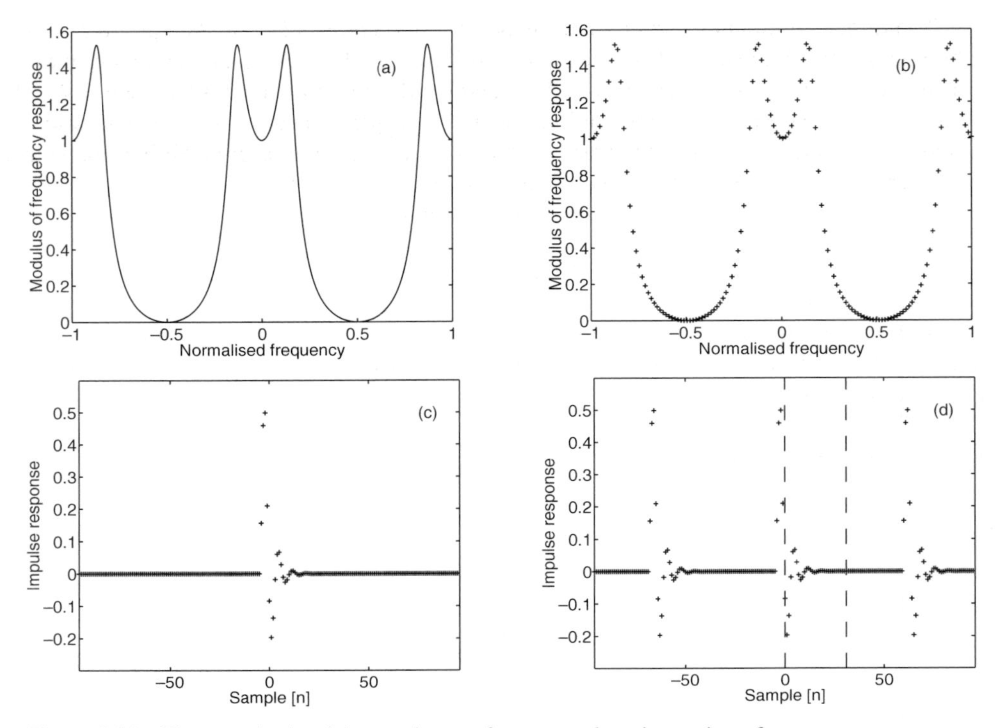

Figure 2.10 The magnitude of the continuous frequency-domain version of

$$A(e^{j\omega T}) = S_{xd}(e^{j\omega T})/F^*(e^{j\omega T}),$$

(a), together with its discrete frequency-domain version $A(k)$, (b), for an example in which this function has a resonant structure and for a DFT size of 64 points. The Wiener filter calculated using the DFT, equation (2.4.28), will be accurate providing the inverse Fourier transform of $A(k)$, shown in (d), is equal to the inverse Fourier transform of $A(e^{j\omega T})$, (c), for $n = 0$ to 32.

desired signals. This is illustrated in Fig. 2.11 in which the vector of K reference signals is defined as

$$\boldsymbol{x}(n) = \left[x_1(n) \quad \cdots \quad x_K(n)\right]^{\mathrm{T}} , \tag{2.5.1}$$

which should not be confused with the vector of delayed versions of the single-channel reference signal in equation (2.3.4). The vector of L desired signals is also defined as

$$\boldsymbol{d}(n) = \left[d_1(n) \quad \cdots \quad d_L(n)\right]^{\mathrm{T}} , \tag{2.5.2}$$

and the vector of L error signals is defined as

$$\boldsymbol{e}(n) = \left[e_1(n) \quad \cdots \quad e_L(n)\right]^{\mathrm{T}} . \tag{2.5.3}$$

The l-th error signal can be written as

$$e_l(n) = d_l(n) - \sum_{k=1}^{K} \sum_{i=0}^{I-1} w_{lki}\, x_k(n-i) , \tag{2.5.4}$$

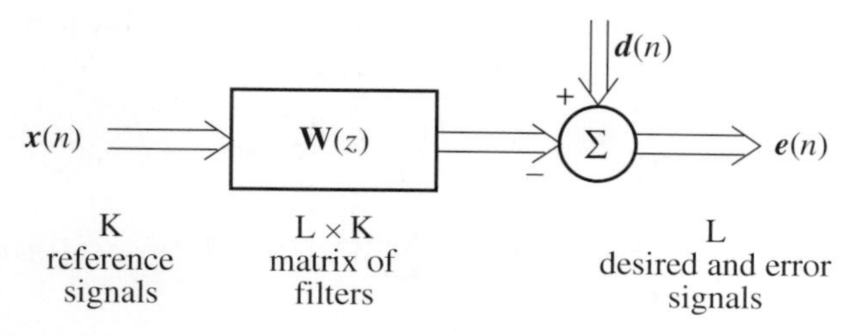

Figure 2.11 Notation for the multichannel electrical filtering problem.

where w_{lki} is the i-th coefficient of the FIR filter contributing to the l-th error signal from the k-th reference signal. Equation (2.5.4) can be written in matrix form as

$$\boldsymbol{e}(n) = \boldsymbol{d}(n) - \sum_{i=0}^{I-1} \mathbf{W}_i \boldsymbol{x}(n-i) , \tag{2.5.5}$$

where the i-th matrix of $L \times K$ filter coefficients is given by

$$\mathbf{W}_i = \begin{bmatrix} w_{11i} & w_{12i} & \cdots w_{1Ki} \\ w_{21i} & w_{22i} & \\ \vdots & \ddots & \\ w_{L1i} & & w_{LKi} \end{bmatrix} . \tag{2.5.6}$$

2.5.1. Time-Domain Solution

The cost function to be minimised can be defined as the sum of the mean-square error signals. Since the l-th error signal is only a function of the KI filter coefficients appearing in equation (2.5.4), however, the problem of finding the optimum value of all LKI filter coefficients can be split into L independent optimisation problems for each set of KI coefficients. Historically, however, this has not been done, and most formulations use a full matrix solution to the complete problem. We will continue with this approach here, since it introduces some of the notation necessary for the analysis of multichannel active control systems in Chapter 5.

The objective of the optimal filtering problem is to minimise the sum of the expectation of the squared errors, which may be written as

$$J = E\left[\sum_{l=1}^{L} e_l^2(n)\right] = \operatorname{trace}\left(E\left[\boldsymbol{e}(n)\boldsymbol{e}^{\mathrm{T}}(n)\right]\right) , \tag{2.5.7}$$

where trace $[\mathbf{M}]$ denotes the trace of the matrix $\mathbf{M}$, which is the sum of its diagonal elements (as described in the Appendix), and $E\left[\boldsymbol{e}(n)\boldsymbol{e}^{\mathrm{T}}(n)\right]$ may be called the correlation matrix since its elements are the auto and cross correlations between each of the elements of $\boldsymbol{e}(n)$.

Using equation (2.5.5) the correlation matrix for the error signals can be written as

$$E\left[\boldsymbol{e}(n)\boldsymbol{e}^{\mathrm{T}}(n)\right] = E\left[\sum_{i=0}^{I-1}\sum_{k=0}^{I-1}\boldsymbol{W}_i\,\boldsymbol{x}(n-i)\boldsymbol{x}^{\mathrm{T}}(n-k)\,\boldsymbol{W}_k^{\mathrm{T}} - \sum_{i=0}^{I-1}\boldsymbol{d}(n)\boldsymbol{x}^{\mathrm{T}}(n-i)\,\boldsymbol{W}_i^{T} - \sum_{i=0}^{I-1}\boldsymbol{W}_i\,\boldsymbol{x}(n-i)\boldsymbol{d}^{\mathrm{T}}(n) + \boldsymbol{d}(n)\boldsymbol{d}^{\mathrm{T}}(n)\right] . \quad (2.5.8)$$

Since the filter coefficients are time invariant this can be written as

$$E\left[\boldsymbol{e}(n)\boldsymbol{e}^{\mathrm{T}}(n)\right] = \sum_{i=0}^{I-1}\sum_{k=0}^{I-1}\boldsymbol{W}_i\,\boldsymbol{R}_{xx}(k-i)\,\boldsymbol{W}_k^{\mathrm{T}} - \sum_{i=0}^{I-1}\boldsymbol{R}_{xd}(i)\,\boldsymbol{W}_i^{\mathrm{T}} - \sum_{i=0}^{I-1}\boldsymbol{W}_i\,\boldsymbol{R}_{xd}^{\mathrm{T}}(i) + \boldsymbol{R}_{dd}(0) , \quad (2.5.9)$$

where it is assumed that the signals are stationary, and the auto- and cross-correlation matrices are defined as

$$\boldsymbol{R}_{xx}(m) = E\left[\boldsymbol{x}(n+m)\boldsymbol{x}^{\mathrm{T}}(n)\right] , \quad (2.5.10)$$

$$\boldsymbol{R}_{xd}(m) = E\left[\boldsymbol{d}(n+m)\boldsymbol{x}^{\mathrm{T}}(n)\right] , \quad (2.5.11)$$

$$\boldsymbol{R}_{dd}(m) = E\left[\boldsymbol{d}(n+m)\boldsymbol{d}^{\mathrm{T}}(n)\right] , \quad (2.5.12)$$

as described in the Appendix, where $\boldsymbol{R}_{xx}(m)$ has the property that

$$\boldsymbol{R}_{xx}^{\mathrm{T}}(m) = \boldsymbol{R}_{xx}(-m) . \quad (2.5.13)$$

Equation (2.5.9) can be written more compactly by defining the overall $KI \times L$ matrix of filter coefficients to be

$$\boldsymbol{W} = \left[\boldsymbol{W}_0\ \boldsymbol{W}_1\ \cdots\ \boldsymbol{W}_{I-1}\right]^{\mathrm{T}} , \quad (2.5.14)$$

the $KI \times L$ matrix of cross-correlation coefficients to be

$$\boldsymbol{R}_{xd} = \left[\boldsymbol{R}_{xd}(0)\ \boldsymbol{R}_{xd}(1)\ \cdots\ \boldsymbol{R}_{xd}(I-1)\right]^{\mathrm{T}} , \quad (2.5.15)$$

and the $KI \times KI$ matrix of auto- and cross-correlation coefficients for the reference signals to be

$$\boldsymbol{R}_{xx} = \begin{bmatrix} \boldsymbol{R}_{xx}(0) & \boldsymbol{R}_{xx}(-1) & \cdots & \boldsymbol{R}_{xx}(1-I) \\ \boldsymbol{R}_{xx}(1) & \boldsymbol{R}_{xx}(0) & & \\ \vdots & & & \\ \boldsymbol{R}_{xx}(I-1) & \cdots & \cdots & \boldsymbol{R}_{xx}(0) \end{bmatrix} , \quad (2.5.16)$$

so that

$$\boldsymbol{R}_{xx}^{\mathrm{T}} = \boldsymbol{R}_{xx} , \quad (2.5.17)$$

which is a block Toeplitz matrix (i.e. its blocks are arranged in a Toeplitz arrangement).

Using these definitions, equation (2.5.9) can be expressed in the matrix form

$$E\left[\boldsymbol{e}(n)\boldsymbol{e}^{\mathrm{T}}(n)\right] = \boldsymbol{W}^{\mathrm{T}}\boldsymbol{R}_{xx}\boldsymbol{W} - \boldsymbol{W}^{\mathrm{T}}\boldsymbol{R}_{xd} - \boldsymbol{R}_{xd}^{\mathrm{T}}\boldsymbol{W} - \boldsymbol{R}_{dd}(0) \ . \qquad (2.5.18)$$

The matrix of derivatives of the trace of this matrix with respect to the elements of $\boldsymbol{W}$ can be calculated, using the rules outlined in Section A.4 of the Appendix, to be

$$\frac{\partial J}{\partial \boldsymbol{W}} = 2\left(\boldsymbol{R}_{xx}\boldsymbol{W} - \boldsymbol{R}_{xd}\right) \ . \qquad (2.5.19)$$

Setting all the elements of this matrix to zero gives

$$\boldsymbol{R}_{xx}\boldsymbol{W}_{\mathrm{opt}} = \boldsymbol{R}_{xd} \ , \qquad (2.5.20)$$

where $\boldsymbol{W}_{\mathrm{opt}}$ is the matrix of optimal filter coefficients.

Assuming $\boldsymbol{R}_{xx}$ is positive definite, the matrix of optimal filter coefficients is thus given by

$$\boldsymbol{W}_{\mathrm{opt}} = \boldsymbol{R}_{xx}^{-1}\boldsymbol{R}_{xd} \ . \qquad (2.5.21)$$

The solution to the multichannel Wiener filtering problem was originally due to Whittle (1963). Wiggins and Robinson (1965) and Robinson (1978) also suggested a generalisation of the Levinson recursion equation to solve equation (2.5.20), which only takes of the order of $(KI)^2$, operations. Notice that each column of $\boldsymbol{W}_{\mathrm{opt}}$ corresponds to a vector that contains all the filter coefficients used to derive a single error signal in equation (2.5.4). Equation (2.5.21) could thus be split up into L separate equations for the L independent minimisation problems discussed above, but the inverse of the same $KI \times KI$ matrix, $\boldsymbol{R}_{xx}$, has to be calculated in each case, so that there is no computational advantage in solving the L individual solutions.

In the single channel case, the autocorrelation matrix was positive definite, provided the reference signal persistently excited the filter, which in practice means that there are at least half as many spectral components in the reference signal as there are coefficients in the filter. In the multichannel case each of the reference signals must also be persistently exciting, but we now have an additional requirement that none of them is perfectly correlated with any other reference signal, or a linear combination of any other reference signals. If two reference signals were exactly the same, for example, the two filters driving each error from these two reference signals would be undetermined, since either could give exactly the same reduction in mean-square level.

Once again the least-squares solution can be regularised by adding a small positive number to the diagonal elements of $\boldsymbol{R}_{xx}$ so that

$$\boldsymbol{W}_{\mathrm{opt}} = \left[\boldsymbol{R}_{xx} + \beta\mathbf{I}\right]^{-1}\boldsymbol{R}_{xd} \ . \qquad (2.5.22)$$

This set of filter coefficients minimises a modified cost function which is similar to equation (2.3.12) and includes β times the sum of squared filter coefficients as well as the sum of squared errors. A physical interpretation of this regularisation procedure in the multichannel case is that uncorrelated white noise signals of mean-square value β have been added to each of the reference signals.

2.5.2. Transform-Domain Solution

An expression for the frequency response of the optimal filters in the multichannel case can also be fairly readily obtained, and this again sets the scene for the multichannel active control problems considered later. The z transform of equation (2.5.5) can be written as

$$\mathbf{e}(z) = \mathbf{d}(z) - \mathbf{W}(z)\,\mathbf{x}(z)\,, \tag{2.5.23}$$

where $\mathbf{e}(z)$, $\mathbf{d}(z)$ and $\mathbf{x}(z)$ are the vector of z transforms of the elements of $\boldsymbol{e}(n)$, $\boldsymbol{d}(n)$ and $\boldsymbol{x}(n)$ in equations (2.5.1, 2, 3) and $\mathbf{W}(z)$ is the matrix of z transforms of the elements of $\boldsymbol{W}_i$ in equation (2.5.6). Note that vectors and matrices of transform-domain quantities are denoted throughout the book by **bold roman** variables, whereas the vectors and matrices of time-domain quantities are denoted by ***bold italic*** variables.

Assume for a moment that each of the reference signals are white noise sequences of unit variance and are all completely uncorrelated. The problem of finding the $L \times K$ matrix of optimal filters now breaks up into $L \times K$ independent problems, since there is no interaction between the filters in this case. Each of these problems can also be readily solved since it reduces to the single channel problem considered in the previous section, whose solution for a single filter can be written as

$$W_{lk,\mathrm{opt}:v}(z) \;=\; \left\{S_{v_k d_l}(z)\right\}_+\,, \tag{2.5.24}$$

where $S_{v_k d_l}$ is the cross spectral density between the k-th uncorrelated reference signal, v_k, and the l-th desired signal, d_l, and $\{\;\}_+$ again denotes the z transform of the causal part of the time domain version of the quantity in brackets.

The reference signals will not, however, generally be uncorrelated or white, but we can use a generalisation of the shaping filter, introduced above, to transform the true reference signals into a new set of reference signals having these properties. In particular we assume that the K observed reference signals, $\boldsymbol{x}(n)$, were originally generated by passing a set of K uncorrelated and white reference signals, $\boldsymbol{v}(n)$, of unit variance, through a matrix of shaping filters so that

$$\mathbf{x}(z) = \mathbf{F}(z)\,\mathbf{v}(z)\,, \tag{2.5.25}$$

where $\mathbf{x}(z)$ and $\mathbf{v}(z)$ are the vectors of z transforms of the observed and uncorrelated reference signals. The spectral properties of a set of signals are often described by the *spectral density matrix*, as discussed in Section A.5 of the Appendix, which for the set of observed reference signals can be written as

$$\mathbf{S}_{xx}(z) \;=\; E\left[\mathbf{x}(z)\mathbf{x}^{\mathrm{T}}(z^{-1})\right]\,, \tag{2.5.26}$$

which is the z transform of $\boldsymbol{R}_{xx}(m)$ in equation (2.5.10), and once again, the expectation is in practice taken over blocks of data. Using the definition of the shaping filter (equation 2.5.25) the cross spectral matrix for the observed reference signals can be written as

$$\mathbf{S}_{xx}(z) \;=\; \mathbf{F}(z)\mathbf{S}_{vv}(z)\mathbf{F}^{\mathrm{T}}(z^{-1})\,. \tag{2.5.27}$$

The elements of $\mathbf{v}(n)$ have been assumed to be mutually uncorrelated, white, and of unit variance. Its cross spectral matrix is thus equal to the identity matrix, and so

$$\mathbf{S}_{xx}(z) \;=\; \mathbf{F}(z)\,\mathbf{F}^{\mathrm{T}}(z^{-1}) \;. \tag{2.5.28}$$

The matrix of shaping filters can thus be considered as part of a *spectral factorisation* of the cross spectral matrix for the multiple reference signals (Youla, 1961), in which the shaping filters in the matrix $\mathbf{F}(z)$ are causal and minimum phase, so that the filters which are the elements of the inverse of this matrix, $\mathbf{F}^{-1}(z)$, are also stable and causal.

The conditions under which such a spectral factorisation exists are discussed for the continuous-time case by Bongiorno (1969), for example, and for the discrete time case here these conditions may be summarised (Kailath *et al.*, 2000), as

(i) $\mathbf{S}_{xx}(z) = \mathbf{S}_{xx}^{\mathrm{T}}(z^{-1})$, which is true from the definition in equation (2.5.26).

(ii) $\mathbf{S}_{xx}\left(\mathrm{e}^{j\omega T}\right)$ must be analytic for all ωT.

(iii) $\mathbf{S}_{xx}\left(\mathrm{e}^{j\omega T}\right)$ must be positive definite at all frequencies.

The final condition is avoided if regularisation is included since the spectral factors then have to be taken of $\mathbf{S}_{xx}(z) + \beta\mathbf{I}$ instead of $\mathbf{S}_{xx}(z)$, which is guaranteed to satisfy condition (iii).

Since the filters corresponding to the elements of the inverse shaping filter matrix are stable and causal, the set of uncorrelated reference signals can be generated from the observed reference signals by operating on them with this inverse filter, so that

$$\mathbf{v}(z) = \mathbf{F}^{-1}(z)\,\mathbf{x}(z) \;. \tag{2.5.29}$$

If the K measured reference signals in $\mathbf{x}(z)$ are, in practice, generated by a smaller number of independent processes, then $\mathbf{S}_{xx}(z)$ will only be positive semi-definite. The vector $\mathbf{x}(z)$ could still be modelled by an equation of the form of (2.5.25), however, if $\mathbf{F}(z)$ were no longer square. In this case, the matrix inverse of $\mathbf{F}(z)$ in equation (2.5.29) would have to be replaced by a pseudo-inverse.

The matrix of optimal filters for the uncorrelated reference signals, whose elements are given by equation (2.5.24), can now be written as

$$\mathbf{W}_{\mathrm{opt:}v}(z) \;=\; \left\{\mathbf{S}_{vd}(z)\right\}_{+} \;, \tag{2.5.30}$$

where

$$\mathbf{S}_{vd}(z) \;=\; E\left[\mathbf{d}(z)\,\mathbf{v}^{\mathrm{T}}(z^{-1})\right] \;. \tag{2.5.31}$$

Using equation (2.5.29), this can thus be written as

$$\mathbf{S}_{vd}(z) \;=\; \mathbf{S}_{xd}(z)\,\mathbf{F}^{-\mathrm{T}}(z^{-1}) \;, \tag{2.5.32}$$

where $\mathbf{S}_{xd}(z)$ is the z transform of $\mathbf{R}_{xd}(m)$ and can also be written as

$$\mathbf{S}_{xd}(z) \;=\; E\left[\mathbf{d}(z)\,\mathbf{x}^{\mathrm{T}}(z^{-1})\right] \;, \tag{2.5.33}$$

which is the cross spectral matrix between the observed reference signals and the desired signals.

The expression for the matrix of optimal filters that operate on the observed reference signals to minimise the sum of the square error signals can now be derived by analogy with the single-channel case. Using the block diagram shown in Fig. 2.12 we can see that this filter must be equal to

$$\mathbf{W}_{\mathrm{opt}}(z) = \left\{\mathbf{S}_{vd}(z)\right\}_{+} \mathbf{F}^{-1}(z) , \tag{2.5.34}$$

and using equation (2.5.32) for $\mathbf{S}_{vd}(z)$, $\mathbf{W}_{\mathrm{opt}}(z)$ can be written as

$$\mathbf{W}_{\mathrm{opt}}(z) = \left\{\mathbf{S}_{xd}(z)\mathbf{F}^{-\mathrm{T}}(z^{-1})\right\}_{+} \mathbf{F}^{-1}(z) . \tag{2.5.35}$$

This reduces to the single-channel result, equation (2.4.23), if there is only a single reference signal and error.

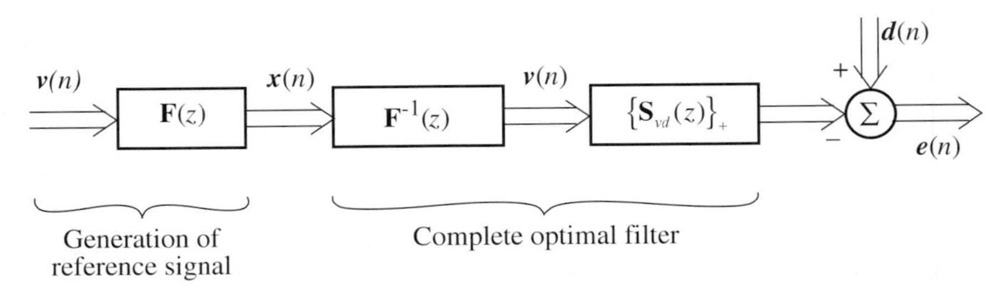

Figure 2.12 Block diagram of the model assumed to generate the reference signals in a multichannel system, together with the elements of the optimal causal filter which minimises the sum of mean-square errors.

If the z transform of the cross-correlation matrix between the reference signals and the error signals is defined to be

$$\mathbf{S}_{xe}(z) = E\left[\mathbf{e}(z)\mathbf{x}^{\mathrm{T}}(z^{-1})\right] , \tag{2.5.36}$$

then using equations (2.5.23, 26 and 33),

$$\mathbf{S}_{xe}(z) = \mathbf{S}_{xd}(z) - \mathbf{W}(z)\mathbf{S}_{xx}(z) . \tag{2.5.37}$$

Equation (2.5.35) can also be derived by setting the causal part of equation (2.5.37) to zero. This is a statement of the principle of orthogonality in this case, which may be written

$$\left\{\mathbf{S}_{xe}(z)\right\}_{+} = \mathbf{0} \quad \text{if} \quad \mathbf{W}(z) = \mathbf{W}_{\mathrm{opt}}(z) , \tag{2.5.38}$$

i.e. each of the error signals is uncorrelated with previous values of each of the reference signals (Davis, 1963). If the constraint of causality is removed from equation (2.5.35) then the matrix of optimal filters is equal to

$$\mathbf{W}_{\mathrm{opt}}(z) = \mathbf{S}_{xd}(z)\mathbf{F}^{-\mathrm{T}}(z^{-1})\mathbf{F}^{-1}(z) , \tag{2.5.39}$$

and using equation (2.5.28) this can be written

$$\mathbf{W}_{\mathrm{opt}}(z) = \mathbf{S}_{xd}(z)\, \mathbf{S}_{xx}^{-1}(z) , \tag{2.5.40}$$

which is a generalisation of the unconstrained single channel case, equation (2.4.2).

The matrix of frequency responses of the optimum causal filters can thus be calculated directly from the matrix of cross spectral densities between the original reference and desired signals, $\mathbf{S}_{xd}(e^{j\omega T})$, and the spectral factorisation of the cross spectral matrix of the original reference signals, $\mathbf{S}_{xx}(e^{j\omega T})$. Methods of calculating the spectral factors based on a polynomial representation of the spectral density matrix have been described by Davis (1963) and Wilson (1972), for example, and a method which uses the discrete-frequency representation has recently been described by Cook and Elliott (1999).

2.6. THE LMS ALGORITHM

In Section 2.3 we saw that the optimum FIR filter, which minimises the mean-square error in the model problem shown in Fig. 2.4, can be directly calculated from a knowledge of the autocorrelation properties of the reference signal and the cross-correlation between the reference and desired signal. In a practical problem, the auto- and cross-correlation functions would have to be estimated from the time histories of these signals, which requires a considerable amount of data to calculate accurately. We have also assumed that the reference and desired signals are stationary in the first place, since otherwise their correlation properties will change with time. The calculation of the optimal filter with I coefficients also involves the inversion of the $I \times I$ autocorrelation matrix. Although this matrix has a special form (it is symmetric and Toeplitz), and so efficient algorithms can be used for its inversion (Levinson, 1947), the computational burden is still proportional to I^2 and so can be significant, particularly for long filters. The matrix inversion may also be numerically unstable with fixed-precision arithmetic if the matrix is ill-conditioned.

Another approach to determining the coefficients of such a filter would be to make them adaptive. Instead of using a set of data to estimate correlation functions, and then using these correlation functions to calculate a single set of 'optimal' filter coefficients, the data is used sequentially to adjust the filter coefficients so that they evolve in a direction which minimises the mean-square error. Generally, all the filter coefficients are adjusted in response to each new set of data, and the algorithms used for this adaptation use a considerably smaller number of calculations per sample than the total number of calculations required to compute the true optimal coefficients. As well as converging towards the optimal filter for stationary signals, an adaptive filter will also automatically re-adjust its coefficients if the correlation properties of these signals change. The adaptive filter may thus be capable of tracking the statistics of non-stationary signals provided the changes in the statistics occur slowly compared with the *convergence time* of the adaptive filter.

2.6.1. Steepest Descent Algorithm

The most widely used algorithm for adapting FIR filters is based on the fact that the error surface for such filters has a quadratic shape, as shown in Fig. 2.5. This suggests that if a filter coefficient is adjusted by a small amount, which is proportional to the negative of the local gradient of the cost function with respect to that filter coefficient, then the coefficient is bound to move towards the global minimum of the error surface. If all the filter coefficients are simultaneously adjusted using this *steepest-descent algorithm*, the adaptation algorithm for the vector of filter coefficients may be written as

$$\mathbf{w}(\text{new}) = \mathbf{w}(\text{old}) - \mu \frac{\partial J}{\partial \mathbf{w}}(\text{old}) , \quad (2.6.1)$$

where μ is a convergence factor and $\partial J/\partial \mathbf{w}$ was defined in equation (2.3.16).

For the model problem shown in Fig. 2.4, the vector of derivatives is given by equation (2.3.17), and using the definitions given in equations (2.3.7) and (2.3.8), this can be written as

$$\frac{\partial J}{\partial \mathbf{w}} = 2E\left[\mathbf{x}(n)\,\mathbf{x}^{\mathrm{T}}(n)\mathbf{w} - \mathbf{x}(n)\,d(n)\right] . \quad (2.6.2)$$

The measured error signal in Fig. 2.4 is given by

$$e(n) = d(n) - \mathbf{x}^{\mathrm{T}}(n)\mathbf{w} , \quad (2.6.3)$$

and so the vector of derivatives may also be written as

$$\frac{\partial J}{\partial \mathbf{w}} = -2E\left[\mathbf{x}(n)\,e(n)\right] . \quad (2.6.4)$$

In order to implement the true gradient descent algorithm, the expectation value of the product of the error signal with the delayed reference signals would need to be estimated to obtain equation (2.6.4), probably by time averaging over a large segment of data, and so the filter coefficients could only be updated rather infrequently.

The suggestion made in the seminal paper by Widrow and Hoff (1960) was that instead of infrequently updating the filter coefficients with an averaged estimate of the gradient, the coefficients be updated at every sample time using an instantaneous estimate of the gradient, which is sometimes called the *stochastic gradient*. This update quantity is equal to the derivative of the instantaneous error with respect to the filter coefficients,

$$\frac{\partial e^2(n)}{\partial \mathbf{w}} = -2\mathbf{x}(n)e(n) . \quad (2.6.5)$$

The adaptation algorithm thus becomes

$$\mathbf{w}(n+1) = \mathbf{w}(n) + \alpha\mathbf{x}(n)e(n) , \quad (2.6.6)$$

where $\alpha = 2\mu$ is the convergence coefficient, which is known as the *LMS algorithm*. The LMS algorithm is simple to implement, is numerically robust and has been very widely used in a variety of practical applications, for example in adaptive electrical noise cancellation, adaptive modelling and inversion and adaptive beam forming (Widrow and Stearns, 1985). A block diagram indicating the operation of the LMS algorithm is shown in Fig. 2.13, in which the error signal and reference signals are multiplied together and their product is used to adapt the coefficients of the filter.

2.6.2. Convergence of the LMS Algorithm

Studies of the convergence properties of the LMS algorithm have had a long and chequered history. The difficulty with this analysis is partly due to the stochastic nature of

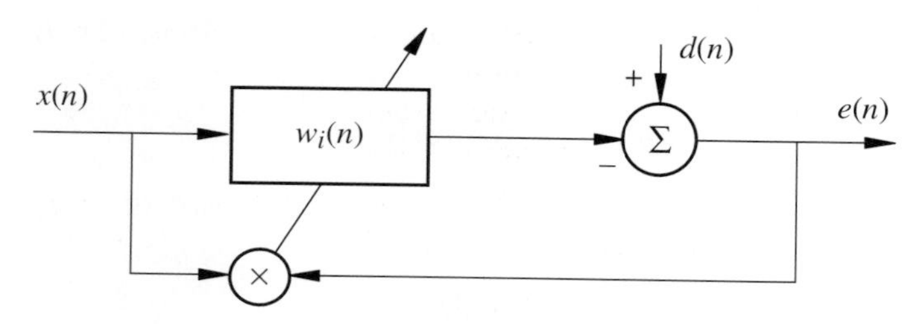

Figure 2.13 Diagrammatic representation of an FIR filter, with coefficients w_i at the n-th sample time, being adapted using the LMS algorithm.

the gradient estimate provided by equation (2.6.5) for random data, which means that the time histories of the filter coefficients themselves will be subject to random variations when applied to one set of data or another. The convergence behaviour is calculated below by considering the average, or mean, behaviour of the filter coefficients. It turns out, however, that this does not provide a complete description of the behaviour of the algorithm, even for a stability analysis, since the variance of the filter coefficient time histories, their mean-square values, can become large even when their average time history converges. Nevertheless, an analysis of behaviour of the filter coefficients in the mean illustrates a number of important properties of the LMS algorithm. We begin such an analysis by taking the expectation value of each term in equation (2.6.6), to calculate the average behaviour of the algorithm over a number of trials with different sets of random data, which are each assumed to have the same statistical properties; thus we have

$$E[\mathbf{w}(n+1)] = E[\mathbf{w}(n)] + \alpha\, E[\mathbf{x}(n)e(n)] \; . \tag{2.6.7}$$

Expanding out the error signal using equation (2.6.3) then gives

$$E[\mathbf{w}(n+1)] = E[\mathbf{w}(n)] + \alpha\left\{E[\mathbf{x}(n)d(n)] - E[\mathbf{x}(n)\mathbf{x}^{\mathrm{T}}(n)\mathbf{w}(n)]\right\} . \tag{2.6.8}$$

We now assume that the filter coefficients change only slightly over I samples, and so the variation of $\mathbf{w}(n)$ is statistically independent of $\mathbf{x}(n)$. This *independence assumption* is difficult to justify for some data sets or for a rapidly adapting filter, but it does lead to a relatively simple model to describe the convergence of the filter, which does give a reasonably good prediction of the observed behaviour of the LMS algorithm in practice. Using the independence assumption, equation (2.6.8) can be written as

$$E[\mathbf{w}(n+1)] = E[\mathbf{w}(n)] + \alpha\left\{E[\mathbf{x}(n)d(n)] - E[\mathbf{x}(n)\mathbf{x}^{\mathrm{T}}(n)]\,E[\mathbf{w}(n)]\right\} . \tag{2.6.9}$$

The cross-correlation vector and autocorrelation matrix are equal to $\boldsymbol{b}$ and $\boldsymbol{A}$, as defined in equations (2.3.8) and (2.3.7). The difference between the expected value of the filter coefficients and their optimal, Wiener, values is defined to be

$$\boldsymbol{\varepsilon}(n) = E[\mathbf{w}(n)] - \mathbf{w}_{\mathrm{opt}} \; , \tag{2.6.10}$$

where $\mathbf{w}_{\mathrm{opt}} = \boldsymbol{A}^{-1}\boldsymbol{b}$, as in equation (2.3.18) so that $\boldsymbol{b} = \boldsymbol{A}\,\mathbf{w}_{\mathrm{opt}}$.

The equation for the average evolution of the filter coefficients, (2.6.9), can now be expressed in the considerably simplified form

$$\boldsymbol{\varepsilon}(n+1) = \left[\mathbf{I} - \alpha\boldsymbol{A}\right]\boldsymbol{\varepsilon}(n) . \tag{2.6.11}$$

The autocorrelation matrix $\boldsymbol{A}$ can be written in terms of the matrix of its eigenvectors, $\mathbf{Q}$, and the diagonal matrix of its eigenvalues, $\boldsymbol{\Lambda}$, so that

$$\boldsymbol{A} = \mathbf{Q}\,\boldsymbol{\Lambda}\,\mathbf{Q}^{\mathrm{T}} , \tag{2.6.12}$$

where the eigenvectors can be assumed to be orthonormal, since $\boldsymbol{A}$ is symmetric, so that

$$\mathbf{Q}\,\mathbf{Q}^{\mathrm{T}} = \mathbf{I} , \tag{2.6.13}$$

and $\mathbf{Q}^{-1} = \mathbf{Q}^{\mathrm{T}}$. The eigenvalues of $\boldsymbol{A}$ are real and non-negative, and are arranged in $\boldsymbol{\Lambda}$ as below

$$\boldsymbol{\Lambda} = \begin{bmatrix} \lambda_1 & & & \\ & \lambda_2 & & 0 \\ & & \ddots & \\ & 0 & & \lambda_I \end{bmatrix} . \tag{2.6.14}$$

Using these properties, equation (2.6.11) can be written as

$$\mathbf{Q}^{\mathrm{T}}\boldsymbol{\varepsilon}(n+1) = \left[\mathbf{I} - \alpha\boldsymbol{\Lambda}\right]\mathbf{Q}^{\mathrm{T}}\boldsymbol{\varepsilon}(n) . \tag{2.6.15}$$

The elements of the vector $\mathbf{Q}^{\mathrm{T}}\boldsymbol{\varepsilon}(n)$ are equal to a transformed set of normalised averaged filter coefficients which correspond to the *principal axes* of the error surface (Widrow and Stearns, 1985). Setting the vector $\mathbf{Q}^{\mathrm{T}}\boldsymbol{\varepsilon}(n)$ equal to $\boldsymbol{\nu}(n)$, equation (2.6.15) becomes

$$\boldsymbol{\nu}(n+1) = \left[\mathbf{I} - \alpha\boldsymbol{\Lambda}\right]\boldsymbol{\nu}(n) , \tag{2.6.16}$$

Since $\boldsymbol{\Lambda}$ is diagonal, equation (2.6.16) represents a set of I independent equations in the elements of $\underline{v}(n) = \left[v_0(n)\ v_1(n)\ \cdots\ v_{I-1}(n)\right]$, which can be written as

$$v_i(n+1) = (1 - \alpha\lambda_i)\,v_i(n) . \tag{2.6.17}$$

The average convergence behaviour of each of the normalised and transformed coefficients of the LMS algorithm is thus independent, and they are said to correspond to the *modes* of the algorithm. The speed of convergence in each of the modes is dependent on the associated eigenvalue of the correlation matrix, λ_i, with the modes associated with the larger eigenvalues having a steeper slope in the error surface and thus more rapid convergence than those associated with smaller eigenvalues. Figure 2.14 shows the error surface for an FIR filtering problem having two eigenvalues, one of which is 100 times larger than the other. The solid line in this figure shows the *average* trajectory of the filter coefficients as the LMS algorithm converges, which is the same as the path of the steepest descent algorithm, equation (2.6.1). For any given initial value, $v_i(n)$ will decay to zero provided

$$\left|1 - \alpha\,\lambda_i\right| < 1 \quad \text{i.e.} \quad 0 < \alpha < 2/\lambda_i , \tag{2.6.18}$$

so that the average behaviour of the LMS algorithm is guaranteed to converge to the optimum Wiener filter under these conditions provided none of the eigenvalues is zero.

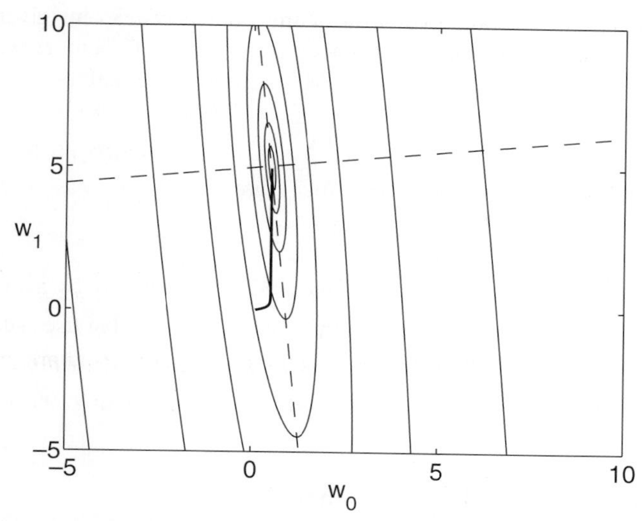

Figure 2.14 Contour map of the error surface for the LMS algorithm with two filter coefficients showing the contours of constant mean square, in 5 dB steps, and the trajectory of the *average* behaviour of the algorithm. Also shown are the principal axes for the error surface, v_i and v_j.

Assuming the adaptive filter is persistently excited, the matrix **A** must be positive definite and so must have positive real eigenvalues. To ensure the convergence of all components of the filter coefficients, equation (2.6.18) must be true for all λ_i, and since the condition is most stringent for the largest eigenvalue, $\lambda_{\max}$, the stability condition derived from the convergence of the mean value of filter coefficients can be written as

$$0 < \alpha < 2/\lambda_{\max} \ . \tag{2.6.19}$$

It is argued by Widrow and Walach (1996), for example, that convergence in the mean-square value of the filter coefficients requires the more restrictive condition on the convergence coefficient, given by

$$0 < \alpha < 2 \Big/ \sum_{i=0}^{I-1} \lambda_i \ . \tag{2.6.20}$$

The sum of the eigenvalues of a matrix is equal to its trace (see Appendix), which is given by the sum of the elements on its main diagonal. The matrix of eigenvectors, **Q**, is also orthonormal, so that trace (**Λ**) is equal to trace (***A***). Since all the diagonal elements of the autocorrelation matrix, ***A*** in equation (2.3.10), are equal to the mean-square value of the reference signal, equation (2.6.20) is equivalent to the condition

$$0 < \alpha < 2/I\overline{x^2} \ , \tag{2.6.21}$$

where $\overline{x^2}$ is equal to $E[x^2(n)]$, i.e. the mean-square value of $x(n)$. The stability condition given by equation (2.6.21) is considerably easier to calculate than equation (2.6.19), and has been found to give a reasonable estimate of the maximum stable value of the convergence coefficient in a number of practical simulations (Clarkson, 1993).

The condition on the upper bound of the convergence coefficient given by equation (2.6.21) leads to the definition of a normalised convergence coefficient, $\tilde{\alpha}$, which is defined such that the convergence coefficient in the LMS algorithm is equal to

$$\alpha = \tilde{\alpha} / I \overline{x^2} \ . \tag{2.6.22}$$

The stability bounds on $\tilde{\alpha}$, from equation (2.6.21), are thus

$$0 < \tilde{\alpha} < 2 \ . \tag{2.6.23}$$

In practice $\overline{x^2}$ must be estimated from the measured data, normally by averaging over the last I data points, and the modified form of the LMS algorithm that uses such an estimate for $\overline{x^2}$ in equation (2.6.22) is called the *normalised LMS algorithm*, an interesting discussion of which is provided by Haykin (1996) and Glentis et al. (1999).

2.6.3. Misadjustment and Convergence Rate

Another consequence of the random variation of the filter coefficients is that, after convergence in a stationary environment, the mean-square error is generally higher than it would be if the filter coefficients were fixed at their optimal value. Unless it happens that the error signal is exactly zero after the adaptive filter converges, there will still be some random changes in the filter coefficients even after convergence. This is because the update term $\boldsymbol{x}(n)\, e(n)$ in equation (2.6.6) will generally have a non-zero value at each time instant, even though its average value must be zero if the filter is adjusted optimally. If the filter coefficients were steady and equal to their optimal value, the mean-square error $J_{\min}$ is given by equation (2.3.21). The mean-square error after convergence of the LMS algorithm, J_∞, is found to be proportional to $J_{\min}$, and $J_\infty - J_{\min}$ is referred to as the *excess* mean-square error for the filter. The ratio of the excess mean-square error to the minimum mean-square error is called the *misadjustment* of the adaptive algorithm, and for the LMS algorithm, Haykin (1996), for example, show that the misadjustment has the form

$$M = \frac{J_\infty - J_{\min}}{J_{\min}} = \sum_{i=0}^{I-1} \frac{\alpha \lambda_i}{2 - \alpha \lambda_i} \ , \tag{2.6.24}$$

which for small α is equal to $\alpha I \overline{x^2}/2$ (Widrow and Stearns, 1985). Clearly the misadjustment can be made arbitrarily small by reducing the convergence coefficient, α, but then the filter will then take longer to converge. Thus with the LMS, and with most other adaptive algorithms, there is a trade-off between the convergence speed and the accuracy of the converged coefficients. The optimum value of the convergence coefficient, α, is thus rather application-specific.

The average convergence behaviour of the individual 'modes' of the LMS algorithm is described by equation (2.6.17). Assuming that all the modes are stable, the average behaviour of the mean-square error associated with each mode will be governed by an exponential decay with a time constant, τ_i, given by

$$\tau_i \approx \frac{1}{2\alpha\lambda_i} \ \text{(samples)} \ . \tag{2.6.25}$$

The mode associated with the largest eigenvalue, λ_{max}, thus decays fastest, with time constant τ_{min}, and that with the smallest eigenvalue, λ_{min}, decays most slowly, with time constant τ_{max}. This is illustrated in Fig. 2.15, which shows the average value of the error for the convergence of the LMS algorithm in the example whose error surface was shown in Fig. 2.14, which has two modes of convergence: one that converges rapidly until the error is attenuated by about 3 dB, and the other that converges more slowly until the error is attenuated by about 25 dB. The ratio of the smallest time constant to the largest can be expressed, using equation (2.6.25) as

$$\frac{\tau_{min}}{\tau_{max}} = \frac{\lambda_{max}}{\lambda_{min}}, \tag{2.6.26}$$

which is independent of α. The range of convergence rates for the modes thus depends on the correlation properties of the reference signal, which determine the eigenvalues of the autocorrelation matrix. It should be noted, however, that slow modes of convergence may not be a problem if these modes are not significantly excited to begin with, i.e. if the values of $\nu_i(0)$ in equation (2.6.17) are very small for these modes, which depends on the properties of the desired signal, $d(n)$.

A physical interpretation of the eigenvalue spread can be obtained from an interesting relationship between power spectral density of a signal and the eigenvalues of the corresponding autocorrelation matrix (Gray, 1972). It can be shown (see, for example, Haykin 1996) that

$$\frac{\lambda_{max}}{\lambda_{min}} \leq \frac{S_{xx}(\max)}{S_{xx}(\min)}, \tag{2.6.27}$$

where $S_{xx}(\max)$ and $S_{xx}(\min)$ are the largest and smallest values of the power spectral density of the reference signal at any frequency. As the size of the autocorrelation matrix becomes large, for many coefficients in the FIR filter, λ_{max} approaches $S_{xx}(\max)$ and λ_{min}

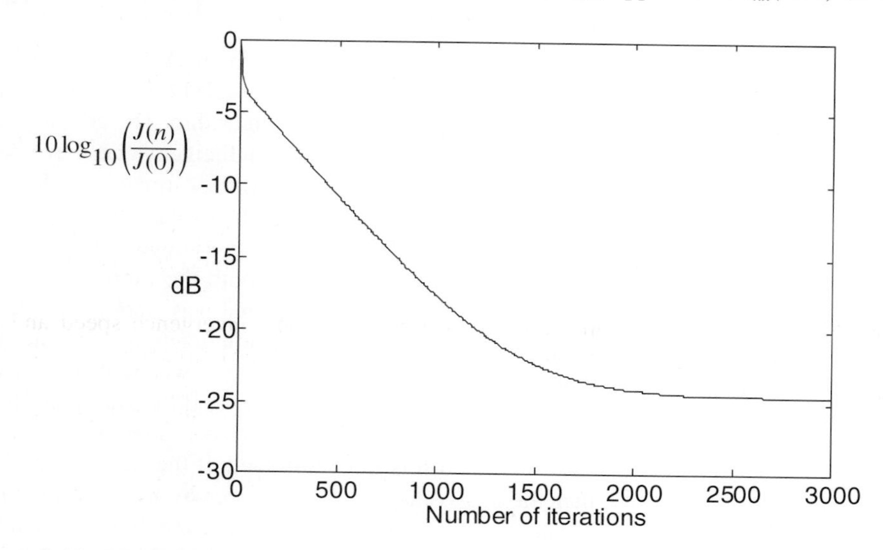

Figure 2.15 The average convergence behaviour of the LMS algorithm in the case in which two modes converge at different rates.

approaches $S_{xx}(\text{min})$ so that $\lambda_{max}/\lambda_{min}$ becomes almost equal to $S_{xx}(\text{max})/S_{xx}(\text{min})$. A potential problem with slow modes of convergence can thus be anticipated if the power spectral density of the reference signal has a large dynamic range.

2.7. THE RLS ALGORITHM

2.7.1. Newton's Method

The slow convergence of the modes associated with small eigenvalues of the Hessian matrix, $\boldsymbol{A}$, is a well-known property of steepest descent adaptive algorithms such as equation (2.6.1). One way of avoiding these problems is to use *Newton's method* to adapt the filter coefficients, which may be written as

$$\boldsymbol{w}(\text{new}) = \boldsymbol{w}(\text{old}) - \mu \boldsymbol{A}^{-1} \frac{\partial J}{\partial \boldsymbol{w}}(\text{old}) , \tag{2.7.1}$$

where the Hessian matrix $\boldsymbol{A}$ is equal to the autocorrelation matrix of the reference signal given by equation (2.3.7).

The convergence properties of this algorithm can be examined by noting from equation (2.3.17), that

$$\frac{\partial J}{\partial \boldsymbol{w}} = 2[\boldsymbol{A}\boldsymbol{w} - \boldsymbol{b}] , \tag{2.7.2}$$

where in this case $\boldsymbol{b}$ is the vector of cross-correlations given by equation (2.3.8). Substituting this into equation (2.7.1) and noting that $\boldsymbol{A}^{-1}\boldsymbol{b} = \boldsymbol{w}_{\text{opt}}$, equation (2.7.1) can be written as

$$\boldsymbol{w}(\text{new}) = (1 - \alpha)\, \boldsymbol{w}(\text{old}) + \alpha\, \boldsymbol{w}_{\text{opt}} , \tag{2.7.3}$$

where $\alpha = 2\mu$. Under the ideal conditions assumed here, in which $\boldsymbol{A}^{-1}$ is known with perfect accuracy and the averaged value of $\partial J/\partial \boldsymbol{w}$ is used, then all the filter coefficients converge independently and at the same rate using Newton's algorithm. The convergence problems of the steepest descent method associated with the small eigenvalues have been resolved by pre-conditioning the gradient with the matrix $\boldsymbol{A}^{-1}$. If the estimates of $\boldsymbol{A}^{-1}$ and $\partial J/\partial \boldsymbol{w}$ in equation (2.7.1) could be accepted with perfect faith then we could set $\mu = \frac{1}{2}$, in which case the algorithm would converge to the optimum result in a single step.

There will generally be practical problems associated with both the estimation of the elements of the matrix $\boldsymbol{A}$ and its inversion. If the reference signal was stationary, a large section of data would need to be used to estimate its autocorrelation function, and hence the elements of $\boldsymbol{A}$, which would be rather time consuming. The matrix $\boldsymbol{A}$ would also still have to be inverted and this would require considerable processing power if the number of filter coefficients, and hence the size of the $\boldsymbol{A}$ matrix, was large. Also there may be numerical problems associated with the inversion of $\boldsymbol{A}$ if it is ill-conditioned. If the eigenvalues of $\boldsymbol{A}$ are equal to λ_i, for $i = 0$ to $I - 1$, then the eigenvalues of $\boldsymbol{A}^{-1}$ are given by $1/\lambda_i$. Thus if some of the eigenvalues of $\boldsymbol{A}$ have very small values and are not very accurately estimated, the calculated value of $\boldsymbol{A}^{-1}$ can be considerably in error. These problems could be overcome to some extent if $\boldsymbol{A}^{-1}$ was carefully pre-computed during an initialisation stage and then used in equation (2.7.1) as a fixed matrix. Such an algorithm would not, however, be able to accommodate significant changes in the statistical properties of the reference signal, and so would not be truly adaptive.

Assuming that $\boldsymbol{A}^{-1}$ is somehow estimated as $\hat{\boldsymbol{A}}^{-1}$ and that the instantaneous estimate of the gradient, equation (2.6.5), is used at every sample time, as in the LMS algorithm, a modified form of Newton's method can be devised, which has the form

$$\boldsymbol{w}(n+1) = \boldsymbol{w}(n) - \alpha \hat{\boldsymbol{A}}^{-1} \boldsymbol{x}(n) e(n) . \tag{2.7.4}$$

2.7.2. Recursive Least-Squares Algorithm

Equation (2.7.4) has many similarities with the *recursive least-squares* (RLS) algorithm, although the RLS algorithm is generally derived from a rather different perspective (see for example, Haykin, 1996). The RLS algorithm performs an exact minimisation of a quadratic cost function at every sample time. The cost function used is often the *exponentially weighted mean-square error*, defined at sample time n to be

$$J(n) = \sum_{l=0}^{n} \lambda^{n-l} e^2(l|n) , \tag{2.7.5}$$

where $e(l|n)$ is the time history of the error that would have been generated if the filter coefficients had always been fixed to be those at the current time, so that

$$e(l|n) = d(l) - \boldsymbol{w}^{\mathrm{T}}(n)\boldsymbol{x}(l) . \tag{2.7.6}$$

Note that all the previous values of the squared error $e^2(l|n)$ used in the definition of $J(n)$ are weighted by progressively higher powers of the *forgetting factor*, λ, which is between 0 and 1, and should not be confused with the eigenvalues referred to above.

The time-dependent cost function given by equation (2.7.5) can be expressed in the quadratic form

$$J(n) = \boldsymbol{w}^{\mathrm{T}}(n)\boldsymbol{A}(n)\boldsymbol{w}(n) - 2\,\boldsymbol{w}^{\mathrm{T}}(n)\boldsymbol{b}(n) + c(n) , \tag{2.7.7}$$

where in this case

$$\boldsymbol{A}(n) = \sum_{l=0}^{n} \lambda^{n-l} \boldsymbol{x}(l)\boldsymbol{x}^{\mathrm{T}}(l) , \tag{2.7.8}$$

$$\boldsymbol{b}(n) = \sum_{l=0}^{n} \lambda^{n-l} \boldsymbol{x}(l) d(l) , \tag{2.7.9}$$

and

$$c(n) = \sum_{l=0}^{n} \lambda^{n-l} d^2(l) . \tag{2.7.10}$$

This quadratic form is minimised at the n-th sample by the vector of filter coefficients given by

$$\boldsymbol{w}_{\mathrm{opt}}(n) = \boldsymbol{A}^{-1}(n)\boldsymbol{b}(n) . \tag{2.7.11}$$

The RLS algorithm is based on an exact implementation of equation (2.7.11) at every sample time.

It should be noted that both $\boldsymbol{A}(n)$ and $\boldsymbol{b}(n)$ contain data up to the n-th sample time and so $\boldsymbol{w}_{\text{opt}}(n)$ cannot be calculated until these samples have been acquired. To be consistent with the form of the LMS algorithm given in equation (2.6.6), in which data from the n-th sample time is used to calculate $\boldsymbol{w}(n+1)$, we must use equation (2.7.11) to calculate $\boldsymbol{w}(n+1)$, and so the aim of the RLS algorithm is to set $\boldsymbol{w}(n+1)$ equal to

$$\boldsymbol{w}(n+1) = \boldsymbol{A}^{-1}(n)\boldsymbol{b}(n) , \qquad (2.7.12)$$

where $\boldsymbol{A}(n)$ and $\boldsymbol{b}(n)$ are defined by equations (2.7.8) and (2.7.9).

The calculation of $\boldsymbol{w}(n+1)$ is based on the previous value of the filter coefficients, $\boldsymbol{w}(n)$, which, from equation (2.7.12), must be given by

$$\boldsymbol{w}(n) = \boldsymbol{A}^{-1}(n-1)\boldsymbol{b}(n-1) . \qquad (2.7.13)$$

From the definition in equation (2.7.9), we can see that $\boldsymbol{b}(n)$ can be calculated from $\boldsymbol{b}(n-1)$ using only data at the n-th sample time, so that

$$\boldsymbol{b}(n) = \lambda\, \boldsymbol{b}(n-1) + \boldsymbol{x}(n)\, d(n) . \qquad (2.7.14)$$

Similarly

$$\boldsymbol{A}(n) = \lambda\, \boldsymbol{A}(n-1) + \boldsymbol{x}(n)\, \boldsymbol{x}^{\mathrm{T}}(n) , \qquad (2.7.15)$$

and the inverse of $\boldsymbol{A}(n)$ could then be taken to compute $\boldsymbol{w}(n+1)$. The calculation of inverse of $\boldsymbol{A}(n)$ at every sample time would be very computationally expensive, however, and it would be preferable to calculate $\boldsymbol{A}^{-1}(n)$ directly from $\boldsymbol{A}^{-1}(n-1)$. This can be achieved by using the *matrix inversion lemma*, which is a special case of the Woodbury inversion formula discussed in the Appendix, to show that

$$\boldsymbol{A}^{-1}(n) = \lambda^{-1}\boldsymbol{A}^{-1}(n-1) - \frac{\lambda^{-2}\boldsymbol{A}^{-1}(n-1)\boldsymbol{x}(n)\boldsymbol{x}^{\mathrm{T}}(n)\boldsymbol{A}^{-1}(n-1)}{1+\lambda^{-1}\boldsymbol{x}^{\mathrm{T}}(n)\boldsymbol{A}^{-1}(n-1)\boldsymbol{x}(n)} . \qquad (2.7.16)$$

Substituting equations (2.7.16) and (2.7.14) into (2.7.12) we obtain

$$\boldsymbol{w}(n+1) = \left[\lambda^{-1}\boldsymbol{A}^{-1}(n-1) - \lambda^{-1}\alpha(n)\,\boldsymbol{A}^{-1}(n-1)\,\boldsymbol{x}(n)\boldsymbol{x}^{\mathrm{T}}(n)\boldsymbol{A}^{-1}(n-1)\right] \times \left[\lambda\, \boldsymbol{b}(n-1) + \boldsymbol{x}(n)d(n)\right] , \qquad (2.7.17)$$

where

$$\alpha(n) = \frac{1}{\lambda + \boldsymbol{x}^{\mathrm{T}}(n)\boldsymbol{A}^{-1}(n-1)\boldsymbol{x}(n)} . \qquad (2.7.18)$$

Expanding equation (2.7.17), using equations (2.7.13) and (2.7.18) and after some manipulation, the new set of filter coefficients can be expressed in terms of the previous set of coefficients as

$$\boldsymbol{w}(n+1) = \boldsymbol{w}(n) + \alpha(n)\,\boldsymbol{A}^{-1}(n-1)\,\boldsymbol{x}(n)\, e(n) , \qquad (2.7.19)$$

where

$$e(n) = d(n) - \boldsymbol{x}^{\mathrm{T}}(n)\, \boldsymbol{w}(n) . \qquad (2.7.20)$$

Equation (2.7.19), together with (2.7.16), (2.7.18) and (2.7.20), constitute the RLS algorithm. The definition of $e(n)$ used in equation (2.7.20) is consistent with that used in the derivation of the LMS algorithm, in that it is calculated using the changing filter coefficients obtained from data up to the previous sample time. This signal is called the *a priori* error. In some algorithms, particularly those used to adapt IIR filters, then the *a posteriori* error is also defined, which is the error signal recalculated using the filter coefficients computed from data up to the new sample time, $\mathbf{w}(n+1)$ in the notation used here.

2.7.3. Fast RLS Algorithms

The RLS algorithm requires a computational effort of $O(I^2)$, i.e. of the order of I^2 operations, every sample, where I is the number of filter coefficients, whereas the LMS algorithm only requires $O(I)$ operations per sample. Considerable effort has gone into developing implementations of the RLS algorithm that exploit the symmetry and redundancy in the update equations to reduce the number of operations to $O(I)$, and the resulting algorithms are called *fast RLS algorithms* (Haykin, 1996). The early versions of these algorithms were numerically unstable due to the build-up of rounding errors, but more robust implementations are now available which use either error feedback or the inherently robust QR lattice architecture (Haykin 1996). Since $\mathbf{A}^{-1}(n)$ is clearly a poor estimate of the autocorrelation matrix for small values of n, the RLS algorithm is usually initialised with $\mathbf{A}^{-1}(0) = \delta^{-1}\,\mathbf{I}$ where δ is a small positive constant. The initial convergence of the RLS algorithm is thus not as simple as the true Newton's method, equation (2.7.1), not only because it effectively uses an instantaneous estimate of the gradient, but also because the autocorrelation matrix is estimated recursively.

The convergence properties of the RLS algorithm can be far superior to those of the LMS algorithm when the signals are stationary and when there is a large spread of eigenvalues in the autocorrelation matrix, as demonstrated by Haykin (1996) and Clarkson (1993), for example. The value of the forgetting factor, λ, controls the trade-off between speed of convergence and misadjustment for the RLS algorithm in a similar way to the convergence coefficient in the LMS algorithm. It should be noted, however, that when the signals are non-stationary, the relative performance of the RLS and LMS algorithms at tracking these non-stationarities depends very much upon the application (Haykin, 1996). For tracking chirp signals, for example, Bershad and Macchi (1989) have shown that the performance of the LMS can be superior to that of the RLS. The RLS algorithm is thus not guaranteed to perform any better than the LMS algorithm except under completely stationary conditions.

2.8. FREQUENCY-DOMAIN ADAPTATION

Another potential method of improving the convergence characteristics of the time-domain LMS algorithm is to use the normalised frequency-domain LMS algorithm. This algorithm can also significantly reduce the computational requirements for the adaptive filter if the filter has a large number of coefficients. There are several other benefits to frequency-domain adaptation when considering feedforward and feedback controllers, as we shall discuss in Chapters 3 and 5, and so this section also serves as a general introduction to the frequency-domain approach. Frequency-domain adaptive filtering has been reviewed by Shynk (1992) and Haykin (1996, Chapter 10).

2.8.1. Block LMS Algorithm

We begin the discussion by returning to equation (2.6.1), which describes the general philosophy of adaptation using the method of steepest descent

$$\boldsymbol{w}(\text{new}) = \boldsymbol{w}(\text{old}) - \mu \frac{\partial J}{\partial \boldsymbol{w}}(\text{old}) , \tag{2.8.1}$$

and the expression for the vector of derivatives, equation (2.6.4),

$$\frac{\partial J}{\partial \boldsymbol{w}} = -2\, E\left[\boldsymbol{x}(n)\, e(n)\right] . \tag{2.8.2}$$

In deriving the conventional LMS algorithm, we used the instantaneous version of these derivatives to update the filter coefficients. Another approach would be to calculate the average value of $\boldsymbol{x}(n)e(n)$ over N samples and use this to update the filter coefficients only every N samples, so that

$$\boldsymbol{w}(n+N) = \boldsymbol{w}(n) + \frac{\alpha}{N} \sum_{l=n}^{n+N-1} \boldsymbol{x}(l)\, e(l) , \tag{2.8.3}$$

where $\alpha = 2\mu$. This is called the *block LMS algorithm*. It has similar convergence properties to the LMS algorithm (Clark et al., 1980, 1981), provided the filter does not converge too quickly compared with the block size, N (Haykin, 1996), since the filter coefficients change less frequently than for the LMS but by larger amounts. The difference between the LMS algorithm and the block LMS algorithm is that the former implicitly uses a recursive averaging of the gradient estimate whereas the latter explicitly uses a finite moving average. The quantity used to update the filter coefficients in equation (2.8.3) can be regarded as an estimate of the *cross-correlation function* between the reference signal, $x(n)$, and the error signal, $e(n)$, which can be written as

$$\hat{R}_{xe}(i) = \frac{1}{N} \sum_{l=n}^{n+N-1} x(l-i)\, e(l) . \tag{2.8.4}$$

This estimate of the cross-correlation function needs to be calculated for $i = 0$ to $I - 1$, where I is the number of filter coefficients being adapted. The most efficient implementation of the block LMS algorithm occurs when $N = I$, and under these conditions the block LMS algorithm requires N^2 multiplications every N samples and thus has about the same computational requirements as the conventional LMS algorithm. If N is large then it would be more computationally efficient to calculate the estimated cross-correlation function from an estimate of the power spectral density, using the fast Fourier transform (FFT) to calculate the discrete Fourier transform (DFT) of blocks of reference and error signals. In order for this estimate of the cross-correlation function to be unbiased, however, some care needs to be taken to prevent circular correlation effects, such as the use of the *overlap-save method* (Rabiner and Gold, 1975). This involves taking $2N$-point FFTs and adding N zeros to the error data block before taking the FFT. Only the causal part of the cross-correlation function is used to update the filter, and so half the length of the $2N$-point cross-correlation function is discarded. This requires an operation similar to that denoted $\{\ \}_+$ in Section 2.4 and in this case involves an inverse FFT of the spectral density, the setting to zero or 'windowing' of the noncausal part of the resulting cross-correlation function, and taking an FFT of the result. The adaptation algorithm for the filter coefficient in the k-th frequency bin at the m-th iteration can thus be written as

$$W_{m+1}(k) = W_m(k) + \alpha \left\{ X_m^*(k) E_m(k) \right\}_+ , \tag{2.8.5}$$

where $X_m(k)$ is the DFT of the m-th $2N$ point segment of $x(n)$, * denotes complex conjugation, and $E_m(k)$ is the DFT of the m-th N point segment of $e(n)$ padded with N zeros, assuming the overlap save method is used to prevent circular correlation. This algorithm was introduced by Clark et al. (1980) and Ferrara (1980), who called it the *fast LMS* (FLMS) *algorithm*, and showed that it was exactly equivalent to the block LMS algorithm.

Once the reference signal and filter response are available in the frequency domain, further computational savings can be made by also performing the convolution required to obtain the filter output in the frequency domain, so that

$$Y_m(k) = W_m(k)\, X_m(k) , \tag{2.8.6}$$

from which the time history must be obtained by taking the inverse FFT (IFFT) and only using the last N points to avoid circular convolution effects. This process inevitably introduces a delay of one block length into the filtering operation. The block diagram of the complete frequency-domain LMS algorithm is shown in Fig. 2.16 (Ferrara, 1985; Shynk, 1992).

The implementation of the frequency-domain adaptation of the filter, and the convolution of the reference signal with the reference signal thus uses 5 $2N$-point FFTs, where $N = I$ and I is the number of adaptive filter coefficients. If each $2N$-point FFT requires $2N \log_2 2N$ operations (Rabiner and Gold, 1975), then frequency-domain adaptive filtering requires about $10N \log_2 2N$ operations every N samples, or $10 \log_2 2N$ operations per sample. This can be compared with N operations per sample to adapt the coefficients using the conventional LMS, and N operations per sample, for the normal time-domain convolution of the reference signal and filter coefficients. The frequency-domain implementation of the LMS algorithm thus requires a factor of about $N/5 \log_2 2N$ fewer computational operations compared with its direct implementation. For a 512-coefficient filter this corresponds to a computational saving of about a factor of 10.

2.8.2. Frequency-dependent Convergence Coefficients

Apart from the computational advantages, it may also be possible to use the frequency-domain implementation to improve the convergence properties of the LMS algorithm. This was originally suggested by Ferrara (1985), who argued that in the frequency domain the error signal in a given frequency bin, $E_m(k)$, is only a function of the filter coefficient in the same bin, $W(k)$, and so each of the frequency-domain filter coefficients converges

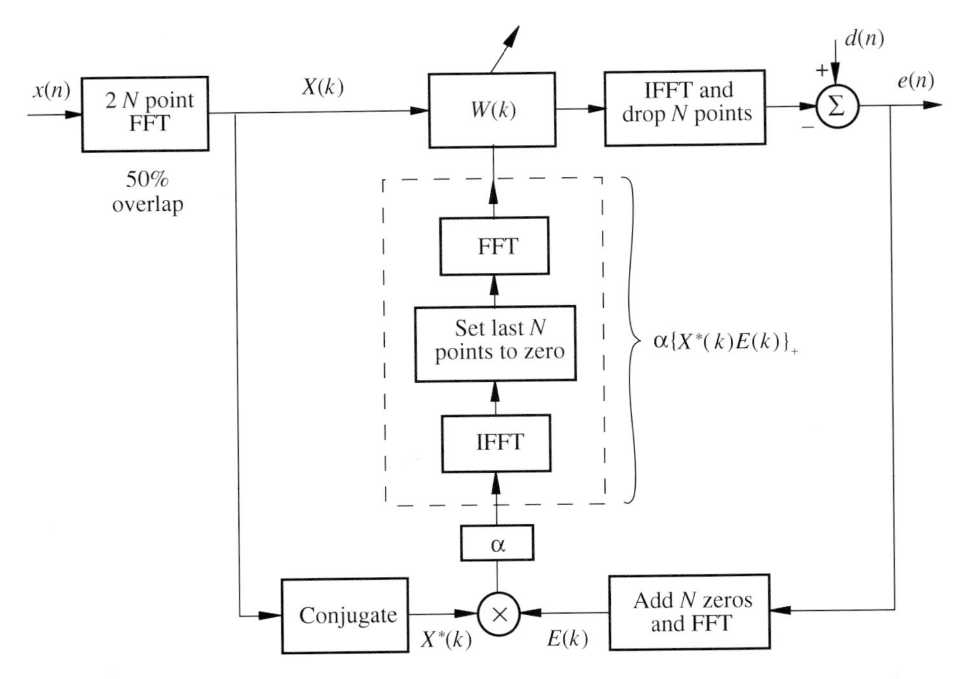

Figure 2.16 Block diagram of a frequency domain implementation of the LMS algorithm.

independently. If this were the case, the convergence coefficient could be selected independently for each bin, so that the adaptation algorithm becomes

$$W_{m+1}(k) = W_m(k) + \left\{\alpha(k)\, X_m^*(k)\, E_m(k)\right\}_+ . \tag{2.8.7}$$

The convergence coefficient used in the adaptation of each individual frequency bin could be normalised by the average power in that bin, for example, so that

$$\alpha(k) = \frac{\tilde{\alpha}}{E\left[|X(k)|^2\right]} , \tag{2.8.8}$$

where $\tilde{\alpha}$ is a single normalised convergence coefficient. For some applications, this normalisation of the frequency-dependent convergence coefficient has been found to considerably improve the convergence rate of the adaptive filter.

Unfortunately, if a frequency-dependent convergence coefficient is used in equation (2.8.7), the adaptive filter is not guaranteed to converge towards the optimum, Wiener, filter (Feuer and Cristi, 1993). This problem is particularly severe when the causally constrained optimum filter is significantly different from the unconstrained optimum filter, as it is in linear prediction problems for example (Elliott and Rafaely, 2000).

A solution to this problem can be obtained by deriving an adaptive algorithm directly based on Newton's method, equation (2.7.3), which can be written in the z domain as

$$W_{m+1}(z) = (1-\alpha)\, W_m(z) + \alpha\, W_{\text{opt}}(z) , \tag{2.8.9}$$

where α is the convergence coefficient and $W_{\text{opt}}(k)$ is the optimum causal filter. Writing the z transform of the cross-correlation function between the reference and error signals at the m-th iteration as

$$S_{xe}(z) \;=\; S_{xd}(z) - W_m(z)\,S_{xx}(z)\;, \tag{2.8.10}$$

allows the transfer function of the optimum Wiener filter, equation (2.4.23), to be expressed as

$$W_{\text{opt}}(z) \;=\; \frac{1}{F(z)}\left\{\frac{S_{xe}(z)}{F\left(z^{-1}\right)}\right\}_+ + \frac{1}{F(z)}\left\{W_m(z)\frac{S_{xx}(z)}{F\left(z^{-1}\right)}\right\}_+ \;. \tag{2.8.11}$$

But $S_{xx}(z) = F(z)\,F\left(z^{-1}\right)$, and since $W(z)$ and $F(z)$ are entirely causal, the final term in equation (2.8.11) is just equal to $W_m(z)$. Substituting the resulting expression for $W_{\text{opt}}(z)$ into equation (2.8.9) allows Newton's algorithm to be written as

$$W_{m+1}(z) \;=\; W_m(z) + \frac{\alpha}{F(z)}\left\{\frac{S_{xe}(z)}{F\left(z^{-1}\right)}\right\}_+ \;. \tag{2.8.12}$$

If we now consider a discrete-frequency version of this algorithm, by approximating $S_{xe}(k)$ with an estimate using the current block of data, $X^*(k)E(k)$, a discrete-frequency approximation to Newton's algorithm can be calculated for each new block of data as

$$W_{m+1}(k) \;=\; W_m(k) \;+\; \alpha_m^+(k)\left\{\alpha_m^-(k)\,X_m^*(k)\,E_m(k)\right\}_+ \;, \tag{2.8.13}$$

where $\alpha_m^+(k) = \sqrt{\alpha}\big/\hat{F}_m(k)$, which corresponds to an entirely causal time sequence, $\alpha_m^-(k) = \left(\alpha_m^+(k)\right)^*$, which corresponds to a noncausal time sequence, and $\hat{F}_m(k)$ is the spectral factor of the practical estimate of the reference signal's power spectral density at the m-th block, $\hat{S}_{xx,m}(k)$, which could in practice be obtained by taking the Hilbert transform of $\sqrt{\hat{S}_{xx,m}(k)}$.

In equation (2.8.13) the bin-normalised convergence coefficient given by equation (2.8.8) has been split into two parts. Since only the noncausal part is used inside the causality constraint in equation (2.8.13) this does not cause noncausal components of the Fourier transform of $X^*(k)\,E(k)$ to affect the adaptation of the filter, which thus converges to the optimal causal filter (Elliott and Rafaely, 2000).

2.8.3. Transform-Domain LMS

Apart from frequency-domain implementations of the LMS algorithm which operate on blocks of data, the data vector could be transformed every sample to give a modified set of signals, which can then be adaptively filtered on a sample-by-sample basis. This approach is called the *transform-domain* LMS (Narayan et al., 1983; Haykin, 1996) or sometimes the sliding frequency-domain LMS.

The transform which generates a completely uncorrelated set of signals is called the Karhunen–Loéve transformation (KLT) but this transform is signal dependent. A number of other, fixed, transforms have this uncorrelating property to some extent. In particular, the DFT is asymptotically equivalent to the KLT for long data vectors. For finite filter lengths, however, different transforms may give a better approximation to the KLT, and the discrete cosine transform (DCT) has been shown to perform well for a variety of low-pass signals (Beaufays, 1995). An advantage of the DFT used above, however, is that some physical insight can be derived into the transform-domain filter coefficients, and this can be very helpful when constraints are imposed on the response of a control filter, as will be described in Chapter 7. Another class of algorithms that pre-process the data to produce a transformed set of signals, which are less correlated than the original reference signal, are those based on the *lattice* structure (see for example Friedlander, 1982) and once again this pre-processing is data dependent.

2.9. ADAPTIVE IIR FILTERS

In active control applications, adaptive filters may be used to cancel a disturbance, or to model the response of a physical system. IIR filters can be considerably more efficient than FIR filters in modelling systems with lightly-damped resonances. For example, an IIR filter may require far fewer coefficients than an FIR filter to accurately model the vibration response of a lightly damped structure. This is because an IIR filter has a transfer function that has non-trivial poles, as well as zeros, as shown in equation (2.2.15). IIR filters can thus be more attractive for modelling the response of some physical systems because their transfer functions are a direct mapping of the differential equation describing the physical system or the system's state-space representation. We will also see that the ideal controller response has a natural IIR structure in some active control applications.

2.9.1. Noise Cancellation and System Identification

The two possible functions of an adaptive filter, to cancel disturbances and to model an unknown system, are illustrated in Fig. 2.17. The first function can be termed adaptive noise cancellation, and is illustrated in Fig. 2.17(a), in which the only objective is to reduce the error signal $e(n)$, usually in the mean-square sense, by cancelling the components of the disturbance, $d(n)$, which are related to the reference signal, $x(n)$. In an active control system, the adaptive canceller has the additional complication of a physical 'plant' response between the output of the adaptive filter and the observation point for the error. Adaptive noise cancellation here, as in the whole of this chapter, is assumed to refer to the cancellation of electrical noise. A second function which an adaptive filter could be used for can be termed adaptive system identification, Fig. 2.17(b), in which the object is to ensure that the coefficients of the adaptive filter $H(z)$ are as close as possible to those of the system to be identified, with transfer function, $H_*(z)$ which generates $d(n)$ from $x(n)$ even though $d(n)$ may contain additional noise, $v(n)$. Although, strictly speaking, $H(z)$ can only denote the transfer function of a time-invariant system, we will use it here to represent the transfer function of an adaptive filter with its coefficients 'frozen' in time.

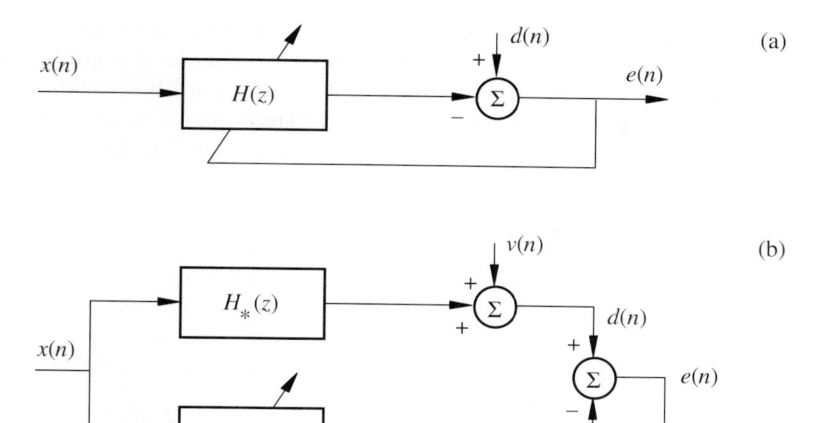

Figure 2.17 An adaptive filter used for (a) adaptive noise canceller, and (b) adaptive system identification.

If $x(n)$ and $v(n)$ were uncorrelated and $x(n)$ was a white-noise signal, the two objectives, of noise cancellation and system identification, would have identical solutions, i.e. the mean-square error would be minimised if the structure and order of $H(z)$ were the same as those of $H*(z)$ and the coefficients of the two filters had the same values. If, however, $x(n)$ were a pure tone, for example, the noise cancellation task could be perfectly accomplished with a variety of filter responses $H(z)$ which did not have the same structure or coefficients as $H*(z)$, but only had the same amplitude and phase at the excitation frequency. System identification can thus be seen to be a more exacting task than noise cancellation, with requirements on the reference signal as well as on the adaptive filter if an accurate model of $H*(z)$ is to be obtained. The two functions of an adaptive filter illustrated in Fig. 2.17 have given rise to two rather different approaches to their design, particularly for adaptive IIR filters. In this section we will briefly review both traditions, so that the applications of adaptive IIR filters in active control can be more clearly understood. Useful reviews of the various approaches to designing and analysing adaptive IIR filters are provided by Treichler (1985), Treichler et al. (1987), Shynk (1989) and Netto et al. (1995).

Unfortunately, it is not nearly as easy to design a reliable and unbiased algorithm to adapt the coefficients of an IIR filter as it is to find such an algorithm for an FIR filter. This is partly because of the problems of guaranteed convergence, but is also because of concerns about the *stability* of the resulting filter. If any of the poles of a fixed IIR filter lie outside the unit circle in the z plane, the filter will be unstable. The stability of an adaptive IIR filter is more complicated to determine than this, however, because the coefficients of the filter, and thus the position of the poles of the equivalent fixed filter, will change with time. Whether an adaptive IIR filter becomes unstable when some of its poles (strictly the poles of the equivalent filter with frozen coefficients) migrate outside the unit circle will depend very much upon which adaptive algorithm is used, as we shall see.

2.9.2. The Output Error Approach

The two basic philosophies used in adapting the coefficients in an IIR filter, are called the output error approach and the equation error approach. The *output error* approach is most widely used in the signal processing literature and is illustrated in Fig. 2.18, which shows the output of the adaptive filter as being of the form

$$y_o(n) = \sum_{j=1}^{J} a_j(n) y_o(n-j) + \sum_{i=0}^{I-1} b_i(n) x(n-i) . \tag{2.9.1}$$

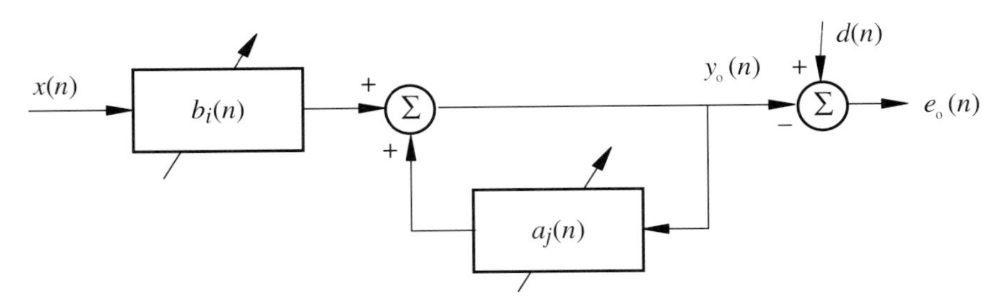

Figure 2.18 Block diagram showing the definition of the output error, $e_o(n)$, for an adaptive IIR filter.

The output error is then defined as

$$e_o(n) = d(n) - y_o(n) , \tag{2.9.2}$$

which is the most obvious definition of the error for an adaptive noise canceller, but unfortunately has a number of problems associated with it. These are principally to do with the recursive nature of equation (2.9.1), which means that the current output is not just a function of the previous inputs, but is also a function of the previous outputs, which, in turn, also depend upon the values of the filter coefficients. Whereas the instantaneous output of an FIR filter is a linear function of the values of its coefficients, $y_o(n)$ has a more complicated dependence on the recursive filter coefficients. This means that the mean-square value of the output error is no longer a quadratic function of the filter coefficients and the error surface may not even be unimodal (Stearns, 1981). Figure 2.19, for example, shows the mean-square error surface for a first-order IIR filter, only having coefficients a_1 and b_0, in equation (2.9.1), when the signal $d(n)$ in Fig. 2.18 is derived by passing a white noise reference signal, $x(n)$, through a second-order IIR filter having coefficients $a_{*1} = 1.1314$, $a_{*2} = -0.25$, $b_{*0} = 0.05$, $b_{*1} = -0.4$ (Johnson and Larimore, 1977). The mean-square error surface clearly has two minima in this case; a local one having a mean-square error of 0.976 at $a_1 = 0.114$, $b_0 = -0.519$, and a global one having a mean-square error of 0.277 at $a_1 = -0.311$, $b_0 = 0.906$. This is an example in which the adaptive filter *undermodels* the system generating the signal $d(n)$, since it is of lower order. The conditions under which no local minima are believed to exist in the error surface are listed by Shynk (1989), and these are (1) that the reference signal is white noise, (2) that the adaptive filter has sufficient order to exactly model the system generating $d(n)$, and (3) that the number of non-recursive coefficients in the adaptive filter exceeds the number of

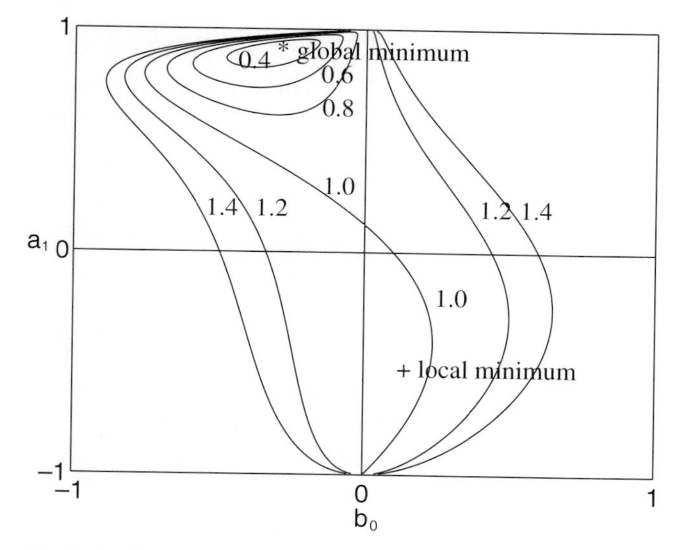

Figure 2.19 Mean-square error surface for an undermodelled adaptive IIR filter, showing contours of constant mean-square error and the existence of two minimum points, one local and one global.

recursive coefficients in the system generating $d(n)$. If the order of the model is greater than that of the system being modelled the system is said to be *overmodelled.* This is difficult to guarantee in practice since the order of the system being modelled is generally unknown. Also, if the adaptive filters have 'extra' poles and zeros, which are not required to model the system, they can create hidden dynamics by pole-zero cancellation, which can be particularly disastrous if the self-cancelling pole-zero pair occurs outside the unit circle.

The arrangement giving rise to the error surface shown in Fig. 2.19 could be viewed as a rather extreme case of undermodelling. It was also originally chosen to demonstrate the danger of gradient descent algorithms for adapting IIR filters, since when the coefficients of the adaptive filter are zero, the local gradient is in the direction of the local rather than the global minimum. Simple gradient descent algorithms are not guaranteed to converge to the global minimum in the error surface if the error surface is not unimodal. Nevertheless, various versions of the gradient descent algorithm are still commonly used for adapting IIR filters, particularly in noise cancellation applications. In these applications it may not be too important if the filter converges to a local rather than a global minimum, provided the residual output error is significantly reduced. The dangers of undermodelling must clearly be born in mind, however, to ensure that the performance obtained by convergence to a local minimum does not significantly degrade the performance.

2.9.3. Gradient Descent Algorithms

We continue the discussion of equation error algorithms by considering the derivatives of the cost function given by the expectation of the squared output error

$$J = E\left[e_o^2(n)\right] , \tag{2.9.3}$$

with respect to the coefficients a_j and b_i, which are given by

$$\frac{\partial J}{\partial a_j}(n) = -2E\left[e_o(n)\frac{\partial y_o(n)}{\partial a_j(n)}\right], \tag{2.9.4}$$

and

$$\frac{\partial J}{\partial b_i}(n) = -2E\left[e_o(n)\frac{\partial y_o(n)}{\partial b_i(n)}\right]. \tag{2.9.5}$$

The derivatives of the output signal with respect to the filter coefficients can be written, using equation (2.9.1), as

$$\frac{\partial y_o(n)}{\partial a_j(n)} = y_o(n-j) + \sum_{k=1}^{J} a_k(n)\frac{\partial y_o(n-k)}{\partial a_j(n)}, \tag{2.9.6}$$

and

$$\frac{\partial y_o(n)}{\partial b_i(n)} = x(n-i) + \sum_{k=1}^{J} a_k(n)\frac{\partial y_o(n-k)}{\partial b_i(n)}. \tag{2.9.7}$$

The evaluation of equations (2.9.6) and (2.9.7) is complicated by their own recursive structure and by the fact that the filter coefficients are varying with time (White, 1975). Strictly speaking, the value of $y_o(n-k)$ in equations (2.9.6) and (2.9.7) should be computed using fixed values of the filter coefficients, equal to the current values $a_j(n)$ and $b_i(n)$, for all past time, which would require the algorithm to have an infinite memory. If we assume that the filter coefficients are only slowly changing with time, then the values of $y_o(n-k)$ computed with fixed values of the filter parameters will not be significantly different from those computed with time-varying parameters, and these will then be equal to the observed values of the filter output.

The variation of the filter coefficients with time implies, for example, that $\partial y_o(n-k)/\partial a_j(n)$ is not just a delayed version of $\partial y_o(n)/\partial a_j(n)$. If, however, we assume that there are only small changes to the filter coefficients at every iteration, i.e. they are *quasi-static*, we can also ignore this complication and write

$$\frac{\partial y_o(n)}{\partial a_j(n)} \approx u_j(n) = y_o(n-j) + \sum_{k=1}^{J} a_k(n)u_j(n-k), \tag{2.9.8}$$

and

$$\frac{\partial y_o(n)}{\partial b_i(n)} \approx v_i(n) = x(n-i) + \sum_{k=1}^{J} a_k(n)v_i(n-k), \tag{2.9.9}$$

where $u_j(n)$ and $v_i(n)$ are signals that are approximately equal to $\partial y_o(n)/\partial a_i(n)$ and $\partial y_o(n)/\partial b_k(n)$ and $a_j(n)$ and $b_i(n)$ are the filter coefficients at the current sample time.

These recursive gradient estimates can then be used as the basis for a steepest-descent algorithm for the adaptation of the filter coefficients using instantaneous gradient estimates given by equations (2.9.4) and (2.9.5), so that

$$a_j(n+1) = a_j(n) + \alpha\, e_0(n)\, u_j(n)\,, \tag{2.9.10}$$

and

$$b_i(n+1) = b_i(n) + \alpha\, e_0(n)\, v_i(n)\,. \tag{2.9.11}$$

This algorithm can be further simplified by again using the fact that the filter coefficients are only slowly-varying to approximate $u_j(n)$ by $u_0(n-j)$ and $v_i(n)$ by $v_0(n-i)$, which means that only a single pair of signals has to be recursively estimated (Shynk, 1989).

2.9.4. The RLMS Algorithm

An algorithm that apparently makes a much cruder approximation to the gradients can be obtained by ignoring the recursive parts of equations (2.9.6) and (2.9.7) altogether, in which case the adaptive equations reduce to (Feintuch, 1976)

$$a_j(n+1) = a_j(n) - \alpha\, e_0(n)\, y_0(n-j)\,, \tag{2.9.12}$$

and

$$b_i(n+1) = b_i(n) - \alpha\, e_0(n)\, x(n-i)\,, \tag{2.9.13}$$

which is known as the recursive LMS (RLMS) algorithm in the signal processing community. The principle is similar to the pseudo linear regression (PLR) algorithm described by Landau (1976) (see also Ljung, 1999 and Harteneck and Stewart, 1996). To maximise the speed of convergence it is generally necessary to have different convergence coefficients in equation (2.9.12) and (2.9.13). Billout et al. (1989) and Crawford et al. (1996) also discuss an alternative 'full feedback' structure in which the recursive output is taken back to the input of the non-recursive part of the filter rather than its output.

Although the RLMS algorithm can converge to a biased solution in some cases, such as for the model problem whose error surface is shown in Fig. 2.19 (Johnson and Larimore, 1977; Widrow and McCool, 1977; Hansen and Snyder, 1997), it is easy to implement and for many systems gives a significant reduction in the output error, and is thus widely used.

For slow adaptation, the difference between the gradient estimates used in the true gradient algorithm, equations (2.9.10) and (2.9.11), and that used in the RLMS algorithm, equations (2.7.12) and (2.9.13), is that in the former case the gradient estimates are filtered by a system which if frozen in time has the transfer function

$$A(z) = 1 - \sum_{k-1}^{J} a_k(n)\, z^{-k}\,. \tag{2.9.14}$$

If the adaptive filter were to converge from having zero coefficients to exactly match the system generating $d(n)$, which is assumed to have a response $H*(z) = B*(z)/A*(z)$, then $A(z)$ would progress from being equal to zero at the start of the convergence to being equal to $A*(z)$ at the end. The stability of the RLMS algorithm has been studied using a variety of methods (see for example, Treichler, 1985; Ren and Kumar, 1992) and stability is generally dependent on the assumption that the transfer function $1/A*(z)$ is *strictly positive real* (SPR), in other words that

$$\mathrm{Re}\left[\frac{1}{A_*(\mathrm{e}^{j\omega T})}\right] > 0 \quad \text{for all } \omega T. \tag{2.9.15}$$

This means that the phase of the frequency response of $1/A*(z)$ must lie within ± 90° at all frequencies, and thus, by implication, that the phase response of the recursive part of the adaptive filter also satisfies this condition during convergence. A very similar phase condition will be derived in the adaptation of feedforward controllers, where it is the difference in phase response between the plant and the internal plant model that is important. Apart from filtering the reference signals, as in equation (2.7.8, 9) for the gradient descent algorithm, the adaptive filter can also be stabilised by filtering the error signal with a filter having response $C(z)$. The condition for stability is then that $C(z)/A*(z)$ is SPR (Treichler, 1985), so that the phase response of $C(z)$ must be within 90° of that of $A*(z)$.

The convergence properties of the RLMS algorithm, and the steepest-descent algorithm that uses the more complete version of the gradients in equations (2.9.8) and (2.9.9), have been examined by Macchi (1995), who shows that despite its apparent crudeness, the RLMS algorithm can have an important *self-stabilising* property. This property is not fully understood, but its effect is that when one of the poles of the equivalent frozen filter migrates outside the unit circle, it is forced back inside by the natural operation of the RLMS algorithm. This periodic drift of the filter coefficients towards instability, followed by recovery, is observed in adaptive echo cancellation applications and in adaptive control systems, and is known as *bursting* (Johnson, 1995). Other adaptive IIR algorithms do not have this self-stabilising property, and become increasingly unstable when a pole migrates outside the unit circle, although the stability could be monitored and various rescue strategies implemented to prevent instability (Clarkson, 1993; Stonick, 1995).

2.9.5. The Equation Error Approach

Another approach to the adaptation of the coefficients in an IIR filter is motivated by the fact that when the output error is small in Fig. 2.18, the output $y_o(n)$ must be nearly equal to the disturbance $d(n)$. This suggests that the filter coefficients could be adjusted to minimise the difference between $d(n)$ and the signal

$$y_e(n) = \sum_{j=1}^{J} a_j(n)\, d(n-j) + \sum_{i=0}^{I-1} b_i(n)\, x(n-i) \,. \tag{2.9.16}$$

This difference signal is called the *equation error*, $e_e(n)$, and is defined as

$$e_e(n) = d(n) - y_e(n) \,. \tag{2.9.17}$$

It has the important property that it is linear in both a_j and b_i. This is because the recursion in Fig. 2.18 has been avoided by using $d(n)$ rather than $y_o(n)$ to drive $a_j(n)$, as shown in Fig. 2.20. Since the equation error is linear in the filter coefficients, the mean-square equation error is a quadratic function of these parameters and adaptive algorithms can be designed which are guaranteed to converge to a unique global minimum of the mean-square equation error. Such algorithms are very widely used in the control literature for system identification (Goodwin and Sin, 1984; Norton, 1986; Ljung, 1999).

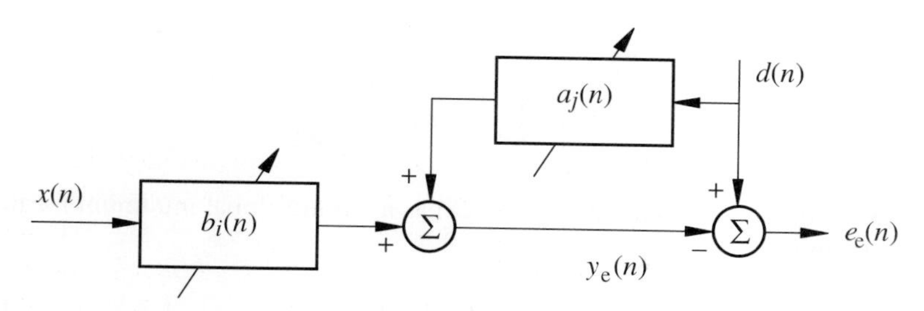

Figure 2.20 Block diagram showing the definition of the equation error, $e_e(n)$, for an adaptive IIR filter.

In the system identification literature, methods that are used to calculate a single estimate of the parameters of the unknown system are known as *off-line* (or batch) algorithms. We will briefly review one such method here, which uses the equation error, before proceeding to consider algorithms that periodically adapt the parameters of the unknown system, which are known as *on-line* identification methods. We begin by writing the output in the equation error system, equation (2.9.16), with fixed filter coefficients in vector form as

$$y_e(n) = \boldsymbol{\theta}^{\mathrm{T}} \boldsymbol{\phi}(n) , \tag{2.9.18}$$

where

$$\boldsymbol{\theta} = \left[a_1 \cdots a_J, b_0 \cdots b_{I-1}\right]^{\mathrm{T}} , \tag{2.9.19}$$

is the vector of filter parameters, and

$$\boldsymbol{\phi}(n) = \left[d(n-1) \cdots d(n-J), x(n) \cdots x(n-I+1)\right]^{\mathrm{T}} , \tag{2.9.20}$$

is the vector of *regressors*, as they are known in the identification literature.

The equation error, given by equation (2.9.16), is thus equal to

$$e_e(n) = d(n) - \boldsymbol{\theta}^{\mathrm{T}} \boldsymbol{\phi}(n) , \tag{2.9.21}$$

which is linear in all of the parameters in $\boldsymbol{\theta}$ since $\boldsymbol{\phi}(n)$ is not a function of $\boldsymbol{\theta}$. The value of the mean-square equation error estimated over a finite time can thus be written in quadratic form as

$$\sum_{n=0}^{N-1} e_e^2(n) = \underline{\theta}^{\mathrm{T}} \boldsymbol{A} \underline{\theta} - 2\boldsymbol{b}^{\mathrm{T}} \underline{\theta} + c , \tag{2.9.22}$$

where in this case

$$\boldsymbol{A} = \sum_{n=0}^{N-1} \boldsymbol{\phi}(n) \boldsymbol{\phi}^{\mathrm{T}}(n) , \tag{2.9.23}$$

$$\boldsymbol{b} = \sum_{n=0}^{N-1} \boldsymbol{\phi}(n) d(n) , \tag{2.9.24}$$

and

$$c = \sum_{n=0}^{N-1} d^2(n) \ . \tag{2.9.25}$$

Assuming $\boldsymbol{A}$ is non-singular, equation (2.9.22) has a unique global minimum for the set of parameters given by

$$\boldsymbol{\theta}_{\text{opt}} = \boldsymbol{A}^{-1}\boldsymbol{b} \ , \tag{2.9.26}$$

which can be calculated directly from a finite set of $N + I$ samples of the input signal, $x(n)$, and $N + J$ samples of the desired signal, $d(n)$. Equation (2.9.26) is called the *ordinary least-squares* estimate in the identification literature (Norton, 1986).

Gradient descent methods such as the LMS could, in principle, also be used to give an adaptive estimate of $\boldsymbol{\theta}$, which in this case would take the form

$$\boldsymbol{\theta}(n+1) = \boldsymbol{\theta}(n) + \alpha\, \boldsymbol{\phi}(n)\, e_{\text{e}}(n) \ . \tag{2.9.27}$$

It is more usual in the identification literature to use an exact least-squares algorithm, however, which is very similar to the recursive least-squares (RLS) method, outlined in Section 2.4 for FIR filters. In the present case the exact least-squares algorithm may be written as

$$\boldsymbol{\theta}(n+1) = \boldsymbol{\theta}(n) + \alpha\, \boldsymbol{A}^{-1}\boldsymbol{\phi}(n)\, e_{\text{e}}(n) \ , \tag{2.9.28}$$

where $\boldsymbol{A}$ is defined here by equation (2.9.23).

2.9.6. Bias Due to Measurement Noise

A significant problem associated with the equation error approach in system identification is the effect of measurement noise. An equivalent block diagram of an equation error algorithm used in the identification of the recursive parameters of an unknown system $H_*(z)$ with output noise $v(n)$ is shown in Fig. 2.21. This is of a rather different form from Fig. 2.20, and is motivated by writing the z transform of the equation error, equation (2.9.17), with the aid of equation (2.9.16), as

$$E_{\text{e}}(z) = A(z)\, D(z) - B(z)\, X(z) \ , \tag{2.9.29}$$

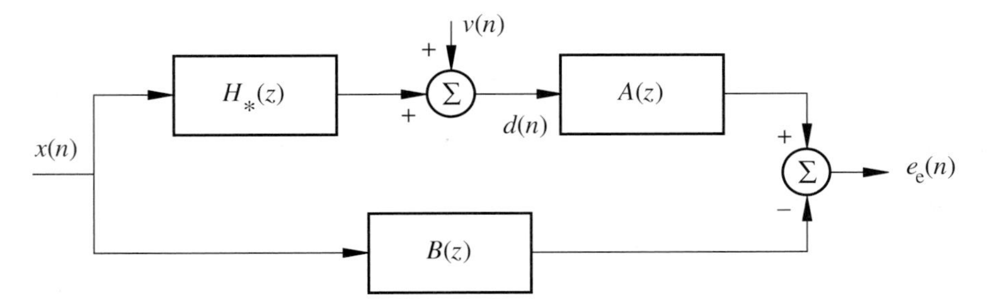

Figure 2.21 Block diagram of the equation error approach used to identify the parameters of an unknown system with output noise.

where, as in equation (2.2.13),

$$A(z) = 1 - a_1 z^{-1} \cdots - a_J z^{-J} , \tag{2.9.30}$$

and

$$B(z) = b_0 + b_1 z^{-1} \cdots + b_{I-1} z^{-I+1} . \tag{2.9.31}$$

It can be seen by inspection of Fig. 2.21 that the values of the coefficients in $A(z)$ that minimise the mean-square value of equation error, $e_e(n)$, will depend not only on the properties of the unknown system, $H*(z)$, but also on the magnitude and characteristics of the measurement noise $v(n)$. This is because $A(z)$ can reduce the mean-square value of $e_e(n)$ in two ways, which are potentially in conflict with each other. On the one hand, $A(z)$ will minimise the mean-square value of $e_e(n)$ by modelling the recursive part of the unknown system and thus reduce the component of $e_e(n)$ correlated with $x(n)$ as far as possible. On the other hand, if $v(n)$ is anything but white noise, $A(z)$ can also reduce the component of $e_e(n)$ that is correlated with $v(n)$.

In minimising the mean-square value of the equation error, the coefficients of $A(z)$ are forced to compromise between these two conflicting roles. These coefficients are thus *biased* away from the values required for system identification by the presence of correlated output noise. Note, however, that $A(z)$ could not reduce the component of $v(n)$ in $e_e(n)$ if $v(n)$ were a white noise signal. This motivates a modification to the ordinary least-squares algorithm in which the characteristics of the noise are identified along with those of the unknown system.

Specifically, we now assume that the noisy output of the unknown system can be expressed as

$$d(n) = \sum_{j=1}^{J} a_{*j}\, d(n-j) + \sum_{i=0}^{I-1} b_{*i}\, x(n-i) + \xi(n) + \sum_{k=1}^{K} c_{*k}\, \xi(n-k) , \tag{2.9.32}$$

in which the subscript $*$ again denotes the values of these coefficients in the system to be identified, and the noise signal, $v(n)$, is assumed to have been generated by filtering a white noise sequence $\xi(n)$, which is uncorrelated with $x(n)$. We also assume that the order of the unknown system is exactly the same as the adaptive filter, equation (2.9.16). We now define an extended set of parameters and input signals such that

$$d(n) = \boldsymbol{\theta}_{*e}^{\mathrm{T}} \boldsymbol{\phi}_e(n) + \xi(n) , \tag{2.9.33}$$

where

$$\boldsymbol{\theta}_{*e} = \begin{bmatrix} a_{*1} & \cdots & a_{*J}, b_{*0} & \cdots & b_{*I-1}, c_{*1} & \cdots & c_{*K} \end{bmatrix}^{\mathrm{T}} , \tag{2.9.34}$$

and

$$\boldsymbol{\phi}_e(n) = \left[d(n) \cdots d(n-J),\, x(n) \cdots x(n-I+1),\, \xi(n-1) \cdots \xi(n-K) \right]^{\mathrm{T}} . \tag{2.9.35}$$

Similarly, we can define the extended set of unknown parameters that must be identified from observations as

$$\boldsymbol{\theta}_e = \left[a_1 \cdots a_{I-1}, b_0 \cdots b_J, c_1 \cdots c_K \right]^{\mathrm{T}} , \tag{2.9.36}$$

so that the equation error is

$$y_e(n) = \boldsymbol{\theta}_e^T \boldsymbol{\phi}_e(n) , \tag{2.9.37}$$

and $\boldsymbol{\phi}_e(n)$ is again given by equation (2.9.35). Note that in forming equation (2.9.36) we are assuming access not to previous values of the observed noise but to previous values of the white noise sequence that is generating the noise. In practice, this is generally obtained by taking the difference between $d(n)$ in equation (2.9.33), and equation (2.9.37), even though this is only equal to $\xi(n)$ when the identification algorithm has exactly matched the true parameters, so that $\boldsymbol{\theta}_e = \boldsymbol{\theta}_{*e}$. The parameters in equation (2.9.37) can now be identified using least-squares techniques, since this equation is linear in all these parameters, and the solution is not biased by the white noise signal $\xi(n)$ in equation (2.9.33). The modified algorithm is called the *extended least-squares* method.

The signal generation model described by equation (2.9.32) is referred to as the *ARMAX* model in system identification (Ljung, 1999), i.e. an ARMA model with eXogenous input, $\xi(n)$. Taking z transforms, equation (2.9.32) can be written as

$$A_*(z)\, D(z) = B_*(z)\, X(z) + C_*(z)\, \Xi(z) , \tag{2.9.38}$$

where

$$A_*(z) = 1 - \sum_{j=1}^{J} a_{*j}\, z^{-j} , \tag{2.9.39}$$

$$B_*(z) = \sum_{i=0}^{I-1} b_{*j}\, z^{-j} , \tag{2.9.40}$$

$$C_*(z) = 1 + \sum_{k=1}^{K} c_{*k}\, z^{-k} , \tag{2.9.41}$$

and $\Xi(z)$ is the z transform of $\xi(n)$, so that

$$D(z) = \frac{B_*(z)}{A_*(z)} X(z) + \frac{C_*(z)}{A_*(z)} \Xi(z) . \tag{2.9.42}$$

It can be seen that the ARMAX model assumes that both the signal model and the noise model share a common pole structure. This assumption is often made in the control literature and its validity in the context of active control will be discussed in Chapter 7.

In conclusion, we have seen that robust algorithms for the adaptation of IIR filters are more difficult to find than robust algorithms for the adaptation of FIR filters. Two different approaches can be used to adapt IIR filters. The output error approach can lead to algorithms that are relatively simple, but may converge to local minima of the error surface. They are often used for adaptive noise cancellation. Algorithms based on the equation error approach are often used for system identification. This approach leads to algorithms that are guaranteed to converge to the global minimum of an error surface, but the coefficients of the resulting filter will not be the same as those of the system being identified if coloured measurement noise is present. This problem can be alleviated by using a modified algorithm which identifies the parameters of the system under control and the system that is assumed to generate the measurement noise, provided these two systems share a common pole structure.

3

Single-Channel Feedforward Control

3.1. INTRODUCTION

We begin the discussion of active control systems by considering the single-channel version of a feedforward controller. Feedforward methods have a long and important history in active control. This dates back to Lueg's original duct controller, as discussed in Chapter 1, in which an upstream reference microphone was used to give advanced information about the waveform of the pressure disturbance propagating down the duct. In this application there is no restriction on the type of disturbance that can be cancelled, provided the reference sensor is placed far enough upstream to overcome any delays in the processor, so that the optimal controller is causal. Thus random or transient signals can be completely cancelled, under noise-free conditions, with a feedforward controller, provided that enough time-advance information is available from the detected signal.

Another class of signal for which feedforward controllers are widely used is that for which the signals are periodic. In this case, the future values of a signal's waveform can, in principle, be perfectly predicted from its waveform in the past. In practice, however, a measured waveform is never perfectly predictable but may be closely synchronised to the process that produces it, for example a rotating or reciprocating machine. In this case, a *reference signal* can generally be obtained from the machine, which can be used instead of that from the reference sensor. This gives 'advanced' warning of the waveform, although the fact that the waveform is substantially unaltered over several cycles is also being used in the controller.

3.1.1. Chapter Outline

Following a discussion of the nature of the feedforward control problem in this section, the control of deterministic disturbances, in particular those which have a periodic waveform, is outlined in Section 3.2. Many of the important issues for an adaptive control system, such as convergence and stability, can be introduced in a straightforward way by considering the superposition of sinusoidal signals. In Section 3.3 the more difficult problem of controlling a random, or stochastic, disturbance is outlined, particularly the need for the controller to be causal.

Section 3.4 describes how such a controller can be made adaptive to track time-varying disturbances, using a variety of algorithms. These algorithms are based on those developed for adaptive digital filters, which were introduced in Chapter 2. In feedforward control applications, the algorithms now have to deal with the dynamic response, delays and uncertainties introduced by having a physical system between the output of the filter and the measurement of the resulting error signal. The computational efficiency, and potentially the speed of convergence, of such adaptive controllers can be improved by adapting them in the frequency domain, as discussed in Section 3.5.

Most of these adaptive algorithms operate by having within them an internal model of the response of the physical system under control. The identification of this response is discussed in Section 3.6, in which it is emphasised that in many practical active control problems the response of the system is relatively well damped, and can be efficiently modelled by a digital FIR filter.

Following an outline of adaptive IIR controllers in Section 3.7, the chapter finishes with two practical applications of single-channel feedforward controllers: the active control of plane sound waves propagating in a duct and the active control of flexural waves propagating on a beam.

3.1.2. Digital Controllers

All the control systems discussed in this chapter will be assumed to be 'digital', i.e. they operate on discrete-time signals. Clearly, the physical signal under control, whether it be the acoustic pressure variation in a duct or the velocity of a structure, is an 'analogue' or continuous-time signal. It is therefore important to establish the relationship between the continuous-time signals present in the physical system under control, and the discrete time signals used by the controller. Although the theory of sampled-data systems is very well developed (Kuo, 1980; Franklin et al., 1990; Åström and Wittenmark, 1997), we will endeavour here to give a physical description of the processes involved and the conditions under which the continuous-time signals can be readily inferred from their discrete-time counterparts.

The block diagram of a linear continuous-time *plant*, subject to a *disturbance* $d_C(t)$, and driven by a discrete-time feedforward controller is shown in Fig. 3.1(a). The name 'plant' is widely used in the control literature to denote the physical system under control which, in the context of active control, is the system between the input to the secondary actuator and the output of the sensor used to measure the residual error signal. The output of this *error sensor* in the absence of active control is termed the disturbance signal, and is caused by the influence of the original, primary, source on this sensor. Note that the disturbance signal plays a similar role to the desired signal in the electrical noise cancelling problem discussed in Section 2.3. It is, however, assumed that the controlled output is added to the disturbance in Fig. 3.1, because of the superposition of physical signals which occur in a linear control system, rather than the controlled output being subtracted from the desired signal to form the error in the electrical cancellation problem of Fig. 2.4, for example. The complete discrete-time control system includes not just the digital controller, $W(z)$, but also the data converters and analogue anti-aliasing and reconstruction filters. The transfer function of the continuous-time plant in the Laplace domain is equal to $G_C(s)$ and it is assumed that an analogue reconstruction filter, with transfer function $T_R(s)$, is used after the digital to analogue converter (DAC). The digital to analogue converter itself generally has an inherent zero-order hold (see Franklin et al., 1990, for example), whose Laplace-domain transfer function is equal to

$$T_Z(s) = \left(1 - e^{-sT}\right)/s \;, \tag{3.1.1}$$

where T is the sampling time.

We will assume that the combined response of the reconstruction filter and the plant provides sufficient attenuation of the frequency components produced by the DAC above half the sample frequency for the output of the plant to be effectively band-limited to below this frequency. Similarly we will assume that the anti-aliasing filter, with transfer function $T_A(s)$, removes any components of the continuous time disturbance above half the sample frequency. Under these conditions the discrete-time error signal is a faithful representation

(a)

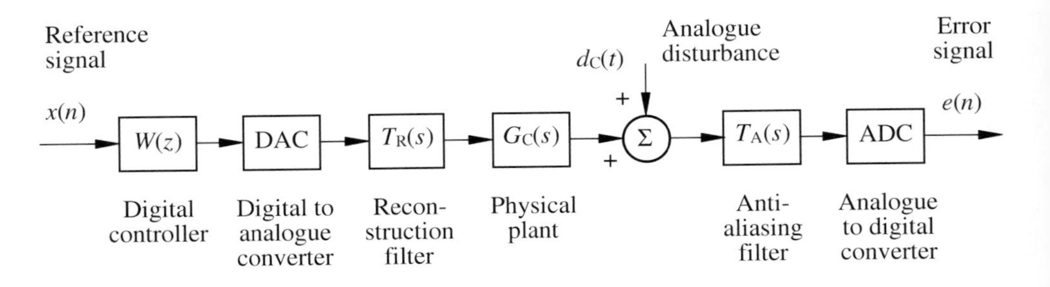

(b)

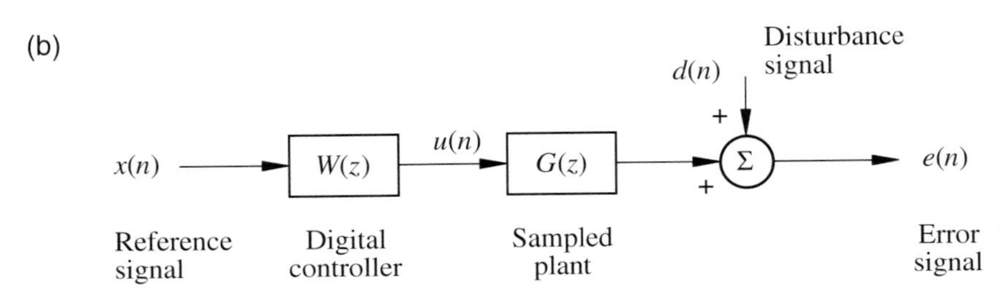

Figure 3.1 Block diagram of a continuous-time 'analogue' plant driven by a sampled-time 'digital' feedforward controller (a) and its equivalent sampled-time representation (b).

of the continuous-time error signal, and, more importantly, the filtered continuous time error signal behaves smoothly between the discrete sampling time.

Assuming no disturbance, the z domain transfer function from the sampled input $u(n)$ to the sampled output $e(n)$ is the effective plant response seen by the digital controller, which is also known as the pulse-transfer function (Franklin et al., 1990). This transfer function, $G(z)$, corresponds to the plant response 'seen' by the digital controller, and includes the response of the continuous-time system under control, the data converters and the analogue anti-aliasing and reconstruction filters. It is also convenient to include any processing delay in the controller within this sampled plant response. This is typically one full sample delay for active control systems that operate synchronously and at relatively high sampling rates, but could also be a fraction of a sample delay for systems that operate at a slower sampling rate, and in which the input and output of the controller are not sampled synchronously. The effective plant response seen by the digital controller can be computed by taking the z transform of the sampled impulse response of the continuous-time signals, which is itself given by the inverse Laplace transform of the s-domain transfer function. Thus

$$G(z) = \mathcal{Z}\mathcal{L}^{-1}\left[T_Z(s)T_R(s)G_C(s)T_A(s)\right], \tag{3.1.2}$$

where $\mathcal{Z}$ denotes taking the z transform and $\mathcal{L}^{-1}$ denotes taking the inverse Laplace transform. The block diagram of the equivalent discrete-time feedforward controller is shown in Fig. 3.1(b), where the discrete-time plant response $G(z)$ includes the responses of all the data converters and analogue filters.

3.1.3. Feedforward Control

In order to describe most clearly the different approaches to feedforward control, we will restrict ourselves in this chapter to a discussion of digital control systems that have a single reference signal, $x(n)$, whether derived from a reference sensor or a synchronising device, a single secondary actuator, with input $u(n)$, and a single error sensor, with output $e(n)$. From the block diagram of such a system shown in Fig. 3.1(b), the z transform of the signal driving the secondary source, $U(z)$, can be written in terms of the reference signal $X(z)$, and the transfer function of the digital controller, $W(z)$, so that

$$U(z) = W(z)\, X(z) \; . \tag{3.1.3}$$

If we assume for the moment that the digital controller is of FIR form, with coefficients w_0, . . ., w_{I-1}, the discrete-time domain version of the control signal that drives the secondary actuator can be written as

$$u(n) \;=\; \sum_{i=0}^{I-1} w_i\, x(n-i) \; . \tag{3.1.4}$$

Note that the current sample of the reference signal $x(n)$ is assumed to be used in calculating the current sample of the control signal and thus the controller is assumed to act instantaneously. Although any practical controller will inevitably have a processing delay, we have assumed that this is contained within the response of the discrete time plant, $G(z)$ in Fig. 3.1(b), as described above, which allows equation (3.1.4) to be written in a way which is consistent with the signal processing literature.

The z transform of the final error signal in Fig. 3.1(b), $E(z)$, is the linear superposition of the disturbance, $D(z)$, and the control input $U(z)$, modified by the response of the physical system under control, the discrete-time plant $G(z)$, so that

$$E(z) = D(z) + G(z)\, U(z) \; . \tag{3.1.5}$$

The plant response has a single input and a single output and is described as being *SISO*. The difference between a feedforward control system and an electrical noise cancellation system is the presence of the plant response between the output of the control filter and the observation point for the error signal. The plant response would represent little problem if it were known with perfect precision and was minimum phase, since an inverse plant response could then be included at the output of the control filter, and the net response between the output of the control filter and the error observation point would be unity, as it is the case for electrical noise cancellation. In practice, however, the plant response is rarely minimum phase, nor is it known with perfect precision, or even completely constant over time. The non-minimum phase characteristics of the plant arise in active control largely because of the distributed nature of the system under control, which allows the disturbances to propagate as waves. In the simplest case, this gives rise to a pure *delay* in the plant response. Such a delay between the filter output and the error observation point would destabilise most of the adaptive filtering algorithms discussed in the previous chapter. The adaptive filtering problem thus has to be re-examined to take the plant response into account. The algorithms used to adjust the control filter in an adaptive feedforward system are based on the same principles as those outlined above for adaptive digital filters, but to account for the presence of the plant they generally require an *internal model* of the plant's response. For the adaptive control algorithm to be stable, this model generally has to represent the plant's response with a certain precision. The response of the plant generally changes with time and/or operating conditions, and so the accuracy with which the internal model must represent the plant's response is an important issue.

3.1.4. Acoustic and Structural Plant Responses

Examples of the plant responses measured for a typical acoustic system and for a typical structural system are shown in Figs 3.2 and 3.3. Figure 3.2 is the response measured from the input of a loudspeaker acting as the secondary actuator, to the output from a pressure microphone, acting as the error sensor, inside a medium-size car. Although the acoustic

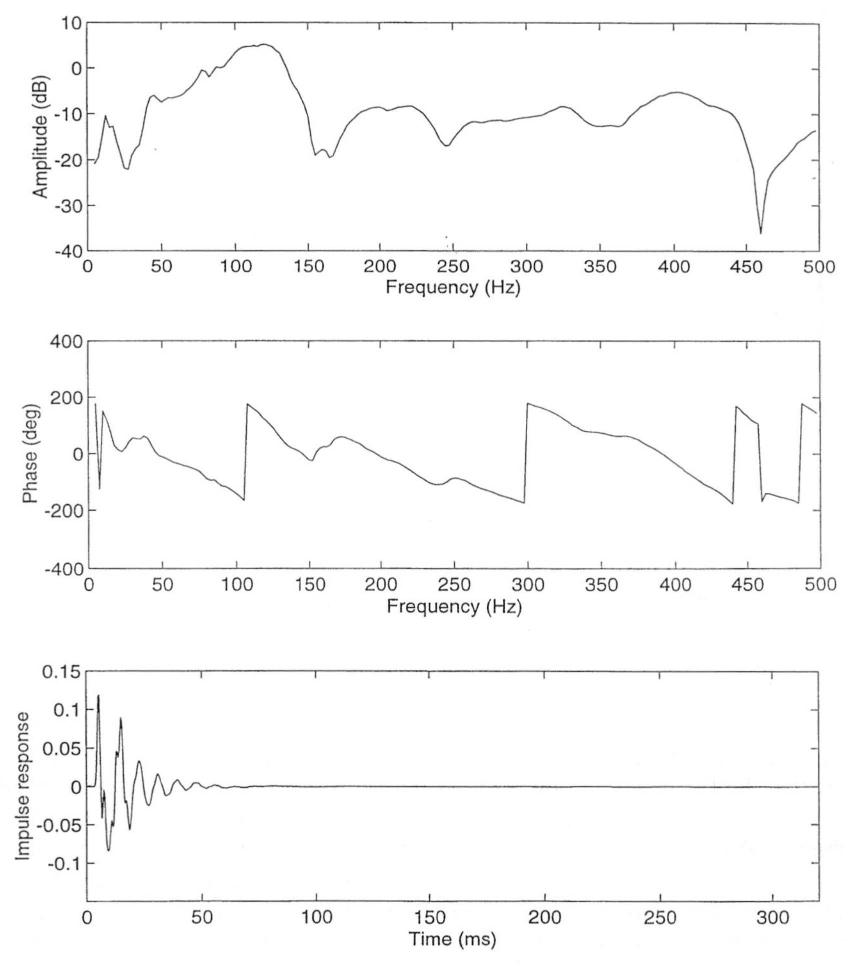

Figure 3.2 The modulus and phase of the frequency response of an acoustic enclosure excited by a loudspeaker and measured using a pressure microphone. The impulse response of the system is also shown.

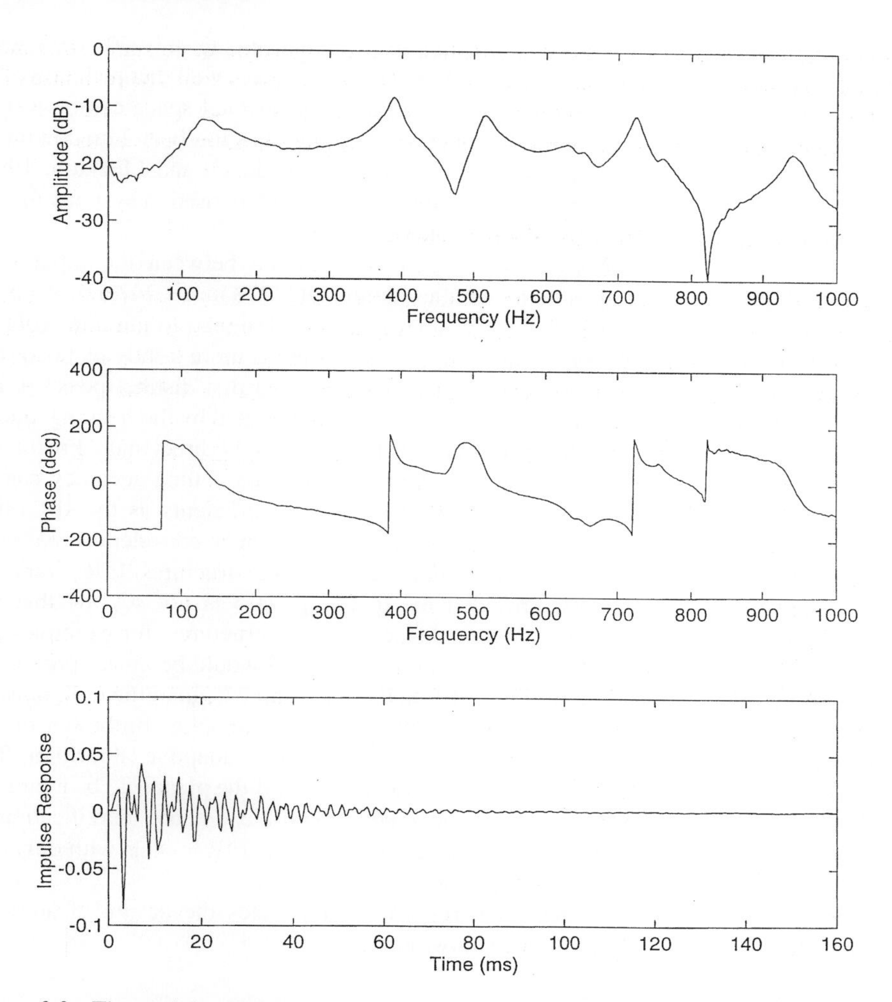

Figure 3.3 The modulus and phase of the frequency response of an aluminium plate excited by a piezoceramic actuator and measured using a distributed piezoelectric film sensor. The impulse response of the system is also shown.

response in such an enclosure can be expressed in terms of a series of modal resonances, as described in Chapter 1, these resonances are very well damped in the car, since the damping ratio (ζ) is typically at least 10%, and the resonances are close together. The result is that the frequency response shows no well-defined peaks, although a number of zeros can be seen in Fig. 3.2 which are due to destructive interference between the contributions from the various acoustic modes to the pressure measured at the microphone. Apart from the discontinuities at $\pm 180°$, the phase response shows the increasing lag with frequency associated with an overall delay of about 5 ms, which can also be seen in the initial part of the impulse response. This delay is partly due to the acoustic propagation time, about 3 ms in this case, and partly due to the processing delay in the digital signal processing system used to measure the response, and its associated converters and anti-aliasing and reconstruction filters. The impulse response in Fig. 3.2 is also very well damped, and can be accurately modelled using an FIR filter having only 50 coefficients operating at a sampling rate of 1 kHz. Even if the enclosure is not well damped, however, an FIR model of the acoustic response in such a three-dimensional space can be just as accurate as an IIR model with a similar number of coefficients, as has been found in many studies of acoustic echo cancellation (see for example Gudvangen and Flockton, 1995; Liavas and Regalia, 1998). FIR plant models are thus very widely used in systems for the active control of sound in three-dimensional enclosures.

The structural response shown in Fig. 3.3 was measured between the input to a piezoceramic (PZT) actuator glued to a thin aluminium plate (278 mm × 247 mm × 1 mm) and the output of a piezoplastic (PVDF) sensor which was designed to measure volume velocity (Johnson and Elliott, 1995a). In this case, the system is more lightly damped ($\zeta \approx$ 3%) and the modes are also more separated in frequency, so that distinct peaks in the frequency response can be seen. The impulse response is dominated by the 'ringing' due to these resonances, and is well modelled with an IIR filter having 24 direct and 24 recursive coefficients, operating at a sampling rate of 2 kHz. A similar modelling accuracy can be achieved with an FIR filter having about twice as many coefficients as the IIR filter, however, and in practice the coefficients of such an FIR filter can be considerably easier to identify than those of the IIR filter. Thus even for engineering structures, FIR filters can provide convenient plant models, provided the damping ratio is not so low that the response is entirely dominated by modal resonances. Space structures, for example, can have very little damping ($\zeta \approx 0.1\%$) and clearly an IIR model would be more appropriate in that case. In the identification of either FIR or IIR plant models, identification noise is injected into the secondary source and the difference between the output of the sensor and the output of the plant model is minimised using a suitable adaptive algorithm. The identification noise must be broadband, typically white noise, if the plant is to be identified over the whole frequency band, but it could be tonal if control, and hence identification, is only required for a single frequency disturbance, in which case FIR models with only two coefficients are needed for accurate identification.

In summary, the presence of the plant response complicates the design of adaptive algorithms for adjusting the control filter in two ways:

1. because of its non-minimum phase properties,
2. because of its potentially time-varying behaviour.

These two issues will be explored more fully in the following sections.

3.2. CONTROL OF DETERMINISTIC DISTURBANCES

A disturbance is entirely deterministic if all its future behaviour can be perfectly predicted from its previous behaviour. In practice, many forms of disturbance can be treated as being almost deterministic over the timescales of importance in active control, so that the signal is almost perfectly predictable. Predictability in a random signal implies that its autocorrelation function is of long duration, which in turn implies that its spectrum must have relatively sharp peaks. Therefore disturbances that are almost deterministic are also referred to as being *narrowband*. The most important example of a deterministic disturbance in active control is that generated by a rotating or reciprocating machine, in which case the resulting sound and vibration can be treated, within the required bounds of accuracy, as being periodic. If the disturbance were perfectly periodic and the plant response did not change, optimal control could be achieved with a fixed, time-invariant, controller. In practice, most disturbance signals encountered in active control are slowly changing, and thus not perfectly periodic, and adaptive controllers are required to maintain good performance. It is possible to implement adaptive control systems that control such periodic disturbances either in the time domain or in the frequency domain. It will be assumed below that the fundamental frequency of the disturbance is known. This information is normally derived from an external periodic reference signal, generated by a tachometer on a rotating or reciprocating engine for example. Methods of estimating this fundamental frequency from the disturbance signal itself, and thus implementing a control system with no external reference signal, have been discussed by Bodson and Douglas (1997) and Meurers and Veres (1999), for example.

3.2.1. Waveform Synthesis

In the time domain, the control signal can be generated by passing a periodic reference signal through an FIR digital controller whose impulse response is as long as the fundamental period of the disturbance. Adjusting the coefficients of this filter can then generate any periodic waveform within the sampling constraints of the digital system. A particularly efficient implementation of a periodic controller is achieved if the sampling rate can be arranged to be an exact integer multiple of the period of the disturbance. For machines running at a constant speed, this could be achieved in practice by using a phase-locked loop to determine the sampling rate, which could be synchronised to a tachometer signal at the rotation rate. The FIR control filter now only has to have as many coefficients as the period of the disturbance divided by the sampling period. Furthermore, if the reference signal is assumed to be a periodic impulse train at the fundamental frequency of the disturbance, the implementation of the controller also becomes extremely efficient (Elliott and Darlington, 1985).

Referring to the block diagram of such a control system shown in Fig. 3.4, the sampled output of the control filter, $u(n)$, is given by the reference signal $x(n)$, filtered by an I-th order FIR controller, so that in general

$$u(n) = \sum_{i=0}^{I-1} w_i \, x(n-i) \, . \tag{3.2.1}$$

In the particular case in which the reference signal is a periodic impulse train, then

$$x(n) = \sum_{k=-\infty}^{\infty} \delta(n-kN) \, , \tag{3.2.2}$$

where the signal has N samples per period and $\delta(n)$ is the Kronecker impulse function, which is equal to 1 if $n = 0$ and is otherwise equal to zero. If the control filter also has N coefficients, its output can be written as

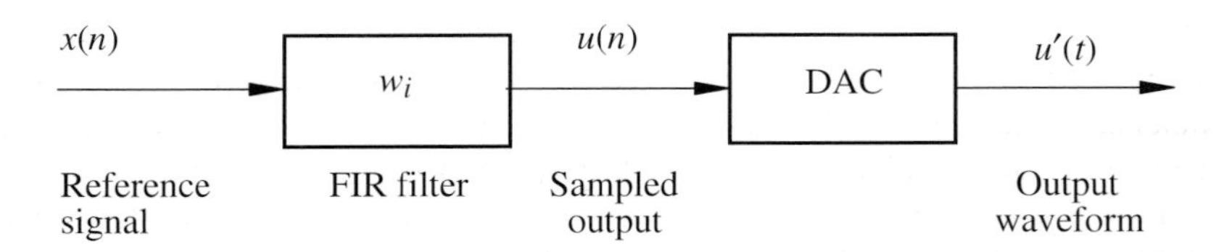

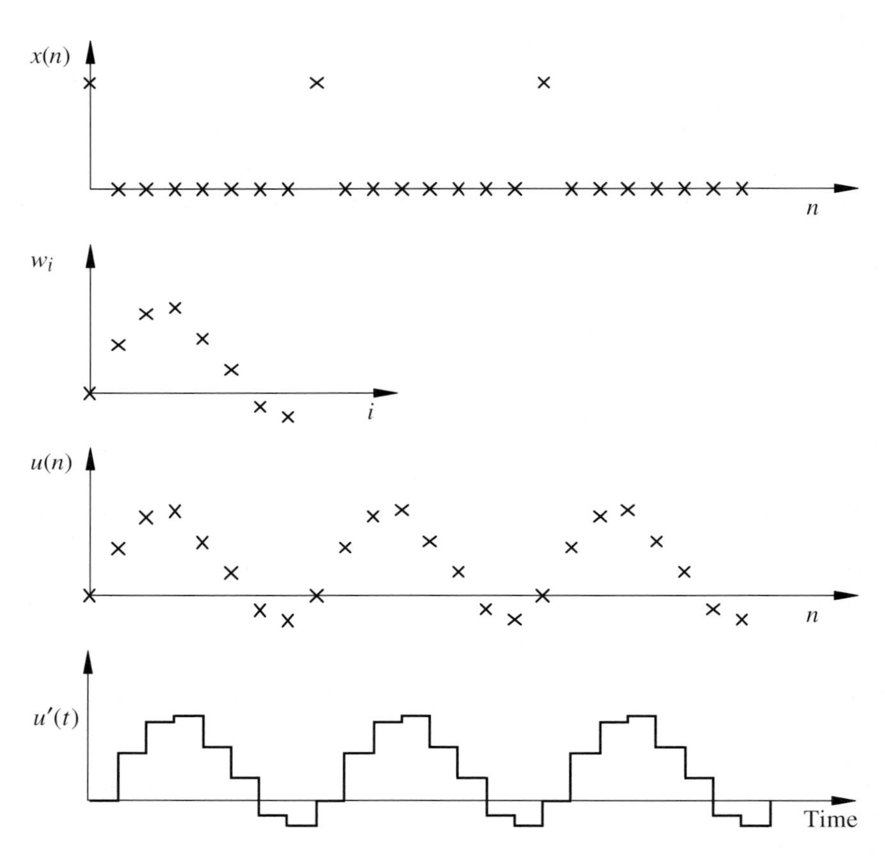

Figure 3.4 Waveform syntheses as FIR digital filtering of a reference signal consisting of a periodic impulse train.

$$u(n) = \sum_{k=-\infty}^{\infty} \sum_{i=0}^{N-1} w_i \, \delta(n - kN) = w_p \, , \tag{3.2.3}$$

where p denotes the smallest value of $(n - kN)$ for any k, and may be interpreted as the 'phase' of $u(n)$ relative to $x(n)$. The output signal is thus a periodic reproduction of the impulse response of the control filter, as illustrated in Fig. 3.4, and the waveform stored in the control filter is effectively 'played out' every cycle after triggering by the impulsive reference signal. The implementation of the control filter thus requires no computation except the sequential retrieval of the N coefficients w_i. This form of controller was originally investigated by Chaplin (1983) who described it as using *waveform synthesis*. Chaplin originally suggested a 'trial and error' or *power sensing* approach to the adaptation of the coefficients of the control filter in which the mean-square value of the error signal is monitored and the coefficients are individually adjusted in turn until this is minimised. Later, more sophisticated algorithms were suggested for adapting all of the filter coefficients simultaneously using the individual samples of the error sequence, which were described as *waveform sensing* algorithms (Smith and Chaplin, 1983). Some of these algorithms can be shown to be variants of the LMS algorithm, which takes on a particularly simple form when the reference signal is a periodic impulse train (Elliott and Darlington, 1985).

3.2.2. Harmonic Control

A more common method of controlling periodic disturbances is to implement a controller that works in the frequency domain, i.e. operates on each harmonic of the disturbance separately. A sampled periodic disturbance can be represented as the finite summation of its harmonics,

$$d(n) = \sum_{k=-K}^{K} D_k \, \mathrm{e}^{\, jk\omega_0 Tn} \, , \tag{3.2.4}$$

where k is the harmonic number, D_k is the complex amplitude of the k-th harmonic, the complex amplitude of the harmonics with negative values of k are the conjugates of the positive-numbered ones, i.e. $D_{-k} = D_k^*$, and there are assumed to be a total of K harmonics. The quantity $\omega_0 T$ is 2π times the normalised fundamental frequency, i.e. the true fundamental frequency, $\omega_0/2\pi$ in hertz, divided by the sampling rate, $1/T$.

If the system under control is linear, the amplitude of the error signal at one harmonic will only depend on the disturbance and the control input at that same harmonic. The control algorithm can then be divided into K *independent* loops, each operating on an individual harmonic, which do not interact in the steady state. The more involved case of a nonlinear system, in which there can be significant interaction between the harmonics, will be discussed in Chapter 8.

The number of harmonics one chooses to control will depend upon the application. The simplest controller, for a disturbance with a relatively small number of harmonics, may have a separate, tonal, reference signal for each harmonic. An individual control filter may then be used for each harmonic, which only has to have its amplitude and phase adjusted in response to the error signal at one frequency. This form of controller was proposed by Conover (1956) and the original control arrangement he used is shown in Fig. 3.5, in which analogue circuits are used to generate the reference signals and for amplitude and phase adjustment. Conover was concerned with the active control of transformer noise. This is most prevalent at the even harmonics of the line frequency, since it is predominantly generated by magnetostriction in the transformer core. The line frequency of Conover's transformer was 60 Hz, or 60 cps in 1956, and he used a full-wave rectifier to double the fundamental frequency and generate a signal rich in harmonics. These harmonics were then individually selected with band-pass filters, to produce tonal reference signals at 120 Hz, 240 Hz and 360 Hz, whose amplitudes and phases could be adjusted before being added together and used to drive the secondary loudspeaker.

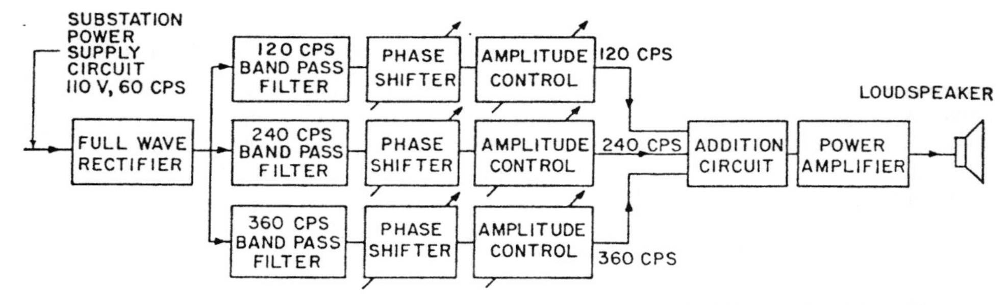

Figure 3.5 Control of a periodic sound source by using individual harmonics (from Conover, 1956).

Conover originally proposed a manual adjustment of the amplitude and phases of the individual harmonics, as illustrated in the cartoon shown in Fig. 3.6, taken from his 1956 paper. Conover does, however, also discuss the need for automatic adaptation and he describes how this was difficult to implement using the analogue technology at his disposal. Using digital techniques, it is far easier to automatically adjust the amplitude and phase of a sinusoidal reference signal, and this has led to the rapid development of frequency-domain controllers in recent years. In practice, the real and imaginary parts of the complex control signal, which correspond to the in-phase and quadrature parts of the physical signal, are generally adjusted in a modern system, rather than the amplitude and phase.

The complex component of the steady state error signal at the k-th harmonic can be written as

$$E_k = D_k + G(\mathrm{e}^{jk\omega_0 T})U_k \ , \tag{3.2.5}$$

where D_k was introduced in equation (3.2.4), $G(\mathrm{e}^{jk\omega_0 T})$ is the complex frequency response of the plant at a normalised angular frequency of $k\omega_0 T$ and U_k is the complex control signal at this frequency.

Clearly the error signal at this harmonic can be driven to zero in this single-channel case if the control input is equal to the optimum value given by

$$U_{k,\mathrm{opt}} = -\frac{D_k}{G(\mathrm{e}^{jk\omega_0 T})} \ . \tag{3.2.6}$$

Note that because this equation only has to be satisfied at a single frequency, there are no problems with the causality of the controller, which can be implemented using any circuit that gives the correct amplitude and phase values at $k\omega_0 T$, even if the plant response has considerable delay.

In practice the plant response and disturbance can never be measured with perfect precision, and an iterative method of adjusting the control signal must generally be used. This has the additional advantage of being able to track non-stationarities in the disturbance, although these must occur on a timescale that is long compared with the period of the disturbance, since otherwise the harmonic series representation of equation (3.2.4) would no longer hold.

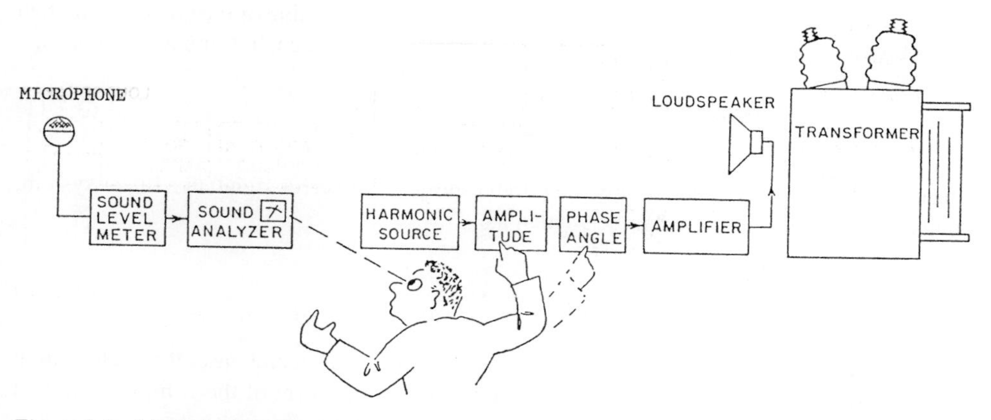

Figure 3.6 Manual adaptation of the amplitude and phase of a feedforward controller for harmonic disturbances (from Conover, 1956).

3.2.3. Adaptive Harmonic Control

One of the simplest iterative adaptation algorithms is the method of steepest descent, in which the real and imaginary parts of the complex control signal are changed in proportion to the negative gradient of the cost function with respect to these control variables. Writing the complex control variable in terms of its real and imaginary parts as

$$U_k = U_{kR} + jU_{kI} , \tag{3.2.7}$$

the steepest-descent algorithm can be written as the pair of equations

$$U_{kR}(n+1) = U_{kR}(n) - \mu \frac{\partial J}{\partial U_{kR}(n)} , \quad U_{kI}(n+1) = U_{kI}(n) - \mu \frac{\partial J}{\partial U_{kI}(n)} , \tag{3.2.8 a,b}$$

where J is the cost function being minimised, n is now the iteration index and μ is a convergence factor, or step size. Alternatively, we can write equations (3.2.8 a and b) as a single equation if we define the *complex gradient* to be

$$g_k(n) = \frac{\partial J}{\partial U_{kR}(n)} + j \frac{\partial J}{\partial U_{kI}(n)} , \tag{3.2.9}$$

in which case the steepest-descent algorithm, equation (3.2.8), can be written as

$$U_k(n+1) = U_k(n) - \mu \, g_k(n) . \tag{3.2.10}$$

Since the cost functions we will consider in this book are restricted to being real variables, we do not need to be concerned about the need to define an analytic gradient of the form discussed by Jolly (1995) or Haykin (1996), for example.

If we now assume a cost function equal to the mean-square value of the error signal, then this is equal to the sum of squared moduli of the error signal at each of the K harmonics,

$$J = \sum_{k=1}^{K} |E_k|^2 . \tag{3.2.11}$$

The control input U_k will affect only the k-th harmonic of the error signal in a linear system, and so the complex gradient can be written as

$$g_k(n) = \frac{\partial |E_k|^2}{\partial U_{kR}(n)} + j \frac{\partial |E_k|^2}{\partial U_{kI}(n)} . \tag{3.2.12}$$

If E_k is given by equation (3.2.5), then the cost function is a special case of the Hermitian quadratic form discussed in the Appendix, and so the scalar form of the complex gradient vector described in the Appendix is equal to

$$g_k(n) = 2\, G^*(\mathrm{e}^{jk\omega_0 T})\, E_k(n) , \tag{3.2.13}$$

where * denotes conjugation. Equation (3.2.13) could also be derived by differentiating the expression for $|E_k|^2$, expressed in terms of the real and imaginary parts of equation (3.2.5), with respect to U_{kR} and U_{kI}. The steepest-descent algorithm can thus be written as

$$U_k(n+1) = U_k(n) - \alpha\, G^*\left(\mathrm{e}^{jk\omega_0 T}\right) E_k(n) , \tag{3.2.14}$$

where $\alpha = 2\mu$ is the convergence coefficient.

If the adaptation proceeds slowly, the error signal will still be in the steady state at each iteration, so that

$$E_k(n) = D_k + G\left(\mathrm{e}^{jk\omega_0 T}\right) U_k(n) , \tag{3.2.15}$$

where the disturbance is assumed to be stationary. Using equations (3.2.15) and (3.2.6), equation (3.2.14) can be written as

$$\left[U_k(n+1) - U_{k,\mathrm{opt}}\right] = \left[1 - \alpha |G_k|^2\right] \left[U_k(n) - U_{k,\mathrm{opt}}\right] , \tag{3.2.16}$$

where $|G_k|^2 = G\left(\mathrm{e}^{jk\omega_0 T}\right) G^*\left(\mathrm{e}^{jk\omega_0 T}\right)$. If $\alpha |G_k|^2$ is considerably smaller than unity, to ensure that the adaptation is slow, then

$$\left(1 - \alpha |G_k|^2\right) \approx \mathrm{e}^{-\alpha |G_k|^2} , \tag{3.2.17}$$

and so if $U_k(0) = 0$ then

$$U_k(n) \approx \left(1 - \mathrm{e}^{-\alpha |G_k|^2 n}\right) U_{k,\mathrm{opt}} . \tag{3.2.18}$$

Substituting this into equation (3.2.15) yields an expression for the convergence of the error signal as a result of implementing the steepest-descent algorithm which is given by

$$E_k(n) = D_k\, \mathrm{e}^{-\alpha |G_k|^2 n} , \tag{3.2.19}$$

so that the real and imaginary components of the error signal decay to zero exponentially with the same time constant, which is equal to $1/\left(\alpha|G_k|^2\right)$ samples.

In a practical implementation, a perfect model of the plant response is not available and the steepest-descent algorithm becomes

$$U_k(n+1) = U_k(n) - \alpha \hat{G}^*(e^{jk\omega_0 T}) E_k(n) \ , \qquad (3.2.20)$$

where $\hat{G}(\mathrm{e}^{jk\omega_0 T})$ is the practical estimate of the plant response, or *plant model* which is used in the control algorithm.

A more ambitious adaptation algorithm would involve an iterative estimate of the optimum solution, equation (3.2.6), which could be written as

$$U_k(n+1) = U_k(n) - \tilde{\alpha} \hat{G}^{-1}(\mathrm{e}^{jk\omega_0 T}) E_k(n) \ , \qquad (3.2.21)$$

where $\tilde{\alpha}$ is another convergence coefficient. Notice, however, that the phase of $\hat{G}^{-1}(\mathrm{e}^{jk\omega_0 T})$ is the same as that of $\hat{G}^*(\mathrm{e}^{jk\omega_0 T})$. The control signal will thus be adjusted in the same direction by both algorithms, and the only difference between them is the amount by which they are adjusted. This will depend on the relative value of the convergence coefficients in the two cases, but it is interesting to note in this case that if

$$\alpha = \frac{\tilde{\alpha}}{\left|\hat{G}(\mathrm{e}^{jk\omega_0 T})\right|^2} \ , \qquad (3.2.22)$$

then the steepest-descent algorithm, equation (3.2.20), is the same as the iterative least-squares algorithm (3.2.21). Thus, in the frequency domain, the method of steepest descent is exactly equivalent to an iterative solution to the least-squares problem for single-channel systems.

3.2.4. Stability Conditions

The conditions for the stability of the adaptive algorithm described by equation (3.2.20) can readily be established. Using equation (3.2.5) for E_k (at the n-th iteration), we can write equation (3.2.20) as

$$U_k(n+1) = U_k(n) - \alpha\left[\hat{G}^*(\mathrm{e}^{jk\omega_0 T}) D_k + \hat{G}^*(\mathrm{e}^{jk\omega_0 T}) G(\mathrm{e}^{jk\omega_0 T}) U_k(n)\right] . \qquad (3.2.23)$$

Equation (3.2.6) can now be used to express D_k as $-G(\mathrm{e}^{jk\omega_0 T}) U_{k,\mathrm{opt}}$, in which case subtracting a factor of $U_{k,\mathrm{opt}}$ from both sides of (3.2.23) allows this equation to be written as

$$\left[U_k(n+1) - U_{k,\mathrm{opt}}\right] = \left[1 - \alpha \hat{G}^*(\mathrm{e}^{jk\omega_0 T}) G(\mathrm{e}^{jk\omega_0 T})\right]\left[U_k(n) - U_{k,\mathrm{opt}}\right] . \qquad (3.2.24)$$

The adaptive algorithm described by equation (3.2.20) is thus guaranteed to converge towards the optimum solution provided

$$\left|1 - \alpha \hat{G}^*(\mathrm{e}^{jk\omega_0 T}) G(\mathrm{e}^{jk\omega_0 T})\right| < 1 . \tag{3.2.25}$$

If the real and imaginary parts of $\hat{G}^*(\mathrm{e}^{jk\omega_0 T}) G(\mathrm{e}^{jk\omega_0 T})$ are written as $R + jX$, then equation (3.2.25) can also be written as

$$\left|1 - \alpha R - j\alpha X\right|^2 < 1 , \tag{3.2.26}$$

or

$$(1 - \alpha R)^2 + (\alpha X)^2 < 1 , \tag{3.2.27}$$

so that the condition for stability is given by

$$0 < \alpha < \frac{2R}{R^2 + X^2} . \tag{3.2.28}$$

This is an important result, which will be used on several occasions below. In the single-channel case, considerable physical insight into the stability condition can be derived by assuming that the estimate of the plant's response at $k\omega_0 T$ can be written as

$$\hat{G}(\mathrm{e}^{jk\omega_0 T}) = M G(\mathrm{e}^{jk\omega_0 T})\, \mathrm{e}^{j\phi} , \tag{3.2.29}$$

where M is a positive real number representing the error in the magnitude, and ϕ represents the error in the phase of the estimated plant response. We can thus write

$$\hat{G}^*(\mathrm{e}^{jk\omega_0 T}) G(\mathrm{e}^{jk\omega_0 T}) = M \left|G(\mathrm{e}^{jk\omega_0 T})\right|^2 \mathrm{e}^{j\phi} , \tag{3.2.30}$$

and so the stability condition given by equation (3.2.28) can be written as

$$0 < \alpha < \frac{2 \cos \phi}{M \left|G(\mathrm{e}^{jk\omega_0 T})\right|^2} . \tag{3.2.31}$$

There are two important points to note about equation (3.2.31). First, the stability condition can never be fulfilled if $\cos\phi$ is negative, so the adaptive algorithm will be unstable unless the phase error in the estimated plant response is less than $\pm 90°$, i.e.

$$|\phi| < 90^\circ \quad \text{for stability} . \tag{3.2.32}$$

This important result will arise repeatedly in the following chapters. Second, provided the phase condition on $\hat{G}^*(\mathrm{e}^{jk\omega_0 T})$ is satisfied, any magnitude error can be compensated for by adjusting the magnitude of the convergence coefficient α.

Clearly the most rapid convergence occurs if there is no phase or magnitude error in $\hat{G}(\mathrm{e}^{jk\omega_0 T})$ and the convergence coefficient is equal to

$$\alpha_{\text{opt}} = \frac{1}{\left|G(\mathrm{e}^{jk\omega_0 T})\right|^2} , \tag{3.2.33}$$

which corresponds to setting the convergence coefficient in the iterative least-squares algorithm of equation (3.2.21) to unity. Under these conditions $\left|1 - \alpha \hat{G}^*(\mathrm{e}^{jk\omega_0 T}) G(\mathrm{e}^{jk\omega_0 T})\right|$

is zero and the adaptive algorithm given by equation (3.2.20) would converge to the optimal solution in one step. Potential errors in $\hat{G}(e^{jk\omega_0 T})$, however, mean that the convergence coefficient is generally adjusted so that it is a fraction of that indicated in equation (3.2.33). With no modelling error, the maximum value of the convergence coefficient for stability is equal to $2\alpha_{\text{opt}}$ according to equation (3.2.31). It is interesting to note from equation (3.2.24) that, provided the adaptive algorithm is stable, it is guaranteed to converge to the optimum result, regardless of the magnitude and phase errors in the estimate of the plant response in this single-channel, single-frequency case. We shall see that the same is not true for broadband controllers, or for multichannel single-frequency controllers.

If control only needs to be implemented at a single frequency, then an efficient implementation can be achieved by arranging for the sampling rate to be exactly four times this frequency (Elliott et al., 1987). The sequence 0, +1, 0, −1, 0, +1 etc., then corresponds to a tonal reference signal at the reference frequency. The time-domain control signal can be very efficiently generated by filtering this reference signal with a two-coefficient FIR filter, whose coefficients now correspond to the real and imaginary parts of the complex control signal used above. A straightforward extension of this technique to three harmonics is described by Boucher (1992), in which case the signals are sampled at eight times the fundamental frequency.

3.2.5. FFT Controller

A particularly efficient implementation of the controller for all harmonics can be achieved if the sampling rate is again assumed to be an integer multiple (N) of the period of the disturbance. The normalised angular fundamental frequency is then $\omega_0 T = 2\pi/N$ and the disturbance signal in equation (3.2.4) can be exactly represented by $N/2$ harmonics. The time-domain control signal can thus be written as

$$u(n) = \sum_{k=-N/2}^{N/2} U_k \, e^{j2\pi kn/N} , \tag{3.2.34}$$

where n is now the sample time rather than the iteration number. In this case the signal $e^{j2\pi kn/N}$ is effectively acting as a complex reference signal. Equation (3.2.34) is the inverse discrete Fourier transform (DFT) of U_k, which may be very efficiently implemented using the fast Fourier transform (FFT), particularly for large N. Similarly, the current estimate of the error signal at the k-th harmonic could be estimated from the last N samples of the error sequence using the equation

$$E_k = \sum_{n=1}^{N} e(n) e^{-j2\pi kn/N} , \tag{3.2.35}$$

which is the DFT of $e(n)$. Thus all $N/2$ harmonics could be controlled without the explicit generation of $N/2$ reference signals by using FFTs to calculate each N-sample block of output signals, according to equation (3.2.33) and using equation (3.2.34) to calculate the complex error signals used in the adaptation algorithm, as shown in Fig. 3.7.

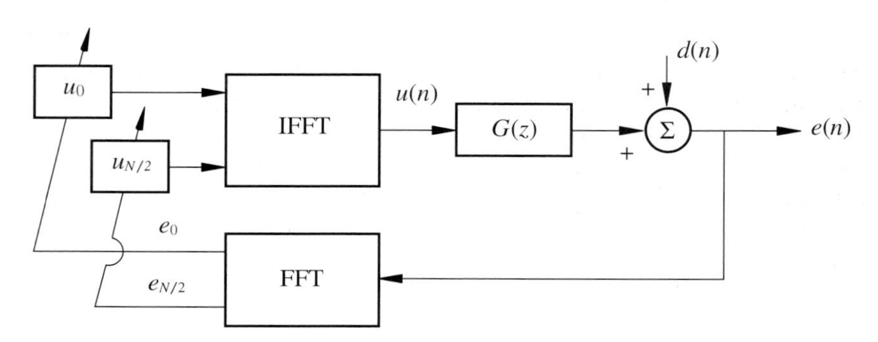

Figure 3.7 An efficient implementation of the frequency-domain harmonic controller, for all harmonics, using the FFT.

3.3. OPTIMAL CONTROL OF STOCHASTIC DISTURBANCES

Stochastic disturbances are characterised as having a random waveform. If they have a relatively well-contained autocorrelation function, then their spectrum is relatively broad, and they are sometimes referred to as being *broadband*. The active feedforward control of stochastic disturbances is considerably more complicated than the control of deterministic disturbances. This is partly because the reference and error signals are no longer perfectly *correlated*, as they were in the deterministic case, and is partly due to the need to control a whole range of frequencies at once. The response of the controller at each frequency cannot be adjusted independently, however, because of the constraint that the controller must be *causal*. In this section, we begin the discussion of feedforward control for such signals by presenting a rather general block diagram, and then deriving the optimal responses for a single-channel controller in both the frequency and the time domains.

3.3.1. Block Diagrams for Feedforward Control

Although the general block diagram derived here has very general application, it is convenient to motivate its form by first considering a specific example. The example we will choose is that of the active control of plane acoustic waves in a uniform duct, as described in Chapter 1. The practical aspects of this application will be more fully described in Section 3.8, but for now we only need to discuss the simplified system shown in Fig. 3.8. In this diagram, the waveform of the incident plane sound wave is measured by a *reference sensor* to give a signal, s, which is passed through a feedforward controller H, whose output, u drives the secondary sound source. The secondary source is indicated as a single loudspeaker in Fig. 3.8 but may in practice be an array of acoustic sources, in the same way that the reference and error sensors may in practice be more complicated than the single microphones shown in Fig. 3.8. The effect of the active control system is measured by a downstream *error sensor*, whose output, e, is not used directly to drive the secondary source but is only used to adjust the response of the feedforward controller H. The physical purpose of each of the various transducers and signals can be clearly appreciated in this example.

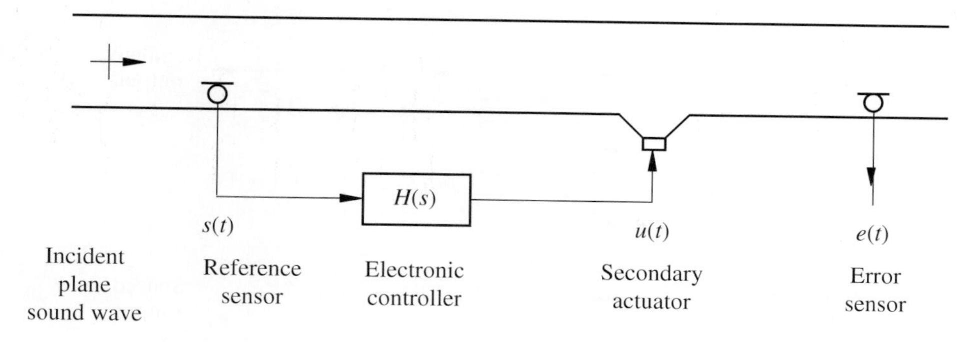

Figure 3.8 Single-channel feedforward control of sound in a duct.

The block diagram of the single-channel feedforward controller is shown in Fig. 3.9 (Ross, 1982; Elliott and Nelson, 1984; Roure, 1985), in which the original primary disturbance in the duct is assumed to be generated by a signal v, which is transmitted to the error sensor by the *disturbance path*, P_e to give d, and to the sensed signal from the reference sensor by the *sensor path* P_s to give z. The output of the secondary actuator affects the error sensor via the *secondary path* or 'error path', G_e, and also affects the sensed signal via the *feedback path*, G_s.

The block diagram shown in Fig. 3.9 can also be put into the form of the generalised plant configuration of Doyle (1983), as in Fig. 3.10, which is used as the starting point for many automatic control problems, as also discussed, for example, by Clark et al. (1998). In this configuration the inputs to the physical system are divided up into those that originate outside the control system, the *exogenous inputs*, of which there is only one, v in this case, and the inputs to the physical system from the actuators, u in this case. The outputs are also divided into the sensed outputs, s in this case, and the regulated outputs, e in this case. The combination of the disturbance, sensor, secondary and feedback paths in Fig. 3.9 are then grouped together as a single matrix which denotes the *generalised plant*. In fact, the block diagram of the control system is not complete as drawn in Figs 3.9 or 3.10 since sensor

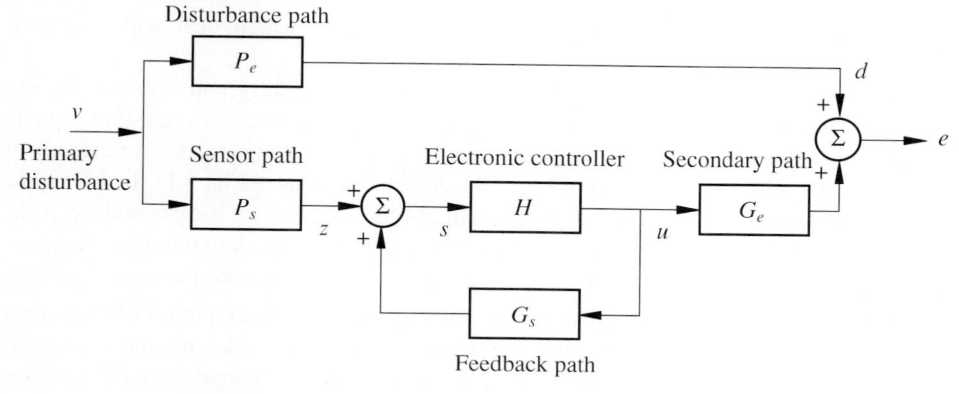

Figure 3.9 General block diagram of a single-channel feedforward active control system.

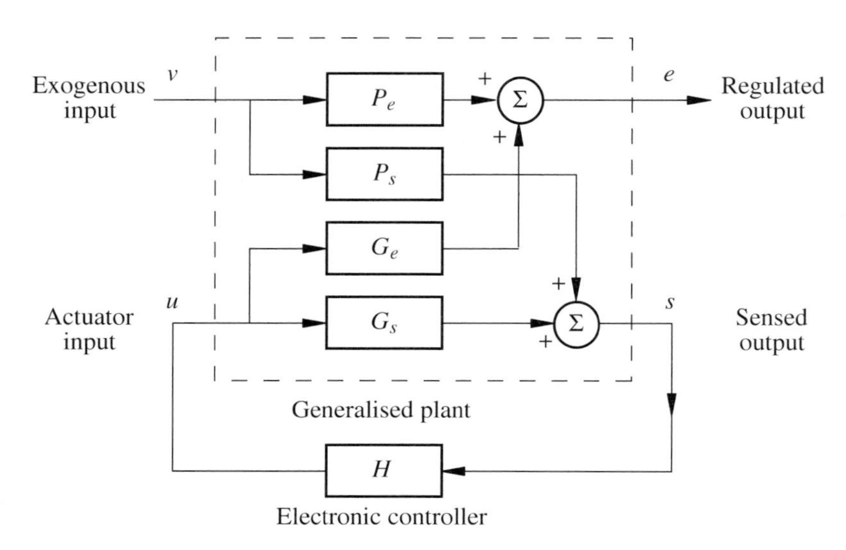

Figure 3.10 Block diagram of the feedforward system drawn in terms of the general control configuration, which is used as the starting point for many automatic control problems.

noise is also generally present at both the reference and error sensors. Sensor noise could be included as additional exogenous inputs in Fig. 3.10 but has been left out for clarity.

In an adaptive feedforward control system, the regulated output is used to adjust the electronic controller and so the part of the generalised plant that most affects the behaviour of the adaptive controller is the secondary path G_e. In this chapter *the plant response* will be used exclusively to refer to the response of the secondary path.

If the reference microphone in Fig. 3.8 were moved to be in the same plane in the duct as the secondary loudspeaker, the control system would become *feedback* rather than feedforward, since no time-advanced information would be available from the reference sensor. Driving the output of the reference sensor, rather than the error sensor, to zero would also block the propagation of the sound waves in the duct in this case, by creating a pressure release surface at the position of the secondary source. The part of the generalised plant that is most important in determining the behaviour of this feedback system would then be the feedback path, G_s, and in Chapters 6 and 7, which deal with feedback controllers, it is G_s that will be referred to as the plant.

If we now assume that the control system is implemented digitally, and that anti-aliasing and reconstruction filters have been included in the system, so that the discrete-time signals are faithful representations of their continuous-time counterparts, as described in Section 3.1, then the block diagram of Fig. 3.9 can be redrawn as in Fig. 3.11. In this figure the plant response, $G_e(z)$, refers to the discrete-time version of the secondary path, together with the data converters and analogue filters required for a digital implementation. The controller, $H(z)$, has also been assumed to have been implemented using a *feedback cancellation* architecture, in which an internal model of the feedback path, $\hat{G}_s(z)$, is used to cancel the effects of the discrete-time version of the true feedback path and associated converters, $G_s(z)$. If this cancellation is perfect, the input to the *control filter* $W(z)$ is then equal to the *reference signal* $x(n)$, which has components due to the original primary source

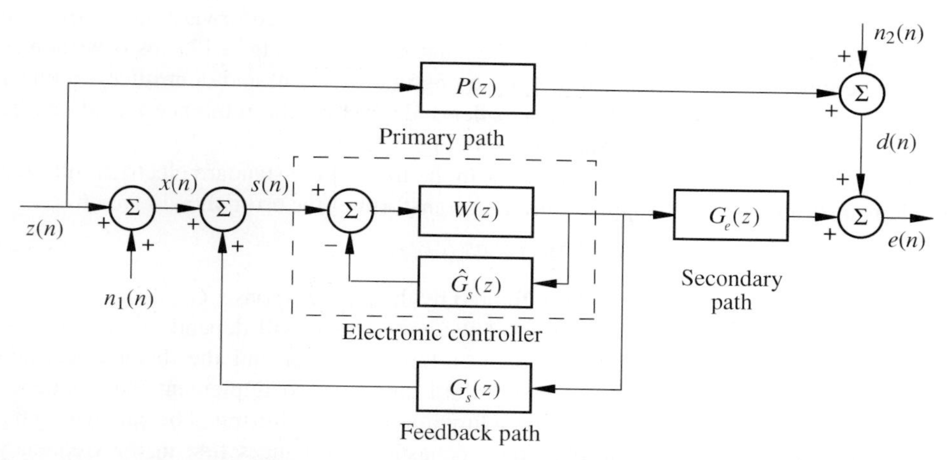

Figure 3.11 Block diagram of the feedforward control system, in which a feedback cancellation architecture has been used within the controller, and $x(n)$ and $d(n)$ are the outputs of the reference sensor and error sensor in the absence of control, i.e. if $W(z) = 0$.

measured at the reference sensor, $z(n)$, and measurement noise at the reference sensor, $n_1(n)$, so that its z transform can be written as

$$X(z) = Z(z) + N_1(z) \, . \tag{3.3.1}$$

The *disturbance* signal measured at the error microphone in the absence of control is denoted $d(n)$ and in general this signal will also contain components due to both the primary source and measurement noise. It is often convenient to assume that the part of the signal due to the primary source at the error sensor is linearly related to that at the reference sensor, so that we can write the z transform of the disturbance signal as

$$D(z) = P(z)\, Z(z) + N_2(z) \, , \tag{3.3.2}$$

where $N_2(z)$ is the z transform of the measurement noise at the error sensor.

In terms of the variables defined in Fig. 3.9, we can see that the transfer function of the *combined primary path*, $P(z)$, must be equal to

$$P(z) = \frac{P_e(z)}{P_s(z)} \, . \tag{3.3.3}$$

Some care must be taken in the interpretation of this combined primary path, however, since although $P(z)$ is stable and causal in many active control applications, as it would be in the anechoic duct example shown in Fig. 3.8 for example, this is not guaranteed to be the case.

3.3.2. Unconstrained Optimisation in the Frequency Domain

The object of this section is to calculate the performance of the optimum controller. We assume that the internal model of the feedbackpath is perfect, so that $\hat{G}_s(z) = G_s(z)$ in

Fig. 3.11, so that this block diagram becomes entirely feedforward, as shown in Fig. 3.12(a). Since there is now only a single plant response in Fig. 3.12, this is written as $G(z)$ and corresponds to $G_e(z)$. Further assuming that the plant and controller are linear and time invariant, the effect of the controller and plant on the reference signal can be transposed to give the block diagram shown in Fig. 3.12(b).

The right-hand side of Fig. 3.12(b) is now in the form of the standard electrical filtering problem, as discussed in Chapter 2, with the z transform of the error signal given by

$$E(z) = D(z) + W(z)\, R(z)\,, \tag{3.3.4}$$

where $R(z)$ is the reference signal, $X(z)$, filtered by the plant response, $G(z)$.

The maximum achievable attenuation of the error signal will depend on two factors. First, the degree of coherence between the reference signal and the disturbance, and second, the ability of the practical feedforward controller to implement the frequency response required for perfect control. These limitations can be illustrated by calculating the performance of the optimal controller for stochastic disturbances first in the frequency domain and then in the time domain. The frequency-domain analysis is used first to determine the optimum controller which is not constrained to be causal or of finite duration, and hence its performance is only limited by the coherence between the reference and disturbance signals. The frequency-domain analysis is then extended by adding the constraint that the controller must be causal, but still of infinite duration. The performance of a practical control filter that is both causal and of finite duration is derived using a time-domain analysis.

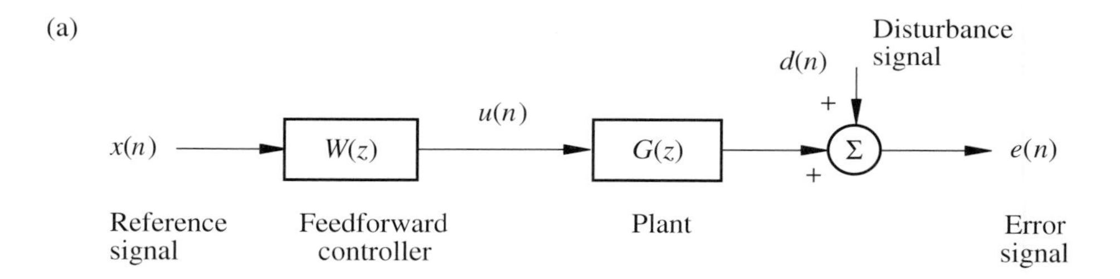

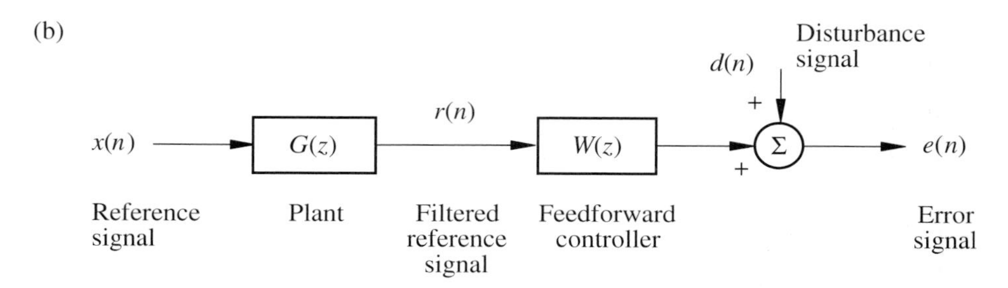

Figure 3.12 Simplified version of the general block diagram for a single-channel feedforward control system assuming perfect feedback cancellation (a), and a rearranged version of this for linear and time invariant systems (b).

In the frequency domain we can write the Fourier transform of $e(n)$, as

$$E(e^{j\omega T}) = D(e^{j\omega T}) + W(e^{j\omega T})R(e^{j\omega T}) . \tag{3.3.5}$$

The object at this stage in the analysis is to adjust $W(e^{j\omega T})$ *at each individual frequency* to minimise the power spectral density of the error signal. There is thus no constraint put on the controller concerning the complexity of the resultant frequency response or, more importantly, its *causality*. Nevertheless, the frequency domain calculation does provide some simple rules concerning the fundamental limits of control. The power spectral density of the error signal is defined to be

$$S_{ee}(e^{j\omega T}) = E\left[E^*(e^{j\omega T})E(e^{j\omega T})\right] , \tag{3.3.6}$$

which can be written as

$$\begin{aligned} S_{ee}(e^{j\omega T}) = {} & W^*(e^{j\omega T})S_{rr}(e^{j\omega T})W(e^{j\omega T}) + W^*(e^{j\omega T})S_{rd}(e^{j\omega T}) \\ & + S_{rd}^*(e^{j\omega T})W(e^{j\omega T}) + S_{dd}(e^{j\omega T}), \end{aligned} \tag{3.3.7}$$

where

$$S_{dd}(e^{j\omega T}) = E\left[D^*(e^{j\omega T})D(e^{j\omega T})\right] , \tag{3.3.8}$$

$$S_{rd}(e^{j\omega T}) = E\left[R^*(e^{j\omega T})D(e^{j\omega T})\right] = G^*(e^{j\omega T})E\left[X^*(e^{j\omega T})D(e^{j\omega T})\right] , \tag{3.3.9}$$

$$S_{rr}(\omega) = E\left[R^*(e^{j\omega T})R(e^{j\omega T})\right] = \left|G(e^{j\omega T})\right|^2 E\left[X^*(e^{j\omega T})X(e^{j\omega T})\right] , \tag{3.3.10}$$

and $E[\ ]$ denotes the expectation operator, which in principle is taken over an ensemble of the stochastic processes and in practice is obtained by averaging the products of the spectra derived from finite time segments of the data over many segments, as described in Chapter 2.

Equation (3.3.7) is a standard Hermitian quadratic form, as described in the Appendix, which is minimised at each frequency by the optimum controller response given by

$$W_{\text{opt}}(e^{j\omega T}) = -\frac{S_{rd}(e^{j\omega T})}{S_{rr}(e^{j\omega T})} = \frac{-S_{xd}(e^{j\omega T})}{G(e^{j\omega T})S_{xx}(e^{j\omega T})} . \tag{3.3.11}$$

Assuming that this optimal controller is implemented at each frequency, the resultant minimum level of the power spectral density of the error signal can be obtained by substituting equation (3.3.11) into equation (3.3.7), which gives

$$S_{ee,\min}(e^{j\omega T}) = S_{dd}(e^{j\omega T}) - \frac{\left|S_{rd}(e^{j\omega T})\right|^2}{S_{rr}(e^{j\omega T})} . \tag{3.3.12}$$

Normalising the minimum error by the power spectral density of the error signal before control, which is equal to $S_{dd}(e^{j\omega T})$, and using equations (3.3.9) and (3.3.10) to express this ratio in terms of the measurable reference signal, we obtain

$$\frac{S_{ee,\min}(e^{j\omega T})}{S_{dd}(e^{j\omega T})} = 1 - \frac{\left|S_{xd}(e^{j\omega T})\right|^2}{S_{xx}(e^{j\omega T})S_{dd}(e^{j\omega T})} = 1 - \gamma_{xd}{}^2(e^{j\omega T}) , \tag{3.3.13}$$

where $\gamma_{xd}^2(e^{j\omega T})$ is the *coherence function* between the signals from the reference sensor and the error sensor in the absence of active control.

The fundamental limits of performance caused by measurement noise can thus be established using equation (3.3.13), by measuring the initial outputs of the reference and error sensors in a practical system, $d(n)$ and $x(n)$, without the need to implement any form of controller. Equation (3.3.13) places stringent requirements on the coherence if significant attenuation in the power spectral density of the error signal is to be achieved. For example, a 10 dB attenuation requires a coherence at this frequency of 90%, $\gamma_{xd}^2(e^{j\omega T}) = 0.9$, and a 20 dB attenuation requires a coherence of 99%, $\gamma_{xd}^2(e^{j\omega T}) = 0.99$. Although equation (3.3.13) provides a simple and powerful method of estimating the maximum attenuation in any feedforward active control system, it does make certain assumptions about both the physical system under control and the nature of the controller. In particular, it assumes that the signal measured at the error sensor is stationary and is a true representation of the physical disturbance that is to be controlled, and that the controller is time invariant and realisable. If the disturbance is non-stationary and the controller is adaptive, then the assumption of time invariance is no longer valid and the performance can be *better* than that predicted by equation (3.3.13). If, on the other hand, the controller response given by equation (3.3.11) is not realisable with the type of filter used in practice, because the frequency response of the required controller is too complicated or, more importantly, because it is not causal, then the performance in practice will be *worse* than that predicted by equation (3.3.13).

3.3.3. Causally Constrained Optimisation in the Transform Domain

The Bode–Shannon formulation for the optimum causal filter, introduced in Section 2.4, can be used here to calculate the optimal *causal* controller in the z domain. The procedure is exactly analogous to that outlined in Section 2.4 except that instead of calculating the filter that operates on the reference signal, $x(n)$, to cancel the disturbance as in Fig. 2.4 we now calculate the controller which operates on the filtered reference signal to cancel the disturbance, as shown in Fig. 3.12(b). We again assume that the reference signal is generated by passing a white noise sequence of unit variance, $V(z)$, through a causal, minimum phase shaping filter, $F(z)$, so that

$$X(z) = F(z)\,V(z)\,, \tag{3.3.14}$$

where

$$S_{xx}(z) = F(z)\,F(z^{-1})\,, \tag{3.3.15}$$

as in Chapter 2.

The filtered reference signal in Fig. 3.12(b) can thus be written as

$$R(z) = G(z)\,F(z)\,V(z)\,. \tag{3.3.16}$$

In order to proceed we now need to define a whitening filter which will act on $r(n)$ to produce a white noise sequence. In Section 2.4, this whitening filter was just the inverse of the reference signal's shaping filter, $F(z)$, which by definition is minimum phase, and so has a stable inverse. In the present case we also have to deal with the plant response, $G(z)$, which will generally not be minimum phase, and will thus not have a stable inverse. We can still operate on $r(n)$ to produce a white noise sequence, however, by using the inverse of the *minimum-phase component* of $G(z)$, together with the inverse of $F(z)$. This white noise sequence will not be the same as that which is assumed to be generating the reference signal, $v(n)$, but still has the important property of being uncorrelated from sample to sample, which is required to derive the optimal causal controller. Specifically, we assume that the response of the plant is split up into a minimum-phase component $G_{\min}(z)$ and an *all-pass component* $G_{\text{all}}(z)$ so that

$$G(z) = G_{\min}(z)\, G_{\text{all}}(z) \ . \tag{3.3.17}$$

Assuming, as we will do throughout this chapter, that the plant is stable, then the poles of $G(z)$ will all lie inside the unit circle. The minimum-phase transfer function, $G_{\min}(z)$, then has the same poles as $G(z)$, except for the poles at the origin associated with a pure delay, and also shares the zeros of $G(z)$ that are inside the unit circle. Any zeros of $G(z)$ which are outside the unit circle appear in conjugate reciprocal locations in $G_{\min}(z)$, i.e. they are reflected about the unit circle. In order for both $G_{\min}(z)$ and $1/G_{\min}(z)$ to have a finite initial value, it was noted in Section 2.1 that $G_{\min}(z)$ must have as many poles as zeros. After the poles and zeros of $G_{\min}(z)$ away from the origin have been determined as above, it may be necessary to include additional poles at the origin to satisfy this condition. Although a digital minimum-phase system is often defined as being one with all its poles and zeros inside the unit circle, by this definition a unit delay would be minimum-phase, even though it does not have a stable causal inverse. We thus assume that the additional poles at the origin, which are associated with pure delays in the plant transfer function, are included in the all-pass component. The minimum-phase plant response, $G_{\min}(z)$, thus has an equal number of poles and zeros which are all inside the unit circle, as does $1/G_{\min}(z)$, which must also represent a stable causal system. The all-pass system has poles at the reflected positions inside the unit circle, as well as the poles at the origin of $G(z)$, and has zeros in the same positions as those of $G(z)$ which lie outside the unit circle (Oppenheim and Shafer, 1975). An example of the decomposition of $G(z)$ into its all-pass and minimum-phase components is shown in Fig. 3.13. Also shown in Fig. 3.13 are the poles and zeros of $G\left(z^{-1}\right)$, which are reflections of those of $G(z)$ about the unit circle with the poles at the origin replaced by zeros at the origin. The poles and zeros of $G_{\min}\left(z^{-1}\right)$ and $G_{\text{all}}\left(z^{-1}\right)$ may be similarly derived, and are also illustrated in Fig. 3.13, and by comparing these with the corresponding pole-zero plots for $G_{\min}(z)$ and $G_{\text{all}}(z)$ we can see that $G_{\min}(z)\, G_{\min}\left(z^{-1}\right) = G(z)\, G\left(z^{-1}\right)$ and that $G_{\text{all}}(z)\, G_{\text{all}}\left(z^{-1}\right) = 1$. The magnitude of the frequency response of the minimum-phase component, $\left|G_{\min}\left(e^{j\omega T}\right)\right|$, is thus the same as that of the plant, $\left|G\left(e^{j\omega T}\right)\right|$, but has the smallest possible phase accumulation consistent with $G_{\min}\left(e^{j\omega T}\right)$ remaining causal. The all-pass component then has the property that

$$\left|G_{\text{all}}(e^{j\omega T})\right|^2 = 1 \qquad \text{for all } \omega T \, . \tag{3.3.18}$$

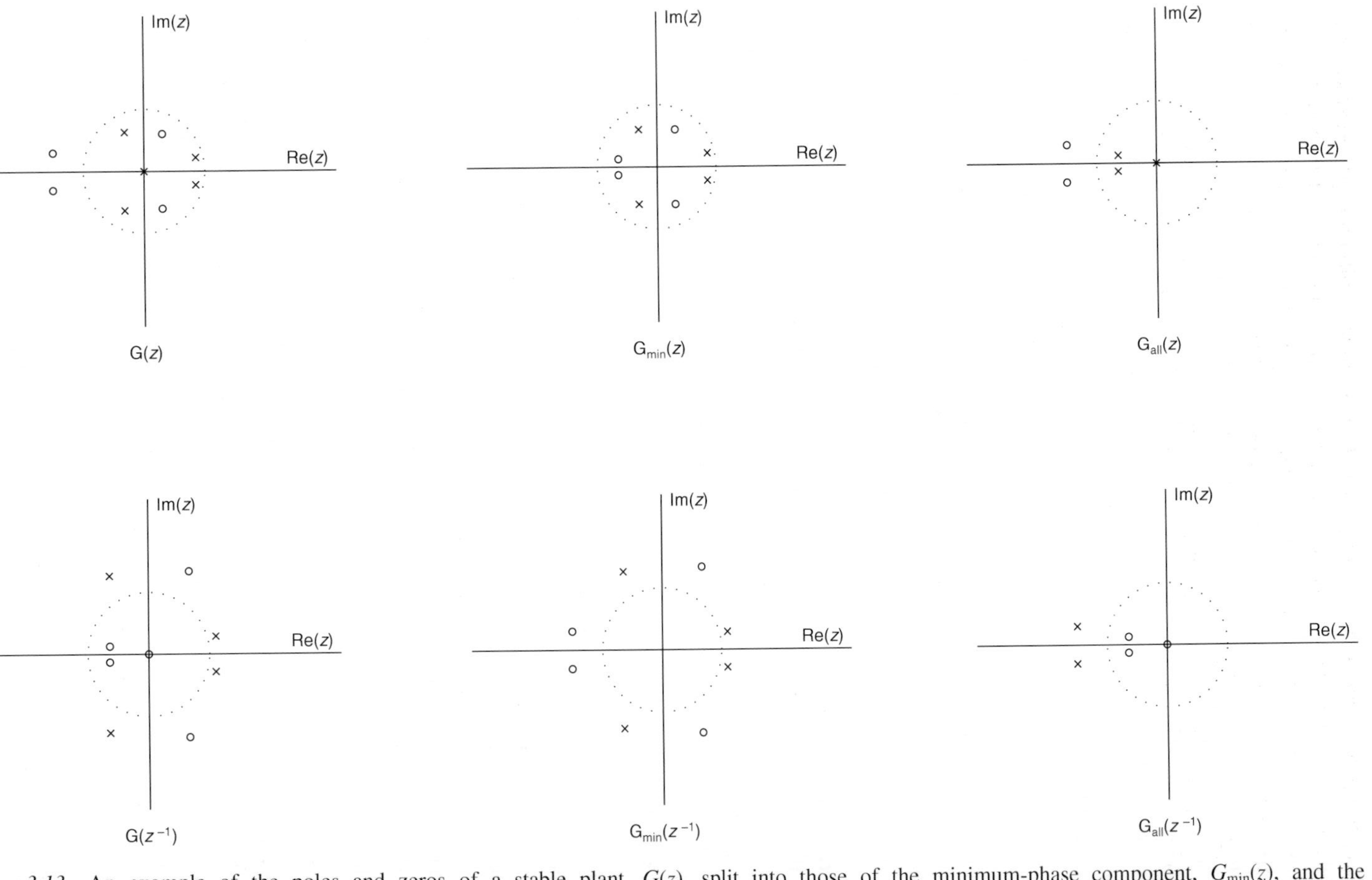

Figure 3.13 An example of the poles and zeros of a stable plant, $G(z)$, split into those of the minimum-phase component, $G_{min}(z)$, and the all-pass component, $G_{all}(z)$. Also shown are the poles of zeros of $G(z^{-1})$, $G_{min}(z^{-1})$ and $G_{all}(z^{-1})$ from which it can be seen that $G_{min}(z)\, G_{min}(z^{-1}) = G(z)\, G(z^{-1})$ and $G_{all}(z)\, G_{all}(z^{-1}) = 1$.

The signal obtained after whitening the filtered reference signal can now be written as

$$V'(z) = \frac{1}{F(z)}\frac{1}{G_{\min}(z)}R(z) \ , \tag{3.3.19}$$

as shown in Fig. 3.14. The optimum filter which operates on $v'(n)$ to best minimise the mean-square error can be written by direct analogy with the results of Section 2.4 as

$$W_{v'\text{opt}}(z) = -\left\{S_{v'd}(z)\right\}_+ \ , \tag{3.3.20}$$

as also shown in Fig. 3.14, where $S_{v'd}(z)$ is the cross spectral density between $v'(n)$ and $d(n)$ and $\{\ \}_+$ denotes that the z transform of the causal part of the time-domain version of the quantity inside the brackets has been taken. The aim of the analysis is to express the optimal causal filter in terms of directly measurable quantities. Using the fact that $R(z) = G(z)\,X(z)$ and equation (3.3.17) for $G(z)$, then $V'(z)$ in equation (3.3.19) can be written as

$$V'(z) = \frac{G_{\text{all}}(z)\,X(z)}{F(z)} \ , \tag{3.3.21}$$

and so we can express the cross spectral density in equation (3.3.20) as

$$S_{v'd}(z) = \frac{G_{\text{all}}\left(z^{-1}\right)S_{xd}(z)}{F\left(z^{-1}\right)} \ . \tag{3.3.22}$$

Since $G_{\text{all}}(z)\,G_{\text{all}}\left(z^{-1}\right) = 1$ then

$$S_{v'd}(z) = \frac{S_{xd}(z)}{F\left(z^{-1}\right)G_{\text{all}}(z)} \ . \tag{3.3.23}$$

The complete expression for the optimal causal feedforward controller is thus

$$W_{\text{opt}}(z) = \frac{-1}{F(z)\,G_{\min}(z)}\left\{\frac{S_{xd}(z)}{F\left(z^{-1}\right)G_{\text{all}}(z)}\right\}_+ \ . \tag{3.3.24}$$

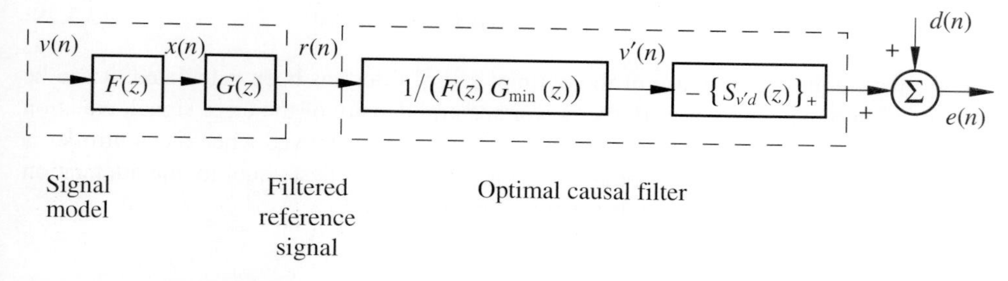

Figure 3.14 Block diagram of the optimal causal filter for the feedforward control problem, in which the filtered reference signal is assumed to be generated by passing a white noise sequence $v(n)$ through a shaping filter $F(z)$ to generate the reference signal, and then through the plant response $G(z)$.

Removing the constraint of causality from equation (3.3.24) and using equations (3.3.15) and (3.3.17), the expression for the frequency response of the optimal controller reduces to

$$W_{\text{opt}}(e^{j\omega T}) = \frac{-S_{xd}(e^{j\omega T})}{G(e^{j\omega T})S_{xx}(e^{j\omega T})}, \tag{3.3.25}$$

which is equal to the unconstrained result derived in equation (3.3.11).

If the poles and zeros of $S_{xx}(z)$, $S_{xd}(z)$ and $G(z)$ are not available to calculate $W_{\text{opt}}(z)$, then the formulation above can also be used to calculate the optimal causal filter in the discrete-frequency domain, directly from measured estimates of the power and cross spectral densities and the plant's frequency response. In order to accurately work in the discrete frequency domain, however, a limit must be observed on the minimum size of the discrete Fourier transform, which is similar to that discussed at the end of Section 2.4. Specifically we require that the DFT size, N, must be large enough for the causal part of the impulse response of $S_{xd}(e^{j\omega T})/F^*(e^{j\omega T})G_{\text{all}}(e^{j\omega T})$ to have decayed to zero by $N/2$ samples. Under these conditions the frequency response of the optimal causal controller is equal to

$$W_{\text{opt}}(k) = \frac{1}{F(k)G_{\min}(k)}\left\{\frac{S_{xd}(k)}{F^*(k)G_{\text{all}}(k)}\right\}_+, \tag{3.3.26}$$

where k is the discrete frequency bin number and

$$\left\{\frac{S_{xd}(k)}{F^*(k)G_{\text{all}}(k)}\right\}_+ = \text{FFT}\left[h(n)\,\text{IFFT}\left(\frac{S_{xd}(k)}{F^*(k)G_{\text{all}}(k)}\right)\right], \tag{3.3.27}$$

in which $h(n)$ is the unit step function, $h(n) = 1$ for $n \geq 0$ and otherwise $h(n) = 0$. The spectral factors can be calculated from the power spectral density of the reference signal in the discrete frequency domain, as in Section 2.4, as

$$F(k) = \exp\left(\text{FFT}\left[c(n)\,\text{IFFT}\,\ln\left(S_{xx}(k)\right)\right]\right), \tag{3.3.28}$$

where ln denotes the natural logarithm and $c(n)$ is given by $c(n) = 0$ for $n < 0$, $c(n) = 1$ for $n > 0$ and $c(0) = \frac{1}{2}$, and the minimum-phase component of the plant response can also be calculated from the cepstral implementation of the discrete Hilbert transform (Oppenheim and Shafer, 1975) so that

$$G_{\min}(k) = \exp\left(\text{FFT}\left[2c(n)\,\text{IFFT}\,\ln\left(|G(k)|\right)\right]\right). \tag{3.3.29}$$

Once the frequency response of the optimal causal filter has been calculated, it can be substituted into the expression for the power spectral density of the error signal, equation (3.3.6), to determine the maximum attenuation that can be achieved when the controller is constrained to be causal. This is bound to be less than, or at best equal to, the attenuation which could be achieved if this constraint was not imposed.

3.3.4. Optimisation in the Time Domain

The optimum performance that can be obtained using a realisable filter that is both causal and of finite length can be calculated by recasting the optimisation problem so that it is formulated in the time domain. Returning to the block diagram of Fig. 3.12(b), we now assume that the feedforward controller is causal and has a finite impulse response (FIR). The error signal is thus given by

$$e(n) \;=\; d(n) + \sum_{i=0}^{I-1} w_i r(n-i) \;=\; d(n) + \boldsymbol{w}^{\mathrm{T}} \boldsymbol{r}(n) \,, \tag{3.3.30}$$

where

$$\boldsymbol{w} \;=\; \left[w_0 \;\cdots\; w_{I-1}\right]^{\mathrm{T}} \,, \tag{3.3.31}$$

is the vector of controller coefficients and

$$\boldsymbol{r}(n) \;=\; \left[r(n) \;\cdots\; r(n-I+1)\right]^{\mathrm{T}} \,, \tag{3.3.32}$$

is the vector of past values of the filtered reference signal. The expectation value of the squared error can thus be written, in a similar way to the mean-square error in the electrical filtering case in Section 2.3, as

$$R_{ee}(0) \;=\; \boldsymbol{w}^{\mathrm{T}} \mathbf{R}_{rr}\, \boldsymbol{w} + 2\boldsymbol{w}^{\mathrm{T}} \mathbf{r}_{rd} + R_{dd}(0) \,, \tag{3.3.33}$$

where

$$R_{ee}(0) \;=\; E\left[e(n)\, e(n)\right] \,, \tag{3.3.34}$$

$$R_{dd}(0) \;=\; E\left[d(n)\, d(n)\right] \,, \tag{3.3.35}$$

$$\mathbf{r}_{rd} \;=\; E\left[\boldsymbol{r}(n)\, d(n)\right] \,, \tag{3.3.36}$$

$$\mathbf{R}_{rr} \;=\; E\left[\boldsymbol{r}(n)\, \boldsymbol{r}^{\mathrm{T}}(n)\right] \,, \tag{3.3.37}$$

so that $\mathbf{r}_{rd}$ is the cross-correlation vector between the filtered reference signal and the disturbance signal and $\mathbf{R}_{rr}$ the autocorrelation matrix of the filtered reference signal. Equation (3.3.33) is of exactly the same form as that used to derive the optimum digital filter in Section 2.3, and assuming that the filtered reference signal is persistently exciting so that $\mathbf{R}_{rr}$ is positive definite, the optimum set of filter coefficients which minimise the mean-square error in this case is given by

$$\boldsymbol{w}_{\mathrm{opt}} \;=\; -\mathbf{R}_{rr}^{-1}\, \mathbf{r}_{rd} \,. \tag{3.3.38}$$

Hence, the expectation of the minimum mean-square error can be written as

$$R_{ee,\min}(0) \;=\; R_{dd}(0) - \mathbf{r}_{rd}^{\mathrm{T}} \mathbf{R}_{rr}^{-1}\, \mathbf{r}_{rd} \,. \tag{3.3.39}$$

The optimum filter and the minimum residual error can thus be calculated from a knowledge of the autocorrelation and cross-correlation properties of the disturbance and filtered reference signals, although these correlation functions are often conveniently calculated from discrete frequency-domain estimates of the power spectral density and cross spectral density.

3.4. ADAPTIVE FIR CONTROLLERS

The coefficients of the optimum causal FIR controller can be calculated directly using equation (3.3.38). There are, however, several problems associated with using this equation to design a real-time controller, particularly the assumption of stationarity used in its derivation and the fact that its solution requires the inversion of a matrix which is potentially large. Thus an adaptive algorithm is sought that would allow a practical controller with a finite length FIR control filter to approach the optimum Wiener filter under stationary conditions, and to track changes in the statistical properties of the disturbance under non-stationary conditions. This is exactly the background that motivated the LMS algorithm for electrical noise cancellation in Chapter 2, and we begin this section with a discussion of a modified form of the LMS algorithm which is widely used in active control systems. In the whole of this section we will continue to assume that the effect of any physical feedback path, $G_s(z)$ in Fig. 3.11, is perfectly cancelled by the internal estimate of the feedback path, $\hat{G}_s(z)$. If $\hat{G}_s(z)$ is not equal to $G_s(z)$, a residual feedback path is set up round the adaptive filter whose effects are small provided the loop gain, which is equal to $W(z)\left(\hat{G}_s(z)-G_s(z)\right)$, is small at all frequencies (Rafaely and Elliott, 1996b). This issue also arises in the use of adaptive filters in feedback controllers and will be more fully discussed in Chapter 7.

3.4.1. Filtered-Reference LMS Algorithm

We will begin by making the assumption that, although the controller is being adapted, its coefficients are changing slowly compared with the timescale of the plant dynamics, so that the error in the arrangement shown in Fig. 3.12(a) can still be accurately represented by the output of Fig. 3.12(b) after the transposition of the filtering operations. The error signal can thus still be represented as

$$e(n) = d(n) + \mathbf{w}^{\mathrm{T}} \mathbf{r}(n) , \tag{3.4.1}$$

where $\mathbf{w}$ and $\mathbf{r}(n)$ are given in equations (3.3.31) and (3.3.32) and $\mathbf{w}$ is nearly time invariant, so that its dependence on the time index n is suppressed.

The philosophy of the LMS algorithm is to adapt the filter coefficients in the opposite direction to the instantaneous gradient of the mean-square error with respect to the coefficients. In this case the gradient can be written as

$$\frac{\partial e^2(n)}{\partial \mathbf{w}} = 2\,e(n)\frac{\partial e(n)}{\partial \mathbf{w}} = 2\,e(n)\mathbf{r}(n), \tag{3.4.2}$$

where equation (3.4.1) and the properties of vector derivations, discussed in the Appendix, have been used to obtain the final part of equation (3.4.2). The appropriate form of the LMS algorithm in this case can thus be written as

$$\mathbf{w}(n+1) = \mathbf{w}(n) - \alpha\,\mathbf{r}(n)\,e(n) , \tag{3.4.3}$$

which is known as the *filtered-reference LMS*, or the *filtered-x LMS* (FXLMS), since the reference signal is generally denoted as x. The algorithm was originally proposed by Morgan (1980) and independently for feedforward control by Widrow et al. (1981) and for the active control of sound in ducts by Burgess (1981).

In practice, the filtered reference signal will be generated using an *estimated* version of the true plant response represented as a *plant model*. This can be implemented as a separate real-time filter, $\hat{G}(z)$, which is used to generate the filtered reference signal, $\hat{r}(n)$, as illustrated in Fig. 3.15. The practical version of the filtered-reference LMS algorithm can thus be written as

$$\mathbf{w}(n+1) = \mathbf{w}(n) - \alpha\, \hat{\mathbf{r}}(n)\, e(n) \ . \tag{3.4.4}$$

A physical interpretation of this algorithm can be obtained by comparing Fig. 3.15 with the corresponding block diagram of the LMS algorithm in Fig. 2.13. In the case of the LMS algorithm, the error signal is multiplied directly by the reference signal to give the cross-correlation estimate used to adapt the filter. If this approach were adapted in the feedforward control case, the error signal would be filtered by the plant response, and this would distort the cross-correlation estimate. The filtered-reference LMS algorithm prefilters the reference signal with the estimated plant response, so the measured error signal and filtered reference signal are again aligned in time to give a valid cross-correlation estimate.

The need for a separate filter to generate the filtered reference signal in equation (3.4.4) clearly increases the computational burden of the filtered-reference LMS algorithm, and so efficient representations of the plant response are very desirable. Although the filtered-reference LMS algorithm has been motivated by the need to control stochastic disturbances, it can also provide a very efficient method of adapting a feedforward controller for single-frequency disturbances. In this case, the reference signal is tonal, and only two coefficients need to be used in the adaptive filter. The plant model also only needs to accurately model the plant at the disturbance frequency in this case, so that it too can be implemented with a two-coefficient FIR filter.

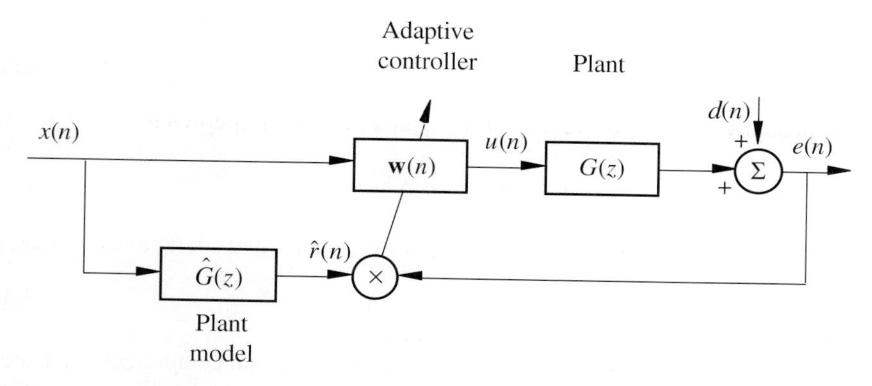

Figure 3.15 Block diagram of the practical implementation of the filtered-reference LMS algorithm, in which a model $\hat{G}(z)$ of the true plant response $G(z)$, is used to generate the filtered reference signal $\hat{r}(n)$.

3.4.2. Conditions for Stability

It is important to establish how well the estimated plant response has to model the true plant in order for the algorithm to behave in a stable and reliable manner. The behaviour of the practical filtered-reference LMS algorithm can be analysed in a similar way to the LMS algorithm in Chapter 2, provided the filter is again assumed to be only slowly adapting. In this case we can substitute equation (3.4.1), for $e(n)$, into equation (3.4.4) to obtain

$$\boldsymbol{w}(n+1) = \boldsymbol{w}(n) - \alpha\left[\hat{\boldsymbol{r}}(n)d(n) + \hat{\boldsymbol{r}}(n)\boldsymbol{r}^{\mathrm{T}}(n)\boldsymbol{w}(n)\right] . \tag{3.4.5}$$

If the algorithm is stable, it will converge to a solution for which the expectation of the update term in square brackets in equation (3.4.5) is equal to zero. Denoting this converged solution as $\boldsymbol{w}_\infty$, we can write

$$\boldsymbol{w}_\infty = -\left[E\left[\hat{\boldsymbol{r}}(n)\boldsymbol{r}^{\mathrm{T}}(n)\right]\right]^{-1} E\left[\hat{\boldsymbol{r}}(n)d(n)\right] . \tag{3.4.6}$$

Thus, even if the practically implemented filtered reference algorithm is stable, it does not generally converge to the truly optimal set of filter coefficients, given by equation (3.3.38), unless the plant model is perfect, so that $\hat{\boldsymbol{r}}(n) = \boldsymbol{r}(n)$.

Since the controller is only changing slowly we can make the usual independence assumption, as discussed in Section 2.6 for the LMS algorithm. Subtracting a factor of $\boldsymbol{w}_\infty$ from both sides of equation (3.4.5) and taking the expectation value, the average behaviour of the adaptive algorithm can then be written in the form

$$E\left[\boldsymbol{w}(n+1) - \boldsymbol{w}_\infty\right] = \left[\mathbf{I} - \alpha\, E\left[\hat{\boldsymbol{r}}(n)\boldsymbol{r}^{\mathrm{T}}(n)\right]\right] E\left[\boldsymbol{w}(n) - \boldsymbol{w}_\infty\right] . \tag{3.4.7}$$

This equation is analogous to equation (2.6.11) for the LMS algorithm, except that in this case the matrix $E\left[\hat{\boldsymbol{r}}(n)\boldsymbol{r}^{\mathrm{T}}(n)\right]$ is not symmetric and so has the more general form of eigenvalue eigenvector decomposition given by equation (A7.6) in the Appendix, rather than the special form of equation (A7.16) used in the analysis of the LMS algorithm. Nevertheless, the generalised correlation matrix in equation (3.4.7) can still be decomposed as

$$E\left[\hat{\boldsymbol{r}}(n)\boldsymbol{r}^{\mathrm{T}}(n)\right] = \mathbf{Q}\,\boldsymbol{\Lambda}\,\mathbf{Q}^{-1} , \tag{3.4.8}$$

so that equation (3.4.7) can be written, as for the LMS algorithm, as

$$\boldsymbol{v}(n+1) = \left[I - \alpha\,\boldsymbol{\Lambda}\right]\boldsymbol{v}(n) , \tag{3.4.9}$$

where in this case the vector of averaged, normalised and rotated coefficients is equal to

$$\boldsymbol{v}(n) = \mathbf{Q}^{-1}\, E\left[\boldsymbol{w}(n) - \boldsymbol{w}_\infty\right] . \tag{3.4.10}$$

In this case, however, the eigenvectors that make up the diagonal components of $\boldsymbol{\Lambda}$ are not necessarily real, and so the magnitude of $1 - \alpha\lambda$, which must be less than unity for the algorithm to be stable, is analogous to equation (3.2.26) in the single frequency analysis above. The conditions for stability of the filtered reference LMS algorithm are thus

$$0 < \alpha < \frac{2\,\mathrm{Re}[\lambda_i]}{|\lambda_i|^2} \quad \text{for all } i \,, \tag{3.4.11}$$

Morgan (1980). The stability of the algorithm thus depends on the sign of the real parts of the eigenvalues of the matrix $E[\hat{\boldsymbol{r}}(n)\boldsymbol{r}^{\mathrm{T}}(n)]$.

If the plant model is perfect then this matrix is positive definite and all the eigenvalues are guaranteed to be real and positive. The convergence modes of the algorithm under these conditions, however, are determined by the eigenvalues of the autocorrelation matrix of the filtered reference signals rather than those of the reference signal itself, and so even if the reference signal is nearly white, the autocorrelation matrix of the filtered reference signal can have a large eigenvalue spread due to the response of the plant. In Chapter 2 it was noted that for the LMS algorithm, the ratio of the maximum to minimum eigenvalue is bounded by the ratios of the maximum and minimum values of the power spectral density of the reference signal. In the present case this condition becomes

$$\frac{\lambda_{\max}}{\lambda_{\min}} \le \frac{\left[|G(\mathrm{e}^{j\omega T})|^2\, S_{xx}(\mathrm{e}^{j\omega T})\right]_{\max}}{\left[|G(\mathrm{e}^{j\omega T})|^2\, S_{xx}(\mathrm{e}^{j\omega T})\right]_{\min}} \,, \tag{3.4.12}$$

so that even if $S_{xx}\left(\mathrm{e}^{j\omega T}\right)$ is constant with ωT, a large eigenvalue spread can be generated by differences between the peaks and troughs in the frequency response of the plant. In the case of a tonal reference signal driving an adaptive filter with only two coefficients, and with the time delay between the responses caused by the two coefficients being one-quarter of the period of the tonal disturbance, the two eigenvalues of the autocorrelation matrix will be equal, which will give the most rapid convergence time. The two coefficients are then operating independently and effectively adjusting the in-phase and quadrature components of the control signal, as in the frequency domain adaptation techniques described in Section 3.2.

If the plant model has a different response from that of the physical plant, then the eigenvalues of the matrix $E[\hat{\boldsymbol{r}}(n)\boldsymbol{r}^{\mathrm{T}}(n)]$ are no longer guaranteed to be entirely real or to have positive real parts. If an eigenvalue is complex then the mode of convergence associated with this eigenvalue will not converge exponentially, on average, but will have an oscillatory component. Provided the real part of the eigenvalue is positive, however, this oscillation will die away and the algorithm will still be stable. If the real part of one of the eigenvalues is negative, then the adaptation will become unstable. It is not straightforward to relate the difference in the response between the plant model and the physical plant to the sign of the real parts of the eigenvalues of the $E[\hat{\boldsymbol{r}}(n)\boldsymbol{r}^{\mathrm{T}}(n)]$ matrix. However, it has been shown by Ren and Kumar (1989), see also Wang and Ren (1999), that a sufficient, but not necessary, condition for the stability of an algorithm that is similar to the filtered reference LMS algorithm, is that the ratio $\hat{G}(\mathrm{e}^{j\omega T})/G(\mathrm{e}^{j\omega T})$ is strictly positive real (SPR). We saw in Chapter 2 that this condition implies that the phase of the quantity under consideration must lie within ±90° at all frequencies. In the current context, the fact that $\hat{G}(\mathrm{e}^{j\omega T})/G(\mathrm{e}^{j\omega T})$ is strictly positive real implies that the phase of the plant model must be within ±90° of the true phase of the plant at all frequencies. This is a neat extension of the phase condition on the plant model derived for the stability of an adaptive controller operating at a single frequency in Section 3.2.

In practice it is possible to implement stable controllers in which the phase condition on the plant model is not satisfied at some frequencies, even though the reference signal is broadband. In general, however, one does not know the spectrum of the reference signal *a priori*, and so it is possible that the spectrum has significant energy at the frequency for which the phase of the plant model was in error by more than 90°. In an extreme case, the reference signal could become almost tonal at this frequency, in which case the adaptive algorithm would clearly become unstable as discussed in Section 3.2. Thus, without a detailed knowledge of the statistics of the reference signal, it would appear that 90° phase accuracy at all frequencies provides the only guaranteed condition on the plant model for the stability of a slowly varying controller. The robustness of the algorithm to modelling error can be improved by using a 'leaky' form of the adaptation algorithm (Widrow and Stearns, 1985), but a detailed discussion of the effects of such a leak are postponed until Section 3.4.7.

3.4.3. Performance Degradation Due to Plant Modelling Errors

It was noted above that when the filtered-reference LMS algorithm is implemented in practice, with an imperfect plant model, then when the controller is stable it converges to equation (3.4.6), which is generally not equal to the optimum solution, given by equation (3.3.38). The value of the residual mean-square error which remains after the algorithm converges is thus necessarily higher than the minimum mean-square error which could be attained if the optimum filter had been implemented. The degree of this degradation in performance with plant modelling error has been investigated in some simulations performed by Saito and Sone (1996).

These authors considered the case of a filtered-reference LMS algorithm applied to a vehicle cabin, in which the impulse response between the input to a loudspeaker and output from a microphone in the cabin lasted for about 150 samples when sampled at a rate of 2 kHz. The performance of the converged filtered-reference LMS algorithm was then calculated for a broadband reference signal when the plant model used in the algorithm was progressively cruder, which was generated by setting to zero the smaller values of the coefficients in the plant's impulse response. In the example considered by Saito and Sone (1996), the maximum attenuation in the disturbance with a perfect plant model was about 9 dB. As the number of non-zero coefficients in the model of the plant's impulse response was reduced, the final attenuation fell, but was not reduced significantly until the plant had only 20 non-zero coefficients. the performance of the algorithm was severely degraded as the number of non-zero coefficients in the impulse response of the plant model was further reduced from 20 to 10, after which the plant model was not accurate enough to maintain the algorithm's stability.

These results suggest that the performance of the filtered-reference LMS algorithm is relatively robust to differences between the response of the physical plant and that of the plant model, but not quite as robust as the algorithm's stability. This observation has practical implications for many applications in which a very efficient implementation of the plant model is required, to reduce the processing requirements of the algorithm, while maintaining an adequate performance from the active control system. The distinction between *robust performance* and *robust stability* will be further discussed for feedback controllers in Chapter 6.

3.4.4. Effect of Plant Delays

The maximum convergence coefficient that can be used in the filtered-reference LMS algorithm with a perfect plant model can be inferred from the analogy between equation (3.4.7) and the corresponding equation for the LMS algorithm. This leads to an estimate of the maximum convergence coefficient which is given by

$$\alpha_{\max} \approx \frac{2}{I\,\overline{r^2}}, \tag{3.4.13}$$

where I is the number of coefficients in the controller and $\overline{r^2}$ is the mean-square value of the filtered reference signal. The fact that the plant response is dynamic, however, means that the analysis outlined above is not valid for rapid adaptation and the value of convergence coefficient given by equation (3.4.13) is often too large to ensure a stable system. In a practical system the most important part of the plant dynamics is often its overall *delay*, Δ, and simulations with a white noise reference signal (Elliott and Nelson, 1989) suggest that a more realistic upper bound in the convergence coefficient in this case may be

$$\alpha_{\max} \approx \frac{2}{(I+\Delta)\overline{r^2}}, \tag{3.4.14}$$

although even smaller values may need to be used if the reference signal is highly coloured or the plant is very resonant. A simplified form of the filtered-reference LMS algorithm, known as the *delayed LMS* (DLMS), is obtained when there is a pure delay between the filter output signal and the observed error signal, as occurs in various telecommunications application (Quershi and Newhall, 1973; Kabal, 1983; Long et al., 1989). Values of the maximum convergence coefficient which are similar to equation (3.4.14) have been obtained analytically in these applications.

The combined effect of delays in the plant, and phase errors in the estimated plant response, on the convergence behaviour of an adaptive controller have been investigated by Boucher et al. (1991). He assumed a tonal reference signal, synchronously sampled at 4 points per cycle, as described in Section 3.2. The value of the convergence coefficient required to give the fastest convergence time for various phase errors in the plant model and various pure delays in the plant are shown in Fig. 3.16(a). The optimum convergence coefficient is clearly largest for no phase error, as expected from equation (3.2.31), and also for smaller plant delays, as expected from equation (3.4.14). The minimum convergence time for the algorithm is obtained using this optimum convergence coefficient, and is plotted in Fig. 3.16(b), which shows that the convergence time is relatively unaffected by phase errors in the plant model of less than about ±40°, but the convergence time then increases significantly until it tends to infinity as the phase error approaches ±90°, and the algorithm becomes unstable for any finite value of the

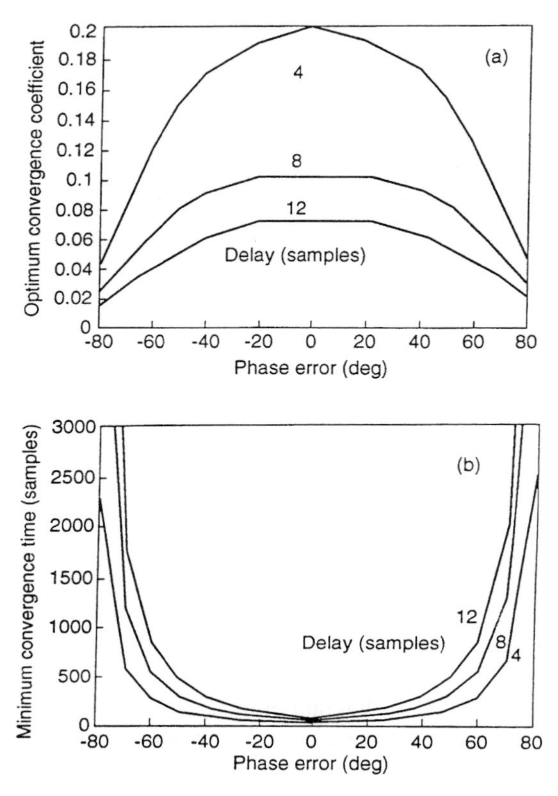

Figure 3.16 The convergence coefficient required to give the fastest convergence time (a) and the convergence time resulting from the use of these optimum convergence coefficients (b) in a simulation of the filtered-x LMS algorithm with a sinusoidal reference signal having four samples per cycle. Various phase errors were assumed between the estimate of the secondary path $\hat{G}(z)$ and the true secondary path $G(z)$. The three graphs correspond to pure delays in the plant $G(z)$, of 4, 8 and 12 samples.

convergence coefficient. Snyder and Hansen (1994) have also considered the effect of phase errors for tonal reference signals sampled at different rates and have shown that in general it is difficult to predict what effect the phase error in the plant model will have, particularly for small plant delays, except that the system cannot be guaranteed stable unless the phase error is less than $\pm 90^\circ$.

In the same way that the maximum convergence coefficient now depends on the filtered reference signal, rather than the reference signal itself, the misadjustment of the filtered-reference LMS algorithm can also be obtained from a generalisation of the expression given for the LMS algorithm in Section 2.6. This is given by

$$M \approx \frac{\alpha I \overline{r^2}}{2} , \tag{3.4.15}$$

where $\overline{r^2}$ is again the mean-square value of the filtered reference signal and it has been assumed that α is small compared with equation (3.4.13). Since the convergence coefficient α must often be made small in active control problems to avoid stability problems, misadjustment is generally less of an issue for adaptive feedforward systems than for electrical cancellation systems using the LMS algorithm.

The filtered-reference LMS algorithm was derived assuming that the coefficients of the adaptive controller change only slowly compared with the timescale of the plant dynamics. In practice it is found that, provided the eigenvalue spread given by equation (3.4.12) is not too large and the convergence coefficient is adjusted as in equation (3.4.14), then the algorithm will generally converge on a timescale which is similar to that of the plant dynamics. Thus, although some care needs to be taken, the filtered-reference LMS algorithm is found to be surprisingly robust to the rather sweeping assumption of 'slowly varying' controller coefficients. Faster-adapting algorithms than those based on steepest descent, such as the RLS algorithm discussed in Chapter 2, could be used in an attempt to increase the speed of convergence, but the assumption of slowly varying controller coefficients, which is at the heart of the filtered-reference approach, would then become much more of an issue. This problem arises because the dynamic response of the plant acts between the output of the controller and the observation point for the error, as shown in Fig. 3.12(a).

3.4.5. Modified Filtered-Reference LMS Algorithm

It is possible to rearrange the block diagram of the adaptive feedforward controller so that the adaptation takes place using the arrangement shown in Fig. 3.12(b), in which the plant response operates on the reference signal before being passed to the adaptive control filter. Adaptation of the control filter is then equivalent to the electrical noise cancellation problem discussed in Chapter 2. The block diagram of such a system is shown in Fig. 3.17, in which the plant model, $\hat{G}(z)$, is used to subtract an estimate of the plant output due to the

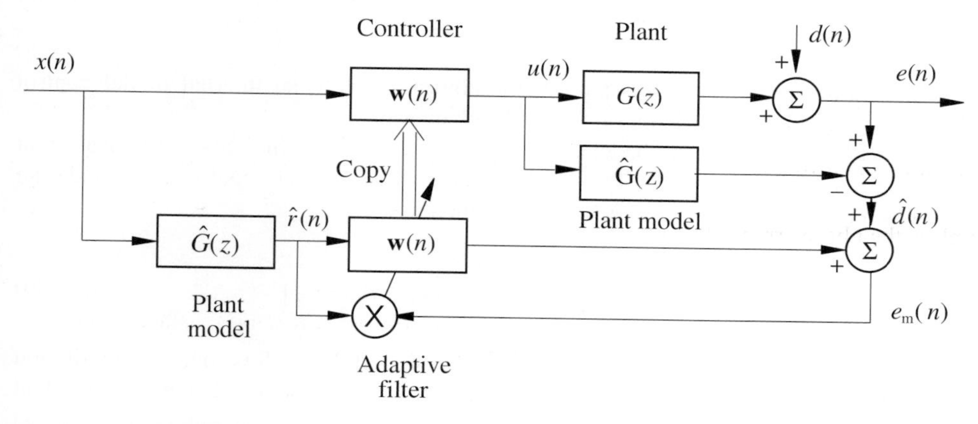

Figure 3.17 Block diagram of the modified filtered-reference LMS algorithm, in which a prediction of the error that would have been obtained if the current controller coefficients had been fixed over all time is obtained by using an internal model of the plant to derive an estimate of the disturbance, $\hat{d}(n)$, which is then added to the reference signal filtered by the plant model and the adaptive controller.

control signal from the error signal, and thus derive an estimate of the disturbance signal, which is denoted as $\hat{d}(n)$. This estimate of the disturbance is then used to generate the new error signal $e_m(n)$, which is used to adapt the filter in the equivalent electrical cancellation task, shown at the bottom of Fig. 3.17. The coefficients of the adaptive filter are then copied into the controller.

This architecture for rapid adaptation of the controller was published by several authors at about the same time (Bjarnason, 1992; Bao et al., 1992; Bronzel, 1993; Doelman, 1993; Flockton, 1993; Kim et al., 1994). The algorithm has been termed the *modified filtered-x LMS* algorithm by Bjarnason (1992). The estimated disturbance signal in Fig. 3.17 is given by

$$\hat{d}(n) = e(n) - \sum_{j=0}^{J-1} \hat{g}_j u(n-j) , \tag{3.4.16}$$

where $\hat{g}_j$ is the J-th order impulse response of the plant model, whose z transform is $\hat{G}(z)$, and

$$u(n) = \sum_{i=0}^{I-1} w_i(n) x(n-i) , \tag{3.4.17}$$

where $w_i(n)$ is the instantaneous impulse response of the I-th order controller. The output error can also be written as

$$e(n) = d(n) + \sum_{j=0}^{J-1} \sum_{i=0}^{I-1} g_j w_i(n-j) x(n-i-j) , \tag{3.4.18}$$

where g_j is the impulse response of the physical plant, whose z transform is $G(z)$, so that the estimated disturbance signal can be written as

$$\hat{d}(n) = d(n) + \sum_{j=0}^{J-1} \sum_{i=0}^{I-1} \left(g_j - \hat{g}_j\right) w_i(n-j) x(n-i-j) . \tag{3.4.19}$$

It is interesting to compare this part of the algorithm with the internal model control arrangement discussed in Chapter 6 for feedback control.

The time-varying nature of the adaptive control filter, with coefficients $w_i(n)$, means that the output error in equation (3.4.18) depends on the values of these coefficients some time ago. The modified error in Fig. 3.17 is now defined to be

$$e_{\mathrm{m}}(n) = \hat{d}(n) + \sum_{i=0}^{I-1} \sum_{j=0}^{J-1} w_i(n) \hat{g}_j x(n-i-j) , \tag{3.4.20}$$

which only depends on the current set of control filter coefficients. It is equal to a prediction of the error signal that would have been produced if the current set of filter coefficients had always been used in the controller.

The control filter in the lower loop of Fig. 3.17 can now be updated using the modified filtered reference LMS algorithm

$$\mathbf{w}(n+1) = \mathbf{w}(n) - \alpha \hat{\mathbf{r}}(n) e_{\mathrm{m}}(n) , \tag{3.4.21}$$

where $\hat{\boldsymbol{r}}(n)$ is again the vector of filtered reference signals. Equation (3.4.21) is a direct form of instantaneous steepest-descent algorithm, which can be derived by analogy with the LMS algorithm without the assumption of slowly varying filter coefficients. Once the coefficients of the filter in this loop have been updated they can be copied into the control filter and are ready to be used to filter the next sample of $x(n)$.

The sensitivity of the modified filtered-reference LMS algorithm to plant modelling errors can be analysed for slowly-varying filter coefficients, since in this case the time-variation in the controller coefficients can be ignored, and equation (3.4.19) reduces to

$$\hat{d}(n) = d(n) + \sum_{j=0}^{J-1} \sum_{i=0}^{I-1} \left(g_j - \hat{g}_j\right) w_i \, x(n-i-j) \, , \tag{3.4.22}$$

and the modified error reduces to

$$e_{\mathrm{m}}(n) = \hat{d}(n) + \sum_{j=0}^{J-1} \hat{g}_j \, w_i \, x(n-i-j) \, . \tag{3.4.23}$$

Substituting equation (3.4.22) for $\hat{d}(n)$ into equation (3.4.23), the modified error for slowly-varying controller coefficients can be seen to be

$$e_{\mathrm{m}}(n) = d(n) + \sum_{j=0}^{J-1} \sum_{i=0}^{I-1} g_j \, w_i \, x(n-i-j) \, , \tag{3.4.24}$$

which is exactly the same as the true error signal, $e(n)$, under the assumed conditions. Thus the behaviour of the modified filtered-reference LMS, equation (3.4.21), is exactly the same as that of the normal filtered-reference LMS with modelling error, equation (3.4.4), under slowly-varying conditions and the stability for small α is again determined by whether the real parts of the eigenvalues of $E\left[\hat{\boldsymbol{r}}(n)\boldsymbol{r}^{\mathrm{T}}(n)\right]$ are entirely real, as in equation (3.4.7).

The maximum value of the convergence coefficient for the modified filtered-reference LMS algorithm, however, is now determined by equation (3.4.13) rather than (3.4.14), since the delay between the adaptation of the control filter and the observation of the error has been removed. Simulation results support the conclusion that the convergence speed of the modified filtered-reference LMS algorithm is indeed similar to that of the LMS algorithm in this case (Bjarnason, 1992; Bao et al., 1992), although a complete convergence analysis of a rapidly changing filter is rather difficult. Further developments of the modified filtered-reference LMS algorithm which reduce the computational load of the algorithm are described by Rupp and Sayed (1998).

Once the need for slowly varying filter coefficients has been removed, more rapid adaptive algorithms could be used to adjust the coefficients of the control filter, $\boldsymbol{w}(n)$, in Fig. 3.17. In particular the use of the RLS algorithm described in Chapter 2 has been explored by Flockton (1993) and Bronzel (1993), and the trade-off between convergence speed and computational load then becomes similar to that for conventional adaptive filters.

3.4.6. Filtered-Error LMS Algorithm

It is possible to account for the presence of the plant response in the feedforward control problem by filtering the error signal instead of filtering the reference signal. In order to derive this *filtered-error LMS algorithm*, however, one must go back to the fundamental block diagram for the feedforward control system, Fig. 3.12(a), and express the error signal explicitly as a function of the impulse response of the controller and the impulse response of the plant.

Assuming that the controller coefficients are fixed, then the error signal can be expressed, following equation (3.4.18), as

$$e(n) = d(n) + \sum_{i=0}^{I-1} \sum_{j=0}^{J-1} g_j \, w_i \, x(n-i-j) \; . \tag{3.4.25}$$

The derivative of the time-averaged square error with respect to the k-th controller coefficient can be written as

$$\begin{aligned} \lim_{N\to\infty} \frac{1}{2N} \sum_{n=-N}^{N} \frac{\partial e^2(n)}{\partial w_k} &= \lim_{N\to\infty} \frac{1}{N} \sum_{n=-N}^{N} e(n) \frac{\partial e(n)}{\partial w_k} \\ &= \lim_{N\to\infty} \frac{1}{N} \sum_{n=-N}^{N} e(n) \sum_{j=0}^{J-1} g_j \, x(n-k-j), \end{aligned} \tag{3.4.26}$$

where the final form of this equation has been obtained by differentiating equation (3.4.25) with respect to w_k to give $\partial e(n)/\partial w_k$. The filtered reference LMS algorithm could be derived from equation (3.4.26) by defining the filtered reference signal to be $x(n)$ convolved with g_j. It is, however, also possible to re-write equation (3.4.26) in an alternative form by defining the dummy time variable

$$n' = n - j, \quad \text{so that} \quad n = n' + j \; . \tag{3.4.27}$$

Equation (3.4.26) can now be expressed as

$$\lim_{N\to\infty} \frac{1}{2N} \sum_{n=-N}^{N} \frac{\partial e^2(n)}{\partial w_k} = \lim_{N\to\infty} \frac{1}{N} \sum_{j=0}^{J-1} \sum_{n'+j=-N}^{N} x(n'-k) \, g_j \, e(n'+j) \; . \tag{3.4.28}$$

The signal that would be obtained by noncausally filtering the error signal with the *time-reversed* impulse response of the plant is now defined to be

$$f(n') = \sum_{j=0}^{J-1} g_j \, e(n'+j) \; . \tag{3.4.29}$$

It is also noted that because j is always finite, then the summation from $-N$ to N as N tends to ∞ on the right-hand side of equation (3.4.28) will be the same for $n' = -\infty$ to ∞ as for $n'+j = -\infty$ to ∞. The time-averaged derivative in equation (3.4.28) can thus be written as

$$\lim_{N\to\infty} \frac{1}{2N} \sum_{n=-N}^{N} \frac{\partial e^2(n)}{\partial w_k} = \lim_{N\to\infty} \frac{1}{N} \sum_{n=-N}^{N} f(n) \, x(n-k) \; . \tag{3.4.30}$$

Provided all the terms in equation (3.4.30) are used over time to slowly update the controller coefficients, the average behaviour of an adaptation algorithm using equation (3.4.30) will be similar to that of a steepest-descent algorithm using the filtered reference signal (Wan, 1993, 1996; see also Elliott, 1998). The problem with using equation (3.4.30) in a real-time algorithm is that an instantaneous estimate of $f(n)$ in equation (3.4.29) cannot be implemented with a causal system. This problem can be overcome by delaying both $f(n)$ and $x(n-k)$ in equation (3.4.30) by $J-1$ samples. The final form of the filtered error LMS algorithm is obtained by adapting the controller coefficients with an instantaneous version of the derivative given by equation (3.4.30) delayed by $J-1$ samples, which can be written

$$\mathbf{w}(n+1) = \mathbf{w}(n) - \alpha\, f(n-J+1)\, \mathbf{x}(n-J+1) \,, \tag{3.4.31}$$

where

$$\mathbf{x}(n-J+1) = \left[x(n-J+1), x(n-J) \cdots x(n-J-I+2)\right]^{\mathrm{T}} . \tag{3.4.32}$$

The delayed filtered error signal, which would be used in practice, is equal to

$$\hat{f}(n-J+1) = \sum_{j=0}^{J-1} \hat{g}_j\, e(n+j-J+1) = \sum_{j'=0}^{J-1} \hat{g}_{J-1-j'}\, e(n-j') \,, \tag{3.4.33}$$

where $\hat{g}_j$ are the coefficients of the impulse response of the FIR plant model, which is assumed for notational simplicity to also have J coefficients. The final form of equation (3.4.33) is obtained by letting $j' = J-1-j$, and emphasises that the error signal is now causally filtered using a *time-reversed* version of the internal model of the plant impulse response. If the z transform of the plant model is written

$$\hat{G}(z) = \sum_{j=0}^{J-1} \hat{g}_j\, z^{-j} \,, \tag{3.4.34}$$

then the transfer function of the filter required to generate the delayed filtered error signal $f(n-J+1)$ from the error signal $e(n)$ can be written as

$$z^{-J+1}\, \hat{G}(z^{-1}) = \sum_{j=0}^{J-1} \hat{g}_j z^{j-1-J} \,. \tag{3.4.35}$$

The block diagram of the complete filtered error LMS algorithm can now be drawn as in Fig. 3.18.

The need to convolve the error signal with all the coefficients of the plant model's impulse response means that the filtered-error LMS algorithm offers no computational advantage over the filtered-reference LMS in the single-channel case. We will see in Chapter 5, however, that this can be very different in multichannel systems, and the main reason for deriving the single-channel filtered-error algorithm here is to prepare the ground for the multichannel case.

In the limit of very slow adaptation, the gradient estimate used by the filtered-error LMS algorithm is the same as that used by the filtered-reference LMS algorithm. The conditions on the plant model to ensure stability will thus be the same for the two algorithms (Elliott, 1998). Simulations of the two algorithms described by Wan (1996) also suggest that the two algorithms can have a similar performance, although it is to be expected that the

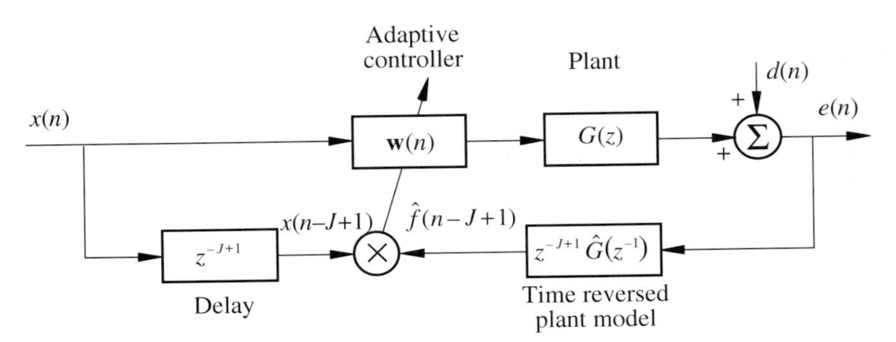

Figure 3.18 Block diagram of the filtered-error LMS algorithm, in which the error signal is filtered by a delayed and time-reversed version of the impulse response of the plant model, $z^{-J+1}\,\hat{G}(z^{-1})$, to generate the filtered error signal $\hat{f}(n-J+1)$.

additional delay of $J-1$ samples, introduced to ensure the filtered error operation is causal, would limit the maximum convergence coefficient and hence reduce the speed of adaptation in some applications.

3.4.7. Leaky LMS Algorithms

The filtered-reference LMS algorithm attempts to minimise the mean-square value of the measured error signal, and thus approach the optimal least-square controller discussed in Section 3.3. Another objective would be to minimise the weighted sum of the mean-square error and the sum of squared control filter coefficients, which we will see has certain practical advantages. The cost function being minimised can then be written as

$$J_2 = E\left[e^2(n)\right] + \beta\,\mathbf{w}^{\mathrm{T}}\mathbf{w} \ , \tag{3.4.36}$$

where β is a positive coefficient-weighting parameter. Expanding this equation out into a quadratic form, as in equation (3.3.33), gives

$$J_2 = \mathbf{w}^{\mathrm{T}}\left(\mathbf{R}_{rr} + \beta\mathbf{I}\right)\mathbf{w} + 2\mathbf{w}^{\mathrm{T}}\mathbf{r}_{rd} + R_{dd}(0) \ , \tag{3.4.37}$$

where $\mathbf{R}_{rr}$, $\mathbf{r}_{rd}$ and $R_{dd}(0)$ are defined in equations (3.3.35–37). The derivative of this new cost function with respect to the elements of $\mathbf{w}$ can be expressed in the form

$$\frac{\partial J_2}{\partial \mathbf{w}} = 2\,E\left[\mathbf{r}(n)e(n)\right] + 2\beta\mathbf{w} \ , \tag{3.4.38}$$

and using the instantaneous version of this derivative to adapt the control filter coefficients at every sample point gives a modified form of the filtered-x LMS algorithm which can be written as

$$\mathbf{w}(n+1) = \left(1-\alpha\beta\right)\mathbf{w}(n) - \alpha\,\mathbf{r}(n)e(n) \ . \tag{3.4.39}$$

If the error signal were to be zero, the coefficients would 'leak' away in this algorithm because of the *leakage term*, $(1 - \alpha\beta)$, and so the algorithm is called the leaky filtered-reference LMS algorithm. Similar modifications can be made to the other single-channel control algorithms discussed above.

When the algorithm is implemented in practice, the filtered reference signal is generated with a model, $\hat{G}(z)$, of the real plant response, as in Fig. 3.12, so that

$$\boldsymbol{w}(n+1) = (1 - \alpha\,\beta)\,\boldsymbol{w}(n) - \alpha\,\hat{\boldsymbol{r}}(n)\,e(n) \; . \qquad (3.4.40)$$

Using a similar analysis to that which leads to equation (3.4.7) above, we find that the leaky version of the algorithm converges provided that the real parts of the eigenvalues of the matrix $E\left[\hat{\boldsymbol{r}}(n)\boldsymbol{r}^{\mathrm{T}} + \beta\,\mathbf{I}\right]$ are positive. The effect of the leakage term is thus to add a term β to each of the eigenvalues of $E\left[\hat{\boldsymbol{r}}(n)\boldsymbol{r}^{\mathrm{T}}\right]$, which improves the robust stability of the algorithm by ensuring that eigenvalues that would otherwise have a small negative real part now have a positive real part. The inclusion of the leakage term in the algorithm inevitably reduces the attenuation which can be achieved in the mean-square error signal if the plant model is perfect, however. The value of β used in practice in equation (3.4.40) thus represents a trade-off between nominal performance and robustness, although in many applications, a small value of β can give significantly improved robustness with only a small degradation in nominal performance.

The condition for stability of the normal filtered reference LMS when slowly adapting is that the ratio $\hat{G}(z)/G(z)$ is strictly positive real (SPR), as discussed above (Ren and Kumar, 1989). This condition can also be written as

$$\mathrm{Re}\left[\hat{G}\left(\mathrm{e}^{j\omega T}\right)G^{*}\left(\mathrm{e}^{j\omega T}\right)\right] > 0 \qquad \text{for all } \omega T \; . \qquad (3.4.41)$$

If the plant model is in error by a magnitude factor $M(\omega T)$ and a phase shift $\phi(\omega T)$ so that

$$\hat{G}\left(\mathrm{e}^{j\omega T}\right) = M(\omega T)\,G\left(\mathrm{e}^{j\omega T}\right)\mathrm{e}^{j\phi(\omega T)} \; , \qquad (3.4.42)$$

then the strictly positive real condition reduces to

$$\cos\left(\phi(\omega T)\right) > 0 \quad \text{for all } \omega T \; . \qquad (3.4.43)$$

Stability of the slowly adapting filtered reference LMS algorithm is thus assured provided the phase error in the plant model is less than 90° at all frequencies, as observed above.

The plant model could alternatively be assumed to be an accurate representation of the plant response under nominal conditions, so that

$$\hat{G}\left(\mathrm{e}^{j\omega T}\right) = G_0\left(\mathrm{e}^{j\omega T}\right) \; , \qquad (3.4.44)$$

but the physical plant response could be subject to changes or *uncertainties* that are modelled by a complex multiplicative factor $\Delta\left(\mathrm{e}^{j\omega T}\right)$, such that

$$G\left(\mathrm{e}^{j\omega T}\right) = G_0\left(\mathrm{e}^{j\omega T}\right)\left(1 + \Delta\left(\mathrm{e}^{j\omega T}\right)\right) \; , \qquad (3.4.45)$$

where

$$\left|\Delta\left(e^{j\omega T}\right)\right| < B\left(e^{j\omega T}\right) \tag{3.4.46}$$

and $B\left(e^{j\omega T}\right)$ is the upper bound on the multiplicative plant uncertainty at the normalised frequency ωT.

If $G\left(e^{j\omega T}\right)$ and $\hat{G}\left(e^{j\omega T}\right)$ are described by equations (3.4.40–44) then the stability condition becomes

$$1 + \mathrm{Re}\left[\Delta\left(e^{j\omega T}\right)\right] > 0 \quad \text{for all } \omega T \ . \tag{3.4.47}$$

The worst-case condition on the multiplicative uncertainty will be when its phase is 180° and its magnitude is equal to its upper bound, equation (3.4.42), in which case the condition for stability of the filtered reference LMS algorithm for slow convergence becomes

$$B\left(e^{j\omega T}\right) < 1 \quad \text{for all } \omega T \tag{3.4.48}$$

(Ren and Kumar, 1989), which can be directly compared with the robust stability condition for a feedback system discussed in Chapter 6. This condition may not be satisfied, however, if $G_0\left(e^{j\omega T}\right)$ takes a particularly small value at some frequency so that $\Delta\left(e^{j\omega T}\right)$ must be considerably greater than unity to describe even a small increase in the absolute value of the plant response.

If a leakage term is included in the filtered-reference LMS algorithm, the stability condition, equation (3.4.41), is modified to

$$\mathrm{Re}\left[\hat{G}\left(e^{j\omega T}\right)G^{*}\left(e^{j\omega T}\right)\right] + \beta > 0 \qquad \text{for all } \omega T \ , \tag{3.4.49}$$

so that if the plant model is subject to phase errors of $\phi(\omega T)$, these errors must be such that (Elliott, 1998b)

$$\cos\left(\phi(\omega T)\right) > \frac{-\beta}{\left|G\left(e^{j\omega T}\right)\right|^{2}} \qquad \text{for all } \omega T \ . \tag{3.4.50}$$

Now if $\left|G\left(e^{j\omega T}\right)\right|^{2}$ is very small at some frequency, in particular less than β, then the right-hand side of equation (3.4.50) is less than –1 and the stability condition is satisfied for any phase error in the plant model. Alternatively, if the plant and plant model are described by equations (3.4.44–48) with the worst-case condition $\Delta\left(e^{j\omega T}\right) = -B\left(e^{j\omega T}\right)$, then the sufficient condition for stability becomes

$$B\left(e^{j\omega T}\right) < 1 + \frac{\beta}{\left|G_0\left(e^{j\omega T}\right)\right|^{2}} \ . \tag{3.4.51}$$

Now if $\left|G_0\left(e^{j\omega T}\right)\right|^2$ is small compared with β at some frequency we can see that B may be considerably greater than unity and yet the slowly-varying algorithm would remain stable (Elliott, 1998b).

As well as providing a useful improvement in robust stability when the filtered reference LMS algorithm is implemented in practice, a leakage term can also help the conditioning of the optimum least-square solution when calculated at the design stage. In the time domain, for example, the filter that minimises equation (3.4.36) is given by

$$\mathbf{w}_{\text{opt},2} = \left[\mathbf{R}_{rr} + \beta\mathbf{I}\right]^{-1}\mathbf{r}_{rd} \ , \tag{3.4.52}$$

which, in contrast to equation (3.3.38), has a unique solution even if the reference signal is not persistently exciting. The modification to the correlation matrix in equation (3.4.52) suggests that an equivalent result would be produced if an uncorrelated white noise signal, of mean-square value equal to β, were to be added to the filtered reference signal in Fig. 3.12(b).

The power spectral density of this modified filtered reference signal can then be written as

$$S_{rr}(z) + \beta = G(z)G\left(z^{-1}\right)S_{xx}(z) + \beta \ . \tag{3.4.53}$$

In order to obtain an expression for the optimal filter in the z domain, equation (3.4.53) can be expressed as the product of $G_{\min}(z)\,G_{\min}\left(z^{-1}\right)$, which is equal to $G(z)\,G\left(z^{-1}\right)$, and $F'(z)\,F'\left(z^{-1}\right)$, where $F'(z)$ is a modified spectral factorisation given by

$$F'(z)\,F'\left(z^{-1}\right) = S_{xx}(z) + \beta \big/ G(z)G\left(z^{-1}\right) \ , \tag{3.4.54}$$

and $F'(z)$ must be used instead of $F(z)$ in the equation for the transfer function of the optimal filter, equation (3.3.24). Alternatively, an equivalent change could be made in the definition of $G_{\min}(z)$.

In the discrete frequency domain the effect of the modified cost function is to change the definition of the spectral factor from equation (3.3.28) to

$$F'(k) = \exp\left(\text{FFT}\left[c(n)\ \text{IFFT}\ \ln\left(S_{xx}(k) + \beta \big/ |G(k)|^2\right)\right]\right) \ , \tag{3.4.55}$$

which has the effect of avoiding numerical problems associated with taking the logarithm of very small numbers if $S_{xx}(k)$ is not very large in some frequency bins.

An alternative cost function to equation (3.4.36), which has an even clearer physical meaning and is also consistent with the optimal feedback control literature, would be the weighted sum of the mean-square error and the mean-square output of the control filter (Darlington, 1995) so that

$$J_3 = E\left[e^2(n)\right] + \rho\, E\left[u^2(n)\right] \ , \tag{3.4.56}$$

where ρ is a positive *effort-weighting parameter*. The mean-square output of the control filter, $E\left[u^2(n)\right]$, is the same as the mean-square input to the secondary actuator in the plant, as shown in Fig. 3.12a, and is called the *control effort*. This output signal can be written as

$$u(n) \;=\; \boldsymbol{w}^{\mathrm{T}}\boldsymbol{x}(n) \;, \tag{3.4.57}$$

where

$$\boldsymbol{x}(n) \;=\; \left[x(n),\, x(n-1)\cdots x(n-I+1)\right]^{\mathrm{T}} \;, \tag{3.4.58}$$

and so equation (3.4.56) can be written in quadratic form as

$$J_3 \;=\; \boldsymbol{w}^{\mathrm{T}}\left(\mathbf{R}_{rr}+\rho\,\mathbf{R}_{xx}\right)\boldsymbol{w}+2\,\boldsymbol{w}^{\mathrm{T}}\boldsymbol{r}_{rd}+R_{dd}(0) \;, \tag{3.4.59}$$

where $\mathbf{R}_{xx}=E\left[\mathbf{x}(n)\,\mathbf{x}^{\mathrm{T}}(n)\right]$.

The vector of optimal filter coefficients which minimise J_3 can thus be written as

$$\boldsymbol{w}_{\mathrm{opt},3} \;=\; \left[\mathbf{R}_{rr}+\rho\,\mathbf{R}_{xx}\right]^{-1}\mathbf{r}_{rd} \;. \tag{3.4.60}$$

Exactly the same result would be obtained if a noise signal, uncorrelated with $x(n)$, were added to the filtered reference signal in Fig. 3.12(b), whose mean-square value was proportional to ρ and whose power spectral density was equal to that of the reference signal. The power spectral density of this modified filtered reference signal is equal to

$$S_{rr}(z)+\rho\,S_{xx}(z) \;=\; \left(G(z)\,G\left(z^{-1}\right)+\rho\right)S_{xx}(z) \;. \tag{3.4.61}$$

In order to perform the spectral factorisations needed to get the transfer function of the optimal filter in this case, it is easier to retain the spectral factorisation of $S_{xx}(z)$ as $F(z)\,F\left(z^{-1}\right)$ but modify the definition of the minimum phase part of the plant so that

$$G'_{\min}(z)\,G'_{\min}\left(z^{-1}\right) \;=\; G(z)\,G\left(z^{-1}\right)+\rho \tag{3.4.62}$$

and use $G'_{\min}(z)$ instead of $G_{\min}(z)$ and $G'_{\mathrm{all}}(z)=G(z)/G'_{\min}(z)$ instead of $G_{\mathrm{all}}(z)=G(z)/G_{\min}(z)$ in equation (3.3.24).

The derivative of the cost function given by equation (3.4.56) with respect to the coefficients of $\boldsymbol{w}$ can be written as

$$\frac{\partial J}{\partial \mathbf{w}} \;=\; 2\,E\left[\boldsymbol{r}(n)\,e(n)\right]+2\,\rho\,E\left[\boldsymbol{x}(n)\,u(n)\right] \tag{3.4.63}$$

and using the instantaneous version of this to adapt the coefficients of the control filter at every sample gives another modified form of the filtered reference LMS algorithm as

$$\boldsymbol{w}(n+1) \;=\; w(n)-\alpha\left[\boldsymbol{r}(n)\,e(n)+\rho\,\boldsymbol{x}(n)\,u(n)\right] \;. \tag{3.4.64}$$

Comparing the quadratic forms of equation (3.4.59) and (3.4.37) suggests that the leaky form of the algorithm is a special case of equation (3.4.64) where the reference signal is assumed to be white noise, in which case $E\left[u^2(n)\right]$ is proportional to $\boldsymbol{w}^{\mathrm{T}}\boldsymbol{w}$. The more general algorithm given by equation (3.4.64) is, however, more computationally expensive to implement than the leaky algorithm and has poorer convergence properties (Darlington, 1995) and so is not widely used in practice.

3.5. FREQUENCY-DOMAIN ADAPTATION OF FIR CONTROLLERS

The adaptation equation for the filtered-reference LMS algorithm, equation (3.4.4), is normally computed at every sample time, and for long control filters this can be a considerable computational burden. Also, if the response of the system under control, the plant $G(z)$, is complicated, then the plant model used in the filtered-reference LMS algorithm has to be rather detailed, which can require long filters for $\hat{G}(z)$ in Fig. 3.15. A considerable computational burden is then associated with the generation of the filtered reference signal. One way of improving this situation for long filters is to update the coefficients of the control filter in the frequency domain so that the time-domain convolution is implemented as a multiplication in the frequency domain, even though the control filter is implemented in the time domain.

Various authors have suggested that both the control filtering and the adaptation be done in the frequency domain (see for example Shen and Spanias, 1992), and this approach, which is widely used for adaptive filters (Shynk, 1992), can lead to very efficient implementations. Unfortunately, performing the convolution associated with the control filter in the frequency domain introduces a delay of at least one FFT block size, which is often unacceptable in the active control of stochastic disturbances, for which any delay in the controller will lead to a degradation in the performance. Therefore, the option of implementing the controller in the time domain, to minimise delays, but performing the adaptation in the frequency domain, to minimise computational loads and improve convergence time, is an attractive one for active control systems (Morgan and Thi, 1995).

In order to derive the frequency-domain version of the filtered-reference LMS algorithm, we first consider the expression for the derivative of the averaged mean-square error with respect to the i-th controller coefficient, which from equation (3.4.2) can be written as

$$\frac{\partial \overline{e^2}}{\partial w_i} = 2\ \overline{e(n)\,r(n-i)} = 2\ \overline{r(n)\,e(n+i)}\ , \tag{3.5.1}$$

This is equal to twice the cross-correlation, $R_{re}(i)$, between the filtered reference signal and the error signal.

The filtered-reference LMS algorithm uses an instantaneous estimate of the gradient term in equation (3.5.1) to update the control filter coefficients at every sample time, as illustrated in Fig. 3.15. Alternatively, a more accurate estimate of the gradient could be found by averaging over the last N samples and updating the coefficients in one step using 'block' adaptation. With a suitable adjustment to the convergence coefficient, these two strategies can have a similar convergence rate, as observed by Elliott and Nelson (1986).

3.5.1. Cross Correlation Estimates via Cross Spectra

The question then arises as to whether there is a more efficient way of estimating the cross-correlation function given by equation (3.5.1). To update all I coefficients of the control filter we need I terms in $R_{re}(i)$. If the control filter does not have too many coefficients, direct averaging in the time domain, as described above, is probably a reasonable method. For larger values of I, it is more efficient to calculate the cross-correlation function in the frequency domain by estimating the Fourier transform of the cross-correlation function, which is equal to the following cross spectral density

$$S_{re}(e^{j\omega T}) = E\left[R^*(e^{j\omega T})E(e^{j\omega T})\right] , \tag{3.5.2}$$

where $E(e^{j\omega T})$ and $R(e^{j\omega T})$ are the Fourier transforms of $e(n)$ and $r(n)$ and E denotes an expectation operator. If $2N$-point discrete Fourier transforms (DFT) are taken of segments of the signals $e(n)$ and $r(n)$, an estimate of the cross spectral density can be computed as

$$\hat{S}_{re}(k) = R^*(k)E(k) , \tag{3.5.3}$$

where k is the discrete frequency index.

To obtain a good estimate of the cross spectral density and to reduce spectral 'leakage', the $2N$ point segments of $e(n)$ and $r(n)$ would generally be smoothly windowed prior to being transformed. In the present case, however, we are interested in obtaining an accurate estimate of a limited portion of the cross-correlation function, which can be obtained without explicit windowing by taking the transforms of N data points for the error signal with N added zeros, and $2N$ data points from the filtered reference signal to implement the overlap–save method outlined in Chapter 2 (Rabiner and Gold, 1975). This ensures that a linear rather than a circular estimate of the correlation is obtained (Shynk, 1992).

Half of the points of the inverse DFT of equation (3.5.3) will then be equal to

$$\hat{R}_{re}(i) = \sum_{n=1}^{N} e(n)r(n-i) \qquad \text{for } 0 < i < N , \tag{3.5.4}$$

i.e. the block-average estimate of the causal part of the cross-correlation function. For large values of N, equation (3.5.4) can be more efficiently calculated for all i from 0 to I via the frequency-domain expression (3.5.3), since this requires 3 $2N$-point FFT calculations, each of which takes of the order of $2N \log_2 2N$ multiplications, and $2N$ complex multiplications ($8N$ real multiplications) per data block, whereas direct calculation of (3.5.4) over N data samples and for $i = 1$ to N requires N^2 multiplications. The computational advantage becomes significant when long control filters are being updated and we can arrange for the control filter length I to be equal to half the DFT block length, N. As discussed in Section 2.6, this method of calculating a block estimate of the gradient was introduced for the LMS algorithm by Clark et al. (1980) and Ferarra (1980), who called it the fast LMS (FLMS) algorithm, and showed that it was exactly equivalent to a block implementation of the gradient descent algorithm.

3.5.2. Frequency-Domain Steepest-Descent Algorithm

Further computational savings for the filtered-reference algorithm used in active control are obtained if $R(k)$ is obtained by directly multiplying the DFT of the reference signal, $X(k)$, by the estimated frequency response of the plant, $\hat{G}(k)$. This can be calculated, for example, from an FIR model of the plant response, $\hat{g}_j$, which is arranged to have the same number of coefficients as half the block length, i.e. $J = N$. Strictly speaking, the effects of the circular convolution generated by multiplying together $X(k)$ and $\hat{G}(k)$ should be removed by taking the inverse FFT of the product, joining the last N points of the result to the last N points of the calculation from the previous block and taking the FFT of the resulting sequence. This process would add two additional FFT calculations to the computational load, although in view of the inherent robustness of the filtered reference LMS algorithm to plant uncertainties, it is not clear whether the circular convolution effects would seriously bias the adaptation, and this complication has been ignored in the discussion below.

Since the calculation of $G(k)$ need only be performed once, the calculation of the frequency-domain filtered-reference signal takes only $2N$ complex multiplications ($8N$ real multiplications) per N point block, whereas calculating N values of this filtered reference signal in the time domain would take N^2 multiplications. Alternatively, it would also be possible to multiply the Fourier transform of the error signal with the conjugate of the plant's frequency response, to implement a frequency-domain filtered-error algorithm, although this would still require the same number of calculations as above in the single-channel case.

A block diagram for the adaptive frequency-domain control filter is illustrated in Fig. 3.19. Every N samples the $2N$-point FFTs of the reference signal and the error signal with zero padding are taken, the former is multiplied at each frequency by the plant frequency response, the latter is conjugated, and the two are multiplied together to give the estimate of the cross spectral density in equation (3.5.3). The first N points of the inverse FFT of this frequency-domain function now can be used as an estimate of the cross-correlation function given by equation (3.5.4). This is used to update the time-domain coefficients of the control filter using data from the m-th block of signals using the equation

$$w_i(m+1) = w_i(m) - \alpha \, \mathrm{IFFT}\left\{R_m^*(k) E_m(k)\right\}_+ , \tag{3.5.5}$$

where the notation $\{\ \}_+$ is again used to denote the fact that only the causal part of the inverse transform is used. The computational cost of the time-domain and frequency-domain adaptation methods are compared in Table 3.1, in which it is assumed that the $2N$-point DFT is implemented with an FFT which requires $2N \log_2 2N$ multiplications.

It has been assumed above that each block of N data points is used to update the control filter. The separation shown in Fig. 3.19, between the functions of generating the control signal, $u(n)$, in the time domain, and adapting the control filter in the frequency domain, suggests that if the control filter need not be adapted very rapidly, the adaptation could be performed using every other data block, for example, and although the control filter would take longer to converge, its final response would be the same. Even more 'sparse' adaptation could be envisaged if the disturbance that was being controlled was nearly stationary. In this case, the most time-consuming part of the processing would be the

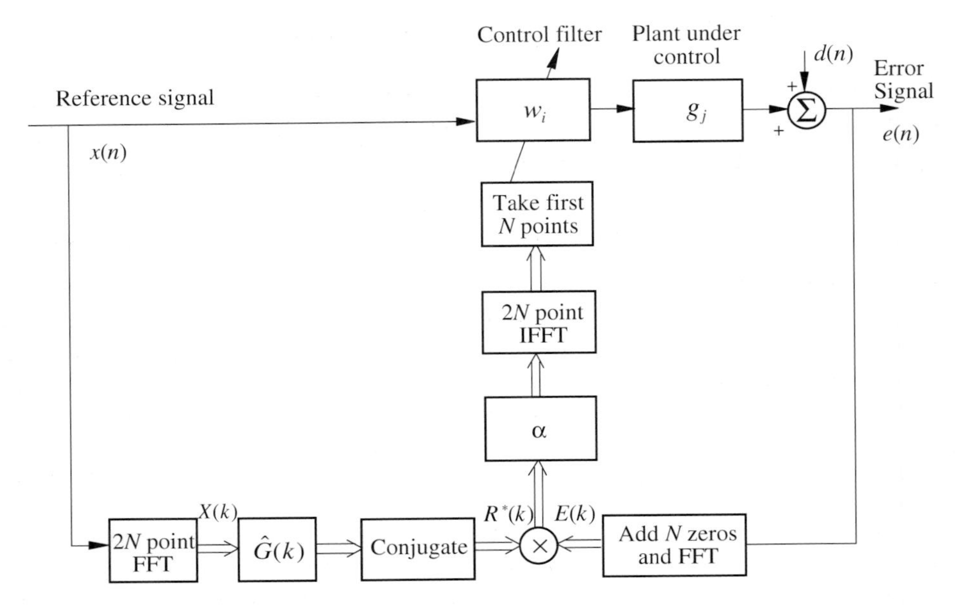

Figure 3.19 Block diagram of a frequency-domain method of adapting the time-domain control filter.

Table 3.1. The number of real multiplications required to implement the filtered-reference adaptation over N samples in the time domain and frequency domain, where it is assumed that the number of coefficients in the control filter, I, and plant model, J, are both equal to N and $2N$ point FFTs, are used to avoid circular correlation. Note that the time-domain implementation of the control filter itself, which takes N^2 multiplications, is common to both systems and has been excluded from the table.

	Filtered reference generation	Control filter update	Total
Time domain	N^2	N^2	$2N^2$
Frequency domain	$8N$	$6N \log_2 2N + 8N$	$(16+6 \log_2 2N)N$

generation of the control signal, which could be implemented with very efficient hardware or software, and the adaptation could be relegated to a background task, with much smaller and less time-critical processing requirements.

3.5.3. Frequency-Dependent Convergence Coefficients

Another potential advantage of adapting the controller response in the frequency domain is that different convergence coefficients could be used in the adaptation of each frequency-domain controller coefficient. This may significantly improve the convergence rate of the slow modes of adaptation if the spectrum of the filtered reference signal has a large dynamic range.

Equation (3.5.5) for the adaptation of the time-domain control filter would thus become

$$w_i(m+1) = w_i(m) - \text{IFFT}\left\{\alpha(k)\, R_m^*(k)\, E_m(k)\right\}_+ . \quad (3.5.6)$$

This modification to the frequency-domain update was suggested by Reichard and Swanson (1993) and Stothers et al. (1995), who both suggested that identification of the plant response could now also be efficiently performed in the frequency domain.

Unfortunately, this algorithm can be shown to converge to a biased solution if the impulse response of the unconstrained controller has significant noncausal components. This is because the causal components of the Fourier transform of $\alpha(k)$ interact with the noncausal part of the Fourier transform of $R^*(k)\, E(k)$, so that they affect the causal part of the product, which is driven to zero by equation (3.5.6). This problem is the same as that encountered in the frequency-domain version of the adaptive electrical filter discussed in Section 2.8. The solution to the problem is also the same, which is to split the frequency-dependent convergence coefficient, $\alpha(k)$, up into 'spectral factors', that have equal amplitude responses but whose impulse responses are either entirely causal, $\alpha^+(k)$ or noncausal, $\alpha^-(k)$, so that

$$\alpha(k) = \alpha^+(k)\alpha^-(k) . \quad (3.5.7)$$

The equation for the adaptation of the time-domain coefficients of the control filter then becomes (Elliott and Rafaely, 2000)

$$w_i(m+1) = w_i(m) - \text{IFFT}\ \alpha_m^+(k)\left\{\alpha_m^-(k)\, R_m^*(k)\, E_m(k)\right\}_+ . \quad (3.5.8)$$

The speed of adaptation of equation (3.5.8) can be considerably greater than is the case where a constant convergence coefficient is used in each frequency bin, if the frequency-dependent convergence coefficient is used to correct for the mean-square level of the filtered reference signal in each frequency bin. In this case

$$\alpha_m(k) = \hat{S}_{rr,m}^{-1}(k) , \quad (3.5.9)$$

where $\hat{S}_{rr,m}(k)$ is the estimated power spectral density of the filtered reference signal for the m-th data block, which may be calculated recursively, for example, using the equation

$$\hat{S}_{rr,m}(k) = (1-\lambda)\hat{S}_{rr,m-1}(k) + \lambda |R(k)|^2 , \quad (3.5.10)$$

where λ is a forgetting factor. Strictly speaking, precautions such as taking a longer FFT size should be used to accommodate the circular effects of multiplying the relevant discrete frequency-domain quantities by $\alpha^+(k)$ and $\alpha^-(k)$, but in practice these effects may not unduly bias the results. Indeed, in many adaptive feedforward control applications, the unconstrained optional controller has only a small noncausal part and the bias involved in using the simpler bin-normalised algorithm of equation (3.5.6) to update the controller can be small. This is not the case if feedback controllers are being adapted in the discrete frequency domain, as discussed in Chapter 7, however.

If the power spectral density of the filtered reference could be estimated perfectly, then

$$\alpha(k) = S_{rr}^{-1}(k) , \quad (3.5.11)$$

where

$$S_{rr}(k) = E\left[R^*(k)R(k)\right] = |G(k)|^2 S_{xx}(k) \; . \tag{3.5.12}$$

Assuming that the power spectral density is divided into its spectral factors, $F(k)$ and $F^*(k)$, and noting that the mean-square response of the minimum phase part of the plant, $G_{\min}(k)$, is equal to that of the plant itself, then the power spectral density of the filtered reference signal is equal to

$$S_{rr}(k) = G_{\min}(k)G^*_{\min}(k)F(k)F^*(k) \; . \tag{3.5.13}$$

Since the Fourier transforms of $G_{\min}(k)$ and $F(k)$, and their inverses, are entirely causal, and those of $G^*_{\min}(k)$ and $F^*(k)$, and their inverses, are entirely noncausal, it is clear that the spectral factorisation of $\alpha(k)$ required to satisfy equation (3.5.9) can be achieved in this case by letting

$$\alpha^+(k) = \left[G_{\min}(k)\,F(k)\right]^{-1} \text{ and } \alpha^-(k) = \left[G^*_{\min}(k)\,F^*(k)\right]^{-1} \; . \tag{3.5.14}$$

Under these conditions, setting the expectation of the update term in equation (3.5.8) to zero results in the optimal causal filter given by equation (3.3.26). A sliding transform domain method of implementing the filtered-reference LMS algorithm using the discrete cosine transform (DCT) has been suggested by Bouchard and Paillard (1996), who show that this can also improve the convergence speed because of the reduction in the eigenvalue spread of the filtered reference signal's autocorrelation matrix. Another method of improving the convergence speed would be to use a lattice structure, as discussed for active noise control applications by Park and Sommerfeldt (1996).

3.6. PLANT IDENTIFICATION

3.6.1. The Need for Plant Identification

The adaptive control algorithms described above require an estimate of the plant response, to act as an internal plant model, for their correct operation. This model must include the entire response between the output of the adaptive controller and the measurement of the residual error. The internal plant model is necessary to rapidly interpret how the changes in the waveform of the error signal have been affected by the changes in the controller coefficients. It is, however, possible to implement algorithms that require no estimate of the plant response. For example, each of the controller coefficients may be slowly adjusted in turn to minimise a measured estimate of the mean-square error (Chaplin, 1983). This sequential adaptation must be performed repeatedly, however, since the error surface is not aligned along the axis of the physical filter coefficients, as explained in Chapter 2, and so the optimum value of one control filter coefficient will depend on the value of all the others. Thus, although this algorithm makes no assumptions about the system under control, beyond its linearity, the price that has to be paid is a very long convergence time, particularly for control filters with a large number of coefficients. Another algorithm which does not require an explicit model of the plant is the time-averaged gradient or *TAG algorithm* described by Kewley et al. (1995), see also Clark et al. (1998). In the TAG algorithm finite difference techniques are used to estimate the elements of the gradient vector and Hessian matrix in the implementation of a version of Newton's algorithm. A variation on the sequential estimation of the elements in the gradient vector is provided by the *simultaneous perturbation method* (Maeda and Yoshida, 1999), in which each of the controller parameters is randomly perturbed at each iteration, but the perturbations introduced into each of the parameters are uncorrelated. Finite difference estimates of the individual elements in the gradient vector will be correct on average using this method, since the perturbations introduced into all of the other parameters will be seen as uncorrelated noise and will average to zero.

It is very unusual in active control applications to have such a sufficiently detailed and accurate knowledge of the system under control that the plant response can be predicted without any direct measurements of the input and output signals. The plant model must thus generally be deduced from measurements taken directly from the system. This problem has been very widely studied in the control literature where it is called *system identification* (see, for example, Norton (1986); Ljung (1999). In many applications, the response of the plant does not change 'significantly' over the timescales on which the control system is operating. In this case, the identification of the plant response can be performed prior to the control system being switched on, which is often called *off-line* system identification. It is clearly important to have a good definition of what constitutes a 'significant' change in the plant response for a particular control system. This will depend on both the magnitude of the change in the plant response and the speed of the change. Generally speaking, the change in the plant should not cause the adaptive control system to become unstable and its performance should also be maintained within reasonable bounds. Therefore, the definition of a 'significant' plant change will depend on the algorithm used to adapt the controller. The filtered-reference LMS algorithm with a tonal reference, as discussed in Section 3.5 for example, is stable and performs well provided the phase response of the internal plant model is within about $\pm 40°$ of that of the true plant response at the disturbance frequency (Boucher et al., 1991). Off-line identification is clearly sufficient provided it is initially accurate and provided the subsequent changes in the plant response are within the class for which the control algorithm being used is robust. If the changes in the plant response are larger than this, then the response must be re-identified. It is common in active control for relatively rapid but small changes to occur in the plant response, which the control algorithm must be made robust to, but rather larger changes in the plant response occur on a longer timescale, which must be re-identified before reliable control can be ensured. The choice between making the algorithm used to adapt the control filter robust to plant variations, or having a method for re-identifying the plant, is even more relevant in the discussion of adaptive feedback systems, but a fuller discussion of this is postponed until Section 7.1.

Off-line system identification can be performed using the adaptive filtering methods described in Chapter 2 with either an FIR model, which is often most appropriate for acoustic systems or engineering structures, or an IIR model, which may be more appropriate for lightly damped structural systems, as described in Section 3.1.

3.6.2. On-line Plant Identification

In some applications of active control, the changes in the plant response that occur while the control system is operating are so large that the control algorithm would become unstable with a fixed plant model. A simple example may be a plant with a lightly damped resonance whose natural frequency changes by more than the bandwidth of the resonance, as shown in Fig. 3.20. The natural frequencies of a lightly damped mechanical system could change by this amount because of a relatively small change in temperature or due to a change in static load, for example. Although the changes in the amplitude and phase are small over most of the frequency range in Fig. 3.20, changes of more than 6 dB in amplitude and 90° in phase can occur at excitation frequencies close to the natural frequency. Other examples of large changes in plant response occur in air ducts for industrial processes, where large changes in temperature or flow-rate can significantly alter the speed of sound propagation.

Under these conditions, the response of the plant must be remeasured over the timescale of the possible changes for control to be maintained. If the control system is kept in operation during this repeated measurement, the process is called *on-line* system identification. In order to perform on-line identification unambiguously without a persistently exciting command signal of the type discussed in Chapter 7, it is generally necessary to add an identification signal to the input signal driving the secondary actuator (Ljung, 1999). There has been some discussion in the active control literature about on-line identification without the use of such an identification signal (Sommerfeldt and Tichy, 1990; Tapia and Kuo, 1990). These techniques identify both the plant response and the primary path response in an 'overall modelling' method. It is shown by Kuo and Morgan (1996) that such methods generally have no unique solution and errors in the estimation of the plant response can be compensated for by errors in the estimation of the primary path. If the response of the control filter changes with time, however, this breaks the symmetry and allows the plant response and primary path to be identified separately under certain conditions (Nowlin et al., 2000).

A block diagram of a feedforward controller with on-line plant identification is shown in Fig. 3.21. Although, in Fig. 3.21, it is assumed that the filtered-reference LMS algorithm is being used to adapt the plant, a similar form of block diagram would be valid for many adaptation algorithms. The identification noise is denoted $v(n)$. The system identification

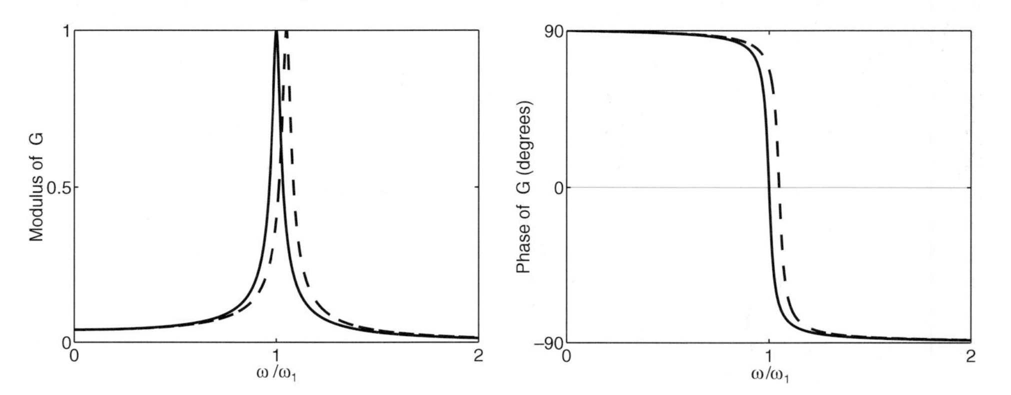

Figure 3.20 The changes in the amplitude and phase response of a plant with a single lightly damped resonance when the resonance frequency changes by the bandwidth of the resonance; solid line before change, dashed line after change.

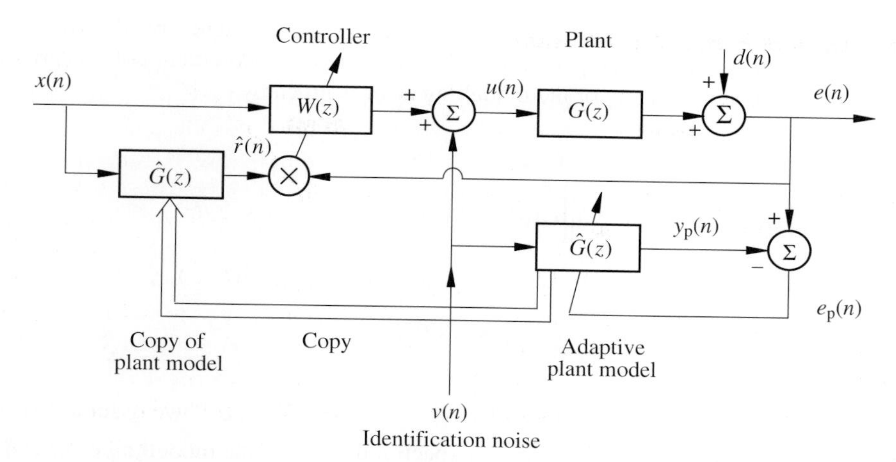

Figure 3.21 Block diagram of an adaptive feedforward active control system with on-line system identification using an identification signal $v(n)$.

part of the control algorithm uses the difference between the measured error signal $e(n)$ and the output of the plant model $y_p(n)$ to generate a modelling error

$$e_p(n) = e(n) - y_p(n) \,, \tag{3.6.1}$$

which is minimised in the system identification procedure. The identified coefficients of the plant model are then passed to the adaptation algorithm for the control filter, to be used in the generation of the filtered reference signal in this case.

If we assume that broad band identification is required, then $v(n)$ will typically be a white noise signal, although there may be advantages in shaping the spectrum of $v(n)$ to be similar to that of the disturbance or the residual error signal on the grounds that the most accurate identification of the plant is required at the frequencies at which the control must be most effective (Coleman and Berkman, 1995). It is important that the identification signal, $v(n)$, is *uncorrelated* with the reference signal, $x(n)$. If this condition is not fulfilled, then the identified model of the plant will be biased compared to the true response. This can be demonstrated by writing the z transform of the measured error signal in this case, from Fig. 3.21, as

$$E(z) \;=\; D(z) + G(z)\big[V(z) + W(z)\,X(z)\big] \;, \tag{3.6.2}$$

and the output of the plant model being identified as

$$Y_p(z) \;=\; \hat{G}(z)\,V(z) \;. \tag{3.6.3}$$

The modelling error, equation (3.6.1), is thus equal to

$$E_p(z) \;=\; \Big[G(z) - \hat{G}(z)\Big]\,V(z) + D(z) + G(z)\,W(z)\,X(z) \;, \tag{3.6.4}$$

so that, in the frequency domain, its spectrum is

$$E_p(e^{j\omega T}) = \Big[G(e^{j\omega T}) - \hat{G}(e^{j\omega T})\Big]V(e^{j\omega T}) + D(e^{j\omega T}) + G(e^{j\omega T})W(e^{j\omega T})X(e^{j\omega T}) \;. \tag{3.6.5}$$

If we assume that the disturbance is related to the reference signal by the frequency response $P(e^{j\omega T})$, as defined in equation (3.3.2), and that no measurement noise is present, then the power spectral density of the modelling error is given by

$$\begin{aligned} S_{pp}(e^{j\omega T}) &= \left|G(e^{j\omega T}) - \hat{G}(e^{j\omega T})\right|^2 S_{vv}(\omega) \\ &+ \left[G^*(e^{j\omega T}) - \hat{G}^*(e^{j\omega T})\right]\left[G(e^{j\omega T})W(e^{j\omega T}) + P(e^{j\omega T})\right] S_{vx}(e^{j\omega T}) \\ &+ S_{vx}^*(e^{j\omega T})\left[G^*(e^{j\omega T})W^*(e^{j\omega T}) + P^*(e^{j\omega T})\right]\left[G(e^{j\omega T}) - \hat{G}(e^{j\omega T})\right] \\ &+ \left|G(e^{j\omega T})W(e^{j\omega T}) + P(e^{j\omega T})\right|^2 S_{xx}(e^{j\omega T}). \end{aligned} \tag{3.6.6}$$

This is a complex Hermitian function of $\left[G(e^{j\omega T}) - \hat{G}(e^{j\omega T})\right]$ and if we assume that the modelling filter is unconstrained, the power spectral density of the modelling error can be minimised at each frequency by the optimum value of this function which is given by

$$G(e^{j\omega T}) - \hat{G}_{\text{opt}}(e^{j\omega T}) = -\frac{\left[G(e^{j\omega T})W(e^{j\omega T}) + P(e^{j\omega T})\right] S_{vx}(e^{j\omega T})}{S_{xx}(e^{j\omega T})}, \tag{3.6.7}$$

so that

$$\hat{G}_{\text{opt}}(e^{j\omega T}) = G(e^{j\omega T}) + \frac{\left[G(e^{j\omega T})W(e^{j\omega T}) + P(e^{j\omega T})\right] S_{vx}(e^{j\omega T})}{S_{xx}(e^{j\omega T})}. \tag{3.6.8}$$

Thus, if there is any cross-correlation between the identification signal $v(n)$ and the reference signal $x(n)$, then $S_{vx}(e^{j\omega T})$ will take a finite value and $\hat{G}_{\text{opt}}(e^{j\omega T})$ will not equal $G(e^{j\omega T})$. The only exception is when the active control system completely cancels the primary disturbance, so that $G(e^{j\omega T})W(e^{j\omega T}) = -P(e^{j\omega T})$, in which case the error signal would be zero if the identification noise were not present. We will see in Chapter 7 that special care needs to be taken in identifying the response of systems with feedback, since the measurement noise is then fed back to the reference signal and so these signals will not be uncorrelated.

This analysis demonstrates the importance of having independent identification noise or 'probe noise' injected into the plant for on-line control, as also discussed by Laugesen (1996). Equation (3.6.2), however, also demonstrates that this identification noise will appear at the output of the active control system. If the contribution of the identification noise to the signal heard at the error sensor is significant compared with the residual disturbance, i.e. after active control, then the performance of the active control system will clearly be degraded. These audible effects can be minimised by adjusting the spectrum of the measured identification noise so that it is similar to the residual disturbance spectrum at the error microphone (Coleman and Berkman, 1995), but will always be present if high levels of identification noise are used. Unfortunately, one must use high levels of identification noise if the plant model is to be able to accurately track *rapid* changes in the plant. This is because even when the reference signal and disturbance are uncorrelated with the identification signal, these signals appear as 'noise' in the modelling error that is being minimised, equation (3.6.4), which would have a large amplitude if the identification signal were small. A considerable amount of averaging must thus be used to extract the signal proportional to the modelling error. If the LMS algorithm is being used for system identification, for example, then to prevent very high levels of misadjustment error, as described in Chapter 2, a small convergence coefficient must be used, giving a correspondingly long convergence time for the adaptive filter. One method of reducing these random errors is to model the disturbance as well as the plant in an 'extended' identification scheme, which could be similar to the extended least-squares method described in Section 2.9. Snyder (1999) also describes how forcing the initial part of the identified impulse response to be zero, for the duration of the known delay in the plant, can significantly improve the estimate of the plant impulse response when extended identification methods are used.

If significant changes occur in the plant response over short timescales, then a slow system identification algorithm may not correct the plant model until the adaptation algorithm has become unstable. One could argue that, theoretically, the identification algorithm will eventually measure the new plant response accurately enough for the control filter adaptation algorithm to be stabilised, and so with on-line modelling the combined system is not truly 'unstable'. In practice, however, the adaptation algorithm for the control filter could increase its coefficients to such an extent during the period in which they are diverging that the data converters and transducers would become saturated and the plant would no longer behave linearly. It is certainly not a pleasant experience to listen to the 'race' between an unstable control algorithm and a stabilising identification algorithm.

The simplest way of preventing the problems caused by the different timescales of the adaptive control algorithm and the adaptive identification algorithm is to make the adaptive control algorithm robust to short-term variations in plant response and the identification algorithm responsive to longer-term trends in the changes in the plant response. Unfortunately, it cannot always be said that short-term plant changes in all applications will be small enough to allow the adaptive controller to be made robust to them. A possible solution under these conditions may be to adjust the level of the identification noise in response to some measurement of the stability of the adaptive controller. A method investigated by Wright (1989), for example, involved scheduling the level of the identification noise on the measured level of the signal at the error sensor. When the mean-square value of this error signal becomes larger, the identification noise is also increased, on the grounds that either the level of the primary disturbance has gone up, and the increased identification noise would not be heard, or that the adaptive controller has become unstable and the plant response should be rapidly re-identified to restore stability. The application of such 'adaptive-identification' methods is very problem specific, however, and they are difficult to analyse in the general case.

In conclusion, one can say that on-line system identification is generally only required, and indeed is generally only possible without significant increases in perceived disturbance, for plants whose response changes slowly compared with the timescales of the algorithm used to adapt the control filter. Under these conditions a variety of on-line identification methods can be used, which employ an identification signal mixed in with the input to the plant, and which will track as wide a variety of plant responses as the structure of the plant model will allow for. Short-term changes in the plant response cannot generally be identified without excessive levels of identification noise, and the algorithm used to adapt the control filter must be designed to be robust to such changes. This section thus provides a framework within which the likely success of an active control system designed for a time-varying plant can be judged. It suggests that when considering the identification of the plant, not only must the form of the nominal plant response be established, in order to ensure a suitable form for the plant model, but the magnitude and form of the possible changes in the plant response must also be determined, together with the likely speed of these changes, to ensure that the control system can either track the longer-term plant changes or be robust to those that occur over the shorter term.

3.7. ADAPTIVE IIR CONTROLLERS

In Chapter 2 we saw that there are two main approaches to the adaptation of digital IIR filters. The equation error approach minimises the difference between the reference signal filtered by the IIR filter denominator, $B(z)$, and the disturbance signal filtered by the IIR filter numerator, $A(z)$, as shown in Fig. 3.22(a), which is similar to Fig. 2.21. This approach has the advantage that the equation error, $e_e(n)$, is linear in the coefficients and so conventional gradient algorithms, based on least-squares methods, can be used to adjust the coefficients, with the guarantee that they will converge to the unique global minimum of the quadratic error surface. The equation error method is widely used for system identification, but is not generally suitable for electrical noise cancellation since the approach is equivalent to minimising a filtered version of the output error $e_o(n)$, as demonstrated in Fig. 3.22(b). For noise cancellation it is the reduction in the magnitude of the output error which is important, rather than identifying the parameters of the system, and it can be seen from Fig. 3.22(a) that even if the equation error was very small, the output error could be large at some frequencies if the response of $A(z)$ was very small. A

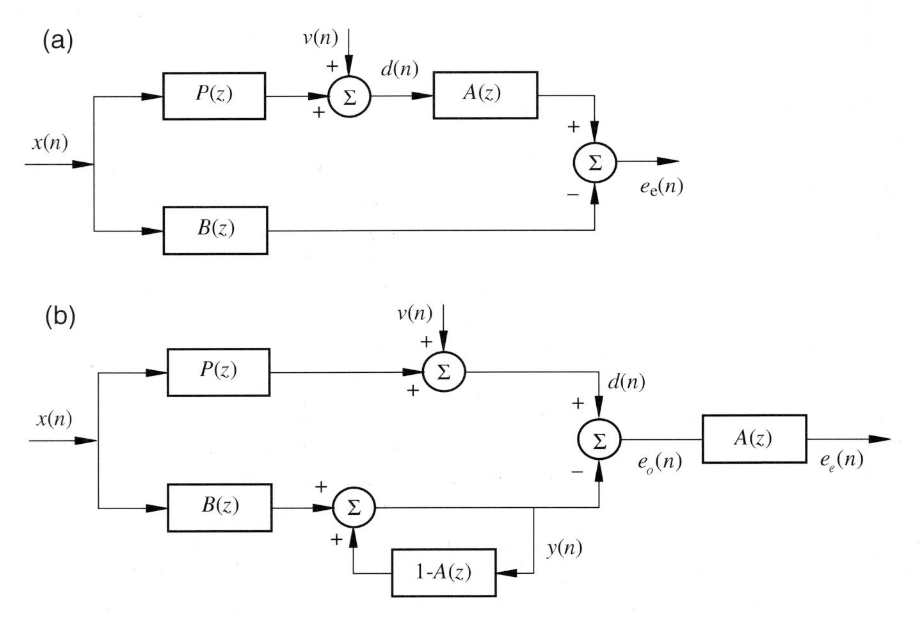

Figure 3.22 Block diagram of the equation error approach to adaptive IIR filters (a), and an equivalent block diagram to demonstrate how the equation error is a filtered form of the output error.

small response of $A(z)$ at certain frequencies corresponds to lightly damped poles of the IIR filter, and so the equation error approach can lead to relatively large output errors at precisely the frequencies at which the disturbance may be largest. It is thus the output error approach, with all its disadvantages, that is most widely used for electrical noise cancellation, and this is also true of adaptive feedforward control.

3.7.1. Form of the Optimum Controller

The block diagram of a single-channel feedforward control system with an IIR controller and no measurement noise is shown in Fig 3.23, which is similar to Fig. 3.11 in that $P(z)$ is the combined primary path, $G_s(z)$ is the physical feedback path from secondary actuator to reference sensor, and $G_e(z)$ is the secondary path from secondary actuator to error sensor. The electronic controller has a transfer function, $H(z)$, which is equal to $B(z)/A(z)$ In contrast to Fig. 3.11, however, there is no explicit feedback cancellation in the controller. The recursive nature of the IIR controller is assumed to be able to provide a complete model of the primary path, secondary path and feedback path. This can be illustrated by writing the transfer function from the primary input to output error in Fig. 3.23 as

$$\frac{E(z)}{X(z)} = P(z) + G_e(z)\left[\frac{B(z)}{A(z) - B(z)\,G_s(z)}\right] . \qquad (3.7.1)$$

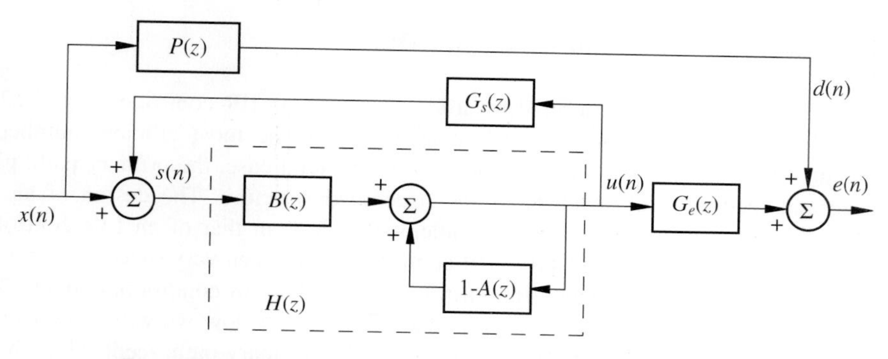

Figure 3.23 Block diagram of an IIR feedforward control system.

With a large number of coefficients, $A(z)$ and $B(z)$ could each accurately approximate any stable transfer function, and equation (3.7.1) could thus be set close to zero with many combinations of the polynomials $A(z)$ and $B(z)$. One of the possible solutions for $A(z)$ and $B(z)$ that sets equation (3.7.1) to zero is

$$A(z) = 1 + B(z)\,G_s(z) \qquad \text{and} \qquad B(z) = -\frac{P(z)}{G_e(z)} . \qquad (3.7.2,3)$$

This has an attractive physical interpretation, since $B(z)$ in this case is the response of the optimum controller if no feedback is present, and $A(z)$ is compensating for the feedback path. Billout et al. (1989) and Crawford et al. (1996) have also considered an arrangement of adaptive filter in which the feedback path, $1 - A(z)$ in Fig. 3.23, is taken back to the input of the non-recursive part of the filter, $B(z)$, rather than back to its output. The adaptation of this *full-feedback* arrangement is shown by Crawford et al. (1996) to be more rapid than that of the conventional IIR arrangement in simulations of an active control system.

It must be emphasised, however, that equations (3.7.2) and (3.7.3) are not the only values of $A(z)$ and $B(z)$ that will set the error to zero in equation (3.7.1), and indeed may not be the most efficient implementation for a practical controller. In equation (3.7.2), for example, the numerator of the IIR filter must model the primary path and the inverse response of the plant, which may require a considerable number of coefficients. The most important question is thus not concerned with the possible values of $A(z)$ and $B(z)$ that will set equation (3.7.1) to a small value, but with the most *efficient* combinations of the two parts of the IIR controller to achieve this task, i.e. how this can be achieved with the smallest number of coefficients. We are also concerned, however, that the error surface will have a unique minimum, so that gradient descent methods can be used to adjust the coefficients of $A(z)$ and $B(z)$, which requires that $A(z)$ and $B(z)$ do not undermodel the required transfer function, as discussed in Section 2.9. This issue is made considerably more complicated, however, by the presence of the feedback path, $G_s(z)$ in Fig. 3.23.

3.7.2. Controller for Sound Waves in a Duct

The general block diagram of a feedforward system with an IIR controller, Fig. 3.23, is not very helpful in determining the IIR controller with the most efficient number of coefficients in $A(z)$ and $B(z)$. This is because, in the general case, the primary path, plant response and feedback path each have an unspecified structure. The design of an IIR controller is thus expected to be more problem-specific than that of an FIR controller. One application area in which adaptive IIR controllers have been very successful is in the active control of plane sound waves in uniform ducts, such as air conditioning ducts. This application is discussed in more detail in Section 3.8, but for now we will be concerned with the form of the optimal controller in this case. The primary path, feedback path and plant response can be very reverberant for an active sound control system in a finite duct. In other words, the impulse responses of $P(z)$, $G_s(z)$ and $G_e(z)$ can last a considerable time if there is little passive damping in the duct. It was shown by Roure (1985) and Elliott and Nelson (1984), however, that when these individual responses are combined together to calculate the frequency response of the optimum controller under noise-free conditions, the resulting controller has a much simpler form than that of the individual responses in the duct. This is because the individual responses share a number of common poles in this case, which cancel out when the optimum controller is calculated, and the zeros of the individual responses have a very special form. The frequency response of the optimum controller in this case can be shown to have the form (Nelson and Elliott, 1992)

$$H_{\text{opt}}(e^{j\omega T}) = \frac{-1}{G_L(e^{j\omega T})G_M(e^{j\omega T})} \frac{e^{-j\omega\Delta T}}{1 - D(e^{j\omega T})e^{-j2\omega\Delta T}} , \tag{3.7.4}$$

where $G_L(e^{j\omega T})$ is the frequency response of the secondary source

$G_M(e^{j\omega T})$ is the frequency response of the reference sensor

$D(e^{j\omega T})$ is the product of the directivity of the secondary source and the reference sensor

and

Δ is the propagation delay between the reference sensor and the secondary source, in samples

One of the most striking things about this expression for the optimal controller is how it only depends on the local properties of the duct The overall reverberant behaviour due to reflections from the ends of the duct have no influence on the optimal controller. A physical interpretation of this result is given in Nelson and Elliott (1992). If we assume that the frequency response of the reference sensor and secondary source are relatively independent of frequency, so that $G_L(e^{j\omega T}) \approx G_L$ and $G_M(e^{j\omega T}) \approx G_M$ and that their directivity is also independent of frequency, so that $D(e^{j\omega T}) = D$, where $D \leq 1$, then the frequency response of the optimum controller is equal to

$$H_{\text{opt}}(e^{j\omega T}) = \frac{1}{G_L G_M} \frac{e^{-j\omega\Delta T}}{1 - De^{-j2\omega\Delta T}} . \tag{3.7.5}$$

Provided the acoustic delay, Δ, is close to being an integer number of samples, the frequency response described by equation (3.7.5) can be very efficiently implemented by a low-order IIR filter since both $A(z)$ and $B(z)$ in Fig. 3.23 would only have to implement gains and simple delays. The structure of a simple IIR controller is thus very well suited to the problem of controlling plane sound waves in ducts with acoustic feedback, and the remainder of this section is devoted to the derivation of an algorithm that is widely used to adapt the coefficients of such a filter.

3.7.3. Filtered-*u* Algorithm

A simplified form of the complete block diagram of Fig. 3.23 is shown in Fig. 3.24, in which a modified form of the reference signal has been used in comparison with the equivalent block diagram for the FIR controller with feedback cancellation shown in Fig. 3.12(a). This modified reference signal, $s(n)$, includes the effect of the feedback path $G_s(z)$ and is the one actually observed as the output of the reference sensor. The z transform of the observed reference signal is given by

$$S(z) = \left[\frac{1}{1 - G_s(z)\,H(z)}\right] X(z) , \tag{3.7.6}$$

in which $X(z)$ is the z transform of the external reference signal, $G_s(z)$ is the transfer function of the physical feedback path, and $H(z)$ is the transfer function of the IIR

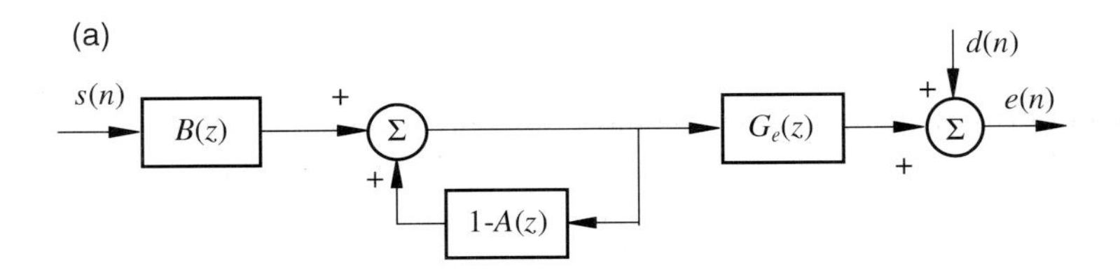

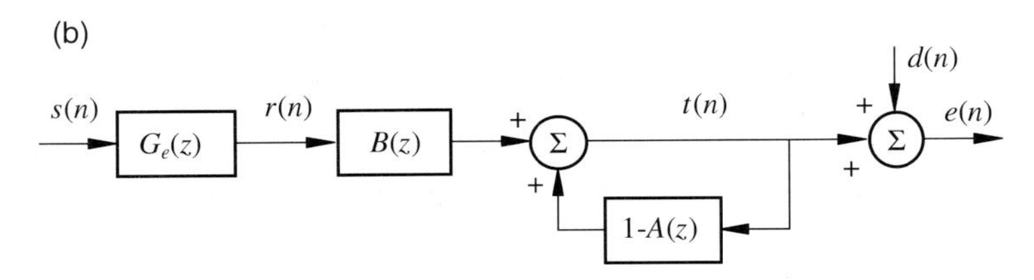

Figure 3.24 Simplified form of the single-channel feedforward system with an IIR controller operating on the observed reference signal (a), and its equivalent form if the controller is almost time-invariant (b).

controller, which is assumed to have coefficients which only change very slowly compared with duration of the impulse response of the system under control, so that at any one instant its transfer function can be written as

$$H(z) = \frac{B(z)}{A(z)} . \tag{3.7.7}$$

As the response of the adaptive controller, $H(z)$, changes as it adapts, the properties of the input signal to the controller, $s(n)$, also change, because of the feedback path $G_s(z)$. The input signal to the adaptive controller is thus *non-stationary* during the adaptation, which complicates the analysis of the adaptation process. Note that the observed reference signal is particularly affected if $G_s(\mathrm{e}^{jwT})H(\mathrm{e}^{jwT})$ becomes close to unity at any frequency, in which case the feedback loop in Fig. 3.23 is close to instability. This is one reason why it is a good idea to reduce the magnitude of the physical feedback path $G_s(z)$ as much as possible, by using a directional reference sensor or secondary source for example.

Although it is the transfer function of the plant response, $G_e(z)$, from secondary actuator to error sensor, which appears in the signal flow diagram of Fig. 3.24(a), the feedback path $G_s(z)$ in Fig. 3.23 cannot be ignored when adaptation of the coefficients of $A(z)$ and $B(z)$ is being considered, as made clear by Flockton (1991) and Crawford and Stewart (1997). Crawford and Stewart (1997) derive expressions for the gradient of the mean-square error with respect to these coefficients in an analogous way to that used in Section 2.9 to derive equations (2.9.10,11) for the IIR filter adapted using the output error. In the case of an IIR control filter, the gradient descent algorithm requires that the input and output signals of the recursive controller are filtered by the modified plant response

$$G'(z) = \frac{G_e(z)}{A(z) - B(z)G_s(z)} , \tag{3.7.8}$$

before being multiplied by the error signal in the calculation of the instantaneous gradient estimates. Crawford and Stewart (1997) also show that if the expression for the gradient is further simplified by assuming that the input and output signals of the recursive controller are filtered by $G_e(z)$ rather than $G'(z)$, this results in the widely used algorithm for the adaptation of recursive controllers called the filtered-u algorithm. This algorithm can also be motivated by assuming that the coefficients of the controller in Fig. 3.24(a) are only changing slowly compared with the duration of the impulse response of the system under control, so that the filtering operations on the observed reference signal can be transposed to give Fig. 3.24(b), and ignoring the implicit feedback path via $G_s(z)$ in Fig. 3.23.

Figure 3.24(b) can then be used to derive a modified form of the RLMS algorithm, as discussed in Chapter 2, which can be used to adapt the coefficients of the controller. It should again be emphasised, however, that, because of the feedback path, the observed reference signal, equation (3.7.6), depends on the response of the controller in this case and will change with time as the controller is adapted. The assumption that the controller is slowly varying is thus even more stringent than was the case with the adaptive IIR electrical filter in Section 2.5, or the equivalent adaptive FIR controller in Section 3.4.

Comparing Fig. 3.24(b) with Fig. 2.18, we can see that the modifications necessary to the RLMS algorithm, equations (2.9.12) and (2.9.13), to account for the presence of the modified reference signal, are that the filtered output signal should now be used to update the coefficients of $A(z)$ and the filtered reference signal should now be used to update the coefficients of $B(z)$. This modified form of the RLMS algorithm can thus be written as

$$a_j(n+1) = \gamma_1 \, a_j(n) - \alpha_1 \, e(n) \, t(n-j) \ , \tag{3.7.9}$$

$$b_i(n+1) = \gamma_2 \, b_i(n) - \alpha_2 \, e(n) \, r(n-i) \ , \tag{3.7.10}$$

where α_1 and α_2 are convergence coefficients, and $t(n)$ and $r(n)$ are the filtered output and filtered reference signals shown in Fig. 3.24(b). The parameters γ_1 and γ_2 in equations (3.7.9) and (3.7.10) are set to be slightly less than unity and allow the coefficients to 'leak' away if the update term is too small. This can have an important stabilising effect on the practical algorithm. This algorithm was originally derived and developed by Eriksson and his co-workers (1987, 1989, 1991a) who called it the *filtered-u algorithm*, since in their derivation both the reference and output signals are gathered into a single vector, u, before filtering by the plant response. The convergence of the filtered-u algorithm has been shown by Wang and Ren (1999a) to depend on a certain transfer function being strictly positive real, which generalises the condition for the convergence of the RLMS algorithm discussed in Section 2.9.

In Section 2.9 we saw that although the RLMS algorithm appeared to be a simplified form of the true gradient descent algorithm for an IIR filter, it had an important self-stabilising property. The reasons why the RLMS algorithm possesses this property are not fully understood (Macchi, 1995) and so it is difficult to say whether it is completely carried over to the case in which a filtered reference signal is used. In practical active sound control applications in ducts, however, the algorithm is found to be self-stabilising under many circumstances (Billet, 1992) and is widely used in practice (Eriksson and Allie, 1989).

A block diagram of the operation of the adaptive recursive controller is shown in Fig. 3.25, in which the feedback path $G_s(z)$ has been retained to emphasise how the

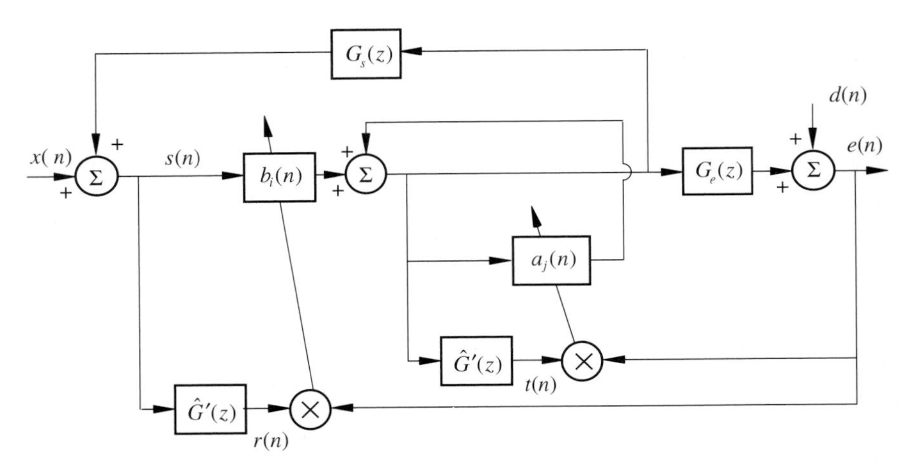

Figure 3.25 Block diagram of the filtered-*u* algorithm for the adaptation of an IIR feedforward controller.

observed reference signal, $s(n)$, is generated, and $\hat{G}'(z)$ is the practical estimate of the modified plant response used to generate the filtered reference and output signals.

The complete control system described by Eriksson et al. (1987, 1989, 1991a) for active control of sound in a duct is shown in Fig. 3.26 and includes the facility for on-line identification of the plant response. This identification is performed on the closed-loop system and so a modified plant response is measured if feedback is present and the controller has non-zero coefficients. If the identification noise is injected into the output of the recursive controller (Eriksson and Allie, 1989), as shown in Fig. 3.26, then the estimated plant response is given by equation (3.7.8), as required to implement the gradient algorithm described by Crawford and Stewart (1997). Eriksson (1991a) also notes, however, that successful control systems can be implemented using plant estimates identified by injecting identification noise at other places in Fig. 3.26, which illustrates that in practice the algorithm given by equations (3.7.9,10) can be quite robust, particularly if appropriate leakage terms are used.

3.8. PRACTICAL APPLICATIONS

In this section, we describe two practical applications of single-channel feedforward control systems: the active control of plane sound waves in ducts and the active control of structural bending waves on a beam. The two applications have different physical problems and use different types of adaptive control algorithms. In the vast majority of feedforward control applications, it is necessary to make the controller adaptive so that good performance can be maintained for changes in the disturbance or small changes in the plant response. A 20 dB attenuation in the error signal at a single frequency requires a controller that is accurate to within ± 0.6 dB in amplitude *and* ± 4° in phase at this frequency. The adaptation of a feedforward controller means that the control system is no longer 'open-loop' as it is often described, rather dismissively, in the control literature.

3.8.1. Control of Plane Sound Waves in Ducts

The active control of plane sound waves in a duct is discussed first. The main elements in such a system have already been shown in Fig. 3.8 and a more detailed diagram of a practical implementation of such a control system is shown in Fig. 3.26. A reference sensor, usually a microphone, is used to measure the waveform of the incident acoustic wave and, after sampling, the resulting signal, $s(n)$, is fed through an electronic controller whose output, $u(n)$, is used to drive the secondary source, which is generally a loudspeaker in this application. Only plane waves are assumed to be propagating in the duct, so that adjusting the controller to cancel the acoustic pressure at the downstream error sensor, $e(n)$, can only be achieved if the acoustic wave travelling downstream from the secondary source is set to zero, as discussed in Chapter 1, thus suppressing acoustic transmission beyond this point.

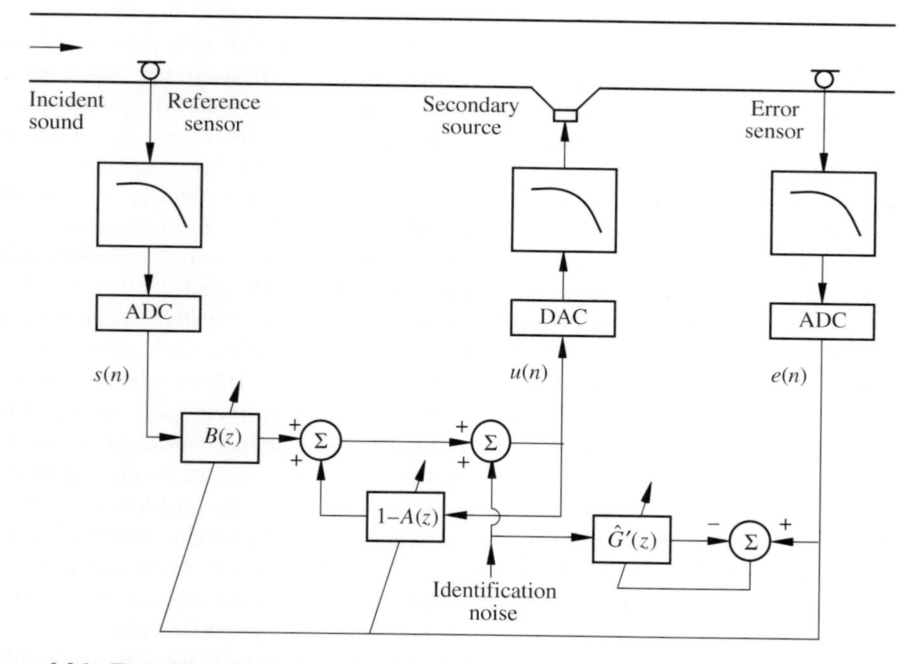

Figure 3.26 The adaptive recursive controller used by Eriksson et al. (1989), for the active control of sound in a duct, including identification noise and on-line identification of the error path.

The physical effect of active control is thus to generate a pressure-release boundary condition at the secondary source position which *reflects* the incident sound wave back upstream towards the primary source. Interference between the original sound wave and this reflected wave generates an acoustic standing wave downstream of the secondary source, as described in Section 1.2, and the pressure at the reference sensor is almost zero at frequencies for which the distance from the secondary source to the reference sensor is an integer number of half wavelengths. The electrical controller thus needs to have a very high gain at these frequencies, as can be deduced from the equation for the frequency response of the ideal controller in this case (Elliott and Nelson, 1984), which can be written in continuous-time form as

$$H_{\text{opt}}(j\omega) = \frac{-1}{G_{\text{L}}(j\omega)G_{\text{M}}(j\omega)} \frac{\mathrm{e}^{-jkl_1}}{1-\mathrm{e}^{-j2kl_1}}, \tag{3.8.1}$$

where it is assumed that neither the reference sensor nor the secondary source is directional, $G_L(j\omega)$ and $G_M(j\omega)$ are the continuous-time frequency responses of the secondary source and reference sensor, $k = \omega/c_0$ is the wavenumber, as described in Chapter 1, with c_0 being the speed of sound, and l_1 is the distance from the reference sensor to the secondary source. This continuous-time form of the frequency response for the ideal controller is equivalent to the discrete-time form of equation (3.7.5) when the general directivity function is included and the delay is equal to $\Delta = f_s\, l_1/c_0$ samples, where f_s is the sampling rate. It is important to note that for the propagation of acoustic waves, terms of the form e^{-jkl_1}, correspond to pure delays, since the speed of sound is almost independent of frequency.

Only plane acoustic waves can propagate in a duct if the excitation frequency is below the first *cut-on frequency* of the duct, which corresponds to the frequency at which half an acoustic wavelength is equal to the larger of the width or height of a rectangular duct, as discussed in Section 1.2. Above this cut-on frequency, higher-order acoustic modes can propagate, which do not have a uniform pressure distribution across the duct cross section. A single reference and error microphone will not be able to accurately measure the amplitude of the acoustic plane wave under these conditions and so the first cut-on frequency sets a *fundamental upper frequency limit* above which a single-channel control system cannot be expected to suppress acoustic propagation in the duct.

The *fundamental low-frequency limit* for such an active control system is generally set by turbulent pressure fluctuations. These are generated by the airflow in the duct, which is inevitably present in an air-conditioning system for example. These turbulent pressure fluctuations decay naturally along the length of the duct and so their contributions at the reference and error sensors are generally uncorrelated. Turbulent pressure fluctuations thus act as a source of measurement noise within the active control system and, since they become increasingly large at lower frequencies, they set a *fundamental lower frequency limit* below which the coherence between the outputs of the reference and error sensors becomes small and thus the control performance becomes poor, in accordance with equation (3.3.13). Although microphones can be designed which are less sensitive to these turbulent pressure fluctuations than normal pressure microphones, as described by Nelson and Elliott (1992), the turbulent pressure fluctuations inevitably increase with the speed of the air flow and with decreasing frequency and so this low-frequency limit cannot be entirely eliminated.

The amplitude and phase response of the feedforward controller must be accurately specified for good attenuation, and so an adaptive digital controller is almost always used. Also, because of the similarity between the form of the optimum controller response in a duct, equation (3.7.4) and that of an IIR filter, adaptive IIR filters make particularly efficient controllers in this application. Figure 3.26 shows the block diagram of a complete control system using an adaptive IIR controller, as described by Eriksson and Allie (1989). As well as an adaptive IIR controller, this system also incorporates on-line system identification using an identification noise signal injected into the secondary source. A separate adaptive filter, $\hat{G}'(z)$, is used to continuously estimate the response of the effective plant, from secondary source input to error sensor output. The signals from the reference and error sensors must be passed through analogue low-pass anti-aliasing filters before being converted into digital form using analogue-to-digital converters (ADCs). Also the digital output from the controller must be smoothed, after conversion into analogue form with a digital-to-analogue converter (DAC), using a similar low-pass analogue filter. The detailed specification of the anti-aliasing and reconstruction filters is discussed in Chapter 10. Although it would appear that the acoustic propagation delay in the equation for the ideal controller, (3.7.4), would always ensure that the optimal digital controller was causal, the delays in the analogue filters and the data converters must also be taken into account if good attenuation is to be achieved using a practical digital system. A formula for the approximate delay through the analogue filters and data converters is derived in Chapter 10 as

$$\tau_A = \left(1.5 + \frac{3n}{8}\right) T , \tag{3.8.2}$$

where T is the sampling time of the digital system and n is the total number of poles in the anti-aliasing filter for the reference sensor *and* in the reconstruction filter for the secondary source. Equation (3.8.2) was calculated assuming a one-sample processing delay, a half sample delay because of the sample and hold circuit in the DAC, that the cut-off frequency of the converters was one-third of the sample rate, and that each pole of the analogue filters contributed a phase shift of 45° at the cut-off frequency.

For a square duct of width 0.5 m, the first acoustic cut-on frequency occurs at about 340 Hz if the duct carries air at normal temperature and pressure, so that the speed of sound is about 340 m s^{-1}. The sample rate of the digital controller might thus be set to 1 kHz, so that T = 1 ms, and perhaps the anti-aliasing and reconstruction filters would each have 6 poles, making n = 12 in equation (3.8.2). The total delay through the analogue parts of the controller is thus about 6 ms. The minimum distance between the reference sensor and the secondary source to ensure that the optimum controller for broadband disturbances is causal ($l = \tau_A c_o$), must thus be at least 2 m, i.e. about 4 times the width of the duct.

The performance of the active control system shown in Fig. 3.26 in attenuating sound propagating in an air conditioning duct with an air flow of about 14 m s^{-1} is shown in Fig. 3.27, as reported by Eriksson and Allie (1989). Figure 3.27 shows the power spectral density of the signal from the error microphone before and after active control. This system had an IIR controller operating at a sample rate of 875 Hz, with 64 recursive coefficients in $A(z)$ and 32 feedforward coefficients in $B(z)$. It was adapted with the filtered-u LMS algorithm described in Section 3.7, using an FIR model of the plant response having 64 coefficients. The control system was placed approximately 12 m from a centrifugal fan in a lined air conditioning duct having cross-sectional dimensions of 0.86 m × 1.12 m. It is clear from Fig. 3.27 that little attenuation has been achieved below 40 Hz, and this is due to the high level of turbulent pressure fluctuation at these frequencies, which causes a significant drop in the coherence between the reference sensor and error sensor. Also, very little attenuation is achieved above about 140 Hz, since the control system was not designed to operate above the first acoustic cut-on frequency of the duct, which was about 150 Hz in this case. Within this bandwidth, however, the random noise propagating down the duct from the fan has been attenuated by up to about 15 dB, which is a typical result for the performance of such a system in an industrial environment.

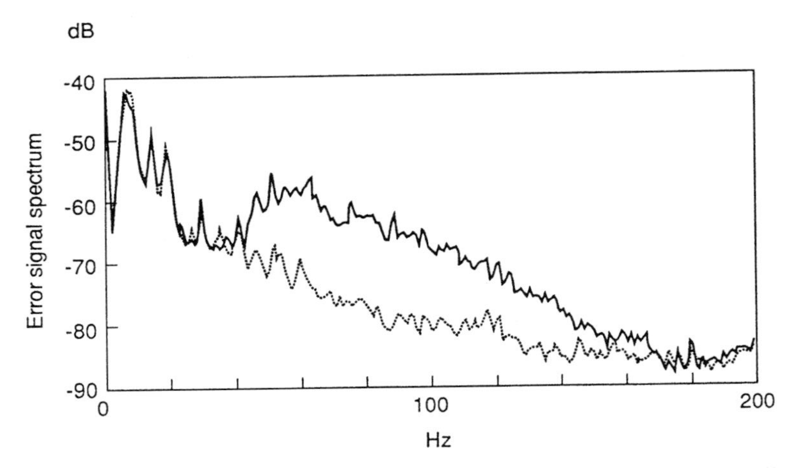

Figure 3.27 The spectrum of the error signal before (——) and after (……) the application of a recursive adaptive controller in an air conditioning duct with an air flow of about 14 m s^{-1} (from Eriksson and Allie, 1989).

3.8.2. Control of Flexural Waves on Beams

The second example of the practical application of a single-channel feedforward active control system used in this section is the suppression of flexural, or bending, waves propagating along a thin beam. Disturbances on a structure often propagate from one point to another as flexural waves, and many structures have beam elements, within which these flexural waves could be controlled. One potential application for such a system would be in isolating a sensitive telescope from the vibrations of a spacecraft, by actively controlling vibration transmission along the telescope boom. Another application could be the control of sound radiated by the hull of a ship whose vibration is excited by a machine mounted to the hull by a series of beams.

Flexural waves on thin beams cause a combination of out-of-plane motion, at right angles to the long axis of the beam, and bending in the beam, and a number of options are available for sensing and actuating such waves, as described for example by Fuller et al. (1996). The system described here uses accelerometers, which are sensitive to out-of-plane motion, to detect the waveform of the incident flexural wave, and to measure the residual wave amplitude at the error sensor, as shown in Fig. 3.28. The secondary actuator in this case is an electromagnetic shaker which applies an out-of-plane force to the beam. This is particularly efficient at low frequencies but does require a stable platform off which to react. At higher frequencies, piezoceramic actuators bonded to the beam become very efficient at generating flexural waves and have the advantage that, because they operate by bending the beam, they require no external reaction point and can be integrated within the structure of the beam.

The frequency limitations set by the physical constraints of this problem are rather different from those of controlling plane acoustic waves in a duct. At low frequencies, the performance can be limited by the fact that as well as detecting the propagating component of the flexural wave, the accelerometer used as the error sensor will also pick up a contribution from the *evanescent* component of the flexural motion generated by the

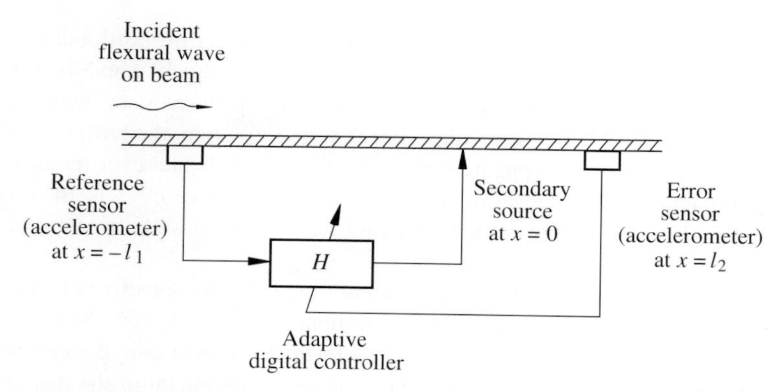

Figure 3.28 An adaptive system for the active control of broadband flexural disturbances on a beam.

secondary source. In contrast to acoustic waves in a duct, which are governed by a second-order wave equation, flexural waves in a thin beam are governed by a fourth-order wave equation, so that a secondary force, F_s, generates two components of flexural motion in both directions along the beam, as shown in Fig. 3.29. The total complex out-of-plane acceleration due to the flexural motion downstream of a secondary force, F_s, operating a single frequency and located at the origin of the coordinate system, can be written as (Fuller et al., 1996),

$$\ddot{w}(x) = \frac{j\omega^2}{4Dk_f^3}\left(e^{-jk_f x} - j e^{-k_f x}\right) F_s , \tag{3.8.3}$$

where D is the bending stiffness of the beam ($D = Ebh^3/12$ where E is the Young's modulus for the beam material, b its width and h its thickness) and k_f is the flexural wavenumber given by

$$k_f = \left(\frac{\omega^2 m}{D}\right)^{1/4} , \tag{3.8.4}$$

where m is the mass per unit length of the uniform beam ($m = \rho bh$ where ρ is the density of the beam material). Note that in contrast to acoustic waves, whose wavenumber is

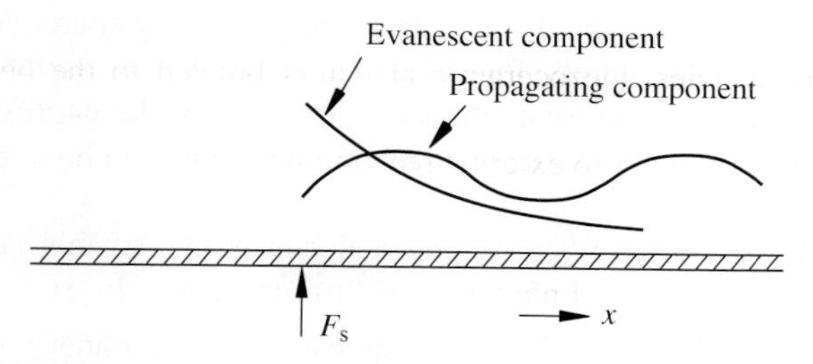

Figure 3.29 Beam excited by a secondary force showing the distribution of out-of-plane acceleration along the beam for evanescent and propagating components of a flexural wave.

directly proportional to frequency, the flexural wavenumber is proportional to $\sqrt{\omega}$. The speed of propagation for flexural waves thus increases with frequency and the propagation is said to be *dispersive*, whereas the speed of propagation of acoustic waves is largely independent of frequency. The out-of-plane acceleration given by the term proportional to $e^{-jk_f x}$ in equation (3.8.3) will propagate with constant amplitude and ever-increasing phase shift along the length of the beam. The term proportional to $e^{-k_f x}$ is the evanescent component, which will decay exponentially and has no phase shift from point to point along the beam.

Returning to the active control system shown in Fig. 3.28, the objective of the controller is to drive the total acceleration at the error sensor position, $x = l_2$, to zero. Since, in general, the acceleration at this position has contributions from the propagating flexural waves due to the primary and secondary sources, *and* the near-field component of the response of the secondary actuator, driving this signal to zero will not necessarily suppress the propagating part of the flexural wave. The attenuation of the *propagating* component can be analysed under these conditions by expressing the net acceleration at the error sensor, assuming it is remote from any discontinuity, as

$$\ddot{w}(l_2) = A e^{-jk_f l_2} - B\left(e^{-jk_f l_2} - j e^{-k_f l_2}\right), \tag{3.8.5}$$

where A is the incident propagating wave amplitude as it passes the secondary actuator, and B the contribution from the secondary source. If the secondary source is adjusted to drive $\ddot{w}(l_2)$ to zero, then

$$B = \frac{-A e^{-jk_f l_2}}{e^{-jk_f l_2} - j e^{-k_f l_2}}. \tag{3.8.6}$$

The residual *propagating* wave amplitude will be equal to $A + B$ and so the ratio of this to the original incident wave amplitude can be deduced, using equation (3.8.6), to be (Elliott and Billet, 1993)

$$\frac{A+B}{A} = \frac{-j e^{-k_f l_2}}{\left(e^{-jk_f l_2} - j e^{-k_f l_2}\right)}. \tag{3.8.7}$$

The attenuation level achieved by driving the acceleration to zero is plotted as a function of frequency in Fig. 3.30, for the case of a 6-mm-thick steel beam with $l_2 = 0.7$ m. It can be seen that in this case, provided attenuations of no more than 20 dB are anticipated, the effect of the near field of the secondary actuator at the error sensor is negligible above about 12 Hz. At this frequency the distance from the secondary source to the error sensor is about 3/8 of a flexural wavelength. Provided the control system operates only above this frequency, we can make the approximation $e^{-k_f l_2} \ll 1$, and the expression for the ideal controller (Elliott and Billet, 1993) then becomes

$$H_{\text{opt}}(j\omega) = \frac{4Dk_f^{\,3}}{j\omega^2} \frac{-e^{-jk_f l}}{1 - e^{-jk_f l_1}\left(e^{-jk_f l_1} - j e^{-k_f l_1}\right)}. \tag{3.8.8}$$

Whereas the near-field components of the response of the secondary actuator provide a fundamental low-frequency limit on the performance of such a control system, the dispersive nature of flexural wave propagation generally leads to a fundamental high-

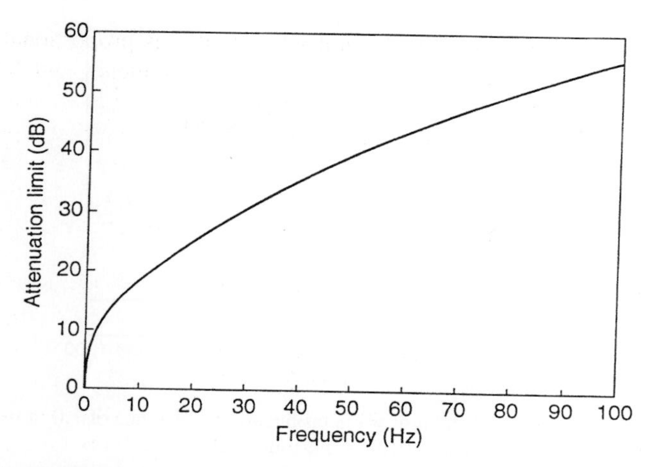

Figure 3.30 Maximum attenuation of a propagating flexural wave in a 6-mm steel beam using a control system with a single error sensor a distance of 0.7 m from the secondary source. The reductions are limited by the near-field of the point secondary force.

frequency limit. This can be demonstrated by calculating the group delay from reference sensor to secondary source, a distance equal to l_1 in Fig. 3.28, which is equal to

$$\tau_g = l_1 \frac{\partial k_f}{\partial \omega} = \frac{l_1}{2\sqrt{\omega}} \left(\frac{m}{D} \right)^{1/4}, \tag{3.8.9}$$

which is plotted as a function of frequency in Fig. 3.31 for the case of a 6-mm-thick steel beam with $l_1 = 1$ m. The electrical controller has again to model a rather complicated frequency response in this application, and the most accurate and flexible way of implementing such a controller is to use digital filters. Such digital controllers have an inherent delay associated with their response, however, due to the processing time of the digital device and the phase lag in the analogue anti-aliasing and reconstruction filters which have to be used, as discussed above. It is clear from the fall in the group delay with frequency seen in Fig. 3.31, that at some upper limiting frequency the group delay along the beam will become less than the inherent delay in the controller, and active suppression of flexural waves with random waveforms will not be possible. In the experiments reported by Elliott and Billet (1993), for example, the delay in the processor was about 2.4 ms and, from Fig. 3.31, broadband control was limited to frequencies below about 800 Hz for the 6-mm steel beam used in the experiments. It is interesting to note that if broadband feedforward control of *longitudinal* waves on a steel beam were being contemplated using a controller with an inherent delay of 2.4 ms, a distance of more than 12 m would be required from reference sensor to secondary actuator. This is because, although longitudinal waves are not dispersive, they have a high wave speed, of about 5000 m s^{-1} in a steel beam. As the beam becomes thinner, the upper frequency limit for the control of flexural waves (due to wave dispersion) is increased and the lower frequency limit (due to the near field of the secondary actuator) is decreased, so that for a given controller geometry, the bandwidth of control is very much greater for a thin beam than it is for a thicker one.

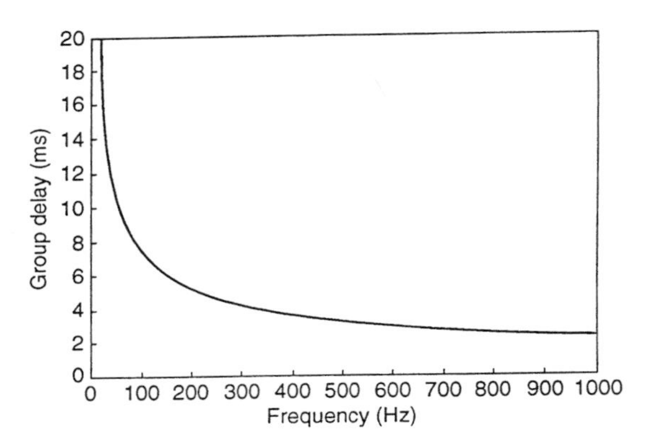

Figure 3.31 The group delay of a flexural wave propagating a distance of 1.0 m from the detection sensor to the secondary source for a 6-mm steel beam.

Figure 3.32 shows the practical results obtained from a feedforward system for the active control of flexural waves on a 6-mm-thick steel beam using an adaptive controller (Elliott and Billet, 1993). In this study, the beam was terminated in sand boxes, which provided anechoic terminations for flexural waves above a few hundred hertz. Figure 3.32 shows the measured power spectral density of the out-of-plane acceleration at the error sensor before and after active control, and it can be seen that reductions in level of up to 30 dB have been achieved from about 100 Hz to 800 Hz. The reductions in the out-of-plane acceleration were also measured at a monitoring sensor position, farther along the beam than the error sensor, and the attenuations observed were similar to those at the error sensor. At frequencies below 100 Hz, external sources of vibration contribute significantly to the output of both reference and error sensors, with a resultant loss in the coherence between their outputs. This limits the performance available from any control system, as was described in Section 3.3. In this experiment the low-frequency limit due to the evanescent

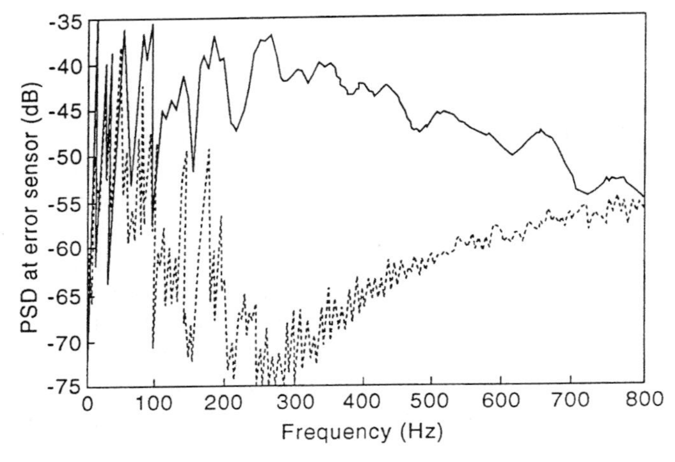

Figure 3.32 The power spectral density at the error sensor on the beam illustrated in Fig. 3.28 before control (——) and after the adaptive controller has converged (- - - - - -).

field, shown in Fig. 3.30, was not approached because of the coherence problems associated with external noise sources. Above about 800 Hz, the delay associated with the propagation of flexural waves from reference sensor to secondary actuator becomes less than the electrical delay in the controller, resulting in the performance limitations due to causality described above.

In this application, the feedforward controller was again digital but used the feedback cancellation arrangement shown in Fig. 3.11. The feedback cancellation filter, $\hat{G}_s(z)$ and the filter used to model the secondary path, $G_e(z)$, were both FIR devices with 256 coefficients, which were adapted to identify these responses prior to control. The FIR control filter was adapted using the filtered-reference LMS algorithm. Although the form of the equation for the ideal controller in this case, equation (3.8.8), is similar to that for acoustic plane waves in a duct, equation (3.8.1), the wavenumber in this case, k_f, is not proportional to frequency, since the propagation is dispersive and in this case terms of the form $e^{-jk_f l_1}$ do *not* correspond to pure delays. Therefore the close correspondence observed in Section 3.7 between the form for the ideal controller and the IIR filter for plane acoustic waves propagating in a duct is no longer true for a flexural wave controller. A discussion of the relative efficiencies of an adaptive IIR controller and an adaptive FIR controller with feedback cancellation in implementing a flexural wave controller are presented by Elliott and Billet (1993). It was found that a considerable additional computational load was involved in implementing on-line identification, which was found necessary to stabilise the adaptive IIR controller in this case, for the reasons discussed in Section 3.7. Using the same signal processing device as was used for the adaptive FIR controller described above, which had 256 coefficients, an IIR controller having only 80 coefficients in $B(z)$ and 100 coefficients in $A(z)$ could be implemented in real time with on-line system identification. The measured performance of this IIR controller was similar to but not quite as good as that of the FIR controller with feedback cancellation in this application.

In both of these applications, successful control has only been possible with a combination of physical understanding, which allows the fundamental limits of control to be determined, and insight into the form of the optimal controller, which guides the structure of the digital control system. Although initial predictions of performance could be made using measurements from the transducers and off-line processing, adaptive controllers were needed to obtain good performance for real-time applications in order to compensate for any slow changes that occurred in the plant response or disturbance spectrum.

Control systems with a single secondary actuator and error sensor are, however, limited to controlling only one form of disturbance propagation in a physical system, whether it be plane sound waves in one direction in a duct, or flexural waves transmitting vibration in one direction in a beam. In many active control problems, there are multiple ways in which the disturbance can propagate from the primary noise source to the point at which control must be achieved, and control may also need to be implemented at a number of positions in space. In order to achieve control under these circumstances, a multichannel control system must be used, with multiple secondary sources and multiple error sensors. The analysis of such multichannel control systems is somewhat more complicated than that of the single-channel systems described in this chapter, and so the treatment is spread over two chapters, with Chapter 4 discussing the control of tonal disturbances and Chapter 5 the control of stochastic disturbances.

4

Multichannel Control of Tonal Disturbances

4.1. INTRODUCTION

In this chapter we will begin the discussion of multichannel feedforward control systems by looking at the control of tonal disturbances. Multichannel systems involve the use of multiple actuators and sensors, and are necessary to control sound or vibration over extended regions of space, as described in Chapter 1. Such systems are often used to control periodic disturbances, which can be broken down into a superposition of sinusoidal components. Control algorithms operating on each of these sinusoids will not interact with one another if the system under control is linear. Thus a control system for periodic disturbances can be implemented with a number of tonal controllers, which can each operate independently, and it is the design of these tonal controllers that we will consider in this chapter. The more complicated case of multichannel systems for the control of stochastic disturbances will be considered in Chapter 5.

A disturbance encountered in practice will never be entirely periodic since its constituent sinusoids will generally be modulated in amplitude and phase to a greater or lesser extent, and will thus have a finite bandwidth. This form of almost-periodic disturbance can thus be described as being *narrowband*. Examples of such signals occur in many applications where the active control of sound and vibration is applied, particularly where the primary source creating the disturbance is a rotating machine or a reciprocating engine. Also, a tachometer signal is generally available from such engines, which can be used to synthesise a suitable reference signal for feedforward control. In this case the inertia of the crankshaft prevents rapid changes in excitation frequency, although the speed of such changes can be somewhat different between a large marine diesel engine and the tuned petrol engine of a sports car.

The majority of this chapter will thus focus on the control of a tonal signal whose amplitude and frequency are assumed to change only slowly with time. In Section 3.2 we saw that the convergence of single-channel control systems for such disturbances could be understood relatively easily. If, for example, the in-phase and quadrature components of the control signal are adapted using the method of steepest descent, then the in-phase and quadrature components of the error signal will decay exponentially and with the same time constant.

The convergence behaviour of a single channel LMS controller operating with a broadband reference signal is determined by a number of modes with different time constants, as explained in Section 3.4. These time constants depend on the eigenvalues of the appropriate autocorrelation matrix and the spread of these eigenvalues is related to the *spectral* properties of the filtered reference signal. In multichannel systems, however, the reference signal can be a pure tone, and yet the matrix that determines the convergence properties of many algorithms can still have a large spread of eigenvalues, which is now due to the *spatial* distribution of the actuators and sensors. As well as their inherent practical interest, it is useful to understand the control of tones in isolation before considering the more complicated problem of multichannel control systems for stochastic disturbances, in which both the spectral properties of the reference signals and the properties of the plant response due to the spatial distribution of the transducers are important.

4.1.1. Chapter Outline

This chapter begins with a discussion of the optimal performance of multichannel control systems for tonal disturbances, towards which the performance of an adaptive algorithm should ideally converge. In Section 4.3 one of the simplest, and most widely used, adaptive algorithms, which uses the method of steepest descent, is analysed in detail to illustrate some of the typical properties and potential problems with such multichannel algorithms. It is important to understand how the convergence properties and performance of any control algorithm are affected by uncertainties in the response of the plant under control, and this becomes a more difficult problem to analyse and understand for multichannel systems because the form of the uncertainties can be difficult to describe, as discussed in Section 4.4. A change in one part of the system, due to a person moving their position within a room in which an active sound control system is operating for example, will change the response from each actuator to each sensor in a very structured manner.

A more general form of adaptive algorithm is discussed in Section 4.5, which reduces to the steepest-descent algorithm in one extreme, and to an algorithm based on Newton's method in another. The two issues of convergence behaviour and final steady-state error are addressed within the context of this algorithm. The analysis of adaptive feedforward systems using the methods usually employed in feedback control systems is considered next, together with a brief discussion of the minimisation of a more general norm of the vector of error signals. Finally, the practical application of a multichannel control system is described, both for the attenuation of the propeller noise in an aircraft and the control of rotor noise inside a helicopter.

4.2. OPTIMAL CONTROL OF TONAL DISTURBANCES

The optimal solution to a control problem will clearly depend on the cost function, or performance index, which one is seeking to minimise. We will begin by considering a cost function proportional to the sum of the mean-square errors, but later the advantages will be discussed of incorporating another term, proportional to mean-square control effort, into the cost function. The disturbance is assumed to be tonal, and of normalised frequency $\omega_0 T$, so that provided the plant is linear, control can be achieved by driving the secondary actuators with tonal signals of the same frequency, whose amplitude and phase must be adjusted to minimise the cost function. The amplitude and phase of the L individual disturbance signals and M individual secondary control signals with respect to some arbitrary reference can be described by complex numbers, which we will group together into the vectors $\mathbf{d}(\mathrm{e}^{j\omega_0 T})$ and $\mathbf{u}(\mathrm{e}^{j\omega_0 T})$. If the $L \times M$ matrix of complex plant responses at the frequency $\omega_0 T$ is denoted $\mathbf{G}(\mathrm{e}^{j\omega_0 T})$, then the vector of complex residual error signals in the steady state can be written as

$$\mathbf{e}(\mathrm{e}^{j\omega_0 T}) = \mathbf{d}(\mathrm{e}^{j\omega_0 T}) + \mathbf{G}(\mathrm{e}^{j\omega_0 T})\mathbf{u}(\mathrm{e}^{j\omega_0 T}) \ . \tag{4.2.1}$$

The block diagram for such a system is shown in Fig. 4.1. For the remainder of this chapter the explicit dependence of the variables on $\omega_0 T$ will be dropped, so that equation (4.2.1) becomes

$$\mathbf{e} = \mathbf{d} + \mathbf{G}\,\mathbf{u} \ . \tag{4.2.2}$$

The cost function we will initially consider is the sum of modulus squared error signals,

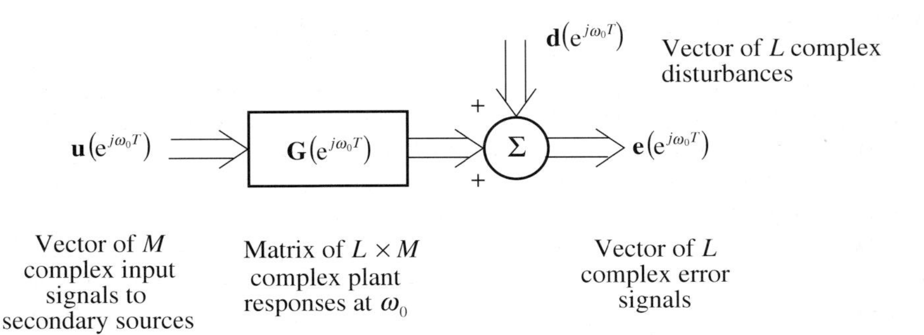

Figure 4.1 Block diagram of a multichannel control system operating at a single frequency of ω_0.

$$J = \sum_{l=1}^{L} |e_l|^2 = \mathbf{e}^{\mathrm{H}}\mathbf{e} , \tag{4.2.3}$$

where H denotes the Hermitian, i.e. complex conjugate, transpose. Substituting equation (4.2.2) into (4.2.3) we see that the cost function has the Hermitian quadratic form:

$$J = \mathbf{u}^{\mathrm{H}}\mathbf{A}\mathbf{u} + \mathbf{u}^{\mathrm{H}}\mathbf{b} + \mathbf{b}^{\mathrm{H}}\mathbf{u} + c , \tag{4.2.4}$$

where $\mathbf{A}$ is the Hessian matrix, and in this case

$$\mathbf{A} = \mathbf{G}^{\mathrm{H}}\mathbf{G} , \qquad \mathbf{b} = \mathbf{G}^{\mathrm{H}}\mathbf{d} , \qquad c = \mathbf{d}^{\mathrm{H}}\mathbf{d} . \tag{4.2.5 a,b,c}$$

The optimal solution we are seeking will depend upon the relative numbers of actuators (M) and sensors (L).

4.2.1. Overdetermined Systems

If there are more sensors than actuators ($L > M$), then the system is described as being *overdetermined* since there are more equations of the form $e_l = d_l + \Sigma\, G_{lm}u_m$ than there are unknowns, u_m. In general the smallest value of the cost function that can be obtained by adjusting the elements of $\mathbf{u}$ under these conditions is not zero. We will assume that the matrix $\mathbf{A} = \mathbf{G}^{\mathrm{H}}\mathbf{G}$ is positive definite in this case, which will almost always be true in practice, as discussed below. Under these conditions the set of secondary signals that minimises the cost function of equation (4.2.3) can be obtained by setting to zero the derivative of equation (4.2.4) with respect to the real and imaginary parts of each component of $\mathbf{u}$, as described in detail in the Appendix. The resulting optimal control vector is

$$\mathbf{u}_{\mathrm{opt}} = -\left[\mathbf{G}^{\mathrm{H}}\mathbf{G}\right]^{-1}\mathbf{G}^{\mathrm{H}}\mathbf{d} , \tag{4.2.6}$$

and substituting this expression back into the general cost function gives the expression for its minimum value in this case as

$$J_{\min} = \mathbf{d}^{\mathrm{H}}\left[\mathbf{I} - \mathbf{G}\left[\mathbf{G}^{\mathrm{H}}\mathbf{G}\right]^{-1}\mathbf{G}^{\mathrm{H}}\right]\mathbf{d} . \tag{4.2.7}$$

4.2.2. Shape of the Error Surface

The optimal control vector has been obtained, for this ideal case, in a single step, and is thus independent of the shape of the multidimensional *error surface* obtained by plotting the cost function, equation (4.2.3), as a function of the real and imaginary parts of each element of $\mathbf{u}$. The behaviour of an adaptive algorithm, however, will generally depend very much on the geometric form of the error surface. The error surface is guaranteed to be quadratic, but its shape will depend strongly on the properties of the matrix $\mathbf{G}^{\mathrm{H}}\mathbf{G}$. To illustrate this, the cost function of equation (4.2.4) can also be expressed, using equations (4.2.6) and (4.2.7), as

$$J = J_{\min} + \left[\mathbf{u} - \mathbf{u}_{\mathrm{opt}}\right]^{\mathrm{H}} \mathbf{A}\left[\mathbf{u} - \mathbf{u}_{\mathrm{opt}}\right] . \tag{4.2.8}$$

The matrix $\mathbf{A} = \mathbf{G}^{\mathrm{H}}\mathbf{G}$ is assumed to be positive definite and so has an eigenvalue/eigenvector decomposition of the form

$$\mathbf{A} = \mathbf{Q}\,\mathbf{\Lambda}\,\mathbf{Q}^{\mathrm{H}} , \tag{4.2.9}$$

where $\mathbf{Q}$ is the unitary matrix of normalised eigenvectors, so that $\mathbf{Q}^{\mathrm{H}}\mathbf{Q} = \mathbf{I}$, and $\mathbf{\Lambda}$ is the diagonal matrix of eigenvalues. Substituting equation (4.2.9) into (4.2.8) and defining the *principal coordinates* for the error surface as the translated and rotated set of secondary signals

$$\boldsymbol{\nu} = \mathbf{Q}^{\mathrm{H}}\left[\mathbf{u} - \mathbf{u}_{\mathrm{opt}}\right] , \tag{4.2.10}$$

then the cost function may be written as

$$J = J_{\min} + \boldsymbol{\nu}^{\mathrm{H}}\,\mathbf{\Lambda}\,\boldsymbol{\nu} = J_{\min} + \sum_{m=1}^{M} \lambda_m \left|\nu_m\right|^2 , \tag{4.2.11}$$

where ν_m is the m-th element of $\boldsymbol{\nu}$ and λ_m is the m-th diagonal element of $\mathbf{\Lambda}$, i.e. the m-th eigenvalue of $\mathbf{G}^{\mathrm{H}}\mathbf{G}$. Equation (4.2.11) clearly shows that the value of the cost function increases quadratically away from the minimum point and varies independently along each of the principal coordinates. The rate of increase of the cost function along each of the principal coordinates depends on the corresponding eigenvalue λ_m.

Since the matrix $\mathbf{G}^{\mathrm{H}}\mathbf{G}$ is assumed to be positive definite, all its eigenvalues are real and positive, and so the error surface is quadratic with a unique global minimum. The eigenvalues do not, however, have equal magnitudes and in many practical situations the *eigenvalue spread*, which is the ratio of the largest to the smallest eigenvalue, can be very large and the control system is then described as being *ill-conditioned*. The error surface under these conditions has long, thin valleys as illustrated by the contour plot shown in Fig. 4.2, which was computed for an example with two principal components and an eigenvalue spread of 100. This problem is referred to in the statistics literature as collinearity, as described for example by Ruckman and Fuller (1995). It is important to note that the *shape* of the error surface is determined only by the eigenvalues of $\mathbf{G}^{\mathrm{H}}\mathbf{G}$, i.e. only by the response of the plant. The *position* of the error surface with respect to the origin of the coordinate system, however, depends on a combination of the disturbance and the plant response, as can be seen from equation (4.2.6) for the coordinates of the minimum point. The eigenvalue spread of $\mathbf{G}^{\mathrm{H}}\mathbf{G}$, and hence the shape of the error surface, will depend on the

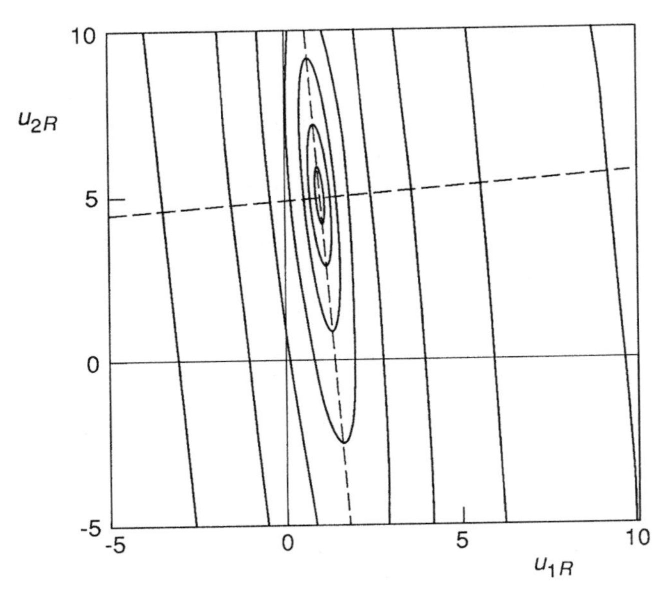

Figure 4.2 Contours of equal cost function, in 5 dB steps, against the real parts of two control signals, u_{1R} and u_{2R}, for an overdetermined control system, with an eigenvalue spread of 100. The dashed lines indicate the principal axes of the error surface.

positioning of the actuators and sensors and their environment. Generally speaking, the eigenvalue spread will get larger the more closely grouped the secondary sources or the error sensors become. An extreme case would be when all the actuators or error sensors are positioned at the same point in space, i.e. they are *collocated*, in which case there would be only one non-zero eigenvalue of $\mathbf{G}^{\mathrm{H}}\mathbf{G}$, which would no longer be positive definite. The eigenvalue spread can also become large if there are more secondary sources than significant modes in a lightly damped system, as explained in Nelson and Elliott (1992, section 12.3). As well as presenting problems for adaptive algorithms, ill-conditioned control problems with large eigenvalue spreads can make the inverse of $\mathbf{G}^{\mathrm{H}}\mathbf{G}$ very sensitive to small perturbations in its elements, which can bring into question the practical usefulness of equation (4.2.7) in determining the optimal reductions when the element $\mathbf{G}$ is determined from measurements.

4.2.3. Fully-Determined Systems

Returning to the general problem, we now consider the case in which there are the same number of actuators as sensors ($L = M$), which is said to be *fully-determined*. In this case the error surface still has a unique global minimum if the matrix $\mathbf{G}^{\mathrm{H}}\mathbf{G}$ is positive definite, but now the optimal control vector takes a rather simple form,

$$\mathbf{u}_{\mathrm{opt}} = -\mathbf{G}^{-1}\mathbf{d} \ , \tag{4.2.12}$$

where it has been assumed that $\mathbf{G}$ is non-singular, which is guaranteed if it is positive definite. The minimum value of the cost function in this case is zero, which implies that all the individual error signals are also zero, as can be verified by substituting equation (4.2.12) into (4.2.2). Although the performance of a fully-determined control system appears to be excellent when evaluated at the error sensors, there is no guarantee that the sound or vibration field at other points in a distributed system would not increase. This danger is also present of course with an overdetermined system, unless a large number of uniformly distributed sensors are used. One indicator that can be used to detect when the secondary sources are exactly cancelling the signals from the error sensors, but may be making matters worse elsewhere, is when the magnitude of the control signals in **u** becomes very large. This aspect of practical control system behaviour will be discussed more fully in the later sections.

4.2.4. Underdetermined Systems

The final case to consider is the *underdetermined* one, in which there are fewer error sensors than secondary actuators ($L < M$). In this case the matrix $\mathbf{G}^{\mathrm{H}}\mathbf{G}$ will not be positive definite and will have at least $M - L$ eigenvalues that are equal to zero. The $\mathbf{G}^{\mathrm{H}}\mathbf{G}$ matrix is thus singular in this case and the solution given by equation (4.2.6) cannot be used. In the underdetermined case there is not a unique solution to the minimisation problem but rather an infinite number of control vectors, **u**, that will set the cost function given by equation (4.2.3) to zero. The error surface in this case is still quadratic but the error surface looks like that shown in Fig. 4.3, in which the contours of equal cost function are parallel straight lines. One of these lines corresponds to the combination of control signals which will set the cost function to zero. Although it is rare to implement a tonal control system with fewer sensors than actuators, a brief consideration of such systems is worthwhile since they form the extreme case of an ill-conditioned overdetermined system, in which one or more of the eigenvalues in equation (4.2.11) are zero and the long valleys shown in Fig. 4.2 have

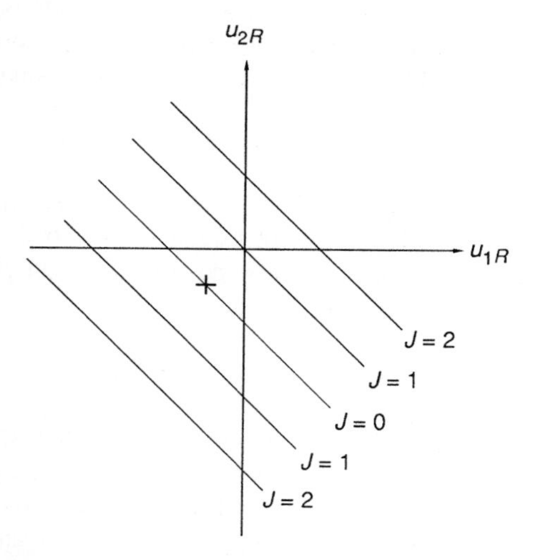

Figure 4.3 Contours of equal cost function for an underdetermined control system. The cross represents the solution which sets the cost function to zero with the least control effort.

stretched out until the contours of equal cost function look parallel over a practical range of control signals.

There are two methods of obtaining a unique solution to an underdetermined problem. The first is an exact mathematical solution, which relies on using the cancellation of the cost function as a constraint and then seeking solutions that minimise the control effort. The *control effort* is defined here to be the sum of the modulus-squared control signals and can be written as

$$P = \sum_{m=1}^{M} |u_m|^2 = \mathbf{u}^{\mathrm{H}}\mathbf{u} , \tag{4.2.13}$$

which can often be related to the power required to drive the actuators in an active control system. The solution of this constrained minimisation problem can be obtained using the method of Lagrange multipliers, as described in general by Golub and Van Loan (1996) and for this application in the appendix of Nelson and Elliott (1992). The optimal set of source strengths is then given by

$$\mathbf{u}_{\mathrm{opt}} = -\mathbf{G}^{\mathrm{H}}\left[\mathbf{G}\mathbf{G}^{\mathrm{H}}\right]^{-1}\mathbf{d} , \tag{4.2.14}$$

where the magnitude of the M real eigenvalues of the matrix $\mathbf{GG}^{\mathrm{H}}$ are assumed to be greater than zero so that this matrix is not singular.

Another method of obtaining an 'engineering' solution to the underdetermined problem is to minimise a cost function which includes both mean-square error and mean-square effort terms, such as

$$J = \mathbf{e}^{\mathrm{H}}\mathbf{e} + \beta\,\mathbf{u}^{\mathrm{H}}\mathbf{u} , \tag{4.2.15}$$

where β is a positive real *effort-weighting parameter*. Note that if adaptive filters were used in the controller, the sum of squared control coefficients, $\mathbf{w}^{\mathrm{T}}\mathbf{w}$, can be made equivalent to the sum of mean-square control signals $\mathbf{u}^{\mathrm{H}}\mathbf{u}$ in this tonal case, and so the distinction drawn between β and ρ in Section 3.4 is not necessary here. Substituting equation (4.2.2) for **e** into equation (4.2.15) results in another Hermitian quadratic form with the same **b** and c parameters as in equation (4.2.5 b and c), but now the Hessian matrix, **A**, takes the form

$$\mathbf{A} = \left[\mathbf{G}^{\mathrm{H}}\mathbf{G} + \beta\mathbf{I}\right] . \tag{4.2.16}$$

The eigenvalues of this matrix are equal to $\lambda'_m = \lambda_m + \beta$, where λ_m are the eigenvalues of $\mathbf{G}^{\mathrm{H}}\mathbf{G}$, which are either positive real or zero. Equation (4.2.16) is thus guaranteed to be positive definite, and so non-singular, for any positive value of β. An optimum control vector can then be calculated which has the form

$$\mathbf{u}_{\mathrm{opt}} = -\left[\mathbf{G}^{\mathrm{H}}\mathbf{G} + \beta\mathbf{I}\right]^{-1}\mathbf{G}^{\mathrm{H}}\mathbf{d} . \tag{4.2.17}$$

It can be shown that equation (4.2.17) tends to (4.2.14) as β tends to zero if the system is underdetermined, and if the system is overdetermined it becomes equal to equation (4.2.6) if $\beta = 0$. For non-zero values of β, however, the minimum mean-square error in an underdetermined system will not be zero using equation (4.2.17), and the solution obtained is thus slightly 'biased' compared to that obtained from equation (4.2.14). The connection between the least-squares problem here and the multiple linear regression technique widely used in the statistical literature was noted by Snyder and Hansen (1990) and further explored by Ruckman and Fuller (1995).

4.2.5. The Pseudo-Inverse

The most general formulation of the exact least-squares solution involves the *pseudo-inverse* of the matrix of plant responses (Golub and Van Loan, 1996), as described in the Appendix, which allows the vector of optimal control signals for all the cases described above to be written as

$$\mathbf{u}_{\text{opt}} = -\mathbf{G}^{\dagger}\mathbf{d} \,. \tag{4.2.18}$$

where $\mathbf{G}^{\dagger}$ is the pseudo-inverse of the matrix $\mathbf{G}$. The analytic form of the pseudo-inverse for each of the cases considered above is shown in Table 4.1. Equation (4.2.18) thus reduces to equation (4.2.6) for the overdetermined case, equation (4.2.12) for the fully-determined case, and equation (4.2.14) for the under-determined case. The pseudo-inverse also provides a solution if the plant matrix is not full rank and will, for example, give the minimum effort solution if $\mathbf{G}^{\mathrm{H}}\mathbf{G}$ is not positive definite in the overdetermined case.

Table 4.1 Form of the pseudo-inverse which is used in the optimal least-squares control problem for various relative numbers of sensors (L) and actuators (M). In all cases the matrix of plant responses at the excitation frequency, $\mathbf{G}$, is assumed to be full rank, so that the matrices that must be inverted in each case are not singular.

Pseudo-inverse	Overdetermined $L > M$	Fully determined $L = M$	Underdetermined $L < M$
$\mathbf{G}^{\dagger}$	$\left[\mathbf{G}^{\mathrm{H}}\mathbf{G}\right]^{-1}\mathbf{G}^{\mathrm{H}}$	$\mathbf{G}^{-1}$	$\mathbf{G}^{\mathrm{H}}\left[\mathbf{G}\mathbf{G}^{\mathrm{H}}\right]^{-1}$

4.2.6. Minimisation of a General Cost Function

Finally in this section we derive the optimal control vector and residual cost function for the general form of cost function given by

$$J = \mathbf{e}^{\mathrm{H}}\,\mathbf{W}_e\,\mathbf{e} + \mathbf{u}^{\mathrm{H}}\,\mathbf{W}_u\,\mathbf{u} \,, \tag{4.2.19}$$

where $\mathbf{W}_e$ and $\mathbf{W}_u$ are weighting matrices for $\mathbf{e}$ and $\mathbf{u}$, which are assumed to be positive definite and Hermitian but are not necessarily diagonal. This form of cost function is widely used in optimal feedback control (Kwakernaak and Sivan, 1972).

The use of the weighting matrix $\mathbf{W}_e$ allows particular aspects of the error signals to be emphasised, such as that corresponding to the sound power radiated from a vibrating system, for example. The weighting matrix $\mathbf{W}_u$ allows certain aspects of the control effort to be emphasised, such as the secondary excitation in the structural modes poorly detected by the error sensors (Elliott and Rex, 1992). Rossetti et al. (1996) have also considered the problem of selecting an effort weighting matrix that does not increase the residual mean-square error and have shown that this is not possible if $\mathbf{G}$ is full rank. When $\mathbf{G}$ is not full rank, however, then Rossetti et al. show that the residual error is unaffected provided $\mathbf{W}_u$

lies entirely in the null space of $\mathbf{G}$. When the system is underdetermined, for example, then an effort weighting matrix of the form

$$\mathbf{W}_u = \beta\left[\mathbf{I} - \mathbf{G}^{\mathrm{H}}\left(\mathbf{G}\mathbf{G}^{\mathrm{H}}\right)^{-1}\mathbf{G}\right], \tag{4.2.20}$$

where β is again a positive real constant, will limit the control effort without increasing the residual mean-square error (Elliott and Rex, 1992).

Substituting equation (4.2.2) for $\mathbf{e}$ into the general cost function given by equation (4.2.19) results in a Hermitian quadratic function of the form of equation (4.2.4) but with

$$\begin{aligned} \mathbf{A} &= \mathbf{G}^{\mathrm{H}}\,\mathbf{W}_e\,\mathbf{G} + \mathbf{W}_u\,, \\ \mathbf{b} &= \mathbf{G}^{\mathrm{H}}\,\mathbf{W}_e\,\mathbf{d}\,, \\ c &= \mathbf{d}^{\mathrm{H}}\,\mathbf{W}_e\,\mathbf{d}\,. \end{aligned} \tag{4.2.21 a,b,c}$$

Assuming that $\mathbf{A}$ is positive definite, the optimal control vector for this case is thus

$$\mathbf{u}_{\mathrm{opt}} = -\left[\mathbf{G}^{\mathrm{H}}\mathbf{W}_e\mathbf{G} + \mathbf{W}_u\right]^{-1}\mathbf{G}^{\mathrm{H}}\mathbf{W}_e\mathbf{d}\,, \tag{4.2.22}$$

which reduces to equation (4.2.17) if $\mathbf{W}_e = \mathbf{I}$ and $\mathbf{W}_u = \beta\mathbf{I}$.

4.3. STEEPEST-DESCENT ALGORITHMS

4.3.1. Complex Gradient Vector

In this section we consider control algorithms that iteratively adjust the control actuator's input signals, $\mathbf{u}$, in order to minimise the sum of squared errors, $J = \mathbf{e}^{\mathrm{H}}\mathbf{e}$, using the method of steepest descent. To do this we need to account for the derivative of the cost function, J, with respect to both the real and the imaginary parts of $\mathbf{u}$, $\mathbf{u}_{\mathrm{R}}$ and $\mathbf{u}_{\mathrm{I}}$, and it is useful to define the *complex gradient vector* as

$$\mathbf{g} = \frac{\partial J}{\partial \mathbf{u}_{\mathrm{R}}} + j\frac{\partial J}{\partial \mathbf{u}_{\mathrm{I}}}\,, \tag{4.3.1}$$

in which $\partial J/\partial \mathbf{u}_{\mathrm{R}}$ and $\partial J/\partial \mathbf{u}_{\mathrm{I}}$ are the vectors of derivatives of J with respect to the elements of $\mathbf{u}_{\mathrm{R}}$ and $\mathbf{u}_{\mathrm{I}}$, which are all entirely real. The steepest descent algorithm for both the real and imaginary parts of the complex vector $\mathbf{u}$ at the $(n+1)$-th iteration can therefore be written as

$$\mathbf{u}(n+1) = \mathbf{u}(n) - \mu\,\mathbf{g}(n)\,, \tag{4.3.2}$$

where μ is the convergence factor and $\mathbf{g}(n)$ is the complex gradient vector at the n-th iteration.

It is shown in the Appendix that for a Hermitian quadratic function, of the form of equation (4.2.4), the complex gradient vector is equal to

$$\mathbf{g} = 2\left[\mathbf{A}\mathbf{u} + \mathbf{b}\right]. \tag{4.3.3}$$

Thus, in the case where the Hessian matrix $\mathbf{A} = \mathbf{G}^{\mathrm{H}}\mathbf{G}$, and the vector $\mathbf{b}$ is $\mathbf{G}^{\mathrm{H}}\mathbf{d}$, then the complex gradient vector at the n-th iteration can be written as

$$\mathbf{g}(n) = 2\left(\mathbf{G}^{\mathrm{H}}\mathbf{G}\,\mathbf{u}(n) + \mathbf{G}^{\mathrm{H}}\mathbf{d}\right) . \tag{4.3.4}$$

The vector of optimal control signals in this overdetermined case, equation (4.2.6), is obtained by setting this vector of complex gradients to zero.

Assuming that the error signals have time to reach their steady-state value at each iteration, then the vector of complex error signals at the n-th iteration can be written from equation (4.2.2) as

$$\mathbf{e}(n) = \mathbf{d} + \mathbf{G}\,\mathbf{u}(n) , \tag{4.3.5}$$

so that

$$\mathbf{g}(n) = 2\,\mathbf{G}^{\mathrm{H}}\,\mathbf{e}(n) , \tag{4.3.6}$$

and the steepest-descent algorithm which minimises the sum of the squared error signals can be written as

$$\mathbf{u}(n+1) = \mathbf{u}(n) - \alpha\,\mathbf{G}^{\mathrm{H}}\,\mathbf{e}(n) , \tag{4.3.7}$$

in which $\alpha = 2\mu$ is now the convergence coefficient.

In a practical implementation of this control algorithm, the true plant response $\mathbf{G}$ would not necessarily be perfectly known and an estimate of this quantity, $\hat{\mathbf{G}}$, must be used in the adaptation algorithm as illustrated in Fig. 4.4. We will postpone a discussion of the effects of this practical complication on the control algorithm until the next section, however, and concentrate here on the stability and convergence properties of equation (4.3.7). In an exactly analogous manner to the analysis of the steepest-descent algorithm for adaptive digital filters given by Widrow and Stearns (1985), the control signal vector can be transformed into its principal components, defined by equation (4.2.10), to derive the stability conditions and convergence properties of the control vector (Elliott et al., 1992). The singular value decomposition of the plant response can then be used to establish the effect of the control algorithm on the error signals. In this section we will use the singular value decomposition at the beginning of the analysis, which allows the response of the system, equation (4.3.5), and the adaptation algorithm, equation (4.3.7), to be written in terms of a set of transformed control signals and error signals. The iterative behaviour of both these signals can then be expressed in a simple and intuitively appealing form.

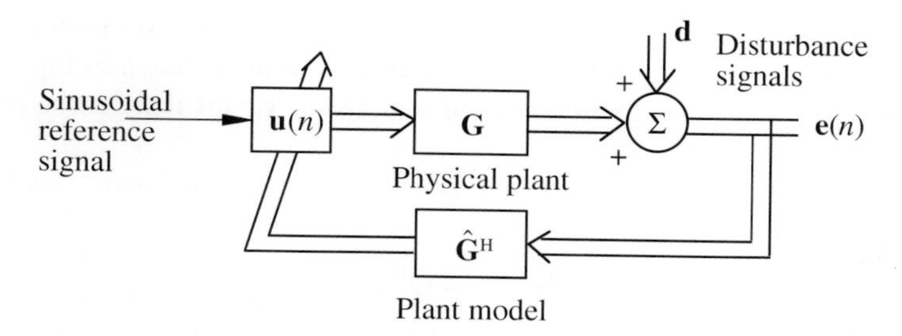

Figure 4.4 Block diagram of the steepest-descent adaptation of the vector $\mathbf{u}(n)$, which defines the sinusoidal input signals to the multichannel plant, whose response is defined at the excitation frequency by the matrix $\mathbf{G}$. The complex error signals are contained in the vector $\mathbf{e}(n)$ which is multiplied by the Hermitian transpose of an estimate of the plant response matrix, $\hat{\mathbf{G}}^{\mathrm{H}}$, before being used to update $\mathbf{u}(n)$.

4.3.2. Transformation into the Principal Coordinates

The *singular value decomposition* (SVD) of the $L \times M$ matrix of complex plant responses is defined to be

$$\mathbf{G} = \mathbf{R}\,\mathbf{\Sigma}\,\mathbf{Q}^{\mathrm{H}}, \tag{4.3.8}$$

where $\mathbf{R}$ is the $L \times L$ unitary matrix of complex eigenvectors of $\mathbf{G}\mathbf{G}^{\mathrm{H}}$, also known as the left singular vectors of $\mathbf{G}$, so that $\mathbf{R}^{\mathrm{H}}\mathbf{R} = \mathbf{R}\mathbf{R}^{\mathrm{H}} = \mathbf{I}$; $\mathbf{Q}$ is the $M \times M$ unitary matrix of complex eigenvectors of $\mathbf{G}^{\mathrm{H}}\mathbf{G}$, also known as the right singular vectors of $\mathbf{G}$, so that $\mathbf{Q}^{\mathrm{H}}\mathbf{Q} = \mathbf{Q}\mathbf{Q}^{\mathrm{H}} = \mathbf{I}$; and the $L \times M$ matrix $\mathbf{\Sigma}$, containing the singular values of $\mathbf{G}$, $\sigma_1, \sigma_2, \ldots, \sigma_M$, is defined as

$$\mathbf{\Sigma} = \begin{bmatrix} \sigma_1 & 0 \cdots\cdots & 0 \\ 0 & \sigma_2 \cdots & \\ \vdots & \ddots & \\ 0 & & \sigma_M \\ 0 & 0 \cdots & 0 \\ \vdots & & \vdots \\ 0 & & 0 \end{bmatrix}, \tag{4.3.9}$$

where the singular values are real and arranged in the order

$$\sigma_1 > \sigma_2 > \cdots > \sigma_M\,. \tag{4.3.10}$$

Note that the full form of the SVD must be used here to describe the behaviour of both $\mathbf{u}$ and $\mathbf{e}$, as opposed to the reduced form, both of which are described in more detail in the Appendix. The SVD of $\mathbf{G}$ can be used to show that

$$\mathbf{G}^{\mathrm{H}}\mathbf{G} = \mathbf{Q}\,\mathbf{\Sigma}^{\mathrm{T}}\mathbf{\Sigma}\,\mathbf{Q}^{\mathrm{H}}, \tag{4.3.11}$$

where $\mathbf{\Sigma}^{\mathrm{T}}\mathbf{\Sigma}$ is a square diagonal matrix with elements σ_m^2. Comparing this expression with equation (4.2.9) we can see that the eigenvalues of the matrix $\mathbf{G}^{\mathrm{H}}\mathbf{G}$ are equal to the squared values of the singular values of $\mathbf{G}$, i.e.

$$\lambda_m = \sigma_m^2\,. \tag{4.3.12}$$

The squared singular values of a transfer response matrix measured from 16 loudspeakers to 32 microphones at 88 Hz in a 6 m × 2 m × 2 m enclosure is shown in Fig. 4.5 (Elliott et al., 1992), which illustrates the wide range of singular values that can be present in a practical implementation.

Substituting equation (4.3.8) into (4.3.5), pre-multiplying by $\mathbf{R}^{\mathrm{H}}$ and using the unitary properties of this matrix yields

$$\mathbf{R}^{\mathrm{H}}\mathbf{e}(n) = \mathbf{R}^{\mathrm{H}}\mathbf{d} + \mathbf{\Sigma}\,\mathbf{Q}^{\mathrm{H}}\,\mathbf{u}(n)\,. \tag{4.3.13}$$

We now define the transformed set of error and disturbance signals as

$$\mathbf{y}(n) = \mathbf{R}^{\mathrm{H}}\mathbf{e}(n)\,, \tag{4.3.14}$$

and

$$\mathbf{p} = \mathbf{R}^{\mathrm{H}}\mathbf{d}\,, \tag{4.3.15}$$

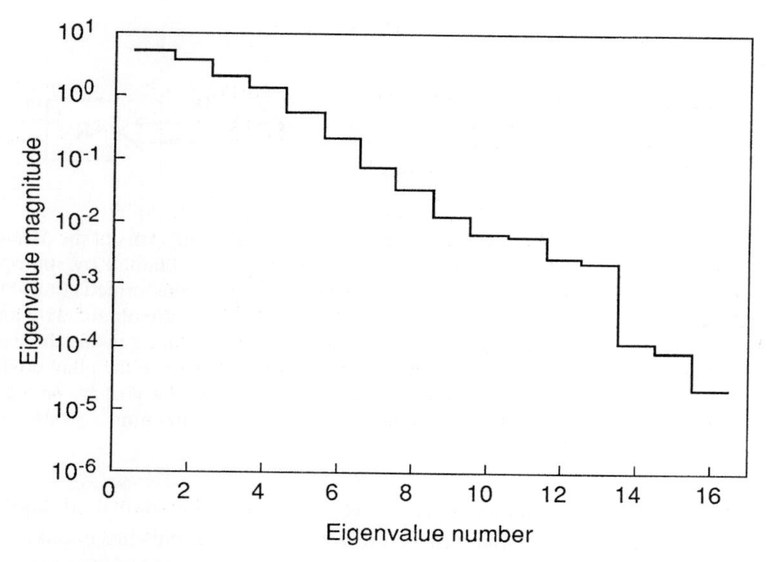

Figure 4.5 The squared singular values of a transfer response matrix measured between 16 loudspeakers and 32 microphones at an excitation frequency of 88 Hz in a 6 m × 2 m × 2 m enclosure (after Elliott et al., 1992).

and the transformed control vector as

$$\mathbf{v}(n) = \mathbf{Q}^{\mathrm{H}}\mathbf{u}(n) . \tag{4.3.16}$$

Equation (4.3.5), describing the response of the physical system under control, can now be written in a particularly simple form in terms of these transformed variables as

$$\mathbf{y}(n) = \mathbf{p} + \mathbf{\Sigma}\, \mathbf{v}(n) . \tag{4.3.17}$$

The diagonal nature of the $\mathbf{\Sigma}$ matrix in equation (4.3.9) means that each of the transformed error signals in equation (4.3.17) is either a function of only a single transformed control signal, so that

$$y_l(n) = p_l + \sigma_l\, v_l(n) \qquad \text{for } 1 \leq l \leq M , \tag{4.3.18}$$

or is completely unaffected by the control signals, so that

$$y_l(n) = p_l \qquad \text{for } M + 1 \leq l \leq L . \tag{4.3.19}$$

The l-th singular value thus quantifies how well the l-th transformed control signal couples into the l-th transformed error signal, and so the singular values are also called the *principal gains*, or principal values, of the multichannel system (Maciejowski, 1989). The block diagram for the multichannel control problem in terms of the transformed inputs and outputs is shown in Fig. 4.6. The use of the singular value decomposition transforms equation (4.3.5), for the physical error signals in the overdetermined case, into a form in which the output signals that are in the rank space of **G**, and can thus be controlled, are clearly divided from those that are in the null space of **G** and thus cannot be controlled.

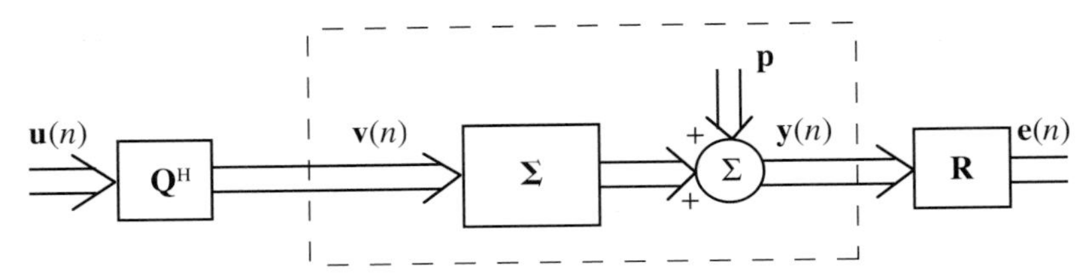

Figure 4.6 Block diagram of the multichannel tonal control problem in terms of the M transformed control signals, which are elements of the vector $\mathbf{v}(n)$ obtained by multiplying the physical control signals $\mathbf{u}(n)$ by the matrix $\mathbf{Q}^{\mathrm{H}}$, and the L transformed control outputs in the vector $\mathbf{y}(n)$, from which the physical output signals $\mathbf{e}(n)$ are obtained by multiplying by the matrix $\mathbf{R}$. The transformed control problem is contained inside the dashed box, within which the $L \times M$ matrix $\boldsymbol{\Sigma}$ contains the singular values of the plant on the leading diagonal and so the first M elements of the transformed output $\mathbf{y}(n)$ are only affected by a single one of the transformed inputs, $\mathbf{v}(n)$, and the remaining $L - M$ transformed outputs are unaffected by these inputs.

The first M elements of the transformed error vector can thus be controlled independently by the transformed control signals, and driven to zero if these signals are equal to

$$v_{m:\mathrm{opt}} = -p_m / \sigma_m \; . \tag{4.3.20}$$

A matrix form of this expression can be obtained, by substituting equation (4.3.8) into equation (4.2.6) for the optimum control vector and rearranging to give

$$\mathbf{v}_{\mathrm{opt}} = -\left[\boldsymbol{\Sigma}^{\mathrm{T}}\boldsymbol{\Sigma}\right]^{-1}\boldsymbol{\Sigma}^{\mathrm{T}}\mathbf{p} \; , \tag{4.3.21}$$

where $\left[\boldsymbol{\Sigma}^{\mathrm{T}}\boldsymbol{\Sigma}\right]^{-1}\boldsymbol{\Sigma}^{\mathrm{T}}$ is equal to the pseudo-inverse of $\boldsymbol{\Sigma}$ in this case, as described in the Appendix.

Using the fact that $\mathbf{R}$ is unitary, the sum of squared errors in the cost function can be written as

$$J = \mathbf{e}^{\mathrm{H}}\mathbf{e} = \mathbf{y}^{\mathrm{H}}\mathbf{y} = \sum_{m=1}^{L} \left|y_m\right|^2 \; . \tag{4.3.22}$$

Substituting equation (4.3.8) into equation (4.2.7) gives the matrix form for the minimum cost function in terms of the transformed primary field,

$$J_{\min} = \mathbf{p}^{\mathrm{H}}\left[I - \boldsymbol{\Sigma}\left[\boldsymbol{\Sigma}^{\mathrm{T}}\boldsymbol{\Sigma}\right]^{-1}\boldsymbol{\Sigma}^{\mathrm{T}}\right]\mathbf{p} \; , \tag{4.3.23}$$

where $\left[I - \boldsymbol{\Sigma}\left[\boldsymbol{\Sigma}^{\mathrm{T}}\boldsymbol{\Sigma}\right]^{-1}\boldsymbol{\Sigma}^{\mathrm{T}}\right]$ is an $L \times L$ diagonal matrix having the first M diagonal elements equal to zero and the remaining $L - M$ diagonal elements equal to unity. The minimum error can thus be expressed as

$$J_{\min} = \sum_{l=M+1}^{L} \left|p_l\right|^2 \; , \tag{4.3.24}$$

and since the cost function without any control can be written as $\sum_{l=1}^{L} \left|p_l\right|^2$, the normalised reduction in the cost function can be readily calculated. In general the cost function will be equal to

$$J = J_{\min} + \sum_{m=1}^{M} |y_m(n)|^2 . \tag{4.3.25}$$

4.3.3. Convergence in the Principal Coordinates

Having expressed the optimal control vector and minimum cost function in terms of the transformed disturbance vector and the singular values of the plant, we can return to the analysis of the steepest-descent algorithm, equation (4.3.7). Using the SVD given by equation (4.3.8) in this equation, the algorithm can be expressed in terms of the transformed variables as

$$\mathbf{v}(n+1) = \mathbf{v}(n) - \alpha\, \boldsymbol{\Sigma}^{\mathrm{T}}\, \mathbf{y}(n) , \tag{4.3.26}$$

and because of the properties of $\boldsymbol{\Sigma}$, each element of $\mathbf{v}$ will only be affected by the corresponding element in $\mathbf{y}(n)$, so that

$$v_m(n+1) = v_m(n) - \alpha\, \sigma_m\, y_m(n) \qquad \text{for } 1 \leq m \leq M . \tag{4.3.27}$$

The elements of both the physical system, equation (4.3.17), and the control algorithm, equation (4.3.26), are thus uncoupled when expressed in terms of the transformed variables, and so the transformed elements of the control signal converge independently of one another.

Substituting equation (4.3.18) for $y_m(n)$ into equation (4.3.27) and using the definition of $v_{m:\mathrm{opt}}$ in equation (4.3.20) we obtain

$$\left(v_m(n+1) - v_{m:\mathrm{opt}}\right) = \left[1 - \alpha\sigma_m^2\right]\left(v_m(n) - v_{m:\mathrm{opt}}\right) , \tag{4.3.28}$$

so that using the fact that $\sigma_m^2 = \lambda_m$, the eigenvalues of the matrix $\mathbf{G}^{\mathrm{H}}\mathbf{G}$, we can write

$$\left(v_m(n) - v_{m:\mathrm{opt}}\right) = \left[1 - \alpha\lambda_m\right]^n \left(v_m(0) - v_{m:\mathrm{opt}}\right) . \tag{4.3.29}$$

The steepest-descent algorithm will converge to the optimum result provided

$$|1 - \alpha\lambda_m| < 1 \qquad \text{for} \quad 1 \leq m \leq M , \tag{4.3.30}$$

so that

$$0 < \alpha < \frac{2}{\lambda_m} \qquad \text{for all } \lambda_m . \tag{4.3.31}$$

The convergence coefficient must thus be positive and its magnitude is limited to

$$\alpha < \frac{2}{\lambda_{\max}} , \tag{4.3.32}$$

where $\lambda_{\max}$ is the largest eigenvalue of $\mathbf{G}^{\mathrm{H}}\mathbf{G}$, although the fastest convergence rate is obtained with $\alpha = 1/\lambda_{\max}$. Some care must be taken with this estimate of the maximum convergence coefficient, however, because it has been derived using equation (4.3.5), which inherently assumes that any transients have died away in the error signals and they have reached their steady state sinusoidal waveforms before being measured. If the time between each iteration of the steepest-descent algorithm is not long compared with the transient response time of the system under control, then high values of the convergence coefficient will invalidate the assumptions inherent in equation (4.3.5), and equation (4.3.32) may give an optimistically high maximum value for the convergence coefficient. A more complete analysis of such transient effects is given in Section 4.6.

If $\alpha\lambda_m \ll 1$ and $v_m(0) = 0$, then equation (4.3.29) can be written as

$$v_m(n) \approx \left(1 - e^{-\alpha \lambda_m n}\right) v_{m:\text{opt}} , \tag{4.3.33}$$

which shows that the transformed control signals converge exponentially towards the optimum solution in the principal coordinates, with time constants determined by $1/(\alpha \lambda_m)$. By substituting $v_m(n)$ in equation (4.3.33) into equation (4.3.18) for $y_l(n)$ and using equation (4.3.20) and (4.3.25), the sum of squared errors can also be expressed for $\alpha\lambda_m \ll 1$ as

$$J(n) \approx J_{\min} + \sum_{m=1}^{M} |p_m|^2 \, e^{-2\alpha \lambda_m n} . \tag{4.3.34}$$

Each $y_m(n)$ in equation (4.3.25) will thus converge independently and exponentially, and can be referred to as the amplitude of a *mode* of the control system.

The convergence of the individual modes and the overall convergence of the cost function are shown in Fig. 4.7 for a simulation of the steepest-descent algorithm operating in the environment discussed above with 16 loudspeakers and 32 microphones at 88 Hz in

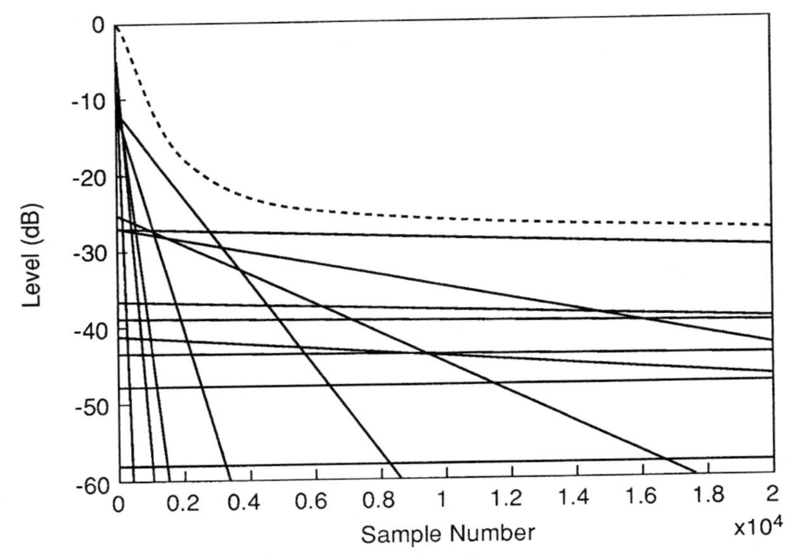

Figure 4.7 The convergence of the sum of squared error signals, normalised by the sum of squared primary disturbances, together with the individual 'modes' of convergence, for a steepest descent control system operating with 16 loudspeakers and 32 microphones in a small enclosure (after Elliott et al., 1992).

a small enclosure (Elliott et al., 1992). The time constant of the slowest mode is $1/(\alpha\lambda_{\min})$ samples, where $\lambda_{\min}$ is the smallest eigenvalue of $\mathbf{G}^{\mathrm{H}}\mathbf{G}$. Since, for rapid convergence, α must be about $1/\lambda_{\max}$, however, the slowest mode of the steepest-descent algorithm must have a time constant which is about $\lambda_{\max}/\lambda_{\min}$ samples, i.e. the eigenvalue spread of the matrix $\mathbf{G}^{\mathrm{H}}\mathbf{G}$. This eigenvalue spread can be very large for some practical multichannel control systems, as we have seen in Fig. 4.5, in which $\lambda_{\max}/\lambda_{\min} \approx 10^5$.

4.3.4. The Well-balanced System

Many active control problems are approximately *well-balanced*, in that $|p_m|^2/\lambda_m$ is reasonably constant with m. This means that the m-th component of the primary field in the transformed coordinates is roughly proportional to the m-th eigenvalue, or principal gain, of the plant matrix. The control system is thus well-balanced if the secondary sources can couple into each of the control modes with a gain that is approximately proportional to the excitation of these modes by the primary source.

The equation for the convergence of the sum of squared errors for a well-balanced control system can be written as

$$J(n) \approx J_{\min} + c\sum_{m=1}^{M} \lambda_m \, \mathrm{e}^{-2\alpha\lambda_m n} , \tag{4.3.35}$$

where c is a constant equal to the average value of $|p_m|^2/\lambda_m$. Note that the slow modes, with small values of λ_m, are also not excited to a great extent in this case, and thus do not play an important role in the initial convergence of the system. The convergence of equations of this form has been analysed by Morgan (1995) in the context of acoustic echo cancellers. Morgan has argued that at any one time only a single mode would dominate the summation in equation (4.3.35), and that the eigenvalue associated with that mode could be determined by differentiating $\lambda\,\mathrm{e}^{-2\alpha\lambda n}$ with respect to λ and equating this to zero, to give

$$\lambda = \frac{1}{2\alpha n} . \tag{4.3.36}$$

Assuming the term proportional to this eigenvalue makes the only contribution to the summation in equation (4.3.35), this equation can be written as

$$J(n) \approx J_{\min} + \frac{c}{2e\alpha n} . \tag{4.3.37}$$

The excess sum of squared errors is thus predicted to converge in proportion to $1/n$, where n is the iteration index. This behaviour can be observed to some extent in Fig. 4.7, in which different modes dominate the convergence behaviour at different times. After a while, however, the convergence behaviour is dominated by a single mode. In order to dominate the long-term convergence properties, this mode must have associated with it both a small eigenvalue, λ_m, and a relatively large value of primary excitation $|p_m|^2$, since it is the level of primary excitation that dictates the initial amplitude of the modal contribution in Fig. 4.7. The final level of the normalised sum of squared errors in Fig. 4.7 is set by equation (4.3.24), and is about –33 dB in this case.

4.3.5. Convergence of the Control Effort

The control effort required to achieve a given reduction in the error can be written, using the fact that $\mathbf{Q}$ is unitary, as

$$\mathbf{u}^{\mathrm{H}}\mathbf{u} = \mathbf{v}^{\mathrm{H}}\mathbf{v} = \sum_{m=1}^{M} |v_m|^2 . \tag{4.3.38}$$

Using equations (4.3.33) and (4.3.20), and the fact that $\sigma_m^2 = \lambda_m$, the control effort at the n-th iteration can thus be written for $\alpha\lambda_m \ll 1$ as

$$\mathbf{u}^{\mathrm{H}}(n)\mathbf{u}(n) \approx \sum_{m=1}^{M} \frac{|p_m|^2}{\lambda_m}\left(1 - \mathrm{e}^{-\alpha\lambda_m n}\right)^2 . \tag{4.3.39}$$

In controlling the m-th component of the transformed disturbance, p_m, the control algorithm thus requires an additional control effort of $|p_m|^2/\lambda_m$. Thus, if the control system is well-balanced, the same control effort is required to control each of the modes. If the primary source particularly excites some control modes that the secondary sources do not couple into efficiently, however, then $|p_m|^2/\lambda_m$ will be large for these modes, and the system will not be well-balanced. This is particularly likely to happen if λ_m is very small, so that the convergence of this mode is also very slow. Also, as the mode converges, the contribution to the control effort from this mode in equation (4.3.39) becomes particularly high, which can lead to excessive steady state actuator driving signals.

The solid line in Fig. 4.8, for example, shows the way in which the control effort changes with the attenuation of the sum of squared error signals during the course of a numerical simulation of the steepest-descent algorithm. This was performed using a set of data measured from 16 loudspeakers to 32 microphones in the enclosure discussed above, in which the primary source was arranged such that the system was *not* well-balanced. In this case, although 20 dB of attenuation in the mean-square error is obtained with only 0.1 units of effort quite early in the convergence process, over 2 units of effort are then required to achieve a further 4 dB or so of attenuation when the algorithm is allowed to converge for some time. The significance of the dashed curve in Fig. 4.8 will be explained in the following section. At a single frequency, the control effort in an active sound control system, in which the electrical impedance of the loudspeakers is relatively insensitive to their acoustic environment, is approximately proportional to the electrical power required to drive the loudspeakers. If the units of control effort in Fig. 4.8 are 100s of electrical watts, for example, then the additional control effort required for the final 4 dB of attenuation at the microphones is probably not justified, particularly since practical experience has shown that this 4 dB of further reduction at the error microphone positions is typically accompanied by an increase in the sound field away from these positions.

4.3.6. Trade-off Between Error Attenuation and Control Effort

In order to prevent large increases in control effort, a simple modification can be made to the steepest-descent algorithm so that it minimises a cost function of the form of equation

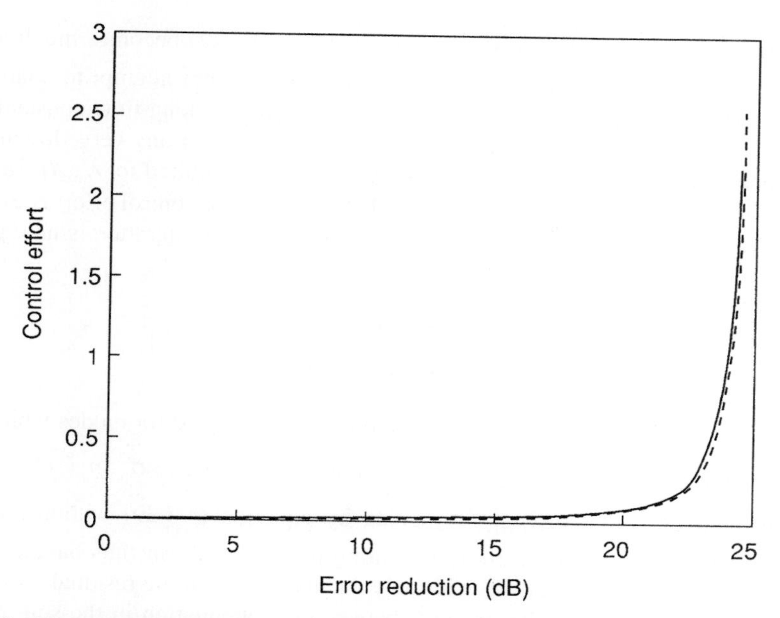

Figure 4.8 The locus of the total control effort with the attenuation in the sum of squared errors for a control system that is not 'well balanced' during the convergence of the steepest-descent algorithm (solid line) and for the optimum steady state solution calculated for various values of the effort weighting parameter β (dashed line).

(4.2.15), which includes the sum of squared errors and a component of control effort proportional to the parameter β. The Hessian matrix in this case is given by equation (4.2.16) and so the complex gradient vector, equation (4.3.3), is now equal to

$$\mathbf{g}(n) \;=\; 2\left(\mathbf{G}^{\mathrm{H}}\mathbf{e}(n) + \beta\,\mathbf{u}(n)\right) . \tag{4.3.40}$$

The steepest-descent algorithm, equation (4.3.2), thus becomes

$$\mathbf{u}(n+1) = \gamma\,\mathbf{u}(n) - \alpha\,\mathbf{G}^{\mathrm{H}}\,\mathbf{e}(n) , \tag{4.3.41}$$

where $\gamma = 1 - \alpha\beta$ provides a level of leakage in the algorithm, as discussed in Chapter 3.

The effort weighting term in the cost function modifies the equation for the convergence of the cost function, with $\alpha\lambda_m \ll 1$, to (Elliott et al., 1992)

$$J(n) \;=\; J_{\min} + \sum_{m=1}^{M} \frac{\sigma_m^2}{\sigma_m^2 + \beta}\left|p_m\right|^2 \mathrm{e}^{-2\alpha\lambda_m n} . \tag{4.3.42}$$

The time constants of convergence are now determined by $1/\lambda_m$, where $\lambda_m = \sigma_m^2 + \beta$ are the eigenvalues of the matrix $\mathbf{G}^{\mathrm{H}}\mathbf{G}+\beta\,\mathbf{I}$, and now

$$J_{\min} = \sum_{m=1}^{M} \frac{\beta}{\sigma_m^2 + \beta}\left|p_m\right|^2 + \sum_{m=M+1}^{L} \left|p_m\right|^2 . \tag{4.3.43}$$

The main effect of the effort weighting parameter β is to discriminate against components of the transformed disturbance for which σ_m^2 is small compared with β, in which case two effects occur. First, the term $\sigma_m^2/(\sigma_m^2+\beta)$ in equation (4.3.42) becomes much less than unity for these singular values, and the control system does not attempt to control these modes. Second, the very small values of σ_m^2 which led to very long time constants above now all become approximately equal to β, so there are no longer any very slow modes of convergence. The time constant of the slowest mode is now limited to λ_{max}/β, assuming $\beta \gg \lambda_{min}$. The effect of the parameter β on the behaviour of the control effort is even more dramatic, as may be expected, and the control effort at the n-th iteration is now given by (Elliott et al., 1992)

$$P(n) = \mathbf{u}(n)^{\mathrm{H}}\mathbf{u}(n) \approx \sum_{m=1}^{M} \frac{\sigma_m^2|p_m|^2}{(\sigma_m^2+\beta)^2}\left(1-\mathrm{e}^{-\alpha\lambda_m n}\right)^2, \tag{4.3.44}$$

where again $\lambda_m = \sigma_m^2 + \beta$. The steady state control effort required for modes which have a value of σ_m^2 that is small compared with β is now proportional to $\sigma_m^2|p_m|^2/\beta^2$, which is much less than the value of $|p_m|^2/\sigma_m^2$ given by equation (4.3.39) without any effort weighting. A small value of the effort weighting parameter β can thus have a dramatic effect on limiting the control effort, with only a small effect on the residual mean-square error. To illustrate this point, the trade-off between the attenuation in the sum of square errors and the control effort has been calculated by computing the optimal error reduction and the resulting control effort with various values of β in the cost function for the active noise control example described above, which was not well-balanced, and these are plotted against each other in the dashed curve of Fig. 4.8. A good practical choice for the value of β would thus be one that ensured that the steady state solution was at the knee of this curve, giving about 20 dB of attenuation for only about 0.1 units of control effort.

It is interesting to note the similarity between the two curves in Fig. 4.8, even though the solid line has been obtained by plotting the instantaneous values of the attenuation and effort as the steepest-descent algorithm evolves in time, whereas the dashed line represents the optimum steady state values of attenuation and effort as the parameter β is varied. This similarity suggests that at every time during its convergence, the steepest-descent algorithm is nearly optimal for a particular value of effort weighting, and this value of the 'effective effort weighting' gradually decreases with time. A similar observation has been made by Sjöberg and Ljung (1992) in connection with the training of neural networks. If a finite value of β is included in the steepest-descent algorithm, the system will converge in a very similar way to the normal steepest-descent algorithm until the value of β in the control algorithm is equal to the gradually decreasing 'effective effort weighting', when the convergence will stop. The convergence for the sum of squared errors as a function of time with a finite value of β thus looks very much like the initial part of Fig. 4.7 except that the attenuation no longer continues to increase once the weighted effort term becomes dominant in the cost function.

4.3.7. Scheduling of Effort Weighting Parameter

It is also possible to change the parameter β while the steepest-descent algorithm is converging, in a way that depends upon the measured control effort. This modification to the steepest-descent algorithm can be used to automatically achieve a good trade-off between performance and control effort, without the analysis required to produce Fig. 4.8. A method of scheduling the effort weighting parameter, $\beta(n)$, on the total control effort $P(n) = \mathbf{u}^{\mathrm{H}}(n)\,\mathbf{u}(n)$ is described by Elliott and Baek (1996) and is illustrated in Fig. 4.9. The effort weighting parameter is set to zero if the measured control effort $P(n)$ is below some threshold P_{T} which is slightly below the desired upper limit on the control effort, P_{L}. If the measured control effort is above this threshold, $\beta(n)$ is increased as a function of $\left(P(n) - P_{\mathrm{T}}\right)$. Figure 4.10 shows the results of a computer simulation of this algorithm in which a linear function was used, so that β was directly proportional to $\left(P(n) - P_{\mathrm{T}}\right)$ for $P(n)$ above P_{T}. A control system with two control inputs and four error signals was simulated using values of the plant matrix and disturbance, measured between two loudspeakers and four microphones at 88 Hz in a small room. The total control effort, $|u_1|^2 + |u_2|^2$ in this case, was limited to 0.8 units on the arbitrary linear scale shown in Fig. 4.10. This limitation is achieved by the increase in the effort weighting parameter $\beta(n)$ which occurs after about 15 iterations, when the total effort begins to exceed the threshold P_{T}. The results obtained from using the unconstrained steepest-descent algorithm, with $\beta(n)$ always equal to zero, are shown as the dashed lines in this figure. With no constraint, the final control effort is about twice that used by the constrained algorithm, but the sum of squared errors, $\mathbf{e}^{\mathrm{H}}\mathbf{e}$, was reduced by 5.5 dB, instead of 4.9 dB in the constrained case. The constant of proportionality used in the linear scheduling of β on $\left(P(n) - P_{\mathrm{T}}\right)$ can be high if the algorithm is only converging slowly when the effort threshold is reached. In this case the steady state solution is close to the optimal solution for the problem of minimising the sum of the squared errors with the *constraint* that the control effort should not be greater than P_{L} (Elliott and Baek, 1996). Various functions could be used to schedule $\beta(n)$ on $P(n)$, and there is a significant literature on this form of constrained optimisation, which is called the *penalty function method* (see for example Luenberger, 1973; Fletcher, 1987).

A similar method can be used to constrain the amplitudes of *individual* elements of the control effort, which can be very desirable if the individual secondary actuators have a maximum power constraint. If individual control efforts are used for each actuator, the control algorithm, generalised from equation (4.3.41), becomes

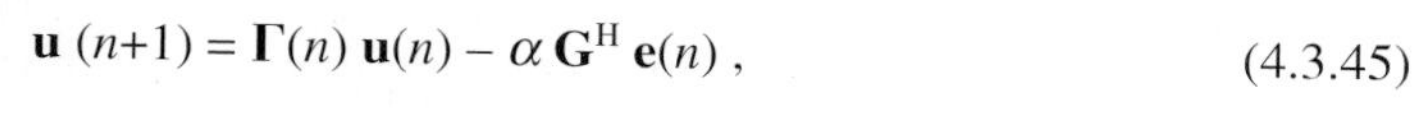

$$\mathbf{u}(n+1) = \mathbf{\Gamma}(n)\,\mathbf{u}(n) - \alpha\,\mathbf{G}^{\mathrm{H}}\,\mathbf{e}(n)\,, \qquad (4.3.45)$$

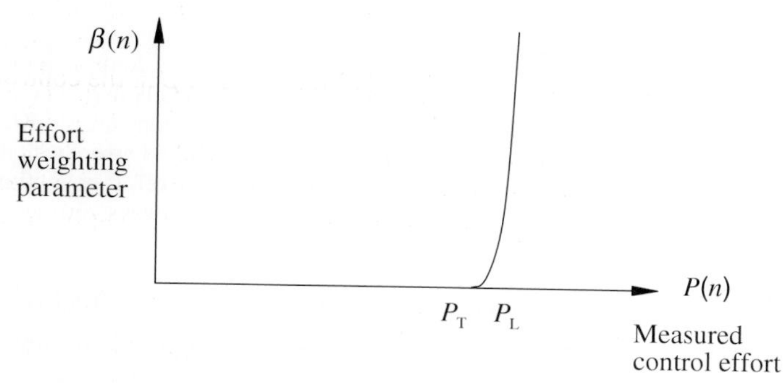

Figure 4.9 A method of scheduling the control effort weighting parameter $\beta(n)$ on the measured value of the control effort, $P(n)$, so that the control effort is maintained below the limited value P_{L}.

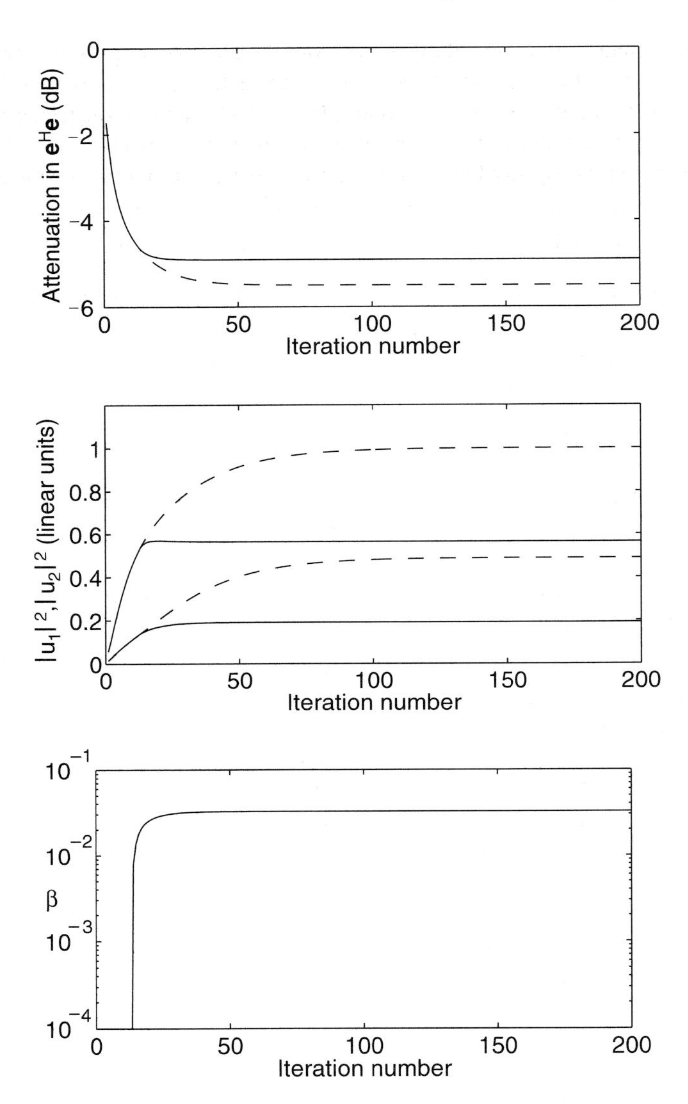

Figure 4.10 Results of a computer simulation of a feedforward adaptive control algorithm operating at a single frequency on a system with two control signals and four error signals in which the *total* control effort ($\mathbf{u}^H\mathbf{u}$) is limited. The variations are shown with iteration number of the level of the attenuation in the sum of squared errors $\mathbf{e}^H\mathbf{e}$, the modulus of the individual control efforts $|u_1|^2$ and $|u_2|^2$, and the single effort weighting parameter β. The dashed graphs are with no effort constraint, i.e. β is always zero.

where $\boldsymbol{\Gamma}(n)$ is a time-varying diagonal matrix with elements $(1 - \alpha\beta_m(n))$ and $\beta_m(n)$ is scheduled on the individual control effort $|u_m(n)|^2$ using the methods outlined above. The results of a simulation of this individually constrained algorithm, using the same plant and disturbance as used above, are shown in Fig. 4.11. The squared modulus of the individual control signals, $|u_1|^2$ and $|u_2|^2$, is now limited to 0.4 units by the individual effort weighting

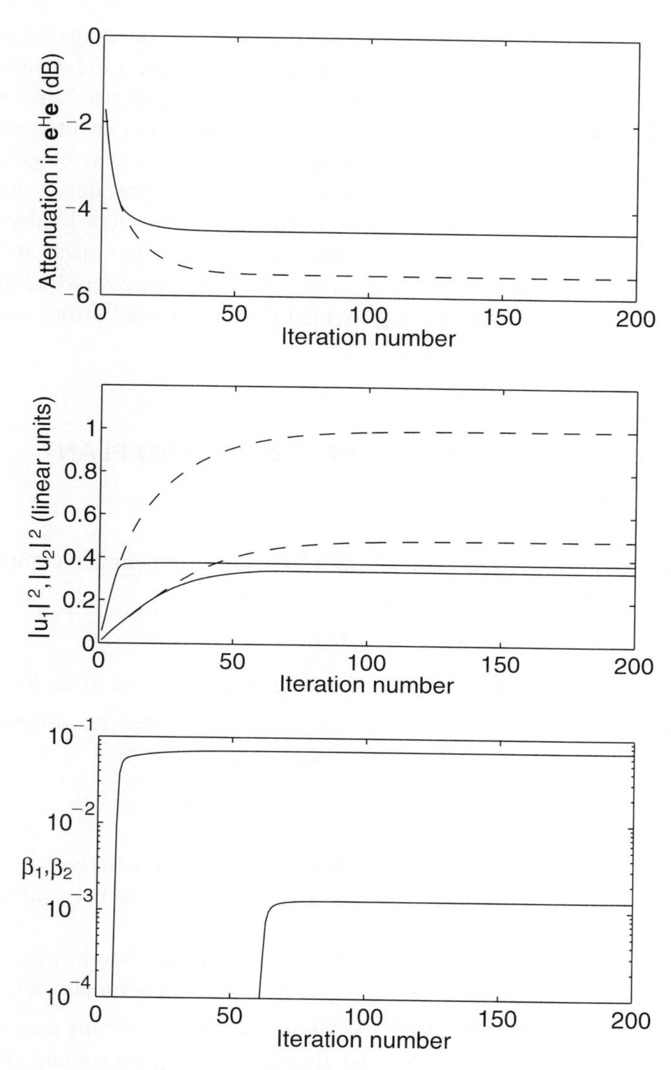

Figure 4.11 Results of a computer simulation of a feedforward adaptive control algorithm operating at a single frequency on a system with two control signals and four error signals in which the *individual* control efforts $|u_m|^2$ are limited. The variations are shown with iteration number of the level of the attenuation in the sum of squared errors $\mathbf{e}^H\mathbf{e}$, the modulus of the individual control efforts $|u_1|^2$ and $|u_2|^2$, and the individual effort weighting parameters β_1 and β_2. The dashed graphs are with no effort constraint, i.e. l is always zero.

parameters which 'cut-in' after about 10 iterations for $\beta_1(n)$ and after about 70 iterations for $\beta_2(n)$, but reach different steady state values. The attenuation in the sum of squared errors is only 4.5 dB in the case of individually constrained secondary actuators, even though the total control effort is about 0.8 units as in Fig. 4.10. This is the inevitable price which has to be paid for restricting the effort expended by one of the actuators to 0.4 units, as opposed to the value of 0.6 units required by constraining the total power to be 0.8 units in Fig. 4.10, or 1 unit if the algorithm was unconstrained.

It can be demonstrated that the adaptive algorithm above once again approaches the optimum solution to the control problem of minimising the sum of the squared errors with the constraint that the individual control efforts are below the prescribed value (Elliott and Baek, 1996). The ability to modify the adaptive control system to converge to near-optimal results, even with effort constraints on the individual actuators, was of great help in the flight tests described in Section 4.8 in which the very low frequency sound inside a helicopter was controlled. This sound was so intense that very large loudspeakers would have had to be used to achieve an unconstrained least-squares minimisation of the internal sound field, but it was found that a useful degree of attenuation could be achieved even with a number of smaller loudspeakers, provided their individual efforts were limited to below the level at which they overloaded.

4.4. ROBUSTNESS TO PLANT UNCERTAINTIES AND PLANT MODEL ERRORS

The steady-state harmonic behaviour of the physical system under control at the n-th iteration was described in Section 4.2 by the equation

$$\mathbf{e}(n) = \mathbf{d} + \mathbf{G}\,\mathbf{u}(n) \tag{4.4.1}$$

where $\mathbf{G}$ is the matrix of transfer responses for the physical plant. If an internal *model* of these responses, $\hat{\mathbf{G}}$, is used to adjust the control vector using the steepest-descent algorithm with effort weighting, then equation (4.3.41) becomes

$$\mathbf{u}(n+1) = (1-\alpha\beta)\mathbf{u}(n) - \alpha\hat{\mathbf{G}}^{\mathrm{H}}\mathbf{e}(n)\ . \tag{4.4.2}$$

In the previous section it was assumed that this plant model was perfect, i.e. $\hat{\mathbf{G}} = \mathbf{G}$. This is clearly a rather sweeping assumption for practical control systems, and one which will be examined in more detail in this section.

There are two important and distinct problems which cause $\hat{\mathbf{G}}$ to be different from $\mathbf{G}$ in practical systems. The first, which has been studied particularly for active control systems (Boucher et al., 1991; Elliott et al., 1992), is when there are modelling errors in $\hat{\mathbf{G}}$, but the plant response itself, $\mathbf{G}$, is assumed to be fixed. The second problem, which is more familiar from the control literature (Morari and Zafiriou, 1989) is when the plant model accurately represents the physical plant in some nominal state, so that $\hat{\mathbf{G}} = \mathbf{G}_0$ say, but the response of the physical plant can then vary about this nominal state. In both cases we are interested in the stability and performance of the control algorithm. Some progress can be made in this analysis using only equations (4.4.1) and (4.4.2), but later in this section more specific assumptions will have to be made about the relationship between $\mathbf{G}$ and $\hat{\mathbf{G}}$.

4.4.1. Convergence Conditions

A general expression for the dynamics of a practical control algorithm is obtained by substituting equation (4.4.1) into equation (4.4.2) to give

$$\mathbf{u}(n+1) = \mathbf{u}(n) - \alpha\left[\hat{\mathbf{G}}^{\mathrm{H}}\mathbf{d} + \left(\hat{\mathbf{G}}^{\mathrm{H}}\mathbf{G} + \beta\mathbf{I}\right)\mathbf{u}(n)\right] . \tag{4.4.3}$$

If the control algorithm is stable, it will reach a steady-state solution when $\mathbf{u}(n+1)$ is equal to $\mathbf{u}(n)$. The term in square brackets in equation (4.4.3) must be zero under these steady state conditions and the control vector will thus be equal to

$$\mathbf{u}_{\infty} = -\left[\hat{\mathbf{G}}^{\mathrm{H}}\mathbf{G} + \beta\mathbf{I}\right]^{-1}\hat{\mathbf{G}}^{\mathrm{H}}\mathbf{d} . \tag{4.4.4}$$

Note that even if the control algorithm is stable it does not generally converge to the true optimal solution in this case, given by equation (4.2.17). Only in the special case in which $\beta = 0$ and there are the same number of actuators as sensors, so that $\mathbf{G}$ and $\hat{\mathbf{G}}$ are square and are assumed to be non-singular, does $\mathbf{u}_{\infty}$ equal the true optimal control vector, which is given by equation (4.2.12) in this fully-determined case.

An iterative mapping of the control vector, offset by the assumed steady state solution, can now be obtained from equation (4.4.4) and equation (4.4.3), as

$$\left(\mathbf{u}(n+1) - \mathbf{u}_{\infty}\right) = \left[\mathbf{I} - \alpha\left(\hat{\mathbf{G}}^{\mathrm{H}}\mathbf{G} + \beta\mathbf{I}\right)\right]\left(\mathbf{u}(n) - \mathbf{u}_{\infty}\right) . \tag{4.4.5}$$

The convergence of the algorithm is determined by the eigenvalues of the matrix $\left[\hat{\mathbf{G}}^{\mathrm{H}}\mathbf{G} + \beta\mathbf{I}\right]$ (Elliott et al., 1992). This matrix is not Hermitian and so, in contrast to the case when $\hat{\mathbf{G}} = \mathbf{G}$, the eigenvalues may be complex and the convergence may have oscillatory as well as exponential modes. Even though the eigenvalues may be complex, the convergence of equation (4.4.5) can still be analysed using the principal coordinates, as in Section 4.3. If the eigenvalues of $\left[\hat{\mathbf{G}}^{\mathrm{H}}\mathbf{G} + \beta\mathbf{I}\right]$ are denoted by λ'_m, then the condition for stability is that

$$\left|1 - \alpha\lambda'_m\right| < 1 \qquad \text{for all } m. \tag{4.4.6}$$

By direct analogy with the single-channel case described in Section 3.2, the convergence condition given by equation (4.4.6) can also be written as a bound on the convergence coefficient, α, as

$$0 < \alpha < \frac{2\,\mathrm{Re}(\lambda'_m)}{\left|\lambda'_m\right|^2} \qquad \text{for all } m , \tag{4.4.7}$$

so that even in the limit of slow convergence, $\alpha \ll 1$, the real parts of all the eigenvalues must be positive for the system to converge. The real parts of the small eigenvalues of $\hat{\mathbf{G}}^{\mathrm{H}}\mathbf{G}$, λ_m say, can rather easily become negative if the eigenvalue spread is large (Boucher et al., 1991). Since the eigenvalues of $\left[\hat{\mathbf{G}}^{\mathrm{H}}\mathbf{G} + \beta\mathbf{I}\right]$ are equal to

$$\lambda'_m = \lambda_m + \beta , \tag{4.4.8}$$

it is clear that a small value of the effort weighting parameter, β, can counteract the effect of eigenvalues with small negative real parts and ensure that the algorithm remains stable.

In general it is the modes with small values of λ_m that must be stabilised and the effect of a small value of β on the convergence of modes with larger values of λ_m will be small, as demonstrated in equations (4.3.42) and (4.3.43).

We now consider the form of the eigenvalues for the matrix $\hat{\mathbf{G}}^{\mathrm{H}}\mathbf{G}$ when there is either plant uncertainty or modelling errors. To begin with the case of plant uncertainty, the response to the physical plant, $\mathbf{G}$, is assumed to be equal to that of a *nominal* plant, $\mathbf{G}_0$, plus a matrix of additive plant uncertainties, $\Delta\mathbf{G}_{\mathrm{p}}$, whereas the plant model, $\hat{\mathbf{G}}$, is assumed to be an accurate model of the nominal plant, so that

$$\mathbf{G} = \mathbf{G}_0 + \Delta\mathbf{G}_{\mathrm{p}} \quad \text{and} \quad \hat{\mathbf{G}} = \mathbf{G}_0 . \tag{4.4.9,10}$$

In the case of modelling uncertainty then the true plant response is assumed to be constant at its nominal value, $\mathbf{G}_0$, and the plant model is subject to additive modelling errors, $\Delta\mathbf{G}_{\mathrm{m}}$, so that

$$\mathbf{G} = \mathbf{G}_0, \quad \hat{\mathbf{G}} = \mathbf{G}_0 + \Delta\mathbf{G}_{\mathrm{m}} . \tag{4.4.11,12}$$

We have seen that the stability of multichannel steepest-descent algorithm depends on the sign of the real parts of the eigenvalues of the matrix $\hat{\mathbf{G}}^{\mathrm{H}}\mathbf{G}$, equation (4.4.7). In the case of additive plant uncertainties, equations (4.4.9,10), the real parts of these eigenvalues are

$$\mathrm{Re}\left[\mathrm{eig}\left(\hat{\mathbf{G}}^{\mathrm{H}}\mathbf{G}\right)\right] = \mathrm{Re}\left[\mathrm{eig}\left(\mathbf{G}_0^{\mathrm{H}}\mathbf{G}_0 + \mathbf{G}_0^{\mathrm{H}}\Delta\mathbf{G}_{\mathrm{p}}\right)\right] . \tag{4.4.13}$$

In the case of additive modelling errors, equations (4.4.11,12), the real parts of the eigenvalues are

$$\mathrm{Re}\left[\mathrm{eig}\left(\hat{\mathbf{G}}^{\mathrm{H}}\mathbf{G}\right)\right] = \mathrm{Re}\left[\mathrm{eig}\left(\mathbf{G}^{\mathrm{H}}\hat{\mathbf{G}}\right)\right] = \mathrm{Re}\left[\mathrm{eig}\left(\mathbf{G}_0^{\mathrm{H}}\mathbf{G}_0 + \mathbf{G}_0^{\mathrm{H}}\Delta\mathbf{G}_{\mathrm{m}}\right)\right] , \tag{4.4.14}$$

in which we have used the fact that the eigenvalues of $\mathbf{A}^{\mathrm{H}}$ are the conjugates of the eigenvalues of $\mathbf{A}$. The additive matrices of plant uncertainties and modelling errors thus play an identical role in determining the *stability* of a multichannel steepest-descent algorithm. The same thing cannot, however, be said of their effect on the performance of the algorithm (Omoto and Elliott, 1996, 1999).

4.4.2. Convergence in the Principal Coordinates

In the rest of this section we will assume that the perturbation is caused by uncertainty in the plant, and examine the effect of this on both stability and performance. We have seen that the stability of the control system depends on whether the real parts of the eigenvalues of $\hat{\mathbf{G}}^{\mathrm{H}}\mathbf{G}$ are positive. For a system with plant uncertainty, $\hat{\mathbf{G}}^{\mathrm{H}}\mathbf{G}$ is given by $\mathbf{G}_0^{\mathrm{H}}\mathbf{G}_0 + \mathbf{G}_0^{\mathrm{H}}\Delta\mathbf{G}_{\mathrm{p}}$, as in equation (4.4.13), and the eigenvalues of this equation can be studied using the singular value decomposition of the *nominal* plant matrix, which is given by

$$\mathbf{G}_0 = \mathbf{R}_0\,\boldsymbol{\Sigma}\,\mathbf{Q}_0^{\mathrm{H}} , \tag{4.4.15}$$

so that
$$\mathbf{G}_0^{\mathrm{H}} = \mathbf{Q}_0\,\mathbf{\Sigma}^{\mathrm{T}}\mathbf{R}_0^{\mathrm{H}}\,, \tag{4.4.16}$$

and
$$\mathbf{G}_0^{\mathrm{H}}\mathbf{G}_0 = \mathbf{Q}_0\,\mathbf{\Sigma}^{\mathrm{T}}\,\mathbf{\Sigma}\,\mathbf{Q}_0^{\mathrm{H}}\,, \tag{4.4.17}$$

since $\mathbf{R}_0$ is unitary.

We now use the eigenvector matrices $\mathbf{R}_0$ and $\mathbf{Q}_0$, obtained for the nominal plant, $\mathbf{G}_0$, to decompose the matrix of perturbations in the plant, so that

$$\Delta\mathbf{G}_{\mathrm{p}} = \mathbf{R}_0\,\Delta\mathbf{\Sigma}\,\mathbf{Q}_0^{\mathrm{H}}\,, \tag{4.4.18}$$

where

$$\Delta\mathbf{\Sigma} = \mathbf{R}_0^{\mathrm{H}}\,\Delta\mathbf{G}_{\mathrm{p}}\,\mathbf{Q}_0\,, \tag{4.4.19}$$

is a matrix of singular value perturbations, which, in general, will *not* be diagonal.

The matrix $\hat{\mathbf{G}}^{\mathrm{H}}\mathbf{G}$ can now be written as

$$\hat{\mathbf{G}}^{\mathrm{H}}\mathbf{G} \;=\; \mathbf{Q}_0\,\mathbf{\Sigma}^{\mathrm{T}}\,\mathbf{\Sigma}\,\mathbf{Q}_0^{\mathrm{H}} + \mathbf{Q}_0\,\mathbf{\Sigma}^{\mathrm{T}}\,\Delta\mathbf{\Sigma}\,\mathbf{Q}_0^{\mathrm{H}} \tag{4.4.20}$$

so that

$$\mathbf{Q}_0^{\mathrm{H}}\,\hat{\mathbf{G}}^{\mathrm{H}}\mathbf{G}\,\mathbf{Q}_0 \;=\; \mathbf{\Sigma}^{\mathrm{T}}\,\mathbf{\Sigma} + \mathbf{\Sigma}^{\mathrm{T}}\,\Delta\mathbf{\Sigma}\,. \tag{4.4.21}$$

The eigenvalues of $\hat{\mathbf{G}}^{\mathrm{H}}\mathbf{G}$ are equal to those of $\mathbf{Q}_0^{\mathrm{H}}\,\hat{\mathbf{G}}^{\mathrm{H}}\mathbf{G}\,\mathbf{Q}_0$ because the two matrices are similar. In order to establish a bound on the eigenvalues of $\mathbf{Q}_0^{\mathrm{H}}\,\hat{\mathbf{G}}^{\mathrm{H}}\mathbf{G}\,\mathbf{Q}_0$ we can use the *Gershgorin circle theorem* (Golub and Van Loan, 1996), which states that if

$$\mathbf{X}^{-1}\,\mathbf{A}\,\mathbf{X} = \mathbf{D} + \mathbf{F} \tag{4.4.22}$$

where $\mathbf{D}$ is a diagonal matrix with elements $d_1, d_2, \ldots, d_M$, and $\mathbf{F}$ has elements f_{mn} but its diagonal elements are zero, then each eigenvalue of $\mathbf{A}$, λ, lies within a disc in the complex plane whose centre is equal to d_m and whose radius is equal to the sum of the moduli of the elements in the corresponding row of $\mathbf{F}$, so that for all λ

$$|\lambda - d_m| \le r_m \qquad \text{where} \qquad r_m \;=\; \sum_{n=1}^{M} |f_{mn}|\,. \tag{4.4.23a,b}$$

Thus each eigenvalue can be written in terms of an 'average' value d_m, if $r_m = 0$, and an 'uncertainty' ε, of maximum magnitude, r_m, so that

$$\lambda_m \;=\; d_m + \varepsilon\,, \tag{4.4.24}$$

where $|\varepsilon| \le r_m$ but the phase of ε is indeterminate. Putting equation (4.4.21) into the form of equation (4.4.22), we have

$$d_m \;=\; \sigma_m^2 + \sigma_m\,\Delta\sigma_{mm}\,, \tag{4.4.25}$$

and

$$f_{mn} = \sigma_m\,\Delta\sigma_{mn} \qquad \text{if } m \ne n \quad \text{and } f_{mm} = 0\,, \tag{4.4.26}$$

where $\Delta\sigma_{mn}$ is the m,n-th element of $\Delta\boldsymbol{\Sigma}$. Equation (4.4.24) for the eigenvalues thus takes the following form in this case:

$$\lambda = \sigma_m^2 + \sigma_m \Delta\sigma_{mm} + \varepsilon \ , \tag{4.4.27}$$

where

$$|\varepsilon| \leq \sigma_m \sum_{\substack{n=1 \\ n \neq m}}^{M} |\Delta\sigma_{mn}| \tag{4.4.28}$$

which is illustrated in Fig. 4.12.

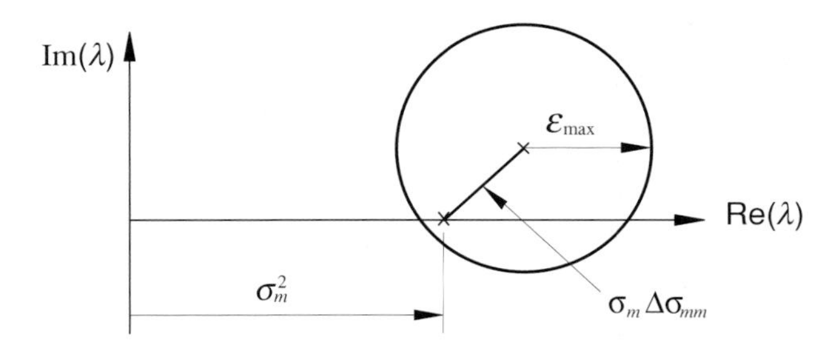

Figure 4.12 The eigenvalues of $\hat{\mathbf{G}}^{\mathrm{H}}\mathbf{G}$, λ, lie in a disc in the complex plane whose centre is $\sigma_m^2 + \sigma_m \Delta\sigma_{mm}$ and whose radius is equal to $\varepsilon_{\max} = \sigma_m \sum_{n \neq m} |\Delta\sigma_{mn}|$.

In order to establish the worst-case conditions for stability we must examine the conditions under which the real parts of any of the eigenvalues of $\hat{\mathbf{G}}^{\mathrm{H}}\mathbf{G}$ become negative. Noting that σ_m is defined to be positive and real, the worst case for stability would be that in which $\Delta\sigma_{mm}$ was real and negative, equation (4.4.28) became an equality, and the phase of ε was such that it too was entirely negative, in which case the eigenvalues would be real and given by

$$\lambda = \sigma_m \left(\sigma_m - \sum_{n=1}^{M} |\Delta\sigma_{mn}| \right) . \tag{4.4.29}$$

The eigenvalues must be positive for stability of the control system, so that a sufficient but not necessary condition for stability is given by

$$\sigma_m > \sum_{n=1}^{M} |\Delta\sigma_{mn}| \ , \qquad \text{for all } m \ . \tag{4.4.30}$$

It will generally be most difficult to satisfy this equation for the smaller singular values for which $m \approx M$. Thus the most critical feature of the matrix $\Delta\boldsymbol{\Sigma}$, from the point of view of stability, will be the magnitude of the elements in rows M, $M-1$ etc., as illustrated in Fig. 4.13. In a number of applications it has been observed that equation (4.4.30) gives an

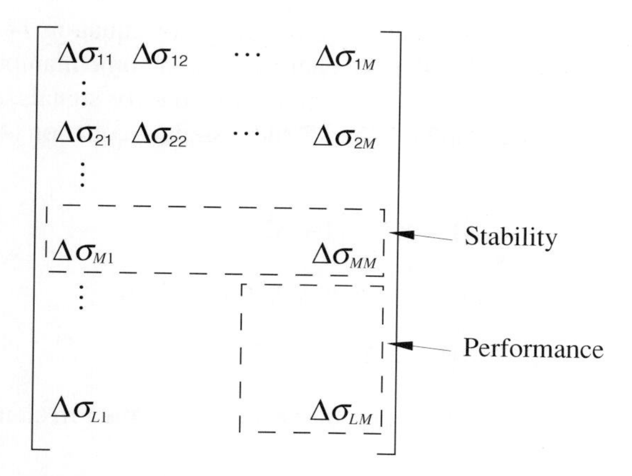

Figure 4.13 The regions of the transformed matrix of plant uncertainties, $\Delta\boldsymbol{\Sigma}$, which are important in determining the stability, and the performance of a multichannel control system.

unduly conservative bound on the level of uncertainty that can be tolerated before instability of an adaptive algorithm. This is because in practical systems with many channels, the effects of the off-diagonal terms in $\Delta\boldsymbol{\Sigma}$ tend to cancel out. Under these conditions it is only the change in the diagonal terms that must be accounted for in equation (4.4.27) and to ensure that the real parts of each eigenvalue is positive we must have

$$\sigma_m > -\mathrm{Re}\left[\Delta\sigma_{mm}\right], \qquad \text{for all } m . \tag{4.4.31}$$

This optimistic criterion has been found to give a reasonably accurate prediction of stability in many practical applications (Omoto and Elliott, 1999).

4.4.3. Effect of Plant Uncertainty on Control Performance

Similar methods can be used to predict the effect of plant uncertainties on the performance of a multichannel control system (Jolly and Rossetti, 1995; Omoto and Elliott, 1996). Assuming that the control system is stable, the steady state solution for the transformed error vector can be obtained using equation (4.4.4) with $\beta = 0$ and the definition of the transformed error signal to give

$$\mathbf{y}_\infty = \left[\mathbf{I} - \boldsymbol{\Sigma}\left(\hat{\boldsymbol{\Sigma}}^{\mathrm{H}}\boldsymbol{\Sigma}\right)^{-1}\hat{\boldsymbol{\Sigma}}^{\mathrm{H}}\right]\mathbf{p} , \tag{4.4.32}$$

where $\hat{\boldsymbol{\Sigma}} = \mathbf{R}^{\mathrm{H}}\hat{\mathbf{G}}\mathbf{Q}$, whereas the minimum value of the transformed error vector, given when $\hat{\mathbf{G}} = \mathbf{G}$, is equal to

$$\mathbf{y}_{\min} = \left[\mathbf{I} - \boldsymbol{\Sigma}\left(\boldsymbol{\Sigma}^{\mathrm{T}}\boldsymbol{\Sigma}\right)^{-1}\boldsymbol{\Sigma}^{\mathrm{T}}\right]\mathbf{p} = \left[\mathbf{I} - \boldsymbol{\Sigma}\boldsymbol{\Sigma}^{\dagger}\right]\mathbf{p} , \tag{4.4.33}$$

where $\boldsymbol{\Sigma}^{\dagger}$ is the pseudo-inverse of $\boldsymbol{\Sigma}$ as described in the Appendix. The diagonal matrix $\mathbf{I} - \boldsymbol{\Sigma}\boldsymbol{\Sigma}^{\dagger}$ has the first M diagonal elements equal to zero and the remaining $L - M$ diagonal elements equal to unity, so the result is consistent with equation (4.3.23). Using the Woodbury formula, as described in the Appendix, and after some manipulation (similar to that in Jolly and Rossetti, 1995), the residual transformed error signals, equation (4.4.32), can be written in terms of the minimum value of these signals, equation (4.4.33), as (Omoto and Elliott, 1999)

$$\mathbf{y}_{\infty} = \left[\mathbf{I} - \boldsymbol{\Sigma}\boldsymbol{\Sigma}^{\dagger}\right]\left[\mathbf{I} + \Delta\boldsymbol{\Sigma}\boldsymbol{\Sigma}^{\dagger}\right]^{-1}\mathbf{p} \; . \tag{4.4.34}$$

For small values of $\Delta\boldsymbol{\Sigma}$ this expression can be approximated by

$$\mathbf{y}_{\infty} \approx \left[\mathbf{I} - \boldsymbol{\Sigma}\boldsymbol{\Sigma}^{\dagger}\right]\left[\mathbf{I} - \Delta\boldsymbol{\Sigma}\boldsymbol{\Sigma}^{\dagger}\right]\mathbf{p} \; . \tag{4.4.35}$$

Using the properties of the $\mathbf{I} - \boldsymbol{\Sigma}\boldsymbol{\Sigma}^{\dagger}$ matrix described above, the l-th element of $\mathbf{y}_{\infty}$ is given by

$$y_l(\infty) = 0 \qquad \text{for } 1 \le l \le M \, , \tag{4.4.36}$$

and

$$y_l(\infty) = p_l - \sum_{m=1}^{M} \frac{\Delta\sigma_{l,m}}{\sigma_m} p_m \qquad \text{for } M+1 \le l \le L \, . \tag{4.4.37}$$

The first M elements of $y_l(\infty)$, equation (4.4.36), are thus the same as those after control with no plant uncertainty, equation (4.3.18) with equation (4.3.20), but the final $L - M$ elements of y_l, equation (4.3.37), have additional terms compared with the case of no plant uncertainties, equation (4.3.19), and these terms thus represent the additional residual error due to the presence of plant uncertainties. A closer examination of equation (4.4.37) reveals that the terms in the summation will tend to be large for values of m close to M, in which case the singular values will be small and the term $1/\sigma_m$ will be large. Thus the performance of the control system is most affected by the elements $\Delta\sigma_{l,m}$ for which $M + 1 \le l \le L$ and $m = M, M - 1$ etc., which corresponds to the elements in the lower right-hand portion of the $\Delta\boldsymbol{\Sigma}$ matrix, as illustrated in Fig. 4.13. Perturbations in the plant matrix that give rise to elements of the $\Delta\boldsymbol{\Sigma}$ matrix having a large magnitude in this region will thus tend to have a very detrimental effect on the performance of the control system.

4.4.4. Examples of the Transformed Matrix of Plant Uncertainties

Examples of the structure of the transformed matrix of plant uncertainties, $\Delta\boldsymbol{\Sigma}$, calculated for various kinds of plant uncertainty between 16 loudspeakers and 32 microphones in a 6 m × 2 m × 2 m enclosure excited at 88 Hz are shown in Figs 4.14 to 4.17 (Omoto and Elliott, 1996, 1999) and 4.18 (Baek and Elliott, 2000). Figure 4.14 corresponds to a random, unstructured, perturbation in the plant, in which each of the elements of $\Delta\mathbf{G}_{\mathrm{p}}$ was generated with independent random real and imaginary parts. Polar plots of each of the elements in the $\Delta\boldsymbol{\Sigma}$ matrix are shown in Fig. 4.14 for 10 such random changes. It can be seen that each of the elements of $\Delta\boldsymbol{\Sigma}$ in this case also appear to have a random phase and approximately equal average amplitude. In particular, the elements in the upper middle rows and the lower right-hand columns, as shown in Fig. 4.13, have a significant magnitude and so only small levels of random uncertainty of this kind can make the control system unstable if no effort weighting is used, or induce poor performance if the system does happen to be stable (Boucher et al., 1991).

In contrast, Fig. 4.15 shows the structure of the $\Delta\mathbf{\Sigma}$ matrix if 10 random changes are made to the positions of the secondary actuators, which are the loudspeakers in the active sound control system described above. Even though the average sum of squared magnitudes of the elements in this matrix is similar to that shown in Fig. 4.14, it now has considerably more structure. In particular, the magnitudes of the elements in both the upper middle rows and the lower right-hand columns are relatively small. Perturbations of the actuator positions in this system thus do not have a significant impact on the stability of the system, unless the distance the actuators are moved becomes substantial, and also do not significantly affect the attenuation achieved at the microphones (Omoto and Elliott, 1999).

If the positions of the error sensors, which are microphones in this case, rather than the loudspeakers are randomly perturbed, the $\Delta\mathbf{\Sigma}$ matrix has a different structure, as shown in Fig. 4.16. In this case, however, the magnitude of the elements in the upper middle rows and the lower right-hand columns is significant. Relatively small changes in the position of the microphones in this control system can thus make the system unstable or perform poorly.

Figure 4.14 The structure of the transformed matrix of plant uncertainties, $\Delta\mathbf{\Sigma}$, in the case of independent random changes in the real and imaginary parts of each of the elements of the plant matrix $\mathbf{G}$.

Figure 4.15 The structure of the transformed matrix of plant uncertainties, $\Delta\mathbf{\Sigma}$, in the case of independent random changes to the positions of the actuators.

Figure 4.17 shows the $\Delta\mathbf{\Sigma}$ matrix if the excitation frequency is changed, so that $\hat{\mathbf{G}}$ is measured at one frequency, but the true response, $\mathbf{G}$, corresponds to a slightly different frequency. The $\Delta\mathbf{\Sigma}$ matrix clearly has a very definite structure in this case with very large elements in the upper diagonal positions, but very small elements in the upper middle rows and lower right-hand columns. Consequently, the stability and the performance of this control system are very insensitive to this kind of uncertainty, even though the magnitude of the elements in the $\Delta\mathbf{\Sigma}$ matrix can be relatively large. Similar results are obtained if the temperature in the enclosure varies, since this changes the natural frequencies of all the acoustic modes in the enclosure, which has a similar effect to changing the excitation frequency (Maurer, 1996).

Another common form of plant uncertainty in a practical active sound control system in a room is that due to the movement of people within the enclosure. The effect of such perturbations on the plant matrix has been studied by Baek and Elliott (2000), and Fig. 4.18 shows the structure of the resulting $\Delta\mathbf{\Sigma}$ matrix if 10 random changes are made to the positions of six spheres inside the enclosure, with the volume of the sphere being similar to that of a person. Once again the largest perturbations occur on the leading diagonal of $\Delta\mathbf{\Sigma}$,

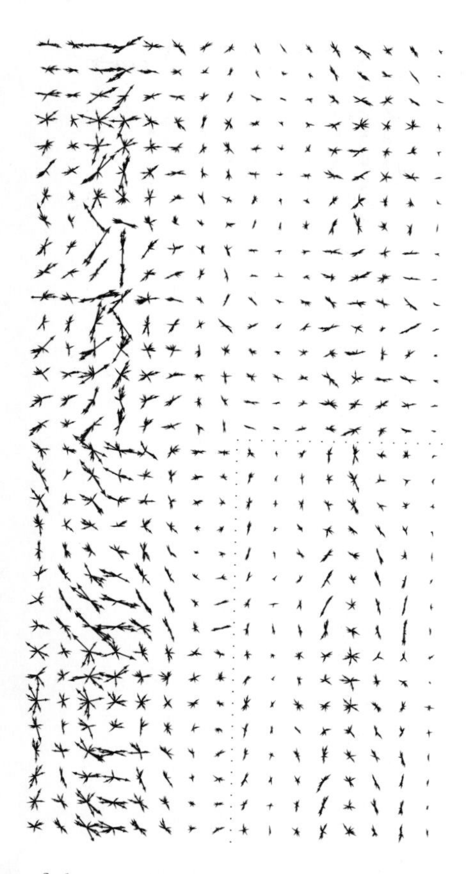

Figure 4.16 The structure of the transformed matrix of plant uncertainties, $\Delta\Sigma$, in the case of independent random changes to the positions of the sensors.

indicating that the values of the principal components themselves are subject to significant change. There are also some perturbations in the upper rows of $\Delta\Sigma$ in this case but these are relatively small in the rows and columns that determine the stability and performance, as indicated in Fig. 4.13.

These examples illustrate the fact that the $\Delta\Sigma$ matrix combines the description of the plant uncertainty with the structure of the nominal plant in a very useful way and so is able to give a straightforward geometric interpretation to the potential effects of such plant uncertainties on the stability and performance of the control system.

4.5. ITERATIVE LEAST-SQUARES ALGORITHMS

The steepest-descent algorithm has been shown to have some useful and robust properties, particularly if a component of effort weighting is included in the cost function. The time constant of the slowest mode of the steepest-descent algorithm, however, is strongly influenced by the eigenvalue spread of the Hessian matrix of the quadratic cost function

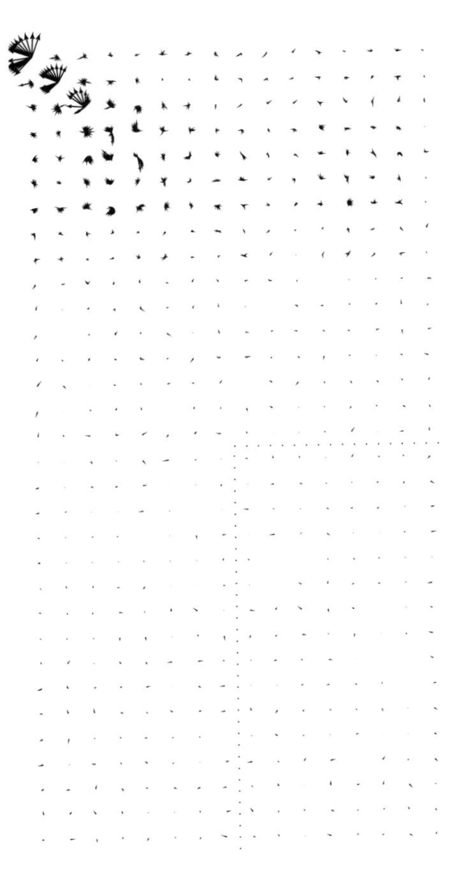

Figure 4.17 The structure of the transformed matrix of plant uncertainties, $\Delta\boldsymbol{\Sigma}$, in the case of changes in the excitation frequency.

being minimised, e.g. **A** in equation (4.2.4). We have seen that this eigenvalue spread can be very large, about 10^5 in the example used to generate Fig. 4.5, but that its effective value can be considerably reduced by including a sensible degree of control effort weighting in the cost function. Nevertheless, in some active control applications the slow modes of convergence associated with the steepest-descent algorithm may well limit the practical performance of the system, since the control system cannot fully track non-stationarities in the disturbance.

It is thus important to investigate adaptive algorithms that hold out the hope of having convergence properties which are less dependent on this eigenvalue spread than the steepest-descent algorithm, such as algorithms that iteratively compute the exact least-squares solution. We will find, however, that the theoretically exact forms of these algorithms are not very robust to implementation errors, particularly those associated with the estimated plant response. A modified form of these exact least-squares algorithms is then introduced, in which a balance is obtained between rapid convergence and good robustness. This balance is determined by a parameter which can be varied to obtain the steepest-descent algorithm in one extreme or the exact least-squares algorithm in the other.

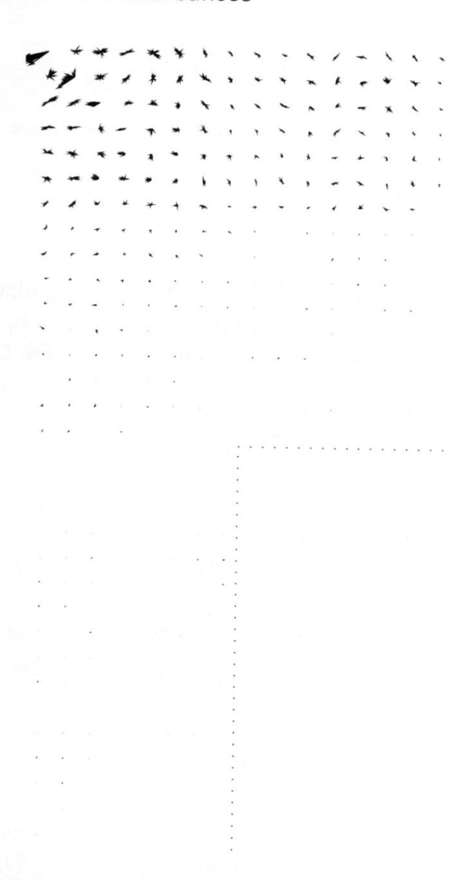

Figure 4.18 The structure of the transformed matrix of plant uncertainties, $\Delta\boldsymbol{\Sigma}$, for the case in which objects are moved to different random locations within the enclosure.

A controller is also discussed which is based on the transformed input and output signals, which were introduced for the analysis of the steepest descent algorithm in Section 4.3. This controller structure can also be arranged to implement a whole range of control algorithms from steepest descent to exact least-squares, but the insight provided by working with the transformed variables also suggests a number of other control strategies. Finally, a decentralised controller is analysed in which only one error signal is used to adjust the input to only one secondary source.

4.5.1. Gauss–Newton Algorithm

We have seen that for the overdetermined multichannel tonal system described by the equation

$$\mathbf{e}(n) = \mathbf{d} + \mathbf{G}\,\mathbf{u}(n)\,, \tag{4.5.1}$$

the optimal solution, which minimises $\mathbf{e}^{\mathrm{H}}\mathbf{e}$, is given by

$$\mathbf{u}_{\text{opt}} = -\left(\mathbf{G}^{\text{H}}\mathbf{G}\right)^{-1}\mathbf{G}^{\text{H}}\mathbf{d} \ , \tag{4.5.2}$$

assuming $\mathbf{G}^{\text{H}}\mathbf{G}$ is not singular. If an estimate of the plant response, $\hat{\mathbf{G}}$, is used to calculate the optimal solution, then we obtain

$$\hat{\mathbf{u}}_{\text{opt}} = -\left(\hat{\mathbf{G}}^{\text{H}}\hat{\mathbf{G}}\right)^{-1}\hat{\mathbf{G}}^{\text{H}}\mathbf{d} \ , \tag{4.5.3}$$

where $\hat{\mathbf{G}}^{\text{H}}\hat{\mathbf{G}}$ is assumed to be positive definite and thus not singular.

An iterative application of this estimate of the optimal solution, which could be used in order to track a time-varying disturbance, leads to a form of the Gauss–Newton iterative algorithm given by

$$\mathbf{u}(n+1) = \mathbf{u}(n) - \left(\hat{\mathbf{G}}^{\text{H}}\hat{\mathbf{G}}\right)^{-1}\hat{\mathbf{G}}^{\text{H}}\mathbf{e}(n) \ . \tag{4.5.4}$$

Substituting equation (4.5.1) for $\mathbf{e}(n)$ into this algorithm gives

$$\mathbf{u}(n+1) = \mathbf{u}(n) - \left(\hat{\mathbf{G}}^{\text{H}}\hat{\mathbf{G}}\right)^{-1}\left(\hat{\mathbf{G}}^{\text{H}}\mathbf{d} + \hat{\mathbf{G}}^{\text{H}}\mathbf{G}\mathbf{u}(n)\right) \ . \tag{4.5.5}$$

If the algorithm is stable, it must converge to a solution that sets the term in the final brackets in equation (4.5.5) to zero, since $\hat{\mathbf{G}}^{\text{H}}\hat{\mathbf{G}}$ is assumed to be positive definite, so that under these conditions the steady state solution is given by

$$\mathbf{u}_{\infty} = -\left(\hat{\mathbf{G}}^{\text{H}}\mathbf{G}\right)^{-1}\hat{\mathbf{G}}^{\text{H}}\mathbf{d} \ . \tag{4.5.6}$$

This is exactly the same steady state control vector as for the practical steepest-descent algorithm, equation (4.4.4) with β = 0. Using equation (4.5.6), the iterative control algorithm given by equation (4.5.4) can be written as

$$\left(\mathbf{u}(n+1) - \mathbf{u}_{\infty}\right) = \left[\mathbf{I} - \left(\hat{\mathbf{G}}^{\text{H}}\hat{\mathbf{G}}\right)^{-1}\hat{\mathbf{G}}^{\text{H}}\mathbf{G}\right]\left(\mathbf{u}(n) - \mathbf{u}_{\infty}\right) \ . \tag{4.5.7}$$

The robustness of this algorithm to plant uncertainties can be analysed by assuming that the plant model exactly represents the true nominal plant, so that $\hat{\mathbf{G}} = \mathbf{G}_0$, and that the plant is subject to some additive uncertainty, so that $\mathbf{G} = \mathbf{G}_0 + \Delta\mathbf{G}_{\text{p}}$, as in the previous section. Again using the SVD of the nominal plant, $\hat{\mathbf{G}} = \mathbf{G}_0 = \mathbf{R}_0\,\boldsymbol{\Sigma}\,\mathbf{Q}_0^{\text{H}}$, then the pseudo-inverse of $\hat{\mathbf{G}}$ can be written as

$$\left(\hat{\mathbf{G}}^{\text{H}}\hat{\mathbf{G}}\right)^{-1}\hat{\mathbf{G}}^{\text{H}} = \mathbf{Q}_0\,\boldsymbol{\Sigma}^{\dagger}\,\mathbf{R}_0^{\text{H}} \ , \tag{4.5.8}$$

where $\boldsymbol{\Sigma}^{\dagger}$ is the pseudo-inverse of $\boldsymbol{\Sigma}$, which can be written in the overdetermined case, as

$$\boldsymbol{\Sigma}^{\dagger} = \begin{bmatrix} 1/\sigma_1 & 0 & & \cdots & \cdots & 0 \\ 0 & 1/\sigma_2 & & & & \\ & & \ddots & & & \\ 0 & & & 1/\sigma_M & \cdots & 0 \end{bmatrix} . \tag{4.5.9}$$

We can also decompose $\Delta\mathbf{G}_\mathrm{p}$ using the eigenvector matrices of $\mathbf{G}_0$ to give the matrix $\Delta\mathbf{\Sigma}$, as in the previous section, so that

$$\Delta\mathbf{G}_\mathrm{p} = \mathbf{R}_0\,\Delta\mathbf{\Sigma}\,\mathbf{Q}_0^\mathrm{H}\,. \tag{4.5.10}$$

Premultiplying equation (4.5.7) by $\mathbf{Q}_0^\mathrm{H}$ and using the definitions of the transformed control variables $\mathbf{v}(n)$ and $\mathbf{v}_\infty$ used in Section 4.3, the iterative least-squares algorithm given by equation (4.5.7) can be written as

$$\left(\mathbf{v}(n+1)-\mathbf{v}_\infty\right) = \left[\mathbf{I}-\mathbf{\Sigma}^\dagger\left(\mathbf{\Sigma}+\Delta\mathbf{\Sigma}\right)\right]\left(\mathbf{v}(n)-\mathbf{v}_\infty\right)\,, \tag{4.5.11}$$

or

$$\left(\mathbf{v}(n+1)-\mathbf{v}_\infty\right) = -\left[\mathbf{\Sigma}^\dagger\Delta\mathbf{\Sigma}\right]\left(\mathbf{v}(n)-\mathbf{v}_\infty\right)\,, \tag{4.5.12}$$

since the matrix $\mathbf{\Sigma}^\dagger\mathbf{\Sigma}=\mathbf{I}$. If $\Delta\mathbf{\Sigma}=\mathbf{0}$ the iterative least-squares algorithm will converge in one step, as expected. In the presence of modelling errors, however, the $M\times M$ matrix $\mathbf{\Sigma}^\dagger\Delta\mathbf{\Sigma}$ takes the form

$$\mathbf{\Sigma}^\dagger\Delta\mathbf{\Sigma} = \begin{bmatrix} \dfrac{\Delta\sigma_{11}}{\sigma_1} & \dfrac{\Delta\sigma_{12}}{\sigma_1} & \cdots & \dfrac{\Delta\sigma_{1M}}{\sigma_1} \\ \vdots & & & \\ \dfrac{\Delta\sigma_{M1}}{\sigma_M} & \dfrac{\Delta\sigma_{M2}}{\sigma_M} & \cdots & \dfrac{\Delta\sigma_{MM}}{\sigma_M} \end{bmatrix}, \tag{4.5.13}$$

where $\Delta\sigma_{mn}$ is the m,n-th element of $\Delta\mathbf{\Sigma}$.

If the plant has some very small singular values, as in the example used to generate Fig. 4.5, then $1/\sigma_M$, $1/\sigma_{M-1}$ etc., will be very large numbers and the lower part of the $\mathbf{\Sigma}^\mathrm{T}\Delta\mathbf{\Sigma}$ matrix will have large components if the magnitudes of the elements $\Delta\sigma_{lm}$ are significant for $l<M$. The lower elements of the vector $\mathbf{v}(n+1)-\mathbf{v}_\infty$ in equation (4.5.12) will then tend to be greatly amplified, compared with the corresponding elements of $\mathbf{v}(n)-\mathbf{v}_\infty$, potentially causing instability. The elements of the $\Delta\mathbf{\Sigma}$ matrix that have the greatest influence on the stability are thus those in the M-th row and directly above, which are the elements in the region of the $\Delta\mathbf{\Sigma}$ matrix associated with stability in Fig. 4.13. The practical implementation of the iterative least-squares algorithm is thus considerably less robust to small perturbations in the plant response than the steepest-descent algorithm. This is because the exact least-squares algorithm causes some elements of the transformed control vector to take very large steps, since they are amplified by a factor of $1/\sigma_M$, $1/\sigma_{M-1}$, etc., in equation (4.5.4), which may not be in the correct direction. This problem can be reduced by including some effort weighting in the cost function being minimised, but will always persist to some extent if the eigenvalue spread of the relevant Hessian matrix is large, which is generally why the Gauss–Newton algorithm was being used in the first place.

4.5.3. A General Adaptive Algorithm

We will now consider a rather general class of algorithm which is inspired by the iterative least-squares method but which overcomes some of its robustness problems. Note that the update term in the iterative least-squares algorithm, equation (4.5.4), is equal to the estimated complex gradient vector, pre-multiplied by the matrix $\left[\hat{\mathbf{G}}^{\mathrm{H}}\hat{\mathbf{G}}\right]^{-1}$. This suggests that a whole family of adaptive algorithms could be produced by premultiplying the estimated gradient vector $\hat{\mathbf{g}}(n) = 2\hat{\mathbf{G}}^{\mathrm{H}}\,\mathbf{e}(n)$ by some more general matrix, $\mathbf{D}$, which is assumed here to be fixed, so that (Boucher, 1992)

$$\mathbf{u}(n+1) \;=\; \mathbf{u}(n) - \mu\,\mathbf{D}\,\hat{\mathbf{g}}(n) \;, \tag{4.5.14}$$

where μ is a convergence factor, which strictly speaking could be absorbed into the definition of $\mathbf{D}$. Note that the steepest-descent algorithm, equation (4.3.7), is obtained if $\mathbf{D} = \mathbf{I}$ and $\hat{\mathbf{G}} = \mathbf{G}$, and the iterative least-squares algorithm, equation (4.5.4), is obtained if $\mathbf{D} = \left[\hat{\mathbf{G}}^{\mathrm{H}}\hat{\mathbf{G}}\right]^{-1}$ and $\mu = 1$. A more general form for $\mathbf{D}$ when minimising the unweighted sum of squared errors is

$$\mathbf{D} \;=\; \left[\hat{\mathbf{G}}^{\mathrm{H}}\hat{\mathbf{G}} + \delta\mathbf{I}\right]^{-1} \;, \tag{4.5.15}$$

where δ is a *regularisation parameter* which is chosen to obtain a good balance between transient performance and robustness. Note that the steady state solution to equation (4.5.14), assuming it is stable, is given by setting $\mu\,\mathbf{D}\,\hat{\mathbf{g}}(n)$ to zero and, provided $\mathbf{D}$ is positive definite, this can only be true if $\hat{\mathbf{g}}(n) = 0$. The steady state solution for this algorithm is thus again given by equation (4.5.6) and is independent of the form of $\mathbf{D}$. The regularisation parameter δ in equation (4.5.15) will thus not affect the potential steady-state performance for the control algorithm, but it will determine its convergence properties and whether the algorithm is robustly stable enough to reach this solution. If an effort weighting parameter is chosen to prevent excessive control effort for ill-conditioned systems, as described above, the estimated complex gradient is equal to

$$\hat{\mathbf{g}} \;=\; 2\left[\hat{\mathbf{G}}^{\mathrm{H}}\mathbf{e}(n) + \beta\,\mathbf{u}(n)\right] . \tag{4.5.16}$$

The complete algorithm to minimise the sum of squared errors is thus given by

$$\mathbf{u}(n+1) \;=\; \mathbf{u}(n) - \alpha\left[\hat{\mathbf{G}}^{\mathrm{H}}\hat{\mathbf{G}} + \delta\,\mathbf{I}\right]^{-1}\left[\hat{\mathbf{G}}^{\mathrm{H}}\mathbf{e}(n) + \beta\,\mathbf{u}(n)\right] \;, \tag{4.5.17}$$

where β is chosen to ensure a sensible steady state solution, as outlined in Section 4.3, and δ is independently chosen to give a good compromise between convergence time and robust stability. For higher values of the convergence coefficient α, however, β will also influence the trade-off between convergence time and robust stability.

In order to ensure that the algorithm given in equation (4.5.14) is as widely applicable as possible, we return to the general form of the cost function given by equation (4.2.19), which is repeated here for convenience,

$$J = \mathbf{e}^{\mathrm{H}}\,\mathbf{W}_e\,\mathbf{e} + \mathbf{u}^{\mathrm{H}}\,\mathbf{W}_u\,\mathbf{u} \;. \tag{4.5.18}$$

Using equation (4.4.1) for $\mathbf{e}$ allows equation (4.5.18) to be written in general Hermitian quadratic form with the $\mathbf{A}$ matrix and $\mathbf{b}$ vector defined in equation (4.2.21). The complex gradient vector for this cost function can thus be written as

$$\mathbf{g}(n) = 2\left[\mathbf{A}\,\mathbf{u}(n) + \mathbf{b}\right] = 2\left[\mathbf{G}^{\mathrm{H}}\,\mathbf{W}_e\,\mathbf{e}(n) + \mathbf{W}_u\,\mathbf{u}(n)\right], \tag{4.5.19}$$

so that the practical estimate of the complex gradient vector is equal to

$$\hat{\mathbf{g}}(n) = 2\left[\hat{\mathbf{G}}^{\mathrm{H}}\,\mathbf{W}_e\,\mathbf{e}(n) + \mathbf{W}_u\,\mathbf{u}(n)\right]. \tag{4.5.20}$$

The pre-conditioning matrix for an iterative least-squares algorithm in this case would be equal to

$$\mathbf{A}^{-1} = \left[\hat{\mathbf{G}}^{\mathrm{H}}\mathbf{W}_e\,\hat{\mathbf{G}} + \mathbf{W}_u\right]^{-1}. \tag{4.5.21}$$

Potential differences between the responses of the plant model and the physical plant mean that a pre-conditioning matrix given by equation (4.5.21) may not give the best trade-off between adaptation speed and robustness, and a more general form would be

$$\mathbf{D} = \left[\hat{\mathbf{G}}^{\mathrm{H}}\mathbf{W}_e\,\hat{\mathbf{G}} + \mathbf{W}_\delta\right]^{-1}, \tag{4.5.22}$$

where $\mathbf{W}_\delta$ is a matrix independently chosen to regularise $\mathbf{D}$. A value of $\mathbf{W}_\delta = \mathbf{W}_u + \delta\mathbf{I}$ may be a sensible first approximation. Assuming $\mathbf{W}_\delta$ and hence $\mathbf{D}$, has been selected the complete adaptive algorithm can be derived by substituting equation (4.5.20) into equation (4.5.14) to give

$$\mathbf{u}(n+1) = \left[\mathbf{I} - \alpha\,\mathbf{D}\,\mathbf{W}_u\right]\mathbf{u}(n) - \alpha\,\mathbf{D}\,\hat{\mathbf{G}}^{\mathrm{H}}\,\mathbf{W}_e\,\mathbf{e}(n) \tag{4.5.23}$$

where α is again equal to 2μ.

4.5.4. Controllers Based on Transformed Signals

Another approach to improving the convergence speed of multichannel tonal control systems is to transform the input and output signals into a form that is easier to work with (Morgan, 1991; Clark, 1995; Popovich, 1996). One such transformation, discussed in Section 4.3, is that based on the singular value decomposition of the plant response matrix

$$\mathbf{G} = \mathbf{R}\,\boldsymbol{\Sigma}\,\mathbf{Q}^{\mathrm{H}}. \tag{4.5.24}$$

The matrices of eigenvectors are used to transform the physical input and output signals into the form

$$\mathbf{v}(n) = \mathbf{Q}^{\mathrm{H}}\,\mathbf{u}(n), \tag{4.5.25}$$

$$\mathbf{y}(n) = \mathbf{R}^{\mathrm{H}}\,\mathbf{e}(n) \tag{4.5.26}$$

so that the transformed output signals are related to the transformed input signals by the equation

$$\mathbf{y}(n) = \mathbf{p} + \boldsymbol{\Sigma}\,\mathbf{v}(n), \tag{4.5.27}$$

where $\mathbf{p} = \mathbf{R}^{\mathrm{H}}\mathbf{d}$ is the transformed disturbance vector, and $\boldsymbol{\Sigma}$ is the $M \times L$ matrix of singular values of $\mathbf{G}$. The first M elements of the transformed output vector, $\mathbf{y}(n)$, are linearly related to a single transformed input, $\mathbf{v}(n)$, via one of the singular values of $\mathbf{G}$, and the final $L - M$ elements of $\mathbf{y}(n)$ are unaffected by any transformed input, as in equations (4.3.18) and (4.3.19). The decoupled structure of equation (4.5.27), and the fact that the singular values are real and thus free from phase shift, suggests that an efficient control algorithm would be to adapt the m-th element of $\mathbf{v}(n)$ to minimise the m-th element of $\mathbf{y}(n)$ using the simple gradient descent algorithm in the transformed or *principal components*, and such an algorithm can be written as

$$v_m(n+1) = v_m(n) - \alpha_m \, y_m(n) \; , \tag{4.5.28}$$

where α_m is the convergence coefficient for the m-th component of $\mathbf{v}(n)$, which is $v_m(n)$. Figure 4.19 shows the block diagram for such a controller architecture.

If the convergence coefficients are made proportional to the corresponding singular values, so that $\alpha_m = \alpha\,\sigma_m$ where α is an overall convergence coefficient, then equation (4.5.28) can be written in matrix terms as

$$\mathbf{v}(n+1) = \mathbf{v}(n) - \alpha\,\boldsymbol{\Sigma}^{\mathrm{T}}\,\mathbf{y}(n) \; . \tag{4.5.29}$$

Using the definitions of $\mathbf{v}(n)$ and $\mathbf{y}(n)$ above, equation (4.5.29) can be written in terms of the physical control signals as

$$\mathbf{u}(n+1) = \mathbf{u}(n) - \alpha\,\mathbf{G}^{\mathrm{H}}\,\mathbf{e}(n) \; , \tag{4.5.30}$$

where $\mathbf{G}^{\mathrm{H}} = \mathbf{Q}\,\boldsymbol{\Sigma}^{\mathrm{T}}\,\mathbf{R}^{\mathrm{H}}$, which is the steepest-descent algorithm for the adaptation of the physical control signals, as in equation (4.3.7).

If the individual convergence coefficients in equation (4.5.28) are instead made inversely proportional to the singular values, so that $\alpha_m = \alpha/\sigma_m$, then the adaptation equation can be written as

$$\mathbf{v}(n+1) = \mathbf{v}(n) - \alpha\left[\boldsymbol{\Sigma}^{\mathrm{T}}\,\boldsymbol{\Sigma}\right]^{-1}\boldsymbol{\Sigma}^{\mathrm{T}}\,\mathbf{y}(n) \; , \tag{4.5.31}$$

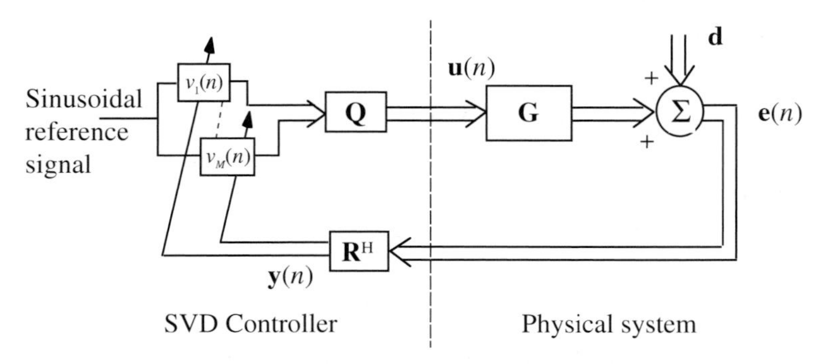

Figure 4.19 Block diagram of a controller in which the transformed input signals, $v_m(n)$, are adapted using the transformed output signals in $\mathbf{y}(n)$. The transformed output signals in $\mathbf{y}(n)$ are obtained from the physical output signals in $\mathbf{e}(n)$ via the matrix $\mathbf{R}^{\mathrm{H}}$, and $\mathbf{v}(n)$ is transformed back into the physical input signals by the matrix $\mathbf{Q}$, where $\mathbf{R}$ and $\mathbf{Q}$ are both obtained from a singular value decomposition (SVD) of the physical plant matrix $\mathbf{G}$.

where $\left[\boldsymbol{\Sigma}^{\mathrm{T}}\boldsymbol{\Sigma}\right]^{-1}\boldsymbol{\Sigma}^{\mathrm{T}}$ is the pseudo-inverse of $\boldsymbol{\Sigma}$ in this case. Again using the definitions of $\mathbf{v}(n)$ and $\mathbf{y}(n)$ and also the definition of the pseudo-inverse in this overdetermined case in Table 4.1, equation (4.5.31) can be written in terms of the physical control signals as

$$\mathbf{u}(n+1) \;=\; \mathbf{u}(n) - \alpha\left[\mathbf{G}^{\mathrm{H}}\mathbf{G}\right]^{-1}\mathbf{G}^{\mathrm{H}}\mathbf{e}(n) \;, \tag{4.5.32}$$

where $\left[\mathbf{G}^{\mathrm{H}}\mathbf{G}\right]^{-1}\mathbf{G}^{\mathrm{H}} = \mathbf{Q}\left[\boldsymbol{\Sigma}^{\mathrm{T}}\boldsymbol{\Sigma}\right]^{-1}\boldsymbol{\Sigma}^{\mathrm{T}}\mathbf{R}^{\mathrm{H}}$, which is Newton's algorithm, as in equation (4.5.4), with a perfect plant model.

Thus both the steepest-descent algorithm and Newton's algorithm are seen to be special cases of the adaptation algorithm in the principal coordinates given by equation (4.5.28) (Maurer, 1996; Cabell, 1998). Cabell (1998) has also suggested that the use of equation (4.5.28) with a constant convergence coefficient, $\alpha_m = \alpha$, can give a good compromise between the slow convergence of the steepest-descent algorithm and the lack of robustness to plant uncertainties experienced by Newton's algorithm. A more general form of equation (4.5.28) that includes individual effort-weighting terms in each of the principal co-ordinates is given by

$$v_m(n+1) \;=\; \left(1-\alpha_m\beta_m\right)v_m(n) - \alpha_m\, y_m(n) \;. \tag{4.5.33}$$

By choosing the control effort weighting and the convergence coefficient independently for each principal component, considerable flexibility can be introduced into this control algorithm, which is called the *PC-LMS algorithm* by Cabell (1998). One possibility would be to use a finite value of control effort weighting for the lower principal components, which are subject to greatest uncertainty, and to set the control effort weighting to zero for the larger principal components, where it would have greatest influence on the reduction in the mean-square error (Rossetti et al., 1996).

Another variant of this algorithm is obtained for the case in which only the values of p_m for larger values of m contribute significantly to the sum of squared values of the errors. In this case only the transformed control signals for these values of m need to be adapted, and the reduced algorithm could be considerably more efficient to implement than when all the transformed control signals are adapted.

4.5.5. Processing Requirements

A general form of linear adaptation algorithm, in which the new control vector, $\mathbf{u}(n+1)$, is a linear function of the current control, $\mathbf{u}(n)$, vector and of the current error, $\mathbf{e}(n)$, is given by

$$\mathbf{u}(n+1) = \mathbf{M}_1\,\mathbf{u}(n) + \mathbf{M}_2\,\mathbf{e}(n) \;. \tag{4.5.34}$$

Note that the algorithms given by equations (4.5.23) and (4.5.33) can both be expressed in this form.

Since we are assuming that $\mathbf{M}_1$ and $\mathbf{M}_2$ are fixed, they can be pre-computed before the control algorithm is implemented in real time. At each iteration this algorithm requires the multiplication of an $M \times M$ matrix with an $M \times 1$ vector and the multiplication of an $M \times L$ matrix with an $L \times 1$ vector, i.e. a total of $M^2 + ML$ complex scalar multiplications.

A computational saving can be made when implementing the steepest-descent algorithm, equation (4.3.7), since in this case $\mathbf{M}_1$ is diagonal. This also has implications for the architecture of the controller, since if $\mathbf{M}_1$ is diagonal the adaptation of any one control signal can proceed independently of the adaptation of all the other control signals. The control algorithm can thus be conveniently partitioned to separately process each of the control signals, and the processors used to implement this may not all be located in one place. Each of the error signals must still be communicated to each of these processors, however, and so the reduction in wiring for a system with independent adaptation of each control signal is not very great compared with a system in which the adaptation of the control signals is coupled.

Another aspect of the general adaptation algorithm in equation (4.5.34) is its behaviour when there are failures in either the secondary actuators or error sensors. In general, either type of failure will cause equation (4.5.34) to converge to a non-optimal solution and may even destabilise the control system. When the steepest-descent algorithm of equation (4.3.7) is implemented, however, equation (4.3.34) has a special form with $\mathbf{M}_1 = \mathbf{I}$ and $\mathbf{M}_2 = -\alpha\mathbf{G}^{\mathrm{H}}$. Because $\mathbf{M}_1$ is diagonal then a failure in one actuator will not affect the adaptation of the other actuators, and the steepest-descent algorithm will still minimise the same cost function, but with a reduced number of actuators. A failure of an error sensor would be equivalent, in the steepest-descent algorithm, to setting a column of $\mathbf{G}^{\mathrm{H}}$ to zero. The resulting algorithm is thus exactly the same as the steepest-descent algorithm that minimises the sum of the squares of the remaining error sensors.

Stability robustness in the face of loop failures, caused for example by faults in the actuators or sensors, is said to determine the *integrity* of a control system (Maciejowski, 1989). The steepest-descent algorithm inherently possesses such integrity, but this cannot be said for all the algorithms described by equation (4.5.34), even though they may converge more quickly than the steepest-descent algorithm under fault-free conditions. In order to reintroduce a tolerance to transducer failures into these other algorithms, it is generally necessary to implement a supervisory program on top of the control program, which monitors the error and control signals and reconfigures the control system if any faults are detected. On-line identification of the matrix of plant responses could be used to detect sensor or actuator failures by monitoring the sum of squared elements in each row and column. Such a procedure inevitably adds to the processing requirements of the control algorithm.

4.5.6. Decentralised Control

Significant savings in the wiring of large control systems can be achieved if the control algorithm is divided into a number of systems, each of which independently adjusts a subset of the actuators' signals to minimise a subset of the sensors' signals. Such a control system is described as *decentralised*. Its stability can be analysed using the approaches outlined above, but in this case the estimated plant response, $\hat{\mathbf{G}}$, only has non-zero terms for the elements that represent the transfer response between the actuators and sensors of each individual control system. An extreme version of this philosophy is when there are as many actuators as sensors, and each actuator is adjusted to minimise the squared error at one sensor only. In this case the order of the transducers can be chosen so that the effective estimate of the plant response used by the whole set of control systems is diagonal.

Physical coupling still exists between the actuators and the other sensors, however, and this could cause the system to become unstable. The stability of such a system can still be established in this case by examining the real parts of the eigenvalues of the matrix $\hat{\mathbf{G}}^{\mathrm{H}}\mathbf{G}$, as in Section 4.4. Simple rules for stability can be established for low-order systems, however, as discussed by Elliott and Boucher (1994). In the case of a symmetric decentralised two-channel system, for example, the physical plant matrix and plant model used in the control algorithm are

$$\mathbf{G} = \begin{bmatrix} G_{11} & G_{12} \\ G_{12} & G_{11} \end{bmatrix} \quad \text{and} \quad \hat{\mathbf{G}} = \begin{bmatrix} G_{11} & 0 \\ 0 & G_{11} \end{bmatrix}. \tag{4.5.35,36}$$

In this case the eigenvalues of the matrix $\hat{\mathbf{G}}^{\mathrm{H}}\mathbf{G}$ are equal to

$$\lambda_{1,2} = |G_{11}|^2 \left[1 \pm G_{12}/G_{11}\right] . \tag{4.5.37}$$

The decentralised control system will be stable provided the real parts of both of these eigenvalues are positive, which is assured if

$$\left|\mathrm{Re}\left(G_{12}/G_{11}\right)\right| < 1 , \tag{4.5.38}$$

which quantifies the intuitive condition for stability, which is that the cross-coupling in the physical system must be smaller than the direct coupling. Elliott and Boucher (1994) considered the specific case of a pair of identical loudspeakers and microphones arranged symmetrically in free space, in which case G_{11} is proportional to e^{-jkl_1}/l_1 where k is the acoustic wavenumber and l_1 is the distance from the loudspeakers to their control microphones, and G_{12} is proportional to e^{-jkl_2}/l_2, where l_2 is the distance from each loudspeaker to the other microphone, as illustrated in Fig. 4.20. In this case the condition for stability, equation (4.5.38) becomes

$$\frac{l_1}{l_2} \cos k\left(l_1 - l_2\right) < 1 , \tag{4.5.39}$$

and when the size of the system is compact compared to the acoustic wavelength, which is typically the case in practice, then $\cos k(l_1 - l_2) \approx 1$, the stability condition can be written in the very simple form

$$l_1 < l_2 . \tag{4.5.40}$$

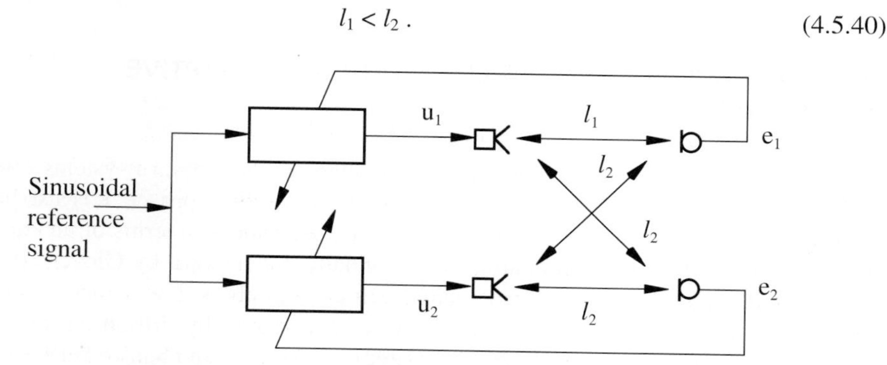

Figure 4.20 Two-channel decentralised active sound control system.

The system is thus stable provided a microphone is closer to the loudspeaker controlling it than it is to the other loudspeaker.

Figure 4.21 shows the results of some simulations of the behaviour of such a two-channel decentralised active sound control system, using the steepest-descent algorithm, for various microphone positions. These results illustrate that, although it may have some slow modes, the system is stable provided the microphones are not actually cross-connected with the loudspeakers used to control their outputs, a condition that would suggest that a mistake had been made in the wiring of the system.

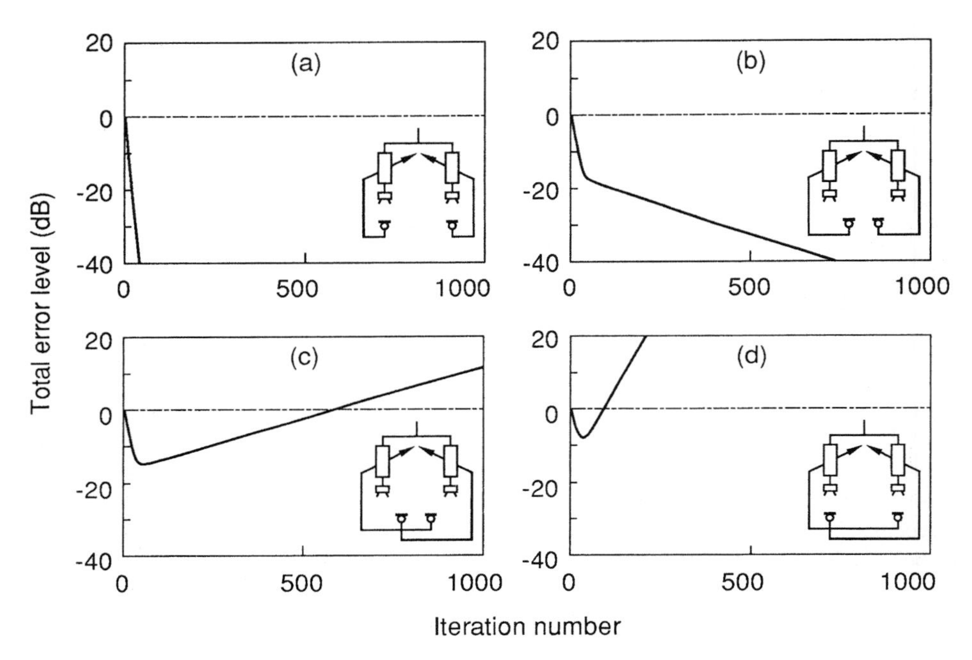

Figure 4.21 Transient behaviour of the sum of squared error signals for two independently implemented controllers in a simulation of an active sound control system in the free field with (a) $kl_1 = 0.1$ and $kl_2 = 0.32$; (b) $kl_1 = 0.17$ and $kl_2 = 0.19$; (c) $kl_1 = 0.19$ and $kl_2 = 0.17$; (d) $kl_1 = 0.32$ and $kl_2 = 0.1$. The approximate geometries of each of the control systems is sketched on each plot.

4.6. FEEDBACK CONTROL INTERPRETATION OF ADAPTIVE FEEDFORWARD SYSTEMS

The stability and convergence properties of the adaptive feedforward systems discussed above have so far been analysed using an iterative optimisation approach. It is also possible to analyse adaptive feedforward systems for tonal disturbances in terms of an equivalent feedback controller. This interpretation was originally put forward by Glover (1977) for algorithms that adaptively cancelled tonal electrical signals and was then used in the analysis of adaptive time-domain feedforward controllers by Elliott et al. (1987), Darlington (1987), Sievers and von Flotow (1992) and Morgan and Sanford (1992). In this section we briefly introduce this interpretation of multichannel adaptive feedforward controllers for tonal disturbances and indicate how it can be used to analyse the dynamic behaviour of a rapidly adapting controller.

Assuming the physical system under control has reached steady state conditions, the vector of complex error signals at the n-th iteration can be written in terms of the disturbance vector at the time of the n-th iteration and the control vector at this time as

$$\mathbf{e}(n) = \mathbf{d}(n) + \mathbf{G}\,\mathbf{u}(n)\,, \tag{4.6.1}$$

where $\mathbf{G}$ is a matrix of complex numbers whose elements are equal to the frequency response of the plant at the excitation frequency, ω_0. The vectors $\mathbf{e}(n)$, $\mathbf{d}(n)$ and $\mathbf{u}(n)$ are thus sequences of complex numbers, whose z-transform can be defined in a similar way to those for the normal (real) sequences of sampled signals in Section 2.2, as discussed by Therrien (1992), for example. If the z-transforms of these sequences are denoted $\tilde{\mathbf{e}}(z)$, $\tilde{\mathbf{d}}(z)$ and $\tilde{\mathbf{u}}(z)$, then the z-transform of equation (4.6.1) can be written as

$$\tilde{\mathbf{e}}(z) \;=\; \tilde{\mathbf{d}}(z) + \mathbf{G}\,\tilde{\mathbf{u}}(z)\;, \tag{4.6.2}$$

where $\mathbf{G}$ is still a matrix of constant complex coefficients.

Similarly, the z-transform of the terms in the equation for the general adaptive algorithm given by equation (4.5.34),

$$\mathbf{u}(n+1) = \mathbf{M}_1\,\mathbf{u}(n) + \mathbf{M}_2\,\mathbf{e}(n)\,, \tag{4.6.3}$$

can be taken to yield

$$\tilde{\mathbf{u}}(z) \;=\; \left[z\mathbf{I} - \mathbf{M}_1\right]^{-1} \mathbf{M}_2\,\tilde{\mathbf{e}}(z)\;. \tag{4.6.4}$$

The matrix relating the z-transform of the error signals to that of the control signals can be interpreted as a feedback controller, $\mathbf{H}(z)$, as shown in Fig. 4.22, so that

$$\tilde{\mathbf{u}}(z) \;=\; -\,\tilde{\mathbf{H}}(z)\,\tilde{\mathbf{e}}(z)\;, \tag{4.6.5}$$

where

$$\tilde{\mathbf{H}}(z) \;=\; -\left[z\,\mathbf{I} - \mathbf{M}_1\right]^{-1} \mathbf{M}_2\;. \tag{4.6.6}$$

Substituting equation (4.6.5) into equation (4.6.2), an equation for the z-transform of the complex error signals in terms of the z-transform of the complex disturbance signals can be obtained in the form

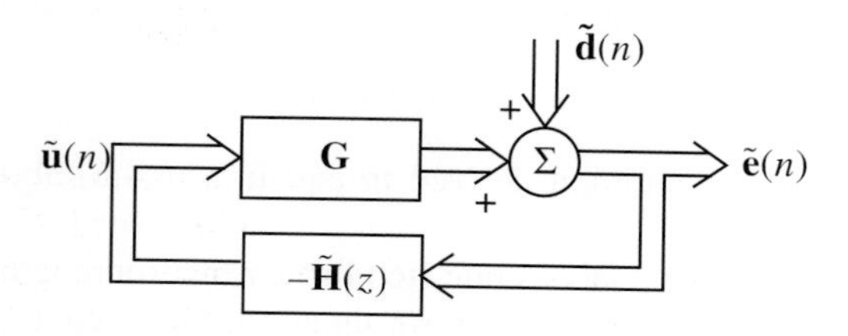

Figure 4.22 Feedback interpretation of an adaptive feedforward controller for tonal disturbances in which the z-transform of the sequence of frequency-domain error signals is transformed by the function $-\tilde{\mathbf{H}}(z)$ to produce the z-transform of the sequence of frequency domain control signals.

$$\tilde{\mathbf{e}}(z) = \left[\mathbf{I} + \mathbf{G}\tilde{\mathbf{H}}(z)\right]^{-1} \tilde{\mathbf{d}}(z) . \qquad (4.6.7)$$

The stability and performance of the adaptive feedforward controller can now be analysed using the tools of multichannel feedback control. If, for example, the simple steepest-descent algorithm is being analysed, then $\mathbf{M}_1$ in equation (4.6.3) is equal to $\mathbf{I}$, and $\mathbf{M}_2$ is equal to $-\alpha\hat{\mathbf{G}}^{\mathrm{H}}$, where $\hat{\mathbf{G}}$ is the complex matrix of estimated plant responses at the excitation frequency, ω_0. The equivalent feedback controller, equation (4.6.6), in this case is thus given by (Elliott, 1998b)

$$\tilde{\mathbf{H}}(z) = \frac{\alpha}{z-1} \hat{\mathbf{G}}^{\mathrm{H}} , \qquad (4.6.8)$$

which can be interpreted as having a phase correcting term, $\hat{\mathbf{G}}^{\mathrm{H}}$, and a digital integrator, $\alpha/(z-1)$. The stability of a feedback control system is governed by the position of its poles, which, as will be explained in more detail in Chapter 6, can be obtained by setting the determinant of $\mathbf{I} + \mathbf{G}\tilde{\mathbf{H}}(z)$ to zero, and for the special case of the steepest-descent algorithm this can be written as

$$\det\left[\mathbf{I} + \frac{\alpha}{z-1}\mathbf{G}\hat{\mathbf{G}}^{\mathrm{H}}\right] = 0 , \qquad (4.6.9)$$

so that

$$\det\left[(z-1)\mathbf{I} + \alpha\mathbf{G}\hat{\mathbf{G}}^{\mathrm{H}}\right] = 0 . \qquad (4.6.10)$$

It is noted in the Appendix that the determinant of a matrix is equal to the product of all its eigenvalues, so that the roots of the characteristic equation are given by the solutions to

$$z - 1 + \alpha\,\lambda_m = 0 , \qquad (4.6.11)$$

where λ_m is the m-th eigenvalue of $\mathbf{G}\hat{\mathbf{G}}^{\mathrm{H}}$, which is equal to the m-th eigenvalue of $\hat{\mathbf{G}}^{\mathrm{H}}\mathbf{G}$ for $1 \le m \le M$ and are otherwise zero, as explained in section A.7 of the Appendix. The position of the m-th pole of the equivalent feedback system is thus given by

$$z_m = 1 - \alpha\lambda_m . \qquad (4.6.12)$$

The stability of the algorithm is assured if none of these poles are outside the unit circle, i.e. $|z_m| \le 1$, for all m, and using equation (4.6.12) this condition is thus equivalent to

$$|1 - \alpha\lambda_m| \le 1 \qquad \text{for all } \lambda_m . \qquad (4.6.13)$$

This is exactly the stability condition derived in equation (4.4.6) above for the case in which $\beta = 0$.

Eatwell (1995) has taken this analysis one step further by dropping the assumption that the response of the plant to the sinusoidal control signals has reached its steady state. Under these more general conditions equation (4.6.1) can be written as

$$\mathbf{e}(n) = \mathbf{d}(n) + \sum_{i=0}^{\infty} \mathbf{G}_i\,\mathbf{u}(n-i) , \qquad (4.6.14)$$

where $\mathbf{G}_i$ is now the complex response of the plant at iteration number i to a sinusoidal input introduced at iteration number 0, and has been termed the *frequency-domain impulse response* by Eatwell (1995). The form of $\mathbf{G}_i$ will depend upon the dynamics of the plant and exactly how the complex error signals are derived from the time-domain error signals, and how the time-domain control signals are derived from their complex equivalent. In the system discussed by Eatwell (1995), complex modulation and demodulation were assumed, followed by low-pass filtering, decimation and interpolation. Since the steady state value of the plant response at ω_0 is equal to $\mathbf{G}$, then the frequency-domain impulse response must have the property that

$$\sum_{i=0}^{\infty} \mathbf{G}_i = \mathbf{G} . \tag{4.6.15}$$

If the iteration rate is so slow that the steady state value of the error is reached after each iteration, as assumed above, then $\mathbf{G}_0 = \mathbf{G}$, $\mathbf{G}_i = 0$ for $i \geq 1$, and equation (4.6.14) becomes equal to equation (4.6.1). More generally, however, equation (4.6.14) allows the response of the plant to be characterised under transient conditions and so allows adaptive feedforward control systems operating at faster iteration rates to be analysed.

Taking the z-transform of equation (4.6.14) we obtain

$$\tilde{\mathbf{e}}(z) = \tilde{\mathbf{d}}(z) + \tilde{\mathbf{G}}(z)\,\tilde{\mathbf{u}}(z) , \tag{4.6.16}$$

where $\tilde{\mathbf{G}}(z)$ is the z-transform of the sequence $\mathbf{G}_i$.

Combining equation (4.6.16) with the z-domain equation for the adaptive feedforward control algorithm, equation (4.6.5), allows the z-transform of the sequence of complex error vectors to now be written as

$$\tilde{\mathbf{e}}(z) = \left[\mathbf{I} + \tilde{\mathbf{G}}(z)\tilde{\mathbf{H}}(z)\right]^{-1} \tilde{\mathbf{d}}(z) , \tag{4.6.17}$$

and so the steady state and transient performance of any adaptive tonal feedforward control system can be analysed under a wide variety of conditions. Eatwell (1995) has used this analysis to show that the iterative Newton's algorithm, for which $\mathbf{M}_1 = \mathbf{I}$ in equation (4.6.3) and $\mathbf{M}_2 = -\alpha\mathbf{G}^{\dagger}$, where $\mathbf{G}^{\dagger}$ is the pseudoinverse of $\mathbf{G}$ which is assumed to have been identified with no error, is only stable for a value of convergence coefficient that satisfies

$$\alpha \leq \frac{1}{\Delta} , \tag{4.6.18}$$

where Δ is the number of samples of pure delay in the response of the physical system under control. This result emphasises the fact that although the analysis in Sections 4.3 and 4.5 can give some guidance as to the convergence properties of the various adaptive feedforward algorithms, the dynamic response of the plant will also limit the speed of convergence if the iteration rate is rapid enough.

4.7. MINIMISATION OF THE MAXIMUM LEVEL AT ANY SENSOR

All of the algorithms described above minimise the sum of the modulus-squared signals from each of the error sensors, although some modifications have been included to improve robustness or limit control effort. The sum of the modulus-squared error signals can be written as the square of the *2-norm* of the vector of error signals, which can be written as

$$\|\mathbf{e}\|_2 = \left(\sum_{l=1}^{L} |e_l|^2 \right)^{1/2} . \tag{4.7.1}$$

In this section we describe an adaptive control algorithm that minimises the maximum level measured at any of the sensors. The cost function that is minimised by such an algorithm can be described as a limiting case of the *p-norm* of the error signal vector, which is described in the Appendix and can be written as

$$J_p = \|\mathbf{e}\|_p = \left[\sum_{l=1}^{L} |e_l|^p \right]^{1/p} , \tag{4.7.2}$$

from which the 2-norm can be derived by setting $p = 2$. As the power p is increased, the values of the error signals that have higher levels will be emphasised to a greater and greater extent. In the limit as p tends to infinity, the only contribution to the summation in equation (4.7.2) that has a significant value will be the error signal that has the highest level. This *∞-norm* of the error signals can thus be written as

$$J_\infty = \|\mathbf{e}\|_\infty = \lim_{p\to\infty} J_p = \max_{1\le l\le L} |e_l| = |e_s| , \tag{4.7.3}$$

where s denotes the value of l for which the error signal has the largest modulus. Note that the value of s will be from 1 to L and will generally change as the control system converges. The value of s may, for example, correspond to the first error signal, $l = s = 1$, at one iteration and the sixth error signal, $l = s = 6$, at the next iteration. The minimisation of the maximum level at any error sensor will result in a controlled field that is more spatially uniform than that obtained by minimising the sum of the squared levels at all error sensors (Elliott et al., 1987; Doelman, 1993; Gonzalez et al., 1998), and this may be important in some applications, particularly if the specification of the control system is written in terms of the maximum measured level.

Before discussing the form of the control algorithm that minimises J_p, it is important to note that the error surface formed by plotting the general cost function, J_p, against the real and imaginary parts of the control signals is guaranteed to be *convex* and thus have a unique global minimum. This property can be demonstrated by showing that the cost function J_p obeys the equation

$$J_p\left(\delta \mathbf{u}_1 + (1-\delta)\mathbf{u}_2\right) \le \delta\, J_p(\mathbf{u}_1) + (1-\delta)\, J_p(\mathbf{u}_2) , \tag{4.7.4}$$

for any δ between 0 and 1 and for any $\mathbf{u}_1$ and $\mathbf{u}_2$, which can be simply proved using the triangle inequality (Gonzalez et al., 1998). Thus the error surface for J_∞ must also be convex and a gradient descent algorithm is guaranteed to converge to a unique global minimum, provided it is stable.

Using the rules for the differentiation of a real cost function with respect to the real and imaginary parts of a complex number outlined in the Appendix, the complex gradient vector of the general cost function, J_p, with respect to the real and imaginary parts of the control signals $\mathbf{u}_R$ and $\mathbf{u}_I$, can be shown to be

$$\mathbf{g}_p = \frac{\partial J_p}{\partial \mathbf{u}_R} + j\frac{\partial J_p}{\partial \mathbf{u}_I} = \left(J_p\right)^{1-p} \sum_{l=1}^{L} \left|e_l\right|^{p-2} \mathbf{r}_l^* e_l \,, \tag{4.7.5}$$

where $\mathbf{r}_l$ is the vector formed by taking the transpose of the l-th row of the $L \times M$ matrix of complex plant responses, $\mathbf{G}$, at the excitation frequency ω_0. Notice that setting $p = 2$ in equation (4.7.5) gives equation (4.3.6), for the gradient of $\mathbf{e}^{\mathrm{H}}\mathbf{e}$ with respect to the real and imaginary parts of $\mathbf{u}$, with an extra term $1/(2\,J_2)$. This extra term arises because J_2 is defined to be the square root of $\mathbf{e}^{\mathrm{H}}\mathbf{e}$. It should be noted, however, that any algorithm which minimises $\mathbf{e}^{\mathrm{H}}\mathbf{e}$ will also minimise J_2.

For the case in which p tends to infinity, the complex gradient vector will only be a function of the error signal that has the highest level, e_s, and can thus be written as

$$\mathbf{g}_\infty = \left|e_s\right|^{-1} \mathbf{r}_s^* e_s \,. \tag{4.7.6}$$

If we assume that the convergence coefficient is normalised by the real number $\left|e_s\right|^{-1}$, a steepest-descent algorithm that minimises J_∞ can be written as

$$\mathbf{u}(n+1) = \mathbf{u}(n) - \alpha\,\mathbf{r}_s^* e_s \,, \tag{4.7.7}$$

which has been termed the *minimax algorithm* by Gonzalez et al. (1995, 1998). At any one instant, the algorithm is the same as that obtained by minimising the modulus-squared value of the single error signal, e_s. Its convergence properties are complicated, however, by the fact that during the course of adaptation the error signal with the highest level will generally switch from one sensor to another. The graph of J_∞ against time, during the course of the convergence of equation (4.7.7), thus takes the form of a series of exponential segments, whose time constants are determined by the values of the scalar quantity $\alpha\mathbf{r}_s^{\mathrm{H}}\mathbf{r}_s$, which changes as the error sensor associated with e_s switches from one sensor to another.

The physical consequences of minimising the maximum level at any microphone in an active sound control system have also been investigated by Gonzalez et al. (1998). They showed that for an active sound control system with 16 loudspeakers and 32 microphones controlling a low-frequency tonal sound field in a room, the maximum level at any microphone, after convergence of the minimax algorithm, was reduced by about 36 dB, as compared with an attenuation of about 32 dB in the maximum level at any microphone for an algorithm that minimised the sum of the mean-squared error signals. This rather modest improvement in J_∞ reflects the fact that the conventional algorithm, which minimises J_2, already tends to discriminate to some extent against microphones with high output levels and thus tends to smooth out the spatial variation in the pressure field. It may be, however, that in other applications the difference in the performance of the two algorithms is more significant and the minimax algorithm may offer worthwhile advantages over the more conventional least-squares approach.

4.8. APPLICATIONS

4.8.1. Controlling Propeller and Rotor Noise in Aircraft

In this final section we describe the application of multichannel feedforward control algorithms to the reduction of tonal components of the sound in the passenger cabins of propeller aircraft and helicopters. The application to propeller aircraft is described first, since in many ways this represents the simpler problem. In this case the amplitude and phase of the tones change only relatively slowly with time during normal cruise conditions. The fact that they change at all is the main reason why the control system must be made adaptive, although there are also relatively small changes in the plant response which must be compensated for, due to people moving around in the cabin for example. The control of sound at harmonics of the blade passage frequencies of the rotors inside a helicopter is rather more challenging. One problem is the potential need for very large loudspeakers to control the sound at the fundamental frequency of the main rotor, which is typically 10–20 Hz on most helicopters. Another serious problem in this application, however, is the fact that the disturbance can have a considerable short-term amplitude and phase modulation, which the control system must track.

4.8.2. Control of Propeller Noise Inside Fixed-wing Aircraft

Many short-haul aircraft, having up to about 50 seats, use propellers instead of jet engines because they can be considerably more efficient at speeds below about 300 mph. The development of prop-fan engines to power larger, faster aircraft has also been suggested because of their improved fuel efficiency. The spectrum of the sound pressure in the passenger cabins of such aircraft contains strong tonal components at the *blade passage frequency* (BPF) of the propellers, which are difficult to attenuate using passive absorption (Wilby et al., 1980; Metzger, 1981). Active control of these tones has been considered since the early 1980s (Chaplin, 1983; Ffowcs-Williams, 1984; Bullmore et al., 1987) and appears to be a good solution to this problem since active sound control works particularly well at low frequencies, where the acoustic modal overlap is relatively small, as described in Chapter 1. Also, with lightweight loudspeakers, active control potentially carries a significantly smaller weight penalty than passive methods of noise control. In this section we will briefly describe the results of some flight trials of a practical active control system operating in the passenger cabin of a British Aerospace 748 turboprop aircraft during a series of flight-trials undertaken in early 1988 (Elliott et al. 1989; Dorling et al., 1989; Elliott et al., 1990) and the subsequent development of this technology.

The 50 seat aircraft used for the trials had a fully trimmed passenger cabin about 9 m long by 2.6 m diameter, and was flown at an altitude of about 10 000 feet under straight and level cruise conditions with the engines running at a nominal 14 200 rpm. As a result of the gearing in the engine and the number of propeller blades, this produces a blade passage frequency (BPF) of 88 Hz. The control system used in the experiments discussed here employed a tachometer on one of the engines to generate reference signals at the BPF and its second and third harmonics, i.e. 88 Hz, 176 Hz and 264 Hz. These reference signals were passed through an array of adaptive digital filters, as described in Chapter 3, and their outputs were used to drive 16 loudspeakers. The coefficients of the digital filters were adjusted to implement the steepest-descent algorithm described above, in order to minimise the sum of the square values of the 32 microphone signals. Many different configurations of loudspeaker and microphone position were investigated. The results presented here are for the distribution illustrated in Fig. 4.23, with the sixteen 200-mm-diameter loudspeakers and thirty two microphones distributed reasonably uniformly on the floor and on the luggage racks throughout the cabin, although there is a greater concentration at the front of the aircraft (seat row 1) since this is close to the plane of the propellers.

The levels of reduction achieved in the sum of the mean-squared pressures at all 32 control microphones, ΔJ(dB), are listed in Table 4.2 at the three control frequencies. These

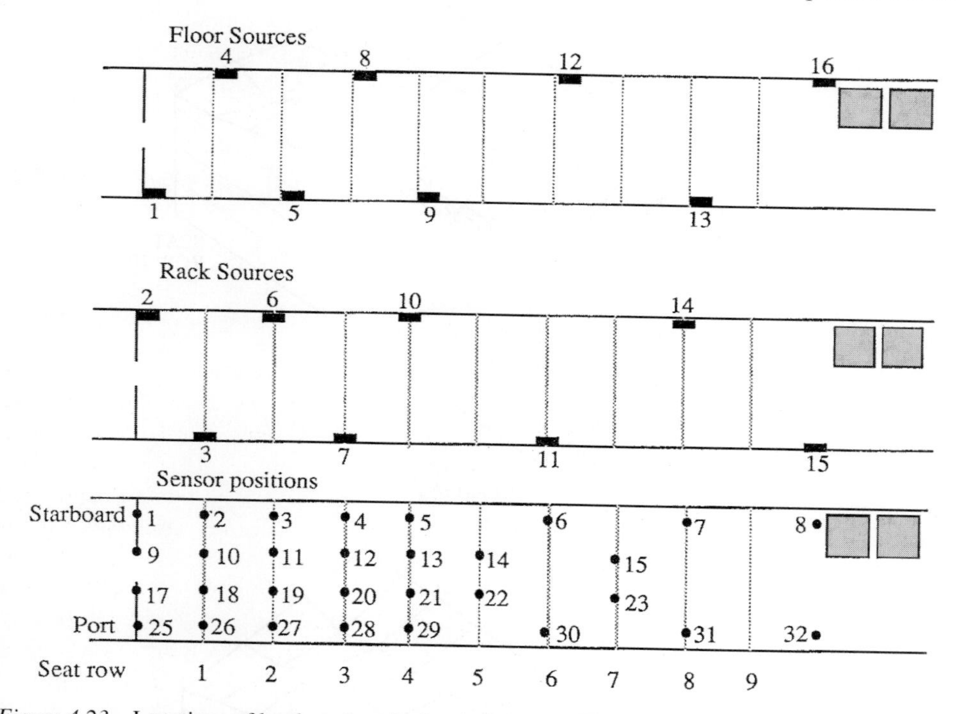

Figure 4.23 Locations of loudspeakers in the aircraft cabin, on the floor (upper) and in the luggage racks (middle), and of the microphones (lower), which are all placed at seated head height. The shaded blocks at the rear of the cabin (right side of the figure), show the position of the control system.

Table 4.2 Changes in the level of the sum of the squared output of 32 control microphones measured during flight trials of a 16-loudspeaker 32-microphone control system in a BAe 748 aircraft (Elliott et al. 1990)

Configuration	Propeller	ΔJ (dB) 88 Hz	176 Hz	264 Hz
Loudspeakers distributed throughout the cabin	Port	−13.8	−7.1	−4.0
	Starboard	−10.9	−4.9	−4.8

were achieved by using the control system with all the loudspeakers operating on the pressure field from either the port or starboard propeller alone. The results are slightly different for the two propellers since they both rotate in the same direction, and so the local stiffening effect on the fuselage due to the floor causes the structural vibration to be asymmetric. It is clear that significant reductions are achieved at the blade passage frequency with smaller reductions at the second and third harmonics. The normalised acoustic pressures measured at the error microphones at 88 Hz, before and after control, are plotted in terms of their physical position in the cabin in Fig. 4.24. Not only has the control

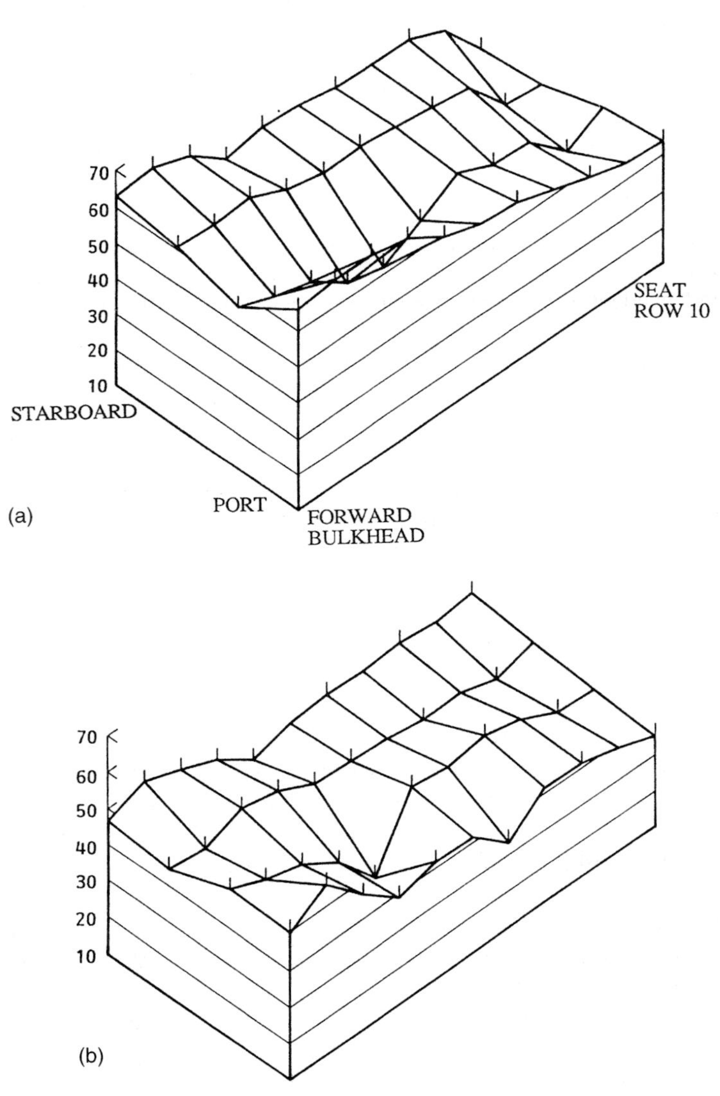

Figure 4.24 Distribution of normalised sound pressure level at 88 Hz, measured at the 32 control microphones illustrated in Figure 4.23 with the control system off (a) and on (b).

system given substantial reductions in the sum of the squared error signals, as listed in Table 4.2, but the individual mean-square pressures at each of the microphones has also been reduced in this case. The maximum pressure level at any one microphone position was reduced by about 11 dB for the results shown in Fig. 4.24.

At the second and third harmonics, the attenuations in the sum of squared microphone signals are somewhat smaller than at the blade passage frequency. This reflects the greater complexity of the acoustic field within the aircraft cabin at these higher frequencies, which makes the sound field more difficult to control, as discussed in Chapter 1. By moving some of the loudspeakers to a circumferential array in the plane of the propellers, at the front of the aircraft, somewhat greater reductions than those shown in Table 4.2 could be achieved at the second and third harmonics. This improvement is achieved by the ability of the secondary loudspeakers to more closely match the fuselage vibration caused by the propellers, and thus control the sound at source, without it being radiated into the cabin. Some care must be taken in the interpretation of the results from the error microphones at these higher frequencies, however, since the greater complexity in the sound field makes it more likely that the pressure level has been substantially reduced only at the locations of the error microphones, and that the average pressure level in the rest of the cabin has remained substantially unaltered. More recent flight trials have included a separate array of monitoring microphones, which are not used in the cost function minimised by the active control system, so that an independent measure of attenuation can be achieved (Borchers et al., 1994).

Although the aircraft control system did not have to respond very quickly under straight and level cruise conditions if the speeds of the two propellers were locked together, this was not possible on the aircraft used for the flight trials. The results presented in Table 4.2 were obtained with only one engine operating at 14 200 rpm and the other detuned to 12 700 rpm, so that its contribution to the sound field was ignored by the control system. As the speeds of the two engines are brought closer together, the sound fields due to the two propellers beat together and the amplitude tracking properties of the adaptive algorithm allow the control system to follow these beats, provided they are not quicker than about 2 beats per second. Reductions in overall A-weighted sound pressure level, of up to 7 dB(A) were measured with the active control system operating at all three harmonics (Elliott et al. 1990). These measurements include a frequency-dependent A-weighting function to account for the response of the ear (Kinsler et al., 1982).

Since this early work, which demonstrated the feasibility of actively controlling the internal propeller noise using loudspeakers, a number of commercial systems have been developed (Emborg and Ross, 1993; Billout et al., 1995) and are now in service on a number of aircraft. Instead of using loudspeakers as secondary sources it is also possible to use structural actuators attached to the fuselage to generate the secondary sound field. These have a number of potential advantages over loudspeakers, since they are capable of reducing both the cabin vibration as well as the sound inside the aircraft. It may also be possible to integrate structural actuators more easily into the aircraft manufacturing process, since no loudspeakers would need to be mounted on the trim panels. Early work in this area using piezoceramic actuators on the aircraft fuselage (Fuller, 1985) demonstrated that relatively few secondary actuators may be needed for efficient acoustic control. More recent systems have used inertial electromagnetic actuators mounted on the aircraft frame (Ross and Purver, 1997) or actively-tuned resonant mechanical systems (Fuller et al., 1995; Fuller, 1997) to achieve a more efficient mechanical excitation of the fuselage at the blade passage frequency and its harmonics.

4.8.3. Control of Rotor Noise Inside Helicopters

The noise inside the passenger cabin of a helicopter also has strong tonal components at the blade passage frequencies of the main rotor(s) and the tail rotor(s). The fundamental frequency of the main rotor is typically about 10–20 Hz on a helicopter, rather than the 80–160 Hz fundamental frequency which is typical for propeller noise in fixed-wing aircraft. Although it could be argued that the ear is very insensitive to these very low frequencies and the rotor tones have little effect on the usual dBA noise rating, they can cause considerable fatigue and significantly contribute to the poor environment that is often perceived inside helicopters. At first sight the active control of helicopter rotor tones would appear to be somewhat simpler than the control of aircraft propeller tones since the fundamental frequencies are about an order of magnitude lower and so the acoustic wavelengths are about an order of magnitude larger. This should mean that a considerably smaller number of acoustic modes will be excited within the cabin, and so a smaller number of loudspeakers would be required to achieve good control. This did not, however, turn out to be the case when a practical control system was flown in a flight trial on a helicopter in 1995 (Boucher et al. 1996; Elliott et al., 1997).

These flight trials were conducted on an EH101 helicopter with the collaboration of GKN Westland Helicopters and Agusta, and again involved an adaptive feedforward system using 16 loudspeakers and 32 microphones. The EH101 is a large helicopter which is capable of carrying 16 passengers in a cabin having dimensions of approximately 6 m × 2 m × 2 m. Although it was anticipated that a considerably smaller number of transducers would ultimately be required for effective control at the lower frequencies, the initial tests were performed with all channels connected, and the results from these tests were then analysed using computer simulations to predict the number and size of the loudspeakers required to produce good performance at each of the frequencies of interest. The control system was tested at the first three harmonics of the blade passage frequency of the main rotor, 17.5 Hz, 35 Hz and 52.5 Hz, and the first two harmonics of the tail rotor, 63.4 Hz and 126.8 Hz. The measured results obtained using all 16 loudspeakers and 32 microphones in a helicopter flight trial at 120 knots are shown in Table 4.3. Also shown in this table are the predicted attenuation that could be achieved with 16 loudspeakers, and an exact least-squares solution implemented at every cycle, with no limitations on the secondary source strength, as calculated from the data recorded during the flight trial.

Table 4.3 Summary of the reductions in the level of the sum of squared pressures at 32 microphones measured using a prototype active control system during flight trials in a large helicopter, and the predicted reductions calculated from the data measured during the flight trials, for a control system with 16 large loudspeakers and 8 practical-sized loudspeakers.

Frequency of tone (Hz)	Attenuation measured in flight (dB)	Predicted attenuation with 16 large loudspeakers (dB)	Attenuation with 8 practical-size loudspeakers (dB)
17.5	3	19	10
35	5.5	20	17
52.5	7	11	10
63	12	19	16.5
126	4	4	4

It can be seen that substantial improvements over the measured results are possible at most frequencies if the exact least-squares solution is implemented. Although this improvement is largely due to the ability of the 16 loudspeakers in the exact least-squares calculations to produce very large volume velocities, it is also partly due to the difficulty that the practical control system has in tracking the non-stationarity of the sound field caused by the rotor. The amplitude and phase of the rotor tones in a helicopter can change much more rapidly than the amplitude and phase of the propeller tones in a fixed-wing aircraft. This is thought to be due to the recirculation of turbulent air round the rotors in the helicopter, which causes a random modulation of the noise-generating mechanism (Elliott et al., 1997). Figure 4.25, for example, shows the short-term average level of the 63 Hz tone from the tail rotor recorded during the flight trial. The control system was switched on about 3 seconds into this recording, but it can be seen that there are large, rapid changes in the level both before and after control.

The control performance was also calculated for a reduced number of loudspeakers and microphones, in which the transducer positions were selected using the guided random search algorithms described in Chapter 8. The results shown in the final column of Table 4.3 are for a configuration with 16 microphones and 8 loudspeakers, with the loudspeakers constrained to give a maximum acoustic volume velocity corresponding to that of a 200-mm-diameter loudspeaker with a 10 mm throw. Useful attenuations are predicted to be obtained at all frequencies considered, but it is surprising that as many as eight loudspeakers are required to give acceptable performance at the lower frequencies. This turns out to be due to the acoustic properties of the walls of the helicopter fuselage, which become almost acoustically transparent at very low frequencies because of their lightweight and planar construction. This has two detrimental effects. First, the primary

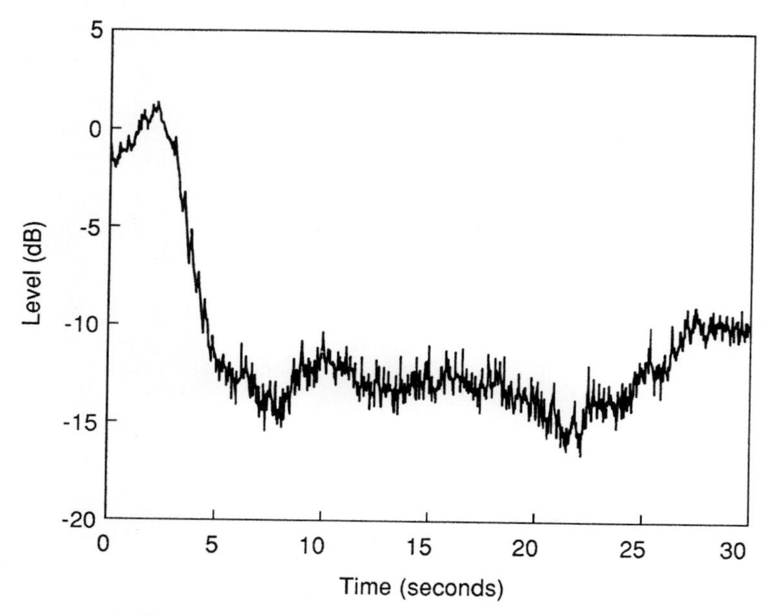

Figure 4.25 Time history of the sum of squared amplitudes of the signals recorded at the microphones during the flight trial of a large helicopter at a frequency of about 63 Hz with the active control system switched on after about 3 seconds.

sound field which the control system is trying to control becomes principally determined by the external sound field, which is a very complicated interference pattern between the nearfields of each of the rotors, instead of being dominated by an enclosed modal sound field with only a few significant modes. Second, the secondary loudspeaker volume velocities required to generate the cancelling field are much larger than they would be for a truly enclosed volume, since the amplification inherent in the acoustic resonances has been lost.

Actively controlling the sound field inside a helicopter thus generally presents a greater technical challenge than controlling the sound field inside a propeller aircraft. By carefully studying the sources of the problems in the helicopter, however, their effect can be quantified and the performance of different kinds of control systems can be predicted. In particular it is possible to include the constraint of finite loudspeaker volume velocity in the calculation of the optimum performance as well as in the adaptive control algorithms, as described above. This prediction of performance can then be made with various combinations of loudspeakers, to quantify the attenuation which could be obtained with different numbers of loudspeakers. Whereas 16 loudspeakers and 32 microphones are necessary to obtain good control of the tonal components of the sound field inside a 50-seat propeller aircraft, only 8 loudspeakers and 16 microphones appear to be necessary to control the tonal components over a similar frequency range in a 16-seat helicopter.

5

Multichannel Control of Stochastic Disturbances

5.1. INTRODUCTION

In this chapter the discussion of multiple-channel feedforward control systems is extended to the case of stochastic, or random, disturbances. Figure 5.1 shows an idealised example of a multichannel system for the active control of stochastic noise generated by a number of potentially incoherent primary sources. Active control at a number of error sensors is achieved by measuring the waveform of the primary sources with an array of reference sensors and feeding these signals through a matrix of control filters to a set of secondary

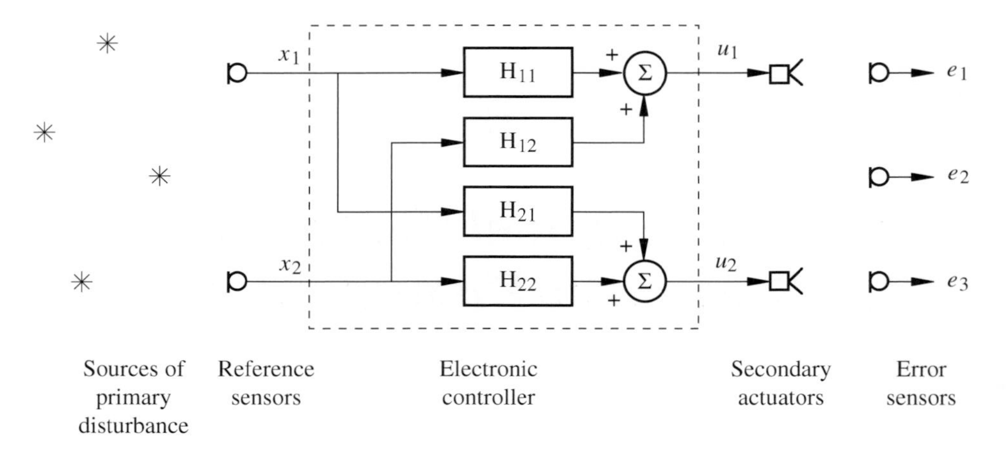

Figure 5.1 An example of a multichannel system for the feedforward control of stochastic noise from a number of primary sources.

actuators. There may be a larger number of incoherent primary sources than reference sensors, and so perfect control at the error sensors cannot be expected since their outputs will generally be only partially correlated with those of the reference sensors.

5.1.1. General Block Diagram

Figure 5.2 shows the general block diagram for the discrete-time multichannel control system of Fig. 5.1. The relationship between the discrete-time signals in Fig. 5.2 and the continuous-time signals in Fig. 5.1 was discussed in Section 3.1. Figure 5.2 includes the potential for acoustic feedback from the secondary sources to the reference sensors via the matrix of responses $\mathbf{G}_s(z)$. These feedback paths can be accounted for by having an internal model of them, $\hat{\mathbf{G}}_s(z)$, acting within the control system. If the model is perfect, i.e.

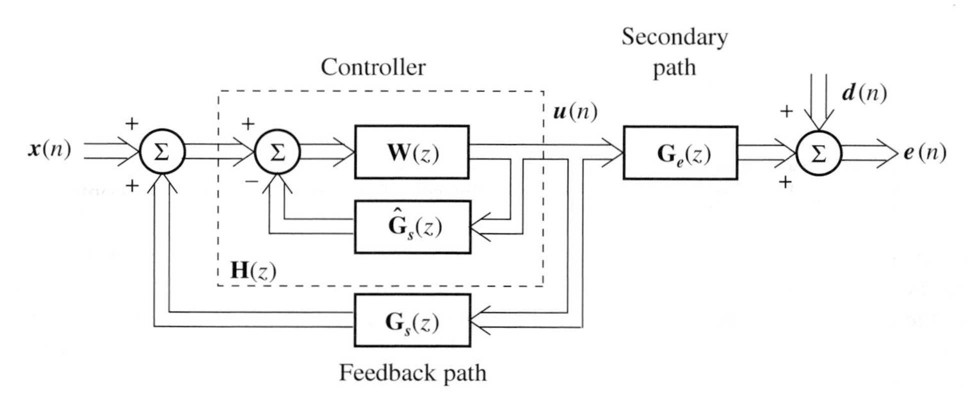

Figure 5.2 Block diagram of multichannel feedforward control system in which the controller is implemented using a feedback cancellation architecture.

$\hat{\mathbf{G}}_s(z) = \mathbf{G}_s(z)$, then the block diagram becomes purely feedforward, as shown in Fig. 5.3. In the calculation of the optimum controller, we will assume that the feedback path has been perfectly cancelled and solve the problem for the purely feedforward controller, $\mathbf{W}(z)$. The matrix of responses for the complete controller can be calculated from this and the feedback cancellation filter using the relationship

$$\mathbf{H}(z) = \left[\mathbf{I} + \mathbf{W}(z)\hat{\mathbf{G}}_s(z)\right]^{-1} \mathbf{W}(z) . \tag{5.1.1}$$

It is assumed for now that this system is stable; but a more detailed discussion of stability in multichannel systems is postponed until Chapter 6. In practice it may not be possible to exactly cancel the feedback path in a real-time controller, particularly if the response of the feedback path is changing with time. The problems of adapting the feedforward controller $\mathbf{W}(z)$ then become very similar to those encountered when adapting a feedback controller, as discussed in Chapter 7.

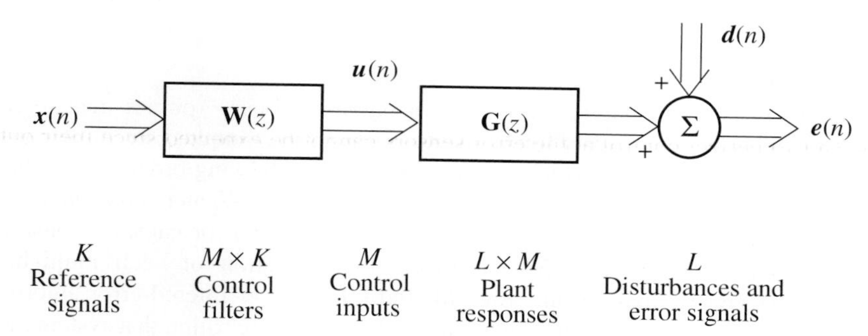

Figure 5.3 Block diagram of a general multichannel feedforward control system with K reference signals, M secondary actuators and L error sensors.

In Fig. 5.3 it is assumed that there are K reference signals, M secondary actuators and L error sensors. It is also assumed in this chapter that $L > M$, so that the system is overdetermined. Since the feedback path is assumed to be perfectly cancelled, only the secondary path, $\mathbf{G}_e(z)$, is needed in Fig. 5.3 and below, and since this is the only plant response required in this chapter it is now denoted $\mathbf{G}(z)$ for notational simplicity. Figure 5.3 could also be used to illustrate a time-domain implementation of a system for the control of deterministic disturbances, in which case the reference signals would typically be sinusoids at harmonic frequencies, as described in Chapter 3. We have seen that in this case only the amplitude and phase of such a signal needs to be manipulated to generate the control inputs. Also, and most importantly, the set of control filters used to control the amplitude and phase of one harmonic reference signal can be adjusted to minimise the sum of squared errors, *independently* of the adjustment of the sets of control filters for the other harmonic reference signals, assuming only that the plant is linear. This independence is due to the fact that the reference signals are *orthogonal*, i.e. their time average products are zero, and it allows the adjustment of the $M \times K$ matrix of control filters to be split into K independent optimisation problems, each for M control filters. In contrast to this deterministic case, stochastic reference signals are often strongly correlated in practical applications, and so the adjustment of the whole set of $M \times K$ control filters must be treated as a single *coupled* optimisation problem. Even if any two of the reference signals are perfectly correlated, both may still need to be used to minimise the error, since one of the reference signals may be providing time-advanced information about the disturbances in a different frequency band from the other.

The nature of the reference signals will depend strongly on the particular application being considered. In the final section of this chapter we will consider the active control of random noise inside a car due to the road excitation. In this application, the reference signals were obtained from accelerometers at carefully selected positions on the body of the car. Each reference signal provides some information about the disturbance measured at the internal microphones, but a number of reference signals have to be used to allow the disturbance to be usefully attenuated. Use of individual reference signals on their own was found not to produce useful attenuations. Multichannel feedforward controllers are also required for the active control of stochastic disturbances propagating as multiple acoustic modes in ducts or multiple vibrational waves in structures. In addition, they form an important building block for multichannel feedback systems, as we shall see in Chapter 6.

5.1.2. Chapter Outline

In order to explain all the effects found in these practical examples, we must consider several aspects of the problem, which are arranged in the following order in this chapter. Sections 5.2 and 5.3 cover the calculation of the optimal, Wiener, controller in the multichannel stochastic case, both for unconstrained filters and for causally constrained and finite-length filters. Section 5.4 discusses the adaptation of such multichannel controllers in the time domain using either filtered-reference or filtered-error algorithms. Section 5.5 describes a modification of the filtered-error LMS algorithm that systematically overcomes many of the causes of slow convergence that can otherwise limit the tracking ability of conventional LMS-based algorithms. Adaptation algorithms can also be efficiently implemented in the transform domain, using the FFT, and this is described in Section 5.6. It also turns out that several methods of improving the convergence behaviour of multichannel controllers can also be implemented in the frequency domain and the final structure of the adaptation algorithm looks very much like that developed for deterministic disturbances in the previous chapter. The difference here, however, is that the controller must be constrained to be causal, whether it is adapted in the time or the frequency domains and some care must be taken in the derivation of the adaptation algorithms to ensure that this condition is satisfied. Finally, in Section 5.7, the use of a multichannel feedforward system is described for the control of the random noise generated inside a vehicle by the tyres travelling over a rough road. The performance measured with a real-time controller is similar to that predicted using the optimal filter calculated off-line from signals previously measured in the vehicle.

5.2. OPTIMAL CONTROL IN THE TIME DOMAIN

In the following two sections we will derive expressions for the array of digital filters in a multichannel controller for stochastic signals that is optimal in the sense of minimising the expectation of the sum of the squared errors, i.e. the H_2 optimal or Wiener filter. This is done using a time-domain formulation in this section and a frequency-domain formulation in the following section. We assume that there are K partially correlated stochastic reference signals, M secondary actuators and L error sensors. Also we assume that any acoustic feedback from the actuators to the sensors has been removed with an internal model, as in Fig. 5.2, so that the general block diagram is as shown in Fig. 5.3.

Two formulations for the vector of time-domain error signals will be presented. The first employs a generalisation of the filtered-reference approach used to derive the optimal controller for the single-channel case in Chapter 3. This formulation leads to an explicit formula for the exact least-squares filter, although the solution can involve some rather large matrices. It also leads to a generalisation of the filtered-reference LMS algorithm for the adaptation of the control filters. The second approach is an extension of that used in the multichannel electrical filtering problem discussed in Chapter 2. Unfortunately, this approach does not lead to an explicit expression for the optimal filter coefficients, but it does suggest a compact form for the frequency-domain H_2 optimal filter, which is discussed in the following section, and a filtered-error algorithm for the adaptation of the filter coefficients, which will be further discussed in Section 5.4.

5.2.1. Formulation Using Filtered Reference Signals

A matrix formulation for the error vector will now be presented, based on the use of filtered-reference signals (Elliott et al., 1987; Nelson and Elliott, 1992). This formulation allows an explicit expression to be derived for the vector of optimal filter coefficients. In general the l-th error sensor signal, $e_l(n)$, can be written as the sum of the l-th disturbance, $d_l(n)$, and the contributions from all M secondary actuator signals, $u_1(n), \ldots, u_M(n)$, filtered by the corresponding elements of the response of the plant, i.e.

$$e_l(n) = d_l(n) + \sum_{m=1}^{M} \sum_{j=0}^{J-1} g_{lmj}\, u_m(n-i) \ , \tag{5.2.1}$$

where the impulse response of the plant from the m-th actuator to the l-th sensor is represented to arbitrary accuracy by the J-th order FIR filter with coefficients g_{lmj}. The signal driving the m-th actuator, $u_m(n)$, is made up of the sum of the contributions from K reference signals, $x_1(n), \ldots, x_K(n)$, each filtered by an I-th order FIR control filter with coefficients w_{mki}, so that

$$u_m(n) = \sum_{k=1}^{K} \sum_{i=0}^{I-1} w_{mki}\, x_k(n-i) \ . \tag{5.2.2}$$

The output of the l-th sensor can thus be expressed as the quadruple summation

$$e_l(n) = d_l(n) + \sum_{m=1}^{M} \sum_{j=0}^{J-1} \sum_{k=1}^{K} \sum_{i=0}^{I-1} g_{lmj}\, w_{mki}\, x_k(n-i-j) \ . \tag{5.2.3}$$

A matrix formulation for the error signal can be obtained by reordering the sequence of filtering for the reference signals, assuming that these filters are time-invariant, so that equation (5.2.3) can be written as

$$e_l(n) = d_l(n) + \sum_{m=1}^{M} \sum_{k=1}^{K} \sum_{i=0}^{I-1} w_{mki} \, r_{lmk}(n-i) \ , \tag{5.2.4}$$

where the *LMK filtered-reference signals* are given by

$$r_{lmk}(n) = \sum_{j=0}^{J-1} g_{lmj} \, x_k(n-j) \ . \tag{5.2.5}$$

The inner product in equation (5.2.4) can be represented in vector form as

$$e_l(n) \ = \ d_l(n) + \sum_{i=0}^{I-1} \boldsymbol{w}_i^{\mathrm{T}} \boldsymbol{r}_l(n-i) \ , \tag{5.2.6}$$

where

$$\boldsymbol{w}_i \ = \ \left[w_{11i} \ w_{12i} \ldots w_{1Ki} \ w_{21i} \ldots w_{MKi} \right]^{\mathrm{T}} \ , \tag{5.2.7}$$

and

$$\boldsymbol{r}_l(n) \ = \ \left[r_{l11}(n) \ r_{l12}(n) \ldots r_{l1K}(n) \ r_{l21}(n) \ldots r_{lMK}(n) \right]^{\mathrm{T}} \ . \tag{5.2.8}$$

The vector of all L error signals,

$$\boldsymbol{e}(n) \ = \ \left[e_1(n) \ \ldots \ e_L(n) \right]^{\mathrm{T}} \ , \tag{5.2.9}$$

can now be written in terms of the vector of disturbances

$$\boldsymbol{d}(n) \ = \ \left[d_1(n) \ \ldots \ d_L(n) \right]^{\mathrm{T}} \ , \tag{5.2.10}$$

and the contributions from all the secondary sources as

$$\boldsymbol{e}(n) \ = \ \boldsymbol{d}(n) + \boldsymbol{R}(n) \boldsymbol{w} \ , \tag{5.2.11}$$

where

$$\boldsymbol{R}(n) \ = \ \begin{bmatrix} \boldsymbol{r}_1^{\mathrm{T}}(n) & \boldsymbol{r}_1^{\mathrm{T}}(n-1) & \ldots & \boldsymbol{r}_1^{\mathrm{T}}(n-I+1) \\ \boldsymbol{r}_2^{\mathrm{T}}(n) & \boldsymbol{r}_2^{\mathrm{T}}(n-1) & & \\ \vdots & & & \\ \boldsymbol{r}_L^{\mathrm{T}}(n) & \boldsymbol{r}_L^{\mathrm{T}}(n-1) & \ldots & \boldsymbol{r}_L^{\mathrm{T}}(n-I+1) \end{bmatrix} , \tag{5.2.12}$$

and the vector containing all *MKI* control filter coefficients is defined as

$$\boldsymbol{w} \ = \ \left[\boldsymbol{w}_0^{\mathrm{T}} \ \boldsymbol{w}_1^{\mathrm{T}} \ \ldots \ \boldsymbol{w}_{I-1}^{\mathrm{T}} \right]^{\mathrm{T}} \ . \tag{5.2.13}$$

With the vector of error signals represented in this matrix form, we can proceed to calculate the optimal, Wiener, set of filter coefficients. This minimises the expectation of the sum of the squared error signals, a cost function which may be written as

$$J = E\left[\boldsymbol{e}^{\mathrm{T}}(n)\boldsymbol{e}(n)\right] = \boldsymbol{w}^{\mathrm{T}}\, E\left[\boldsymbol{R}^{\mathrm{T}}(n)\boldsymbol{R}(n)\right]\boldsymbol{w} + 2\boldsymbol{w}^{\mathrm{T}}\, E\left[\boldsymbol{R}^{\mathrm{T}}(n)\boldsymbol{d}(n)\right] + E\left[\boldsymbol{d}^{\mathrm{T}}(n)\boldsymbol{d}(n)\right] . \tag{5.2.14}$$

This quadratic function of the control filter coefficients, $\boldsymbol{w}$, has a unique global minimum if $E\left[\boldsymbol{R}^{\mathrm{T}}(n)\boldsymbol{R}(n)\right]$ is positive definite. The conditions for this matrix being positive definite correspond to a generalised form of the persistently exciting condition described in Chapter 2. The minimum value of the cost function is obtained for the vector of control filter coefficients given by

$$\boldsymbol{w}_{\mathrm{opt}} = -\left\{E\left[\boldsymbol{R}^{\mathrm{T}}(n)\boldsymbol{R}(n)\right]\right\}^{-1} E\left[\boldsymbol{R}^{\mathrm{T}}(n)\boldsymbol{d}(n)\right] , \tag{5.2.15}$$

and has the value

$$J_{\min} = E\left[\boldsymbol{d}^{\mathrm{T}}(n)\boldsymbol{d}(n)\right] - E\left[\boldsymbol{d}^{\mathrm{T}}(n)\boldsymbol{R}(n)\right]\left\{E\left[\boldsymbol{R}^{\mathrm{T}}(n)\boldsymbol{R}(n)\right]\right\}^{-1} E\left[\boldsymbol{R}^{\mathrm{T}}(n)\boldsymbol{d}(n)\right] . \tag{5.2.16}$$

The matrix $E\left[\boldsymbol{R}^{\mathrm{T}}(n)\boldsymbol{R}(n)\right]$ which must be inverted to calculate $\boldsymbol{w}_{\mathrm{opt}}$ or $J_{\min}$ has dimensions $MKI \times MKI$, which can be of very high order if a large number of secondary actuators are being driven by a large number of reference signals via control filters with many coefficients. If, for example, there are 6 reference signals, 4 secondary sources and 128 coefficients in each filter, the matrix which has to be inverted has 3072×3072 elements. Fortunately, if the elements of $\boldsymbol{r}_l(n)$ are ordered as in equation (5.2.8), then $E\left[\boldsymbol{R}^{\mathrm{T}}(n)\boldsymbol{R}(n)\right]$ has a block Toeplitz form (Nelson et al., 1990) as for the multichannel electrical filtering problem in equation (2.5.16), and iterative methods are available for its efficient inversion (see for example Robinson, 1978).

5.2.2. Formulation Using Matrices of Impulse Responses

The vector of time histories of the L error signals in Fig. 5.3 can also be expressed as

$$\boldsymbol{e}(n) = \boldsymbol{d}(n) + \sum_{j=0}^{J-1} \boldsymbol{G}_j\, \boldsymbol{u}(n-j) , \tag{5.2.17}$$

where $\boldsymbol{G}_j$ is the $L \times M$ matrix of the j-th coefficients of the plant impulse response functions, between each actuator and each sensor, which are all assumed to be stable, and $\boldsymbol{u}(n)$ is the vector of M control signals which are the input signals to the actuators. The vector of control signals can also be written as

$$\boldsymbol{u}(n) = \sum_{i=0}^{I-1} \boldsymbol{W}_i\, \boldsymbol{x}(n-i) , \tag{5.2.18}$$

where $\boldsymbol{W}_i$ is the $M \times K$ matrix of the i-th coefficients of the FIR controller matrix and $\boldsymbol{x}(n)$ is the vector of K reference signals. The vector of error signals is thus equal to

$$\mathbf{e}(n) = \mathbf{d}(n) + \sum_{j=0}^{J-1} \sum_{i=0}^{I-1} \mathbf{G}_j \mathbf{W}_i \mathbf{x}(n-i-j) . \tag{5.2.19}$$

The cost function to be minimised is again assumed to be equal to the expectation of the sum of the squared error signals and, following the approach adopted for the multichannel electrical filtering problem in Section 2.4, this can be written as

$$J = \operatorname{trace}\left[E\left(\mathbf{e}(n)\mathbf{e}^{\mathrm{T}}(n)\right)\right] . \tag{5.2.20}$$

Using equation (5.2.19) the expectation of the outer product in equation (5.2.20) can be written as

$$E\left[\mathbf{e}(n)\mathbf{e}^{\mathrm{T}}(n)\right] = \mathbf{R}_{dd}(0) + \sum_{j=0}^{J-1} \sum_{i=0}^{I-1} \mathbf{G}_j \mathbf{W}_i \mathbf{R}_{xd}^{\mathrm{T}}(i+j)$$
$$+ \sum_{j=0}^{J-1} \sum_{i=0}^{I-1} \mathbf{R}_{xd}(i+j)\ \mathbf{W}_i^{\mathrm{T}} \mathbf{G}_j^{\mathrm{T}} + \sum_{j=0}^{J-1} \sum_{i=0}^{I-1} \sum_{i'=0}^{I-1} \sum_{j'=0}^{J-1} \mathbf{G}_j \mathbf{W}_i \mathbf{R}_{xx}(i'+j'-i-j) \mathbf{W}_{i'}^{\mathrm{T}} \mathbf{G}_{j'}^{\mathrm{T}} , \tag{5.2.21}$$

where the $L \times K$ cross-correlation matrix between the reference signals and disturbances is defined as

$$\mathbf{R}_{xd}(m) = E\left[\mathbf{d}(n+m)\mathbf{x}^{\mathrm{T}}(n)\right] , \tag{5.2.22}$$

the $K \times K$ matrix of auto- and cross-correlations between the reference signals is defined as

$$\mathbf{R}_{xx}(m) = E\left[\mathbf{x}(n+m)\mathbf{x}^{\mathrm{T}}(n)\right] , \tag{5.2.23}$$

and $\mathbf{R}_{dd}(0)$ is similarly defined.

Using the rules outlined in the Appendix for differentiating the trace of a matrix function with respect to the elements of one of the matrices, the derivative of the cost function with respect to the elements of the i-th matrix of controller coefficients, $\mathbf{W}_i$, can be written as

$$\frac{\partial J}{\partial \mathbf{W}_i} = 2\left[\sum_{j=0}^{J-1} \mathbf{G}_j^{\mathrm{T}} \mathbf{R}_{xd}(i+j) + \sum_{j=0}^{J-1} \sum_{j'=0}^{J-1} \sum_{i'=0}^{I-1} \mathbf{G}_j^{\mathrm{T}} \mathbf{G}_{j'} \mathbf{W}_{i'} \mathbf{R}_{xx}(i+j-i'-j')\right] . \tag{5.2.24}$$

The set of optimal control filters is given by setting equation (5.2.24) to zero for $i = 0$ to $I - 1$. Whereas in the multichannel electrical filtering case the resulting equations could be represented in a single matrix form, there appears to be no simple way of achieving this with equation (5.2.24), in which case this equation cannot be used to derive the optimal controller in the time domain.

An important result is obtained if the matrix of cross-correlation functions between each error signal and each reference signal is defined to be

$$\mathbf{R}_{xe}(m) = E\left[\mathbf{e}(n+m)\mathbf{x}^{\mathrm{T}}(n)\right] , \tag{5.2.25}$$

which can be written using equation (5.2.19) as

$$\mathbf{R}_{xe}(m) = \mathbf{R}_{xd}(m) + \sum_{j=0}^{J-1} \sum_{i=0}^{I-1} \mathbf{G}_j \mathbf{W}_i \mathbf{R}_{xx}(m-i-j) . \tag{5.2.26}$$

The matrix of derivatives of the cost function with respect to the elements of the i-th matrix of controller coefficients (equation 5.2.24) can thus be written as

$$\frac{\partial J}{\partial \mathbf{W}_i} = 2\sum_{j=0}^{J-1} \mathbf{G}_j^{\mathrm{T}} \mathbf{R}_{xe}(i+j) . \tag{5.2.27}$$

The conditions that must be satisfied by an H_2 optimal causal filter of *infinite* length are thus that

$$\sum_{j=0}^{\infty} \mathbf{G}_j^{\mathrm{T}} \mathbf{R}_{xe}(i+j) = \mathbf{0} \qquad \text{for } i \geq 0 . \tag{5.2.28}$$

Note that equation (5.2.28) does not represent the convolution of $\mathbf{R}_{xe}(m)$ with $\mathbf{G}_j^{\mathrm{T}}$ but the convolution of $\mathbf{R}_{xe}(m)$ with a *time-reversed* version of this impulse response.

Setting the z- transform of the causal part of equation (5.2.28) to zero, using an extension of the notation introduced in Chapter 2, we obtain the z-domain optimality condition,

$$\left\{\mathbf{G}^{\mathrm{T}}(z^{-1})\mathbf{S}_{xe}(z)\right\}_{+} = \mathbf{0} , \tag{5.2.29}$$

where $\mathbf{G}(z)$ is the matrix of transfer functions for the plant, $\mathbf{S}_{xe}(z)$ is the z-transform of $\mathbf{R}_{xe}(m)$, and $\{\ \}_{+}$ again denotes that the causal part of the time domain version of the function has been taken. Equation (5.2.29) is a convenient starting point for the derivation in the following section of the H_2 optimal causal filter in the frequency domain.

5.3. OPTIMAL CONTROL IN THE TRANSFORM DOMAIN

Expressions for the optimal controller can also be derived in the transform domain. The optimal controller is first obtained without any constraints in the frequency domain, in which case it is not necessarily causal. The resulting form for the controller is relatively simple, however, and some insight can be gained by comparing this controller to that for deterministic signals considered in the previous section. The causally constrained matrix of filters is then derived in the z-domain using multichannel spectral factorisation.

5.3.1. Unconstrained Controller

Referring to Fig. 5.3, the Fourier transform of the vector of L error signals, $\mathbf{e}$, can be written in terms of the vector of transformed disturbances $\mathbf{d}$, the $L \times M$ matrix of frequency response functions of the plant, $\mathbf{G}$, the $M \times K$ matrix of frequency response functions of the controller, $\mathbf{W}$, and the Fourier transform of the vector of K stochastic reference signals, $\mathbf{x}$, as

$$\mathbf{e}(\mathrm{e}^{j\omega T}) = \mathbf{d}(\mathrm{e}^{j\omega T}) + \mathbf{G}(\mathrm{e}^{j\omega T})\mathbf{W}(\mathrm{e}^{j\omega T})\mathbf{x}(\mathrm{e}^{j\omega T}) , \tag{5.3.1}$$

where the explicit dependence of all the variables on ($e^{j\omega T}$) will be suppressed for notational convenience in the remainder of this section.

In the first part of this section we find the matrix of controller responses that minimises a cost function, equal to the expectation of the sum of the squared errors, independently at each frequency. The controller is thus not constrained to be causal. The cost function at each frequency can be written in the alternative forms

$$J = E\left(\mathbf{e}^{\mathrm{H}}\mathbf{e}\right) = \operatorname{trace} E\left(\mathbf{e}\,\mathbf{e}^{\mathrm{H}}\right) , \qquad (5.3.2, 5.3.3)$$

where E denotes the expectation operator.

The outer product and trace form of the cost function, equation (5.3.3), was used for the multichannel electrical filtering case in Section 2.4, and turns out to provide the most convenient method of analysis here (Elliott and Rafaely, 1997). If the inner product form of the cost function, equation (5.3.2), is used in the analysis, exactly the same result can be derived, but it is less easy to generalise to the causal case.

Using equation (5.3.1), the cost function given by equation (5.3.3) can be written as

$$J = \operatorname{trace} E\left[\mathbf{GWxx}^{\mathrm{H}}\mathbf{W}^{\mathrm{H}}\mathbf{G}^{\mathrm{H}} + \mathbf{GWxd}^{\mathrm{H}} + \mathbf{dx}^{\mathrm{H}}\mathbf{W}^{\mathrm{H}}\mathbf{G}^{\mathrm{H}} + \mathbf{dd}^{\mathrm{H}}\right] . \qquad (5.3.4)$$

The expectation only needs to be taken over the stochastic parts of equation (5.3.4), which involve $\mathbf{x}$ and $\mathbf{d}$. It is convenient to define the $K \times K$ matrix of power and cross spectral densities for the reference signals as

$$\mathbf{S}_{xx} = E\left[\mathbf{x}\,\mathbf{x}^{\mathrm{H}}\right] , \qquad (5.3.5)$$

the $L \times K$ matrix of cross spectral densities between the disturbance and reference signals as

$$\mathbf{S}_{xd} = E\left[\mathbf{d}\,\mathbf{x}^{\mathrm{H}}\right] , \qquad (5.3.6)$$

and the $L \times L$ matrix of power plus cross spectral densities for the disturbance signals as

$$\mathbf{S}_{dd} = E\left[\mathbf{d}\,\mathbf{d}^{\mathrm{H}}\right] , \qquad (5.3.7)$$

all of which are referred to as *spectral density matrices*. Notice that the definition of $\mathbf{S}_{xd}$ is slightly different from that used by Bendat and Piersol (1986), for example, as discussed in Section A.5 of the Appendix.

The cost function in equation (5.3.4) can now be written as

$$J = \operatorname{trace}\left[\mathbf{G}\,\mathbf{W}\,\mathbf{S}_{xx}\,\mathbf{W}^{\mathrm{H}}\,\mathbf{G}^{\mathrm{H}} + \mathbf{G}\,\mathbf{W}\,\mathbf{S}_{xd}^{\mathrm{H}} + \mathbf{S}_{xd}\,\mathbf{W}^{\mathrm{H}}\,\mathbf{G}^{\mathrm{H}} + \mathbf{S}_{dd}\right] . \qquad (5.3.8)$$

In order to derive the optimum controller response at each frequency, we differentiate equation (5.3.8) with respect to the real and imaginary parts of $\mathbf{W}$, which are equal to $\mathbf{W}_{\mathrm{R}}$ and $\mathbf{W}_{\mathrm{I}}$, using the rules derived in the Appendix, to give

$$\frac{\partial J}{\partial \mathbf{W}_{\mathrm{R}}} + j\frac{\partial J}{\partial \mathbf{W}_{\mathrm{I}}} = 2\left(\mathbf{G}^{\mathrm{H}}\,\mathbf{G}\,\mathbf{W}\,\mathbf{S}_{xx} + \mathbf{G}^{\mathrm{H}}\,\mathbf{S}_{xd}\right) . \qquad (5.3.9)$$

The frequency responses of the unconstrained optimum matrix of controllers can now be derived by setting the matrix of complex derivatives in equation (5.3.9) to zero. The cost function has a unique global minimum, provided both $\mathbf{G}^{\mathrm{H}}\mathbf{G}$ and $\mathbf{S}_{xx}$ are positive definite, and are hence non-singular, under which conditions the optimal unconstrained controller can be written as

$$\mathbf{W}_{\mathrm{opt}} = -\left[\mathbf{G}^{\mathrm{H}}\mathbf{G}\right]^{-1}\mathbf{G}^{\mathrm{H}}\mathbf{S}_{xd}\,\mathbf{S}_{xx}^{-1} . \tag{5.3.10}$$

This equation is the main result that we seek. If the frequency response matrix of the plant, $\mathbf{G}$, and the spectral density matrices $\mathbf{S}_{xd}$ and $\mathbf{S}_{xx}$ in equations (5.3.5) and (5.3.6) can be measured or are known, then the matrix of optimum controllers can be calculated at each frequency. A very similar expression can be derived for the controller which minimises a cost function that includes the mean-square control effort as well as the mean-square error, in which case the matrix $\mathbf{G}^{\mathrm{H}}\mathbf{G}$ is 'regularised' by a diagonal matrix before being inverted. Apart from making the optimal controller easier to calculate numerically, such control effort weighting can also shorten the duration of the impulse responses of the optimal filters, which can be important in reducing circular convolution effects when manipulating responses in the discrete-frequency domain (Kirkeby et al., 1998). The formulation can also be extended to include frequency-dependent error weighting matrices (Minkoff, 1997). A more general form of equation (5.3.10) could also be written using the pseudo-inverse of $\mathbf{G}$ instead of $\left[\mathbf{G}^{\mathrm{H}}\mathbf{G}\right]^{-1}\mathbf{G}^{\mathrm{H}}$, which would allow for poor conditioning of $\mathbf{G}^{\mathrm{H}}\mathbf{G}$ at some frequencies, and similarly the pseudo-inverse of $\mathbf{S}_{xx}$ could be used instead of $\mathbf{S}_{xx}^{-1}$, which would also allow for poor conditioning of the reference signal matrix at some frequencies.

If we consider the special case of a scalar reference signal of unit amplitude at a single frequency of ω_0, then equation (5.3.10) is only defined at ω_0, $\mathbf{S}_{xx}$ is then a scalar equal to unity, and $\mathbf{S}_{dx}$ is a vector, equal to $\mathbf{d}(\omega_0)$. Equation (5.3.10) then reduces to the optimum overdetermined solution for the vector of control signals in the case of a tonal reference, as given by equation (4.2.6).

Also, we could define the least-squares estimate of the primary path from the reference signals to the error signals to be

$$\mathbf{P} = \mathbf{S}_{xd}\,\mathbf{S}_{xx}^{-1} . \tag{5.3.11}$$

The complete optimum feedforward controller illustrated in Fig. 5.2 including a perfect feedback cancellation path, $\hat{\mathbf{G}}_s = \mathbf{G}_s$, has the matrix of frequency responses

$$\mathbf{H}_{\mathrm{opt}} = \left[\mathbf{I} + \mathbf{W}_{\mathrm{opt}}\mathbf{G}_s\right]^{-1}\mathbf{W}_{\mathrm{opt}} , \tag{5.3.12}$$

and using equation (5.3.10) for $\mathbf{W}_{\mathrm{opt}}$ with equation (5.3.11), $\mathbf{H}_{\mathrm{opt}}$ can also be written as

$$\mathbf{H}_{\mathrm{opt}} = \left[\mathbf{G}_e^{\mathrm{H}}\mathbf{P}\mathbf{G}_s - \mathbf{G}_e^{\mathrm{H}}\mathbf{G}_e\right]^{-1}\mathbf{G}_e^{\mathrm{H}}\mathbf{P} , \tag{5.3.13}$$

where $\mathbf{G}_e$ in Fig. 5.2 is equal to $\mathbf{G}$ in equation (5.3.10). Equation (5.3.13) is identical to the result derived some time ago for deterministic disturbances by Elliott and Nelson (1985b).

An expression for the minimum possible value of the cost function can be derived by substituting equation (5.3.10) into equation (5.3.8) and using the fact that trace (**AB**) = trace (**BA**), to give

$$J_{\min} = \operatorname{trace}\left[\mathbf{S}_{dd} - \mathbf{S}_{xd}\,\mathbf{S}_{xx}^{-1}\,\mathbf{S}_{xd}^{\mathrm{H}}\,\mathbf{G}\left(\mathbf{G}^{\mathrm{H}}\mathbf{G}\right)^{-1}\mathbf{G}^{\mathrm{H}}\right] . \tag{5.3.14}$$

Note that if there are as many secondary actuators as error sensors, then $\mathbf{G}$ is a square matrix and $\mathbf{G}\left(\mathbf{G}^{\mathrm{H}}\mathbf{G}\right)^{-1}\mathbf{G}^{\mathrm{H}}$ is equal to the identity matrix. Under these conditions, the expression for the minimum value of the cost function becomes independent of the plant response and only depends on the correlation properties of the reference and disturbance signals (Minkoff, 1997).

In the special case of a single secondary actuator and error sensor (L=M=1), but with multiple reference signals (K>1), then $\mathbf{G}$ becomes a scalar and $\mathbf{G}\left(\mathbf{G}^{\mathrm{H}}\mathbf{G}\right)^{-1}\mathbf{G}^{\mathrm{H}}$ is equal to unity. $\mathbf{S}_{dd}$ and $\mathbf{S}_{xd}\,\mathbf{S}_{xx}^{-1}\,\mathbf{S}_{xd}^{\mathrm{H}}$ are also scalars in this case, and so the trace operation in equation (5.3.14) can be omitted and the fractional reduction in the cost function is given by

$$\frac{J_{\min}}{J_{\mathrm{p}}} = 1 - \frac{\mathbf{S}_{xd}\,\mathbf{S}_{xx}^{-1}\,\mathbf{S}_{xd}^{\mathrm{H}}}{S_{dd}} = 1 - \gamma_{\mathbf{x},d}^{2} , \tag{5.3.15}$$

where $J_{\mathrm{p}} = S_{dd}$ is the cost function before control and $\gamma_{\mathbf{x},d}^{2}$ is the *multiple coherence function* between the K reference signals and the single disturbance signal (see for example Newland, 1993).

5.3.2. Causally Constrained Controller

We now consider the problem of constraining the controller to be causal. We have seen that the causal controller must satisfy the time-domain equation given by equation (5.2.28). In the z-domain this condition can be written, as in equation (5.2.29), as

$$\left\{\mathbf{G}^{\mathrm{T}}\left(z^{-1}\right)\mathbf{S}_{xe}(z)\right\}_{+} = \mathbf{0} , \tag{5.3.16}$$

where $\mathbf{S}_{xe}(z)$ is the z transform of $\mathbf{R}_{xe}(m)$ and which, subject to the conditions outlined in Section A.5 of the Appendix, can be written as

$$\mathbf{S}_{xe}(z) = E\left[\mathbf{e}(z)\,\mathbf{x}^{\mathrm{T}}\left(z^{-1}\right)\right] . \tag{5.3.17}$$

Note that $\mathbf{G}^{\mathrm{T}}\left(z^{-1}\right)$ represents the z transform of the transposed and time-reversed matrix of stable plant impulse responses and is sometimes said to be the *adjoint* of the matrix $\mathbf{G}(z)$ (Grimble and Johnson, 1988; Wan, 1996), although this should not be confused with the classical adjoint of the constant matrix defined in Section A.3 of the Appendix. Since

$$\mathbf{e}(z) = \mathbf{d}(z) + \mathbf{G}(z)\,\mathbf{W}(z)\,\mathbf{x}(z) . \tag{5.3.18}$$

Then

$$\mathbf{S}_{xe}(z) = \mathbf{S}_{xd}(z) + \mathbf{G}(z)\,\mathbf{W}(z)\,\mathbf{S}_{xx}(z) , \tag{5.3.19}$$

where

$$\mathbf{S}_{xd}(z) = E\left[\mathbf{d}(z)\mathbf{x}^{\mathrm{T}}\left(x^{-1}\right)\right] , \tag{5.3.20}$$

which is the z-transform of $\boldsymbol{R}_{xd}(m)$, and

$$\mathbf{S}_{xx}(z) = E\left[\mathbf{x}(z)\mathbf{x}^{\mathrm{T}}\left(z^{-1}\right)\right] , \tag{5.3.21}$$

which is the z-transform of $\boldsymbol{R}_{xx}(m)$, and the expectation is again taken across the z-transforms of sections of the time history, as explained in Section A.5 of the Appendix.

The condition for the optimal matrix of causal filters $\mathbf{W}_{\mathrm{opt}}(z)$ can thus be written using equations (5.3.16) and (5.3.19) as

$$\left\{\mathbf{G}^{\mathrm{T}}\left(z^{-1}\right)\mathbf{S}_{xd}(z) + \mathbf{G}^{\mathrm{T}}\left(z^{-1}\right)\mathbf{G}(z)\,\mathbf{W}_{\mathrm{opt}}(z)\mathbf{S}_{xx}(z)\right\}_{+} = \mathbf{0} . \tag{5.3.22}$$

The spectral density matrix for the reference signals is now expressed in terms of its spectral factors, as described in Section 2.4, so that

$$\mathbf{S}_{xx}(z) = \mathbf{F}(z)\mathbf{F}^{\mathrm{T}}\left(z^{-1}\right) , \tag{5.3.23}$$

where both $\mathbf{F}(z)$ and $\mathbf{F}^{-1}(z)$ are causal and stable.

We also separate the $L \times M$ matrix of plant responses, $\mathbf{G}(z)$, into its all-pass and minimum phase components (Morari and Zafiriou, 1989) so that

$$\mathbf{G}(z) = \mathbf{G}_{\mathrm{all}}(z)\mathbf{G}_{\mathrm{min}}(z) , \tag{5.3.24}$$

where the $L{\times}M$ matrix of all-pass components is assumed to have the property that

$$\mathbf{G}_{\mathrm{all}}^{\mathrm{T}}\left(z^{-1}\right)\mathbf{G}_{\mathrm{all}}(z) = \mathbf{I} , \tag{5.3.25}$$

so that

$$\mathbf{G}^{\mathrm{T}}\left(z^{-1}\right)\mathbf{G}(z) = \mathbf{G}_{\mathrm{min}}^{\mathrm{T}}\left(z^{-1}\right)\mathbf{G}_{\mathrm{min}}(z) , \tag{5.3.26}$$

where $\mathbf{G}_{\mathrm{min}}(z)$ is the $M \times M$ matrix that corresponds to the minimum-phase component of the plant response and thus has a stable causal inverse. Assuming that the matrix $\mathbf{G}^{\mathrm{T}}\left(z^{-1}\right)\mathbf{G}(z)$ satisfies the conditions under which a spectral factorisation exists, as described in Section 2.4, then the decomposition in equation (5.3.26) has a similar form to the factorisation in equation (5.3.23). The all-pass component can then be calculated from $\mathbf{G}(z)$ and $\mathbf{G}_{\mathrm{min}}(z)$ as

$$\mathbf{G}_{\mathrm{all}}(z) = \mathbf{G}(z)\mathbf{G}_{\mathrm{min}}^{-1}(z) , \tag{5.3.27}$$

so that

$$\mathbf{G}_{\mathrm{all}}^{\mathrm{T}}\left(z^{-1}\right) = \mathbf{G}_{\mathrm{min}}^{-\mathrm{T}}\left(z^{-1}\right)\mathbf{G}^{\mathrm{T}}\left(z^{-1}\right) , \tag{5.3.28}$$

where $\left[\mathbf{G}_{\mathrm{min}}^{-1}\left(z^{-1}\right)\right]^{\mathrm{T}}$ has been written as $\mathbf{G}_{\mathrm{min}}^{-\mathrm{T}}\left(z^{-1}\right)$ to keep the notation compact.

The condition for the optimal causal filter, equation (5.3.22), can now be written, using similar notation, as

$$\left\{\mathbf{G}_{\min}^{\mathrm{T}}\left(z^{-1}\right)\left[\mathbf{G}_{\min}^{-\mathrm{T}}\left(z^{-1}\right)\mathbf{G}^{\mathrm{T}}\left(z^{-1}\right)\mathbf{S}_{xd}(z)\mathbf{F}^{-\mathrm{T}}\left(z^{-1}\right)+\mathbf{G}_{\min}(z)\mathbf{W}_{\mathrm{opt}}(z)\mathbf{F}(z)\right]\mathbf{F}^{\mathrm{T}}\left(z^{-1}\right)\right\}_{+}=0\ , \tag{5.3.29}$$

where $\mathbf{G}_{\min}^{\mathrm{T}}\left(z^{-1}\right)$ and $\mathbf{F}^{\mathrm{T}}\left(z^{-1}\right)$ correspond to matrices of time-reversed causal sequences and, since these are both minimum phase, the optimality condition reduces to (Davis, 1963)

$$\left\{\mathbf{G}_{\mathrm{all}}^{\mathrm{T}}\left(z^{-1}\right)\mathbf{S}_{xd}(z)\mathbf{F}^{-\mathrm{T}}\left(z^{-1}\right)+\mathbf{G}_{\min}(z)\mathbf{W}_{\mathrm{opt}}(z)\mathbf{F}(z)\right\}_{+} = 0\ , \tag{5.3.30}$$

where equation (5.3.28) has been used to simplify the term $\mathbf{G}_{\min}^{-\mathrm{T}}\left(z^{-1}\right)\mathbf{G}^{\mathrm{T}}\left(z^{-1}\right)$ in equation (5.3.29).

Since $\mathbf{W}_{\mathrm{opt}}(z)$, as well as $\mathbf{G}_{\min}(z)$ and $\mathbf{F}(z)$, correspond to matrices of entirely causal sequences, the causality brackets can be removed from the second term in equation (5.3.30), which can then be solved to give

$$\mathbf{W}_{\mathrm{opt}}(z) = -\mathbf{G}_{\min}^{-1}(z)\left\{\mathbf{G}_{\mathrm{all}}^{\mathrm{T}}\left(z^{-1}\right)\mathbf{S}_{xd}(z)\mathbf{F}^{-\mathrm{T}}\left(z^{-1}\right)\right\}_{+}\mathbf{F}^{-1}(z)\ . \tag{5.3.31}$$

Equation (5.3.31) provides an explicit method for calculating the matrix of optimal causal filters that minimise the sum of squared errors in terms of the cross spectral matrix between the reference and desired signals, the minimum-phase and all-pass components of the plant response, and the spectral factors of the spectral density matrix for the reference signals. If the plant has as many outputs as inputs, so that $\mathbf{G}(z)$ is a square matrix, then $\mathbf{G}_{\mathrm{all}}^{\mathrm{T}}\left(z^{-1}\right)$ is equal to $\mathbf{G}_{\mathrm{all}}^{-1}\left(z^{-1}\right)$ and equation (5.3.31) becomes equivalent to the corresponding result in Morari and Zafiriou (1989),

$$\mathbf{W}_{\mathrm{opt}}(z) = -\mathbf{G}_{\min}^{-1}(z)\left\{\mathbf{G}_{\mathrm{all}}^{-1}\left(z^{-1}\right)\mathbf{S}_{xd}(z)\mathbf{F}^{-\mathrm{T}}\left(z^{-1}\right)\right\}_{+}\mathbf{F}^{-1}(z)\ , \tag{5.3.32}$$

from which it is clear that the multichannel result reduces to equation (3.3.24) in the single channel case. Also, if the constraint of causality is removed and z is set equal to $\mathrm{e}^{j\omega T}$, then equation (5.3.31) becomes equal to the unconstrained result in the frequency domain, equation (5.3.10), since $\mathbf{G}_{\min}^{-1}\mathbf{G}_{\mathrm{all}}^{\mathrm{H}}=\left[\mathbf{G}^{\mathrm{H}}\mathbf{G}\right]^{-1}\mathbf{G}^{\mathrm{H}}$ and $\mathbf{F}^{-\mathrm{H}}\mathbf{F}^{-1}=\mathbf{S}_{xx}^{-1}$. The connection between the Wiener solution to the multichannel control problem, as derived here, and the more recent polynomial approach is discussed by Casavola and Mosca (1996).

More general cost functions than the sum of squared errors can also be considered using minor modifications to the formulation above, using a weighted sum of squared errors and weighted sum of squared control efforts for example. Simple forms of this general cost function can also act to regularise the least-squares solution. If, for example, the cost function to be minimised has the form

$$J = \operatorname{trace} E\left(\boldsymbol{e}(n)\boldsymbol{e}^{\mathrm{T}}(n)+\rho\boldsymbol{u}(n)\boldsymbol{u}^{\mathrm{T}}(n)\right)\ , \tag{5.3.33}$$

then the optimal causal control filter has exactly the same form as equation (5.3.31) except that the spectral factorisation of equation (5.3.26) becomes

$$\mathbf{G}'^{\mathrm{T}}_{\min}\left(z^{-1}\right)\mathbf{G}'_{\min}(z) = \mathbf{G}^{\mathrm{T}}\left(z^{-1}\right)\mathbf{G}(z) + \rho\mathbf{I} . \tag{5.3.34}$$

Also, if each of the reference signals is assumed to contain additive white sensor noise of variance β, which is uncorrelated from one reference sensor to another, then the spectral factorisation in equation (5.3.23) becomes

$$\mathbf{F}'(z)\mathbf{F}'^{\mathrm{T}}\left(z^{-1}\right) = \mathbf{S}_{xx}(z) + \beta\mathbf{I} . \tag{5.3.35}$$

Although both control effort weighting and sensor noise will to some extent reduce the attenuation which can be obtained under the assumed, nominal, conditions, using small values of ρ and β need only have a small influence on the nominal attenuation, but can make the optimal filter easier to calculate and more robust to small changes in plant response and reference signal statistics.

5.4. ADAPTIVE ALGORITHMS IN THE TIME DOMAIN

In this section two generalisations of the LMS algorithm will be described for a controller composed of an array of FIR filters, which is adapted to minimise the sum of the squared outputs of a number of error sensors. As in the previous section it will be assumed that the FIR control filters each have I coefficients, are driven by K reference signals and, in turn, drive M secondary actuators. There are thus a total of MKI coefficients to be adapted to minimise the sum of the squared outputs from L sensors. Both of the adaptation algorithms described here operate directly in the time domain on the coefficients of the control filter using each new set of data samples.

5.4.1. Filtered-Reference LMS Algorithm

The first algorithm to be described is the multichannel generalisation of the filtered-reference LMS algorithm introduced in Chapter 3. The algorithm relies on the generation of a set of LMK filtered-reference signals, obtained by filtering each of the K reference signals by each of the $L \times M$ paths in the plant response. These filtered-reference signals can be used to obtain a simple expression for the vector of L error signals, as in equation (5.2.11). In Section 5.2, the control filter coefficients were assumed to be time invariant. We now allow them to change with time, but assume that they are only varying slowly compared with the timescales of the plant dynamics. The vector of error signals is now approximated by

$$\boldsymbol{e}(n) = \boldsymbol{d}(n) + \boldsymbol{R}(n)\,\boldsymbol{w}(n) , \tag{5.4.1}$$

where $\boldsymbol{w}(n)$ is a vector containing all MKI control filter coefficients at the n-th sample time.

In common with the single-channel filtered-reference algorithm, the multichannel version adjusts each of the control filter coefficients to minimise an instantaneous cost function, in this case given by the instantaneous sum of squared outputs of the error signals, which can be written as

$$\boldsymbol{e}^{\mathrm{T}}(n)\,\boldsymbol{e}(n) = \boldsymbol{w}^{\mathrm{T}}(n)\,\boldsymbol{R}^{\mathrm{T}}(n)\,\boldsymbol{R}(n)\,\boldsymbol{w}(n) + 2\,\boldsymbol{w}^{\mathrm{T}}(n)\,\boldsymbol{R}^{\mathrm{T}}(n)\,\boldsymbol{d}(n) + \boldsymbol{d}^{\mathrm{T}}(n)\,\boldsymbol{d}(n) . \tag{5.4.2}$$

The derivatives of this cost function with respect to the vector of filter coefficients at the same sample time can be written as

$$\frac{\partial \boldsymbol{e}^{\mathrm{T}}(n)\boldsymbol{e}(n)}{\partial \boldsymbol{w}(n)} = 2\left[\boldsymbol{R}^{\mathrm{T}}(n)\,\boldsymbol{R}(n)\boldsymbol{w}(n) + \boldsymbol{R}^{\mathrm{T}}(n)\boldsymbol{d}(n)\right] . \tag{5.4.3}$$

Using equation (5.4.1) this vector of derivatives can also be written as

$$\frac{\partial \boldsymbol{e}^{\mathrm{T}}(n)\boldsymbol{e}(n)}{\partial \boldsymbol{w}(n)} = 2\,\boldsymbol{R}^{\mathrm{T}}(n)\boldsymbol{e}(n) . \tag{5.4.4}$$

The vector of derivatives of the instantaneous squared errors with respect to the control filter coefficients are now used to adapt the control filter coefficients using the method of steepest descent, which is sometimes called the stochastic gradient algorithm, to give the multichannel generalisation of the filtered-reference LMS algorithm:

$$\boldsymbol{w}(n+1) \;=\; \boldsymbol{w}(n) - \alpha\,\boldsymbol{R}^{\mathrm{T}}(n)\boldsymbol{e}(n) , \tag{5.4.5}$$

where α is a convergence coefficient. This algorithm has been called the *multiple error LMS* algorithm (Elliott and Nelson, 1985a; Elliott et al., 1987). A very similar algorithm was derived for adapting control filters implemented using tapped delay lines with analogue weights by Arzamasov and Mal'tsev (1985). In practice, the true matrix of plant responses is not usually available to generate the filtered-reference signals and so a model of the plant responses must be used. The matrix of filtered-reference signals obtained with these estimated plant responses is denoted as $\hat{\boldsymbol{R}}(n)$, and the practical form of the multichannel filtered-reference LMS algorithm is thus

$$\boldsymbol{w}(n+1) \;=\; \boldsymbol{w}(n) \;-\alpha\,\hat{\boldsymbol{R}}^{\mathrm{T}}(n)\boldsymbol{e}(n) . \tag{5.4.6}$$

The convergence properties of this algorithm can be analysed in a similar way to that used for the single channel algorithm in Chapter 3. Equation (5.4.1) for $\boldsymbol{e}(n)$ is first substituted into equation (5.4.6) to give

$$\boldsymbol{w}(n+1) \;=\; \boldsymbol{w}(n) - \alpha\left[\hat{\boldsymbol{R}}^{\mathrm{T}}(n)\boldsymbol{d}(n) + \hat{\boldsymbol{R}}^{\mathrm{T}}(n)\,\boldsymbol{R}(n)\boldsymbol{w}(n)\right] . \tag{5.4.7}$$

Providing the algorithm is stable, it will converge to a set of control filter coefficients that sets the expectation value of the term in square brackets in equation (5.4.7) to zero. The steady-state vector of control filter coefficients after convergence is thus given by

$$\boldsymbol{w}_{\infty} \;=\; -\left\{E\left[\hat{\boldsymbol{R}}^{\mathrm{T}}(n)\,\boldsymbol{R}(n)\right]\right\}^{-1}\; E\left[\hat{\boldsymbol{R}}^{\mathrm{T}}(n)\boldsymbol{d}(n)\right] . \tag{5.4.8}$$

If the response of the plant model used to generate $\hat{\boldsymbol{R}}(n)$ is not equal to the true plant response, then the steady-state control vector will not, in general, be equal to the optimum set of control coefficients which minimise the sum of the squared errors, given by equation (5.2.15).

Equation (5.4.8) can be used with the usual independence assumption to write the expectation behaviour of the adaptive algorithm in equation (5.4.7) as

$$E\left[\boldsymbol{w}(n+1) - \boldsymbol{w}_{\infty}\right] \;=\; \left[I - \alpha\,E\left[\hat{\boldsymbol{R}}^{\mathrm{T}}(n)\,\boldsymbol{R}(n)\right]\right]\; E\left[\boldsymbol{w}(n) - \boldsymbol{w}_{\infty}\right] , \tag{5.4.9}$$

from which it is clear that the algorithm will converge only if

$$0 < \alpha < \frac{2\,\mathrm{Re}(\lambda_m)}{|\lambda_m|^2} \qquad \text{for all } \lambda_m\,, \tag{5.4.10}$$

where λ_m are the potentially complex eigenvalues of the matrix $E\left[\hat{\boldsymbol{R}}^{\mathrm{T}}(n)\,\boldsymbol{R}(n)\right]$. Care must again be exercised when using equation (5.4.10) to calculate the maximum convergence coefficient, since it has been derived assuming that the vector of error signals is given by equation (5.4.1), which itself assumes that the filter coefficients are changing only slowly compared with the dynamic response of the plant. Clearly, no value of α can be found that satisfies (5.4.10), however, if the real part of any of the eigenvalues of $E\left[\hat{\boldsymbol{R}}^{\mathrm{T}}(n)\,\boldsymbol{R}(n)\right]$ is negative, which defines the conditions for stability of equation (5.4.7) for very small values of the convergence coefficient. It has been shown by Wang and Ren (1999b) that a sufficient condition for stability of the multichannel filtered-reference algorithm with a persistently exciting reference signal is that the matrix $\hat{\mathbf{G}}^{\mathrm{H}}(\mathrm{e}^{j\omega T})\mathbf{G}(\mathrm{e}^{j\omega T}) + \mathbf{G}^{\mathrm{H}}(\mathrm{e}^{j\omega T})\hat{\mathbf{G}}(\mathrm{e}^{j\omega T})$ must be positive definite for all ωT, i.e.

$$\mathrm{eig}\left[\hat{\mathbf{G}}^{\mathrm{H}}(\mathrm{e}^{j\omega T})\mathbf{G}(\mathrm{e}^{j\omega T}) + \mathbf{G}^{\mathrm{H}}(\mathrm{e}^{j\omega T})\hat{\mathbf{G}}(\mathrm{e}^{j\omega T})\right] > 0 \qquad \text{for all } \omega T\,, \tag{5.4.11}$$

where eig[ ] denotes the eigenvalues of the matrix in the brackets, which is similar to the single-frequency result derived in Chapter 4. It should be emphasised that although this condition is sufficient for stability of the adaptive algorithm, it is not necessary: it is possible for some reference signals that the condition is violated, even though the system is stable. In general, however, equation (5.4.11) provides a useful guide to the stability limits of the adaptive algorithm when the plant response is different from that of the plant model.

The robustness of the stability to plant uncertainty can be improved, in a similar way to that used in the frequency-domain case considered in Section 4.4, by including a small term proportional to the sum of the squared controller coefficients in the instantaneous cost function being minimised, which thus becomes

$$J(n) = \boldsymbol{e}^{\mathrm{T}}(n)\boldsymbol{e}(n) + \beta\,\boldsymbol{w}^{\mathrm{T}}(n)\boldsymbol{w}(n)\,, \tag{5.4.12}$$

where β is a positive coefficient-weighting parameter. If this cost function is now expressed as a quadratic function of the filter coefficients, the resultant equation is of exactly the same form as equation (5.2.14) except that the Hessian matrix is now $\boldsymbol{R}^{\mathrm{T}}(n)\boldsymbol{R}(n) + \beta\mathbf{I}$ instead of $\boldsymbol{R}^{\mathrm{T}}(n)\boldsymbol{R}(n)$. The optimal controller thus becomes

$$\mathbf{w}_{\mathrm{opt}} = -\left\{E\left[\boldsymbol{R}^{\mathrm{T}}(n)\,\boldsymbol{R}(n) + \beta\boldsymbol{I}\right]\right\}^{-1} E\left[\boldsymbol{R}^{\mathrm{T}}(n)\boldsymbol{d}(n)\right]\,, \tag{5.4.13}$$

and the gradient vector becomes

$$\frac{\partial J}{\partial \boldsymbol{w}(n)} = 2\left[\boldsymbol{R}^{\mathrm{T}}(n)\boldsymbol{e}(n) + \beta\,\boldsymbol{w}(n)\right]\,. \tag{5.4.14}$$

Note that in the broadband case, adding a term proportional to the sum of the squared controller coefficients to the cost function is not, in general, the same as adding a term proportional to the control effort, i.e. the sum of the squared control signals, as discussed in Section 3.4. Although the incorporation of a quadratic effort weighting term allows a direct comparison with the control literature, the use of a quadratic controller coefficient-weighting term gives rise to a computationally simple modification of the adaptation algorithm, which can have better convergence properties than an algorithm that accounts for control effort (Darlington, 1995). The addition of a coefficient-weighting term in the cost function being minimised introduces a leakage factor into the adaptation algorithm which thus becomes

$$\boldsymbol{w}(n+1) = \gamma \boldsymbol{w}(n) - \alpha \hat{\boldsymbol{R}}^{\mathrm{T}}(n)\boldsymbol{e}(n) , \tag{5.4.15}$$

where $\hat{\boldsymbol{R}}(n)$ has been used as a practical estimate of $\boldsymbol{R}(n)$, and $\gamma = 1 - \alpha\beta$. The convergence condition is then changed from equation (5.4.10) to

$$0 < \alpha < \frac{\mathrm{Re}(\lambda_m) + \beta}{|\lambda_m + \beta|^2} \quad \text{for all } \lambda_m , \tag{5.4.16}$$

where λ_m is again an eigenvalue of $E\left[\hat{\boldsymbol{R}}(n)\,\boldsymbol{R}(n)\right]$. If one of these eigenvalues has a small negative real part, it can easily be seen how a small value of β can stabilise an otherwise unstable system. The effect of β on the frequency-domain stability condition in equation (5.4.11) is to require that the matrix $\hat{\mathbf{G}}^{\mathrm{H}}\left(\mathrm{e}^{j\omega T}\right)\mathbf{G}\left(\mathrm{e}^{j\omega T}\right) + \mathbf{G}^{\mathrm{H}}\left(\mathrm{e}^{j\omega T}\right)\hat{\mathbf{G}}\left(\mathrm{e}^{j\omega T}\right) + 2\beta\mathbf{I}$ is positive definite at all frequencies. It is found in practice that small values of β can once again improve the robust stability of the algorithm without unduly degrading the performance, so that robust performance can often be achieved.

Once the stability of the adaptive system has been established, the analysis of the convergence properties (Elliott and Nelson, 1993) closely follows the convergence analysis of the tonal control system in Chapter 4, in that the cost function converges in a series of modes, whose time constants are determined by the eigenvalues of the Hessian matrix, which in this case is equal to β plus the eigenvalues of $E\left[\hat{\boldsymbol{R}}^{\mathrm{T}}(n)\,\boldsymbol{R}(n)\right]$. A more general case of the multichannel filtered-reference algorithm in equation (5.4.15) has been described by Elliott and Nelson (1993), in which the error signals and control signals are weighted in the cost function by matrices analogous to $\boldsymbol{W}_e$ and $\boldsymbol{W}_u$ in equation (4.2.19). Even if the plant modelling is perfect so that the matrix $E\left[\hat{\boldsymbol{R}}^{\mathrm{T}}(n)\,\boldsymbol{R}(n)\right]$ becomes equal to the generalised autocorrelation matrix $E\left[\boldsymbol{R}^{\mathrm{T}}(n)\,\boldsymbol{R}(n)\right]$, the eigenvalues of this matrix can have a large range. In the multichannel control of stochastic disturbances this eigenvalue spread can be attributed to four distinct causes.

First, as in the analysis of the single-channel LMS algorithm, the spread of eigenvalues in the autocorrelation matrix is partly due to the spectral range in the reference signal.

Second, in a system with multiple reference signals the range of the eigenvalues is also affected by the correlation between the various reference signals.

Third, the range of eigenvalues is affected by the spatial arrangement of the actuators and sensors in the physical system, in the same manner as that for the single-frequency case considered in Chapter 4, in which the convergence was limited by the spread of the eigenvalues in the matrix $\mathbf{G}^{\mathrm{H}}(\mathrm{e}^{j\omega_0 T})\mathbf{G}(\mathrm{e}^{j\omega_0 T})$.

Finally, the convergence rate is limited by the dynamic response of each of the paths in the plant response, particularly any delay.

All of these effects can compound together in the general case to generate a very large eigenvalue spread, which can slow down the convergence speed considerably, particularly when many coefficients are used in each of the control filters. These problems with eigenvalue spread can be significantly reduced if the preconditioned LMS algorithm is used or if the adaptive algorithm is cast in the frequency domain, as discussed in the following sections. Also in this case, the effects of cross-correlation between the reference signals can be separated from those due to the spatial distribution of actuators and sensors in the plant, which simplifies the design of more rapidly-adapting control algorithms.

It would, in principle, also be possible to improve the adaptation time of the time domain algorithm by using an iterative exact least-squares, or Gauss–Newton, method. If the matrix $\boldsymbol{D} = E\left[\hat{\boldsymbol{R}}^{\mathrm{T}}(n)\,\hat{\boldsymbol{R}}(n) + \delta\,\boldsymbol{I}\right]^{-1}$ is precomputed, where δ is a regularisation factor which can be used to improve the robustness of the algorithm, as in Chapter 4, then an iterative least-squares algorithm using the instantaneous gradient estimate can be written as

$$\boldsymbol{w}(n+1) \;=\; \boldsymbol{w}(n) - \alpha\,\boldsymbol{D}\,\hat{\boldsymbol{R}}^{\mathrm{T}}(n)\boldsymbol{e}(n)\ . \tag{5.4.17}$$

This algorithm has not received a great deal of attention in the literature, perhaps because of the potentially large size of the matrix $\hat{\boldsymbol{R}}^{\mathrm{T}}(n)\,\hat{\boldsymbol{R}}(n)$, as mentioned in the previous section, which can create problems in estimating the inverse matrix required for $\boldsymbol{D}$. A different normalisation procedure has been used by Douglas and Olkin (1993) to improve the convergence rate of the multiple-channel filtered-reference LMS algorithm. These authors emphasise the importance of using a normalised algorithm when the reference signals are not stationary, in which case a fixed convergence coefficient can cause the algorithm to become unstable if the level of the reference signal unexpectedly increases.

5.4.2. Filtered-Error LMS Algorithm

A multichannel form of the filtered-error LMS algorithm introduced in Chapter 3 can be derived by direct manipulation of the quadruple summation formulae for the time-domain error signal, equation (5.2.3) (Popovich, 1994; Elliott, 1998c). A more direct and elegant derivation can be obtained, however, by examination of the derivative of the cost function with respect to the matrix of i-th control filter coefficients in equation (5.2.27), which can be written using equation (5.2.25) as

$$\frac{\partial J}{\partial \boldsymbol{W}_i} \;=\; 2E\left[\sum_{j=0}^{J-1} \boldsymbol{G}_j^{\mathrm{T}}\,\boldsymbol{e}(n+j)\boldsymbol{x}^{\mathrm{T}}(n-i)\right]\ . \tag{5.4.18}$$

The vector of M filtered-error signals can now be defined as

$$\boldsymbol{f}(n) = \sum_{j=0}^{J-1} \boldsymbol{G}_j^{\mathrm{T}}\,\boldsymbol{e}(n+j)\ , \tag{5.4.19}$$

which allows equation (5.4.18) to be written as

$$\frac{\partial J}{\partial \mathbf{W}_i} = 2E\left[\boldsymbol{f}(n)\boldsymbol{x}^{\mathrm{T}}(n-i)\right] . \tag{5.4.20}$$

Using the instantaneous version of this derivative to adapt the matrix of i-th control filter coefficients at the n-th time, the multichannel version of the filtered-error LMS algorithm can be written as

$$\mathbf{W}_i(n+1) = \mathbf{W}_i(n) - \alpha\, \boldsymbol{f}(n)\boldsymbol{x}^{\mathrm{T}}(n-i) , \tag{5.4.21}$$

which has been termed the *adjoint LMS algorithm* by Wan (1996).

As for the single-channel case, however, it is not possible to implement this algorithm with a causal system, because equation (5.4.19) requires time-advanced error signals. Delaying both the filtered error signals and the reference signals by J samples, however, gives a form of equation (5.4.21) that can be causally implemented,

$$\mathbf{W}_i(n+1) = \mathbf{W}_i(n) - \alpha\, \boldsymbol{f}(n-J)\boldsymbol{x}^{\mathrm{T}}(n-i-J) . \tag{5.4.22}$$

The vector of delayed filtered-error signals can be written as

$$\boldsymbol{f}(n-J) = \sum_{j'=1}^{J} \boldsymbol{G}_{J-j'}^{\mathrm{T}}\, \boldsymbol{e}(n-j') , \tag{5.4.23}$$

where $j' = J - j$, which emphasises the fact that $\boldsymbol{e}(n)$ is causally filtered by a time-reversed version of the transposed matrix of plant impulse responses. Strictly speaking, the filtered-error and reference signals only need to be delayed by $J-1$ samples, but an extra sample delay has been included to make the equations simpler to follow.

One advantage of the multichannel filtered-error algorithm over the multichannel filtered-reference algorithm is in terms of its computational efficiency. The multichannel filtered-error LMS algorithm involves the M filtered error signals in equation (5.4.19), each of which requires the filtering of the L individual error signals, and so only LM individual filtered-error signals need to be generated. In contrast, the multichannel filtered-reference algorithm in equation (5.4.5) requires the generation of KLM filtered-reference signals. In many practical implementations of the algorithm, it is the generation of the filtered error or filtered reference signals that takes up the majority of the processing time. The number of multiplication operations required per sample to implement the filtered-reference and filtered-error LMS algorithms in the general case with K reference signals, M secondary sources and L error sensors is shown in Table 5.1, in which there are assumed to be I coefficients in every control filter and each path in the plant response is modelled as

Table 5.1 Number of operations required per sample to implement the multichannel filtered-reference and filtered-error LMS algorithms in the time domain with K reference signals, L error sensors, M secondary actuators, I coefficients in each control filter and each element of the matrix of plant responses modelled with an FIR filter having J coefficients

Algorithm	Filtered signal generation	Control filter update	Total number of operations
Filtered reference	$JKLM$	$IKLM$	$(I+J)KLM$
Filtered error	JLM	IKM	$(IK+JL)M$

an FIR filter with J coefficients. For a control system with many reference or error signals, so that K or L is large, then the filtered-error formulation can represent a significantly more efficient method of implementing an adaptive controller than the filtered-error approach. If, for example, $K = 6$, $M = 4$, $L = 8$ and $I = J = 128$, then the filtered-reference LMS algorithm requires about 49,000 multiplications per sample for its implementation, whereas the filtered-error LMS requires about 7,000 multiplications per sample.

The block diagram of the multichannel filtered-error LMS algorithm is shown in Fig. 5.4, in which the vector of error signals is passed through the transposed matrix of time reversed plant responses, $\boldsymbol{G}^{\mathrm{T}}\left(z^{-1}\right)$, with a delay of z^{-J} to ensure causality, which is also present in the reference signal path. If the multichannel filtered-error LMS algorithm is used to minimise a cost function that also includes a term proportional to the sum of squared coefficients in all the control filters, as in equation (5.4.12), the algorithm is modified by introducing a leakage factor into the coefficient update equation in exactly the same way as for the filtered-reference LMS algorithm in equation (5.4.15).

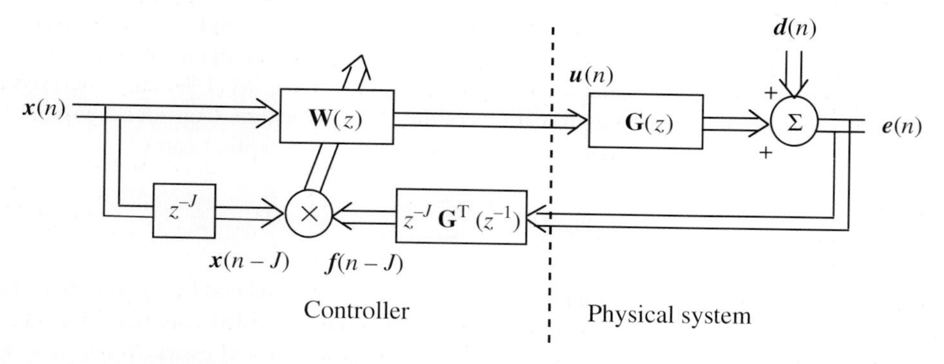

Figure 5.4 Block diagram of the filtered-error LMS algorithm for adapting the feedforward controller that uses the outer product (⊗) of the vector of delayed reference signals $\boldsymbol{x}(n-J)$, and the vector of delayed filtered error signals, $\boldsymbol{f}(n-J)$.

The multichannel filtered-error LMS algorithm can be derived from the multichannel filtered-reference LMS algorithm by only reordering and delaying the terms used to update the controller coefficients (Wan, 1996; Elliott, 1998c). This would also be true if estimated plant impulse responses were used to generate these derivatives instead of the true ones. Therefore, we can conclude that for small values of α, and hence slow adaptation rates, the convergence behaviour of the multichannel filtered-error LMS algorithm must be the same as that of the multichannel filtered-reference LMS algorithm. In particular, the stability condition for the eigenvalues of the generalised autocorrelation matrix, equation (5.4.10) must still apply and so the filtered-error algorithm is equally sensitive, or insensitive, to plant modelling error as the filtered-reference algorithm.

As in the single-channel case discussed in Section 3.4, however, the maximum convergence speed of the filtered-error LMS algorithm may be slower than that of the filtered-reference LMS algorithm, because of the additional delays which have been introduced into the error signals to ensure that $\mathbf{G}^{\mathrm{T}}\left(z^{-1}\right)$ is causal. Douglas (1999) has introduced a fast implementation of the multichannel LMS algorithm which avoids these delays by reordering the terms in the adaptation equation. The speed of convergence is then the same as that of the filtered-reference LMS algorithm, but the computational complexity scales like the filtered-error LMS algorithm. Douglas (1999) also extends this technique to develop a fast, multichannel, implementation of the modified filtered-reference LMS algorithm, the single channel version of which was described in Section 3.4, which effectively removes the delays in the plant response from the adaptation of the controller. Even in this case, however, it should be noted that the speed of the multichannel LMS algorithm is still limited by the auto- and cross-correlation functions of the reference signals, and by the spatial distribution of the actuators and sensors in the plant response, as listed at the end of Section 5.4.1.

5.4.3. Sparse Adaptation of the Filter Coefficients

The computational burden of the multichannel LMS algorithms described above can be very high if there are a large number of reference signals, secondary sources or error sensors, or if the sample rate is high. Some computational savings can be made by moving away from the original LMS philosophy of updating every filter coefficient at every sample time and adopting an approach that uses *sparse adaptation*. Douglas (1997) has analysed a number of such 'partial update' modifications of the LMS algorithm. Consider the filtered-reference LMS algorithm of equation (5.4.5) written out in the explicit form,

$$w_{mki}(n+1) = w_{mki}(n) - \alpha \sum_{l=1}^{L} e_l(n)\, r_{lmk}(n-i) \ , \qquad (5.4.24)$$

from which it is clear that each of the MKI filter coefficients is updated every sample using every error signal. An early observation (Elliott and Nelson, 1986) was that instead of updating the filter coefficients every sample, they could be adapted every N samples, so that

$$w_{mki}(n+N) = w_{mki}(n) - \alpha \sum_{l=1}^{L} e_l(n)\, r_{lmk}(n-i) \ , \qquad (5.4.25)$$

in which case the value of α could be increased and the adaptation speed could to some extent be maintained for values of N that were similar to the delays in the plant (see also Snyder and Tanaka, 1997). Although some computational saving can be achieved using this technique, each of the filtered-reference signals for I samples in the past must be calculated every N samples. A more efficient approach, which only involves the calculation of the current filtered-reference signal every N samples, is described by Orduña-Bustamante (1995) and involves the sequential adaptation of the MK filter coefficients with $i = 0$ at one time instant, those with $i = 1$ at the second time instant, etc. This can be written as

$$w_{mk0}(n+1) = w_{mk0}(n) - \alpha \sum_{l=1}^{L} e_l(n)\, r_{lmk}(n) \ , \qquad (5.4.26)$$

$$w_{mk1}(n+2) = w_{mk1}(n) - \alpha \sum_{l=1}^{L} e_l(n+1)\, r_{lmk}(n) \ , \qquad (5.4.27)$$

and so on until

$$w_{mkI}(n+I+1) \;=\; w_{mkI}(n) - \alpha \sum_{l=1}^{L} e_l(n+I)\, r_{lmk}(n) \;. \tag{5.4.28}$$

Note that only one value of each of the filtered-reference signals need be calculated every N samples using this approach.

Another method of reducing the computational load of the filtered-reference LMS algorithm is to use only a single error signal to adapt all the filter coefficients at each sample as in the *scanning error* algorithm (Hamada, 1991), which can be written as

$$w_{mki}(n+1) \;=\; w_{mki}(n) - \alpha\, e_1(n)\, r_{1mk}(n-i) \;, \tag{5.4.29}$$

$$w_{mki}(n+2) \;=\; w_{mki}(n+1) - \alpha\, e_2(n+1)\, r_{2mk}(n-i+1) \;, \tag{5.4.30}$$

and so on until

$$w_{mki}(n+L) \;=\; w_{mki}(n+L-1) - \alpha\, e_L(n+L-1)\, r_{Lmk}(n-i+L-1) \;. \tag{5.4.31}$$

Some care must be taken with the sparse adaptation algorithms described by equation (5.4.25) and equations (5.4.29–31), since each error signal is only sampled every N data points and the convergence of these algorithms is potentially affected by the aliasing of these signals.

The filtered-error LMS algorithm described by equation (5.4.22) can be explicitly written out as

$$w_{mki}(n+1) = w_{mki}(n) - \alpha\, f_m(n-J)\, x_k(n-i-J) \;, \tag{5.4.32}$$

and this could also be updated using a similarly sparse approach. Sparse adaptation could be achieved in this case either by scanning the filtered errors, to give

$$w_{1ki}(n+1) = w_{1ki}(n) - \alpha\, f_1(n-J)\, x_k(n-i-J) \;, \tag{5.4.33}$$

$$w_{2ki}(n+2) = w_{2ki}(n) - \alpha\, f_2(n-J+1)\, x_k(n-i-J+1) \;, \tag{5.4.34}$$

and so on until

$$w_{Mki}(n+M) = w_{Mki}(n) - \alpha\, f_M(n-J+M-1)\, x_k(n-i-J+M-1) \;, \tag{5.4.35}$$

or by sequentially using each reference signal to give

$$w_{m1i}(n+1) = w_{m1i}(n) - \alpha\, f_m(n-J)\, x_1(n-i-J) \;, \tag{5.4.36}$$

$$w_{m2i}(n+2) = w_{m2i}(n) - \alpha\, f_m(n-J+1)\, x_2(n-i-J+1) \;, \tag{5.4.37}$$

and so on until

$$w_{mKi}(n+K) = w_{mKi}(n) - \alpha\, f_m(n-J+K-1)\, x_K(n-i-J+K-1) \;, \tag{5.4.38}$$

or potentially by sequentially using each filtered-error signal with each individual reference signal in turn. The convergence and misadjustment properties of these sparse adaptation algorithms have not been explored.

5.5. THE PRECONDITIONED LMS ALGORITHM

Rather than adapting the control filter matrix $\mathbf{W}(z)$ directly, using the filtered-error algorithm whose block diagram is shown in Fig. 5.4 for example, the form of the optimum matrix of control filters can be used to motivate an alternative controller architecture, which can overcome many of the problems associated with steepest-descent algorithms such as the LMS. These problems were spelt out in Section 5.4, and are mainly caused by two distinct effects in multichannel systems: first the fact that the reference signals are correlated, and second the fact that the plant response is coupled. Returning to equation (5.3.31) for the matrix of optimum control filters, we note that if this matrix is written as

$$\mathbf{W}(z) = \mathbf{G}_{\min}^{-1}(z)\mathbf{C}(z)\mathbf{F}^{-1}(z) , \tag{5.5.1}$$

where $\mathbf{G}_{\min}^{-1}(z)$ is the $M \times M$ inverse matrix of the minimum-phase plant response, $\mathbf{F}^{-1}(z)$ is the inverse of the spectral factor of the reference signal's spectral density matrix, and $\mathbf{C}(z)$ is an $M \times K$ matrix of transformed controllers, then the optimal responses for the transformed controller $\mathbf{C}(z)$ can be seen from equation (5.3.31) to take on the particularly simple form

$$\mathbf{C}_{\text{opt}}(z) = -\left\{\mathbf{G}_{\text{all}}^{\text{T}}\left(z^{-1}\right)\mathbf{S}_{xd}(z)\mathbf{F}^{-\text{T}}\left(z^{-1}\right)\right\}_{+} . \tag{5.5.2}$$

The transformed control matrix $\mathbf{C}(z)$ could now be made adaptive, and adjusted at the m-th iteration according to Newton's algorithm,

$$\mathbf{C}_{m+1}(z) = \mathbf{C}_m(z) + \alpha\Delta\mathbf{C}_m(z) , \tag{5.5.3}$$

where α is a convergence coefficient and $\Delta\mathbf{C}_m(z)$ is the difference between the optimum matrix of controller responses and the current controller, at the m-th iteration. This is equal to

$$\Delta\mathbf{C}_m(z) = -\left\{\mathbf{G}_{\text{all}}^{\text{T}}\left(z^{-1}\right)\mathbf{S}_{xe,m}(z)\mathbf{F}^{-\text{T}}\left(z^{-1}\right)\right\}_{+} , \tag{5.5.4}$$

where

$$\mathbf{S}_{xe,m}(z) = E\left[\mathbf{e}_m(z)\mathbf{x}_m^{\text{T}}\left(z^{-1}\right)\right] , \tag{5.5.5}$$

and $\mathbf{e}_m(z)$ and $\mathbf{x}_m(z)$ are the vectors of errors and reference signals at the m-th iteration. Equation (5.5.4) can also be written as

$$\Delta\mathbf{C}_m(z) = -\left\{E\left[\mathbf{a}_m(z)\mathbf{v}_m^{\text{T}}\left(z^{-1}\right)\right]\right\}_{+} , \tag{5.5.6}$$

where

$$\mathbf{a}_m(z) = \mathbf{G}_{\text{all}}^{\text{T}}\left(z^{-1}\right)\mathbf{e}_m(z) , \tag{5.5.7}$$

which is a vector of M filtered-error signals, and

$$\mathbf{v}_m(z) = \mathbf{F}^{-1}(z)\mathbf{x}_m(z) , \tag{5.5.8}$$

is the vector of uncorrelated innovation signals assumed to generate the reference signals, as in Section 2.4.

Taking the inverse z-transform of the spectral density matrix in equation (5.5.6) gives the required adjustment for the i-th element of the causal impulse response of $\mathbf{C}(z)$ at the n-th sample time. This is given by

$$\Delta \boldsymbol{C}_i(n) = E\left[\boldsymbol{a}(n)\boldsymbol{v}^{\mathrm{T}}(n-i)\right] , \tag{5.5.9}$$

where $\boldsymbol{a}(n)$ and $\boldsymbol{v}(n)$ are the time-domain versions of the signals defined by equations (5.5.7) and (5.5.8). In moving from the z domain to the time domain, we can describe control filters with a finite number of coefficients, although clearly a matrix of control filters with a limited number of coefficients cannot converge to the optimal solution in the z domain, which assumes a matrix of control filters which are causal but of infinite duration.

If an instantaneous version of this matrix of cross-correlation functions is used to adapt the matrix of the i-th coefficients of an FIR transformed controller matrix at every sample time, a new algorithm that is similar in form to the filtered-error LMS algorithm is derived, which can be written as (Elliott, 2000)

$$\boldsymbol{C}_i(n+1) = \boldsymbol{C}_i(n) - \alpha\, \boldsymbol{a}(n)\boldsymbol{v}^{\mathrm{T}}(n-i) . \tag{5.5.10}$$

The generation of the signals $\boldsymbol{a}(n)$ and $\boldsymbol{v}(n)$ is shown in the block diagram in Fig. 5.5.

Comparing this block diagram to that shown in Fig. 5.4, we notice that $\mathbf{W}(z)$ is now implemented according to equation (5.5.1), with the reference signals, $\boldsymbol{x}(n)$, being pre-processed by $\mathbf{F}^{-1}(z)$ to give $\boldsymbol{v}(n)$, which drives the transformed controller matrix $\mathbf{C}(z)$, and the output of $\mathbf{C}(z)$ is then multiplied by $\mathbf{G}_{\min}^{-1}(z)$ to generate the control signals, $\boldsymbol{u}(n)$, for the plant. The vector of innovation signals $\boldsymbol{v}(n)$ is also required in the adaptation equation (5.5.10), together with the vector of filtered-error signals, which are generated in this case by passing $\boldsymbol{e}(n)$ through the transposed and time-reversed version of the all-pass component of the plant response, $\mathbf{G}_{\mathrm{all}}^{\mathrm{T}}\left(z^{-1}\right)$.

In order to implement equation (5.5.10) causally, however, delays must be introduced

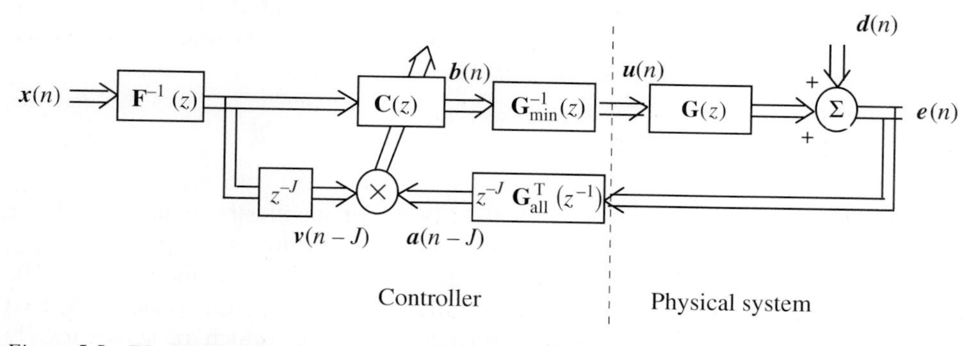

Figure 5.5 Block diagram of the preconditioned LMS algorithm for adapting the transformed matrix of controllers $\mathbf{C}(z)$, which uses the outer product (⊗) of the vector of delayed innovations for the reference signals, $\boldsymbol{v}(n-J)$, and the vector of delayed filtered error signal $\boldsymbol{a}(n-J)$.

into $\boldsymbol{v}(n)$ and $\boldsymbol{a}(n)$, as for the filtered-error LMS algorithm described above, which are also shown in Fig. 5.5. Alternatively the adaptive algorithm could be implemented in the discrete-frequency domain, provided precautions were taken to avoid circular convolution and correlation effects, in which case the one block delay in the adaptation would ensure causality.

The adaptive algorithm implemented with the block diagram shown in Fig. 5.5 avoids many of the convergence problems of the filtered-error LMS algorithm. The reference signals are whitened and decorrelated by preconditioning with the inverse spectral factor $\mathbf{F}^{-1}(z)$, in a similar manner to the virtual reference signals described by Dehandschutter et al. (1998) and Akiho et al. (1998), for example. Also, the transfer function from the output of the transformed controller, $\boldsymbol{b}(n)$, to the filtered-error signals used to update this controller, $\boldsymbol{a}(n)$, can be deduced by setting the disturbance to zero in Fig. 5.5, to give

$$\mathbf{a}(z) = \mathbf{G}_{\text{all}}^{\text{T}}\left(z^{-1}\right)\mathbf{G}(z)\mathbf{G}_{\text{min}}^{-1}(z)\mathbf{b}(z) \; . \tag{5.5.11}$$

Using the properties of $\mathbf{G}_{\text{all}}(z)$ and $\mathbf{G}_{\text{min}}(z)$ in equations (5.3.24) and (5.3.25) we find that $\mathbf{a}(z)$ is equal to $\mathbf{b}(z)$, without any cross coupling, and preconditioning the plant response with its minimum phase component means that the transfer function in the adaptation loop consists only of the delay required to make $\mathbf{G}_{\text{all}}^{\text{T}}\left(z^{-1}\right)$ causal in a practical system. The algorithm defined by equation (5.5.10) thus overcomes many of the causes of slow convergence in traditional filtered-error and filtered-reference LMS algorithms by separately preconditioning the reference signals and plant response, and has been termed the *preconditioned LMS (PLMS) algorithm* (Elliott, 2000).

5.5.1. Simulation Example

A simulation is presented to illustrate the properties of the algorithm described by equation (5.5.10) for a control system with two reference signals, two secondary sources and two error sensors ($K = L = M = 2$). The physical arrangement is illustrated in Fig. 5.6. Two independent Gaussian white noise signals, n_1 and n_2, are used to generate the disturbance signals via band-pass filters, with a bandwidth between normalised frequencies of 0.1 and 0.4, and also to generate the reference signals available to the control algorithm, x_1 and x_2, via filters which give the reference signals a pink noise spectrum (3 dB/octave slope), and a mixing matrix of real numbers $\mathbf{M}$ which in this case is equal to

$$\mathbf{M} = \begin{bmatrix} 0.75 & 0.25 \\ 0.25 & 0.75 \end{bmatrix} . \tag{5.5.12}$$

In the arrangement for the conventional LMS algorithm illustrated in Fig. 5.6 the reference signals are fed to a matrix of 128-point FIR control filters, W_{11}, W_{21}, W_{12}, W_{22}, operating at a sampling rate of 2.5 kHz, which drive the two secondary loudspeakers. The secondary loudspeakers are assumed to operate under free-field conditions and are spaced 0.5 m apart and are 1 m away from the two error microphones which are symmetrically positioned 1 m apart. The disturbance signals at the two microphones are assumed to be generated by two primary loudspeakers symmetrically positioned 3.6 m apart in the plane of the error sensors. The secondary loudspeakers are farther away from the error

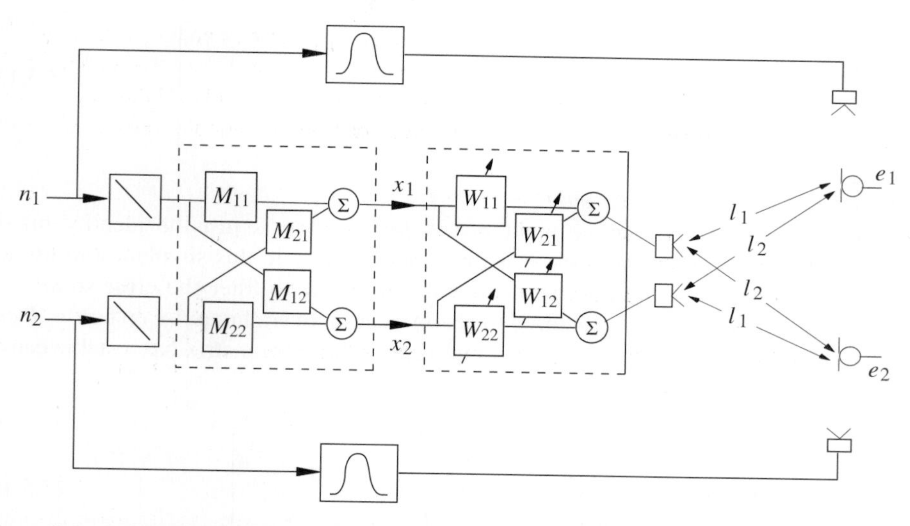

Figure 5.6 Block diagram of a simulation in which two uncorrelated white noise signals, n_1 and n_2, are used to generate both the disturbance signals at the error microphones via band-pass filters, and the observed reference signals, x_1 and x_2, via filters with a slope of 3 dB/octave and a mixing matrix $\boldsymbol{M}$. The acoustic plant is a symmetric arrangement of two secondary loudspeakers and two error microphones in free space.

microphones than the primary loudspeakers, and so perfect cancellation at the error sensors is not possible and the exact least-square solution for the control filters gives a reduction in the sum of squared errors in this case of about 18 dB.

The secondary loudspeakers and error microphones are symmetrically arranged in free space and so the continuous-time plant response can be written in this case as

$$\mathbf{G}(j\omega) = \begin{bmatrix} \frac{A}{l_1} e^{-jkl_1} & \frac{A}{l_2} e^{-jkl_2} \\ \frac{A}{l_2} e^{-jkl_2} & \frac{A}{l_1} e^{-jkl_1} \end{bmatrix}, \tag{5.5.13}$$

where A is an arbitrary constant, l_1 is the distance from the upper secondary loudspeaker to the upper error sensor (1.03 m in this simulation), l_2 is the distance from the upper secondary loudspeaker to the lower error microphone (1.25 m in this simulation) and k is the acoustic wavenumber, which is equal to ω/c_0 where c_0 is the speed of sound.

If the signals are sampled at a rate of f_s, the normalised plant response matrix can be written in the z domain as

$$\mathbf{G}(z) = \begin{bmatrix} z^{-N_1} & \frac{l_1}{l_2} z^{-N_2} \\ \frac{l_1}{l_2} z^{-N_2} & z^{-N_1} \end{bmatrix}, \tag{5.5.14}$$

where N_1 is the integer value of $l_1 f_s/c_0$, N_2 is the integer value of $l_2 f_s/c_0$ and $l_1/l_2 < 1$ by definition. For this plant matrix, the all-pass and minimum-phase decomposition can be derived by inspection to give

$$\mathbf{G}(z) = \mathbf{G}_{\text{all}}(z)\mathbf{G}_{\min}(z) = \begin{bmatrix} z^{-N_l} & 0 \\ 0 & z^{-N_1} \end{bmatrix} \begin{bmatrix} 1 & \dfrac{l_1}{l_2} z^{-\Delta N} \\ \dfrac{l_1}{l_2} z^{-\Delta N} & 1 \end{bmatrix}, \tag{5.5.15}$$

where $\Delta N = N_2 - N_1$, which is positive.

The all-pass component of the plant matrix in this case is just the identity matrix multiplied by a delay of N_1 samples. It is thus not necessary in this simulation to use the transpose of the all-pass component of the plant response to filter the error signals, as shown in Fig. 5.6, and the delay required to obtain a causal overall impulse response, J, can be set equal to N_1. The minimum-phase component of the plant matrix has a stable causal inverse which is equal in this case to

$$\mathbf{G}_{\min}^{-1} = \frac{1}{1-\left(l_1/l_2\right)^2 z^{-2\Delta N}} \begin{bmatrix} 1 & -\dfrac{l_1}{l_2} z^{-\Delta N} \\ -\dfrac{l_1}{l_2} z^{-\Delta N} & 1 \end{bmatrix}. \tag{5.5.16}$$

It is also possible in this simple example to analytically calculate the inverse of the spectral factors of the reference signal's spectral density matrix, $\mathbf{F}^{-1}(z)$, which will convert the measured reference signals back into the noise signals n_1 and n_2. In this case, $\mathbf{F}^{-1}(z)$ consists of the inverse of the mixing matrix, $\mathbf{M}$ in equation (5.5.12), and a pair of minimum-phase filters which will whiten the resulting pink noise signals. A method of performing this spectral factorisation in the general case using only the discrete-frequency version of the power spectral matrix has been presented by Cook and Elliott (1999) and the use of similar techniques to perform the all-pass and minimum-phase decomposition of a matrix of discrete-frequency plant responses is being explored at the time of writing.

The upper curve in Fig. 5.7 shows the time history of the sum of squared errors when the filtered-error LMS algorithm is used to adapt the coefficients of the control filters in the arrangement shown in Fig. 5.6. In all the simulations shown in Fig. 5.7 the convergence coefficient was set to half the lowest value that resulted in instability. The sum of squared errors has been normalised by the value before control and is plotted as a reduction in Fig. 5.7. The multiple time constants associated with the LMS algorithm are clearly seen and the reduction has only reached about 17 dB after 10,000 samples. If the inverse spectral factors are used to reconstruct the original noise signals, n_1 and n_2, and these are used as the reference signals in the normal filtered-error algorithm, the filter converges in about 8,000 samples, as is also shown in Fig. 5.7 as LMS with $\mathbf{F}^{-1}(z)$. Keeping the original reference signals but using the inverse of the minimum-phase part of the plant response to reduce the overall plant response to the all-pass components results in a similar convergence time to that of the LMS with $\mathbf{F}^{-1}(z)$ and is not shown in Fig. 5.7. The lowest curve in Fig. 5.7 corresponds to the case in which the inverse spectral factors and the minimum-phase plant response are used to implement the algorithm given by equation (5.5.10), which is denoted LMS with $\mathbf{F}^{-1}(z)$ and $\mathbf{G}_{\min}^{-1}(z)$. This results in the fastest convergence rate, for which 18 dB reduction is achieved in about 3,000 samples.

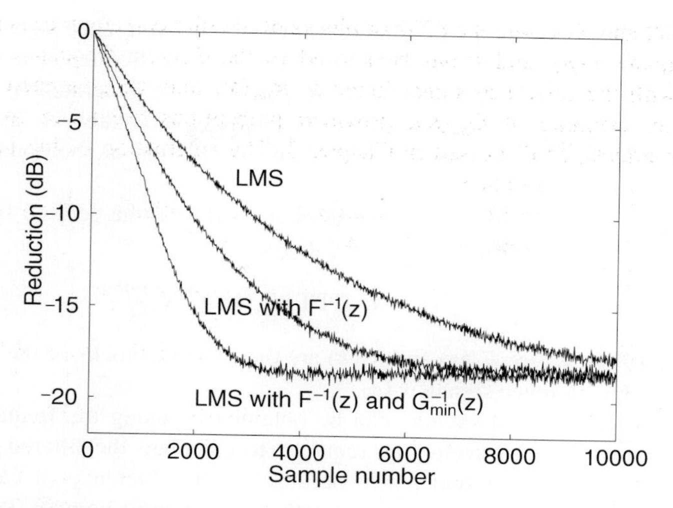

Figure 5.7 Time histories of the reduction in the sum of squared error signals for the simulation whose block diagram is shown in Fig. 5.6 with three adaptation algorithms. Upper graph: LMS algorithm using reference signals x_1 and x_2 and error signals e_1 and e_2; middle graph: LMS algorithm using uncorrelated reference signals n_1 and n_2 obtained by filtering x_1 and x_2 with $\boldsymbol{F}^{-1}(z)$. Lower graph: the preconditioned LMS algorithm which uses $\boldsymbol{F}^{-1}(z)$ to generate uncorrelated reference signals and $\boldsymbol{G}_{\min}^{-1}(z)$ to diagnose the overall plant response.

5.6. Adaptive Algorithms in the Frequency Domain

Having discussed in the previous section the direct implementation of steepest-descent algorithms in the time domain, we will now consider a more indirect implementation in the discrete-frequency domain. There are several reasons for introducing this extra layer of complexity. Initially, these algorithms can be justified on the grounds of computational efficiency for control filters with many coefficients. Although this was true in the single-channel case discussed in Chapter 3, we will see that the computational savings are even more significant in the multichannel case. Moreover, once the signals are in the frequency domain it is again possible to compensate for some of the shortcomings of the conventional time-domain algorithms, in particular the limited convergence speed due to eigenvalue spread of the generalised autocorrelation matrix of the filtered-reference signals. The frequency-domain formulation also allows the eigenvalue spread problems due to the intercorrelations between the reference signals, and the spatial distribution of actuators and sensors, to be considered separately.

Both the multichannel filtered-reference and filtered-error LMS algorithms can be readily implemented in the frequency domain. The frequency-domain version of the multichannel filtered-reference LMS algorithm, equation (5.4.6), can be written as a generalisation of equation (3.5.5) as

$$w_{mki}(n+N) = w_{mki}(n) - \alpha\,\mathrm{IFFT}\left\{\sum_{l=1}^{L} R_{lmk}^{*}(\kappa)\,E_l(\kappa)\right\}_{+} , \tag{5.6.1}$$

where $R_{lmk}(\kappa)$ and $E_l(\kappa)$, are the FFTs of blocks of the filtered-reference signals $r_{lmk}(n)$ and the error signals $e_l(n)$, and κ has been used as the discrete frequency index to avoid confusion with the reference signal index k. $R_{lmk}(\kappa)$ may be generated by multiplying $X_k(\kappa)$ by an estimate of $G_{lm}(\kappa)$, provided precautions are taken to avoid circular convolution effects, as discussed in Chapter 3. The calculation of the update term thus takes *LMK* complex multiplications.

The frequency domain filtered-error algorithm can similarly be derived from the time domain version, as given in Equation (5.4.32), to give

$$w_{mki}(n+N) = w_{mki}(n) - \alpha \ \text{IFFT}\left\{F_m(\kappa)\, X_k^*(\kappa)\right\}_+ , \tag{5.6.2}$$

(Stothers et al., 1995), where $F_m(\kappa)$ and $X_k(\kappa)$ are the FFTs of blocks of the filtered-error signals $f_m(n)$ and the reference signals $x_k(n)$.

Considerable computational savings can be obtained by using the frequency-domain update, partly because the convolutions required to calculate the filtered-error signals become simple multiplications and partly because of the efficiency of calculating the required correlation functions in the frequency rather than the time domain. The algorithms are most efficient if the number of controller coefficients is equal to the data block size N, although $2N$-point FFTs still need to be used to avoid circular correlation effects, as in the single-channel case described in Chapter 3. Table 5.2 shows the total number of real multiplications required to implement N samples of adaptation for the time-domain and frequency-domain versions of the multichannel filtered-error LMS algorithm. If we continue with the example discussed in Section 5.4 for which $K = 6$, $M = 4$, $L = 8$ and $I = J = 128$, and assume that the block size (N) is also 128, then the time-domain algorithm requires about 900,000 real multiplications to adapt to N samples of data, whereas the frequency-domain algorithm only requires about 135,000 real multiplications.

A general matrix formulation for the frequency-domain steepest-descent algorithm can be obtained by using the results derived in Section 5.4. The gradient of the sum of the mean-square errors with respect to each control filter coefficient can be written as in equation (5.4.18), whose value for one block of data can be expressed in the frequency domain and used to adapt the time-domain coefficients of the controller. The practical frequency-domain steepest-descent algorithm can then be written entirely in the frequency domain as

$$\mathbf{W}_{\text{new}}(\kappa) \ = \ \mathbf{W}_{\text{old}}(\kappa) - \alpha\left\{\hat{\mathbf{G}}^{\text{H}}(\kappa)\,\mathbf{e}(\kappa)\,\mathbf{x}^{\text{H}}(\kappa)\right\}_+ \tag{5.6.3}$$

Table 5.2 The number of real multiplications required to implement the filtered-error adaptation over N samples in the time domain and the frequency domain for a system with K reference signals, M control signals and L error signals. The number of coefficients in the filter, I, and the plant model, J, are assumed equal to N and $2N$-point FFTs are used to avoid circular convolution. Note that the time-domain implementation of the control filter itself is common to both systems and the computation required for this has been excluded from the table

Algorithm	Filtered error generation	Control filter update	Total
Time domain	LMN^2	KMN^2	$(K+L)MN^2$
Frequency domain	$8LMN$	$(K+L+MK)\,2N\log_2(2N)$ $+\,8KMN$	$(K+L+MK)\,2N\log_2(2N)$ $+\,8(K+L)MN$

where $\hat{\mathbf{G}}(\kappa)$ is the matrix of responses of the internal model of the physical plant at the discrete frequency κ. The frequency-domain versions of both the filtered-reference and the filtered-error algorithms, equations (5.6.1) and (5.6.2), can be derived from this general expression.

In order to improve the convergence speed of the frequency-domain algorithms it has been suggested that separate convergence coefficients could be used in each frequency bin, $\alpha(\kappa)$, to compensate for the mean-square value of the filtered-reference signal in each bin (Reichard and Swanson, 1993; Stothers et al., 1995). In general, however, such an algorithm may converge to a biased solution because the constraint of causality is applied to the product of the term within the causality brackets in equation (5.6.3) and $\alpha(\kappa)$, as discussed in Section 3.5.

A version of Newton's algorithm in the frequency domain can be obtained by using a practical approximation to the expression for the optimum matrix of causal filters in the frequency domain derived in Section 5.3, $\hat{\mathbf{W}}_{\text{opt}}(\kappa)$, and writing Newton's algorithm in the form

$$\mathbf{W}_{\text{new}}(\kappa) = (1-\alpha)\,\mathbf{W}_{\text{old}}(\kappa) + \alpha\,\hat{\mathbf{W}}_{\text{opt}}(\kappa) \;. \tag{5.6.4}$$

The matrix of optimal filter coefficients can be written by expressing equation (5.3.31) in the discrete-frequency domain as

$$\mathbf{W}_{\text{opt}}(\kappa) = -\,\mathbf{G}_{\min}^{-1}(\kappa)\left\{\mathbf{G}_{\text{all}}^{\text{H}}(\kappa)\,\mathbf{S}_{xd}(\kappa)\,\mathbf{F}^{-\text{H}}(\kappa)\right\}_{+}\mathbf{F}^{-1}(\kappa) \;, \tag{5.6.5}$$

where $\mathbf{G}_{\min}^{-1}(\kappa)$ is the inverse of the minimum phase parts of the plant response, equation (5.3.24), and $\mathbf{F}(\kappa)$ and $\mathbf{F}^{\text{H}}(\kappa)$ are the spectral factors of $\mathbf{S}_{xx}(\kappa)$, equation (5.3.23).

Suppressing the frequency variable for a moment, then since $\mathbf{e}=\mathbf{d}+\mathbf{G}\mathbf{W}\mathbf{x}$, the cross spectral matrix between the reference and disturbance signals can be written as

$$\mathbf{S}_{xd} = \mathbf{S}_{xe} - \mathbf{G}\mathbf{W}_{\text{old}}\mathbf{S}_{xx} \;, \tag{5.6.6}$$

so that

$$\begin{aligned}\mathbf{W}_{\text{opt}} &= -\,\mathbf{G}_{\min}^{-1}\left\{\mathbf{G}_{\text{all}}^{\text{H}}\,\mathbf{S}_{xe}\,\mathbf{F}^{-\text{H}}\right\}_{+}\mathbf{F}^{-1} \\ &\quad +\mathbf{G}_{\min}^{-1}\left\{\mathbf{G}_{\text{all}}^{\text{H}}\,\mathbf{G}\,\mathbf{W}_{\text{old}}\,\mathbf{S}_{xx}\,\mathbf{F}^{-\text{H}}\right\}_{+}\mathbf{F}^{-1} \;.\end{aligned} \tag{5.6.7}$$

Using the definition of the spectral factors $\mathbf{F}$ and $\mathbf{F}^{\text{H}}$ we can write

$$\mathbf{S}_{xx}\,\mathbf{F}^{-\text{H}} = \mathbf{F} \;, \tag{5.6.8}$$

which is a matrix of entirely causal responses. Also, since $\mathbf{G} = \mathbf{G}_{\text{all}}\,\mathbf{G}_{\min}$, and using equation (5.3.25), then we can also write

$$\mathbf{G}_{\text{all}}^{\text{H}}\,\mathbf{G} = \mathbf{G}_{\text{all}}^{\text{H}}\,\mathbf{G}_{\text{all}}\,\mathbf{G}_{\min} = \mathbf{G}_{\min} \;, \tag{5.6.9}$$

which is another matrix of entirely causal responses. Since all the terms within the causality brackets for the second term in equation (5.6.7) are already causal, the causality bracket can be removed and the optimum matrix of controllers can be written as

$$\mathbf{W}_{\text{opt}} = -\,\mathbf{G}_{\min}^{-1}\left\{\mathbf{G}_{\text{all}}^{\text{H}}\,\mathbf{S}_{xe}\,\mathbf{F}^{-\text{H}}\right\}_{+}\mathbf{F}^{-1} + \mathbf{W}_{\text{old}} \;. \tag{5.6.10}$$

An estimate of this matrix of optimal filters is obtained if the cross spectral density $\mathbf{S}_{xe}$ is approximated by its estimate based only on the current block of data, i.e. $\mathbf{ex}^{\mathrm{H}}$, to give

$$\hat{\mathbf{W}}_{\mathrm{opt}} = -\mathbf{G}_{\min}^{-1}\left\{\mathbf{G}_{\mathrm{all}}^{\mathrm{H}}\,\mathbf{e}\,\mathbf{x}^{\mathrm{H}}\,\mathbf{F}^{-\mathrm{H}}\right\}_{+}\mathbf{F}^{-1}+\mathbf{W}_{\mathrm{old}}\ . \tag{5.6.11}$$

Using this expression as the estimate of the matrix of optimal controller responses in equation (5.6.4), a multichannel version of Newton's algorithm in the discrete-frequency domain may be written using equation (5.3.28) as

$$\mathbf{W}_{\mathrm{new}}(\kappa) = \mathbf{W}_{\mathrm{old}}(\kappa)-\boldsymbol{\alpha}^{+}(\kappa)\left\{\boldsymbol{\alpha}^{-}(\kappa)\hat{\mathbf{G}}^{\mathrm{H}}(\kappa)\mathbf{e}(\kappa)\mathbf{x}^{\mathrm{H}}(\kappa)\mathbf{F}^{-\mathrm{H}}(\kappa)\right\}_{+}\mathbf{F}^{-1}(\kappa)\ , \tag{5.6.12}$$

where

$$\boldsymbol{\alpha}^{+}(\kappa) = \sqrt{\alpha}\,\hat{\mathbf{G}}_{\min}^{-1}(\kappa)\ \text{ and }\ \boldsymbol{\alpha}^{-}(\kappa) = \sqrt{\alpha}\,\hat{\mathbf{G}}_{\min}^{-\mathrm{H}}(\kappa)\ . \tag{5.6.13,14}$$

The block diagram for this algorithm is shown in Fig. 5.8. In comparison with the steepest-descent algorithm, equation (5.6.3), the term within the causality constraint has been pre- and post-multiplied by the matrices of noncausal responses given by $\boldsymbol{\alpha}^{-}$ and $\mathbf{F}^{-\mathrm{H}}$, and outside the causality constraint, the update term has been pre- and post-multiplied by the matrices of causal responses given by $\boldsymbol{\alpha}^{+}$ and $\mathbf{F}^{-1}$. This form of Newton's algorithm thus compensates separately for the matrix of plant responses, via $\boldsymbol{\alpha}^{+}(\kappa)$ and $\boldsymbol{\alpha}^{-}(\kappa)$ in equations (5.6.13,14), and the correlation structure of the reference signals, via the spectral factors of their cross spectral matrix $\mathbf{F}(\kappa)$ and $\mathbf{F}^{\mathrm{H}}(\kappa)$.

If the matrix of plant responses does not change significantly during the operation of the control system, then $\mathbf{G}_{\min}^{-1}$, and hence $\boldsymbol{\alpha}^{+}$ and $\boldsymbol{\alpha}^{-}$, can be computed off-line. In general,

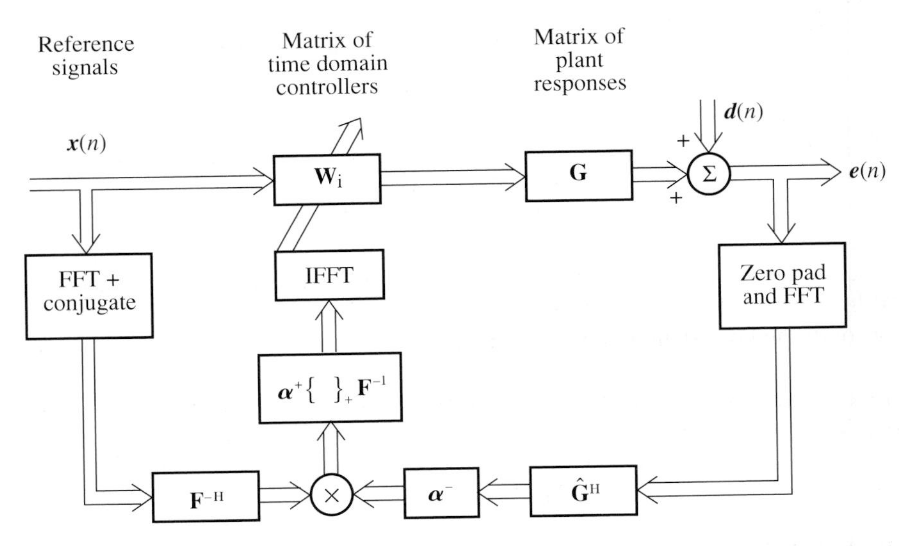

Figure 5.8 Frequency-domain adaptation of the matrix of control filters in which the spectral factors $\boldsymbol{\alpha}^{+}$, $\boldsymbol{\alpha}^{-}$, and $\mathbf{F}^{-1}$, $\mathbf{F}^{-\mathrm{H}}$ compensate separately for the spatial properties of the plant and correlation properties of the reference signals. The conventional frequency-domain form of the filtered-error adaptation algorithm is obtained if the spectral factors are set equal to the relevant identity matrices.

however, the spectral factors $\mathbf{F}^+$ and $\mathbf{F}^-$ will have to be calculated on-line, from a running estimate of $\mathbf{S}_{xx}$, since the reference signals will usually be non-stationary.

Although equation (5.6.12) has been motivated in a rather different way from the algorithm described in Section 5.5, it is interesting to compare the forms of the two algorithms. If the transformed matrix of frequency-domain controllers is defined to be

$$\mathbf{C}(\kappa) = \hat{\mathbf{G}}_{\min}(\kappa)\,\mathbf{W}(\kappa)\,\mathbf{F}(\kappa) \qquad (5.6.15)$$

then equation (5.6.12) can be written as

$$\mathbf{C}_{\text{new}}(\kappa) = \mathbf{C}_{\text{old}}(\kappa) - \alpha\left\{\hat{\mathbf{G}}_{\text{all}}^{\text{H}}(\kappa)\,\mathbf{e}(\kappa)\,\mathbf{x}^{\text{H}}(\kappa)\,\mathbf{F}^{-\text{H}}(\kappa)\right\}_+ \,. \qquad (5.6.16)$$

This is of exactly the same form as equation (5.5.10) if the vector of filtered-error signals in the frequency domain is defined in a similar way to equation (5.5.7) as

$$\mathbf{a}(\kappa) = \hat{\mathbf{G}}_{\text{all}}^{H}(\kappa)\,\mathbf{e}(\kappa) \qquad (5.6.17)$$

and a vector of frequency-domain innovation signals is defined in a similar way to equation (5.5.8) to give

$$\mathbf{v}(\kappa) = \mathbf{F}^{-1}(\kappa)\,\mathbf{x}(\kappa)\,. \qquad (5.6.18)$$

Although it is anticipated that the convergence properties of the time-domain and frequency-domain versions of this algorithm will be similar, there may be numerical and computational advantages to implementing it in the frequency domain using equation (5.6.16).

5.7. APPLICATION: CONTROLLING ROAD NOISE IN VEHICLES

The low-frequency noise in many vehicles is dominated by the sound caused by the vehicle travelling over uneven road surfaces. This noise can hinder our perception of warning sounds, spoil our enjoyment of other sounds which we would like to hear, such as the radio, and can also be very wearing on a long journey. The current trend for more fuel-efficient vehicles, having lighter bodies, tends to exacerbate this low-frequency noise problem, since there is then very little passive isolation or absorption in the vehicle at low frequencies. Active control appears to be a good solution to this problem since it works best at low frequencies and can potentially use the loudspeakers and amplifiers already fitted to the car for the audio system. In this section we describe the design of such an active control system for road noise using a feedforward approach with reference signals derived from accelerometers on the suspension and car body. The physical arrangement for such a system is shown in Fig. 5.9, in which the secondary actuators are loudspeakers inside the vehicle, a car in this case, and the error sensors are microphones arranged to measure the acoustic pressure at discrete points inside the car.

5.7.1. Selection of Reference Signals

At first sight, it would appear that only two accelerometers on the car body would be needed to measure reference signals that were well correlated with the pressures measured

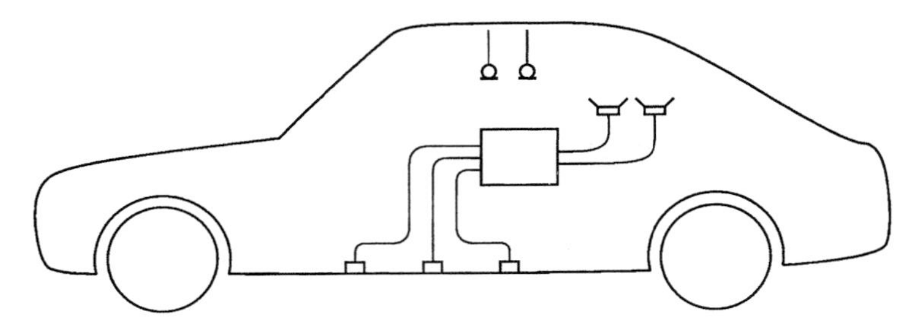

Figure 5.9 Physical arrangement for feedforward of road noise in a vehicle, in which reference signals derived from accelerometers on the car body are used to drive the secondary loudspeakers via a digital controller, which is adapted to minimise the sum of the squared signals at the error microphones.

at the microphones, one on either side of the car measuring the vertical motion of the front wheels. This turns out to be woefully inadequate on most vehicles, and one of the main engineering tasks in getting acceptable performance from a practical system turns out to be the selection of a suitable set of accelerometers as reference sensors.

We have seen that the processing requirements necessary to implement an adaptive controller increase significantly as the number of reference signals becomes large. Also, much of the cost of an active system in this application will be associated with the accelerometers used to generate the reference signals. In practice, then, we not only want to select a set of accelerometer positions that give acceptable performance, but we also want to select the smallest possible number of such accelerometers.

Although some engineering judgement can be used in the selection of suitable reference accelerometer positions, in practice there are generally many more possible accelerometer positions than the number that can be used by a practical real-time controller. An exhaustive search through all combinations of the potential reference signals could take a considerable amount of time. There are about 10^9 ways of selecting 6 reference signals from a potential set of 32 accelerometers, for example. The guided random search algorithms described in Chapter 9 have proved to be a very efficient way of conducting the search for good accelerometer positions in practice. Even using these guided random search algorithms, however, the potential performance of several thousand sets of reference signals has to be assessed, and efficient methods of calculating the optimum performance are very useful. Typically, an initial selection of potential reference signals is made on the basis of their mutual coherence, and using results based on unconstrained frequency domain optimisation, such as equation (5.3.15). The final selection is then made using a guided random search, taking the attenuation of the optimum causally constrained controller as the objective function.

As an example of the use of unconstrained frequency-domain methods in the initial selection of reference signals, Fig. 5.10 shows the coherence measured between the outputs of two accelerometers measuring the vertical motion on the front and rear suspension of a small car when driven in a straight line over a rough road at about 100 km h^{-1} (Sutton et al., 1994). One would intuitively expect these signals to be well correlated since the front and rear tyres are travelling over the same piece of road, but at different times. In fact the two signals are only well correlated at very low frequencies. This is because the irregularities in the road that give rise to the motion at frequencies above about 150 Hz are smaller than

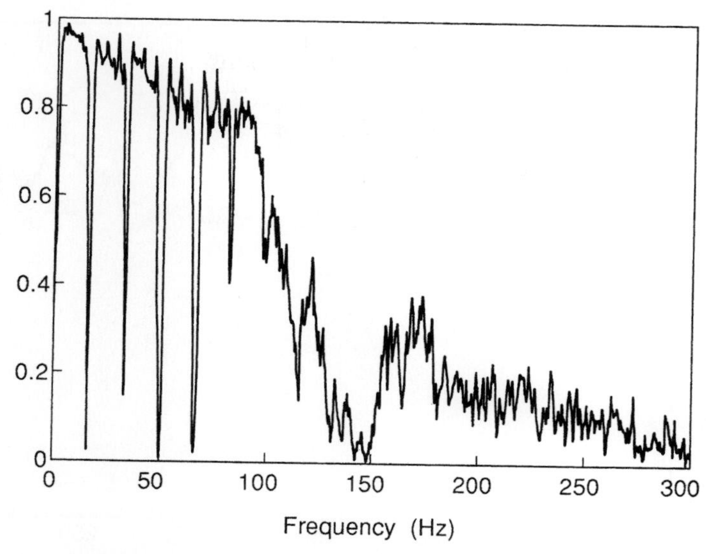

Figure 5.10 Coherence measured between the vertical acceleration at the front and rear suspensions on the nearside of a 1.3-litre front-wheel-drive car. The coherence falls sharply when the road surface wavelength is less than the length of the tyre contact patch.

the size of the patch of rubber in contact with the road, which has a length of about 200 mm (Moore, 1975). Thus the piece of road affecting the front tyre above 150 Hz is somewhat different from that affecting the rear tyre, and the vertical motions of the two wheels become uncorrelated. The dips in the coherence at about 16 Hz and its harmonics are due to slight differences in the rotation frequencies of the two wheels. The other components of the motion at the suspension points, due to fore-and-aft, side-to-side and angular motion, are also found to have as much effect on the internal soundfield as the vertical motion. This is because the vertical motion tends to be rather well isolated from the body of the car due to the design of the suspension system. One is thus left with a system that is influenced by multiple forms of motion at each wheel, which are only partially uncorrelated. It is thus not surprising that numerical search algorithms have to be used to find a good set of reference signals.

5.7.2. Predicted and Measured Performance

The time-domain predictions of performance can be used to calculate the attenuation resulting from using an optimal causal set of control filters having finite length, and can also be useful in assessing the sensitivity of the control system to various other shortcomings of a practical controller. Figure 5.11, for example, shows the power spectral density of the A-weighted sound pressure inside a small car when driven over a rough road at 60 km h^{-1}, and the predicted level of the power spectral density after active control with a controller using six reference signals and constrained to be causal, operating on a plant that was assumed to have a delay of either 1 ms or 5 ms (Elliott and Sutton, 1996). The

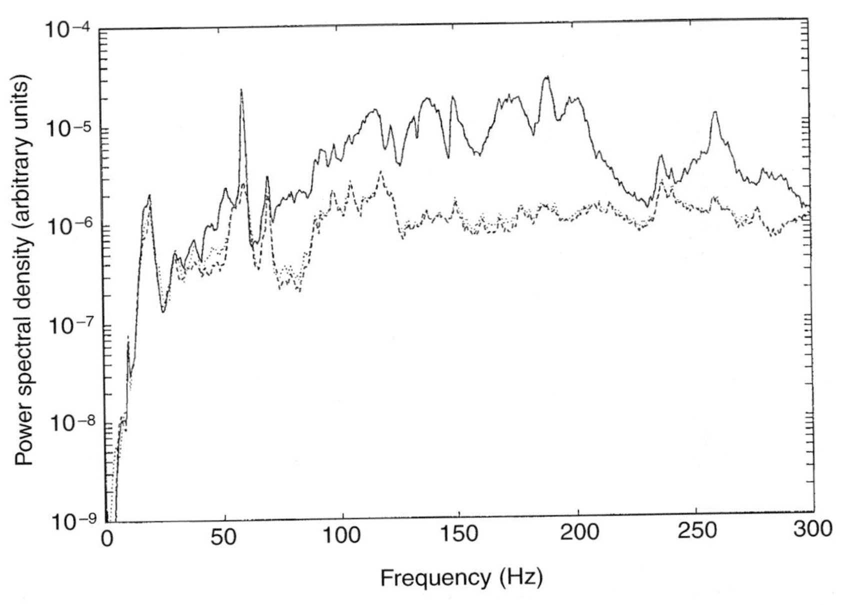

Figure 5.11 A-weighted power spectral density of the pressure measured in a small car (solid line) and predictions of the residual spectrum with a feedforward control system operating on a plant with a delay of 1 ms (dashed line) or 5 ms (dotted line).

predicted reductions in the overall level of the power spectral density using such a control system is about 7 dB, but this is relatively unaffected by the 1 ms or 5 ms delay in the plant. If an unconstrained frequency-domain method had been used to calculate the performance, however, a reduction of up to 12 dB would have been predicted, which could not be achieved in a practical implementation (Sutton et al., 1994). The 5 ms plant delay is typical in practice and is partly due to the acoustic propagation time between the secondary loudspeaker and the error microphone, which is about 3 ms per metre of separation, and partly due to the delays in the analogue anti-aliasing and reconstruction filters and the processing delays in the digital system. The fact that the performance of the control system is relatively insensitive to plant delays up to 5 ms is thought to be due to the relatively slow speed of flexural wave transmission through the vehicle body at these low frequencies (Sutton et al., 1994).

A real-time control system was also implemented and tested on the vehicle used to obtain the data described above (Saunders et al., 1992). The control system used six reference signals, two 110-mm-diameter loudspeakers placed in the front doors of the vehicle, and two microphones in the headrests of the front seats. It operated at a sampling rate of 1.2 kHz, using 12 individual control filters, each having 128 coefficients. The control filters were adapted using a time-domain implementation of the multichannel filtered-reference algorithm described in Section 5.4. The plant response, between the secondary loudspeakers and error microphones, was measured in an initial identification phase by applying white noise to each secondary loudspeaker in turn. The best locations for the accelerometers were determined by simultaneously measuring the vehicle vibration at many positions, and the interior acoustic pressures, during a series of road tests. Likely

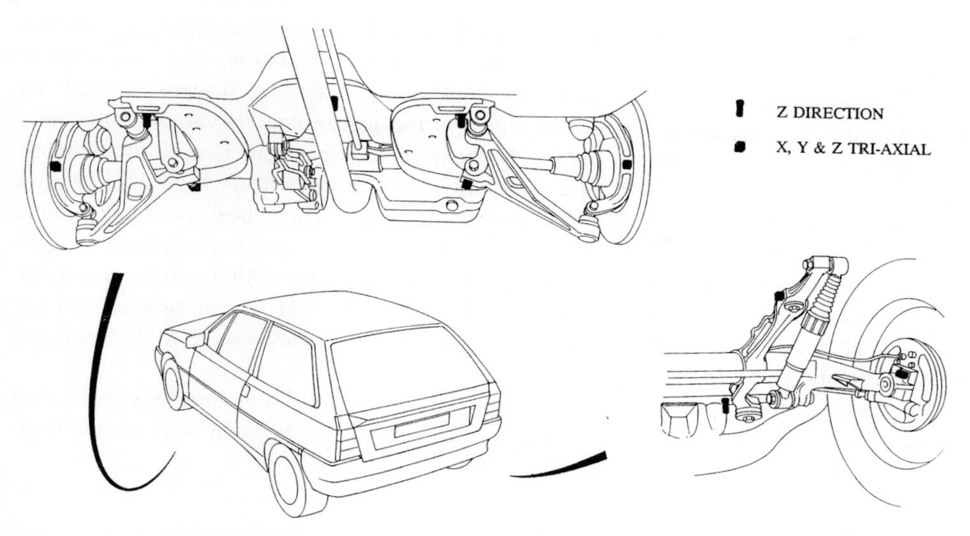

Figure 5.12 Typical locations for the reference accelerometers in an active control system for road noise. The six positions used in all the measurements reported here were: floor (vertical), wishbone (vertical) and hub (lateral) on left-hand and right-hand front of the vehicle, as shown in the top part of the sketch.

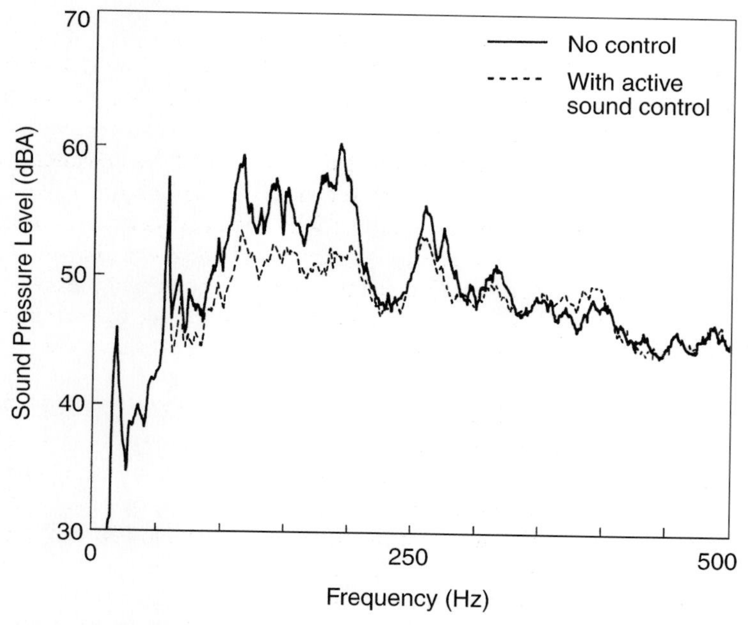

Figure 5.13 Spectrum of A-weighted sound pressure level in a small car, using a real-time active control system, measured at the driver's ear position. Vehicle speed was 60 km h^{-1} over an asphalt surface impregnated with coarse chippings. Solid line with no active control; dashed line with active control.

positions for the accelerometers were deduced from multiple-coherence calculations, but the final selection was made on the basis of time-domain calculations to ensure that the controller was causal. The accelerometer positions considered on the suspension and body are shown in Fig. 5.12. The optimum placement of the accelerometers varies significantly between different vehicle models, but for the front-wheel-drive vehicle used by Saunders et al. all six accelerometers were finally placed close to the front wheels. Figure 5.13 shows the measured A-weighted sound pressure level at the driver's head position in the vehicle measured when driving at 60 km h^{-1} along an asphalt road with coarse chippings, with the active control system switched off (solid line) and on (dashed line). Although the frequency range shown in Fig. 5.13 is larger than that shown in Fig. 5.11, it can be seen that the measured attenuations in internal noise from 100 Hz to 200 Hz are similar to the predicted attenuations shown in Fig. 5.11.

These results show that useful attenuations in interior road noise are possible using active sound control, and that the attenuations achieved in practice can be predicted from off-line calculations of the optimal controllers.

6

Design and Performance of Feedback Controllers

6.1. INTRODUCTION

In this chapter we begin the discussion of feedback control systems by discussing the design of fixed controllers, and their performance. Feedback control systems are distinguished from feedforward systems in that they have no reference sensor to give time-advanced information about the disturbance being controlled. They are used in a variety of applications for the active control of sound and vibration, particularly where the primary sources which generate the disturbance cannot be directly observed, or in which there are too many primary sources to economically obtain reference signals from each one. Examples of feedback active sound control systems include active headsets and active headrests, the latter being designed to create a fixed zone of quiet for broadband disturbances. Feedback systems for the control of vibration are also widely used, particularly to control lightly damped structures, for which the disturbance at each resonance peak can be relatively narrowband. Control of such structures can be achieved with direct velocity feedback, which increases the effective damping in the system, although more complicated strategies may need to be employed if, for example, the ultimate aim of the control system is the reduction of radiated sound.

In our discussion of feedback control we will continue to characterise the plant under control using an input–output approach, rather than using a state variable model. State variable methods for active feedback control have been reviewed, for example, by Fuller et al. (1996), Preumont (1997) and Clark et al. (1998), and such methods can provide a complete model of the global response of the system under control. They are particularly applicable to the control of the first few modes of a structure, and are widely used for flight control systems in aircraft, the control of low-frequency vibration in distributed structures (see for example Meirovich, 1990; Rubenstein et al., 1991) and the control of sound radiated by a lightly damped structures (see for example Baumann et al., 1991; Petitjean and Legrain, 1994). When sound or vibration is controlled over a broad bandwidth with an active system, the plant can be subject to significant delays and be of high order, with a response that is characterised by many well-damped and overlapping modes, whose natural frequencies can change over rather short timescales. A state variable model for such a system may be very complicated to identify, whereas an input–output model can be determined from direct measurement of the plant's frequency response, for example. A frequency response approach also allows some of the design methods and design compromises which are inherent in a feedback control system to be illustrated in a straightforward fashion, and one that is consistent with conventional signal processing formulations.

In the control literature, single-channel control systems are referred to as having single-input single-output (SISO) plants and multichannel systems as having multiple-input multiple-output (MIMO) plants. Also, analogue systems are more accurately referred to in the control literature as continuous-time systems and digital systems as sampled-time systems, although we will generally retain the less formal titles here, since they are more widely used by the signal processing community. Active systems are generally designed for *disturbance rejection*, and most of the discussion in the following two chapters will concentrate on this application of feedback control. In the conventional control literature, however, more emphasis is generally placed on feedback controllers designed to ensure that the plant output follows some command signal, which are called *servo-control* systems.

6.1.1. Chapter Outline

The general characteristics of feedback control systems will be described in this section, followed, in the remainder of the chapter, by a discussion of the design of fixed, time-invariant, controllers, both for single-channel and multiple-channel systems, implemented using either analogue or digital techniques. The adaptation of feedback controllers will be discussed in the following chapter. There is a large and well-developed literature on the design of feedback control systems and the author has found the following general textbooks particularly helpful: Franklin et al. (1994), Morari and Zafiriou (1989), Skogestad and Postlethwaite (1996).

The design of analogue feedback controllers is described in Section 6.2, in which frequency response methods such as those of Bode and Nyquist are particularly emphasised. The complications involved in sampling the plant output and using a digital feedback controller are outlined in Section 6.3 and, using this sampled-time example, the advantages of one particular architecture of feedback controller are emphasised in Section 6.4. This architecture uses an internal model of the plant response to transform the feedback control problem into a feedforward control problem, and thus allows the insight and analytical tools developed in the previous chapters to be brought to bear on the design of feedback controllers.

One important practical condition which must then be introduced is that of robust stability, so that the control loop remains stable for realistic changes or uncertainties in the response of the plant. Section 6.5 describes the design of optimal controllers in the time domain, which minimise a cost function that includes the mean-square error and a mean-square effort term, where the latter is shown to improve the robust stability of the system. Section 6.6 discusses the corresponding transform-domain design of optimal feedback controllers, of which the well-known minimum-variance controller is shown to be a special case. Some of the complications involved in applying feedback control to multichannel plants are described in Section 6.7 and it is shown in Section 6.8 how the calculation of robust stability becomes rather more difficult in a multichannel system because of the difficulties involved in describing realistic uncertainties in each channel of the plant response. Section 6.9 discusses how the optimal controller for such multichannel systems could be calculated.

Finally, in Section 6.10, the design methods introduced in the previous section are applied to the specific example of actively controlling the sound in a headrest. The need for a measurement of the plant uncertainty as well as the plant response is emphasised, so that a robustly stable controller can be designed. A discrete-frequency method is then used for the design of an H_2/H_∞ feedback controller which minimises the mean-square value of the error signal, an H_2 norm, but maintains robust stability, an H_∞ constraint. The predicted performance of such a feedback controller is then compared with the performance measured in an experiment with a real-time controller.

6.1.2. Disturbance Rejection

A physical representation of an active sound control system using an analogue feedback loop is shown in Fig. 6.1(a), in which the signal from a single microphone, $e(t)$, is fed back to the secondary loudspeaker, with input signal $u(t)$, via an analogue feedback controller with the Laplace-domain transfer function $-H(s)$. The phase inversion in the feedback controller reminds us that we are hoping to design a negative feedback system. The equivalent block diagram for this feedback control system is shown in Fig. 6.1(b), in which the Laplace-domain transfer function of the plant, from the loudspeaker input to the microphone output, is denoted $G(s)$, and the output of the microphone in the absence of control is denoted $d(t)$, which is the disturbance in this case.

The Laplace transform of the error signal can be written as

$$E(s) = D(s) - G(s)H(s)E(s), \tag{6.1.1}$$

so that the transfer function from disturbance to error signal is equal to

$$\frac{E(s)}{D(s)} = \frac{1}{1+G(s)H(s)} = S(s) \ . \tag{6.1.2}$$

This transfer function is also known as the *sensitivity function*, $S(s)$, of the system. Since the objective of this control system is good rejection of the disturbances, we would like the sensitivity function to be small. The conditions required for this to happen are most clearly seen in the frequency domain, and letting $s = j\omega$ in equation (6.1.2), the frequency response of the sensitivity function can be written as

$$S(j\omega) = \frac{1}{1+G(j\omega)H(j\omega)} , \tag{6.1.3}$$

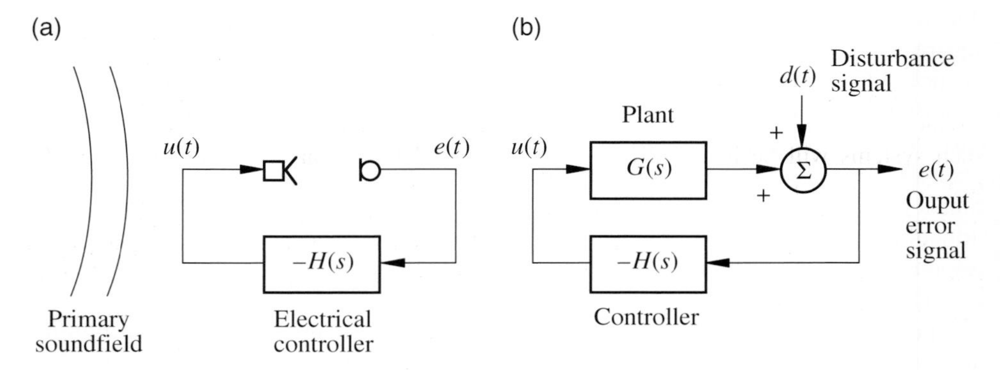

Figure 6.1 Physical block diagram of a feedback control system for the suppression of an acoustic disturbance (a) and its equivalent block diagram (b).

where $G(j\omega)$ is the frequency response of the plant and $H(j\omega)$ is the frequency response of the controller.

If the frequency response of the plant were relatively flat and free from phase shift below some frequency ω_0, and the electrical controller were a high gain amplifier, so that $G(j\omega)H(j\omega)$, the *loop gain*, was large at all frequencies below ω_0, then

$$|1+G(j\omega)H(j\omega)| \gg 1, \qquad \omega < \omega_0 , \tag{6.1.4}$$

in which case

$$S(j\omega) \ll 1, \qquad \omega < \omega_0 . \tag{6.1.5}$$

The action of the control system at these frequencies can be described in simple terms by imagining a small perturbation being introduced as the disturbance in Fig. 6.1(b). As the perturbation begins to increase, it is immediately amplified and its waveform is inverted before being fed back to the summing junction. The waveform of the perturbation is thus very nearly balanced by the signal fed back through the plant.

The condition described by equation (6.1.4) can be closely fulfilled in purely electrical systems for which negative feedback circuits were first invented (Black, 1934). For operational amplifier circuits such a feedback loop is said to have a *virtual earth*, since the input signal is driven down to earth potential by the feedback loop, and this term is sometimes carried over to describe acoustic feedback systems. In acoustic systems, however, the plant generally exhibits considerably more dynamic response than an electronic amplifier. The electroacoustic response of a moving-coil loudspeaker, for example, introduces significant phase shift near its mechanical resonance frequency, which is typically between about 10 Hz and 100 Hz. The acoustic path from loudspeaker to microphone will also inevitably involve some delay, due to the acoustic propagation time, which will introduce an increasing phase shift with frequency. As the net phase shift in the plant approaches 180°, the negative feedback action described above becomes positive feedback and the control system can become unstable.

6.1.3. Following a Command Signal

Many applications of feedback control are not concerned with disturbance rejection, but with the manipulation of the plant output to follow some prescribed set point or *command signal*. Such a feedback control system was described above as having a servo action, and the block diagram of a servo-control system is illustrated in Fig. 6.2. A brief description of such systems will be included here both for completeness, and also because once the disturbances are under control, the tracking of a set point can be an important objective for active vibration control systems, e.g. in the position control of flexible systems, or active sound control systems, e.g. in sound reproduction. No disturbance has been included in Fig. 6.2 and the plant output has been denoted $y(t)$ in this case, to conform with most of the control literature. The objective of the controller is to drive the plant so that its output follows the command signal, $c(t)$, as closely as possible and to this end the controller is driven by the difference between the required output and the actual output,

$$\varepsilon(t) = c(t) - y(t) , \tag{6.1.6}$$

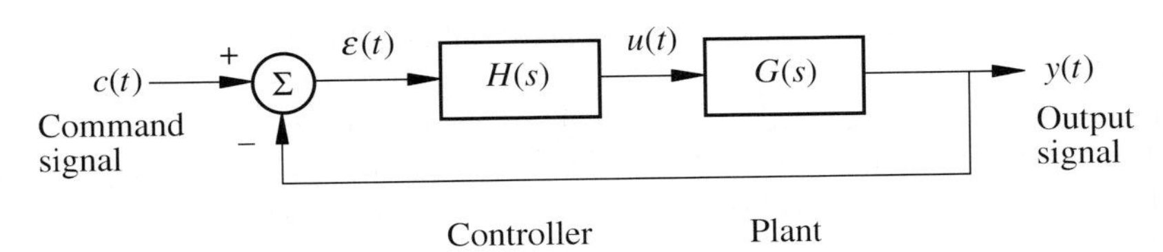

Figure 6.2 Block diagram of a negative feedback system having a servo action, in which the output of the plant $y(t)$ is arranged to follow the command signal $c(t)$.

The explicit phase inversion of the controller in Fig. 6.1(b) has now been dropped since it is included in the differencing operation required to generate $\varepsilon(t)$. The Laplace transform of the plant output in Fig. 6.2 is equal to

$$Y(s) = G(s)H(s)\left[C(s) - Y(s)\right] , \tag{6.1.7}$$

so that the transfer function from required output to actual output is equal to

$$\frac{Y(s)}{C(s)} = \frac{G(s)H(s)}{1+G(s)H(s)} = T(s) . \tag{6.1.8}$$

This transfer function is called the *complementary sensitivity function*, $T(s)$, because the sum of this and the sensitivity function, equation (6.1.2), is equal to unity,

$$S(s) + T(s) = 1 . \tag{6.1.9}$$

The frequency response of the complementary sensitivity function is equal to

$$T(j\omega) = \frac{G(j\omega)H(j\omega)}{1+G(j\omega)H(j\omega)} . \tag{6.1.10}$$

If we again assume that below some frequency, ω_0, the loop gain is large, so that equation (6.1.4) again holds, we find that the complementary sensitivity function will be close to unity in this case, i.e. for excitation frequencies below those where the plant dynamics become significant, a high-gain controller can achieve good servo control. The complementary sensitivity function also affects the output of a feedback control system designed to reject disturbances if *sensor noise* is present. Sensor noise can be modelled as an independent input to the controller, $n(t)$, which does not directly affect the physical output, $e(t)$, but is caused by the process of measuring this output, as shown in Fig. 6.3. Notice that the sensor noise $n(t)$ in Fig. 6.3 enters the control loop at exactly the same point as the command signal in Fig. 6.2, apart from a phase inversion, and so the transfer function from sensor noise to output is the same as that given in equation (6.1.8). The physical output signal due to both the disturbance and sensor noise can thus be written as

$$E(s) = S(s)D(s) - T(s)N(s) . \tag{6.1.11}$$

The complementary sensitivity function plays an important role in determining the robustness of the control system to variations in the plant response, as will be explained in more detail in the next section. Such variations have an effect at the output of the plant which is similar to that of sensor noise, and so the compromise between performance and robustness can be viewed as a trade-off between minimising the two terms in equation (6.1.11), subject to the condition $S(s) + T(s) = 1$.

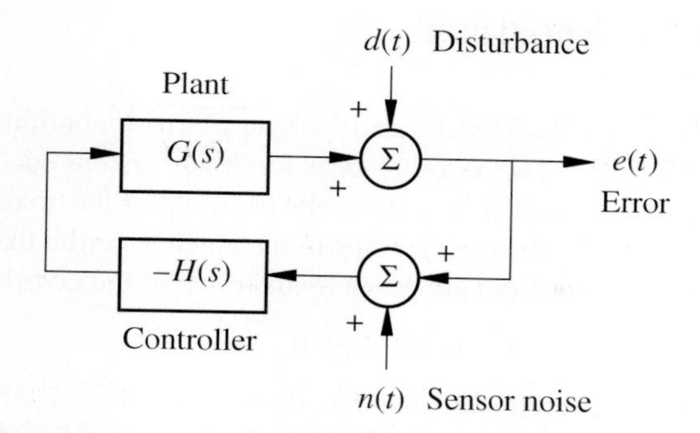

Figure 6.3 Disturbance-rejecting feedback control system with sensor noise, $n(t)$.

6.1.4. Bandwidth Limitations Due to a Delay

The bandwidth over which feedback control can be applied is partly determined by the *delay* in the plant response. Assume for example, that the plant response corresponds to that of a pure delay of τ seconds, so that

$$G(j\omega) = e^{-j\omega\tau} . \tag{6.1.12}$$

The best a servo-control system could then achieve would be to track the required signal with the same delay (Skogestad and Postlethwaite, 1996), so that

$$T(j\omega) = e^{-j\omega\tau} . \tag{6.1.13}$$

If we now consider the rejection of disturbances in such a control system, which is determined by the sensitivity function $S(j\omega)$ of equation (6.1.2), then since $S(j\omega) + T(j\omega) = 1$, in this case we have

$$S(j\omega) = 1 - e^{-j\omega\tau} . \tag{6.1.14}$$

The disturbances will be attenuated at all frequencies for which $|S(j\omega)| < 1$, which corresponds to

$$\omega\tau < \frac{\pi}{3} . \tag{6.1.15}$$

The bandwidth, in hertz, over which the disturbances can be attenuated for a plant with a delay of τ seconds can thus be written as

$$\text{Bandwidth (Hz)} < \frac{1}{6 \times \text{Delay (sec)}} . \tag{6.1.16}$$

A more detailed description of the performance of feedback control systems will be presented in the next section, but equation (6.1.16) is a useful 'rule of thumb' for calculating the bandwidth of control in many active control applications, in which the plant response has significant delay.

6.2. ANALOGUE CONTROLLERS

In this section, the stability and design of single-channel continuous-time feedback controllers will be discussed. The *stability* of a feedback system such as that shown in Fig. 6.1(b) is most easily established in terms of the positions of the poles of the closed-loop transfer function. The poles of the transfer functions which describe the systems shown in Fig. 6.1(b) and 6.2 are the same, and are given by the roots of the *characteristic equation*

$$1 + G(s)H(s) = 0 . \tag{6.2.1}$$

If the output of the control system is to remain bounded (finite) for any bounded input, then all of the poles of the closed-loop transfer function must be in the open left-hand half of the s plane, i.e. with none on the imaginary axis (see for example Franklin et al., 1994). Strictly speaking, this definition of stability is only valid provided the plant and controller are themselves stable, since otherwise 'hidden' instabilities could occur within the control loop, which are not observed at the output, $e(t)$ in Fig. 6.3. A stronger stability condition is provided by the concept of *internal stability* (see for example Morari and Zafiriou, 1989), which is only satisfied if bounded inputs injected at *any* point in the control system generate bounded responses at any other point.

In the discussion of active control systems below, we will thus assume that both the plant and the controller are stable, since this considerably simplifies the analysis. The more general analysis is described, for example, in Morari and Zafiriou (1989).

6.2.1. Nyquist Stability Criterion

In many applications of active control, an explicit pole/zero model of the plant, which is needed to calculate the roots of equation (6.2.1), is not available. This is either because the model would be of very high order, or because the plant response involves delays, or because of the considerable variability that can exist in such systems. It is thus more appropriate to continue with the frequency-response approach adopted in other parts of this book and consider the *Nyquist* definition of stability for a feedback system (as also described, for example, in Franklin et al., 1994). This involves an examination of the polar plot of the open-loop frequency response $G(j\omega)H(j\omega)$. If, at some frequency, $G(j\omega)H(j\omega)$ were equal to exactly -1, then it is clear from equation (6.1.3) that the response of the feedback control system at this frequency would become unbounded. Since we are assuming that the plant under control and the controller are themselves stable, then the definition of stability provided by the Nyquist criterion is that the polar plot of the open-loop frequency response $G(j\omega)H(j\omega)$ must not enclose the *Nyquist point* $(-1,0)$ as ω varies from $-\infty$ to ∞ (Nyquist, 1932), as illustrated in Fig. 6.4. The locus of $G(j\omega)H(j\omega)$ for negative ω is shown as a dashed line in this figure, which is the mirror image of the locus for positive ω about the real axis since

$$G(-j\omega)H(-j\omega) = G^*(j\omega)H^*(j\omega) , \tag{6.2.2}$$

where * denotes complex conjugation.

Note that the magnitude of the frequency response of most physical systems tends to zero at high frequencies (they are said to be *strictly proper* in the control literature), in which case the polar plots of the open-loop frequency response converge on the origin as

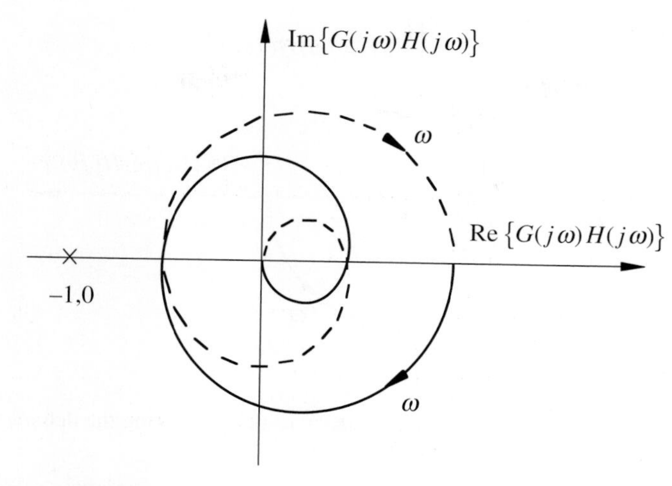

Figure 6.4 The polar plot of a stable open-loop frequency response, for $\omega > 0$ (solid line) and $\omega < 0$ (dashed line), which does not encircle the Nyquist point (−1,0) and so has a stable closed-loop response.

ω or $-\omega$ tends to ∞. As ω or $-\omega$ tends to zero, the open-loop frequency response also tends to the same, real, value equal to the open-loop 'dc gain' of the system. By adopting a bounded-input, bounded-output definition of stability and assuming that both the plant and controller are stable according to this definition, we have excluded the possibility of there being an integrator in the loop, which would cause the response of the system to become progressively larger as the frequency tends to zero. The lack of an integrator would limit the steady-state performance of a servo system, but we are primarily interested here in the rejection of relatively high frequency disturbances and so such an assumption is not overly restrictive for our purposes. The complete locus in Fig. 6.4 can be deduced from frequency response *measurements* made on the open-loop system, and the stability of the closed-loop system can thus be predicted from these measurements before the loop is closed.

6.2.2. Gain and Phase Margins

Apart from determining the absolute stability of a feedback system, the Nyquist plot of the open-loop frequency response can also provide useful information about the *relative stability* of such a system, i.e. its stability margins. Two parameters have commonly been used to quantify the relative stability of a feedback control system, and they both give information about the changes that can be allowed in the response of the plant whilst maintaining closed-loop stability. The first of these parameters is called the *gain margin* and is defined to be the increase that can be tolerated in the overall gain of the plant without the feedback loop becoming unstable. In Fig. 6.5 it is clear that if the magnitude of the open-loop frequency response was increased by a factor greater than $1/g_c$, then the locus would enclose the Nyquist point and become unstable. The gain margin in dB can thus be defined as

$$\text{Gain margin} = -20 \log_{10}(g_c) \ . \tag{6.2.3}$$

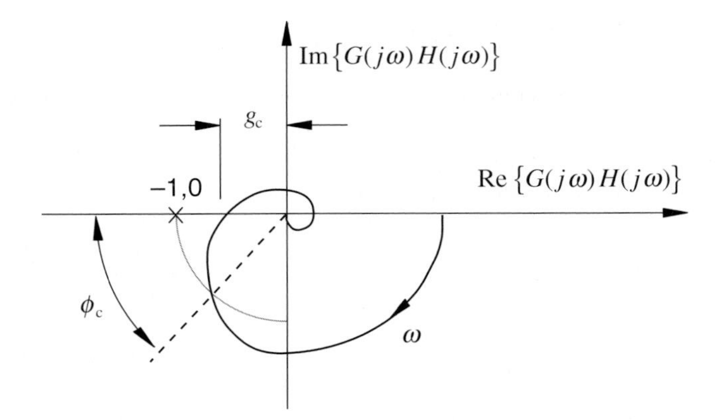

Figure 6.5 Nyquist plot of the open-loop frequency response showing the definition of the phase margin, ϕ_c, and gain margin, $-20\ \log_{10}(g_c)$.

The other widely used relative stability parameter is the *phase margin*, defined to be the additional phase shift that could be tolerated in the plant, without the closed-loop system becoming unstable. The phase margin must thus be equal to the phase shift of the plant at the frequency for which the modulus of the open-loop gain is unity, which is illustrated as ϕ_c in Fig. 6.5.

One problem with describing the relative stability using the gain and phase margins is that they do not indicate whether the system will be unstable if there is a change in *both* gain and phase. A rather contrived example is shown in Fig. 6.6, for which the gain and phase margins are both very large, but the stability of the system is sensitive to small increases in both gain and phase lag. The causes of the changes in the open-loop frequency response which can give rise to instability are generally perturbations in the response of the plant. These perturbations can be due to a multitude of reasons, for example: a change in temperature or operating condition, or a change in the amplitude of the signal driving the plant if the system is weakly nonlinear. If the control system remains stable despite such changes it is described as being *robustly stable*.

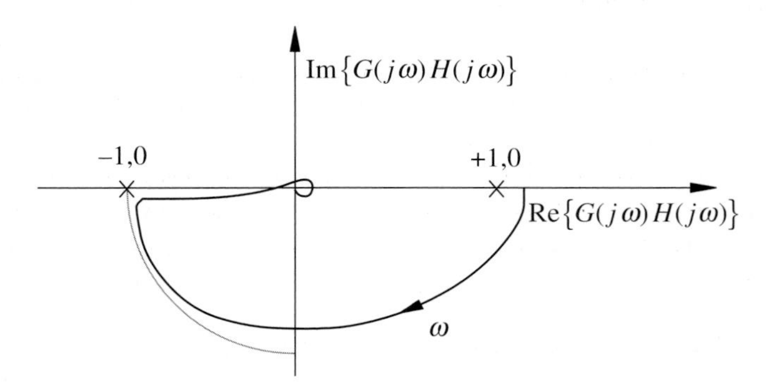

Figure 6.6 An example of an open-loop frequency response that has a very large gain and phase margin, but whose closed-loop stability is still sensitive to small changes in the plant response.

6.2.3. Unstructured Plant Uncertainties

If the robust stability of a system is to be determined, a model of the perturbations, or *uncertainties*, in the plant response is required. The response of the plant under normal operating conditions, i.e. with no uncertainty, is called the *nominal* plant response. The effect of changes in the plant response on the Nyquist plot of the open-loop frequency response is to introduce regions of uncertainty round the nominal open-loop frequency response at each frequency. The shapes of these regions of uncertainty are determined by the types of change that occur in the plant. If the uncertainty is due to a fixed range of gain and phase variations, the region is confined to the segment shown in Fig. 6.7(a). If the uncertainty were due to different operating conditions, the region of uncertainty might take the form shown in Fig. 6.7(b), for example. These shapes of uncertainty region are difficult to deal with analytically, but can always be enclosed by a *disc of uncertainty*. We will be concerned below with designing controllers whose stability is robust to plant uncertainties that are described by such discs, and although the design may be rather conservative if the true uncertainty region is not well described by a disc, a robustly stable controller can always be designed by making this assumption.

Such a disc of uncertainty is generated by an *unstructured* model of uncertainty in which we assume that the true plant response at a given frequency is given by

$$G(j\omega) = G_0(j\omega)\left[1 + \Delta_G(j\omega)\right] , \tag{6.2.4}$$

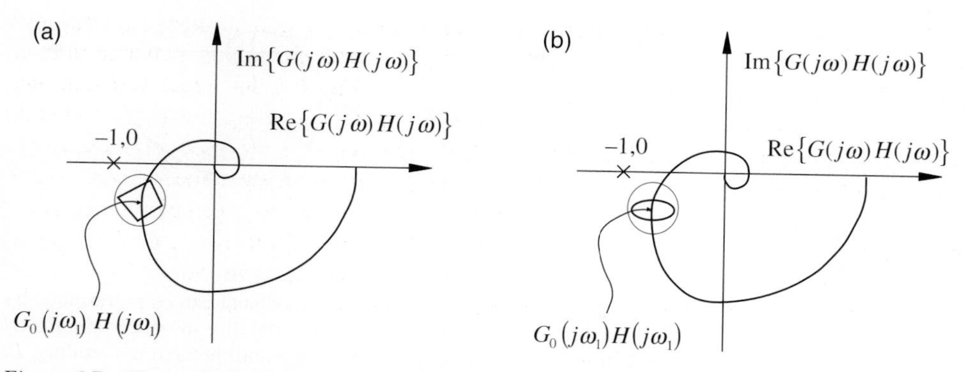

Figure 6.7 The regions of uncertainty in the Nyquist plot about the nominal open-loop response at a given frequency (ω_1) due to (a), a given phase and gain margin and (b), variation over a number of operating points, both of which are enclosed by a disc of uncertainty.

where $G_0(j\omega)$ is the nominal plant response, $\Delta_G(j\omega)$ is the multiplicative plant uncertainty and it is assumed that $G(j\omega)$ is stable for all $\Delta_G(j\omega)$. The uncertainty, $\Delta_G(j\omega)$, is generally complex, but it is assumed that its magnitude is bounded at each frequency to be no greater than $B(\omega)$, so that

$$\left|\Delta_G(j\omega)\right| \le B(\omega) \qquad \text{for all } \omega \ . \tag{6.2.5}$$

At a given frequency, all plant responses that are described by equations (6.2.4) and (6.2.5) lie within a disc in the complex plane whose centre is equal to the nominal plant response, $G_0(j\omega)$, and whose radius is $B(\omega)\left|G_0(j\omega)\right|$.

6.2.4. Condition for Robust Stability

Referring to Fig. 6.8(a), the open-loop frequency response of a system with unstructured uncertainty could, at a given frequency, be located anywhere in a disc in the Nyquist plane, whose centre is $G_0(j\omega)H(j\omega)$ and whose radius is $B(\omega)\left|G_0(j\omega)H(j\omega)\right|$. For robust stability, the closed-loop system cannot be unstable for any plant whose response satisfies equations (6.2.4) and (6.2.5), so that the disc of uncertainty must not enclose the Nyquist point. This means that the distance from the centre of the circle to the Nyquist point, $\left|1+G_0(j\omega)H(j\omega)\right|$, must be greater than the radius of the circle, so that

$$\left|1+G_0(j\omega)H(j\omega)\right| > B(\omega)\left|G_0(j\omega)H(j\omega)\right| , \tag{6.2.6}$$

and so

$$\frac{\left|G_0(j\omega)H(j\omega)\right|}{\left|1+G_0(j\omega)H(j\omega)\right|} < \frac{1}{B(\omega)} . \tag{6.2.7}$$

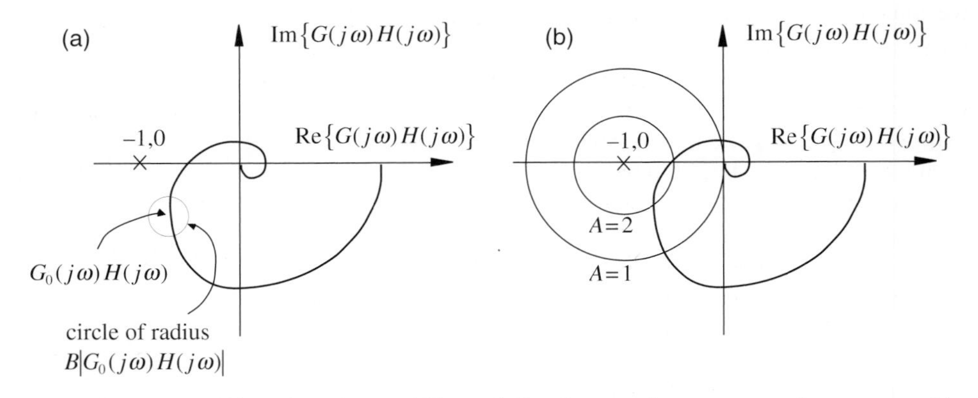

Figure 6.8 The conditions for robust stability and disturbance enhancement can be represented by geometric constraints in the Nyquist plot. (a) For robust stability the uncertainty disc, whose radius is proportional to the upper bound on the multiplicative uncertainty, B, must not encircle the (−1,0) point. (b) In order for the system not to enhance the disturbance by more than a factor of A, the Nyquist plot must not pass through a circle drawn about the (−1,0) point having a radius of $1/A$.

The left-hand side of equation (6.2.7) is equal to the modulus of the complementary sensitivity function for the nominal plant, $T_0(j\omega)$ defined in equation (6.1.8). The condition for robust stability can thus be written as

$$\left|T_0(j\omega)\right| < \frac{1}{B(\omega)} \quad \text{for all } \omega , \tag{6.2.8}$$

or

$$\left|T_0(j\omega)B(\omega)\right| < 1 \quad \text{for all } \omega . \tag{6.2.9}$$

If a control system were required to be robustly stable to changes in the plant response described by a multiplicative uncertainty bounded by $B(\omega) = 0.5$, for example, the control system would have a gain margin of about 3.5 dB and a phase margin of about 30°, as can be confirmed by a simple geometric construction in the Nyquist plane.

The maximum value of $|T_0(j\omega)B(\omega)|$ over frequency (or, strictly speaking, the least upper bound of this function) is known as its *supremum* and the condition for robust stability can also be written as

$$\sup_{\omega} |T_0(j\omega)B(\omega)| = \|T_0 B\|_{\infty} < 1 , \tag{6.2.10}$$

where $\|T_0 B\|_{\infty}$ denotes the ∞-norm of T_0B, the weighted complementary sensitivity function, and this form of the robust stability condition is widely used in the H_∞ control literature (Doyle et al., 1992; Morari and Zafiriou, 1989; Skogestad and Postlethwaite, 1996). Thus we see that robust stability involves a *constraint* on the complementary sensitivity function, which must be satisfied at each frequency. The symbol H_∞ is used to denote that the norm is defined for the Hardy space of transfer functions that are stable and proper, i.e. their response is finite as $\omega \rightarrow \infty$. Later we will also use the H_2 norm of a transfer function, which is proportional to the square root of the integrated modulus squared response over all frequencies, and this is defined for the Hardy space of transfer functions which are stable and strictly proper, i.e. their response tends to zero as $\omega \rightarrow \infty$ (see also, for example, Skogestad and Postlethwaite, 1996, or the Appendix). In active control the cost function that we generally want the control system to minimise is proportional to the mean-square value of a function, or the square of its H_2 norm, and the controller which minimises such a cost function is said to be H_2-optimal. It is also possible to design controllers that minimise the H_∞ norm of a function, or the maximum value of the frequency response (Zames, 1981; Doyle et al., 1992), which are called H_∞-optimal controllers. We shall see that practical feedback systems for active control should typically be designed to minimise an H_2 cost function while maintaining multiple H_∞ constraints, and such controllers are said to perform H_2/H_∞ control.

If the multiplicative uncertainty in the plant is too great, it may not be possible to obtain good suppression of disturbances while ensuring robust stability. The attenuation of disturbances is determined by the magnitude of the sensitivity function, $S(j\omega)$ in equation (6.1.3). When the magnitude of $S(j\omega)$ is less than unity, so that the disturbance is attenuated, then the relationship between the sensitivity function and complementary sensitivity function, equation (6.1.9), can be used to relate their moduli via the triangle inequality to give

$$|S(j\omega)| + |T(j\omega)| \geq 1 , \tag{6.2.11}$$

so that

$$|T(j\omega)| \geq 1 - |S(j\omega)| . \tag{6.2.12}$$

But for the control system to be robustly stable, $1/B(\omega)$ must be greater than $|T(j\omega)|$ under nominal conditions, equation (6.2.8), and so

$$\frac{1}{B(\omega)} > 1 - |S(j\omega)| , \tag{6.2.13}$$

which gives the limit on the smallest value of the sensitivity function, and hence the greatest attenuation of disturbances, which can be obtained with a given plant uncertainty, as

$$|S(j\omega)| > 1 - 1/B(\omega) \; . \tag{6.2.14}$$

If the multiplicative uncertainty is greater than unity at some frequency, then $1 - 1/B(\omega)$ must be greater than zero, and for a robustly stable control system $|S(j\omega)|$ must also be greater than zero. Perfect suppression of disturbances is therefore impossible at frequencies where $|B(\omega)| > 1$, but the maximum disturbance attenuation that could be achieved under these conditions can be calculated from equation (6.2.14). The practical design of feedback controllers often involves a compromise between robust stability and good disturbance rejection.

6.2.5. Disturbance Enhancement

We can also make some rather general statements about the conditions under which *enhancement* or amplification of the disturbance can take place using the Nyquist plot. Suppose we are interested in the conditions under which the modulus of the sensitivity function, and hence the amplification of the disturbance, is greater than some value A. Since $S(j\omega) \;=\; [1 + G(j\omega)H(j\omega)]^{-1}$, we can see that if

$$|S(j\omega)| > A \; , \tag{6.2.15}$$

then

$$|1 + G(j\omega)H(j\omega)| < \frac{1}{A} \; . \tag{6.2.16}$$

Geometrically, equation (6.2.16) implies that the disturbance will be amplified by a factor of at least A if the open-loop frequency response falls within a disc round the Nyquist point of radius $1/A$. This is illustrated in Fig. 6.8(b) for A = 1 and 2, i.e. amplifications of 0 dB, or no disturbance enhancement, and 6 dB enhancement of the disturbance. If the open-loop frequency response of a practical system has a phase shift of greater than 180° at any frequency, the polar plot of the open-loop frequency response will be likely to pass into the disc indicating some enhancement (Berkman et al., 1992). The more robustly stable the control system is made, and hence the farther the open-loop frequency response remains from the Nyquist point, the smaller will be the maximum amplification of the disturbance. This dependence can be quantified by again using the triangle inequality on the sum of the sensitivity and complementary sensitivity functions, equation (6.1.9), but now under the conditions that the magnitude of $S(j\omega)$ is greater than unity, in which case

$$|S(j\omega)| - |T(j\omega)| \le 1 \; , \tag{6.2.17}$$

so that

$$|T(j\omega)| \ge |S(j\omega)| - 1 \; . \tag{6.2.18}$$

For the control system to be robustly stable $1/B(\omega)$ must be greater than $|T(j\omega)|$ under nominal conditions, equation (6.2.8), and so

$$1/B(\omega) > |S(j\omega)| - 1 \;, \tag{6.2.19}$$

which gives the limit on the largest value of the sensitivity function, and hence the greatest enhancement of disturbances as

$$|S(j\omega)| < 1 + \frac{1}{B(\omega)} \;. \tag{6.2.20}$$

Both this limit on disturbance enhancement and the limit on disturbance attenuation for a robustly stable controller derived above, equation (6.2.14), can be described by the overall limit of $\left|B(\omega)\left[1 - |S(j\omega)|\right]\right| < 1$ (Rafaely, 1997), whose consequences are illustrated in Fig. 6.9. This shows the maximum level of the disturbance enhancement for a robustly stable controller, calculated from equation (6.2.20), and the maximum level of the disturbance attenuation for a robustly stable controller, calculated from equation (6.2.14), against the bound on the unstructured multiplicative uncertainty $B(\omega)$. For a robustly stable control system, the magnitude of the sensitivity function in dB must be outside the shaded area of Fig. 6.9, i.e. it must fall between these two limits. Provided the bound on the multiplicative uncertainty, $B(\omega)$, is less than unity, there is no limit to the disturbance attenuation that can be achieved with a robustly stable controller, but for $B(\omega) = 1$ the disturbance enhancement is limited to 6 dB for such a controller. If the bound on the multiplicative uncertainty is greater than 1, equations (6.2.14) and (6.2.20) limit both the attenuation achievable and the enhancement possible with a robustly stable controller,

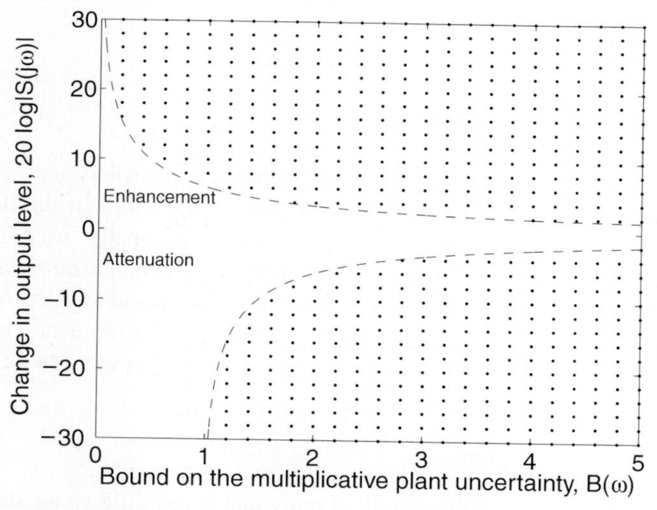

Figure 6.9 The limits on the disturbance enhancement and disturbance attenuation for a robustly stable control system due to multiplicative uncertainty in the plant response. For a given magnitude of multiplicative uncertainty, $B(\omega)$, the enhancement or attenuation of the disturbance must fall between the two curves if the control system is to be robustly stable.

so that with $B(\omega) = 2$, for example, the attenuation must be less than 6 dB, but the enhancement can only be about 4 dB at most.

A phase lag of 180° will always occur at high frequencies if the open-loop frequency response falls off at least as fast as 40 dB per decade, in which case a very interesting general result can be demonstrated. If this condition holds and the plant and controller are stable, then Bode (1945) showed that the integral of the log sensitivity function, which is linearly proportional to the disturbance attenuation in dB, over the whole frequency range must be zero, i.e.

$$\int_0^{\infty} 20 \log_{10} |S(j\omega)| \, d\omega = 0 , \tag{6.2.21}$$

which is called *Bode's sensitivity integral.*

This profound limitation on performance appears at first sight to impose a severe constraint on all controllers. On closer inspection, however, we realise that good attenuation, $|S(j\omega)| \ll 1$, over a small bandwidth, where the disturbance has significant energy, can be traded for small enhancements, $|S(j\omega)| > 1$, over large ranges of frequency, where the disturbance has little energy. Unfortunately, this disturbance enhancement can only be made arbitrarily small for minimum-phase plants, and for non-minimum-phase plants a disturbance attenuation over a finite range of frequencies inevitably gives rise to a certain minimum level of disturbance enhancement at another frequency (Freudenberg and Looze, 1985). If the disturbance is 'pushed down' at some frequencies, then for a non-minimum-phase plant the disturbance at other frequencies will inevitably 'pop up', which is described as the *waterbed effect* (see for example Doyle et al., 1992). One of the advantages of a feedforward control system using a time-advanced reference signal is that the performance is *not* then limited by the Bode sensitivity integral, as discussed by Hon and Bernstein (1998). In this case broadband attenuation of a disturbance can be achieved without any out-of-band enhancements.

6.2.6. Analogue Compensators

Apart from analysing the potential stability of a feedback control system, the frequency-domain methods introduced above can also provide a useful tool in the design of such systems. Analogue feedback control systems have traditionally used rather simple controllers. First- and second-order RC circuits, known as *compensators*, are commonly used to improve the performance of analogue feedback systems and to increase their relative stability. The circuit of one such compensator, together with its frequency response when $R_1 = 9R_2$, is shown in Fig. 6.10. The frequency response of this circuit can be written as

$$\frac{V_{\text{out}}(j\omega)}{V_{\text{in}}(j\omega)} = \frac{1 + j\omega R_2 C}{1 + j\omega (R_1 + R_2) C} . \tag{6.2.22}$$

At low frequencies, the gain of this circuit is unity and it has little phase shift. The useful aspect of the circuit is that at high frequencies the phase shift is again small but the gain is now $R_2/(R_1 + R_2)$, and so the loop gain can be considerably attenuated at frequencies around the Nyquist point, for example, with little phase shift. In the frequency region from $\omega = 1/(R_1 + R_2)C$ to $\omega = 1/R_2C$, the phase shift is negative and can be significant if $R_1 \gg R_2$.

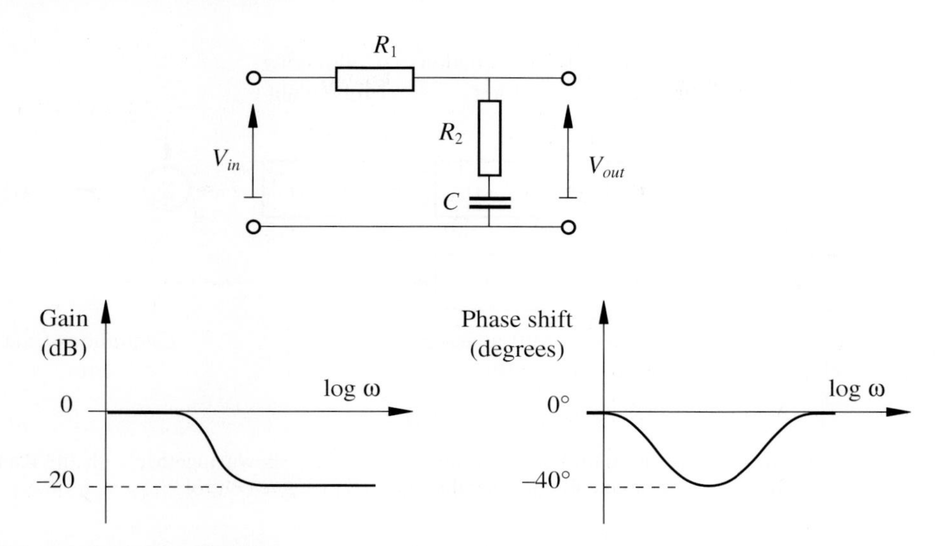

Figure 6.10 Circuit diagram and frequency response of a phase lag compensator, with $R_2/(R_1 + R_2) = 0.1$.

This phase lag can destabilise a feedback system by reducing its phase margin if it occurs too near the Nyquist point. The best design of this *phase-lag compensator* thus involves choosing R_1 and R_2 relative to C and the overall gain of the controller such that the loop gain is reduced near the Nyquist point, without excessive phase lag being introduced in this vicinity.

The use of phase lag compensators in increasing the gain and phase margins in servo systems is fully described by Dorf (1990), for example. Its use in improving the robust stability of a feedback control system used in an active headset will be described in Section 7.6.

6.3. DIGITAL CONTROLLERS

In this section we will discuss the limitations of implementing the controller 'digitally', i.e. with a sampled-time system. Much of this material is similar to that covered for digital feedforward control in Section 3.1, but is reiterated here for completeness. There are a number of excellent text books on digital control systems, including Kuo (1980), Åström and Wittenmark (1997) and Franklin et al. (1990), from which most of the material in this section has been abstracted. Figure 6.11 shows the general block diagram of a single-channel continuous-time plant, $G_C(s)$, subject to a disturbance $d_C(t)$ and controlled with a sampled-time controller, $H(z)$. It has been assumed that an analogue reconstruction filter, with response $T_R(s)$, has been used to smooth the waveform from the digital-to-analogue converter (DAC) driven by the digital controller. This itself includes a *zero-order hold*, with a continuous-time impulse response that is given by a pulse of duration equal to one sample time, T, and so its transfer function is given by

$$T_Z(s) = \frac{1 - e^{-sT}}{s} . \tag{6.3.1}$$

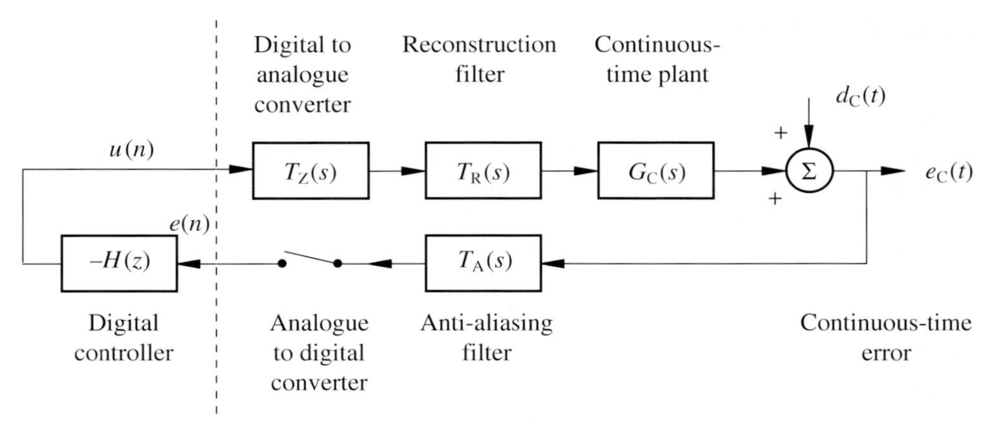

Figure 6.11 An analogue plant controlled by a digital controller, shown together with the data converters and analogue filters generally required in active control.

Also, an anti-aliasing filter, with response $T_A(s)$, has been included before the sampling of the error signal by the analogue-to-digital converter (ADC) to help prevent frequencies above half the sample rate from contaminating the sampled error signal. The analogue-to-digital converter is assumed to act perfectly and is thus represented by an intermittently closed switch, as in the control literature.

The signals to the right of the DAC and ADC in Fig. 6.11 thus exist in continuous time and those to the left exist in sampled time. The behaviour of this control system can be viewed in two rather different ways depending on whether one concentrates on the continuous-time signals, or the sampled-time ones. It is more difficult to analyse the continuous-time response of this system because, in general, there is no unique linear relationship between, for example, the Laplace transforms of the continuous-time output from the sampled system and the continuous time input to the sampled system. This is because frequency components of the continuous-time input signal above half the sampling rate will generally be aliased after sampling by the ADC, and the zero-order hold in the DAC will generate a weighted and folded version of the input spectrum above half the sampling frequency. These problems are alleviated if we assume, as in Section 3.1, that the combination of the reconstruction filter and the plant provides sufficient attenuation above half the sampling rate for any higher-frequency components of the DAC output to be insignificant at the plant output.

The continuous-time output of the plant will thus be a smooth function of time, over timescales comparable with the sample time, and we can be assured that there will be no significant 'intersample behaviour' (Morari and Zafiriou, 1989). If we also assume that the anti-aliasing filter effectively removes all the frequency components of the continuous-time disturbance, $d_C(t)$, above half the sampling frequency, then the sampled-time error signal will also be a faithful representation of the filtered continuous-time error signal. In general the behaviour of the complete control system shown in Fig. 6.11 can be most easily analysed in the sampled-time domain, since all the aliasing effects noted above can be contained in the definition of the sampled-time variables. If suitable antialiasing and reconstruction filters are used, we can also be assured that such an analysis will offer a complete description of the behaviour of the continuous-time section of Fig. 6.11.

6.3.1. Pulse Transfer Function

The *z-domain* transfer function, $G(z)$, between the input to the DAC, $u(n)$, and the output of the ADC, $e(n)$, is the plant response 'seen' by the digital controller, which is also called the *pulse transfer function* (Kuo, 1980). This can be derived by taking the z-transform of the sampled version of the continuous-time impulse response for the complete analogue system under control, which is given by the inverse Laplace transform of its s-domain transfer function. Thus $G(z)$ can be written as

$$G(z) = \mathcal{Z}\mathcal{L}^{-1}\left[T_Z(s)T_R(s)G_C(s)T_A(s)\right], \tag{6.3.2}$$

where $T_Z(s)$ is defined by equation (6.3.1), $\mathcal{Z}$ denotes the z-transform and $\mathcal{L}^{-1}$ the inverse Laplace transform.

The z-transform of the sampled disturbance signal can also be written as

$$d(n) = \mathcal{Z}\mathcal{L}^{-1}\left[T_A(s)D_C(s)\right], \tag{6.3.3}$$

and so the block diagram for the sampled-time signals in the digital controller can be drawn in the considerably simplified form shown in Fig. 6.12.

The sensitivity function of the digital feedback controller can thus be written as

$$S(z) = \frac{E(z)}{D(z)} = \frac{1}{1+G(z)H(z)}. \tag{6.3.4}$$

Although the stability of this transfer function is now determined by whether its poles are inside the unit circle in the z plane, rather than whether they were on the left-hand side of the s plane, the Nyquist criterion for stability remains the same. In the sampled-time case, the open-loop frequency response need only be plotted in the Nyquist plane for frequencies up to half the sample rate, since the frequency response of a sampled system is periodic. All of the results regarding robust stability in the previous section, for example, can thus be carried directly over into the analysis of digital controllers (Morari and Zafiriou, 1989), on which we will concentrate for the remainder of this chapter.

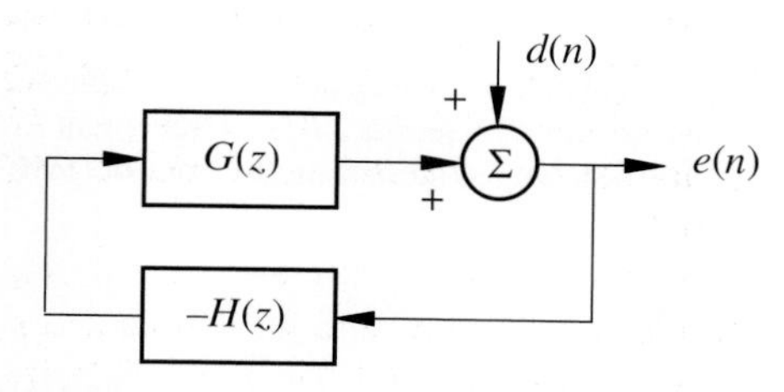

Figure 6.12 Block diagram that describes the behaviour of the sampled-time signals in a digital feedback controller.

6.4. INTERNAL MODEL CONTROL (IMC)

In this section, a reformulation of the feedback control problem will be outlined that allows a strong link to be made with the work described in the previous chapters. This is achieved by transforming the feedback design problem into an equivalent feedforward one.

We will assume that the plant is stable, which is most often the case in active control problems, but provided an unstable plant is first stabilised by any stable feedback controller, exactly the same methods of design as those described below can then be used on the stabilised plant (Morari and Zafiriou, 1989; Maciejowski, 1989). Consider the internal structure of a feedback controller shown in Fig. 6.13, in which the complete negative feedback controller, $-H(z)$, is contained within the dashed lines shown in this figure. The feedback controller contains an *internal model*, $\hat{G}(z)$, of the plant response, $G(z)$, which is fed by the input to the plant and whose output is subtracted from the observed error, $e(n)$. The resulting output, $\hat{d}(n)$, acts as the input to the *control filter*, $W(z)$, whose output, $u(n)$, drives the plant. The transfer function of the complete feedback controller is thus

$$H(z) = \frac{-W(z)}{1+\hat{G}(z)W(z)} . \tag{6.4.1}$$

Substituting this expression into equation (6.3.4) for the sensitivity function of the complete feedback control system gives

$$S(z) = \frac{E(z)}{D(z)} = \frac{1+\hat{G}(z)W(z)}{1-\left[G(z)-\hat{G}(z)\right]W(z)} . \tag{6.4.2}$$

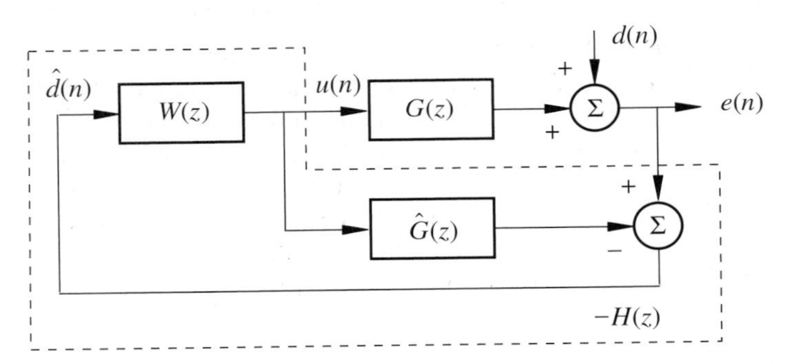

Figure 6.13 Block diagram of a feedback control system in which the feedback controller, $H(z)$, is implemented using an internal model $\hat{G}(z)$ of the plant $G(z)$. This arrangement of feedback controller is known as internal model control (IMC).

6.4.1. Perfect Plant Model

We assume for the time being that the internal plant model perfectly represents the response of the plant, so that $\hat{G}(z) = G(z)$. Morari and Zafiriou (1989) show that with a perfect plant model, the control system shown in Fig. 6.13 is internally stable, that is, any bounded input will only produce a bounded output, provided both $G(z)$ and $W(z)$ are stable. If the plant model is perfect, then the contribution to the observed error from the plant output is completely cancelled by that from the internal model, and so the signal fed to the control filter is equal to the disturbance, i.e. $\hat{d}(n) = d(n)$ in Fig 6.13. The equivalent block diagram of the feedback control system is thus given by Fig. 6.14, which has an entirely

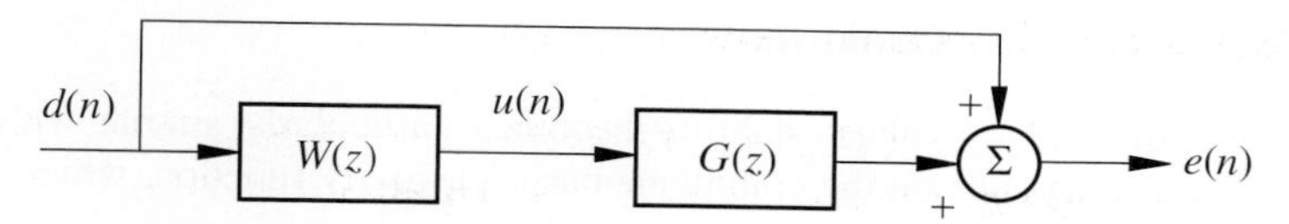

Figure 6.14 Equivalent block diagram for the internal model control arrangement of feedback controller in the case when the response of the plant model exactly matches that of the plant.

feedforward structure. The feedforward structure in this case is confirmed by substituting $\hat{G}(z) = G(z)$ into equation (6.4.2) to give the overall transfer function for the complete feedback control system as

$$S(z) = \frac{E(z)}{D(z)} = 1 + G(z)W(z) \ . \tag{6.4.3}$$

There does seem to be something very strange about replacing the 'closed-loop' feedback control system in Fig. 6.12 by an equivalent 'open-loop' feedforward one in Fig. 6.14. We must remember, however, that at this stage we are assuming complete knowledge of the plant, and so in principle there is no need for a feedback path, since any feedback controller could be replaced by an 'equivalent feedforward compensator' (as originally described by Newton, Gould and Kaiser, 1957). The real power of this controller arrangement comes, however, when we consider what happens if the plant is not known with perfect precision due to uncertainties, as discussed below.

The immediate advantage of the internal model structure for the feedback controller is that the control filter that minimises the mean-square error can now be designed using the standard Wiener techniques discussed in Chapter 3. This was the original motivation for such a controller, which was proposed for analogue controllers by Newton, Gould and Kaiser (1957). These authors showed that the optimum least-squares controllers could then be designed 'analytically', rather than using the rather ad-hoc methods outlined in Section 6.2. This approach to the design of the controller was extended by Youla et al. (1976a,b), to the case in which the plant is unstable. For a stable plant, it should be emphasised that *any* stabilising controller can be put into the form of Fig. 6.13, and so the control filter, $W(z)$, 'parametrises all stabilising controllers'. This structure of feedback controller is sometimes known as *Youla parametrisation* or *Q parametrisation*. In process control, this arrangement of feedback controller is widely used for both the design and implementation of feedback controllers and is known as *internal model control* (IMC), which is how it will be described here. An excellent review of the properties of IMC systems is given by Morari and Zafiriou (1989). The simple physical interpretation of the IMC architecture allows it to be readily extended to deal with nonlinear systems, as discussed by Hunt and Sbarbaro (1991) for example, and multichannel systems, as discussed in Section 6.9 below.

If the control filter, $W(z)$, is implemented as an FIR filter, the mean-square error in Fig. 6.14 is a quadratic function of the coefficients of this filter, exactly as in Chapter 3. More generally, though, the attenuation of other norms of the error signal will also be *convex functions* of these filter coefficients with a unique global minimum, even if certain practical constraints are put on the control system (Boyd and Barratt, 1991; Boyd and Vandenberghe, 1995). This parametrisation of feedback controllers has thus become widely used in H_∞ control (Zames, 1981).

6.4.2. Robust Stability Constraint

The constraint imposed by robust stability becomes particularly simple with an IMC architecture since it depends on the complementary sensitivity function, which following equation (6.1.9) is equal to

$$T(z) = 1 - S(z) . \tag{6.4.4}$$

For an IMC system with perfect plant modelling, the sensitivity function is given by equation (6.4.3), and so the complementary sensitivity function takes the simple form

$$T(z) = -G(z)W(z) . \tag{6.4.5}$$

The complementary sensitivity function is thus also a linear function of the coefficients of an FIR control filter. The condition for robust stability, equation (6.2.8), can now be written as

$$\left|G_0\left(e^{j\omega T}\right)W\left(e^{j\omega T}\right)\right| < \frac{1}{B\left(e^{j\omega T}\right)} , \qquad \text{for all } \omega T , \tag{6.4.6}$$

where $G_0\left(e^{j\omega T}\right)$ is the frequency response of the nominal plant, which can also be written as

$$\left\|G_0 W B\right\|_\infty < 1 . \tag{6.4.7}$$

In the following section, we will calculate the *nominal performance* of a feedback control system, which can be achieved if the plant response is fixed at its nominal value. In practice, we are also interested in the *robust performance* of the system, which is the performance that can be achieved for all plants in a certain class, subject to the controller being stable, i.e. the robust stability condition must also apply. It is difficult to analytically obtain an efficient solution to the optimum robust performance problem. A two-stage approach, which is not guaranteed to be optimal but should provide a good engineering solution, is described by Morari and Zafiriou (1989). The two steps are

(1) Design the controller for nominal performance, as discussed in the following section.
(2) Reduce the 'gain' of the control filter so that the robust stability criteria, equation (6.4.6), is satisfied.

In the process control examples considered by Morari and Zafiriou, the plant uncertainty occurs mainly at higher frequencies and so the gain of the control filter is reduced at these frequencies by augmentation with a low-pass filter, whose cut-off frequency is reduced until the robust stability criterion is just satisfied. Elliott and Sutton (1996) have discussed adding an extra control effort-weighting term to the cost function being minimised by the control filter, which will have the effect of reducing the gain of the control filter in frequency regions where the uncertainty is large. In Chapter 7 we also discuss an adaptive approach in which the response of the control filter can be adjusted to minimise the error signal in the frequency domain, with the constraint that the robust stability condition is satisfied.

Our initial assumption, that the plant model was perfect in the IMC system, at first sight appears to oversimplify the feedback design problem, by throwing away the feedback. It can be seen above, however, that the reason why feedback is needed in the first place is because of uncertainties in the system under control, and that these uncertainties can be accounted for in a very direct way using the IMC controller architecture, by making the stability of the system robust to such uncertainty. The IMC architecture for a feedback controller is similar to the 'feedback cancellation' architecture for the general feedforward controller discussed in Chapter 3, as emphasised by Elliott and Sutton (1996). Similar formulations for active feedback control systems have been suggested by Eriksson (1991b), Forsythe et al. (1991), Oppenheim et al. (1994), Elliott (1993) and Elliott et al. (1995). The architecture was also used for adaptive control by Walach and Widrow (1983), as described in detail in their more recent book (Widrow and Walach, 1996).

Considerable insight can be gained into the performance of feedback controllers from the equivalent block diagram of Fig. 6.14. If, for example, the plant were a pure delay, the control filter would have to act as a predictor in driving the plant with an estimate of the future behaviour of the disturbance, so that it could be cancelled at the error sensor. The block diagram shown in Fig. 6.14 would then be equivalent to the 'line enhancer' used in signal processing to enhance, or suppress, harmonic lines in the spectrum of the signal, as described in Section 2.3.3. In the special case of the plant being a pure delay, the optimum response of the control filter depends entirely on the statistics of the disturbance. In general, however, the optimum control filter will depend on both the plant dynamics and the properties of the disturbance, as discussed in Section 6.5.

6.4.3. Disturbance Rejection at Remote Error Sensors

In some active control applications, local feedback control is used to attenuate the disturbance measured at a number of remote error sensors. An example of such an application is the active control of sound radiation from a structure using feedback from structural sensors to structural actuators, in which case the remote error sensors may be microphones placed in the acoustic far field of the plate in order to control an approximation to the total radiated sound power. A general form of the block diagram for such a system is shown in Fig. 6.15, in which the disturbance signals at both the sensed output, s, which is used for feedback, and the regulated output, e, which is used to assess the performance of the controller, are assumed to be generated from an external or exogenous input, v. This block diagram is of exactly the same type as that used for feedforward controllers in Fig. 3.10, which emphasises the general form of this configuration (Doyle, 1983). In general, the block diagram could include multiple inputs and outputs, but the scalar case is considered here for simplicity.

It is now assumed that an IMC architecture is used in the feedback controller H, so that the block diagram can be drawn as in Fig. 6.16(a), in which $d_e(n)$ is the disturbance at the regulated output, generated by passing the exogenous signal $v(n)$ through $P_e(z)$, and $d_s(n)$ is the disturbance signal measured at the sensed output, generated by passing the exogenous signal through $P_s(z)$. If the model of the 'local' plant $G_s(z)$ is assumed to be perfect for the time being, then this block diagram is considerably simplified, as in Fig. 6.16(b). The block diagram is now of exactly the same form as a feedforward controller with a 'reference signal' equal to the disturbance signal at the sensed output and a 'disturbance signal' equal to the disturbance signal at the regulated output (Elliott et al., 1995).

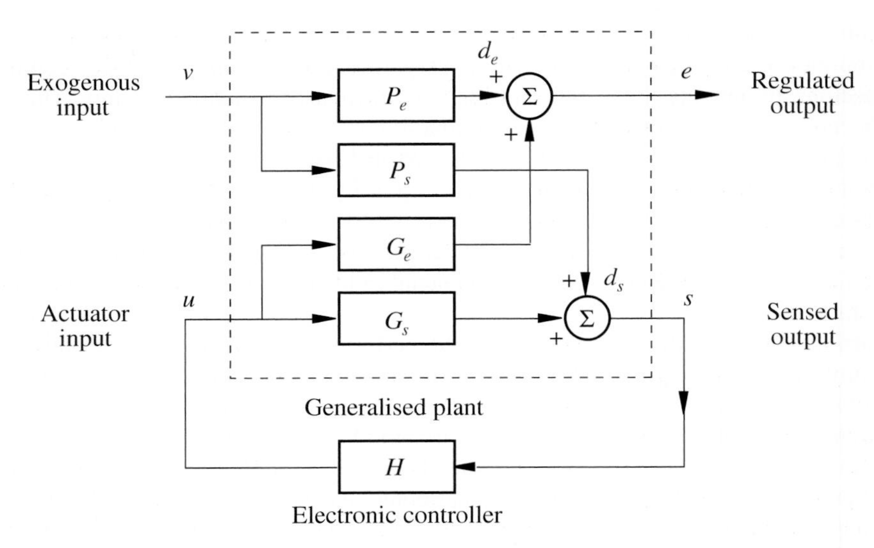

Figure 6.15 Block diagram of a feedback system with a remote error sensor, drawn in terms of the general control configuration which is used as the starting point for many automatic control problems.

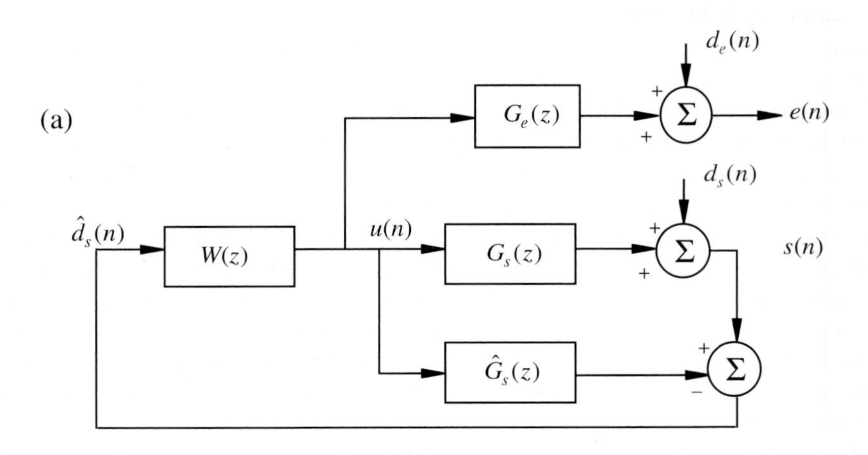

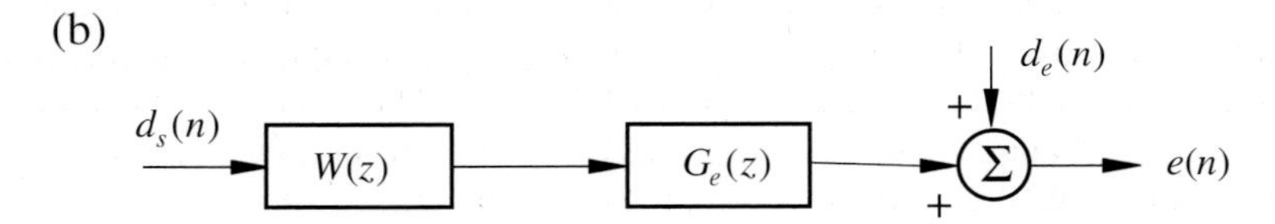

Figure 6.16 Block diagram of a feedback system with a remote error sensor (a) when the controller is implemented with IMC, and (b) the simplified form of this block diagram when the plant model in the IMC controller perfectly matches the plant, so that $\hat{G}_s(z) = G_s(z)$.

The design of a feedback controller to minimise the mean-square output of a remote sensor can now proceed as below except that the remote plant response is used to calculate the error signal,

$$e(z) = d_e(z) + G_e(z)\,W(z)\,d_s(z)\,, \tag{6.4.8}$$

whereas the local plant response must be used to determine the robust stability constraint, which now becomes

$$\left\| G_s\, W\, B \right\|_\infty < 1\,. \tag{6.4.9}$$

6.5. OPTIMAL CONTROL IN THE TIME DOMAIN

Using an IMC arrangement, the optimum least-squares feedback controller can be calculated fairly readily in either the time domain, as discussed in this section, or the frequency domain, as discussed in the following section. Before going into the details of this analysis, however, a number of other potential criteria for optimality of the feedback controller will be discussed. To begin with, we will assume that the plant response, $G(z)$, is known perfectly and does not change from its nominal value, $G_0(z)$. The performance that can be achieved with the controller under these conditions is known as the *nominal performance*. We have seen that under these conditions, an internal model can be used whose response perfectly matches that of the plant, so that the sensitivity function, which is a measure of disturbance rejection, is equal to

$$S_0(z) = 1 + G_0(z)W(z)\,. \tag{6.5.1}$$

If the nominal plant response were minimum phase, it would have a stable inverse and so the sensitivity function could be set to zero at all frequencies for a control filter given by

$$W(z) = -\frac{1}{G_0(z)}\,. \tag{6.5.2}$$

Some care must be taken even in this simple case, however, since if the response of the plant tends to zero at dc or half the sampling rate, for example, the response of the ideal controller would have to tend to infinity at this frequency. The stability of such a controller would then be very sensitive to plant uncertainties, but we will see that as soon as the controller is made robust to plant uncertainty, this problem does not arise.

In active control, the plant response is very rarely minimum phase and so ways of approximating the 'ideal' controller response given by equation (6.5.2) must be developed. What we will find, however, is that whereas the controller that minimises the mean-square error for a minimum-phase plant is independent of the disturbance being controlled, the optimal control filter is, in general, a function of both the plant response and the disturbance.

6.5.1. Optimal Least-Squares Control

Optimum least-square feedback control has been studied extensively using state-space representations of the plant, in what has been called linear quadratic gaussian, or *LQG control* (see, for example, Kwakernaak and Sivan, 1972; Anderson and Moore, 1989). This theory has been very successful in solving problems, such as those in the aerospace field, in which simple state-space models can accurately represent the dynamics of the system under control. In other, more industrial, applications, however, accurate plant models are not available and LQG designs have sometimes been found not to be robust enough to be used in practice. We will continue with our philosophy of avoiding state-space formulations and consider the optimum least-squares problems using a purely frequency-domain approach and H_2 norms. The H_2 norm of a stable, sampled-time, scalar transfer function can be defined (subject to the reservations discussed by Morari and Zafiriou, 1989) to be

$$\|F(z)\|_2 = \left(\frac{1}{2\pi} \int_0^{2\pi} \left| F\left(e^{j\omega T}\right) \right|^2 d\omega T \right)^{1/2} , \tag{6.5.3}$$

so that by Parseval's theorem, the sum of the squares of all the samples of the time domain version of $F(z)$ is equal to $\|F(z)\|_2^2$.

Returning to the active control problem and assuming that the spectrum of the disturbance is known, the cost function to be minimised for stochastic signals is given by the average mean-square value of the error signal, which is equal to the frequency integral of its power spectral density,

$$J_1 = \int_0^{2\pi} S_{ee}\left(e^{j\omega T}\right) d\omega T . \tag{6.5.4}$$

If the disturbance is assumed to have been generated by passing a white noise signal, having unit power spectral density, through a shaping filter $F\left(e^{j\omega T}\right)$, then using the equivalent block diagram in Fig. 6.14, the power spectral density of the error can be written as

$$S_{ee}\left(e^{j\omega T}\right) = \left| F\left(e^{j\omega T}\right)\left[1 + G\left(e^{j\omega T}\right) W\left(e^{j\omega T}\right)\right] \right|^2 , \tag{6.5.5}$$

so that the cost function in equation (6.5.4) is proportional to the square of the H_2 norm of the frequency response whose modulus is taken in equation (6.5.5). To calculate the nominal performance, with a perfect internal model and an FIR control filter, the optimisation problem is thus quadratic, since the equivalent block diagram is entirely feedforward. Although a parametrisation of the control filter as an FIR filter provides a convenient method of solving the least-squares problem in this section, the use of an FIR filter may also be particularly appropriate if the plant is strongly resonant and the control filter has to compensate for the poles of the plant with its own zeros. A particularly efficient implementation of the controller for such a resonant plant would then use IMC with an FIR control filter and an IIR plant model.

The coefficients of the optimal FIR control filter can be found using the formulation for the Wiener filter presented in Chapter 2, which was used in Chapter 3 for scalar feedforward controllers. The difference is that now the reference signal is equal to the disturbance and so the filtered reference signal used in Chapter 3 becomes the filtered disturbance signals, whose z transform is defined to be

$$R(z) = G(z)F(z) . \tag{6.5.6}$$

If the cross-correlation vector between the disturbance and the filtered disturbance is now written, as in Section 3.3, as

$$\mathbf{r}_{rd} = E[\mathbf{r}(n)d(n)] , \tag{6.5.7}$$

where

$$\mathbf{r}(n) = \left[r(n) \cdots r(n-I+1)^{\mathrm{T}}\right] , \tag{6.5.8}$$

and I is the number of coefficients in the FIR control filter, and the autocorrelation matrix of the filtered disturbance is equal to

$$\mathbf{R}_{rr} = E\left[\mathbf{r}(n)\mathbf{r}^{\mathrm{T}}(n)\right] , \tag{6.5.9}$$

then the coefficients of the H_2-optimal FIR control filter can be calculated, following the discussion of Chapter 3, as

$$\mathbf{w}_{\mathrm{opt}} = -\mathbf{R}_{rr}^{-1}\,\mathbf{r}_{rd} . \tag{6.5.10}$$

The performance achieved by this optimal control filter in attenuating the disturbance can then be directly calculated, as in equation (3.3.39). For a feedback controller the performance thus depends on the cross-correlation between the disturbance signal and the disturbance signal filtered by the nominal plant response, whereas for a feedforward controller the performance depends on the cross-correlation between the disturbance and the external reference signal filtered by the plant response.

6.5.2. Road Noise Example

The close connection between the formulation outlined above for the H_2-optimal feedback controller, and that for the feedforward controller discussed in Chapter 3 allows a consistent comparison to be made between the performance of the two control approaches in a given application (Elliott and Sutton, 1996). Returning to the example of road noise in a car, discussed in Section 5.7, the performance of a feedback controller can be calculated from the same data as was used to produce the predictions shown in Fig. 5.11 for the feedforward system. The feedback system has the significant advantage of not requiring any of the six accelerometers used to generate reference signals for the feedforward system, although there is a significant price to pay for this advantage, as we shall see.

The A-weighted power spectral density of the error signal measured using a pressure microphone in a car travelling at 60 km h^{-1} over a rough road is shown as the solid line in Fig. 6.17. The predicted residual pressure spectrum at the microphone is also shown in Fig. 6.17 for an optimal feedback control system designed using IMC, with a perfect plant model and a 128-coefficient FIR control filter at a sampling frequency of 1 kHz, for the two cases when the plant is a pure delay of either 1 ms or 5 ms. Whereas this increase in plant delay had little effect on the performance of a feedforward control system, Fig. 5.11, it has a profound effect on the performance of the feedback system. In the case of a 1 ms plant delay, the residual spectrum of the error signal is almost flat, indicating that the residual error is almost white noise. Indeed, the residual spectrum is entirely flat if a long enough control filter is used, since the equivalent block diagram for the feedback controller,

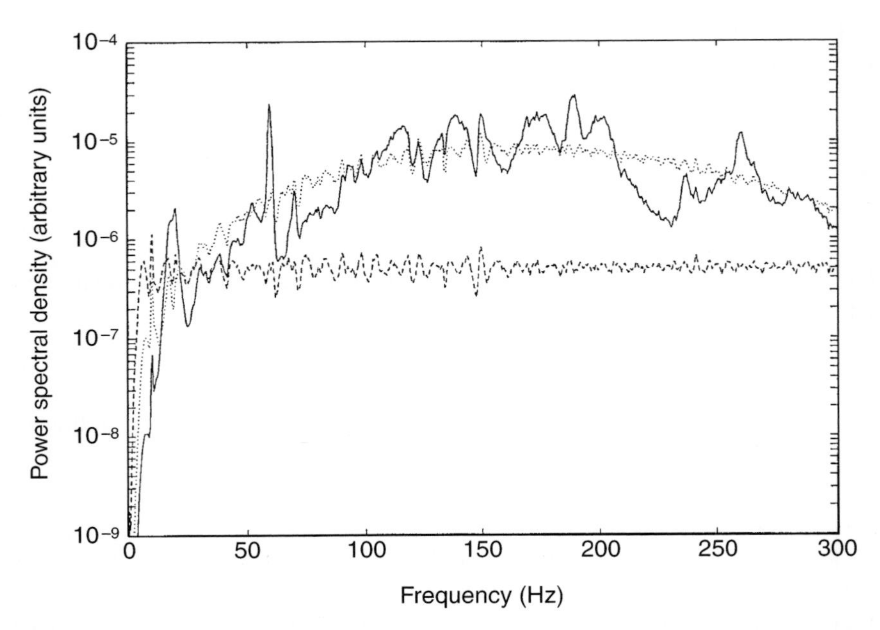

Figure 6.17 A-weighted power spectral density of the pressure measured in a small car (solid line) and predictions of the residual spectrum with feedback control system operating on a plant with a delay of 1 ms (dashed line) or 5 ms (dotted line).

Fig. 6.14, is then similar to that of the optimal predictor discussed in Section 2.3.3, Fig. 2.6, with a delay of $\Delta = 1$ sample. A long control filter will thus cancel all the predictable components of the disturbance, leaving only that part which is uncorrelated from one sample to the next, which is white noise. With a 5 ms plant delay, the residual spectrum is more 'coloured', but the peaks in the spectrum of the original disturbance, which correspond to the more predictable components of the pressure signal, have been largely eliminated. The feedback controller in this case is thus exploiting the predictability of the disturbance over the timescale of the plant delay to achieve control.

Interesting parametric studies can be carried out on the relative performance of feedback and feedforward controllers using this formulation. Figure 6.18, for example, shows the attenuation predicted in the overall level of the A-weighted noise in the car discussed above as a function of the delay in the plant, for both feedforward and feedback controllers (Elliott and Sutton, 1996). It is clear that the performance of the feedback control system is much more dependent on plant delays than that of the feedforward control system. This is to be expected, since the reference signals in the feedforward controller should, by definition, be providing time-advance information to the controller. The feedforward controller also has the advantage that it only controls components of the disturbance which are correlated with the reference signals. In the road-noise application, a feedforward system would not interfere with speech or warning sounds inside the car, whereas a feedback system could not discriminate between road noise and these other disturbances, and would attenuate them all.

For very short plant delays, below about 1.5 ms, the performance of the feedback controller can exceed that of the feedforward controller, since the performance of the latter

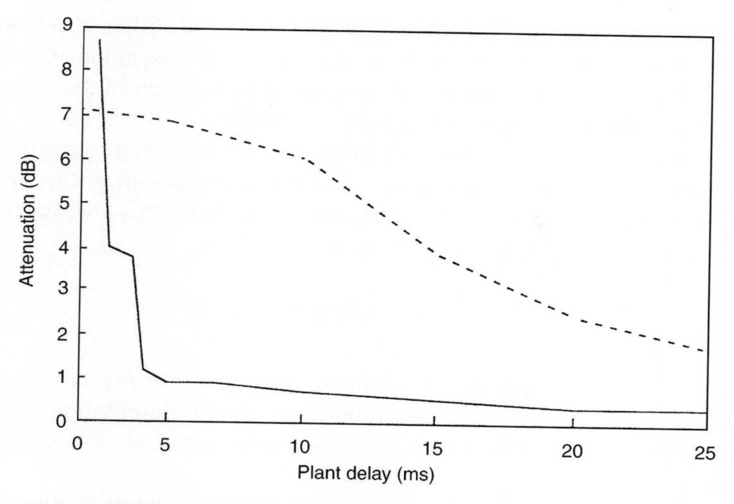

Figure 6.18 Variation of the attenuation in the mean-square value of the *A*-weighted pressure with delay in the plant for a feedforward control system (dashed line) and a feedback control system (solid line).

is limited by the coherence between the reference signals and the disturbance, as discussed in Section 3.3, whereas the feedback controller does not suffer from this limitation. In an active sound control system, the plant delay is partly due to delays in the analogue filters, data converters, signal processing and transducers, but is also dependent on the acoustic propagation delay between the loudspeaker and microphone, which is about 3 ms per metre in air at normal temperature and pressure. Provided all the other sources of delay can be minimised, the plant delay in an active control system is thus limited by the distance from loudspeaker to microphone. For the road noise disturbance used to generate Fig. 6.18, very good performance can be achieved only with plant delays of less than about 1 ms, which implies that the error microphone cannot be placed further than about 0.3 m from the secondary loudspeaker and in practice may need to be placed considerably closer. The spatial extent of the zone of quiet created by such a feedback controller is going to be limited because of this distance, as discussed in Chapter 1, and so we find that the control performance at the error microphone and the physical effect of the complete active control system are intimately linked. This issue has been explored in more detail by Rafaely and Elliott (1999), as is further discussed in Section 6.10.

6.5.3. Robust Controllers

All of the example results above have been calculated in terms of the nominal performance of the feedback controller, by minimising the mean-square error while assuming the plant response is equal to its nominal value $G_0(e^{j\omega T})$, which is perfectly known. There is thus no guarantee that the controller will be robust to changes in the plant response. In general, the H_2-optimal controllers designed above will not be very robust, and they have the reputation of being too 'aggressive' (Morari and Zafiriou, 1989). In Section 6.4 we saw that the robust stability of any IMC controller could be readily determined, given the bound on the multiplicative plant uncertainty, $B(\omega)$. It is also clear from equation (6.4.6) that the feedback controller can be made more robust by reducing the gain of the control filter at frequencies where there is a danger of instability.

The most straightforward way of reducing the gain of the controller is to minimise a cost function that not only includes mean-square error but also the sum of the squared filter coefficients, as in equation (3.4.38). Using Parseval's theorem, this cost function can be expressed entirely in the frequency domain, as

$$J_2 = \int_0^{2\pi} \left(S_{ee}\left(e^{j\omega T}\right) + \beta \left| W\left(e^{j\omega T}\right) \right|^2 \right) d\omega T \ . \tag{6.5.11}$$

In practice it has been found that minimising equation (6.5.11) can give a better compromise between performance and robustness than minimisation of the more conventional LQG cost function, which includes control effort as well as mean-square error (Mørkholt et al., 1997).

The impulse response of the optimum FIR control filter that minimises equation (6.5.11) can be written, as in equation (3.4.52), as

$$\mathbf{w}_{\text{opt}} = -\left[\mathbf{R}_{rr} + \beta \mathbf{I}\right]^{-1} \mathbf{r}_{rd} \ . \tag{6.5.12}$$

Minimisation of equation (6.5.11) will tend to reduce the gain of the control filter only at frequencies where the modulus of its frequency response is large, regardless of the requirement for robust stability at different frequencies. A way of modifying the H_2 optimisation problem to reduce the gain more precisely at the frequencies where the robust stability condition, equation (6.4.6), is most stringent would be to minimise the quadratic cost function given by

$$J_3 = \int_0^{2\pi} \left(S_{ee}\left(e^{j\omega T}\right) + \beta \left| G_0\left(e^{j\omega T}\right) W\left(e^{j\omega T}\right) B\left(e^{j\omega T}\right) \right|^2 \right) d\omega T \ . \tag{6.5.13}$$

Elliott and Sutton (1996) have noted that if 'sensor noise' is added to the disturbance estimate used as the reference signal in Fig. 6.14, which is generated by a passing white noise signal, $v(n)$, of variance β through a shaping filter having frequency response $B\left(e^{j\omega T}\right)$ as shown in Fig. 6.19, then minimising the resultant output error is exactly equivalent to minimising J_3 in equation (6.5.13). This method of improving the robustness of the controller is similar to the *fictitious sensor noise* approach proposed by Doyle and Stein (1979). If the disturbance signal is assumed to be generated by passing white noise of unit variance through a shaping filter of frequency response $F\left(e^{j\omega T}\right)$, then the cost function given by equation (6.5.13) can also be written as

$$J_3 = \int_0^{2\pi} \left[\left| S\left(e^{j\omega T}\right) F\left(e^{j\omega T}\right) \right|^2 + \beta \left| T\left(e^{j\omega T}\right) B\left(e^{j\omega T}\right) \right|^2 \right] d\omega T \ , \tag{6.5.14}$$

where $S\left(e^{j\omega T}\right)$ is the sensitivity function and $T\left(e^{j\omega T}\right)$ is the complementary sensitivity function. Equation (6.5.14) emphasises the compromise in feedback controller design between minimising the mean-square error, proportional to $\left| S\left(e^{j\omega T}\right) F\left(e^{j\omega T}\right) \right|^2$, while also

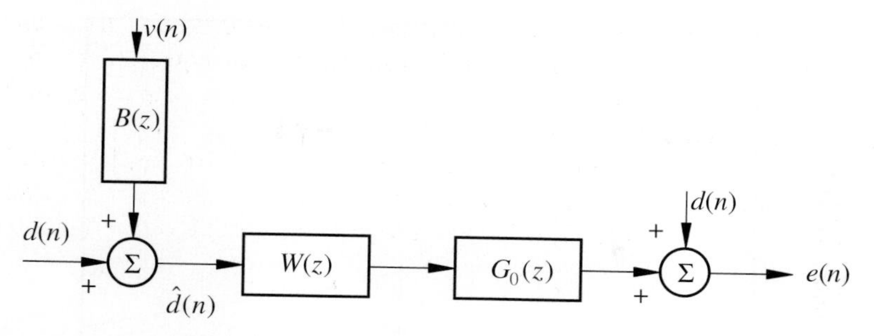

Figure 6.19 Block diagram of the internal model controller shown in Fig. 6.14 with a perfect model of the nominal plant, $\hat{G}(z) = G(z) = G_0(z)$, and 'sensor noise' injected into the estimated disturbance signal to make the controller more robust.

maintaining robustness, which is dependent on the magnitude of $\left|T\left(e^{j\omega T}\right)B\left(e^{j\omega T}\right)\right|^2$. The impulse response of the optimal FIR control filter which minimises equation (6.5.13) is given by

$$\mathbf{w}_{\text{opt}} = -\left[\mathbf{R}_{\hat{r}\hat{r}}\right]^{-1}\mathbf{r}_{\hat{r}d} , \tag{6.5.15}$$

where $\mathbf{R}_{\hat{r}\hat{r}}$ is the autocorrelation matrix of the noisy disturbance estimate, $\hat{d}$ in Fig. 6.19, filtered by the nominal plant, and $\mathbf{r}_{\hat{r}d}$ is the cross-correlation vector between this signal and the disturbance. If the plant response is a pure delay and the bound on the multiplicative uncertainty is constant at all frequencies, then the cost function given by equation (6.5.13) reduces to the simpler form in equation (6.5.11), but in the more general case, minimising equation (6.5.13) provides a more flexible method of designing a robust controller. In order to design a robustly stable controller, the value of the parameter β is gradually increased until the controller that results from minimising equation (6.5.13) satisfies the robust stability criterion, equation (6.4.6). It should be noted that such a controller is only 'optimal' in that it minimises equation (6.5.13). It does not optimally minimise the mean-square error, given by equation (6.5.4), which is an H_2 criterion, while imposing the constraint of robust stability given by the H_∞ condition of equation (6.4.7). The solution to this mixed H_2/H_∞ control problem can be obtained by iterative programming methods (Boyd and Barratt, 1991; Rafaely, 1997) or the use of discrete frequency-domain optimisation, as discussed in the final section of this chapter.

As the controller is designed to be more robust, the potential performance of the control system will be reduced. Using the method outlined above, this important trade-off can be explored by calculating the overall attenuation in the disturbance and the maximum allowable bound on the plant uncertainty for controllers designed with different balances between performance and robustness in the cost function being minimised, i.e., different values of β in equation (6.5.13). The results of such a calculation for a feedback system operating on a plant with a 1 ms delay, controlling the road noise disturbance used above, are shown in Fig. 6.20 (Elliott and Sutton, 1996). If the controller is designed with no regard to robustness, an attenuation of nearly 9 dB can be achieved, but the control system is only stable for multiplicative plant uncertainties of less than about 6% at low frequencies. This corresponds to a change in the amplitude of the plant response of about

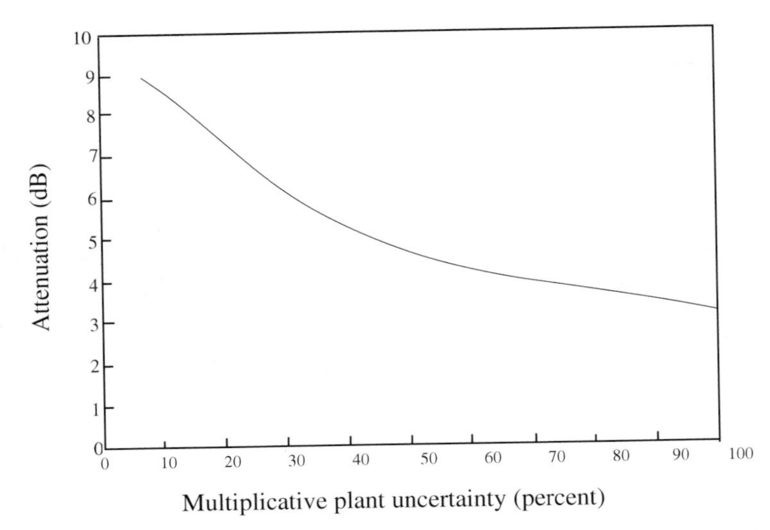

Figure 6.20 The variation in the attenuation of the mean-square error signal, against the maximum fractional plant uncertainty ($1/T_{max}$), as the parameter β is varied for the road noise feedback controller.

0.5 dB or a change in the phase of about 3.5° . In a practical vehicle installation, the change in the plant response between loudspeaker and microphone is likely to be greater than these limits if anyone in the vehicle moves significantly, and the control system would then become unstable.

If the controller is made robust to multiplicative plant uncertainties of less than about 33%, which corresponds to variations in the plant's amplitude and phase response of about 2.5 dB and 20°, as is typical of the changes which may occur in practice, then the attenuation in the disturbance falls to about 5 dB. Although the details of this trade-off between robustness and performance are clearly very application specific, the methods outlined in this section do allow a range of achievable specifications to be explored.

6.6. OPTIMAL CONTROL IN THE TRANSFORM DOMAIN

In this section the optimal H_2 controller will be formulated in the z domain, again assuming an IMC architecture. This architecture allows the design of an optimal feedback controller to be cast as a feedforward control optimisation problem, as in Fig. 6.14, and so the frequency-domain least-squares solutions found for feedforward controllers in Chapter 3 can again be used. The H_2-optimum feedforward controller in the z domain was shown in Section 3.3 to have the form

$$W_{\text{opt}}(z) = \frac{-1}{F(z)G_{\min}(z)}\left\{\frac{S_{xd}(z)}{F(z^{-1})G_{\text{all}}(z)}\right\}_+ , \tag{6.6.1}$$

where the power spectral density of the reference signal has been factorised into the response of a minimum-phase shaping filter $F(z)$, and its time-reversed form $F(z^{-1})$,

$$S_{xx}(z) = F(z)F(z^{-1}) , \tag{6.6.2}$$

the transfer function of the plant, which is assumed to be stable, has been split into its minimum-phase, $G_{\min}(z)$ and all-pass, $G_{\text{all}}(z)$, components, so that

$$G(z) = G_{\min}(z)G_{\text{all}}(z) , \tag{6.6.3}$$

and $S_{xd}(z)$ is the z-transform of the cross-correlation function between the reference and disturbance signals.

Figure 6.14 shows the equivalent feedforward system for the IMC arrangement with a perfect model, and in this case, the reference signal is equal to the disturbance signal. The cross spectral density function $S_{xd}(z)$ is then equal to the power spectral density of the disturbance in this case, $S_{xx}(z)$, whose factorisation is shown in equation (6.6.2). The H_2-optimum feedback control filter in the IMC arrangement can thus be written as

$$W_{\text{opt}}(z) = \frac{-1}{F(z)G_{\min}(z)} \left\{ \frac{F(z)}{G_{\text{all}}(z)} \right\}_+ . \tag{6.6.4}$$

The expression for the frequency response of the complete H_2-optimal feedback controller can be deduced from equation (6.4.1) as being

$$H_{\text{opt}}(z) = \frac{-W_{\text{opt}}(z)}{1+G_0(z)\, W_{\text{opt}}(z)} . \tag{6.6.5}$$

A method of solving equations of the form of equation (6.6.4) by manipulating the poles and zeros of $G(z)$ and $S_{xx}(z)$ was discussed in Chapter 3.

Another elegant method of obtaining solutions to equation (6.6.5) for the optimum feedback controller is to express the plant response and the spectral factors of the disturbance spectrum as the ratio of polynomials in z^{-1}, and using the procedures described by Kučera (1993b) for example. The polynomial approach, together with that of Youla (1976a), also has the advantage of being able to explicitly design the optimum feedback controller for an unstable plant response. The equivalence of the Wiener method of deriving the optimal H_2 feedback controller, equation (6.6.4), and the conventional state-space approach to the LQG control problem has been demonstrated by Safonov and Sideris (1985).

The discrete-frequency response of the optimum feedback controller can be calculated in practice from directly measured data by using the cepstral method of calculating the spectral factors from the power spectral density of the disturbance and the minimum phase component of the plant's frequency response, as described in detail in Section 3.3.3. The causality constraint in equation (6.6.4) can also be imposed in the discrete-frequency domain, although care needs to be taken to ensure that the number of points in the discrete Fourier transform is sufficiently large to contain the impulse response of $F(z)/G_{\text{all}}(z)$, as discussed at the end of Section 2.4.

It is interesting to note that if the constraint of causality is removed from equation (6.4.4), the expression for the frequency response of the optimum controller becomes independent of the disturbance spectrum, and can be written as

$$W_{\text{opt}}(z) = \frac{-1}{G(z)} , \tag{6.6.6}$$

which implies that the complete feedback controller, equation (6.6.5), has an infinite gain at all frequencies. This clearly emphasises the importance of the causality constraint on the design of feedback controllers. The complete feedback controller, $H_{opt}(z)$ in equation (6.6.5), may also be inherently unstable, even with a stable plant and a stable control filter, and when the complete closed-loop system, whose sensitivity function is given by equation (6.3.4), is stable. This instability of the 'open-loop' controller can occur particularly if it has been designed with very little robustness, and although, in principle, it does not create any problems for the closed-loop system, it is generally avoided in practice because of the danger of instability with a sensor failure or during saturation, and the difficulties with commissioning such a system, as described, for example, by Arelhi et al. (1996).

6.6.1. Robust Control

We have seen in the previous section that minimising the modified H_2 cost function given by equation (6.5.13) can make the stability of the controller more robust to plant uncertainties. It was also shown that minimising the output error of a control system with frequency-shaped 'sensor noise' added to the disturbance, as shown in Fig. 6.19, would also have the effect of minimising this cost function. In order to use this method of making the IMC controller more robust we must thus continue to distinguish between the disturbance signal, $d(n)$, and the disturbance estimate, $\hat{d}(n)$ in Fig. 6.19, even though the effect of the plant is assumed to be perfectly cancelled in calculating $d(n)$. The optimal control filter given by equation (6.6.1) cannot then be simplified by assuming that the reference signal is equal to the disturbance in this case, but the full form must be retained. The spectral factorisation involved in the calculation for the robust controller must now take into account the power spectral density of the disturbance and that of the sensor noise, so that, if the mean-square value of $v(n)$ in Fig. 6.19 is β, the required spectral factorisation is

$$F(z)\,F(z^{-1}) \;=\; S_{dd}(z) + \beta\, B(z)\, B\left(z^{-1}\right) \;, \qquad (6.6.7)$$

which is similar to the form of spectral factorisation used by Bongiorno (1969) and Youla et al. (1976a). In this case the reference signal is equal to $\hat{d}(n)$, but since $d(n)$ and $v(n)$ in Fig. 6.19 are uncorrelated, then $S_{xd}(z)$ in equation (6.6.1) is equal to $S_{dd}(z)$, and hence the optimal control filter can be calculated by solving equation (6.6.7) for $S_{dd}(z)$.

6.6.2. Minimum-Variance Control

Finally in this section we consider the optimal controller for a very special case of the plant, which is where the plant response is equal to a minimum-phase system and a pure delay of k samples, so that

$$G(z) \;=\; G_{\min}(z)\, z^{-k} \;. \qquad (6.6.8)$$

It is interesting to consider this form of response since it is widely used as a model of the plant response in the control literature, particularly in the calculation of the so-called minimum-variance controller (as described, for example, by Wellstead and Zarrop, 1991).

The all-pass component of the plant response in equation (6.6.8) is clearly equal to the delay and the optimal control filter, equation (6.6.4), can be written in the z domain in this case as

$$W_{\mathrm{opt}}(z) = \frac{-\{z^k F(z)\}_+}{G_{\min}(z) F(z)} . \tag{6.6.9}$$

An example of the impulse response of the minimum phase disturbance shaping filter, $F(z)$, is shown in Fig. 6.21 and its transfer function may be written as

$$F(z) = f_0 + f_1 z^{-1} + f_2 z^{-2} + \cdots \tag{6.6.10}$$

where f_0, f_1, etc. are the samples of the impulse response. The transfer function $\{z^k F(z)\}_+$ can now be identified as that of the causal part of the time-advanced impulse response shown in Fig. 6.21, so that

$$\{z^k F(z)\}_+ = f_k + f_{k+1} z^{-1} + \cdots \tag{6.6.11}$$

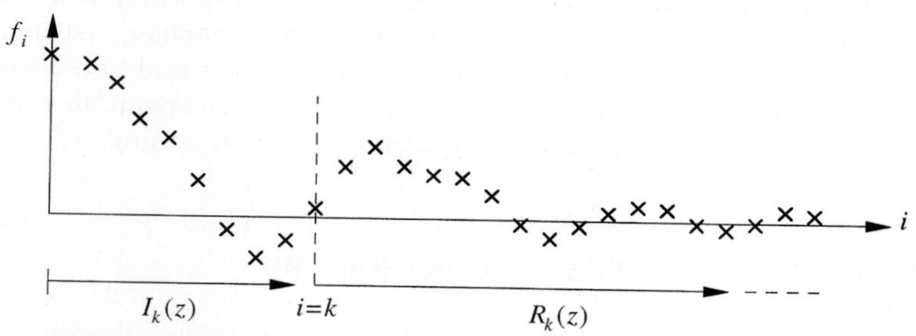

Figure 6.21 The impulse response of the disturbance shaping filter $F(z)$ is divided up into an initial part $I_k(z)$ for $i = 0$ to $k - 1$, and a later, residual, part $R_k(z)$ for $i \geq k$.

which we will call the 'remainder' of the plant response after k samples and denote as $R_k(z)$. Thus the optimum H_2 control filter can be written as

$$W_{\mathrm{opt}}(z) = \frac{-R_k(z)}{G_{\min}(z) F(z)} . \tag{6.6.12}$$

If $G_{\min}(z) = 1$, equation (6.6.12) becomes the transfer function of the optimal least-squares prediction filter discussed in Section 2.3.3.

The complete feedback controller is given by equation (6.6.5), and in this case its transfer function is equal to

$$H_{\mathrm{opt}}(z) = \frac{1}{G_{\min}(z)} \left[\frac{R_k(z)}{F(z) - z^{-k} R_k(z)} \right] . \tag{6.6.13}$$

It is clear from the definitions of $F(z)$, equation (6.6.10), and $R_k(z)$, equation (6.6.11), that the denominator of the term in square brackets in equation (6.6.13) has a response given by the first k points of the impulse response of $F(z)$. This will be denoted as the 'initial' part of the plant response, $I_k(z)$, where

$$I_k(z) = F(z) - z^{-k} R_k(z) , \tag{6.6.14}$$

whose impulse response is also shown in Fig. 6.21. The complete feedback controller for the particular class of plant given by equation (6.6.8) can thus be expressed as

$$H_{\mathrm{opt}}(z) = \frac{1}{G_{\min}(z)}\left[\frac{R_k(z)}{I_k(z)}\right], \qquad (6.6.15)$$

which is equal to the inverse of the minimum-phase part of the plant multiplied by the ratio of the transfer functions corresponding to the remainder of the impulse response of the disturbance shaping filter after k samples, to that of its initial part. This is called the *minimum-variance controller* and various techniques for making such controllers adaptive or 'self-tuning' have been developed (Wellstead and Zarrop, 1991). The controller in equation (6.6.15) has the interesting property that it can be split into two parts, one of which is a function only of the plant response and the other of which is a function only of the disturbance. This has led some researchers to propose a feedback control structure that consists of a fixed filter that corresponds to the optimum least-squares inverse of the plant, or 'compensator', and a separate filter that allows the controller to be tuned to a particular disturbance (Berkman et al., 1992). It should be emphasised, however, that although the response of a plant can always be approximated by a minimum-phase system and a pure delay, as long as the pure delay is long enough, this may not lead to a controller that performs as well as one designed without this underlying assumption. In general, then, equation (6.6.4) is still to be preferred for calculating optimal H_2 controllers.

6.7. MULTICHANNEL FEEDBACK CONTROLLERS

If a sampled-time plant has M inputs and L outputs, its response can be described by an $L \times M$ matrix of responses, $\mathbf{G}(z)$. The feedback controller will thus generally have M outputs and L inputs and can be described by the $M \times L$ matrix of responses $\mathbf{H}(z)$ as illustrated in Fig. 6.22 for a disturbance rejection problem. The algebraic manipulation necessary to analyse such a multichannel system is rather more complicated than for the single-channel case, since the performance and stability are now governed by matrix norms rather than the scalar norms used above. The properties of matrix norms are discussed in the Appendix. In this section, the performance and stability of multichannel feedback

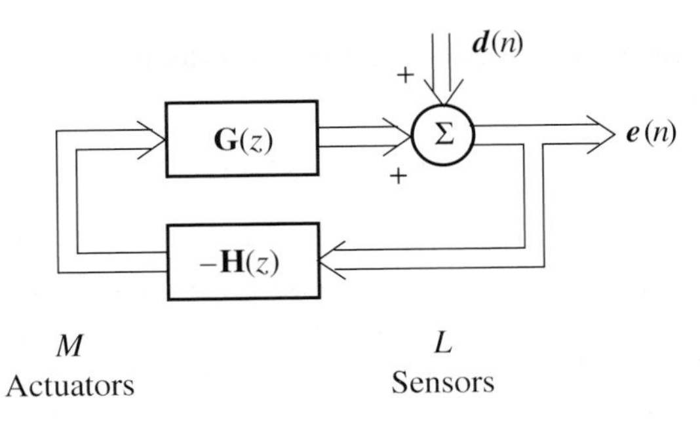

Figure 6.22 Block diagram of a multichannel feedback control system for the suppression of disturbances in a MIMO plant that has M inputs and L outputs.

systems will be briefly considered using methods analogous to those used above for single-channel systems. For a more detailed and complete description of these issues the reader is referred to the textbooks by Maciejowski (1989) and Skogestad and Postlethwaite (1996).

6.7.1. Stability

We begin by considering the conditions for such a feedback control system to be stable. The z-transform of the vector of output, or error signals in Fig. 6.22, can be written as

$$\mathbf{e}(z) = \mathbf{d}(z) - \mathbf{G}(z)\mathbf{H}(z)\mathbf{e}(z) , \tag{6.7.1}$$

so that

$$[\mathbf{I} + \mathbf{G}(z)\mathbf{H}(z)]\mathbf{e}(z) = \mathbf{d}(z) . \tag{6.7.2}$$

The matrix $\mathbf{I} + \mathbf{G}(z)\mathbf{H}(z)$ is called the *return difference matrix*, and provided it is not singular, an explicit expression can be found for the vector of error signals, which is given by

$$\mathbf{e}(z) = [\mathbf{I} + \mathbf{G}(z)\mathbf{H}(z)]^{-1}\mathbf{d}(z) . \tag{6.7.3}$$

The $L \times L$ matrix $[\mathbf{I} + \mathbf{G}(z)\mathbf{H}(z)]^{-1}$ is the matrix sensitivity function, and can be written as

$$[\mathbf{I} + \mathbf{G}(z)\mathbf{H}(z)]^{-1} = \frac{\operatorname{adj}[\mathbf{I} + \mathbf{G}(z)\mathbf{H}(z)]}{\det[\mathbf{I} + \mathbf{G}(z)\mathbf{H}(z)]} , \tag{6.7.4}$$

where adj[ ] and det[ ] refer to the adjugate, or classical adjoint, and the determinant of the matrix in square brackets, as discussed in the Appendix. The closed-loop system will be stable provided both the plant and controller are stable, i.e. each of the elements in the matrices $\mathbf{G}(z)$ and $\mathbf{H}(z)$ is stable, and the roots of

$$\det[\mathbf{I} + \mathbf{G}(z)\mathbf{H}(z)] = 0 , \tag{6.7.5}$$

all lie inside the unit circle in the z plane.

The stability of a multichannel feedback control system can also be determined from the open-loop frequency response using a generalisation of the Nyquist criterion (Maciejowski, 1989). Assuming that both the plant and controller are individually stable, the generalised Nyquist criterion states that the closed-loop system is stable provided that the locus of the function

$$\det[\mathbf{I} + \mathbf{G}(e^{j\omega T})\mathbf{H}(e^{j\omega T})] , \tag{6.7.6}$$

does not encircle the origin as ωT goes from $-\pi$ to π. Such a locus is illustrated in Fig. 6.23(a). This criterion is considerably less useful than that for a single-channel system since the locus can be very complicated for systems with many channels, and it is not clear how the shape of the locus changes if the gain of all the elements in the controller is altered. It is thus difficult to obtain a clear geometric guide to the relative stability of the system.

We can now use the fact that the determinant of a matrix is equal to the product of its eigenvalues to express the single polar plot defined by the locus of (6.7.6) as a series of

(a)

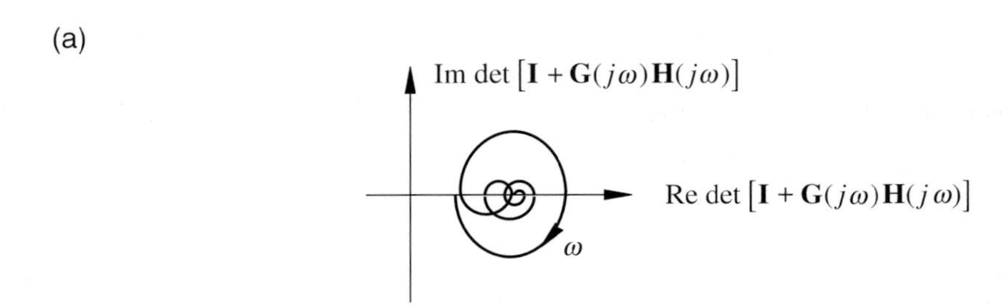

(b)

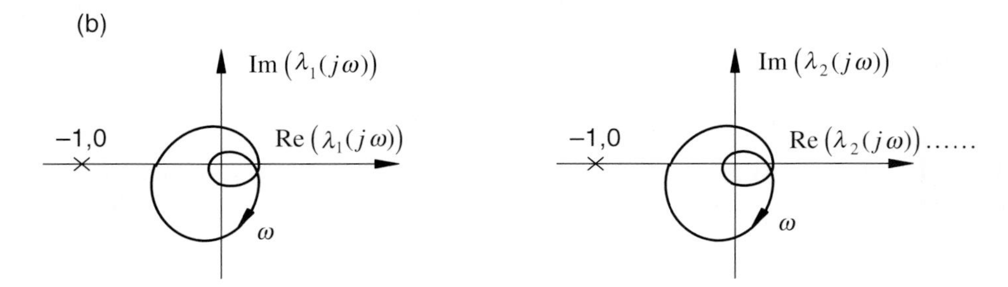

Figure 6.23 Multichannel generalisations of the Nyquist stability criterion for (a) the complete characteristic equation, and (b) the individual eigenvalues of the return difference matrix.

more simple polar plots, which are analogous to the single-channel Nyquist criterion. Specifically we note that

$$\det\left[\mathbf{I}+\mathbf{G}\left(\mathrm{e}^{j\omega T}\right)\mathbf{H}\left(\mathrm{e}^{j\omega T}\right)\right] = \left[1 + \lambda_1\left(\mathrm{e}^{j\omega T}\right)\right]\left[1 + \lambda_2\left(\mathrm{e}^{j\omega T}\right)\right]\cdots, \tag{6.7.7}$$

where $\lambda_i\left(\mathrm{e}^{j\omega T}\right)$ are the eigenvalues of the matrix $\mathbf{G}\left(\mathrm{e}^{j\omega T}\right)\mathbf{H}\left(\mathrm{e}^{j\omega T}\right)$. The locus of equation (6.7.7) does not enclose the origin, provided that the locus of *none* of the eigenvalues, which are called the *characteristic loci*, encircle the (–1,0) point as ωT goes from $-\pi$ to π, as illustrated in Fig. 6.23(b). The practical usefulness of this construction is somewhat limited, however, since the concepts of gain and phase margin only carry over to the multichannel case if there is a simultaneous change in the gain or phase of all elements in the matrix of plant response (Skogestad and Postlethwaite, 1996). An example of the physical interpretation of these eigenvalues in a simple two-channel control system is described by Serrand and Elliott (2000) for an active isolation experiment that consisted of a rigid beam with an active mount at either end. In this case the two eigenvalues could be associated with the resonant responses of the heave (vertical) and pitching (rocking) modes of the beam.

6.7.2. Small-Gain Theorem

A potentially rather conservative, i.e. sufficient but not necessary, condition for stability is that the modulus of all the eigenvalues of $\mathbf{G}\left(\mathrm{e}^{j\omega T}\right)\mathbf{H}\left(\mathrm{e}^{j\omega T}\right)$ is less than unity for all frequencies, i.e.

$$\left|\lambda_i\left(e^{j\omega T}\right)\right| < 1 \qquad \text{for all } i \text{ and } \omega T. \tag{6.7.8}$$

The eigenvalue of a matrix with the largest modulus is called its *spectral radius* and for the matrix $\mathbf{G}\left(e^{j\omega T}\right)\mathbf{H}\left(e^{j\omega T}\right)$ can be written as

$$\rho\left[\mathbf{G}\left(e^{j\omega T}\right)\mathbf{H}\left(e^{j\omega T}\right)\right] = \max_i \left|\lambda_i\left(e^{j\omega T}\right)\right|. \tag{6.7.9}$$

The spectral radius of a matrix product has the important property that

$$\rho\left[\mathbf{G}\left(e^{j\omega T}\right)\mathbf{H}\left(e^{j\omega T}\right)\right] \le \bar{\sigma}\left[\mathbf{G}\left(e^{j\omega T}\right)\right] \bar{\sigma}\left[\mathbf{H}\left(e^{j\omega T}\right)\right], \tag{6.7.10}$$

where $\bar{\sigma}\left(\mathbf{G}\left(e^{j\omega T}\right)\right)$ and $\bar{\sigma}\left(\mathbf{H}\left(e^{j\omega T}\right)\right)$ are the largest *singular values* of the matrices $\mathbf{G}\left(e^{j\omega T}\right)$ and $\mathbf{H}\left(e^{j\omega T}\right)$, which are real by definition.

Thus, an even more conservative condition for stability than equation (6.7.8) is that

$$\bar{\sigma}\left[\mathbf{G}\left(e^{j\omega T}\right)\right] \bar{\sigma}\left[\mathbf{H}\left(e^{j\omega T}\right)\right] < 1 \qquad \text{for all } \omega T. \tag{6.7.11}$$

This condition for stability is known as the *small-gain theorem*. The singular values of the plant and controller matrices are also known as the *principal gains* of these elements and the small-gain theorem states that the stability of a multichannel feedback system is assured if the largest principal gain of the plant multiplied by the largest principal gain of the controller is less than unity at all frequencies. Although this condition for stability may be very conservative in general, we shall see in the next section that it can provide a rather tight bound for robust stability in the multichannel case.

6.8. ROBUST STABILITY FOR MULTICHANNEL SYSTEMS

Robust stability involves the ability of a feedback system to remain stable despite uncertainties in the plant response. For single-channel plants, we saw in Section 6.2 that such uncertainty could be represented by discs in the Nyquist plane, but for multichannel systems things are more complicated because of the potential interrelationships between the uncertainties in different elements of the $L \times M$ matrix of plant frequency response functions, $\mathbf{G}\left(e^{j\omega T}\right)$.

6.8.1. Uncertainty Descriptions

Even if we assume that these uncertainties are independent from element to element, or *unstructured*, then there are two different ways of expressing multiplicative uncertainty in the multichannel case. The first assumes that the family of possible plant responses can be described by

$$\mathbf{G}\left(e^{j\omega T}\right) = \mathbf{G}_0\left(e^{j\omega T}\right)\left[\mathbf{I} + \boldsymbol{\Delta}_1\left(e^{j\omega T}\right)\right] \tag{6.8.1}$$

where $\mathbf{G}_0\left(e^{j\omega T}\right)$ is the matrix of frequency responses for the nominal plant. A block diagram corresponding to equation (6.8.1) is shown in Fig. 6.24(a), from which we can see why $\boldsymbol{\Delta}_I\left(e^{j\omega T}\right)$ is described as the *multiplicative input uncertainty*.

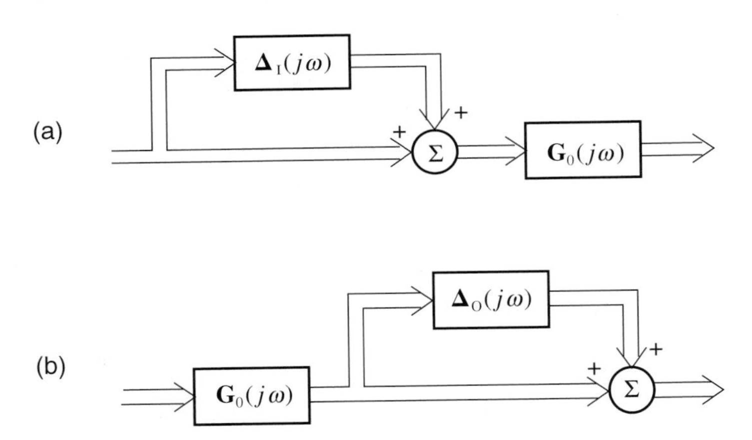

Figure 6.24 Uncertainty in multichannel plants represented by perturbations on the input (a) or output (b).

Expanding out equation (6.8.1), the fractional uncertainty can be written as

$$\mathbf{G}_0(e^{j\omega T})\ \boldsymbol{\Delta}_I(e^{j\omega T}) = \mathbf{G}(e^{j\omega T}) - \mathbf{G}_0(e^{j\omega T}) \ . \tag{6.8.2}$$

If we now assume that the plant has more outputs than inputs, so that $L > M$, an explicit expression for the input uncertainty matrix can be obtained as a function of the plant response and nominal plant response. This is derived by pre-multiplying both sides of equation (6.8.2) by $\mathbf{G}_0^H\left(e^{j\omega T}\right)$ and assuming that $\mathbf{G}_0^H\left(e^{j\omega T}\right)\mathbf{G}_0\left(e^{j\omega T}\right)$ is non-singular, so that

$$\boldsymbol{\Delta}_I\left(e^{j\omega T}\right) = \left[\mathbf{G}_0^H\left(e^{j\omega T}\right)\mathbf{G}_0\left(e^{j\omega T}\right)\right]^{-1}\mathbf{G}_0^H\left(e^{j\omega T}\right)\left[\mathbf{G}\left(e^{j\omega T}\right) - \mathbf{G}_0\left(e^{j\omega T}\right)\right] , \tag{6.8.3}$$

which can be useful in obtaining examples of $\boldsymbol{\Delta}_I\left(e^{j\omega T}\right)$ from measured sets of data. The quantity $\boldsymbol{\Delta}_I\left(e^{j\omega T}\right)$ is the $M \times M$ matrix of multiplicative input uncertainties, and if the uncertainties are unstructured, the elements of this matrix will be independent random variables. The only thing that is then known about $\boldsymbol{\Delta}_I\left(e^{j\omega T}\right)$ is its magnitude, which must be specified by a matrix norm in the multichannel case. A choice of norm that is convenient for subsequent analysis is the matrix 2-norm, which is equal to the largest singular value of the matrix. We will assume that at each frequency this norm is always less than some upper bound $B_I\left(e^{j\omega T}\right)$, so that

$$\left\|\boldsymbol{\Delta}_I\left(e^{j\omega T}\right)\right\|_2 = \bar{\sigma}\left[\boldsymbol{\Delta}_I\left(e^{j\omega T}\right)\right] \le B_I\left(e^{j\omega T}\right) \ . \tag{6.8.4}$$

Alternatively, we could assume a family of possible plant responses described by

$$\mathbf{G}\left(e^{j\omega T}\right) = \left[\mathbf{I} + \boldsymbol{\Delta}_O\left(e^{j\omega T}\right)\right]\ \mathbf{G}_0\left(e^{j\omega T}\right) , \tag{6.8.5}$$

whose block diagram is shown in Fig. 6.24(b) and from which it is clear why this is described as the *multiplicative output uncertainty*. Equation (6.8.5) can also be written as

$$\boldsymbol{\Delta}_{\mathrm{O}}\left(\mathrm{e}^{j\omega T}\right)\mathbf{G}_0\left(\mathrm{e}^{j\omega T}\right) = \mathbf{G}\left(\mathrm{e}^{j\omega T}\right) - \mathbf{G}_0\left(\mathrm{e}^{j\omega T}\right) . \tag{6.8.6}$$

If the plant has more outputs than inputs, $L > M$, then it is generally not possible to obtain a unique expression for the $L \times L$ matrix $\boldsymbol{\Delta}_{\mathrm{O}}\left(\mathrm{e}^{j\omega T}\right)$ in terms of the $L \times M$ matrices of plant and nominal plant responses.

If we again assume that the uncertainty is unstructured, then all we know about $\boldsymbol{\Delta}_{\mathrm{O}}\left(\mathrm{e}^{j\omega T}\right)$ is that its 2-norm is bounded at each frequency by

$$\overline{\sigma}\left(\boldsymbol{\Delta}_{\mathrm{O}}\left(\mathrm{e}^{j\omega T}\right)\right) \leq B_{\mathrm{O}}\left(\mathrm{e}^{j\omega T}\right) . \tag{6.8.7}$$

Even if we leave aside the differences in the dimensions of the input and output uncertainty matrices, then we should not view them as equivalent representations of the same effect, as a simple example will demonstrate. If we assume that the plant has as many inputs as outputs, $M = L$, so that the matrices $\mathbf{G}\left(\mathrm{e}^{j\omega T}\right)$ and $\mathbf{G}_0\left(\mathrm{e}^{j\omega T}\right)$ are square and the latter is assumed to be invertible, then the input and output uncertainty matrices are also square and can be written in this case as

$$\boldsymbol{\Delta}_{\mathrm{O}}\left(\mathrm{e}^{j\omega T}\right) = \left[\mathbf{G}\left(\mathrm{e}^{j\omega T}\right) - \mathbf{G}_0\left(\mathrm{e}^{j\omega T}\right)\right]\mathbf{G}_0^{-1}\left(\mathrm{e}^{j\omega T}\right) , \tag{6.8.8}$$

$$\boldsymbol{\Delta}_{\mathrm{I}}\left(\mathrm{e}^{j\omega T}\right) = \mathbf{G}_0^{-1}\left(\mathrm{e}^{j\omega T}\right)\left[\mathbf{G}\left(\mathrm{e}^{j\omega T}\right) - \mathbf{G}_0\left(\mathrm{e}^{j\omega T}\right)\right] . \tag{6.8.9}$$

We can thus express the output uncertainty matrix in terms of the input uncertainty matrix in this case as

$$\boldsymbol{\Delta}_{\mathrm{O}}\left(\mathrm{e}^{j\omega T}\right) = \mathbf{G}_0\left(\mathrm{e}^{j\omega T}\right)\boldsymbol{\Delta}_{\mathrm{I}}\left(\mathrm{e}^{j\omega T}\right)\mathbf{G}_0^{-1}\left(\mathrm{e}^{j\omega T}\right) . \tag{6.8.10}$$

To get an idea of the relative sizes of $\boldsymbol{\Delta}_{\mathrm{O}}\left(\mathrm{e}^{j\omega T}\right)$ and $\boldsymbol{\Delta}_{\mathrm{I}}\left(\mathrm{e}^{j\omega T}\right)$, we take the 2-norm of equation (6.8.10), where the 2-norm has the property that $\overline{\sigma}(\mathbf{AB}) \leq \overline{\sigma}(\mathbf{A})\overline{\sigma}(\mathbf{B})$ so that

$$\overline{\sigma}\left[\boldsymbol{\Delta}_{\mathrm{O}}\left(\mathrm{e}^{j\omega T}\right)\right] \leq \overline{\sigma}\left[\mathbf{G}_0\left(\mathrm{e}^{j\omega T}\right)\right] \overline{\sigma}\left[\boldsymbol{\Delta}_{\mathrm{I}}\left(\mathrm{e}^{j\omega T}\right)\right] \overline{\sigma}\left[\mathbf{G}_0^{-1}\left(\mathrm{e}^{j\omega T}\right)\right] . \tag{6.8.11}$$

But the largest singular value of $\mathbf{G}_0^{-1}\left(\mathrm{e}^{j\omega T}\right)$ is equal to the reciprocal of smallest singular value of $\mathbf{G}_0\left(\mathrm{e}^{j\omega T}\right)$, which can be written as

$$\overline{\sigma}\left[\mathbf{G}_0^{-1}\left(\mathrm{e}^{j\omega T}\right)\right] = 1/\underline{\sigma}\left[\mathbf{G}_0\left(\mathrm{e}^{j\omega T}\right)\right] , \tag{6.8.12}$$

and the ratio of the largest to the smallest singular values of a matrix is called its *condition number,* which is written as

$$\kappa\left[\mathbf{G}_0\left(\mathrm{e}^{j\omega T}\right)\right] = \frac{\overline{\sigma}\left[\mathbf{G}_0\left(\mathrm{e}^{j\omega T}\right)\right]}{\underline{\sigma}\left[\mathbf{G}_0\left(\mathrm{e}^{j\omega T}\right)\right]} . \tag{6.8.13}$$

Since the largest singular value of the input uncertainty matrix is bounded by $B_{\mathrm{I}}\left(\mathrm{e}^{j\omega T}\right)$, as in equation (6.8.4), then from equation (6.8.11) we have

$$\bar{\sigma}\left[\mathbf{\Delta}_{\mathrm{O}}\left(\mathrm{e}^{j\omega T}\right)\right] \leq \kappa\left[\mathbf{G}_0\left(\mathrm{e}^{j\omega T}\right)\right] B_{\mathrm{I}}\left(\mathrm{e}^{j\omega T}\right) . \tag{6.8.14}$$

If we assume that the largest singular value of the output uncertainty matrix now takes its maximum upper bound, then it is possible in the worst case that

$$B_{\mathrm{O}}\left(\mathrm{e}^{j\omega T}\right) \leq \kappa\left[\mathbf{G}_0\left(\mathrm{e}^{j\omega T}\right)\right] B_{\mathrm{I}}\left(\mathrm{e}^{j\omega T}\right) , \tag{6.8.15}$$

so that at each frequency

$$\frac{B_{\mathrm{O}}\left(\mathrm{e}^{j\omega T}\right)}{B_{\mathrm{I}}\left(\mathrm{e}^{j\omega T}\right)} \leq \kappa\left[\mathbf{G}_0\left(\mathrm{e}^{j\omega T}\right)\right] . \tag{6.8.16}$$

We have seen in Chapter 4 that the plant matrix of a multichannel active control system can be very ill-conditioned at certain frequencies. The largest singular value of the plant matrix can then be many orders of magnitude greater than the smallest singular value, resulting in a huge condition number. If uncertainty at the input to a plant were represented as output uncertainty, then equation (6.8.15) shows that the upper bound on the norm of this equivalent output uncertainty would potentially have to be much larger than that of the original input uncertainty. This example (taken from Morari and Zafiriou, 1989) demonstrates that in order to get a tight bound on the uncertainty for multichannel systems, it is important to model the plant uncertainty where it physically occurs, rather than where it is mathematically convenient.

In this context it is interesting to return to the form of the perturbed singular value matrices used in Section 4.4 to describe the uncertainty of a multichannel plant response at a single frequency. Equation (6.8.2) describes the change in the plant response if the uncertainty is on the input to the system, which can be compared with equation (4.4.18) for the perturbed singular value matrix $\Delta\mathbf{\Sigma}$, to obtain

$$\mathbf{G}_0 \mathbf{\Delta}_{\mathrm{I}} = \mathbf{R}_0 \Delta\mathbf{\Sigma}\, \mathbf{Q}_0^{\mathrm{H}} , \tag{6.8.17}$$

where $\mathbf{G}_0 = \mathbf{R}_0\, \mathbf{\Sigma}_0\, \mathbf{Q}_0^{\mathrm{H}}$ is the singular value decomposition of the nominal plant matrix at the frequency of interest and the explicit dependence on frequency has been dropped for notational convenience. Since $\mathbf{R}_0^{\mathrm{H}}\, \mathbf{R}_0 = \mathbf{I}$ and $\mathbf{Q}_0^{\mathrm{H}}\, \mathbf{Q}_0 = \mathbf{I}$, however, the perturbed singular value matrix can be written as

$$\Delta\mathbf{\Sigma} = \mathbf{\Sigma}_0\, \mathbf{Q}_0^{\mathrm{H}} \mathbf{\Delta}_{\mathrm{I}} \mathbf{Q}_0 . \tag{6.8.18}$$

If the real and imaginary parts of the elements of $\mathbf{\Delta}_{\mathrm{I}}$ are randomly distributed, i.e. unstructured, then the elements of $\mathbf{Q}_0^{\mathrm{H}} \mathbf{\Delta}_{\mathrm{I}} \mathbf{Q}_0$ will also be unstructured. Pre-multiplying this matrix by $\mathbf{\Sigma}_{\mathrm{O}}$ in equation (6.8.18) will produce a matrix $\Delta\mathbf{\Sigma}$ that has random elements of equal variances along its *rows*, with the largest elements in the top row and elements of decreasing magnitude in subsequent rows, since they are multiplied by the smaller singular values in $\mathbf{\Sigma}_{\mathrm{O}}$. This is exactly the form of the $\Delta\mathbf{\Sigma}$ matrix observed in practice when the positions of the actuators, i.e. the inputs to the plant, are perturbed, as shown in Fig. 4.15 of Chapter 4.

Similarly, if unstructured uncertainty is assumed in the output of the system, the change in the plant response is given from equation (6.8.5) as

$$\boldsymbol{\Delta}_O \mathbf{G}_0 = \mathbf{R}_0 \, \Delta\boldsymbol{\Sigma} \, \mathbf{Q}_0^{\mathrm{H}} , \tag{6.8.19}$$

which has also been set equal to the change in the plant output in terms of the perturbed singular value matrix, equation (4.4.18). Using the unitary properties of $\mathbf{R}_0$ and $\mathbf{Q}_0$ again, we obtain

$$\Delta\boldsymbol{\Sigma} = \mathbf{R}_0^{\mathrm{H}} \boldsymbol{\Delta}_O \, \mathbf{R}_0 \, \boldsymbol{\Sigma}_0 \tag{6.8.20}$$

in this case. If the real and imaginary parts of $\boldsymbol{\Delta}_O$ are random numbers, then $\mathbf{R}_0^{\mathrm{H}} \boldsymbol{\Delta}_O \mathbf{R}_0$ will be similarly unstructured, and post-multiplying by $\boldsymbol{\Sigma}_0$ will give a $\Delta\boldsymbol{\Sigma}$ matrix that has random elements of equal variance in its *columns* in this case, with the largest elements in the left-hand column and elements of decreasing size in subsequent columns, again according to the relative size of the singular values in $\boldsymbol{\Sigma}_0$. This is exactly the form of the $\Delta\boldsymbol{\Sigma}$ matrix observed when the position of the sensors, i.e. the outputs from the system, are perturbed, as shown in Fig. 4.16 of Chapter 4. The structure of the $\Delta\boldsymbol{\Sigma}$ matrix may thus be useful in determining the most parsimonious form of the unstructured uncertainty for matrices of measured plant responses.

6.8.2. Structured Uncertainty

Apart from unstructured uncertainties, there may also be perturbations in the plant response due to uncertainties in the internal parameters of the plant, such as temperature, operating condition, etc. These parametric changes will lead to correlated changes in the elements of the plant matrix, which are described as *structured uncertainties*, and these are considerably more difficult to measure than the unstructured uncertainties discussed above. It is important to get an accurate model for such uncertainties, however, in order to get a condition for robust stability which is as tight as possible.

The general case of plant uncertainties could be represented as in Fig. 6.25 in which $\boldsymbol{\Delta}$ represents the matrix of uncertainties, which we want to make as small and unstructured as possible. If $\boldsymbol{\Delta} = \mathbf{0}$ then the plant response is by definition equal to its nominal value $\mathbf{G}_0(e^{j\omega T})$, but in general we can only write

$$\mathbf{G}(e^{j\omega T}) = \text{function}\left[\mathbf{G}_0(e^{j\omega T}), \boldsymbol{\Delta}(e^{j\omega T})\right] . \tag{6.8.21}$$

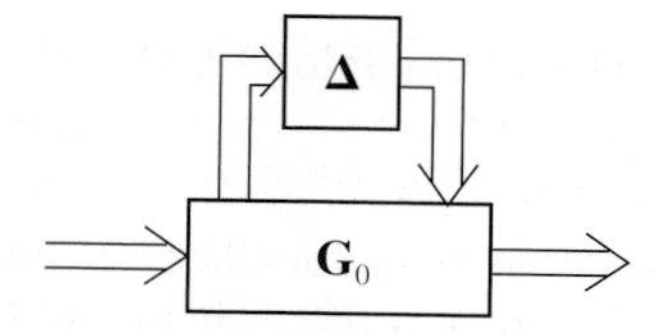

Figure 6.25 General form of uncertainty in a multichannel plant **G**, in which the internal parameters are perturbed by the elements of the matrix **Δ**.

If the perturbations are relatively small, then we can linearise equation (6.8.21) and express it in the form

$$\mathbf{G}(e^{j\omega T}) = \mathbf{G}_0(e^{j\omega T}) + \mathbf{P}_2(e^{j\omega T})\boldsymbol{\Delta}(e^{j\omega T})\mathbf{P}_1(e^{j\omega T}) , \tag{6.8.22}$$

where $\mathbf{P}_1(e^{j\omega T})$ is an $L \times N_2$ matrix and $\mathbf{P}_2(e^{j\omega T})$ is an $N_2 \times M$ matrix of transformed plant responses and $\boldsymbol{\Delta}(e^{j\omega T})$ is the $N_1 \times N_2$ uncertainty matrix.

It can be seen that input and output uncertainty are special cases of equation (6.8.22), but in general the dimensions of the matrix $\boldsymbol{\Delta}(e^{j\omega T})$ may be considerably smaller than those of the plant. Once again the assumption is that the only thing that is known about $\boldsymbol{\Delta}(e^{j\omega T})$ is the maximum upper bound of its largest singular value, $B(e^{j\omega T})$.

If we assume that a feedback loop has been connected around the plant in Fig. 6.25, the stability of the system can be determined in the absence of any disturbances, and the complete feedback control system can be split into two parts as shown in Fig. 6.26. All the uncertainties are gathered into the matrix with transfer response **Δ**, of which we only know that $\bar{\sigma}(\boldsymbol{\Delta}) < B$, and the matrix **M** represents the response of all the other parts of the plant and controller from the 'output' from the uncertainty matrix, **Δ**, to the 'input' of the uncertainty matrix.

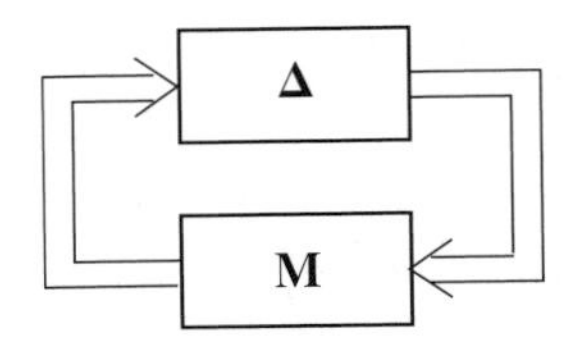

Figure 6.26 Block diagram of a complete feedback control system, **M**, which is subject to uncertainties contained in the matrix **Δ**, used to derive the robust stability condition.

6.8.3. Robust Stability

The system, **M**, is assumed to be stable if the uncertainty **Δ** is zero, i.e. to be *nominally stable,* and using the small-gain theorem we can say that the system with uncertainty will also be stable provided that

$$\bar{\sigma}\left[\boldsymbol{\Delta}(e^{j\omega T})\right] \bar{\sigma}\left[\mathbf{M}(e^{j\omega T})\right] < 1 \qquad \text{for all } \omega T . \tag{6.8.23}$$

Normally this is a rather conservative condition for stability, because the direction of the gain associated with $\bar{\sigma}\left[\boldsymbol{\Delta}(e^{j\omega T})\right]$ and that associated with $\bar{\sigma}\left[\mathbf{M}(e^{j\omega T})\right]$ do not generally both tend to destabilise the system. In the particular case when **Δ** is an uncertainty matrix, however, the direction of the gain in **Δ** is completely unknown, and in the worst case it must be taken to be such that the system will be destabilised. Under these conditions, equation (6.8.23) is no longer conservative and substituting in the upper bound for

$\bar{\sigma}\left[\mathbf{\Delta}\left(e^{j\omega T}\right)\right]$ as $B\left(e^{j\omega T}\right)$, we find that the *necessary and sufficient* condition for robust stability in a multichannel system can be written as

$$\bar{\sigma}\left[\mathbf{M}\left(e^{j\omega T}\right)\right] < \frac{1}{B\left(e^{j\omega T}\right)} \qquad \text{for all } \omega . \tag{6.8.24}$$

or

$$\|\mathbf{M}\, B\|_\infty < 1 , \tag{6.8.25}$$

where $\| \ \|_\infty$ denotes the operator H_∞ norm which is equal to the largest of any of the singular values at any frequency, as described in Section A.9 of the Appendix.

The importance of this condition will be illustrated by a few examples. If the plant is subject to unstructured multiplicative output uncertainty, the block diagram of the complete feedback controller will be as shown in Fig. 6.27. The transfer response of the control system as 'seen' by the uncertainty is shown inside the dashed lines and corresponds to the matrix $\mathbf{M}$ in Fig. 6.26. An equation for $\mathbf{M}$ can be obtained fairly readily in this case. If the 'input' to $\mathbf{M}$ is denoted $\mathbf{x}$ and the 'output' from $\mathbf{M}$ is denoted $\mathbf{y}$, then from Fig. 6.27 we can see that

$$\mathbf{y} = -\left[\mathbf{I} + \mathbf{G}_0\mathbf{H}\right]^{-1}\mathbf{G}_0\mathbf{H}\ \mathbf{x} , \tag{6.8.26}$$

so that in this case we can write $\mathbf{M}$ as

$$\mathbf{M} = -\left[\mathbf{I}+\mathbf{G}_0\mathbf{H}\right]^{-1}\mathbf{G}_0\mathbf{H} , \tag{6.8.27}$$

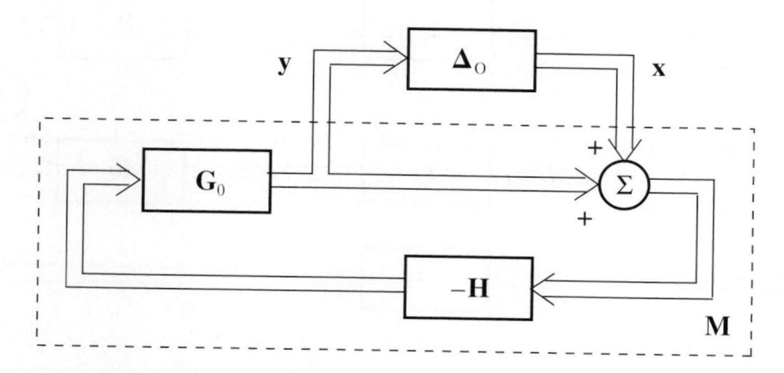

Figure 6.27 Feedback control system with output uncertainty in the plant, arranged in the form of the block diagram shown in Fig. 6.26.

which is the multichannel version of the complementary sensitivity function, $\mathbf{T}_0$, for the nominal plant. The condition for robust stability in the case of unstructured output uncertainty can thus be written, from equation (6.8.24), as

$$\bar{\sigma}\left(\mathbf{T}_0\left(e^{j\omega T}\right)\right) < \frac{1}{B\left(e^{j\omega T}\right)} \qquad \text{for all } \omega T , \tag{6.8.28}$$

or

$$\|\mathbf{T}_0 B\|_\infty < 1 . \tag{6.8.29}$$

To illustrate the flexibility of the small-gain theorem, we now assume that only some elements of the singular value matrix of the plant are subject to uncertainties, as discussed in Section 4.4. The block diagram of the complete control system with uncertainty in $\mathbf{\Sigma}_1$ is shown in Fig. 6.28, where the nominal plant response has been decomposed into the form

$$\mathbf{G}_0 = \mathbf{R}\Sigma\mathbf{Q}^{\mathrm{H}} = [\mathbf{R}_1\,\mathbf{R}_2]\begin{bmatrix}\mathbf{\Sigma}_1 & 0\\ 0 & \mathbf{\Sigma}_2\end{bmatrix}\begin{bmatrix}\mathbf{Q}_1^{\mathrm{H}}\\ \mathbf{Q}_2^{\mathrm{H}}\end{bmatrix} . \tag{6.8.30}$$

The matrix $\mathbf{\Sigma}_1$ is diagonal, but does not necessarily contain the largest singular values of $\mathbf{G}_0$. The uncertainty matrix $\mathbf{\Delta}$ in Fig. 6.28 is not assumed to be diagonal, however. In this case Fig. 6.28 can be used to show that the transfer function of the complete control system as seen by the uncertainty is equal to

$$\mathbf{M} = -\mathbf{Q}_1^{\mathrm{H}}\,\mathbf{H}[\mathbf{I}+\mathbf{G}_0\,\mathbf{H}]^{-1}\,\mathbf{R}_1 . \tag{6.8.31}$$

The robust stability of this control system is thus determined by whether the largest singular value of equation (6.8.31) multiplied by the upper bound on the largest singular value of $\mathbf{\Delta}$ in this case is less than unity for all frequencies.

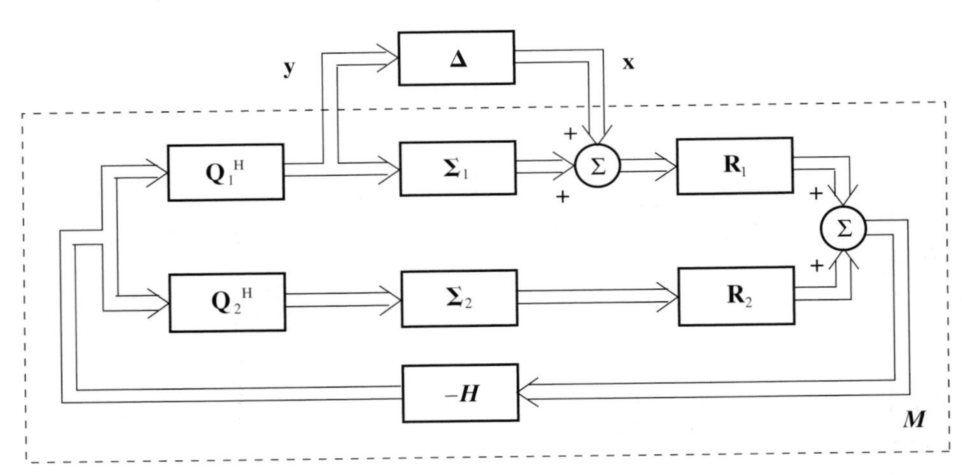

Figure 6.28 Feedback control system with uncertainty in some of the singular values of the plant, arranged in the form of the block diagram shown in Fig. 6.26.

6.9. OPTIMAL MULTICHANNEL CONTROL

The direct design of the matrix of feedback controllers, $\mathbf{H}(z)$ in equation (6.7.3), to minimise the norm of the error signal vector, $\mathbf{e}$, presents more problems for multichannel systems than single-channel systems. Fortunately, the use of an internal model control structure for the feedforward controller carries across directly to the multichannel case, as illustrated in Fig. 6.29(a). The plant model, $\hat{\mathbf{G}}(z)$ has the same dimensions,

$L \times M$, as the plant, $\mathbf{G}(z)$, and the control filter matrix $\mathbf{W}(z)$ is of dimensions $M \times L$. The matrix of responses for the complete feedback controller can be written, using Fig. 6.29(a), as

$$\mathbf{H}(z) = -\left[\mathbf{I} + \mathbf{W}(z)\hat{\mathbf{G}}(z)\right]^{-1} \mathbf{W}(z) . \tag{6.9.1}$$

We can therefore adopt the same strategy for calculating the optimum performance as was used above, in which $\mathbf{W}(z)$ is initially designed to give the optimum nominal performance for a fixed, nominal plant, and the question of robustness is then dealt with as a modification to this design.

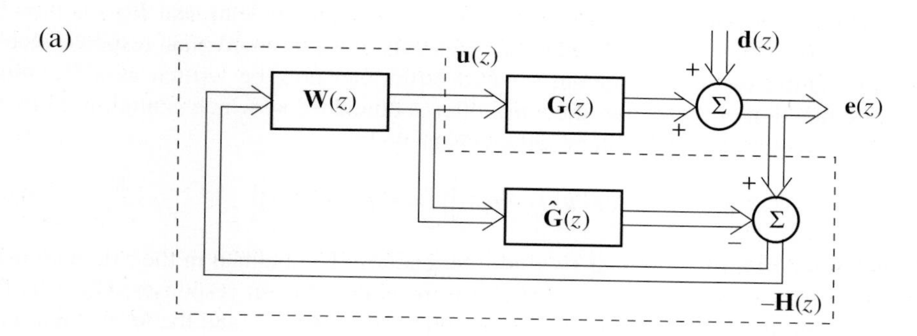

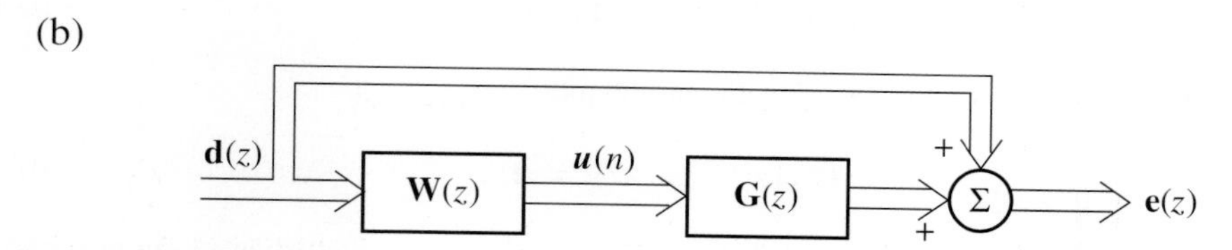

Figure 6.29 Multichannel feedback control system in which the controller is implemented using internal model control (a), and its equivalent block diagram if the plant model is perfect (b).

6.9.1. Nominal Performance

To calculate the nominal performance, we first assume that the response of the plant model exactly matches that of the nominal plant, $\hat{\mathbf{G}}(z) = \mathbf{G}_0(z)$, so that the equivalent block diagram is as shown in Fig. 6.29(b). The vector of error signals can now be written in terms of the vector of disturbance signals as

$$\mathbf{e}(z) = \left[\mathbf{I} + \mathbf{G}(z)\mathbf{W}(z)\right]\mathbf{d}(z) , \tag{6.9.2}$$

so that comparing equation (6.9.2) with equation (6.7.3), which defines the matrix sensitivity function, we find that with the IMC arrangement, the matrix sensitivity function becomes

$$\mathbf{S}(z) = \mathbf{I} + \mathbf{G}(z)\mathbf{W}(z) . \tag{6.9.3}$$

The entirely feedforward arrangement shown in Fig. 6.29(b) is a special case of the multichannel optimisation problem considered in Section 5.2, in which the vector of reference signals is now equal to the vector of disturbances. In the case of a feedback controller, however, we are not faced with the limitations of performance imposed by coherence, since the reference signals are now, by definition, completely coherent with the disturbance signals, but we must still account for the constraints of causality.

If each of the individual filters in the matrix $\mathbf{W}(z)$ is an FIR device, with I coefficients, the analysis used in Section 5.2 can be used to show that the vector of time-domain error signals can be written as

$$\boldsymbol{e}(n) = \boldsymbol{d}(n) + \boldsymbol{R}(n)\boldsymbol{w} \,, \tag{6.9.4}$$

where $\boldsymbol{w}$ is the concatenated vector of $M \times L \times I$ filter coefficients and $\boldsymbol{R}(n)$ is now the matrix of *disturbance* signal filtered by the elements of the nominal plant response matrix. The expectation of the sum of the squared errors can now be written as a Hermitian quadratic function of the vector of control filter coefficients, $\boldsymbol{w}$, which is minimised by the H_2-optimal vector of control filter coefficients given by

$$\boldsymbol{w}_{\text{opt}} = -\left\{E\left[\boldsymbol{R}^{\text{T}}(n)\boldsymbol{R}(n)\right]\right\}^{-1} E\left[\boldsymbol{R}^{\text{T}}(n)\boldsymbol{d}(n)\right] . \tag{6.9.5}$$

An expression for the optimal H_2 controller can also be obtained in the z domain using the results of Section 5.3. The $L \times M$ matrix of nominal plant responses, $\mathbf{G}_0(z)$, is first factorised into the $L \times M$ matrix of all-pass components, $\mathbf{G}_{\text{all}}(z)$, and the $M \times M$ matrix of minimum-phase components, $\mathbf{G}_{\min}(z)$, so that

$$\mathbf{G}_0(z) = \mathbf{G}_{\text{all}}(z)\mathbf{G}_{\min}(z) \,, \tag{6.9.6}$$

where $\mathbf{G}_{\text{all}}(z)$, $\mathbf{G}_{\min}(z)$ and $\mathbf{G}_{\min}^{-1}(z)$ are stable and

$$\mathbf{G}_{\text{all}}^{\text{T}}\left(z^{-1}\right) \mathbf{G}_{\text{all}}(z) = \mathbf{I} \,. \tag{6.9.7}$$

We also assume that the spectral density matrix for the disturbance signals can be factorised so that

$$\mathbf{S}_{dd}(z) = \mathbf{F}(z)\mathbf{F}^{\text{T}}(z^{-1}) \,, \tag{6.9.8}$$

where both $\mathbf{F}(z)$ and $\mathbf{F}^{-1}(z)$ are stable and causal. Since for the equivalent feedforward arrangement shown in Fig. 6.29(b), the reference signals, normally $\boldsymbol{x}(n)$, are equal to the disturbance signals, $\boldsymbol{d}(n)$, then in this case the matrix $\mathbf{S}_{xd}(z)$ in equation (5.3.20) becomes equal to $\mathbf{S}_{xx}(z)$, which is itself equal to $\mathbf{S}_{dd}(z)$.

The matrix of transfer functions for the H_2-optimal causal control filters can now be written as a special case of equation (5.3.31) as

$$\mathbf{W}_{\text{opt}}(z) = \mathbf{G}_{\min}^{-1}(z)\left\{\mathbf{G}_{\text{all}}^{\text{T}}(z^{-1})\mathbf{F}(z)\right\}_{+} \mathbf{F}^{-1}(z) \,, \tag{6.9.9}$$

from which the transfer function of the matrix of complete feedback controllers can be calculated using equation (6.9.1)

Simple methods of improving the robust performance by reducing the gain of the control filter can be derived as a generalisation of the single-channel case discussed in Sections 6.5 and 6.6. The controller could be made more robust to output uncertainty, for example, by including some frequency-weighted sensor noise in the measured disturbance signal before taking the spectral factors of its spectral density matrix. We have seen, however, that the matrix whose norm has to be limited for robust stability, in equation (6.8.24), depends very much on the structure of the plant uncertainty in the multichannel case. It should finally be mentioned that the H_∞ optimal controller would minimise the ∞-norm of the matrix sensitivity function

$$\|\mathbf{S}\|_\infty = \sup_\omega \bar{\sigma}\left[\mathbf{S}\left(e^{j\omega T}\right)\right] . \tag{6.9.10}$$

Even with the simplified form of the matrix sensitivity function given by equation (6.9.3) the problem of calculating a set of control filters $\mathbf{W}(z)$ to minimise equation (6.9.10) while maintaining the robust stability constraint, equation (6.8.25), is not straightforward and the interested reader is referred to Morari and Zafiriou (1989), Skogestad and Postlethwaite (1996) or Zhou et al. (1996) for further discussion.

6.10. APPLICATION: ACTIVE HEADREST

The most widespread use of feedback systems for active sound control is probably in active headsets. Current designs generally use analogue feedback loops, to minimise the loop delay and thus maximise the control bandwidth, but there is also potential for using a combination of fixed analogue and adaptive digital controllers. The issues involved in the design of adaptive feedback systems are not described until Chapter 7, however, and so a discussion of active headsets is postponed until the end of that chapter.

As an example of the more analytical design of feedback controllers, the use of such controllers in an active headrest, of the type shown in Fig. 6.30, is described here. The aim of such a system is to attenuate the acoustic disturbance at the microphone and hence generate a 'zone of quiet' around the microphone, which would include the ear positions of a listener whose head is also shown in Fig. 6.30. Such an active headrest was originally proposed by Olsen and May in 1953, and the acoustical principles of its operation were discussed in Chapter 1. The design of a practical controller for such an application involves

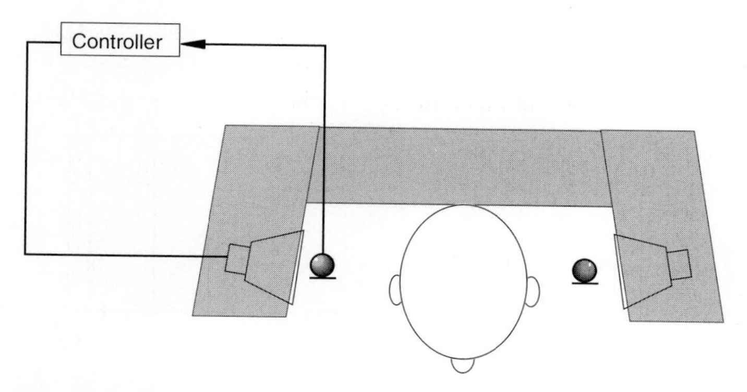

Figure 6.30 Plan view of one channel of an active headrest system showing the secondary loudspeaker, microphone, and the listener's head in the central position.

a detailed and interactive analysis of both the acoustic and control aspects of the problem, as described for example by Rafaely and Elliott (1999) and Rafaely et al. (1999). In principle, such a control system is also multichannel, since it has two sensors and two actuators on each seat. In practice, however, the response of a microphone to the closest loudspeaker is so much larger than its response to the other loudspeaker that the cross-coupling has very little effect on the stability and performance of the individual control systems, which can generally be designed as two independent single-channel systems. In this section, we will consider the design of a fixed single-channel controller for this application as described by Rafaely (1997) and Elliott and Rafaely (1997), which gives an optimum trade-off between performance, defined here to be attenuation of the microphone output, and robustness, particularly the robust stability of the system to changes in the plant response due to movements of the listener's head.

6.10.1. Response of the Plant and Plant Uncertainty

Figure 6.31 shows the frequency response of the nominal plant in this case, measured from the input of the loudspeaker to the output of the microphone, with the listener's head in the central position. The microphone was placed 2 cm from the front of the loudspeaker, which was arranged to be at the side of the listener's head, as shown in Fig. 6.30. The frequency response falls off at low frequencies and has a broad peak at about 250 Hz due to the loudspeaker response. It also displays a series of well-damped peaks between 1 kHz and 10 kHz, which are due to acoustic resonances between the headrest and the head. The most pronounced feature of the phase response is its linearly increasing lag with frequency, which is mainly due to the acoustic propagation delay between the loudspeaker and microphone.

The frequency response between the loudspeaker and microphone was also measured with the listener's head in a number of other positions, in order to assess the uncertainty in

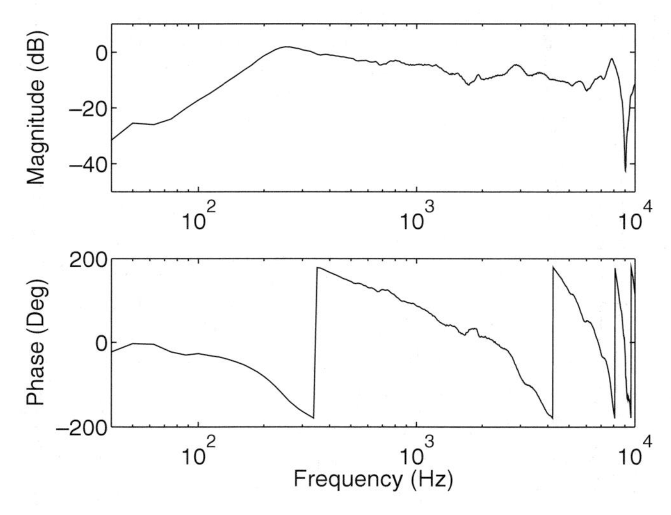

Figure 6.31 Frequency response of the plant, between secondary loudspeaker and microphone, for the headrest system.

the plant response due to head movement. Figure 6.32 shows the Nyquist plot of the nominal plant response, together with the scatter of responses at particular frequencies, obtained from the set of plant responses with various head positions. If this set of plant responses is described by a multiplicative uncertainty model,

$$G(j\omega) = G_0(j\omega)\left(1 + \Delta_G(j\omega)\right) \tag{6.10.1}$$

and the multiplicative uncertainty is assumed to be bounded, so that

$$\left|\Delta_G(j\omega)\right| \leq B(\omega) , \tag{6.10.2}$$

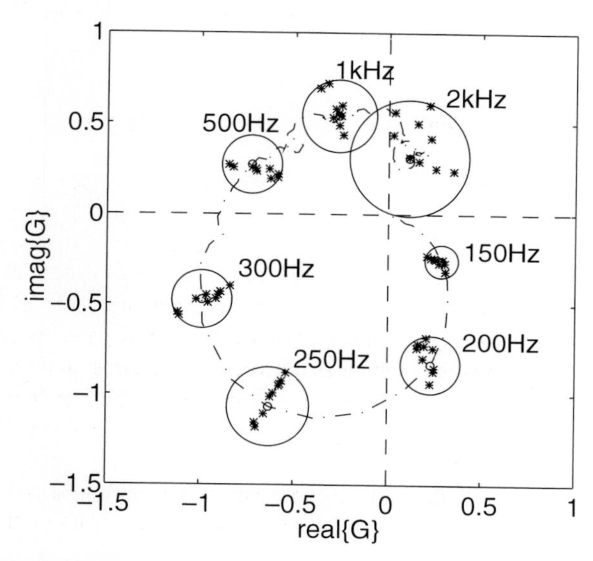

Figure 6.32 Nyquist plot of the nominal plant response for the headrest, with the listener's head in the central position (dot-dash line), and seven other measurements of plant response at various discrete frequencies, measured with the listener's head in various other positions (marked as *), together with the discs describing the smallest values of multiplicative plant uncertainty that describe these variations at each frequency.

at each frequency, then the bound can be represented as a circle in the Nyquist plot about the nominal response, as described in Section 6.2. Such a set of circles has been drawn in Fig. 6.32 to represent the bounds required to model the uncertainty in the observed set of plant responses using equations (6.10.1) and (6.10.2). The diameter of the circles is proportional to the size of the bound on the multiplicative uncertainty. This increases at higher frequencies because the details of the acoustic resonances that occur in this frequency region depend on the head position. It is the relatively large uncertainties at high frequencies which have the greatest effect on limiting the gain of the controller for robust stability. Although it could be argued that the uncertainty in the mid-frequency region is not as unstructured as is assumed by equation (6.10.2)—the uncertainty at 250 Hz, for example, is almost entirely due to gain changes—the high frequency uncertainties do have both gain and phase changes associated with them and so the unstructured multiplicative model for the uncertainty is a reasonable one in the most important frequency range for stability. A smooth curve was fitted to the uncertainty bounds obtained from Fig. 6.32 at a number of discrete frequencies, as shown in Fig. 6.33, which was used in the controller design outlined below.

Figure 6.34 shows the power spectral density of the assumed disturbance for this design, although in practice a measured disturbance spectrum could be used in the design, in which case all the parameters required for the optimisation described below could be obtained from direct measurements.

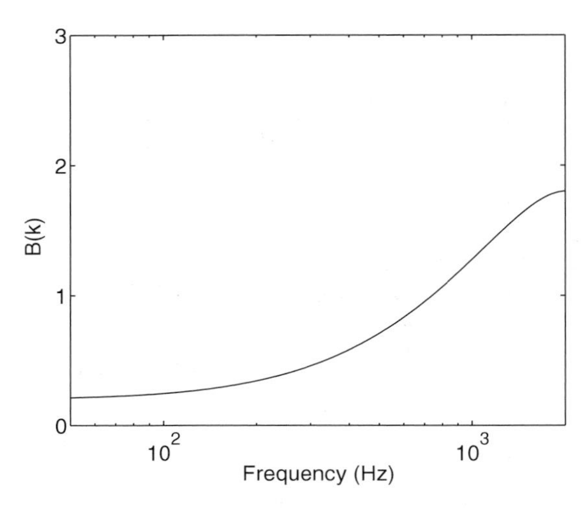

Figure 6.33 Assumed bound on the multiplicative plant uncertainty as a function of frequency, $B(k)$.

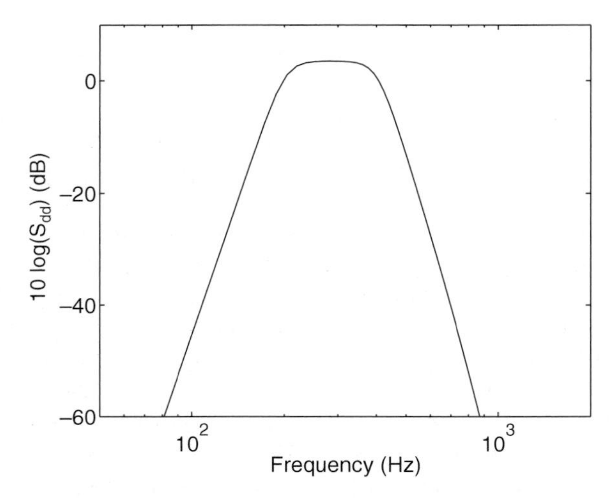

Figure 6.34 Power spectral density of the disturbance signal assumed for the controller design.

6.10.2. H_2/H_∞ Controller Design

The controller is designed by assuming an internal model control architecture with the control filter parametrised as a long FIR digital filter. The control filter is calculated that minimises the mean-square output of the microphone, an H_2 performance criterion, while maintaining robust stability and limiting the enhancement of the disturbance out of the control bandwidth, both of which are H_∞ constraints. The control problem thus has a mixed H_2/H_∞ form, but still remains a convex problem with respect to the coefficients of the FIR control filter. It is insightful to set up the optimisation problem in the discrete frequency domain (Boyd et al., 1988), in which case it can be expressed as seeking to minimise the mean-square output of the microphone under nominal conditions, which can be written using equation (6.5.5) as

$$J = \sum_{k=0}^{N-1} S_{dd}(k) \left|1 + G_0(k) W(k)\right|^2 , \tag{6.10.3}$$

where $S_{dd}(k)$ is the power spectral density of the disturbance, $G_0(k)$ is the frequency response of the nominal plant and $W(k)$ is the frequency response of the controller, all at the discrete frequency k, which ranges from 0 to $N-1$. This minimisation is subject to the constraint imposed by robust stability, which can be written following equation (6.4.6) as

$$\left|G_0(k) W(k)\right| < 1/B(k) \qquad \text{for all } k , \tag{6.10.4}$$

or

$$\left|G_0(k) W(k)\right| B(k) < 1 \qquad \text{for all } k , \tag{6.10.5}$$

and the constraint that the disturbance is not amplified by more than the factor $A(k)$, which can be written, using equation (6.4.3), as

$$\left|1 + G_0(k) W(k)\right| < A(k) \qquad \text{for all } k , \tag{6.10.6}$$

or

$$\left|1 + G_0(k) W(k)\right| / A(k) < 1 \qquad \text{for all } k . \tag{6.10.7}$$

In the design reported below, the disturbance enhancement was limited to 3 dB, so that $A \approx 1.4$. This convex optimisation problem was solved using an iterative gradient-descent method by Elliott and Rafaely (1997) and a sequential quadratic programming technique by Rafaely and Elliott (1999), which yielded essentially the same results. A similar approach to the design of H_2/H_∞ controllers has been taken by Titterton and Olkin (1995).

The modulus of the frequency response of the control filter, $|G(k)|$, the power spectral density of the residual error spectrum, $S_{ee}(k)$, the robust stability constraint, equation (6.10.5), and the disturbance enhancement constraints, equation (6.10.7), are all plotted in Fig. 6.35 after various numbers of iterations of the gradient-descent design process. After a large number of iterations, the control filter has converged to a reasonable approximation of the inverse of the plant response over the bandwidth of the disturbance, but its response outside of this bandwidth has been limited to prevent enhancement of the disturbance at low frequencies and to ensure robust stability at high frequencies. It can be seen from Fig. 6.35(c) and (d) that the constraints are just met in these two frequency

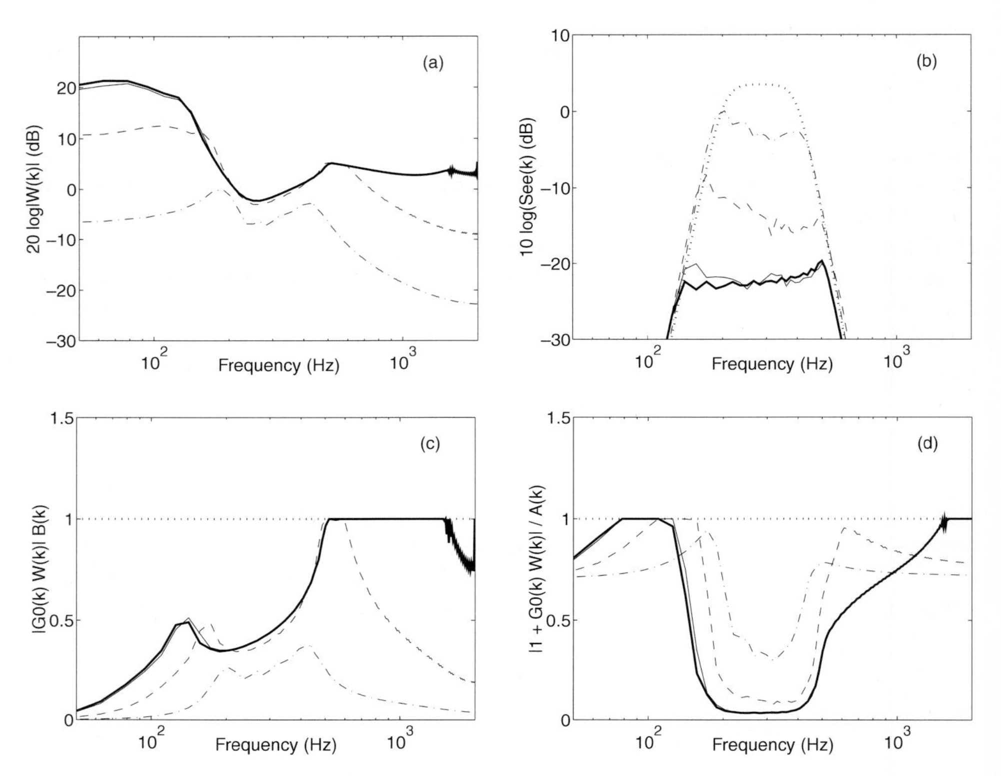

Figure 6.35 The results of the H_2/H_∞ controller design calculated using sequential quadratic programming (solid lines), before control (dotted line) and after 100 (dash-dotted line), 5,000 (dashed lines) and 150,000 (solid thin lines) iterations of the discrete-frequency domain iterative algorithm. The various graphs show: (a) the magnitude of the frequency response of the control filter in dB $\left(20 \log_{10} |W(k)|\right)$; (b) the level of the power spectral density of the residual error $\left(10 \log_{10} S_{ee}(k)\right)$; (c) the robust stability measure $\left(|G_0(k)W(k)|B(k)\right)$, which should be less than 1 at all frequencies for the robust stability constraint to be satisfied with a plant uncertainty of $B(k)$; and (d) the disturbance enhancement measure $\left(|1+G_0(k)W(k)|/A(k)\right)$, which should be less than 1 at all frequencies if the disturbance is not to be enhanced by more than a factor of $A(k)$.

regions. The rms output of the microphone was reduced by about 22 dB in this simulation.

6.10.3. Other Controller Designs

If no robustness constraint is imposed, slightly higher levels of attenuation are possible, as shown by the dashed line in Fig. 6.36(b), but then the controller is not robustly stable at frequencies of about 800 Hz, and enhances the disturbance by more than 6 dB at frequencies of about 100 Hz. If the controller had been designed to minimise a conventional LQG cost function, with a single effort weighting parameter, adjusted so that the constraints are approximately maintained, then the results, as shown by the thin solid

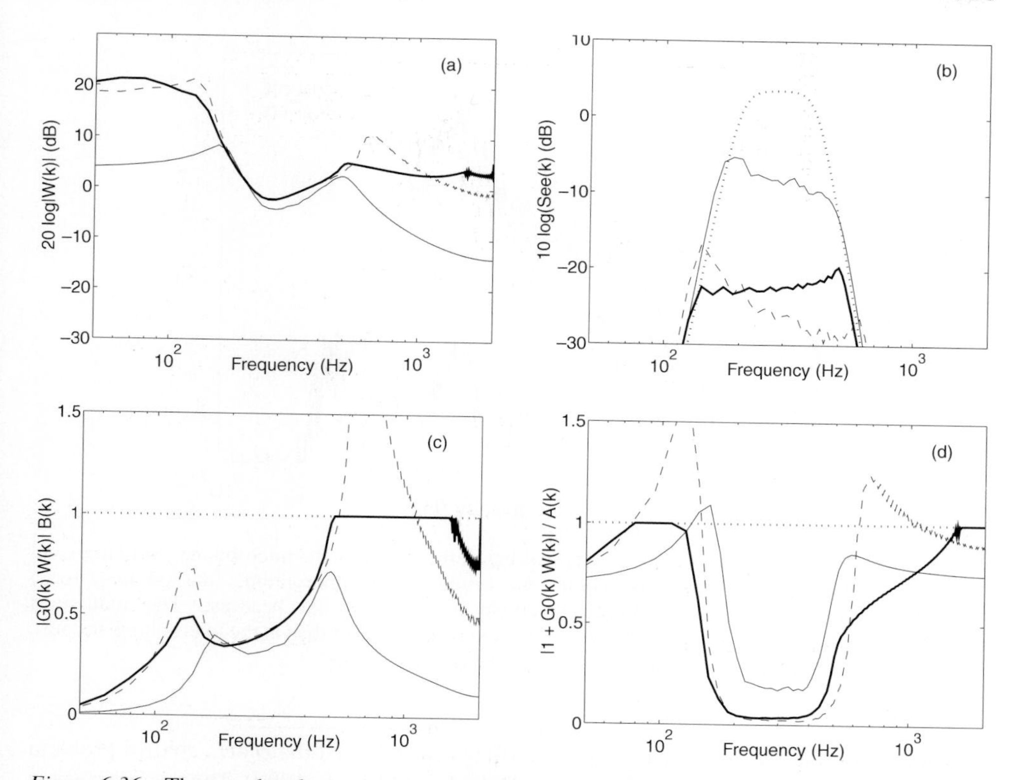

Figure 6.36 The results from the H_2 controller design with no robustness or disturbance enhancement constraints (dashed lines), the H_2 controller design with a frequency-independent effort weighting term, as in LQG theory (thin solid lines), and the H_2/H_∞ controller from above for reference (thick solid line). The various graphs shown are (a) the magnitude of the frequency response of the control filter in dB $\left(20 \log_{10} |W(k)|\right)$; (b) the level of the power spectral density of the residual error $\left(10 \log_{10} S_{ee}(k)\right)$; (c) the robust stability measure $\left(|G_0(k)W(k)|B(k)\right)$, which should be less than 1 at all frequencies for the robust stability constraint to be satisfied with a plant uncertainty of $B(k)$; and (d) the disturbance enhancement measure $\left(|1+G_0(k)W(k)|/A(k)\right)$, which should be less than 1 at all frequencies if the disturbance is not to be enhanced by more than a factor of $A(k)$.

lines in Fig. 6.36, indicate that only about 10 dB reduction in the mean-square error signal can be obtained. These results demonstrate that controllers designed only to minimise the mean-square error are, indeed, rather 'aggressive', with potentially poor robustness properties, and that the use of conventional LQG cost functions can significantly degrade the performance of the control system if the controller is to be robust. Although there are a number of methods of calculating optimal H_2/H_∞ controllers apart from those outlined above, this approach does give an optimal trade-off between the characteristics which are important in many active control applications, which are minimum mean-square error, robust stability, and limited out-of-band disturbance enhancement.

Finally, Fig. 6.37 shows the power spectral density of the signal measured at the microphone output when a digital IIR controller was implemented, whose frequency

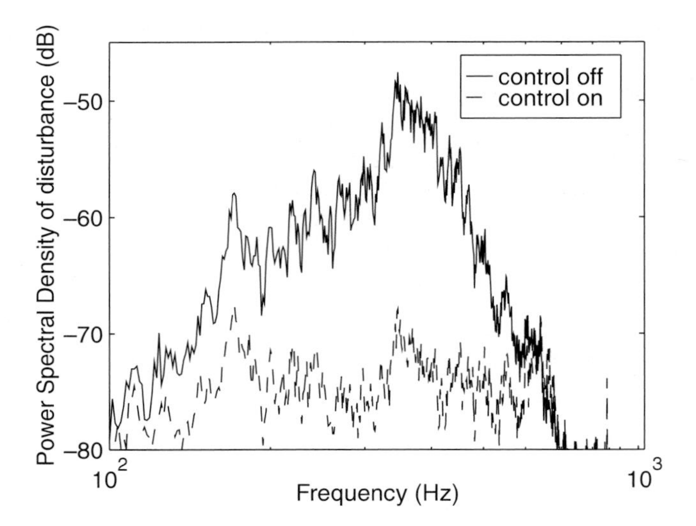

Figure 6.37 Power spectral density of the disturbance measured by the microphone (solid line) and the residual error after control measured by the microphone (dashed line) for a feedback controller implemented in an experimental headrest. The reductions experienced at the listener's ears are somewhat smaller than at the microphone because of the limitations imposed by the acoustic zone of quiet.

response was adjusted to fit that of the controller which solved the H_2/H_∞ control problem outlined above in an experimental headrest (Rafaely and Elliott, 1999). The performance is similar to that predicted in Fig. 6.35(b) except for the slightly different disturbance spectrum used in this case. It should be emphasised that these results represent the changes to the acoustic pressure at the microphone position next to the loudspeaker in the headrest. The reductions in pressure measured at other positions, particularly at the ears of the listener, are somewhat less than those shown above, because of acoustic effects due to the limited spatial extent of the zone of quiet around the control microphone, as described in Chapter 1. Better results can be obtained at the listener's ear by using a virtual microphone arrangement (Elliott and David, 1992; Garcia Bonito et al., 1997; Rafaely and Elliott, 1999) in which case the feedback loop is designed to minimise the output from a remote error microphone, which is used in the design of the controller but is removed during its operation. The block diagram for such a system was shown in Figs. 6.15 and 6.16 and the design of a suitable controller follows directly from the discussion at the end of Section 6.4. There is then a conflict in the design of such a system between placing the virtual microphone close to the listener's ear position, so that the zone of quiet is centred in the right place but the delay in the remote plant, $G_e(z)$ in Fig. 6.16, is rather large, and placing the virtual microphone close to the secondary loudspeaker, in which case the control performance is improved because the delay in the remote plant is reduced, but the zone of quiet will limit the acoustic performance at the listener's ear. The interaction between the design of the controller and the acoustic performance is further complicated by the fact that, as the listener's head moves, the plant response changes, but good attenuation is only ever required at the position of the listener's ears. These issues have been discussed in more detail by Rafaely et al. (1997, 1999).

As a result of all of these issues, the design of a practical active headrest is not as straightforward as outlined above. Particular attention must be paid to the interaction between the fundamental performance limitations imposed by the feedback control loop and those imposed by the behaviour of the acoustic field. This interaction depends critically on the physical arrangement of the headrest, the position of the loudspeaker and microphone with respect to that of the head, the disturbance spectrum and the objectives set for the control system. Finally, it should be reiterated that the performance shown in Fig. 6.37 was measured at the control microphone, and should not be taken as an indication of the attenuation actually experienced by a listener in a practical active headrest. The sensitivity of the performance to the factors outlined above suggests that it may not be practical to develop a general-purpose active headrest, but that individual designs may be more appropriate for particular applications.

7

Adaptive Feedback Controllers

7.1. INTRODUCTION

Having described the trade-offs and performance limitations inherent to any feedback controller in the previous chapter, this chapter is concerned with methods of making the feedback controller adaptive, in an attempt to maintain good performance under changing conditions. Even if the controller is not adaptive, we have seen that it is more difficult to specify the optimum controller for a feedback system than it is for a feedforward one, mainly because of the requirement that the system must be stable for all practical plant responses, i.e. it must be *robustly stable*. One of the aims of a fully-adaptive controller could be to make the feedback system stable over a greater range of plant responses by compensating for any changes in this response. An important general question is thus whether the requirement of robust stability can be dropped for a fully-adaptive controller, on the grounds that the adaptive algorithm could always re-identify the plant response after any change. Although the answer to this question depends on the details of the particular application, we will find that in active control, some level of robustness must still be maintained even if the feedback controller is adaptive.

In many active control applications, changes in the plant response can occur on a timescale that is short compared with the typical adaptation time of the controller, and this can severely limit the ability of the system to track such changes. An example of such a change in the plant response would be that caused by the motion of the passengers in a vehicle that was fitted with an active noise control system. In order to accurately track such changes in the plant response, the level of identification noise injected into the system would have to be so large that any benefit of active control in attenuating the original acoustic disturbance might be completely lost, as described in more detail in Section 3.6. In general then, even an adaptive feedback controller must be robustly stable, and methods of ensuring that this condition is satisfied are discussed in the following two sections, which rely heavily on the use of the internal model control architecture for the feedback controller.

7.1.1. Chapter Outline

Following a general discussion in the remainder of this section on the need for adaptation, Section 7.2 describes the use of the time-domain LMS algorithm in the adaptation of the control filter in an internal model control architecture for feedback systems. Differences between the response of the physical plant and the internal plant model are seen to provide a residual feedback loop around the adapting control filter, which can interfere with its convergence and even make it unstable under certain conditions. Some methods of avoiding such instability by adapting the filter in the frequency domain are presented in Section 7.3. The most promising approach involves the indirect adaptation of the control filter in the frequency domain, and then the on-line design of an efficient IIR filter to implement the complete feedback controller in the time domain. Such an approach begins to look more like the conventional adaptive control strategy of iterative controller redesign than the sample-by-sample adaptation that is more familiar in signal processing, and this is because of the overriding need to ensure closed-loop stability at all times.

After a brief diversion into the use of combined feedforward and feedback control in servo systems, Section 7.4 then considers the use of such combined systems in active control, where the advantages of the two systems can complement each other. The effect of an analogue feedback controller on the plant can be to make the overall system easier to control digitally, since sharp resonances can be damped to give a shorter transient response. This theme is developed in Section 7.5, in which the combination of an inner analogue feedback loop and an outer digital feedback loop is discussed. The analogue controller can reduce stationary broadband components in the disturbance, while the digital controller can be made adaptive to track nonstationary narrowband components. The application of such a combined analogue and digital feedback controller to an active headset is described at the end of Section 7.6, after a general discussion of the need for active control in headsets and the performance of fixed analogue feedback systems.

7.1.2. Feedback Loop and Adaptation Loop

We have already analysed the stability of an adaptive feedforward system in the face of plant uncertainty, and have also seen that most of the adaptive algorithms used in such applications require an internal model of the plant. This discussion was centred around the stability of the adaptive algorithm, whereas now we also have the possibility that the feedback loop itself might become unstable, even when the adaptation algorithm does not. If the feedback system is made slowly adaptive, there are thus two distinct mechanisms for instability, first due to the 'inner loop' with the feedback controller, and second due to the 'outer loop' which includes the adaptive control algorithm feeding back information from the error signal to update the controller, as illustrated in Fig. 7.1. Interaction between these two loops can make the analysis of adaptive feedback systems very complicated. Even if the feedback controller is adapted slowly, it is not easy to obtain general conditions under which this adaptation will not destabilise the feedback loop. No attempt will be made in this chapter formally to prove stability under such general conditions. We will restrict ourselves to discussing some architectures for adaptive feedback controllers that appear to be well suited to the active control problem and attempt to explain their behaviour in an intuitive manner.

The adaptation of feedback systems, or *adaptive control*, has been extensively studied for at least 50 years (Åström, 1987; Åström and Wittenmark, 1995), although there were some spectacular failures of these techniques applied to aircraft control in the 1960s, which were attributed to a lack of theoretical understanding of the algorithms being used. To quote Åström (1987) 'early work on adaptive flight control . . . was characterised by a lot of enthusiasm, poor hardware, and non-existent theory'. Considerable strides have been made in the theoretical and practical understanding of adaptive control since that time. Most examples of adaptive control in the literature are concerned with servo systems, for

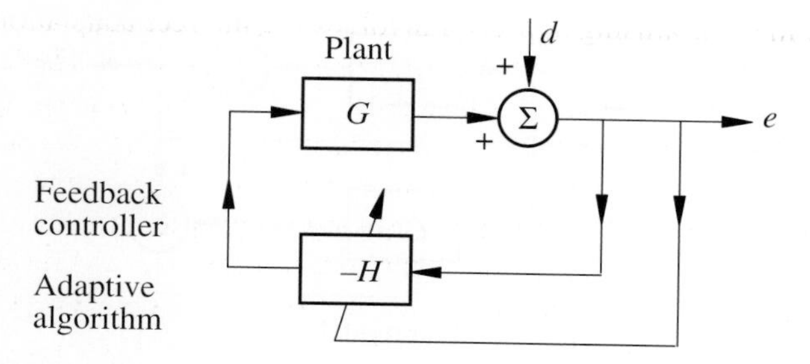

Figure 7.1 Block diagram of an adaptive feedback controller for disturbance rejection showing the inner feedback loop, due to the feedback controller, and the outer feedback loop, due to the adaptive algorithm.

which changes in the plant response are responsible for most of the requirement to adapt the controller. 'Adaptive control' in the control literature thus generally implies that the controller is being adjusted in response to a change in the *plant*. This is either achieved 'directly' without an explicit model of the plant or 'indirectly' by identifying a model of the plant and then using this to calculate the new controller.

7.1.3. Adaptation to Non-Stationary Disturbances

The emphasis on disturbance rejection in active control means that there is a need for feedback controllers that are adaptive, even if the plant response does not change. This is because the response of the optimum controller depends on the spectral properties of the disturbance, as well as on the plant response. If the disturbance were tonal, for example, very good attenuation could be achieved with a negative feedback controller having a high gain at the frequency of the tone but low gain at other frequencies. If the frequency of the tonal disturbance were to change, however, this controller would not give any attenuation, and the peak in its frequency response would have to be moved to again ensure good rejection of the disturbance. This form of adaptive feedback controller thus does not explicitly adapt to changes in the plant but is designed to maintain performance in the face of changes in the disturbance.

It is very difficult to directly adapt the parameters of a feedback controller as suggested in Fig. 7.1. This is particularly true if the controller is implemented using analogue components. In this chapter we will thus concentrate on the adaptation of digital feedback controllers, which we will restrict to being single channel (SISO). We have seen in Section 6.4 how the feedback control problem can be transformed into an equivalent feedforward control structure using internal model control (IMC) and a perfect plant model. If the control filter in the IMC controller is implemented as an FIR filter, the mean-square error will thus be a quadratic function of the coefficients of such a filter, and the filter can be adapted using simple gradient descent algorithms, exactly as described in Chapter 3. One of the simplest forms of digital adaptive feedback controller is thus as illustrated in Fig. 7.2, in which only the control filter is adapted to minimise the mean-square output

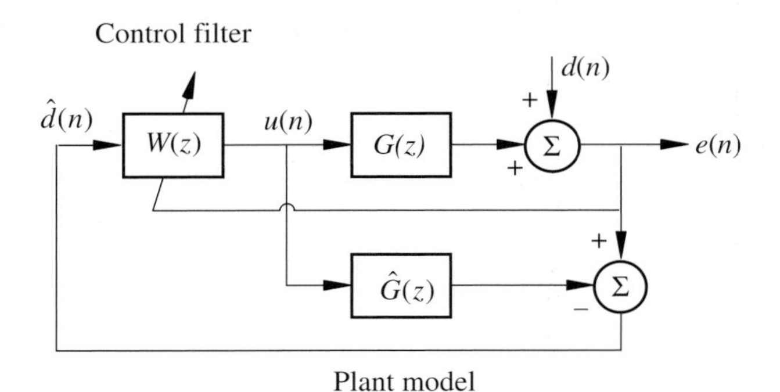

Figure 7.2 The simplest form of adaptive feedback controller for non-stationary disturbances using an internal model control (IMC) architecture, with only the control filter $W(z)$ adapted to minimise the output error.

error. The use of adaptive techniques using this feedback controller architecture was originally suggested by Walach and Widrow (1983) and Widrow (1986).

A convenient and simple algorithm for the adaptation of the control filter would be the filtered-reference LMS algorithm described in Chapter 3. The use of time-domain and frequency-domain versions of this algorithm in a feedback controller is described in the following two sections. Their behaviour is not the same as for the entirely feedforward controllers described in Chapter 3, however, because of the effects of differences between the responses of the plant and plant model. This introduces a level of residual feedback around the adaptive filter that complicates its convergence properties, as we shall see. The difference between the responses of the plant and the plant model generally arises because of changes in the response of the plant. Provided that both the feedback loop and the adaptive algorithm remain stable for such changes, which we will see is true provided that the changes in the plant are small in some sense and suitable modifications are made in the adaptive algorithm, then the simple adaptive controller will not only compensate for changes in the properties of the disturbance, but will also compensate to some extent for small changes in the plant response.

7.1.4. Adaptation to Changes in Plant Response

If the changes in the plant response are too large to allow the simple system shown in Fig. 7.2 to be stabilised, it is possible to use the more complete adaptive feedback controller shown in Fig. 7.3. In this arrangement the control filter is adapted to minimise the output error, $e(n)$, as before, and the plant model is adapted to minimise the modelling error, which in this case happens to be equal to the estimated disturbance, $\hat{d}(n)$. Such fully-adaptive IMC methods have been described by Datta and Ochoa (1996, 1998) and Datta (1998).

Since we are concerned with disturbance cancellation and have no independent estimate of the disturbance, then identification noise, which is shown as $v(n)$ in Fig. 7.3, must be

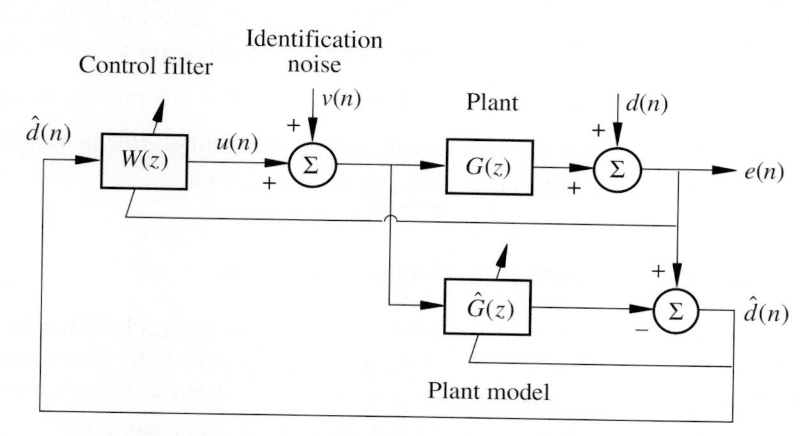

Figure 7.3 A possible arrangement for a complete form of adaptive feedback controller using IMC, in which the control filter is adapted to minimise the output error and the plant model is adapted to minimise the modelling error. Note that with this arrangement, the plant model $\hat{G}(z)$ is not guaranteed to converge to the true plant $G(z)$.

injected into the system for identification. The identification noise is assumed to be uncorrelated with the disturbance and to be persistently exciting. If the control loop were being used for servo-control, and the command signal were sufficiently broadband, then the command signal could be used for identification, as described for example by Widrow and Walach (1996). The adaptation of the plant model is complicated by the feedback path, however, as described by Gustavsson et al. (1977), for example. This can be illustrated by calculating an expression for the modelling error in this case.

Assuming that the control filter and plant model are fixed, the z transform of the modelling error can be written, using Fig. 7.3, as

$$\hat{D}(z) = D(z) + \left[G(z) - \hat{G}(z)\right]\left[V(z) + U(z)\right] , \tag{7.1.1}$$

where $U(z)$ is the output of the control filter and can be written as

$$U(z) = W(z)\hat{D}(z) . \tag{7.1.2}$$

Substituting equation (7.1.2) into (7.1.1) we obtain

$$\hat{D}(z) = \frac{1}{1-\left[G(z)-\hat{G}(z)\right]W(z)} D(z) + \frac{G(z)-\hat{G}(z)}{1-\left[G(z)-\hat{G}(z)\right]W(z)} V(z) . \tag{7.1.3}$$

Clearly in this case, $\hat{D}(z)$ is not a linear function of $\hat{G}(z)$ and so there may be problems with adapting an FIR plant model to minimise its mean-square value, particularly if $\hat{G}(z)$ is not initially similar to $G(z)$. Even if the adaptation process is stable, however, the plant model $\hat{G}(z)$ will not in general converge to $G(z)$, partly because the disturbance is now filtered by a function of $\hat{G}(z)$, as discussed by Widrow and Walach (1996). The identified plant model will thus be biased because the adaptive algorithm is able to minimise the mean-square value of the contribution from the disturbance in equation (7.1.3), as well as that from the identification noise. These problems do not occur in the identification of purely feedforward systems, as discussed in Chapter 3, since in this case the reference signal is provided externally, rather than being derived from the error as the disturbance estimate. The recursive term $1/\left(1-\left[G(z)-\hat{G}(z)\right]W(z)\right)$ is thus not present in equation (7.1.3) when identifying a feedforward signal, and the modelling error in Chapter 3 is linearly proportional to the coefficients of the FIR plant model $\hat{G}(z)$.

7.1.5. Closed-Loop Identification of Plant Response

One approach to the unbiased identification of the plant, when the feedback loop is closed with a fixed controller, is described as the joint input–output method by Gustavsson et al. (1977). In this method, the response of the measured plant output to identification noise inputs is first estimated, which from Fig. 7.3 and with a fixed $\hat{G}(z)$ will give

$$F_1(z) = \frac{E(z)}{V(z)} = \frac{G(z)}{1-\left[G(z)-\hat{G}(z)\right]W(z)} . \tag{7.1.4}$$

Since the identification noise is assumed to be uncorrelated with the disturbance, the disturbance signal will not bias a least-squares estimate of $F_1(z)$, although considerable averaging may have to be used to get an estimate with small random errors if the level of the identification noise is low compared with that of the disturbance. Iterative adjustment of F_1, and copying of F_1 to $\hat{G}$ is suggested as a practical identification scheme by Widrow and Walach (1996), but it is not clear whether the convergence of such a process can always be guaranteed.

The response of the plant input signal to identification noise inputs is then separately estimated, which from Fig. 7.3 will give

$$F_2(z) = \frac{U(z)}{V(z)} = \frac{1}{1-\left[G(z)-\hat{G}(z)\right]W(z)} . \tag{7.1.5}$$

The true plant response is thus equal to the ratio

$$\frac{F_1(z)}{F_2(z)} = G(z) . \tag{7.1.6}$$

Rafaely and Elliott (1996a) have described a way in which each of these identification processes can be performed with separate adaptive filters.

When both the control filter and the plant model are adapted at the same time, the best choice of the relative timescales for the adaptation of the two filters will depend strongly on the particular application, as discussed in Section 3.6. If the plant response can change rapidly, the timescale of adaptation for the plant model should also be made fast, to ensure that the algorithm adapting the control filter does not become unstable. The feedback system implemented using IMC is guaranteed to be stable provided that the plant model is an accurate one. Thus, both the feedback loop and the adaptation of the control filter can be made stable if the plant identification is performed rapidly. Unfortunately, such rapid identification requires high levels of identification noise, which are then fed to the secondary actuator, and can increase the mean-square value of the signal at the error sensor, in spite of any attenuation of the disturbance which the feedback signal may achieve.

In some active control applications, significant changes in the plant response only occur over timescales that are long compared with the adaption time required of a typical controller. It is possible to 'track' these slow plant changes with an identification algorithm using very low levels of identification noise, as described in Section 3.6, and thus achieve good long-term performance. In other applications the changes in the plant response can be rapid but not very large, in which case some level of performance can still be maintained by making the adaptation of the control filter robust to such changes, as discussed below. Considerable difficulties remain in the design of a reliable adaptive system for the active control of a system whose changes in response are both large and rapid.

7.2. TIME-DOMAIN ADAPTATION

We have seen in Chapter 3 that the dynamics of a feedforward control system that simultaneously identifies and controls a plant can be very complicated. These dynamics are further complicated by the feedback loop in a feedback control system. The combined behaviour of two adaptive processes, for plant identification and control filter adjustment, and a feedback loop is thus too involved to be easily modelled in any general way. The approach adopted in this section is to consider the behaviour of the filtered-reference LMS algorithm in adapting only the control filter in an IMC system, assuming that the plant model is time-invariant.

In practice, the adaptation processes of plant identification, described above, and control filter adjustment, discussed here, could always be performed sequentially. If each of these processes can be shown to be independently stable and to converge to the optimum result, the complete system would clearly perform reliably, even if it were not as fast as a simultaneously adapting system.

7.2.1. Effect of Plant Modelling Error on an Adaptive Control Filter

The adaptation of the FIR control filter $W(z)$ in Fig. 7.2 will now be considered. We will assume that the changes in the response of the control filter occur on a timescale that is long compared to that of the plant dynamics, so that the changes are quasi-static. From Fig. 7.2, we can see that the estimated disturbance signal can be written as

$$\hat{D}(z) = D(z)+\left[G(z)-\hat{G}(z)\right]U(z) , \tag{7.2.1}$$

where $U(z)$ is the output of the control filter $W(z)$, and the error signal can be written as

$$E(z) = D(z)+G(z)\,W(z)\,\hat{D}(z) . \tag{7.2.2}$$

The internal model control arrangement shown in Fig. 7.2 can thus be redrawn as in Fig. 7.4. If the control filter is completely static, the ratio of the error signal to the disturbance, which is equal to the sensitivity function, can be deduced from this block diagram as being equal to

$$S(z) = \frac{E(z)}{D(z)} = 1+\frac{G(z)\,W(z)}{1-\left[G(z)-\hat{G}(z)\right]W(z)} , \tag{7.2.3}$$

so that

$$S(z) = \frac{1+\hat{G}(z)\,W(z)}{1-\left[G(z)-\hat{G}(z)\right]W(z)} , \tag{7.2.4}$$

as given for the usual IMC arrangement in equation (6.4.2).

The block diagram of the controller shown in Fig. 7.4 emphasises that adaptation of the control filter can be viewed as if it were in a feedforward configuration, although some level of residual feedback is present round the control filter due to the imperfect plant model. This is exactly the same as would occur if an imperfect feedback cancellation filter had been used in the feedforward system of Fig. 3.11 in Section 3.3. Clearly, if the plant model is perfect, so that $\hat{G}(z) = G(z)$, then the system is entirely feedforward. The system would then be stable for all possible values of the coefficients of the FIR control filter, and

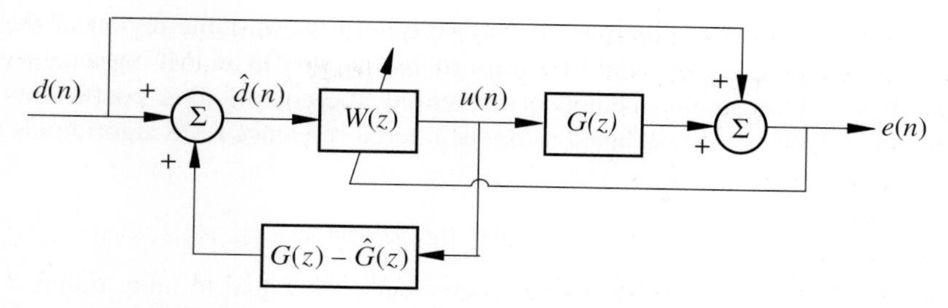

Figure 7.4 Another form of the block diagram for the IMC system with an adaptive control filter, as shown in Fig. 7.2.

the error surface obtained by plotting the mean-square error against the control filter coefficients, would be entirely quadratic. We now imagine what would happen to this error surface if the response of the plant model is not exactly equal to that of the plant. The 'loop gain' around the control filter can be written as

$$L(z) = \left[\hat{G}(z) - G(z)\right] W(z) \,, \tag{7.2.5}$$

so that the spectrum of the error signal is equal to

$$E(e^{j\omega T}) = \frac{1 + \hat{G}(e^{j\omega T}) W(e^{j\omega T})}{1 + L(e^{j\omega T})} D(e^{j\omega T}) \,. \tag{7.2.6}$$

Provided the loop gain is much less than unity at all frequencies, then the feedback path around the control filter will have little effect on the error and thus little effect on the convergence of the control filter. Even if the plant model is very accurate, however, but its response is not exactly equal to that of the plant, the loop gain could still become comparable to unity if the control filter had very large coefficients, and there would then be a danger of instability. The error surface is thus relatively unaffected by small levels of plant modelling error provided that the filter coefficients are not too large, but at the extreme edges of the error surfaces, where the filter coefficients are very large, the error surface is no longer quadratic. The mean-square error will suddenly become infinite when the filter coefficients are so large as to make the system become unstable. As the degree of mismatch between the plant and plant model increases, this unstable region of the error surface will come closer and closer to the origin and also to the position of the optimum filter coefficients.

This region of instability on the error surface would be of concern, but not necessarily a disaster, provided that the algorithm used to adjust the control filter always converged directly to the minimum of the error surface. Unfortunately, when the filtered-reference LMS algorithm is used with random data, the trajectory of the coefficients along the error surface is subject to noise, and with plant modelling errors the average values of the filter coefficients after convergence will not exactly correspond to the optimum ones. We have seen in Chapter 3 that the filtered-reference algorithm may even become unstable if the response of the plant model is sufficiently different from that of the true plant. Thus, when using a filtered-reference LMS algorithm to adjust the control filter, care must be taken to ensure that the response of the filter does not become any greater than it needs to be to achieve reasonable control, otherwise it may stray into the unstable region of the error surface. This requirement is similar to that for robust stability in an IMC system, given by equation (6.4.6). One simple method of preventing the control filter coefficients from becoming too large when it is adapted using the filtered-reference LMS algorithm is to use a leakage term, so that

$$\mathbf{w}(n+1) = \gamma \mathbf{w}(n) - \alpha \hat{\mathbf{r}}(n) e(n) , \qquad (7.2.7)$$

where $\mathbf{w}(n)$ is the vector of control filter coefficients, γ is equal to unity minus a small leakage term, α is the convergence coefficient, and $\hat{\mathbf{r}}$ is the vector of filtered reference signals used in practice, which are obtained in this case by filtering the estimated disturbance signal by the estimated plant response.

7.2.2. The Error Surface with an Imperfect Plant Model

The fact that the filtered reference signal is generated by a model of the plant response, rather than by the plant response itself, means that the average gradient estimate used by equation (7.2.7) is not equal to the true one. Thus even if a true gradient descent algorithm would be stable in the case described above, the filtered-reference algorithm might not. In practice, however, it is generally found that errors in the plant model cause the feedback controller to become unstable because the adaptive control filter is pushed into the unstable part of the error surface, some time before the adaptive algorithm itself becomes unstable. This behaviour is illustrated in Fig. 7.5, which shows the error surfaces obtained in a series of simulations performed by Rafaely and Elliott (1996b). In this case the plant model was that measured for an active headset under nominal conditions, as described in Section 7.6, and the response of the plant itself was taken to be that measured on the headset under different conditions, when the fit of the headset against the ear was tighter, and by also including a number of samples of extra delay in the plant. The disturbance was a pure tone with 20 samples per cycle and the control filter had only two coefficients so that the error surface could be readily visualised. Figure 7.5(a) shows the error surface for this problem and the convergence of the filtered-reference LMS algorithm when the plant model is perfect. In this case the error surface is quadratic and the LMS algorithm converges to the global minimum.

If the plant response is equal to that of the headset pushed against the ear, then the new error surface and the behaviour of the adaptive algorithm are as shown in Fig. 7.5(b). If the coefficients of the control filter are in the region of the error surface marked by dots, then the feedback system is unstable. The difference between the phase response of the plant and that of the plant model in this case is 20° at the disturbance frequency, and although this does not cause the LMS algorithm to become unstable, it does converge to a sub-optimal solution.

If two samples of delay are added to the plant response, then the phase error between the plant and the plant model at the disturbance frequency increases to 56°, and the stable region of the error surface has decreased considerably, as shown in Fig. 7.5(c). Within the stable region, the error surface in this case is relatively flat, and so the transition from the stable to the unstable region of the error surface is very abrupt. The optimum values of the filter coefficients also lie very close to this boundary and so, as the LMS algorithm drives the filter

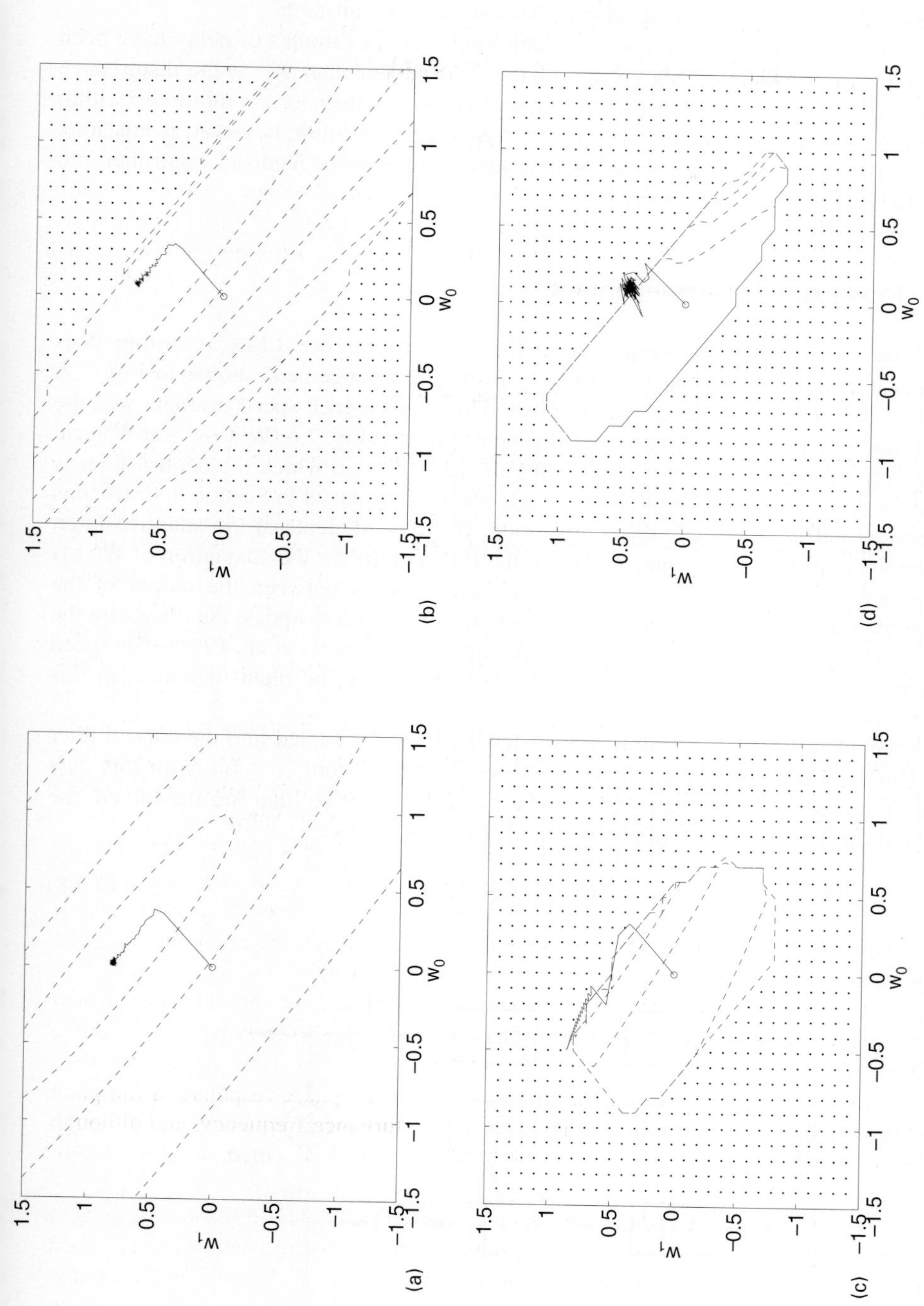

Figure 7.5 Error surfaces for an adaptive FIR control filter in an IMC feedback system with a fixed plant model and various plant responses. The dashed lines are contours of constant mean-square error, the dotted area denotes the part of the error surface corresponding to an unstable system, and the solid line indicates the convergence path of the filtered-reference LMS algorithm initialised at the coefficient values indicated by the circle. (a) Plant response equal to model. (b) Plant response changes to a tight headset fit. (c) Plant delay increased by 2 samples. (d) Plant delay increased by 6 samples.

coefficients towards their optimum value, the noise in the adaptation process pushes them over the stability boundary, and the adaptive system becomes unstable.

Finally, Fig. 7.5(d) shows the case in which a total of six samples of delay have been added to the plant model, which gives a phase error greater than 90° at the disturbance frequency. In this case the error surface is rather shallow but does have a minimum within the stable region, at about $w_0 = -0.3$, $w_1 = 0$. The LMS algorithm, however, is unstable, because of the large phase shifts in the plant model, and drives the feedback controller into an unstable part of the error surface.

7.2.3. Modified Error Arrangement

Even if the plant model is adequate and the filtered-reference LMS algorithm does converge, it may not converge very quickly using the arrangement shown in Fig. 7.2, because of the presence of delays in the plant. This convergence speed problem was the original motivation for an alternative arrangement shown in Fig. 7.6 (Widrow and Walach, 1996; Auspitzer et al., 1995; Bouchard and Paillard, 1998), in which the control filter is adapted using a form of the 'modified error' algorithm, introduced in Section 3.4, which is implemented outside the feedback control loop. If the coefficients of the adaptive filter, $W(z)$, are not copied into the control filter in the IMC controller, the adaptation of $W(z)$ is now guaranteed to be stable, since there are no dynamics between the output of the adaptive filter and the modified error, $e_m(n)$, which can be used to update the filter with the normal LMS algorithm, or potentially faster algorithms (Auspitzer et al., 1995). The speed of convergence of the adaptive filter is thus not limited by the plant dynamics in this arrangement.

After the adaptive filter has converged, its coefficients are copied into the control filter of the IMC feedback controller. If the feedback control system is stable with this new control filter, and assuming that the coefficients of the adaptive filter are then fixed, the expression for the z-transform of the modified error is given by

$$E_m(z) = \left[1 + \hat{G}(z)\,W(z)\right]\hat{D}(z) \;, \tag{7.2.8}$$

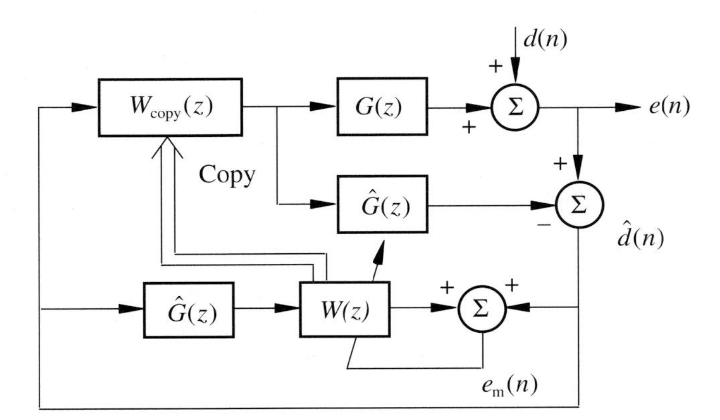

Figure 7.6 Alternative adaptation method for the control filter in an IMC arrangement, which is similar to the modified error method used for adaptive feedforward control.

where

$$\hat{D}(z) = \frac{D(z)}{1-\left[G(z)-\hat{G}(z)\right]W_{\text{copy}}(z)} . \tag{7.2.9}$$

But if $W_{\text{copy}}(z) = W(z)$, then the modified error is equal to

$$E_{\text{m}}(z) = \frac{1+\hat{G}(z)\,W(z)}{1-\left[G(z)-\hat{G}(z)\right]W(z)}\; D(z) , \tag{7.2.10}$$

which is exactly the same as the expression for the output error that can be obtained from equation (7.2.4). Thus, under steady state conditions the modified error is equal to the output error and any algorithm that minimises the modified error clearly also minimises the output error.

The modified error is equal to the output error if $W_{\text{copy}}(z) = W(z)$, but if this condition is imposed by copying the coefficients at every sample time, then the error surface seen by the adaptive filter reduces to that described in the previous section, with the danger of instability caused by the coefficients straying into the unstable region. If, on the other hand, the adaptive filter is allowed to converge completely before being copied, the convergence is guaranteed to be stable, but the modified error will no longer be equal to the output error after convergence. An iterative adaptation scheme is thus suggested, in which the adaptive filter is allowed to converge, its coefficients are copied into the control filter, and then it is allowed to converge again. There is, however, no guarantee that the feedback loop will be stable when each new set of control filter coefficients are copied across. The possibility of instability will be reduced, however, if the magnitude of the control filter response is not allowed to get too large, so that the magnitude of the 'loop gain' $L\left(e^{j\omega T}\right)$, in equation (7.2.5) does not get too close to unity at any frequency. This is very similar to the condition imposed on the controller by robust stability, as discussed in more detail below. Provided that the iterative procedure is stable and unbiased, however, then the difference between $W(z)$ and $W_{\text{copy}}(z)$ will decay to zero, and so the modified error minimised at each iteration becomes closer and closer to the output error, as indicated in equation (7.2.10).

Another approach, which would be applicable if the plant response were subject to slow but significant changes, is described by Datta and Ochoa (1998). They describe an IMC controller in which the plant model is periodically re-identified and then the control filter is designed to minimise the mean-square tracking error for a servo-control system, assuming that the current plant model is correct. This approach, in which the optimum controller is designed assuming that the current plant model perfectly represents the true plant, is an example of what is called the *certainty equivalence principle* in the control literature (Åström and Wittenmark, 1995).

7.3. FREQUENCY-DOMAIN ADAPTATION

We have seen in Chapters 3 and 4 that there are several advantages to adapting feedforward controllers using frequency-domain methods. When the control filter has many coefficients, frequency-domain adaptation is more computationally efficient than time-domain adaptation, and also some of the slow convergence problems associated with the time-domain LMS algorithm can be overcome by having individual convergence coefficients in each frequency bin, provided the convergence coefficient is split into its spectral factors for use inside and outside the causality constraint, so that the convergence is unbiased.

In this section, we briefly consider the controller architectures required for frequency-domain adaptation of a single-channel feedback controller. As well as having the advantages we have seen in feedforward controllers, frequency-domain adaptation also allows us to impose some important *constraints* on the adaptation of the control filter. Perhaps the most important of these is the constraint of robust stability, which, as we have seen in Chapter 6, can be naturally expressed as an inequality in the frequency domain and this inequality takes a particularly simple form for a feedback controller implemented using IMC.

We have also seen in Chapter 6 that any delay in the controller has an even more dramatic impact on the performance of a feedback control system than it does on a feedforward one. It is thus important that we continue to use the structure introduced in Chapter 3, in which the control filter is implemented directly in the time domain, but the adaptation algorithm is implemented in the frequency domain. We will also restrict ourselves to the adaptation of only the control filter, and assume that the plant model is fixed, even if the plant response does vary slowly and within specified limits.

7.3.1. Direct Adaptation of the Control Filter

The block diagram of a single-channel digital IMC feedback controller is shown in Fig. 7.7, in which the FIR control filter, whose coefficients are w_i, is implemented in the time-domain but is then adapted using the filtered-reference LMS algorithm in the frequency-domain. In the feedforward case, shown in Fig. 3.19, the updates for the time-domain FIR controller coefficients were obtained directly by taking the causal part of the inverse FFT of the estimated cross spectral density, $R^*(k)\,E(k)$. We will find that in the feedback case it is important to maintain a frequency-domain representation of the control filter that has been implemented. To do this, one could take the causal part of the update quantity in the frequency domain using the operation denoted $\{\ \}_+$ in Fig. 7.7, which involves an inverse FFT, windowing of the impulse response and another FFT, and update the control filter, $W(k)$, in the frequency domain. The time-domain controller is then obtained by taking the inverse FFT of $W(k)$, as illustrated in Fig. 7.7. A slightly more efficient method of maintaining a frequency-domain representation of the control filter would be to update the time-domain control filter directly, as shown for the feedforward controller in Fig. 3.19, which requires a single inverse FFT, and then to take the FFT of the impulse response of the time-domain control filter at every iteration.

The frequency-domain filtered-reference LMS algorithm is used in Fig. 7.7 to adjust the control filter to minimise the mean-square value of the error. The update equations follow directly from the feedforward case, equation (3.5.5), and can be written as

$$W_{\text{new}}(k) \;=\; W_{\text{old}}(k) \;-\; \alpha\left\{\hat{R}^*(k)\,E(k)\right\}_+ \;, \qquad (7.3.1)$$

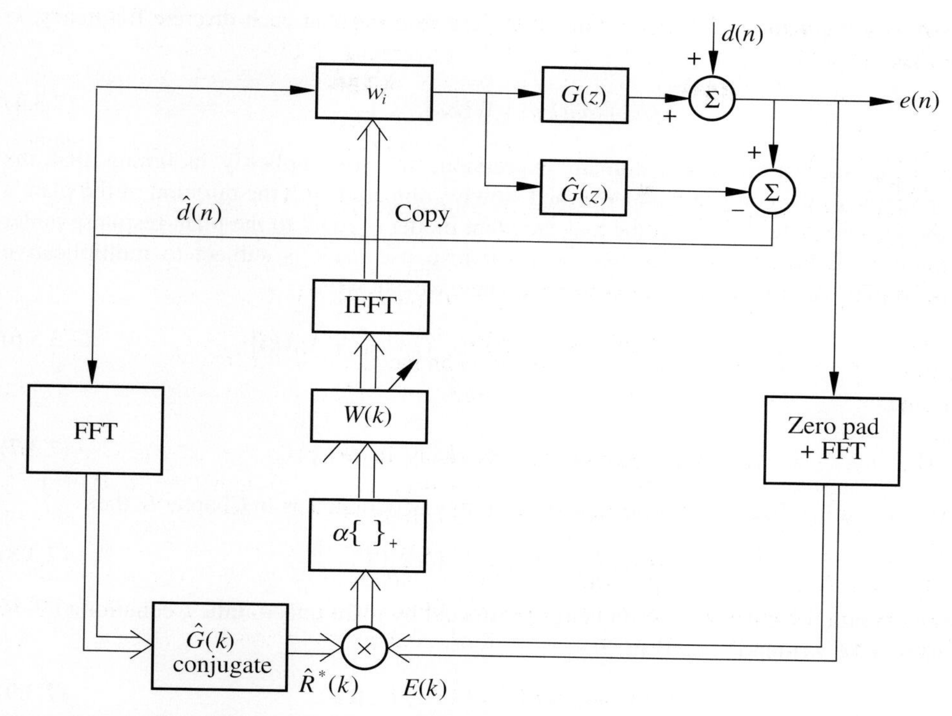

Figure 7.7 Block diagram of a feedback control system implemented using IMC in which the time-domain control filter w_i in the feedback loop is adapted using a frequency-domain technique to minimise the output error.

and

$$w_{i,\text{new}} = \text{IFFT}\left[W_{\text{new}}(k)\right] , \tag{7.3.2}$$

where $\hat{R}^*(k)$ is the conjugate of the FFT of the practical filtered reference signal, which in this case is equal to the estimate of the disturbance signal, filtered by a frequency-domain version of the plant model, so that

$$\hat{R}(k) = \hat{G}(k)\,\hat{D}(k) . \tag{7.3.3}$$

It has been assumed that the discrete-frequency response of the plant model, $\hat{G}(k)$, provides a good model of the plant, so that no resonance in the plant response will be missed by the frequency discretisation, for example. This will be assured provided that the block size of the FFT is at least twice as long as the number of samples needed to represent the significant part of the plant's impulse response. Precautions must also be taken to avoid circular convolution effects, as described in Chapter 3.

The effect of the residual feedback path, left by the differences between $G(z)$ and $\hat{G}(z)$ in the IMC arrangement, on the adaptation algorithm will be similar in the frequency domain to those described in Section 7.2 for the time-domain algorithm. In broad terms, the adaptation will not be affected by the residual feedback, provided that the residual loop gain, equation (7.2.5), is considerably less than unity at each discrete frequency, k, so that

$$\left| G(k) - \hat{G}(k) \right| \left| W(k) \right| \ll 1 . \tag{7.3.4}$$

In using such a frequency-domain expression, we are implicitly assuming that the adaptation of the controller is performed slowly compared with the duration of the plant's transient response. If we assume that the plant model is equal to the plant response under nominal conditions, G_0, and that the plant response itself is subject to multiplicative uncertainty, Δ_G, then at each discrete frequency, we can write

$$\hat{G}(k) = G_0(k) \quad \text{and} \quad G(k) = G_0(k)\left[1 + \Delta_G(k)\right] , \tag{7.3.5,6}$$

so that

$$\left| G(k) - \hat{G}(k) \right| = \left| G_0(k) \right| \left| \Delta_G(k) \right| . \tag{7.3.7}$$

Also assuming that the multiplicative uncertainty is bounded as in Chapter 6, then

$$\left|\Delta_G(k)\right| \leq B(k) \qquad \text{for all } k . \tag{7.3.8}$$

The condition for the adaptation being unaffected by plant uncertainties, equation (7.3.4), can then be written in the form

$$\left|G_0(k)\, W(k)\, B(k)\right| \ll 1 \qquad \text{for all } k . \tag{7.3.9}$$

This condition is of exactly the same form as that for robust stability in a fixed feedback controller, equation (6.10.5), except that instead of requiring that $\left|G_0\, W\, B\right|$ be less than unity, equation (7.3.9), requires that $\left|G_0\, W\, B\right|$ now be much less than unity. The requirement for reliable convergence of the direct adaptive feedback controller is thus seen to be more restrictive than that for robust stability with a fixed feedback controller. In other words, guaranteeing the stability of a directly adapted feedback controller requires significantly smaller plant uncertainty than that required to guarantee the stability of a fixed feedback controller.

7.3.2. Indirect Adaptation of the Control Filter

In order to avoid the problems associated with residual feedback in the adaptation loop, the control filter could be adapted indirectly, using a frequency-domain version of the modified error arrangement, shown in Fig. 7.6 for time-domain adaptation. The frequency-domain version is shown in Fig. 7.8. In this case only a single signal, $\hat{d}(n)$, needs to be Fourier transformed in order to implement the modified error algorithm. The adaptation of the control filter is now of exactly the same form as for electrical noise cancellation and so the filter coefficients can be adjusted rapidly, without any danger of the adaptation being destabilised by the plant dynamics. The control filter is again constrained to be causal during the adaptation, but now the Fourier transforms of the coefficients are only copied into the time-domain feedback controller after the filter has converged.

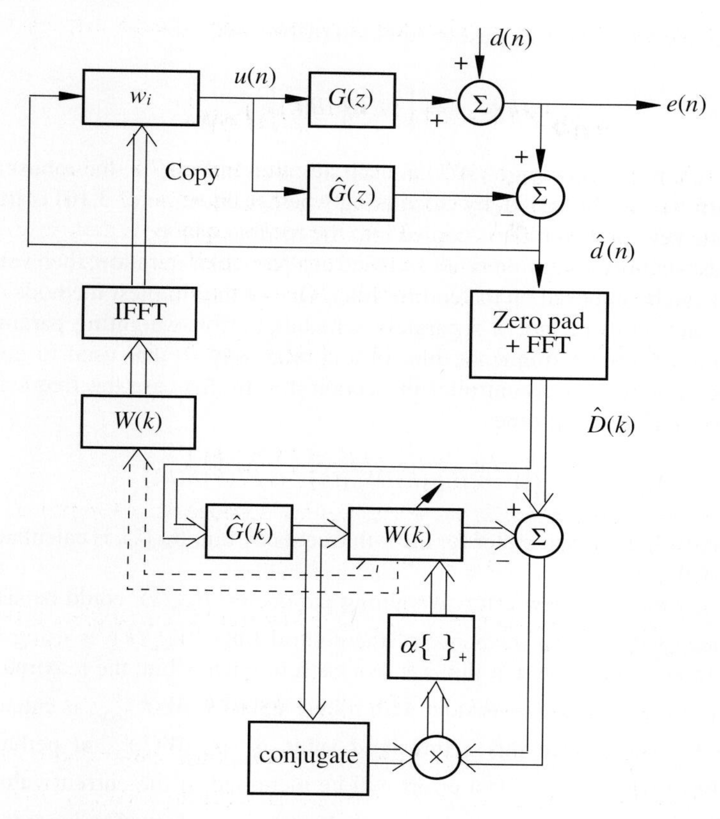

Figure 7.8 Indirect frequency-domain adaptation of the control filter in an IMC arrangement using the modified error method.

Even though the stability of the adaptation may now be assured, we still have to deal with the problem of the stability of the feedback loop once the time-domain control filter has been updated. We will assume that since the time-domain control filter is updated only infrequently, the stability of the closed-loop system can be analysed as if the feedback controller were time-invariant. In this case, the conditions for the robust stability of a fixed controller derived in the previous chapter can be used to ensure the stability of the feedback loop, even if the plant response changes with time.

7.3.3. Constraints on the Control Filter

If we again assume that the plant model, which we already use for the IMC controller and for generating the filtered reference signal in the LMS algorithm, is a fair measure of the nominal plant response, so that $\hat{G}(k) = G_0(k)$, then the condition for robust stability can be written as a *constraint* on the magnitude of the control filter response at each frequency bin of the form

$$|W(k)| < |W(k)|_{\max} \qquad \text{for all } k \,, \tag{7.3.10}$$

where we can calculate $|W(k)|_{max}$ using equation (6.10.5) as

$$|W(k)|_{max} = \left[\left|\hat{G}(k)\right| \; B(k) \right]^{-1} . \quad (7.3.11)$$

Since we have a representation of $W(k)$ at each iteration in Fig. 7.8, the robust stability of the new controller can be tested, by calculating whether equation (7.3.10) is true for each bin, before the new control filter is copied into the feedback loop.

If the robust stability condition is not satisfied at a particular iteration, then various forms of constraint can be imposed on the control filter. One of the simplest methods of applying such a constraint is to include a separately scheduled effort-weighting parameter in the update equation for each frequency bin, in a similar way to that used to constrain the control effort in a pure tone controller in Section 4.3. In this case the frequency-domain update, equation (7.3.1), becomes

$$W_{new}(k) = \left\{ \left(1 - \alpha \, \beta_{old}(k)\right) W_{old}(k) - \alpha \, \hat{R}^*(k) E(k) \right\}_+ , \quad (7.3.12)$$

and the effort-weighting parameter for the k-th frequency bin, $\beta_{old}(k)$, is calculated from the magnitude of $W_{old}(k)$.

One way in which the new effort-weighting parameter, $\beta_{new}(k)$, could be scheduled on the magnitude of the current response of the control filter $|W_{new}(k)|$ is using the penalty function method, as illustrated in Fig. 7.9. For each frequency bin, the maximum modulus of the control filter response consistent with robust stability, $|W(k)|_{max}$, is calculated using equation (7.3.11) and from this a threshold value is set, $|W(k)|_T$, at perhaps 90% of $|W(k)|_{max}$, above which the control effort will be increased. If the current value of $|W(k)|$ is below this threshold, the effort weighting in this frequency bin, $\beta(k)$, is set to zero. If, as the control filter is slowly adapted, $|W(k)|_{new}$ rises above $|W(k)|_T$, however, $\beta(k)$ is gradually increased, as shown in Fig. 7.9. If $|W(k)|$ approaches $|W(k)|_{max}$ the value of $\beta(k)$ becomes very high in equation (7.3.12) and any further increase in $|W(k)|$ is prevented.

Another important constraint in a practical feedback controller is that while minimising the components of the disturbance that have a high level, the disturbance is not unduly

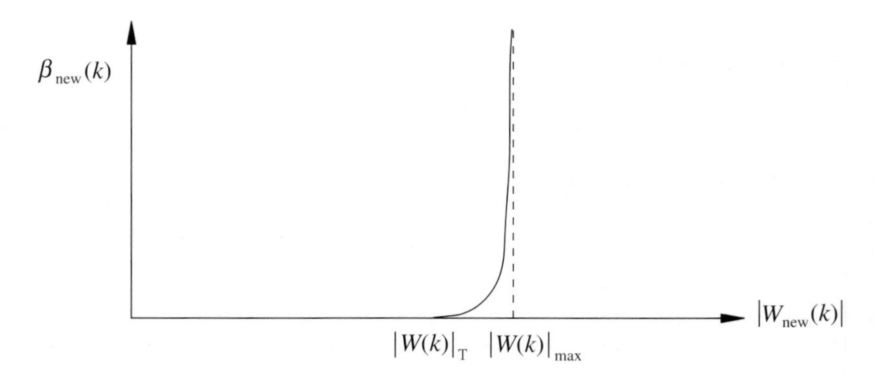

Figure 7.9 Scheduling of the effort weighting parameter on the magnitude of the control filter response at each bin to improve the robust stability of the feedback controller.

amplified at any other frequency. We know from Bode's sensitivity integral that some amplification is usually inevitable, particularly for non-minimum-phase systems, as discussed in Section 6.1, but it is desirable that the degree of this amplification be limited in any frequency band, as far as possible. The amplification of the disturbance is equal to the modulus of the sensitivity function, which for an IMC system can be written in the rather simple form of equation (6.4.3). In terms of discrete-frequency variables, the sensitivity function is equal to

$$S(k) = 1 + G(k)\, W(k)\,. \tag{7.3.13}$$

If we wish to constrain the amplification to be less than A, where $A > 1$, then this puts a practical constraint on the control filter of the form

$$\left|1+\hat{G}(k)\,W(k)\right| \;<\; A \qquad \text{for all } k\;, \tag{7.3.14}$$

where the plant model has been used as the best estimate of the plant response available within the controller. The condition given by equation (7.3.14) can thus be checked at every iteration, or every few iterations, of the controller adaptation. If it is found to be violated for a particular frequency bin, one method of limiting the amplification would be to reduce the magnitude of the controller response in this frequency bin using an individual effort weighting, as outlined above.

The frequency-domain control filter could also be constrained so that the total control effort is limited, to prevent overloading the actuators or power amplifiers. This constraint is similar to the one considered in Section 4.3, but in this case the control effort must be calculated over the whole frequency spectrum. The control effort in this case can be written as

$$E\left[u^2(n)\right] \;\approx\; \sum_{k=0}^{N} \left|\hat{D}(k)\,W(k)\right|^2\,, \tag{7.3.15}$$

where $\hat{D}(k)$ is the discrete frequency version of the estimated disturbance signal in Fig. 7.8. If a single effort weighting factor were now applied to the adaptation of the control filter in *all* the frequency bins, this could be scheduled on the control effort to limit its maximum value in a manner analogous to that used above.

7.3.4. Controller Implementation

The adaptation procedure outlined above provides a way of adjusting the frequency response of the control filter to minimise the mean-square error, while maintaining important practical constraints. The method of then implementing the controller suggested by Fig. 7.8 is to take the inverse Fourier transform of this frequency response and implement the control filter as an FIR filter in the time domain. The optimal form of the least-squares controller in the frequency domain was discussed in Section 6.6, and equation (6.6.4) suggests that the optimal filter must compensate for the frequency response of the plant and the disturbance shaping filter. If the plant response had lightly-damped resonances and an all-pole structure, the natural form of the control filter would be an FIR one. Under other circumstances, however, the frequency response of the optimum control filter could be required to have sharp peaks, and its impulse response could only be accurately implemented with an FIR filter having many coefficients. Similarly, the obvious method of implementing the plant model in the IMC arrangement shown in Fig. 7.8 would be an FIR filter, and this would have to have many coefficients if the plant were particularly resonant. The implementation of the complete feedback controller in Fig. 7.8 could thus require two FIR filters with many coefficients, even though the complete feedback controller may be adequately implemented using a low-order IIR filter.

Another approach to the real-time implementation of the complete controller would be to regard the adjustment of the frequency response of the control filter as part of an iterative design procedure. The next stage in this design procedure would be to calculate the discrete-frequency response of the complete feedback controller which has to be implemented in real time. This can be written as

$$H(k) = \frac{-W(k)}{1+\hat{G}(k)\,W(k)} \,. \tag{7.3.16}$$

Finally, an IIR filter is designed that best approximates this frequency response, and it is this IIR filter that is actually implemented in real time to achieve feedback control. This was the approach used to implement the headrest controller described in Section 6.10. Repeated application of such a design procedure is a form of indirect adaptive control. The last stage in the design procedure begs many questions, since under general conditions it is not clear how well the frequency response of the resulting IIR controller will approximate the required frequency response given by equation (7.3.16). Also the structure of the IIR filter, i.e. the number of recursive and non-recursive coefficients, would probably have to be fixed *a priori* because of limitations on the hardware for the digital controller, and this may not be optimum under all circumstances. The extent to which the resulting IIR controller attenuated the mean-square error, and perhaps more importantly fulfilled the practical constraints, could be checked before implementation, however, in order to identify any significant problems when compared to the ideal controller designed in the frequency domain. If such problems were identified, there would be the option of either redesigning the IIR filter in some way, or even of not adjusting the controller at all in this iteration, but maintaining the response to be that designed in the previous iteration.

It is interesting to note that if the controller is designed so that the stability of the feedback loop is robust to uncertainties in the plant response, it will also tend to be robust to changes in the controller response. It is thus expected that for a robust controller design, the subsequent fitting of an IIR filter response to the frequency response produced by this design will not be too critical.

This philosophy of iterative redesign of the controller is getting further and further away from the approach generally used in adaptive signal processing, of having a simple filter structure, with coefficients which are adapted on a sample-by-sample basis to minimise a well-defined cost function. In feedback control, it is the twin requirements of computational efficiency and the maintenance of important constraints, especially that of robust stability, that have driven us down this less direct path to adapting the controller. The approach ultimately leads to two separate processes: a real-time implementation of the controller in which the control signal, $u(n)$, is calculated from the error signal, $e(n)$, at every sample time, and a much less frequent adjustment of the coefficients in the controller, which could either be calculated 'off-line' on a separate processor, or as a background task on the same processor as is being used to implement the controller.

7.4. COMBINED FEEDBACK AND FEEDFORWARD CONTROL

We start this section by digressing somewhat from the main theme of the chapter, which is disturbance cancellation, to briefly describe some interesting architectures for servo-control systems which employ both fixed feedback and adaptive feedforward techniques. These architectures have potential application to active control systems that attempt to follow some required signal, as in sound reproduction or motion tracking, for example.

7.4.1. Feedforward Interpretation of Servo Control

The direct form of the feedback servo-control system was discussed in Section 6.1, and the block diagram for such a system is shown again in Fig. 7.10(a) for a discrete-time implementation. The disturbance signal has been omitted in this figure to allow us to concentrate on the servo action. The controller is driven by the command error signal $\varepsilon(n)$, which is the difference between the command signal, $c(n)$, and the measured output of the plant, $y(n)$. The total transfer function from the command signal input to the plant output is given by

$$\frac{Y(z)}{C(z)} = \frac{G(z)\,H(z)}{1+G(z)\,H(z)} \,. \tag{7.4.1}$$

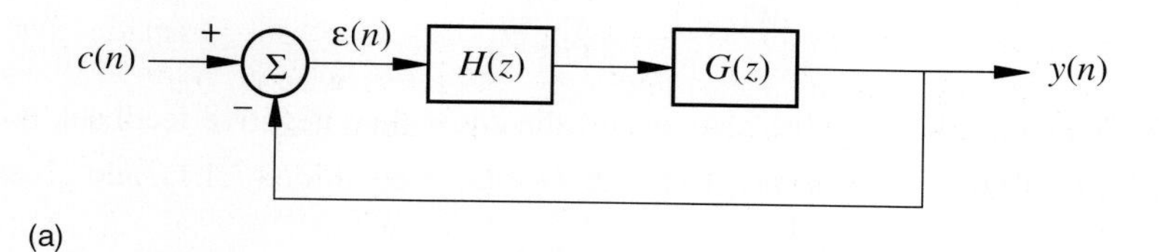

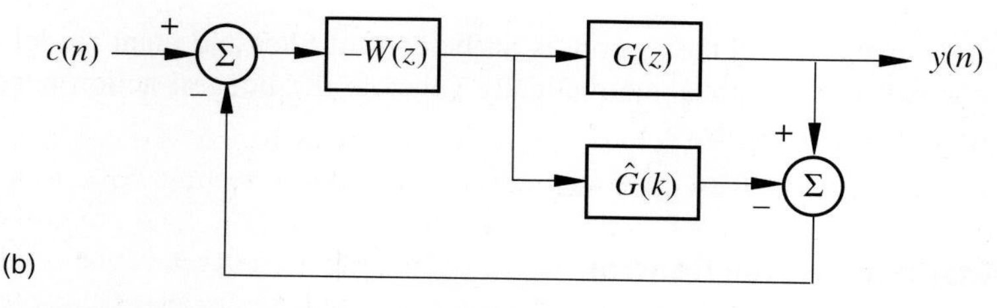

Figure 7.10 Block diagram of a servo-control system in direct form (a), in which the object is for the output of the plant, $y(n)$, to follow the command signal, $c(n)$, so that the error $\varepsilon(n)$ is small. Also shown (b) is the servo controller implemented using IMC, which becomes purely feedforward if the plant model is accurate, i.e. when $\hat{G}(z) = G(z)$.

If the loop gain, $G(z)H(z)$, can be made large in the working frequency region, without causing instability in other frequencies, then the plant output will faithfully follow the command input, since the frequency response corresponding to equation (7.4.1) will be nearly equal to unity over the working range of frequencies. In many applications it is most important that accurate servo action occurs for constant or slowly-varying command signals, and so the loop gain must be large at low frequencies. If the plant does not have an inherent integral action, as would a dc motor with voltage input and angular displacement output, for example, then perfect steady state control can only be achieved with an integrator in the controller, which would have an infinite gain at dc.

We now consider the servo controller implemented using IMC, as shown in Fig. 7.10(b), where a sign change has been introduced into the control filter W to compensate for the differencing operation in the feedback loop and so ensure that the notation is consistent with that used above. If the plant model is perfect, then the signal that is fed back in this arrangement, $z(n)$, is zero. The control filter, W, is thus implementing purely *feedforward* control under these conditions. Morari and Zafiriou (1989) explain this by emphasising that the only role for $z(n)$ in Fig. 7.10(b) is to compensate for uncertainties in the process being controlled. If the process is known perfectly then a feedforward system can perform as well as a feedback one. The IMC architecture still retains a level of feedback control if the plant model is not perfect, however, since then $z(n)$ will not be zero.

The frequency response of the complete control system using IMC can be written from above as

$$\frac{Y(\mathrm{e}^{j\omega T})}{C(\mathrm{e}^{j\omega T})} = \frac{-G(\mathrm{e}^{j\omega T})\,W(\mathrm{e}^{j\omega T})}{1-\left[G(\mathrm{e}^{j\omega T})-\hat{G}(\mathrm{e}^{j\omega T})\right]W(\mathrm{e}^{j\omega T})} . \tag{7.4.2}$$

If we require the low-frequency response of the system to be unity, then with a perfect plant model the dc response of the control filter must be given by

$$W(\mathrm{e}^{j0}) = \frac{-1}{\hat{G}(\mathrm{e}^{j0})} . \tag{7.4.3}$$

Equation (7.4.3) implies that the response of the equivalent negative feedback controller, obtained by comparing equation (7.4.2) with $G = \hat{G}$, to equation (7.4.1), and given by

$$H(z) = \frac{-W(z)}{1+\hat{G}(z)\,W(z)} , \tag{7.4.4}$$

is infinite at dc, even though the responses of the control filter and plant model are finite. The IMC controller would thus automatically generate the integral action necessary for good low-frequency performance.

7.4.2. Adaptive Inverse Control

Another way in which the uncertainty in the plant response can be compensated for in a mainly feedforward controller is to make it adaptive, as shown in Fig. 7.11(a). The difference between the command signal and plant output, $\varepsilon(n)$, is now used only as an error

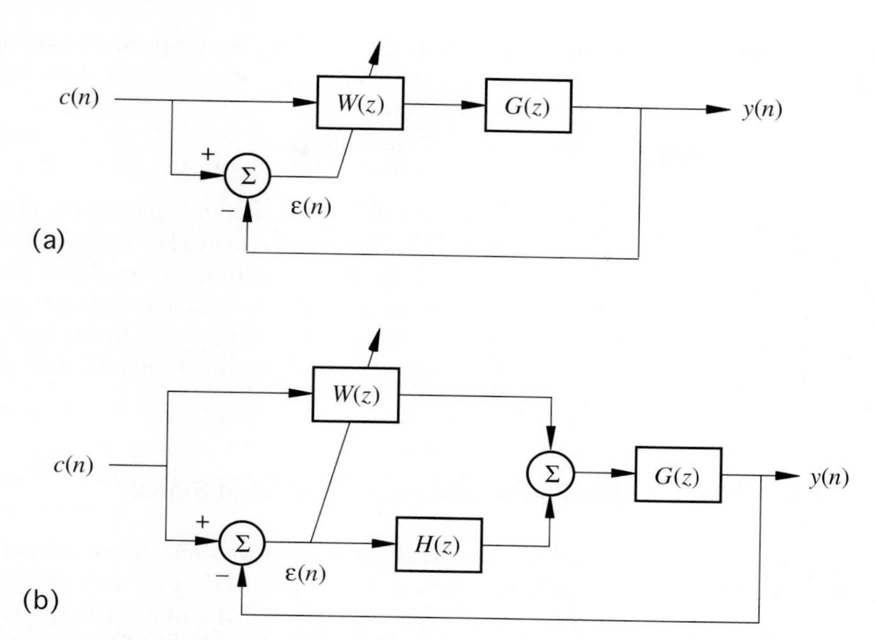

Figure 7.11 Adaptive feedforward system for servo control of the plant output (a) and a servo system that uses both feedback control and adaptive feedforward control (b).

signal, whose mean-square value is minimised by the adaptive feedforward controller. This approach to servo control has been studied extensively by Widrow and Walach (1996), who have called it *adaptive inverse control*. If the control filter was an FIR digital filter, then its coefficients could be adapted using the filtered-reference LMS algorithm, for example. We have seen that this algorithm requires an internal model of the plant to generate the filtered reference signal, and the adaptive controller shown in Fig. 7.11(a) uses this model in a rather different way to the IMC feedback controller shown in Fig. 7.10(b).

7.4.3. Combined Feedback and Feedforward Servo-Control

A combination of fixed feedback control and adaptive feedforward control is shown in Fig. 7.11(b). This block diagram is the same as the adaptive feedforward controller Fig. 7.11(a), except that the error signal is also fed back directly through the fixed controller H, as in Fig. 7.10(a). This arrangement was originally suggested in the context of neural control, i.e. the control of nonlinear systems using neural network controllers, by Kawato et al. (1988), and Psaltis et al. (1988). Kawato et al. describe how a low-bandwidth feedback controller could provide slow but reliable servo action while the adaptive feedforward system gradually learnt the inverse dynamics of the plant. As the action of the feedforward controller is improved by adaptation, the error signal, $\varepsilon(n)$ in Fig. 7.11(b), becomes smaller, and so the need for feedback control is reduced. Kawato et al. (1988) compare this gradual transition, from slow feedback control to rapid feedforward control, to the way in which we develop our own motor skills. The complete system being controlled by the feedforward system in Fig. 7.11(b) comprises both the plant G and the feedback controller, H. The response of the system as 'seen' by the feedforward controller will thus be

$$G'(z) = \frac{G(z)}{1+G(z)H(z)}, \tag{7.4.5}$$

and it is an estimate of this response that would have to be used to generate the filtered reference signal if the filtered-reference LMS algorithm were used to adapt the feedforward controller. It is not of course necessary for the feedback controller to be digital, and a particularly efficient implementation may be to use an analogue feedback controller round the plant, and then only sample the output from the whole analogue loop. In this case, the block diagram would revert to Fig. 7.11(a) with a suitably modified sampled-time plant response.

7.4.4. Servo Control with a Time-Advanced Command Signal

We now consider the case in which the command signal is known some time in advance of the time at which the plant is required to follow it. This would be the case, for example, in the repetitive motion of a robot arm, where the path that the robot arm will be expected to follow is the same from cycle to cycle, or where the required trajectory of a machine tool is known in advance. Such time-advanced information cannot be easily used in purely feedback controllers, such as that shown in Fig. 7.10(a), but can be readily incorporated into an adaptive feedforward system, such as that shown in Fig. 7.12(a). In this block

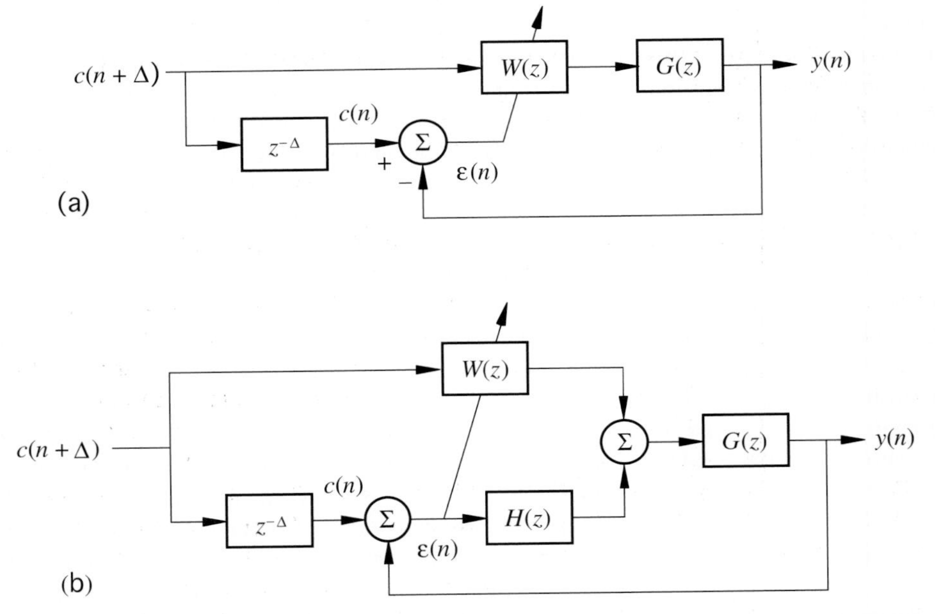

Figure 7.12 Adaptive feedforward control system that makes use of time-advanced information about the required signal (a), and a system that combines adaptive feedforward control using a time advanced signal, with a fixed feedback control (b).

diagram, it is assumed that the command signal is known Δ samples before the plant is required to follow it. This signal is denoted $c(n + \Delta)$, and is used to drive the feedforward controller, W. The error signal required to adapt the feedforward controller is obtained by delaying the time-advanced command signal by Δ samples, to give $c(n)$, and taking the difference between this and the plant output as before. In general, the delay could be replaced by a *reference model* whose response the plant is required to replicate (Widrow and Walach, 1996).

The advantage of having this time-advanced information is that, for perfect control, the feedforward controller now has to approximate a delayed inverse of the plant response, rather than an inverse whose impulse response is zero for all negative time. The time-advance in the command signal thus plays the part of a *modelling delay* in inverse modelling, and much more accurate inverses of non-minimum-phase plants can be obtained with such a modelling delay. The ability of the controller to invert the dynamic response of the plant, and so implement accurate servo-control action, is significantly improved by using this time-advanced information. The application of such an adaptive inverse control system in controlling the flow of anaesthetic drugs is described by Widrow and Walach (1996).

If it is necessary to maintain servo-control action, even if it is slow, while the feedforward controller with the time-advanced command signal is being adapted, then the combined feedback/feedforward system shown in Fig. 7.12(b) could be used. This is a straightforward development of the system shown in Fig. 7.11(b) and described above. Once again, the behaviour of the complete control system would gradually change from slow feedback control to rapid feedforward control as the adaptive feedforward controller learnt the inverse dynamics of the plant.

7.4.5. Combined Feedback and Feedforward Disturbance Control

There may also be advantages in using a combination of feedback and feedforward controllers when the problem is one of disturbance cancellation, particularly when the feedforward controller is adaptive (Doelman, 1991; Imai and Hamada, 1995; Saunders et al., 1996; Clark and Bernstein, 1998). The block diagram of such a system is shown in Fig. 7.13. In this case, the feedback controller will modify the response of the system being controlled by the adaptive feedforward system, so that its effective response is given by equation (7.4.5). This modification to the effective plant response by the feedback controller can make the job of the adaptive feedforward algorithm easier in a

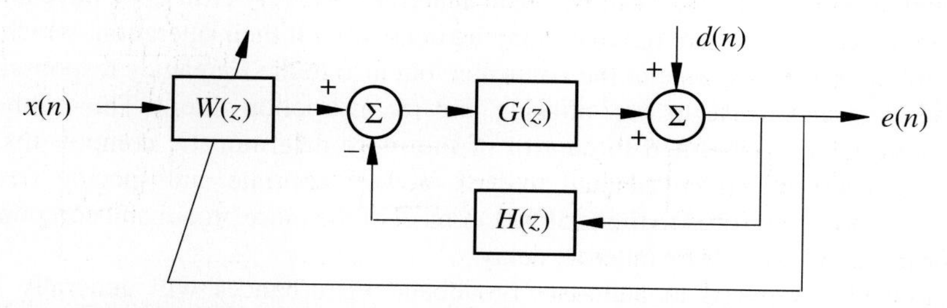

Figure 7.13 Combined feedback and adaptive feedforward system for control of the disturbance, $d(n)$.

number of ways. First, it may reduce the variability of the effective plant response when there is uncertainty in the physical plant, as will be discussed more fully in the following section. Second, the feedback controller could reduce the amplitude of any resonances in the physical plant. This would make the effective plant response easier to model, and thus improve the computation efficiency, but also reduce the transient response of the effective plant, and thus reduce the convergence time of the adaptive feedforward system.

The performance of the feedforward and feedback systems in controlling disturbances can also be complementary. The performance of the feedforward system is limited if there is not perfect coherence between the reference signal and disturbance, as discussed in Chapter 3. When controlling broadband noise in practical environments, this loss of coherence is likely to occur at relatively narrow frequency bands, due, for example, to the reference sensor being at a node of the sound field being controlled. By providing some level of control within these narrow frequency bands, the feedback control can give useful improvements in the overall performance of the combined control system (Tseng et al., 1998).

7.5. COMBINED ANALOGUE AND DIGITAL CONTROLLERS

In this section, we stay with the problem of disturbance rejection and consider a system having both analogue and digital feedback paths.

7.5.1. Advantages and Disadvantages of Digital Controllers and Analogue Controllers

The performance of a feedback controller when attenuating broadband disturbances is greatly affected by delays in the control loop. The reason for this lies in the need for the controller to predict the future behaviour of the disturbance, as discussed in Section 6.4. The attenuation of narrowband disturbances is less affected by such delays, because these signals are more predictable. The controllers required for the control of narrowband disturbances, however, are generally required to be of greater complexity than for broadband disturbances, since their frequency responses must be designed to attenuate a specific disturbance spectrum.

Although a digital controller can be made adaptive relatively easily, we have also seen that such controllers have an inherent delay associated with their operation, which is due not only to the processing time of the controller, but also to the frequency responses of the data converters and analogue anti-aliasing and reconstruction filters. Thus, although a digital controller is well suited to control of the more deterministic components of the disturbance, since it can be adapted to have a very accurate and specific frequency response, it is not well suited to the feedback control of the more broadband components of the disturbance, because of its inherent delay.

The controller required to attenuate broadband disturbances will generally have a relatively smooth frequency response and so could generally be implemented with a simple filter. If this broadband component in the disturbance is also reasonably stationary, then only a fixed controller is required, which can be implemented using analogue components. This will minimise the loop delay and hence maximise the attenuation of the broadband disturbance. Even some narrowband disturbances can be controlled by simple analogue controllers, provided that they are caused by system resonances, and the plant response is also large at these frequencies, giving a high loop gain. The disturbance frequencies would not change significantly over time for a fixed plant response in this case, and so could again be assumed to be stationary.

There are thus complementary advantages in using an analogue feedback controller, to attenuate stationary and broadband disturbances, and in using a digital feedback controller, to attenuate non-stationary deterministic disturbances. In order to get the best overall performance when the disturbance contains both stationary broadband and non-stationary deterministic disturbances, it is thus important to consider a *combination* of both analogue and digital controllers. The design of such a combined system involves some interaction between the design of the digital and analogue parts, and this interaction will be discussed in this section. The performance of a combined analogue and digital controller for an active headset will be discussed in the following section.

7.5.2. Effective Plant Response

As well as the complementary properties described above, there may also be advantages in using an analogue feedback loop to make the response of the plant seen by the digital system more manageable for an adaptive feedback control system. The use of an analogue feedback loop round a very resonant plant can decrease the duration of its impulse response, making it easier to model and to control, as discussed at the end of Section 7.4. The variability of the plant response may also be reduced by feedback.

The block diagram of a combined analogue and digital feedback controller is shown in Fig. 7.14. The Laplace-domain transfer function of the original continuous-time plant is denoted $G_C(s)$, and it is assumed that a fixed analogue controller with transfer function $H_C(s)$ has been designed and implemented to attenuate the broadband components of the disturbance $d_C(t)$ and that this analogue feedback loop is stable. The residual error from this analogue feedback loop is then passed through an anti-aliasing filter, with transfer function $T_A(s)$, to an analogue-to-digital converter before being operated on by an adaptive digital controller. The digital controller uses the IMC architecture with a fixed plant model, $\hat{G}(z)$, and an adaptive control filter, whose 'frozen' transfer function is denoted as $W(z)$ in Fig. 7.14. The output of the digital controller is passed to a digital-to-analogue converter, which has an inherent zero-order hold with transfer function $T_Z(s)$, and then through a reconstruction filter, with transfer function $T_R(s)$, before being added to the input to the continuous-time plant, together with the output of the analogue controller. A similar combination of fixed and adaptive feedback controllers has been described by Tay and Moore (1991), although their objective was to maintain performance for changes in plant response rather than disturbance non-stationarity, and so their digital plant model was also adaptive.

The complete sampled-time plant which is 'seen' by the digital controller thus consists of the data converters and associated analogue filters, together with the analogue control loop which has a transfer function, from plant input to error output, given by

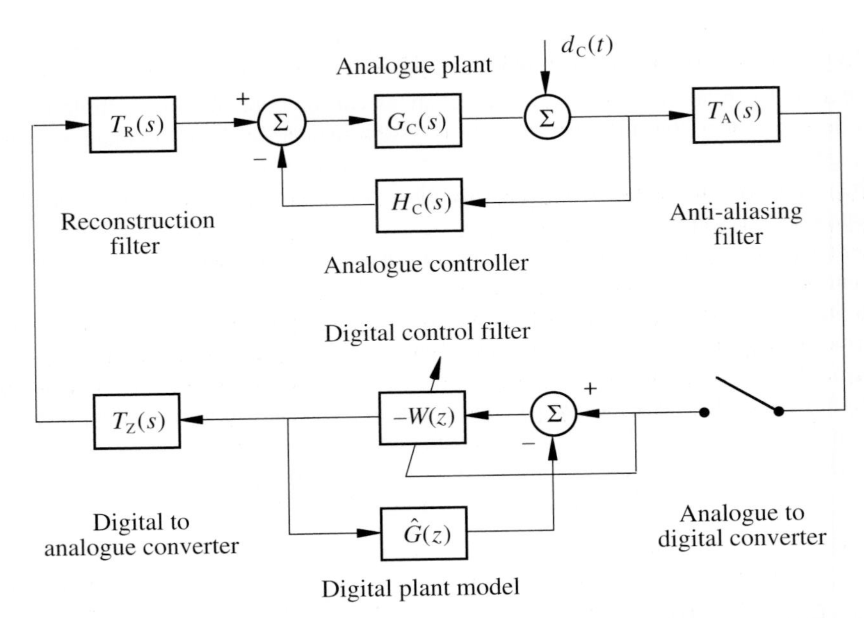

Figure 7.14 Block diagram of a combined feedback control system that has a fixed analogue inner loop, via the analogue controller $H_a(s)$, and an adaptive digital outer loop, via the adaptive digital IMS system with control filter $W(z)$.

$$T_C(s) = \frac{G_C(s)}{1+G_C(s)\,H_C(s)} . \tag{7.5.1}$$

The z-domain transfer function of the plant 'seen' by the digital controller is given by the sampled version of the continuous-time impulse response of this feedback loop, the data converters and the filters. Following on from Section 6.3, the transfer function for this total sampled-time plant response can be written as

$$G(z) = \mathcal{Z}\mathcal{L}^{-1}\left[\frac{T_Z(s)\,T_R(s)G_C(s)\,T_A(s)}{1+G_C(s)\,H_C(s)}\right], \tag{7.5.2}$$

where $\mathcal{Z}$ denotes the z-transform and $\mathcal{L}^{-1}$ denotes the inverse Laplace transform.

The design of the analogue controller could be performed independently of the design of the digital controller. The design of the digital system is influenced by the analogue controller, however, since it affects the response of the effective plant that the digital system has to control. Because the analogue controller is fixed, its effect on the functioning of the digital feedback controller can be readily accounted for, provided that the nominal sampled-time plant response has been identified with the analogue feedback loop closed.

7.5.3. The Uncertainty of the Effective Plant Response

The uncertainty associated with the sampled-time plant response, $G(z)$, is also affected by the analogue feedback loop. At frequencies where the analogue loop gain, $G_C(j\omega)H_C(j\omega)$, is large, the closed-loop transfer function of the analogue system will tend to

$$T_C(j\omega) \approx \frac{1}{H_C(j\omega)} \quad \text{if} \quad G_C(j\omega)H_C(j\omega) \gg 1 , \tag{7.5.3}$$

provided that the analogue system remains stable. Under these conditions, the closed-loop transfer function is largely independent of the physical plant response, $G_C(j\omega)$, and so the uncertainty associated with the sampled-time plant response, $G(z)$ in equation (7.5.2), is less than it would have been if no analogue feedback loop had been present.

The multiplicative uncertainty of a typical plant generally increases with frequency, and becomes greater than unity at high frequencies. The loop gain of the analogue system must therefore be small at high frequencies to ensure robust stability, and so the closed-loop frequency response of the analogue feedback system will tend to that of the plant alone, so that

$$T_C(j\omega) \approx G_C(j\omega) \quad \text{if} \quad G_C(j\omega)H_C(j\omega) \ll 1 . \tag{7.5.4}$$

At frequencies where equation (7.5.4) holds, the uncertainty of the closed-loop analogue system is clearly equal to that of the analogue plant itself. Thus, the presence of the analogue feedback system will not, in general, reduce the *maximum* uncertainty of the closed-loop analogue transfer function, which tends to be reached at high frequencies, but can reduce its multiplicative uncertainty at lower frequencies.

We have seen that a feedback control system using both fixed analogue and adaptive digital control could be particularly effective at controlling disturbances with both broadband and deterministic components. The analogue control loop could be designed independently of the digital one, but the performance of the digital control loop may well be improved by the presence of the analogue loop, since the closed-loop analogue system could be more heavily damped than the open-loop analogue system, and may also be less variable, thus making it easier to control.

7.6. APPLICATION: ACTIVE HEADSETS

There are two main types of headset that are used for rather different acoustical purposes. The 'circumaural', or 'closed-backed' headset, or *ear defender* is illustrated in Fig. 7.15 and has a rigid earshell, which is tensioned against the ear by a band over the head, and a soft cushion which is designed to seal the headset against the side of the face. The main purpose of such headsets is generally hearing protection, although they may be fitted with small loudspeakers for communication if the headset is to be worn by air crew for example. The other type of headset is called 'superaural' or 'open-backed', and has an open foam element between a small loudspeaker and the ear. The main purpose of such headsets is generally sound reproduction, as in portable tape and CD players.

It is possible to use active control in either of these headsets to reduce the external noise heard when wearing the headset. The application of active feedback control to the ear defender type of headset was historically investigated first, and we will initially consider the design and performance of fixed feedback control systems for these headsets.

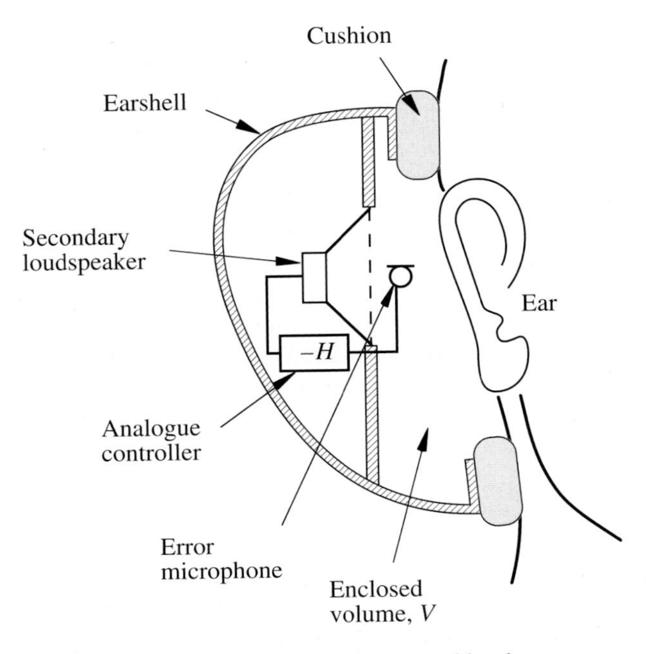

Figure 7.15 An active circumaural headset.

7.6.1. Passive Performance of a Closed-Back Headset

The passive performance of a typical circumaural headset can be derived using a simplified form of the analysis presented by Shaw and Thiessen (1962). We assume that the headset shell is rigid, of mass M, and is mounted on a cushion having a mechanical stiffness of K_c which perfectly seals the headset against the head. The force acting on the headset shell due to a uniform external pressure of p_{ext} will equal

$$f = p_{ext} A \,, \tag{7.6.1}$$

where A is the area of the headset shell where it is joined to the cushion. The dynamic response of the headset is governed by its mass and the total stiffness that acts on it, which is equal to that of the cushion, K_c, plus the mechanical stiffness of the air inside the headset, which is equal to

$$K_a = A^2 \frac{\gamma p_0}{V} \,, \tag{7.6.2}$$

where γ is the ratio of principal specific heats of air, p_0 is the ambient pressure and V is the volume of air in the headset. The complex displacement of the shell when subject to a sinusoidal force $f(j\omega)$ is thus equal to

$$x(j\omega) = \frac{f(j\omega)}{K_t - \omega^2 M + j\omega R} \,, \tag{7.6.3}$$

where K_t is the total stiffness, given by $K_c + K_a$, and R is the mechanical resistance provided by the damping in the cushion. The pressure created inside the headset by this displacement is given by

$$p_{\text{int}}(j\omega) = \frac{K_{\text{a}}}{A} x(j\omega) , \tag{7.6.4}$$

so that using equations (7.6.3) and (7.6.1), the ratio of the internal to the external acoustic pressure can be written as

$$\frac{p_{\text{int}}(j\omega)}{p_{\text{ext}}(j\omega)} = \frac{K_{\text{a}}}{K_{\text{a}} + K_{\text{c}} - \omega^2 M + j\omega R} , \tag{7.6.5}$$

which is called the *transmission ratio* of the headset. At frequencies well below the natural frequency of the headset, given by

$$\omega_0 = \sqrt{\frac{K_{\text{a}}+K_{\text{c}}}{M}} , \tag{7.6.6}$$

the transmission ratio is equal to the constant value given by $K_{\text{a}}/(K_{\text{a}} + K_{\text{c}})$. In most cases, such as the one discussed below, K_{c} is considerably larger than K_{a} and so the low-frequency transmission ratio is approximately equal to $K_{\text{a}}/K_{\text{c}}$. Since K_{a} is inversely proportional to the enclosed volume inside the headset, as in equation (7.6.2), this volume must be made as large as possible for good passive performance.

Shaw and Thiessen (1962) quote the typical parameters of a well-designed headset as being (in SI units) $M \approx 0.13$ kg, $K_{\text{c}} \approx 10^5$ N m^{-1} and $V \approx 170 \times 10^{-6}$ m^3, so that $K_{\text{a}} \approx 1.3 \times 10^4$ N m^{-1}, which gives a natural frequency, $\omega_0/2\pi$, of about 140 Hz, and a low-frequency transmission ratio of about –20 dB. The modulus of the transmission ratio for such a headset is plotted in Fig. 7.16 as a function of frequency. The mechanical resistance of the cushion has been assumed to be $R = 60$ N s m^{-1} in this figure, which is somewhat lower than the typical value suggested by Shaw and Thiessen (1962), in order to illustrate the resonance effect. Although the passive transmission ratio is very good at higher frequencies, more than 40 dB above 500 Hz in Fig. 7.16, below this frequency the

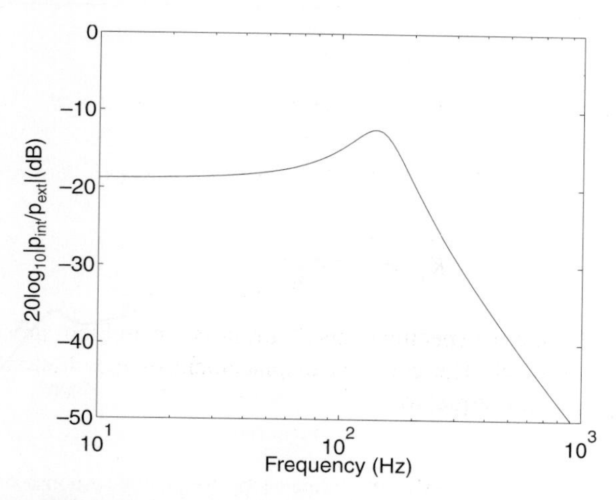

Figure 7.16 The modulus of the ratio of internal to external pressure, in dB, for a typical circumaural headset without active attenuation.

performance is limited by the need to provide a good seal around the ear, so that the cushion cannot be too stiff. In a number of applications, the headset is required to give good noise reductions at low frequencies as well as high frequencies and the application of active sound control at these low frequencies is appealing. The use of a feedback loop to reduce the noise in a headset was originally suggested in the 1950s (Simshauser and Hawley, 1955; Olson, 1956). A rather complete theoretical and experimental investigation of an active headset using a fixed analogue controller was reported soon afterwards (Meeker, 1958, 1959). The simplicity and lack of delay inherent in analogue controllers has made them widely used in commercial active headset systems up to the present day.

The frequency response of the plant, between the input to the loudspeaker and the output of the microphone, in an active closed-back headset fitted to a subject's head, as measured by Wheeler (1986), is shown in Fig. 7.17. The physical proximity of the secondary source

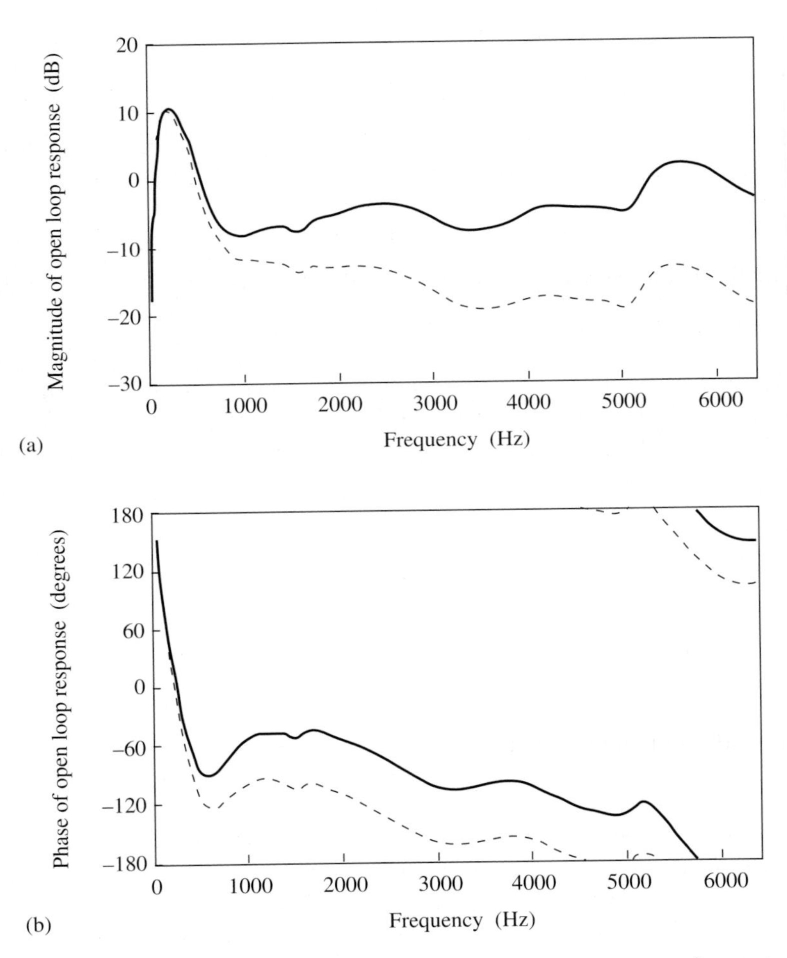

Figure 7.17 Modulus (a) and phase (b) of the open-loop frequency response of an active headset, measured by Wheeler (1986). The response is shown both before (solid line) and after (dashed line) the insertion of a phase lag compensating network in the loop.

and feedback microphone, which are within about 1 cm of each other, means that there is very little acoustic delay in the plant response. The dynamics are dominated by the mechanical resonance of the loudspeaker at low frequencies (about 200 Hz) and resonances associated with acoustic modes inside the earshell at higher frequencies. It is important to have an appropriate amount of foam filler in the earshell so that these acoustic resonances are well damped and do not cause excessive phase shifts at high frequencies, which can lead to instability. It is also important to ensure that the plant response is as close to being minimum-phase as possible, since this will reduce the disturbance enhancement that may occur due to the 'waterbed effect', described in Section 6.2.

7.6.2. Analogue Phase Lag Compensation

Also shown in Fig. 7.17 is the open-loop frequency response of the active headset when a phase lag compensator, of the type shown in Fig. 6.10, is included as the analogue controller. The compensator makes little difference to the response at low frequencies, but at high frequencies gives an extra 15 dB or so of attenuation at the price of only about 40° extra phase lag at 6 kHz. The gain margin of the system can be calculated from the open-loop gain at the frequency where the phase shift falls through 180°, which is at about 5.7 kHz without the compensator and 4.5 kHz and 5.2 kHz after the compensator has been included. It can be seen from Fig. 7.17 that the gain margin of the system has been improved by about 15 dB by the incorporation of the phase-lag compensator.

One of the main causes of uncertainty in the response of the active headset is the variation observed when the headset is worn by different subjects. This is due to differences in the fit and sealing of the headset from person to person, and also because the dynamic response of the ear, and hence the acoustic impedance presented to the headset by the ear, varies from subject to subject. Wheeler (1986) also conducted a series of experiments in which he measured the response of an active headset on a number of different subjects and found that the variation in the amplitude of the plant response from 1 kHz to 6 kHz was about ± 3 dB, and the variation in the phase of the plant response was about ± 20° . The active headset thus has to be designed with gain and phase margins of at least these figures. It is interesting to note, in view of the discussion Section 6.2, that both the amplitude and phase variation could be fairly accurately accounted for in this case by assuming an unstructured multiplicative uncertainty of the form of equations (6.2.4) and (6.2.5) with an upper bound of $B(\omega) \approx 0.3$ over this frequency range.

The additional attenuation of noise inside the headset when this active control system was operating is shown in third-octave bands in Fig. 7.18 (Wheeler, 1986). It can be seen that nearly 20 dB of additional attenuation has been provided in the third-octave bands at about 200 Hz, although a small degree of amplification of the noise occurs in the third-octave bands about 4 kHz. It should be remembered, however, that the frequency range spanned by a third octave band at 4 kHz is considerably greater than a third octave band at 200 Hz. The attenuation integrated over a linear frequency range thus does not violate the Bode limit discussed in Section 6.1. Crabtree and Rylands (1992) have measured the performance of five different types of active headset and found that they all show a maximum attenuation of about 20 dB at 100–200 Hz, falling to zero below about 30 Hz and above about 1 kHz, so the result shown in Fig. 7.18 can be considered as typical.

Finally, Fig. 7.19 shows the noise levels in a typical military vehicle in third-octave

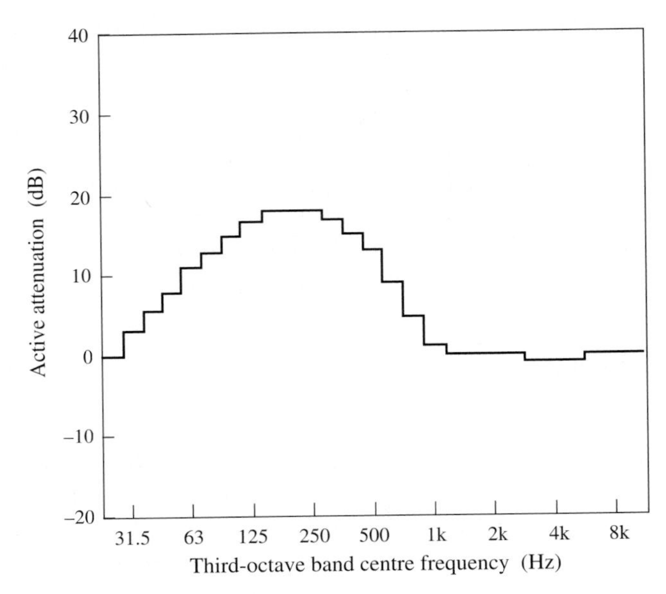

Figure 7.18 Additional headset attenuation provided by the feedback control loop, after Wheeler (1986), in third-octave bands.

bands, together with the noise levels experienced at the ear of a person wearing either a conventional passive ear defender or an active one, as quoted by Wheeler (1987). The range of levels and frequencies important in speech communication are also shown. The passive attenuation provided by the headset is better above about 125 Hz than it is below this frequency, as predicted by the passive transmission ratio curve in Fig. 7.16. The active control system clearly reduces the external noise levels at low frequencies so they are below the typical speech levels in this important frequency range.

7.6.3. Disturbance-dependent Compensation

A more analytical method of designing the analogue feedback controller for an active headset using H_∞ methods has been set out by Bai and Lee (1997). The bound on the multiplicative uncertainty $B(\omega)$, assumed by these authors rose smoothly from slightly less than unity at low frequencies to about 2 at high frequencies, and they assumed a disturbance spectrum that had a broad peak at about 300 Hz. The resulting controller had 5 poles and 5 zeros, but apart from a peak at about 300 Hz to best control the assumed disturbance, the overall frequency response characteristics of the controller were similar to the phase-lag network discussed above.

Higher performance can be obtained over a narrow frequency band by using sharply tuned compensators (see for example the work of Carme, 1987, which is described in Nelson and Elliott, 1992). It has also been noted by Veight (1988) that the design of a compensator required to minimise noise inside the headset depends on the spectrum of the primary acoustic disturbance, and he describes headsets with several different compensators, designed for different noise environments. An interesting design for a

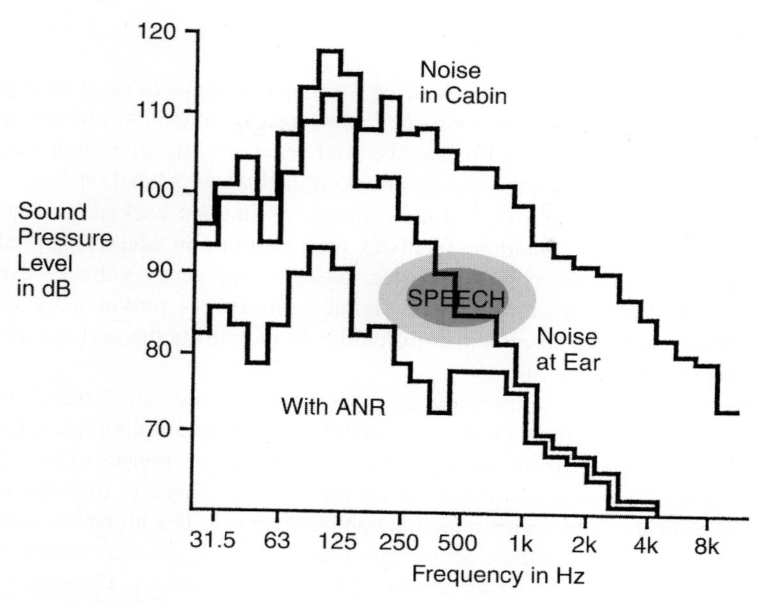

Figure 7.19 Noise levels in the cabin of a typical military vehicle, in third-octave bands, together with the noise levels experienced at the ear of a person wearing a conventional headset (noise at ear) and a headset using feedback active control (with ANR). The range of levels and frequencies important in speech communication are also shown. Reproduced from Wheeler (1987).

general-purpose active headset would thus have a compensator that could adapt itself to different disturbance spectra. It is difficult and expensive to implement adaptive analogue circuits, and so the most practical method of designing an adaptive controller may be to use the digital techniques described in this chapter. We have already noted, however, that such controllers have an inevitable delay. This delay cannot be reduced to below that of the analogue system without using an uneconomically high sampling rate and so a practical digital controller may have problems achieving control over a broad band of frequencies. Adaptive feedforward digital controllers have been successfully implemented for headsets by Brammer et al. (1997), for example. In this section we will concentrate on the design of entirely feedback control systems. Even with a significant delay in the loop, we have seen that digital feedback controllers can control more predictable disturbances, such as narrowband noise, which is the type of noise that tends to vary from one environment to another. A combined analogue/digital controller is described below for an open-backed headset, in which the fixed analogue controller is designed to attenuate broadband disturbances, and the adaptive digital controller can vary its response to attenuate the different narrowband disturbances that are found in different noise environments. It is important that the adaptive digital filter does not degrade the robustness of the analogue controller, and so the robustness conditions described in Section 6.5 place an important constraint on the response of the digital filter in this case.

7.6.4. Active Open-Backed Headsets

In an open-backed headset, such as those used for music reproduction, an analogue active noise control can still attenuate environmental noise to some extent, but its performance is more limited than that of the closed-backed headset because of the very high variability of the plant response as the headset is moved about on the listener's head (Rafaely, 1997). In this section we consider the design and performance of an open-backed headset which in addition to having a simple analogue feedback loop also has an adaptive digital one. An interesting discussion of the control of siren noise in emergency vehicles using open-backed active headsets with an adaptive digital controller is provided by Casali and Robinson (1994), who also describe the difficulties of measuring the performance of such a system using current standards.

One of the main decisions that must be made in such a design is the choice of the sampling rate. Assuming a fixed processing capacity, a slower sampling rate allows longer digital filters to be implemented, and thus sharper frequency responses to be synthesised, but will inevitably involve longer delays in the digital loop. This will limit the bandwidth over which significant disturbance rejection can be achieved. If a higher sampling rate is used, the loop delay is smaller, allowing disturbances of greater bandwidth to be attenuated, but the complexity of the control filter will be reduced, allowing only a few spectral peaks in the disturbance to be attenuated. Clearly the choice of sampling rate will depend on the spectrum of a typical disturbance.

7.6.5. Adaptive Digital Feedback System

In the example described here (Rafaely, 1997), a commercial open-backed headset with an analogue controller was used (NCT, 1995). An additional digital feedback loop was implemented around the analogue system using IMC, as shown in Fig. 7.14. The sampling rate of the digital system was 6 kHz, with anti-aliasing and reconstruction filters having a cut-off frequency of 2.2 kHz (Rafaely, 1997). The control filter was adapted in the time domain using the filtered-reference LMS algorithm, which included a leak in the update equation. The leakage factor was set to ensure that the stability of the adaptive feedback loop was robust to reasonable levels of plant uncertainty. In order to retain good performance, however, this leak could not be set high enough to maintain stability under extreme conditions, when the headset was pressed very close to the head for example, and instability was prevented in these cases by using a monitoring program in the digital controller which reset the coefficients if the error signal from the microphone became too large. This 'rescue routine' was also found to be important in providing robust performance under conditions of high transient disturbances, such as when the headset is buffeted against the head (Twiney et al., 1985), in which case the headset tends to overload, and the plant model will no longer match the response of the true plant. The FIR plant model used in the digital feedback controller had 100 coefficients, and after identification had a response that was very close to that of the nominal plant. The adaptive FIR control filter also had 100 coefficients.

The performance of the system with and without the adaptive digital control loop was tested in a sound field that had both broadband and narrowband components, as shown in Fig. 7.20 (Rafaely, 1997). Figure 7.20 shows the power spectral density of the pressure

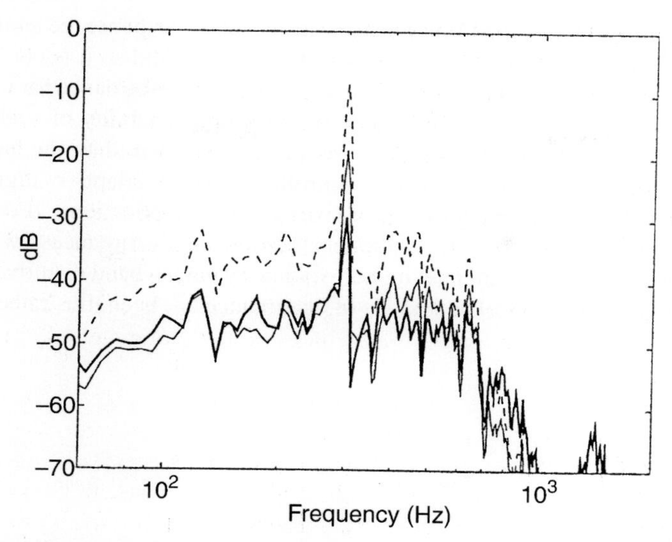

Figure 7.20 Power spectral density of the signal from the microphone near the ear in an active headset, which has both a fixed analogue feedback controller and an adaptive digital feedback controller, when subject to a noise field having both broadband and narrowband components. The dashed line is with no control, the light curve shows the measured spectrum with only the analogue controller operating, and the dark curve shows the measured spectrum once the adaptive digital controller has been allowed to converge (Rafaely, 1997).

measured at the ear position of a manikin on which the active headset was tested, which gives a better measure of the subjective performance of the control system than the response measured at the error microphone in the control loop. The analogue feedback loop alone provides about 10 dB of attenuation up to 500 Hz, so that the tonal component is still prominent in the residual spectrum seen in Fig. 7.20. If the adaptive digital controller is now switched on, the tonal component of the signal at the error microphone in the control loop is further attenuated by over 20 dB, but because of the distance between this microphone and the ear, the tonal component of the pressure measured in the manikin's ear position, as shown in Fig. 7.20, is only attenuated by a further 10 dB, although this still gives a worthwhile reduction in the subjective annoyance of the tone.

In this chapter, we have seen that adaptive controllers can be considerably more difficult to implement in feedback systems than they are in feedforward systems. The main reason for this is that adaptive algorithms may adjust the feedback controller to give a response for which the feedback system is not stable. Although this is guaranteed not to happen when using an IMC controller with a perfect plant model, uncertainties in the physical plant response always have to be taken into account and this introduces a level of 'residual' feedback round the loop which can still destabilise the system. The danger of instability becomes particularly significant if the control filter in the IMC controller has a large response at a frequency where the plant response is especially uncertain. The possibility of instability can be reduced by using a leakage term in the algorithm that adapts the control filter. This method of ensuring robust stability for the controller can be made particularly precise if the adaptive algorithm is implemented in the discrete frequency domain.

In active control applications, the changes in the plant response generally occur too quickly to be accurately tracked by a fully adaptive system, and so it is common to use a fixed plant model, but to adapt the control filter to maintain performance for non-stationary disturbances. Using the methods outlined above, the robust stability of such an adaptive feedback controller can be maintained. We have also seen that the advantages of conventional, fixed, analogue feedback controllers and of adaptive digital feedback controllers can be complementary in active control applications. Fixed analogue controllers are good at attenuating stationary broadband disturbances, while adaptive digital controllers are good at attenuating non-stationary narrowband disturbances. The use of a combined system to control both kinds of disturbance has been illustrated for an active headset.

8

Active Control of Nonlinear Systems

8.1. INTRODUCTION

Nonlinear systems are distinguished from linear ones in that they do not obey the principle of *superposition*: the response of a nonlinear system to two inputs applied simultaneously is not the same as the sum of the responses to the two inputs applied individually. At first sight it would appear that the active control of such systems would not be appropriate, since active control conventionally relies on the superposition of a secondary field and a primary field to attenuate the disturbance. There are, however, a number of areas in which the behaviour of a nonlinear system can beneficially be modified using active control and some of these will be briefly reviewed in this chapter. We concentrate on the active control of two broad classes of nonlinear system: systems in which the nonlinearity is sufficiently 'weak' for linear systems theory to give a reasonable first-order description of their behaviour, and systems with a stronger nonlinearity that exhibit chaotic behaviour, which is very different from the behaviour of a linear system.

It is far more difficult to make general statements about the behaviour of nonlinear systems than about the behaviour of linear systems. For linear systems the principle of superposition provides a very powerful array of analysis techniques, such as the Laplace and Fourier transforms, which allow the system's response to be described in terms of a single function, e.g. the frequency response or the impulse response. Nonlinear systems do not possess such a simple descriptor of their behaviour, and different nonlinear systems can behave in dramatically different ways. A nonlinear system can also behave in very different ways when subjected to different kinds of input signals. It is therefore not reasonable to try to describe the behaviour of some types of nonlinear systems without specifying how they are excited.

8.1.1. Weakly Nonlinear Systems

We begin by considering systems in which the nonlinearity is only relatively mild or *weak*. The fundamental behaviour of such systems can be understood using linear theory, but with modifications that are due to the nonlinearity. A simple example may be the saturation of an amplifier or transducer, as illustrated in Fig. 8.1. The output is linearly proportional to the input for small signals, but the output then saturates for inputs above or below certain limits. If a low-level sinusoidal input is applied to such a system, the output will also be a sinusoid, but if the input amplitude is increased, the output waveform will be clipped.

The output of the nonlinear system shown in Fig. 8.1 depends only on the instantaneous value of the input, and so only a single value of output is associated with each input. For other simple forms of nonlinearity, such as hysteresis, the output depends on not just the

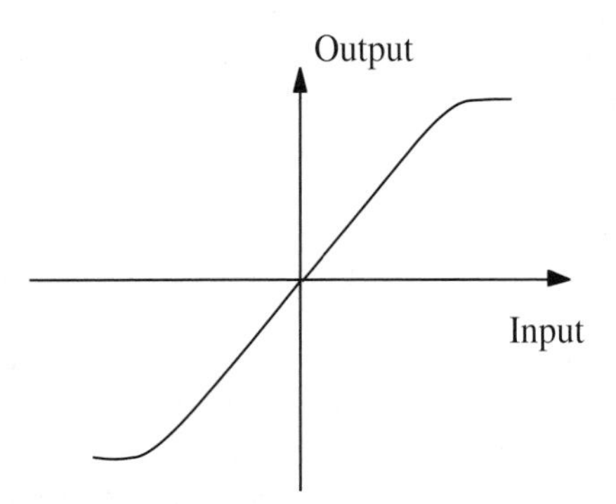

Figure 8.1 A simple example of a single-valued nonlinearity in which the output level smoothly saturates for input amplitudes above a certain limit.

current value of the input but also on previous values of the input and/or output, and so the output can have multiple values for a given input, depending on the previous output.

A simple example of a multiple-valued nonlinearity is exhibited by a system with backlash, or lost motion, whose behaviour is illustrated in Fig. 8.2. If a sinusoidal input were applied to this system whose peak to peak amplitude was less than the lost motion, which is equal to the horizontal distance between the two response lines in Fig. 8.2, there would be no output. If the amplitude of the sinusoidal input is larger than the lost motion, the output will remain constant once the input has reached its maximum value, until the decreasing input has taken up the lost motion and then the output will follow the waveform of the input until it reaches its minimum value when the lost motion would again have to be taken up before the output started to rise. The resulting path around the input–output characteristic is shown by the arrows. The output waveform will thus be distorted, and its peak-to-peak amplitude will be equal to the peak-to-peak amplitude of the input minus the lost motion.

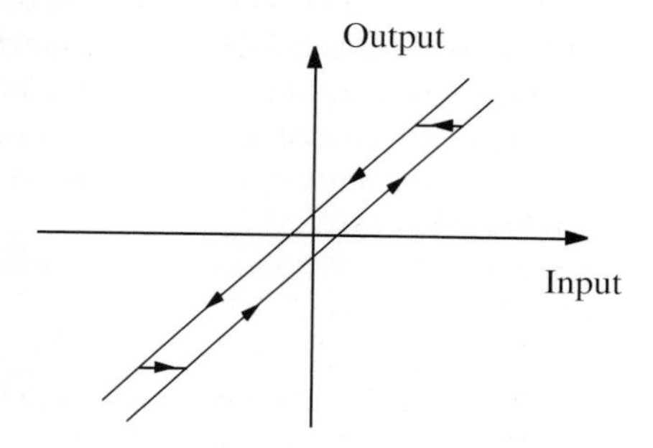

Figure 8.2 The behaviour of a system with backlash, which is an example of a multiple-valued nonlinear system.

Although, when driven by sinusoidal inputs, the outputs of the systems illustrated in Figs. 8.1 and 8.2 are both distorted, they still have the same fundamental frequency as the input. One way in which the behaviour of such nonlinearities could be characterised is by the harmonic structure of the output when driven by a sinusoidal input of a specified amplitude. The nonlinear behaviour of amplifiers, for example, is often quoted in terms of their *harmonic distortion*. If the system is driven by an input that has two frequency components, the output will also generally contain frequency components at the sum and difference of these frequencies, which are referred to as *intermodulation products*.

Weak nonlinearities are encountered in active control systems due to the nonlinearity of the physical system under control (Klippel, 1995), or nonlinearity in the transducers. They are particularly prevalent in high-amplitude active vibration control systems, in which the secondary actuators tend to exhibit both saturation and hysteresis. If we assume that the disturbance is sinusoidal, then the control problem is typically to adjust the input waveform so that a cancelling sinusoidal response is produced at the output of the system, as shown in Fig. 8.3. This waveform must thus be pre-distorted to allow for the distorting effect of the nonlinear system. Note that although the system under control may be nonlinear, its output is assumed to be linearly added to the disturbance signal in Fig. 8.3. The principles of a frequency-domain controller, which will automatically synthesise such an input waveform, will be described in Section 8.3.

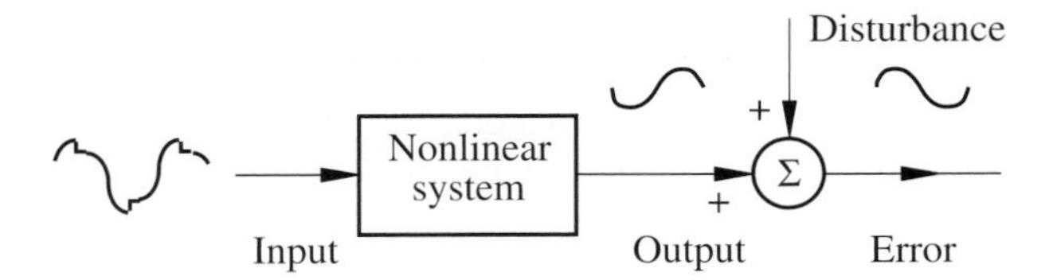

Figure 8.3 In order to cancel a sinusoidal disturbance at the output of a weakly nonlinear system, the input waveform must be pre-distorted so that after having passed through the nonlinear system, the resultant output is sinusoidal.

8.1.2. Chaotic Systems

Nonlinear systems can exhibit much more exotic behaviour than harmonic distortion. Chaotic outputs can be produced by nonlinear systems with very simple describing equations. The output waveform of such a chaotic system may appear almost random, even though it is produced by a simple deterministic equation. Instead of the output getting progressively more distorted as the input amplitude is increased, as would be the case for the systems shown in Figs. 8.1 and 8.2 for example, chaotic systems typically display a rather characteristic progression of significantly different behaviour patterns as a parameter in the defining nonlinear equation is increased.

Consider, for example, the behaviour of a *Duffing oscillator*, which is determined by the nonlinear differential equation

$$\frac{\mathrm{d}^2 y(t)}{\mathrm{d}t^2} + c\frac{\mathrm{d}y(t)}{\mathrm{d}t} - \tfrac{1}{2}\,y(t) + \tfrac{1}{2}\,y^3(t) \;=\; x(t)\;, \tag{8.1.1}$$

when driven by the harmonic input

$$x(t) \;=\; A\cos(\omega_d\, t)\;. \tag{8.1.2}$$

A physical interpretation of the output signal, $y(t)$, in equation (8.1.1) is that it corresponds to the normalised displacement of a second-order mechanical system with a nonlinear stiffness, when driven by a normalised force of $x(t)$. An example of such a nonlinear mechanical system is a thin steel beam, which hangs down so that it is suspended between two powerful magnets, as sketched in the upper part of Fig. 8.4, and is driven close to its fundamental natural frequency. The dynamics of such a system have been described by Moon and Holmes (1979) and the arrangement is sometimes known as a 'Moon-beam'. When it is not driven, so that $x(t) = 0$, equation (8.1.1) has two stable equilibrium positions at $y(t) = +1$ and -1, which correspond to the steel beam being statically deflected towards one or the other of the magnets. The system also has an equilibrium point at $y(t) = 0$ that is unstable, because the local spring constant is negative at this point. The potential energy function for equation (8.1.1) thus has two minima at $y = \pm 1$ and a local maximum at $y = 0$, as shown in the lower graph of Fig. 8.4. The parameter c in equation (8.1.1) corresponds to the normalised damping factor of such a mechanical system and ω_d in equation (8.1.2) to the normalised driving frequency.

In order to demonstrate the range of behaviour which can be exhibited by this relatively simple nonlinear system, the steady state output waveform, $y(t)$ in equation (8.1.1), which would correspond to the displacement of the beam's tip, is plotted in Fig. 8.5 for a range of

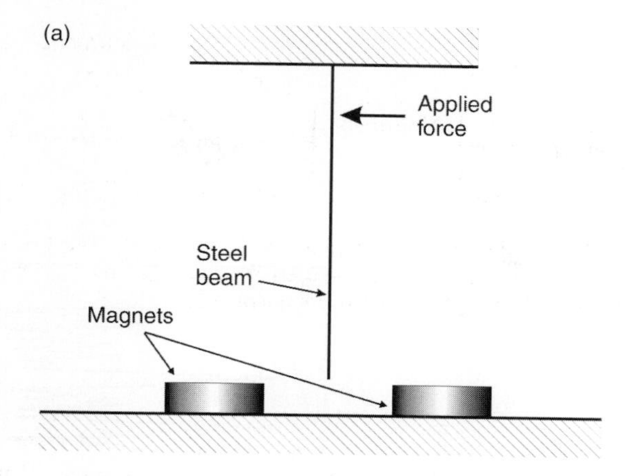

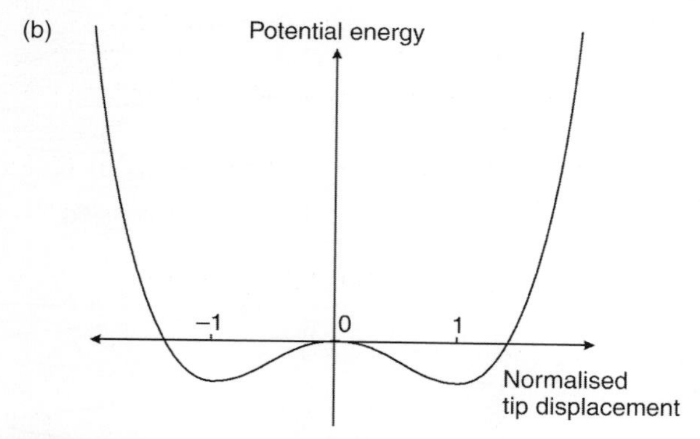

Figure 8.4 (a) Physical arrangement of a cantilevered steel beam hung between two magnets, and (b) the potential energy of the beam as a function of the normalised tip displacement, which has a double potential well due to the nonlinearity in the system's stiffness function.

amplitudes of the harmonic input, A in equation (8.1.2), with c = 0.168 and ω_d = 1. Figure 8.5(a) shows the waveform of the response when the sinusoidal input amplitude is very small, A = 0.0025, in which case the response is centred around the equilibrium position $y(t)$ = 1 but has an amplitude which is much less than unity, and has an almost sinusoidal waveform. For small perturbations, the local behaviour of the Duffing oscillator about the stable equilibrium points is almost linear, and so a sinusoidal input produces an almost sinusoidal response. As the sinusoidal input level is increased, the nonlinearity of the spring response in equation (8.1.1) becomes more evident and Fig. 8.5(b) shows the response when A = 0.174, in which case the output waveform is still centred on $y(t)$ = 1 and has the same fundamental frequency as the input, but is slightly distorted.

If the driving amplitude in equation (8.1.2) is increased still further, the output waveform only repeats itself every two periods of the driving signal, as shown in Fig.8.5(c) for A = 0.194, which is a subharmonic response. A further increase in the driving amplitude causes the beam to move from one equilibrium position to another in an apparently random

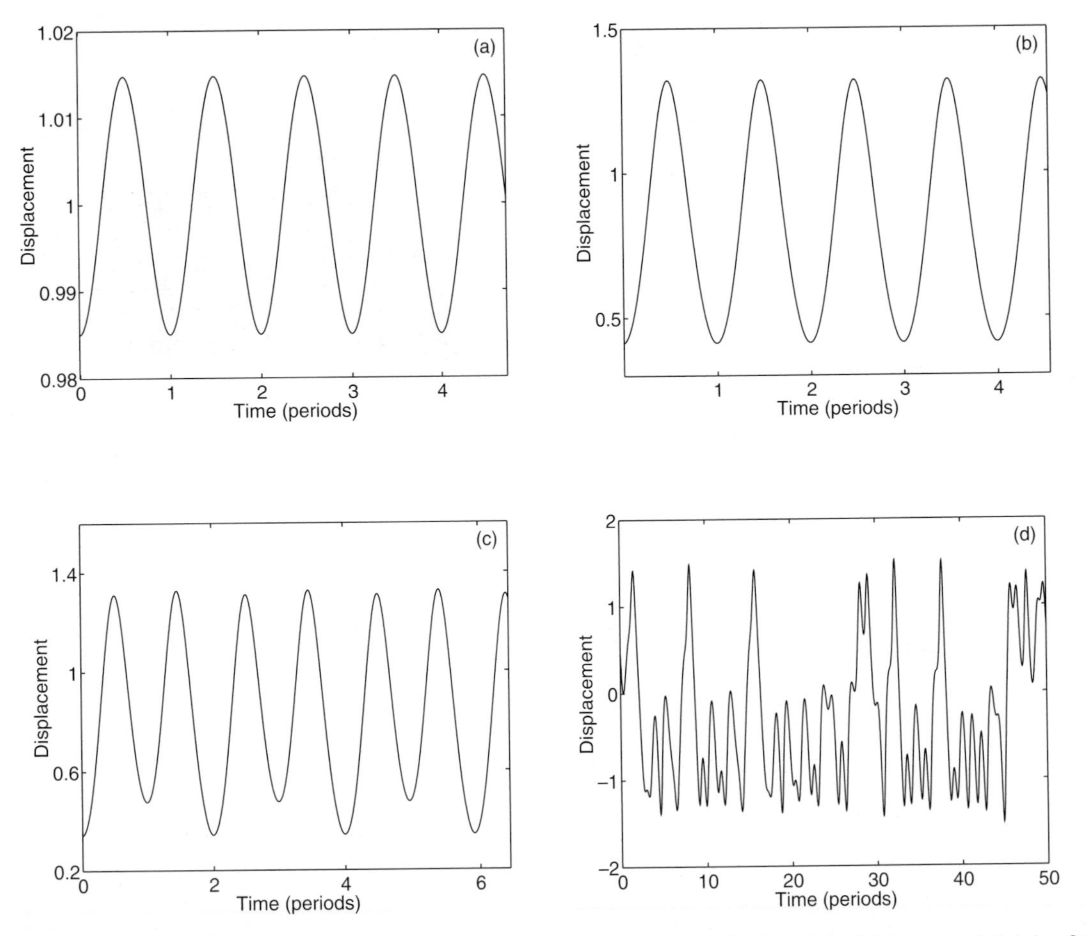

Figure 8.5 Waveforms of the output of the nonlinear Duffing oscillator for four different amplitudes of the driving signal: (a) $A = 0.00025$, almost linear response; (b) $A = 0.174$, response with harmonic distortion; (c) $A = 0.194$, subharmonic response; (d) $A = 0.2348$, chaotic response.

fashion, as shown in Fig. 8.5(d) for $A = 0.2348$. This waveform also has a continuous power spectral density, instead of the line spectra characteristic of the other waveforms shown in Fig. 8.5, which would suggest that the signal was stochastic if it were analysed using conventional Fourier methods. In fact the waveform is completely predictable, although very sensitive to perturbations, and is an example of *deterministic chaos*. A much more complete description of the Duffing oscillator, and chaotic systems in general, can be found in numerous books such as those by Moon (1992) and Ott (1993) and brief introductions are provided in the articles by Kadanoff (1983) and Aguirre and Billings (1995a), for example.

If the output waveform of the Duffing oscillator is sampled once every cycle of the driving frequency, a single constant value is produced if the output waveform has the same fundamental frequency as the input, as in Fig. 8.5(a) and (b) and the output is said to be a period-1 oscillation. If the output waveform has a fundamental frequency which is half that of the input, such as in Fig. 8.5(c), then its period is twice that of the input and sampling at the input frequency will produce two constant values, which is called a period-2 oscillation. The transition from a period-1 oscillation to a period-2 oscillation as the input amplitude is increased is called a *bifurcation* and occurs at an excitation level of $A \approx 0.205$, for the Duffing oscillator whose response is shown in Fig. 8.5, as illustrated in Fig. 8.6. If the input amplitude, A in equation (8.7.2), is increased slightly further then sampling the output waveform at the frequency of the input produces 4 levels, and the period has again doubled. This period doubling occurs repeatedly as the input amplitude is increased still further, until, when the input amplitude is sufficient to cause chaotic motion of the system, this gives rise to a continuous distribution of sampled levels in Fig. 8.6.

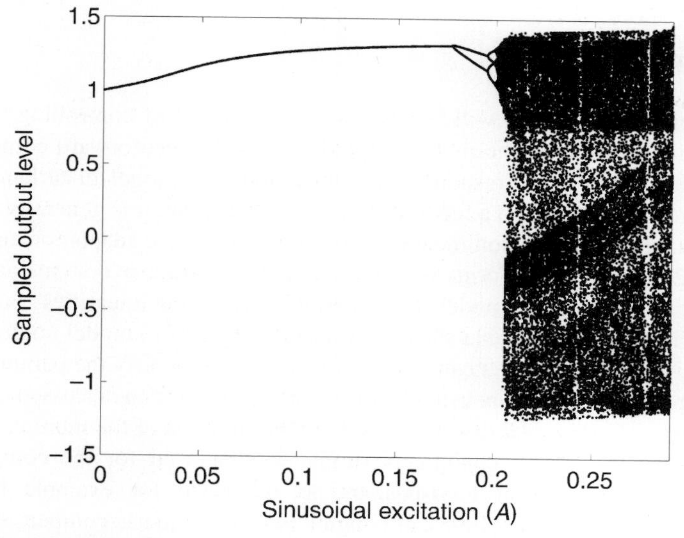

Figure 8.6 Levels of the steady state output waveform of the Duffing oscillator, $y(n)$, when sampled at the frequency of the sinusoidal input signal, $x(t) = A\cos(\omega_d t)$, for different values of the sinusoidal excitation, A. The bifurcations from period-1 to period-2 and to period-4 oscillations can be seen clearly, together with the transition to chaotic motion, when the distribution of levels becomes continuous.

The series of bifurcations which occurs before the input amplitude has reached this level is said to be the period-doubling route to chaos, and is characteristic of many nonlinear systems. The Duffing oscillator can revert back to having a periodic solution if the driving amplitude is greater than the maximum value shown in Fig. 8.6 (Moon, 1992). It can also be seen in Fig. 8.6 that even when the input amplitude is larger than that required for the system to become chaotic, there are small ranges of input amplitudes for which the behaviour of the system once again becomes periodic. These are shown by the thin vertical stripes at $A \approx 0.242$ and 0.276 for example, although the fine structure of Fig. 8.6 is very complicated and there are many other, more narrow ranges of A for which the behaviour also becomes periodic (Ott, 1993). This suggests that rather small changes of the input parameters in a chaotic system can produce profound changes in the character of the response. The extreme sensitivity of chaotic systems to perturbations will be discussed in Section 8.5, and Section 8.6 describes ways in which it is possible to exploit this sensitivity; to change the behaviour of chaotic systems in a very significant way with only very small control signals. The practical application of such control systems is still in its infancy, although the possibility of controlling aerodynamic phenomena, which may be chaotic, using very small perturbations has been put forward by Ffowcs-Williams (1987).

The purpose of this brief diversion into the behaviour of the Duffing oscillator has been to demonstrate the rich variety of behaviour that such a system can exhibit, and to distinguish between the harmonic response of a weakly nonlinear system, which may be distorted but still has the same fundamental frequency as the excitation, as in Figs. 8.5(a) and (b), and the more extreme behaviour of a strongly nonlinear system shown in Figs. 8.5(c) and (d).

8.1.3. Chapter Outline

In this chapter, the discussion will be restricted to examples of controlling two different forms of nonlinear system. The first example involves the feedforward compensation of weak nonlinearities in plant responses, as illustrated for a tonal disturbance signal in Fig. 8.3. In order to make such a feedforward system adaptive, it is generally necessary to have an internal model of the nonlinear plant response within the adaptation algorithm, and in Section 8.2 the analytical forms of some models of nonlinear systems are discussed. The need for a recursive model for multiple-valued nonlinearities is particularly emphasised. Section 8.3 describes the use of neural networks to model nonlinear systems and the adaptive algorithms that can be used to adaptively identify the parameters in such a model. Adaptive compensation using feedforward control is then discussed in Section 8.4 initially in terms of sinusoidal disturbances, but then in terms of the more general case of random disturbances. This technique is similar to that used for the compensation of nonlinearities in moving-coil loudspeakers, as described for example by Gao and Snelgrove (1991) and Klippel (1998), although it is possible to use compensators having a simple and well-defined structure in this case because of the rather low-order dynamics of the loudspeaker. Another method of linearising the response of loudspeakers and hydraulic actuators is the use of local feedback, although this requires a sensor to measure the local response of the actuator. Loudspeaker compensation and local feedback are briefly discussed by Elliott (1999), but will not be considered further in this chapter. Finally, a brief outline of the defining features of chaotic systems is provided in Section 8.5 before some potential strategies for controlling such systems are discussed in Section 8.6.

8.2. ANALYTICAL DESCRIPTIONS OF NONLINEAR SYSTEMS

In this section, we briefly review a number of mathematical models for nonlinear systems. The study of nonlinear systems has a long history and a considerable body of literature, some of which is mathematically very sophisticated. We can therefore only give an indication here of the properties and limitations of the more important models. The interested reader is referred to the review articles and textbooks by Billings (1980), Billings and Chen (1998) and Priestley (1988), for example.

We begin with a model called the Volterra series, which is perhaps the easiest to understand because it reduces so readily to the single-valued nonlinearity of the type shown in Fig. 8.1. All models will be written in discrete-time form to be consistent with the representation used in the rest of the book. There are, however, a number of fundamental problems associated with using discrete-time representations for nonlinear systems that physically operate in continuous time. These are principally caused by a difficulty in defining the bandwidth of a nonlinear system's response, as is necessary to prevent aliasing when the response is sampled. Even the response of weakly nonlinear systems will have many harmonics if driven by a sinusoidal input at, say, one quarter of the sample rate. If these are filtered out before the response is sampled, then part of the nonlinear behaviour is lost and the response of the filter is added to that of the system being measured. If they are not filtered out, they will alias down and corrupt the measured response at lower frequencies. In this introductory treatment, such difficulties will, however, be ignored and the sampled signals will be taken as accurate and complete representations of the response of the nonlinear system.

8.2.1. The Volterra Series

The *Volterra series* describing the discrete-time output, $y(n)$, of a causal nonlinear system in terms of the input sequence, $x(n)$, is given by

$$
\begin{aligned}
y(n) = {} & h_0 + \sum_{k_1=0}^{\infty} h_1(k_1)x(n-k_1) \\
& + \sum_{k_1=0}^{\infty}\sum_{k_2=0}^{\infty} h_2(k_1,k_2)x(n-k_1)x(n-k_2) \qquad (8.2.1) \\
& + \cdots ,
\end{aligned}
$$

where the coefficients $h_1(k_1)$ and $h_2(k_1,k_2)$ are called the first- and second-order *Volterra kernels* and the series can in principle be extended to any arbitrary order.

If h_0 and the second- and higher-order kernels are zero, the Volterra series reduces to a linear convolution, with $h_1(k_1)$ then being the linear impulse response. If the system is memoryless, so that all kernels are zero for k_1, k_2, etc. > 0, then the Volterra series reduces to a power series, which describes the instantaneous output in terms of the instantaneous input. The Volterra series is often described as a Taylor series with memory. In order for the series to be convergent, the nonlinear system must be stable and have a finite memory. Even if these conditions are fulfilled, however, the series may require a very large number of kernels. If the memory is restricted to 100 samples, for example, so that the linear impulse response is specified by 100 first-order kernels, the quadratic part of the Volterra series requires 10,000 second-order kernels. The cubic part of this Volterra series, which would be needed to describe even the simplest symmetric nonlinearity, require 1,000,000 third-order kernels. There are only half as many independent second-order kernels and an eighth as many independent third-order kernels as quoted above due to symmetry, but it can be seen that the number of kernels rises rapidly with the order of the nonlinearity. This rapid rise in the number of kernels required to describe systems with even a modest memory size and degree of nonlinearity is sometime referred to as the *curse of dimensionality*. The problem of identifying such a large number of kernels in practice has led to approximate methods in which only the most 'important' kernels are identified.

Apart from the curse of dimensionality, a major drawback of the Volterra series is that it is not able to describe nonlinear systems in which the output is a function of the previous value of the output itself. A simple example of such a system would be the backlash function, whose behaviour is illustrated in Fig. 8.2. Unless the Volterra series had an infinitely long memory and knowledge of the initial conditions, it would not be possible to use it to predict the position of the system within the deadband of the backlash function. For a sinusoidal input, a system described by a Volterra series will produce an output that has only higher harmonic components. The Volterra series cannot produce the sub-harmonics that can be generated by multiple-valued nonlinear systems (Billings, 1980) or recursive nonlinear systems, such as Duffing's equation.

8.2.2. The NARMAX Model

A rather general model of a nonlinear system, in which the output is a potentially nonlinear function of previous output as well as previous inputs, is the 'nonlinear autoregressive moving average with exogenous input' or *NARMAX model* (Chen and Billings, 1989). It was originally introduced in an attempt to avoid the difficulty of an excessive number of parameters associated with the Volterra series and for its resemblance to the linear recursive model,

$$y(n) = \sum_{j=1}^{J} a(j)y(n-j) + \sum_{i=0}^{I-1} b(i)x(n-i) \ . \tag{8.2.2}$$

The general form of the discrete-time NARMAX model includes exogenous, i.e. externally introduced, noise terms, but if these are ignored for now, the model becomes a NARMA one, which can be written as

$$y(n) = F\left[y(n-1)\cdots y(n-J), x(n)\cdots x(n-I+1)\right] \ , \tag{8.2.3}$$

where $F(\ )$ is some nonlinear function. The exclusion of the exogenous terms from equation (8.2.3), which would generally include the disturbances in the description of a plant response, is consistent with block diagrams of the form of Fig. 8.3, in which the output of the nonlinear system is linearly added to the disturbance. It should, however, be emphasised that such a block diagram does not represent all forms of nonlinear control problem, because in general the disturbance may also affect the operating point of the plant, and in this case the interference terms between the disturbance and the plant's input and output must be included in equation (8.2.3). It would be straightforward to include such terms as exogenous inputs in the formulation considered below and in Section 8.4, but the equations then become significantly more complicated. The assumption that the plant output is linearly added to the disturbance will therefore be retained, since it still demonstrates the main features of the active compensation of nonlinear systems. The polynomial form of the NARMA model can be written as

$$\begin{aligned} y(n) = & \sum_{j=1}^{J} a_1(j)\, y(n-j) + \sum_{j=1}^{J} \sum_{k=1}^{K} a_2(j,k)\, y(n-j)\, y(n-k) + \cdots \\ & + b_0 + \sum_{i=0}^{I-1} b_1(i)\, x(n-i) + \sum_{i=0}^{I-1} \sum_{l=0}^{L-1} b_2(i,l)\, x(n-i)\, x(n-l) + \cdots \\ & + \sum_{m=0}^{M-1} \sum_{p=1}^{P} c_2(m,p)\, x(n-m)\, y(n-p) + \cdots \quad . \end{aligned} \tag{8.2.4}$$

If the Volterra series is viewed as the nonlinear generalisation of a linear FIR filter, the polynomial NARMA model could be viewed as the nonlinear generalisation of linear IIR filter. For linear systems, an IIR filter has advantages over an FIR filter in terms of the smaller number of coefficients that have to be specified to give a desired response, but it is always possible to approximate any stable response with an arbitrarily long FIR filter. The behaviour of nonlinear systems is more profoundly affected by the addition of the feedback term and the higher-order cross terms in equation (8.2.4), since this equation can describe multiple-valued nonlinearities, such as hysteresis and backlash, which cannot be described by the Volterra series.

One problem with such a general model is finding a reliable method of identifying the coefficients for practical systems. If the underlying difference equation for the system were known, this could be recast into the form of equation (8.2.4) and the coefficients could be found analytically. Unfortunately, this is not often the case in practice and a general identification method would ideally provide estimates of the coefficients from the input and output signals of the system, which is treated as a 'black box'. An interesting historical perspective on this important problem is provided by Gabor et al. (1961). The identification of recursive coefficients, even for stable linear systems, is not straightforward, especially if output noise is present, as described in Section 2.9. With the additional complexity of the nonlinearities, it is not surprising that it is also difficult to derive an exact identification method for the NARMAX coefficients that would work under general circumstances. Progress has been made by Billings and Voon (1984) in developing a suboptimal least-squares algorithm, which is similar to the equation error approach discussed below, and has been shown to be successful for identifying several cases of input and output nonlinearity, as further discussed by Billings and Chen (1998).

8.3. NEURAL NETWORKS

Neural networks are systems whose architecture is inspired by the arrangement of nerves in biological systems and by their operation. Neural networks are widely used both in the modelling and control of nonlinear systems. Strictly speaking, the neural networks being discussed here should be described as *artificial neural networks*, since they can only mimic a small fraction of the behaviour of a real biological system, but the generic term 'neural network' is very widely used. There is a wide variety of types of neural networks, as described for example in the textbook by Haykin (1999). We will principally be concerned here with *feedforward networks*, i.e. those in which a series of input signals is passed forward through a network to produce an output signal. In this section, various different architectures for feedforward neural networks will be described, together with their use in modelling nonlinear systems. An interesting discussion of neural networks from a system identification viewpoint is provided by Ljung and Sjöberg (1992). The use of neural networks for the feedforward control of nonlinear systems will be discussed in the following section, and some of the material covered here has been included in preparation for this discussion. There has been an explosion in the literature on neural networks in the last decade or so, whose beginning was perhaps marked by the first IEEE International Conference on Neural Networks in 1987. Since then, several specialist journals have begun and many excellent review articles and textbooks have been published, for example Lippmann (1987), Hush and Horne (1993), Widrow and Lehr (1990) and Haykin (1999). In this brief section we can thus only hope to give an indication of the main areas of work which appear to be directly relevant to the active control of nonlinear systems.

8.3.1. Multilayer Perceptrons

Perhaps the most well-known type of neural network is the *multilayer perceptron*, whose typical structure is illustrated in Fig. 8.7. This shows a set of signals $x_k^{(0)}$ at the 'input layer' being fed forward via two 'hidden layers' to the final 'output layer', which gives the signals $y_n^{(3)}$. The instantaneous input signals, $x_1^{(0)}, \ldots, x_K^{(0)}$, are passed to the processing elements in the first hidden layer, which are called *neurons* (again, strictly speaking, artificial neurons), whose outputs can be written as

$$y_l^{(1)} = f(x_l^{(1)}) \,, \tag{8.3.1}$$

where $x_l^{(1)}$ is a weighted sum of the input signals,

$$x_l^{(1)} = \sum_{k=0}^{K} w_{kl}^{(1)} \, x_k^{(0)} \,, \tag{8.3.2}$$

and $f(\;)$ is a nonlinear function. The block diagram of a single neuron is shown in Fig. 8.8. The coefficients in the nodal summations, e.g. $w_{kl}^{(1)}$, are generally referred to as *weights* in the neural network literature. A dc bias term is also generally included in the summation for each neuron, which adjusts the 'operating point' along the nonlinear function. This can be assumed to be included in equation (8.3.2) if we let $x_0^{(0)} = 1$, for example, in which case $w_{0l}^{(1)}$ is the bias term for this neuron.

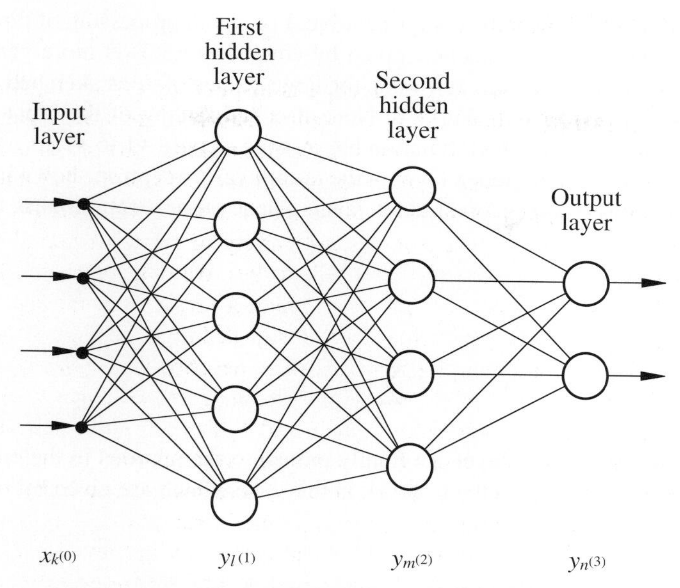

Figure 8.7 The architecture of a multilayer perceptron.

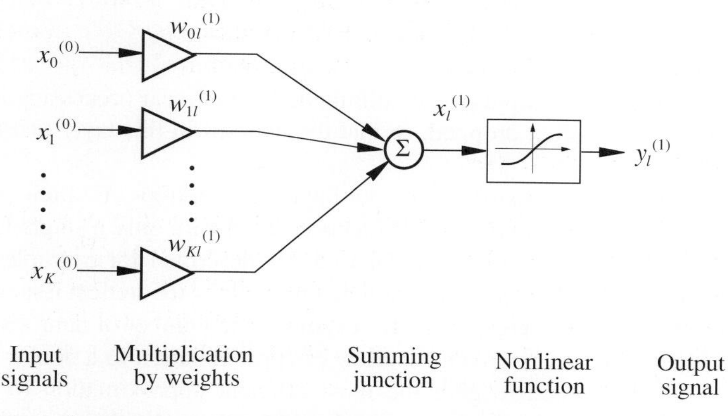

Figure 8.8 Block diagram of an individual neuron with a sigmoid nonlinearity, in this case in the first hidden layer of a multilayer perceptron.

In the original perceptron arrangement, the nonlinear function was a saturating *signum* function (Widrow and Hoff 1960) but, for reasons that will become clear when we discuss the adaptation of the network's weights, a smooth nonlinear function has considerable advantages. One widely used nonlinearity in signal processing applications is the *sigmoid* or tanhfunction, given by

$$y = f(x) = \tanh(x) = \frac{e^x - e^{-x}}{e^x + e^{-x}}. \tag{8.3.3}$$

Neural networks are also widely used for pattern recognition problems, in which an asymmetric nonlinearity, such as $\left(1+e^{-x}\right)^{-1}$, is often preferred, since the output represents pixel levels, which are always positive. For signal processing of bipolar signals, however, the symmetric tanh function given by equation (8.3.3) is more generally used, whose output smoothly saturates at $y = \pm 1$ for large values of x, as sketched in the block diagram of the single neuron in Fig. 8.8. Note also that scaling of the input signals now becomes important, since the tanh function has a range of only +1 to −1.

The outputs from the first hidden layer in the multilayer perceptron shown in Fig. 8.7 are passed to the neurons in the second hidden layer, whose outputs can, in turn, be written as

$$y_m^{(2)} = f(x_m^{(2)}) \,, \tag{8.3.4}$$

where

$$x_m^{(2)} = \sum_{l=0}^{L} w_{ml}^{(2)} y_l^{(1)} \,. \tag{8.3.5}$$

The outputs from the second layer are finally passed to the neurons in the output layer to produce the output signal from the network in this case, which are given by

$$y_n^{(3)} = f(x_n^{(3)}) \,, \tag{8.3.6}$$

where

$$x_n^{(3)} = \sum_{m=1}^{M} w_{nm}^{(3)} y_m^{(2)} \,. \tag{8.3.7}$$

In order for the range of the outputs to be unlimited, the nonlinear processing in the output layer, equation (8.3.6) is often dropped, so that the neurons in this layer just act as linear combiners.

One of the powerful properties of multilayer perceptrons is their potential to approximate any continuous single-valued function, even with only a single hidden layer, as stated in the *universal approximation theorem* (as described for example by Haykin, 1999). In spite of the universal approximation theorem, many theoretical issues are still not fully resolved for multilayer perceptrons: for example, the number of neurons needed in a single hidden layer to achieve a given accuracy of approximation to a nonlinear function, the number of layers needed to give the most efficient approximation to a nonlinear function, or how the weights in these neurons should be optimally determined.

8.3.2. Backpropagation Algorithm

An important advantage of the multilayer perceptron is that the coefficients can easily be adapted using a method that has been found to be very successful in practice, called the *backpropagation algorithm*. In the language used to describe neural networks, this is a *supervised learning method* in that, during a learning phase, the output of the network is compared with a known desired signal to give an indication of how well the network is doing.

The backpropagation algorithm is a form of steepest-descent algorithm in which the error signal, which is the difference between the current output of the neural network and the desired output signal, is used to adjust the weights in the output layer, and is then used to adjust the weights in the hidden layers, always going back through the network towards the inputs. Thus, although the neural network operates on the input signals to give an output in an entirely feedforward way, during learning, the resulting error is propagated back from the output to the input of the network to adjust the weights. The formulation of the backpropagation algorithm will be illustrated here using the simplified network shown in Fig. 8.9, which has a single hidden layer and a single output signal, obtained from a *linear* output neuron. The error between the desired signal and the output of this network is defined to be

$$e = d - y^{(2)} , \tag{8.3.8}$$

where the time dependence of the signals has been suppressed for clarity.

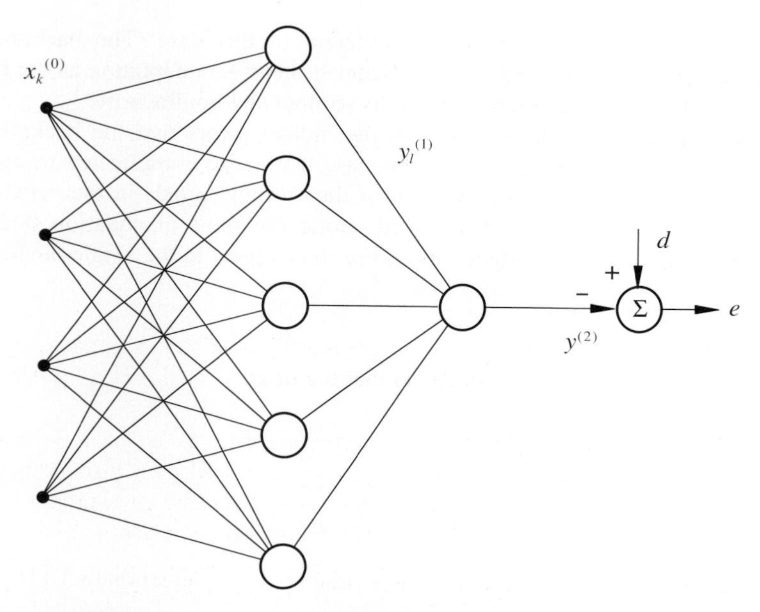

Figure 8.9 A multilayer perceptron with a single hidden layer, whose output is compared with a desired signal for supervised learning using the backpropagation algorithm.

The backpropagation algorithm adjusts the weights in each of the neurons in proportion to the gradient of the squared error with respect to this weight, i.e. for the *i,j*-th weight in the *h*-th layer,

$$w_{ij}^{(h)}(\text{new}) = w_{ij}^{(h)}(\text{old}) - \mu \frac{\partial e^2}{\partial w_{ij}^{(h)}}(\text{old}) , \tag{8.3.9}$$

where μ is a convergence factor, which may be different for each weight.

For the output layer in Fig. 8.9, with weights $w_l^{(2)}$, this gradient is equal to

$$\frac{\partial e^2}{\partial w_l^{(2)}} = 2e \frac{\partial e}{\partial w_l^{(2)}} = -2\, e \frac{\partial y^{(2)}}{\partial w_l^{(2)}} . \tag{8.3.10}$$

In this case the output layer has a linear neuron,

$$y^{(2)} = \sum_{l=1}^{L} w_l^{(2)} y_l^{(1)} , \tag{8.3.11}$$

where $y_l^{(1)}$ are the outputs from the neurons in the hidden layer, so that

$$\frac{\partial y^{(2)}}{\partial w_l^{(2)}} = y_l^{(1)} , \tag{8.3.12}$$

and the backpropagation algorithm for the weights in the linear output layer becomes

$$w_l^{(2)}(\text{new}) = w_l^{(2)}(\text{old}) + \alpha_l^{(2)} e\, y_l^{(1)} , \tag{8.3.13}$$

where $\alpha_l^{(2)} = 2\mu$ is the convergence coefficient in this case. The backpropagation algorithm uses exactly the same gradient descent strategy for adaptation as the LMS, and so it reduces to the LMS algorithm for neurons without any nonlinearity.

It is in the adaptation of the weights in the hidden layers that the backpropagation algorithm really comes into its own. In this case it is more complicated to express the derivative in equation (8.3.9) as a function of the various signals in the network, but a straightforward procedure can still be used, along the lines of equation (8.3.10), but employing the chain rule for differentiation. For the weights in the single hidden layer in Fig. 8.9

$$\frac{\partial e^2}{\partial w_{kl}^{(1)}} = 2e \frac{\partial e}{\partial w_{kl}^{(1)}} = -2e \frac{\partial y^{(2)}}{\partial w_{kl}^{(1)}} . \tag{8.3.14}$$

We can now use the chain rule to write

$$\frac{\partial y^{(2)}}{\partial w_{kl}^{(1)}} = \frac{\partial y^{(2)}}{\partial y_l^{(1)}} \frac{\partial y_l^{(1)}}{\partial x_l^{(1)}} \frac{\partial x_l^{(1)}}{\partial w_{kl}^{(1)}} . \tag{8.3.15}$$

Each of these derivatives can be evaluated individually. Using equation (8.3.11) we can see that

$$\frac{\partial y^{(2)}}{\partial y_l^{(1)}} = w_l^{(2)} . \tag{8.3.16}$$

Assuming that the nonlinearity in the hidden layer is a tanh function (equation (8.3.3)), we can express the middle derivative in equation (8.3.15), after some manipulation, as

$$\frac{\partial y_l^{(1)}}{\partial x_l^{(1)}} = \frac{4}{\left(e^{x_l^{(1)}} + e^{-x_l^{(1)}}\right)^2} = 1 - y_l^{(1)^2} . \tag{8.3.17}$$

Finally, using equation (8.3.2) for the weighted summation in the hidden layer, we can see that

$$\frac{\partial x_l^{(1)}}{\partial w_{kl}^{(1)}} = x_k^{(0)} , \tag{8.3.18}$$

i.e. the k-th input signal. The adaptation equation for the weights in the hidden layer can be derived by substituting each of these individual expressions into equation (8.3.15) and using this in equation (8.3.9) to give

$$w_{kl}^{(1)}(\text{new}) = w_{kl}^{(1)}(\text{old}) + \alpha_{kl}^{(1)}\, e\, w_l^{(2)}\left(1 - y_l^{(1)^2}\right) x_k^{(0)} , \qquad (8.3.19)$$

where $\alpha_{kl}^{(1)}$ is the convergence coefficient in this case.

The adaptation equation for the weights in the hidden layer thus depends on the weights in the output layer, and in general the adaptation equation for the weights in any given layer will depend on the weights in all the other layers between the given layer and the output. For a neuron in any layer of the network, the derivative of the output with respect to a weight in this neuron can always be expanded in the form of equation (8.3.15), i.e. as a product of an *influence coefficient* (the rate of change of the network output with the output of this neuron), and terms which only depend on the operation of the neuron under consideration, equations (8.3.17) and (8.3.18). This general structure will be useful when considering feedforward control.

Finally we should note the role of the gradient of the neuron's nonlinearity in the adaptation process. For the sigmoid nonlinearity this gradient term, equation (8.3.17), is a smooth function which is always positive, but can have a very small magnitude if the nonlinearity is near saturation, i.e. $y_l^{(1)} \approx \pm 1$. The binary signum function does not have an analytic derivative, and it is this which prevented the backpropagation algorithm from being used to implement learning on the early versions of multilayer perceptrons.

The fact that the backpropagation algorithm uses the method of gradient descent means that its convergence properties depend on the shape of the multidimensional *error surface* formed by plotting the squared error as a function of the weights in all the layers. This surface is impossible to visualise in all of its dimensions, but some idea of its properties can be obtained by plotting segments of the surface, for example the variation of the squared error as two of the weights are varied, with all the other weights kept constant. For the simple network considered above (Fig. 8.9), with no nonlinearity in the output neuron, the error surface obtained by varying the weights on the output layer is quadratic. This is to be expected since the output layer is just a linear combiner in this case.

The error surface obtained by varying the weights in the hidden layer is more interesting because of the nonlinearity in the neurons. Figure 8.10, for example, shows the error surfaces obtained by Widrow and Lehr (1990) when varying two weights in a hidden layer, firstly with the network untrained (upper graph) and secondly after all the other weights in the network had been adjusted using the backpropagation algorithm (lower graph). Clearly the error surfaces are not quadratic, but are characterised in this case by relatively flat planes and steep valleys. The surfaces before and after training are similar except that the minimum is deeper after all the other weights in the network have been trained. If the convergence coefficient in a conventional steepest-descent algorithm were adjusted so that the algorithm was stable while descending the steep valleys, the convergence rate would be very slow while traversing the flat planes in the error surface. The backpropagation algorithm does have a reputation for being very slow to converge. The shape of the error surface means that the convergence rate can be rather variable, with no apparent reduction in the error for long periods of time, followed by rapid convergence. The convergence properties of the backpropagation algorithm can be significantly improved by using a

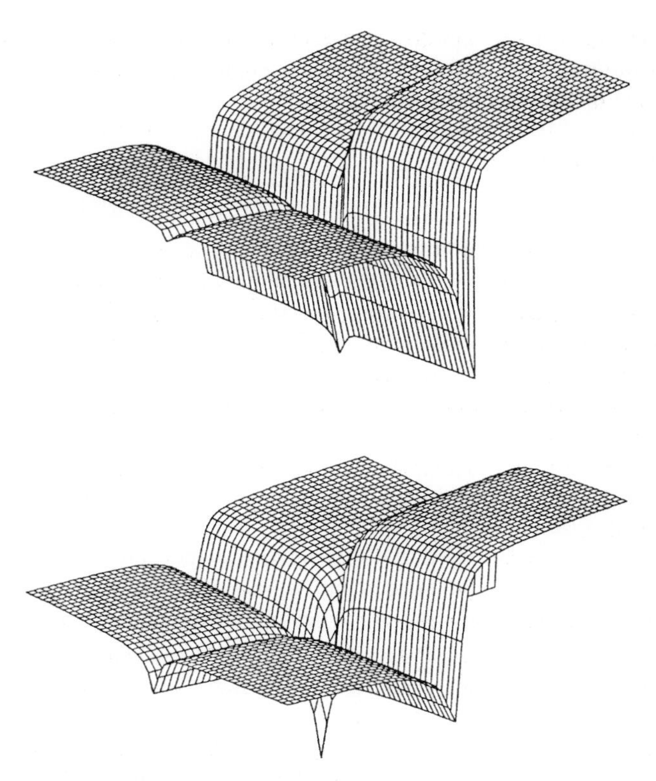

Figure 8.10 Error surfaces obtained when two weights in the first hidden layer are varied in a multilayer perceptron before training (above), and after training (below). After Widrow and Lehr (1990) © 1990 IEEE.

number of modifications to the basic algorithm, as discussed for example by Haykin (1999). One modification which appears to be particularly important is to have individual convergence parameters for each weight, and to adjust the values of these convergence coefficients during adaptation.

Another important implication of the neuron nonlinearities on the shape of the error surface is that it is no longer guaranteed to be convex, or unimodal, i.e. to have a single unique minimum. It is possible for the error surface to have local minima, which the backpropagation algorithm may converge to, at which the squared error is higher than that at another, deeper, minimum some distance away on the error surface. Also, because of the symmetry in the network architecture, even if the minimum found is global it will not be unique. In practice this symmetry is often broken by initialising the weights with different random values before training. Although the possibility of local minima is acknowledged as being an issue in the literature, it does not appear to be a significant problem in practical applications of multilayer perceptrons with sigmoid nonlinearities (Haykin, 1999). Widrow and Lehr (1990), for example, rather pragmatically remark that even though the solution found by the backpropagation algorithm may not be globally optimal, it often gives satisfactory performance in practical use.

The multilayer perceptron shown in Fig. 8.9 will process any ordered set of input data. In active control we are particularly interested in using such neural networks to perform two tasks. The first is the identification of nonlinear systems, which will be briefly described below, and the second is the control of nonlinear systems, which will be addressed in the following section.

8.3.3. Application to Modelling Dynamic Systems

We are generally interested in the identification of systems that are not only nonlinear but also have some dynamics. Their current output will therefore not just depend on the current input, but also on previous inputs. For a single-input single-output system it may be possible to use the multilayer perceptron shown in Fig. 8.9 to model such a system if the inputs to the network are arranged to be past values of a single input signal, $x(n)$, as shown in Fig. 8.11, which is called a *time-delay neural network*. The network weights could then be adapted using the backpropagation algorithm at each sample time using the instantaneous difference, $e(n)$, between the output of the unknown nonlinear system, $d(n)$, and that of the neural network, $y(n)$, when driven by $x(n)$. This arrangement is rather reminiscent of the Volterra series, in that only past input values are used to calculate the output, and it shares some of the drawbacks of the Volterra series, since it is only able to represent a rather limited class of nonlinear systems. If the nonlinear system had hysteresis or backlash, for example, the output would not be a single-valued function of the input, and such a system could not be modelled using the network shown in Fig. 8.11.

What is required to model such a system is a network whose output depends not just on previous inputs, but also on previous outputs. This could be achieved by feeding the output of the neural network, $y(n)$, back to another set of inputs via another tapped delay line, thus

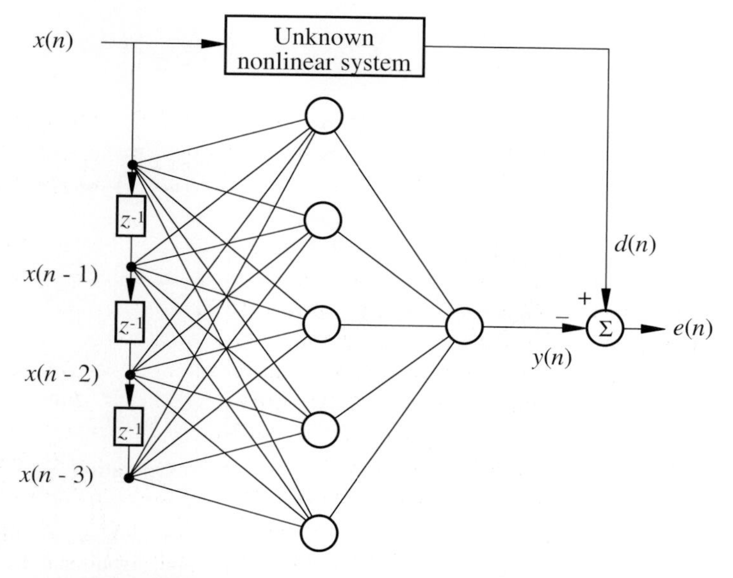

Figure 8.11 A simple multilayer perceptron driven by an input time sequence via a tapped delay line, and used for the identification of an unknown nonlinear system.

generating a *recursive neural network*. The operation of this network would be a special case of the general NARMA model, defined by equation (8.2.3). This modification to the neural network would mean that it was not entirely feedforward, and in general it could not be trained using the usual backpropagation algorithm. An alternative architecture for training would be to take additional inputs to the neural network not from the output of the neural network itself, but from the output of the unknown system, i.e. $d(n)$ (Narendra and Parthasarathy, 1990). If the network did converge to be a good model of the system being identified, then $d(n) \approx y(n)$ and the operation of the network would be similar to the recursive network. This identification strategy is illustrated for a simplified case in Fig. 8.12, and is similar to the *equation error* methods used for recursive linear system identification, as described in Section 2.9.5. One danger of this approach is that if any significant noise is present on the output of the unknown nonlinear system, the weights in the recursive part of the network will be biased. This is because they have the potentially conflicting tasks of reproducing the recursive behaviour of the unknown system and attenuating the measurement noise. This identification strategy is thus most useful when the system being identified is relatively noise-free, or when the noise generation process itself can be modelled and thus the noise waveform can be predicted and subtracted from the observed output. The advantage of the arrangement shown in Fig. 8.12 is that it is still entirely feedforward during training, since the current output depends only on the past input values, assuming slow adaptation, and so the backpropagation algorithm can be used to train the network weights. Once the weights have been trained, they are fixed, and

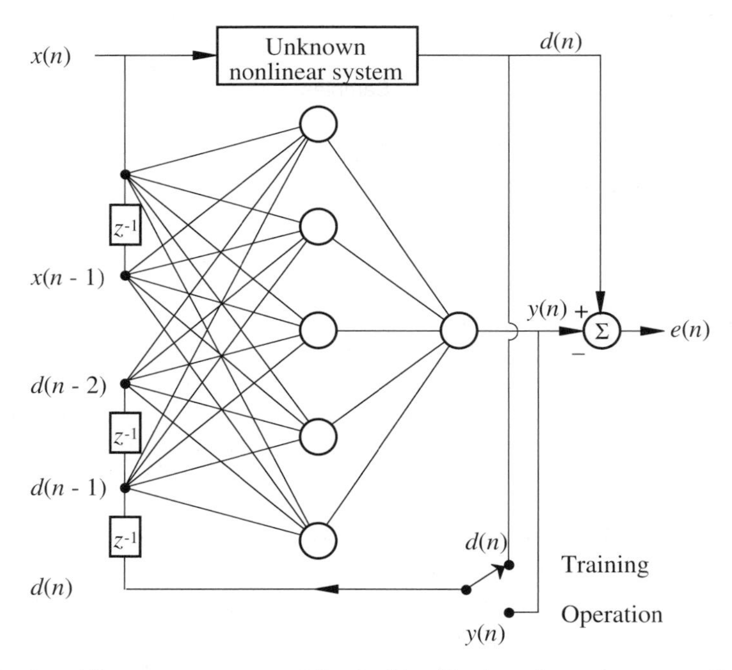

Figure 8.12 A multilayer perceptron used for the identification of an unknown nonlinear system, whose inputs during training are the current and past input signals and the past desired signals, and during operation are the current and past input signals and the past output signals.

assuming that the resulting system is stable, the nonlinear model can be implemented by reconnecting the second delay line to the output of the neural network instead of the output of the unknown system, as shown in Fig. 8.12.

Figure 8.13 shows the results of a simulation of this identification method, when applied to a backlash function (Sutton and Elliott, 1993). The input signal to the backlash function and to the neural network, $x(n)$, was narrowband random noise. The neural network had two inputs, which were $x(n)$ and $d(n-1)$, one hidden layer with two neurons having sigmoid nonlinearities, and a linear output layer. The output of the backlash function $d(n)$, appears to be slightly delayed compared with its input, because of the time taken to traverse the deadband, and is also somewhat smoothed. The neural network, in 'training' mode, was adapted using the backpropagation algorithm. After the weights had converged it was then switched to the 'operating' mode and its output, $y(n)$, is also plotted in Fig. 8.13. Although this neural network does not perfectly model the operation of the backlash function, the outputs of the neural network and the backlash are very similar. It was found that such a degree of similarity could not be obtained using a neural network model driven only by past input values, such as in Fig. 8.11.

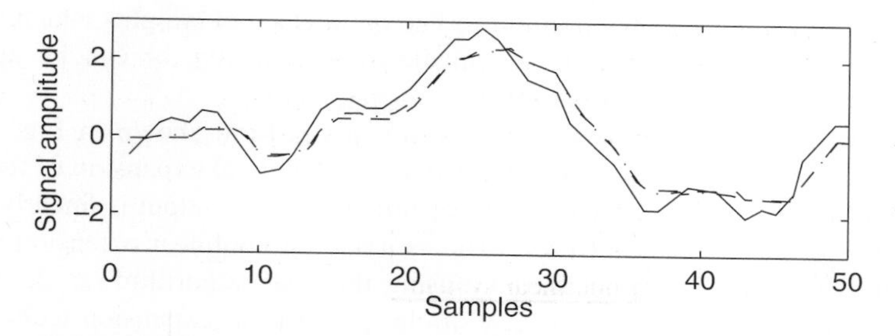

Figure 8.13 The input $x(n)$ (solid line) and output $d(n)$ (dashed line) of a backlash function when driven by narrowband random noise and the output, $y(n)$ (dot dashed line) of a recursive neural model of the backlash function.

Although, from a system identification perspective, a neural network could be regarded as 'just another model structure', the multilayer perceptron does possess a number of useful features, as discussed, for example, by Ljung and Sjöberg (1992). One important property is that the signum function naturally models the kind of saturating nonlinearities that are often found in practice, and can do so far more easily than a polynomial expansion, for example. One criticism of the general multilayer perceptron structure is that it can involve far more parameters for adaptation than are needed for a parsimonious model. This over-parametrisation can give rise to the possibility of *over-fitting* the observed data. A common method of preventing such over-fitting, and thus improving the *generalisation* of the network, is to divide the observed data into two sets, the first set being used to train the network, using backpropagation for example, and the second set being used to test the network, or *cross-validate* the observed performance (Haykin, 1999). Although the mean-square error obtained from the training data set will generally continue to decrease as the adaptive algorithm converges, the mean-square error for the test data is typically reduced as the algorithm converges initially, but then begins to increase as the algorithm over-fits the training data. The best model is obtained by stopping the adaptation when the mean-square error for the test data is at a minimum. It was noted in Section 4.3.6 that stopping the adaptation of a steepest-descent algorithm before convergence is complete has an effect which is similar to minimising a cost function that includes a term proportional to the sum of squared weights as well as the mean-squared error, and this serves as a method of regularising the resultant solution. Ljung and Sjöberg (1992) show that such regularisation reduces the effective number of parameters fitted by the network and effectively stops over-fitting by automatically adjusting the effective order of the resulting model.

8.3.4. Radial Basis Function Networks

Finally in this section, we mention another important class of nonlinear network which is illustrated in Fig. 8.14. This is different from the multilayer perceptron illustrated in Fig. 8.7 in that instead of the nonlinearities being distributed throughout the network, all of whose weights can be adjusted, the nonlinearity is now confined to only the first stage of the network, and is fixed. The outputs from this fixed nonlinear stage are then fed to a linear combiner, whose weights can be adjusted. The advantage of such an architecture is that the output is always a linear function of the weights and so the error surface is guaranteed to be quadratic with a unique global minimum. The optimal set of weights which minimises a quadratic error criterion can thus be calculated analytically, or can be approached adaptively using the LMS or RLS algorithms.

A number of different nonlinear networks are of the form shown in Fig. 8.14. The Volterra series, for example, could be regarded as a polynomial expansion of the past and previous input data, which is a fixed nonlinear process whose output is linearly weighted by the Volterra kernels, equation (8.2.1). This is a natural nonlinear extension to the FIR filter, which can readily be made adaptive using the LMS algorithm (as described, for example, by Rayner and Lynch, 1989). A similar polynomial expansion technique for a pattern-matching application is shown in fig. 4 of Widrow et al. (1988).

If the input signal to a network was a pure tone and the output was required to be a general periodic signal at the same frequency, a convenient nonlinear network to use, with

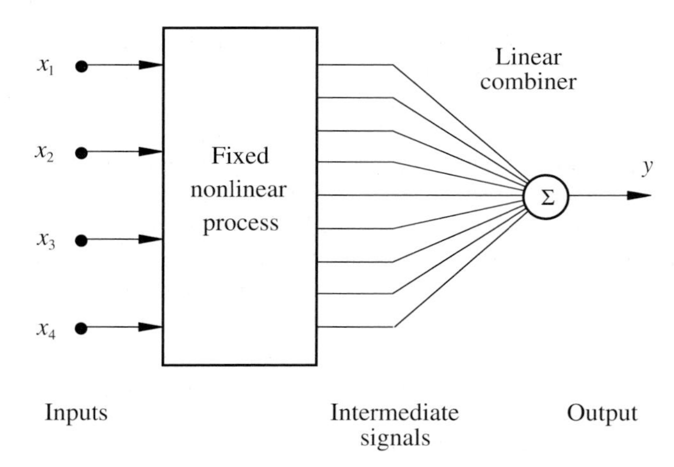

Figure 8.14 Architecture of a nonlinear network that has a fixed nonlinear processing stage which transforms the inputs into a set of intermediate signals which are linearly combined to produce the output.

the form shown in Fig. 8.14, would have a harmonic generator as the fixed nonlinear process. The harmonic outputs of the fixed nonlinear process, which would in general need to have both sine and cosine components, could then be weighted and summed to give a periodic signal as an output that has the same period as the input sinusoid. The use of such a network as a harmonic controller will be described in the next section.

Perhaps the most widely known type of nonlinear network with the architecture shown in Fig. 8.14 is the *radial-basis function network* (Broomhead and Lowe, 1988) whose properties are also described by Haykin (1999) and Brown and Harris (1995), for example. If the set of input signals to the network are described by the vector $\mathbf{x} = \left[x_1\ x_2 \cdots x_K\right]^{\mathrm{T}}$, the output of the network can be written as

$$y = \sum_{i=1}^{I} w_i\, \phi\left(\left\|\mathbf{x} - \mathbf{x}_i\right\|\right) , \qquad (8.3.20)$$

where w_i is the weight for the i-th intermediate output in Fig. 8.14. These outputs are formed by the set of I *radial-basis functions*, $\phi\left(\left\|\mathbf{x} - \mathbf{x}_i\right\|\right)$, where $\|\ \|$ denotes a vector norm, which is normally taken to be Euclidean (as described in the Appendix), and $\mathbf{x}_i$ are the *centres* of the radial basis functions. A common choice for the basis function is the Gaussian curve, so that

$$\phi(r) = \exp(-r^2/c^2) , \qquad (8.3.21)$$

where $r = \left\|\mathbf{x} - \mathbf{x}_i\right\|$ and c is a constant.

To continue our discussion of nonlinear system identification, we could imagine that the input vector $\mathbf{x}$ contains the current and past values of an input sequence to the network. The centres of the radial basis functions, $\mathbf{x}_i$, could then describe the waveforms of signals that were to be emphasised by the network, and these could, for example, be terms in a Fourier series. The output of the radial basis functions would then be a measure of how different the input waveform in the immediate past, $\mathbf{x}$, was from the signal defined by $\mathbf{x}_i$. The final output of the network is then the weighted sum of these functions, which will be significant when particular combinations of the signals $\mathbf{x}_i$ are present. The behaviour of the network is thus significantly influenced, but not completely determined, by the choice of the centres $\mathbf{x}_i$. Numerous ways of choosing the positions of the centres in radial basis function networks have been proposed, including the use of vectors representing individual input signals that are considered particularly representative of a training set, or by using automatic clustering algorithms, so that the distribution of the centres approximates that of the training data (as described for example by Chen et al., 1992). Brown and Harris (1995) have also investigated triangular basis functions and demonstrated the relationship between such systems and those using fuzzy logic.

Finally, it is interesting to note that a multilayer perceptron with a linear output layer and with fixed weights in the hidden layers could also be represented by the block diagram shown in Fig. 8.14. In the multilayer perceptron, however, the nonlinear process in the hidden layers has normally been defined by adapting the weights using backpropagation and a set of training data, whereas for the radial-basis function network, the nonlinear process is defined by the choice of the centres, although this is also normally based on a set of training data.

8.4. ADAPTIVE FEEDFORWARD CONTROL

The principles of adaptive feedforward control for nonlinear plants are similar to those used in the control of linear plants. The block diagram of a single-channel version of the arrangement is shown in Fig. 8.15. This is not such a general block diagram as was the case with a linear plant, because of the assumption that the disturbance acts at the output of the plant. If the plant is linear, the effect of any disturbances that act within the plant can be exactly represented by an equivalent output disturbance. This is not the case for a nonlinear plant, however, as noted above, since an internal disturbance will, in general, also affect the operating point and hence the response of the system to the input signal. Although Fig. 8.15 is not a completely general representation for nonlinear systems, it does serve as a convenient model problem in the present discussion.

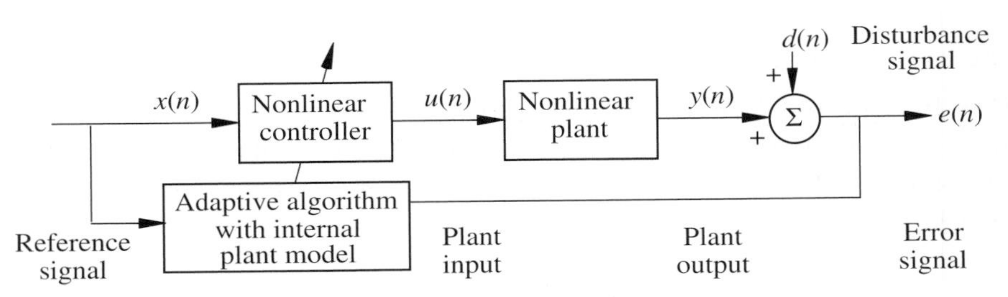

Figure 8.15 The block diagram of an adaptive feedforward control system for a nonlinear plant using a nonlinear controller.

Perhaps the greatest problem with the design of a feedforward controller for nonlinear systems is in the prediction of its optimal performance. This can be calculated using Wiener filtering theory for linear systems, but as yet, no such general theory has emerged for nonlinear systems. Some of the conditions for the invertibility of a nonlinear system must be similar to those for linear systems, such as causality, but other conditions must also apply. Consider for example the simple saturating nonlinearity shown in Fig. 8.1. Clearly, no input to this system could produce an output whose amplitude was greater than the saturation limit. The lack of a general method of predicting the optimum performance of a control system with a nonlinear plant means that there is no way of knowing whether the residual error, which is left after the adaptation of a practical controller, is due to fundamental difficulties in inverting the plant response, or non-ideal performance of the adaptation algorithm. Although quantifying the performance of a model of a nonlinear system is not straightforward, as discussed by Aguirre and Billings (1995a) for example, this does not present the fundamental difficulties encountered in quantifying the performance of an inverse model of such a system, as is required for feedforward control in Fig. 8.15.

Having acknowledged this limitation, however, it is interesting to note that using the internal model control (IMC) architecture shown in Fig. 8.16, the problem of designing a feedback system for a nonlinear system can again be reduced to that of designing a model and of designing a feedforward controller. Both the model and the controller could conveniently be implemented with neural networks, as discussed by Hunt and Sbarbaro (1991) for example.

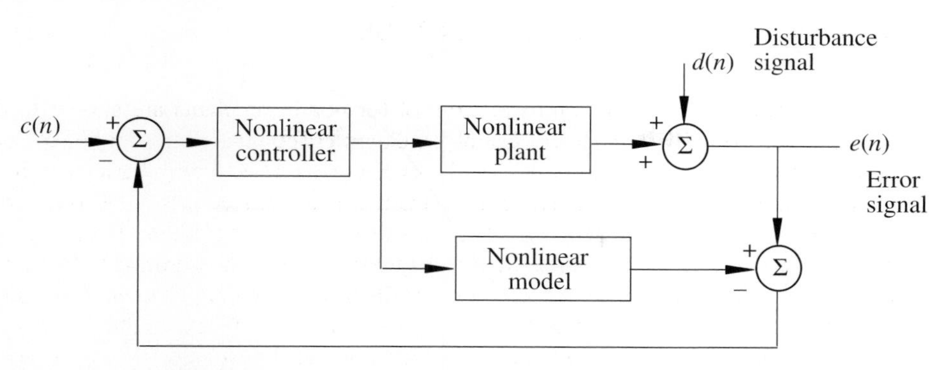

Figure 8.16 Internal model control (IMC) architecture of a feedback controller for a nonlinear plant with additive output disturbances *d*, and the command signal *c*, which uses a model of the nonlinear plant and a nonlinear feedforward controller.

8.4.1. Inversion of a Static Nonlinearity

We begin this section by considering the adaptive feedforward control of a plant that has a single-valued differentiable nonlinearity and has no dynamics, using the arrangement shown in Fig. 8.15. With these assumptions on the behaviour of the plant, its output, y, will only depend on the current value of the plant input, u. If we make no assumptions about the internal structure of the controller, but only assume that it contains a set of adjustable parameters $w_0, \ldots, w_I$, we can formulate a steepest-descent adaptation algorithm for these coefficients in a similar way to that used in the derivation of the backpropagation algorithm. The rate of change of the squared instantaneous error with respect to one of the controller parameters can be written as

$$\frac{\partial e^2}{\partial w_i} = 2e\frac{\partial e}{\partial w_i} = 2e\frac{\partial y}{\partial w_i}. \tag{8.4.1}$$

It is assumed that the plant output, y, only depends on the current plant input, u, and so the rate of change of plant output with a controller coefficient can be written as

$$\frac{\partial y}{\partial w_i} = \frac{\partial y}{\partial u}\frac{\partial u}{\partial w_i}. \tag{8.4.2}$$

The term $\partial u/\partial w_i$ in equation (8.4.2) depends only on the form of the controller and denotes the rate of change of controller output with this controller coefficient. If the controller were a neural network, for example, this term could be readily calculated using the backpropagation algorithm and is a form of reference signal. The term $\partial y/\partial u$ in equation (8.4.2) is the rate of change of plant output with input for the nonlinear plant about the current operating point, u_0, say. Since the plant is assumed to have a single-valued nonlinearity with no dynamics, this term is just the slope of the input–output characteristic at the current operating point, as shown in Fig. 8.17.

In order to calculate the gradient of the squared error with respect to the controller coefficients, the differential behaviour of the plant, i.e. the derivative of its input–output characteristic, about the known operating point must thus be calculated or identified. If the

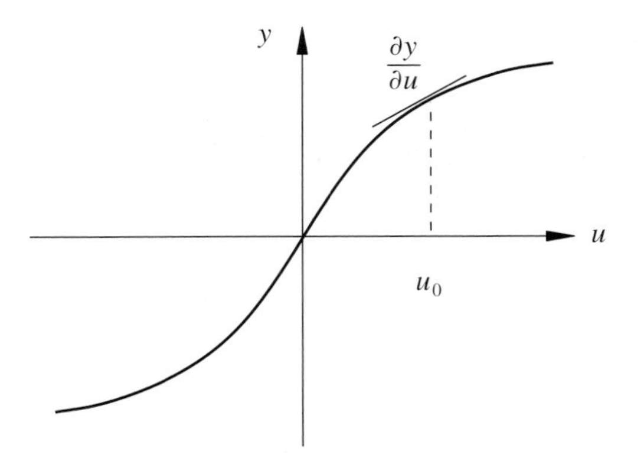

Figure 8.17 The differential response of a single-input single-output plant about an operating point u_0 is given by the slope of the input–output characteristic at this point. For plants whose outputs depend on multiple input signals, the differential response is the gradient of the corresponding multidimensional surface about the operating point.

plant has only a single input and output and is only exercised over a limited range of operating points, the derivatives of the plant input–output characteristic could be measured at a number of operating points prior to control and stored away for use by the control algorithm. Alternatively, a nonlinear model of the plant could be identified, and the derivatives at the current operating point could be deduced from this model. If the plant has many inputs and outputs, the input–output characteristic becomes a series of high-dimensional surfaces. Although it may be more difficult to identify the plant's behaviour in this case, the control approach outlined above can still be applied, provided that the rate of change of each output with every input is accounted for. There is, however, no guarantee that the method of steepest descents will converge to the global minimum of the multidimensional error surface, which may be very complicated in the general case of a nonlinear controller and a nonlinear plant.

8.4.2. Harmonic Control of Periodic Disturbances

As a first example of the adaptive feedforward control of a dynamic nonlinear system, we consider a frequency-domain approach to the control of periodic disturbances in weakly nonlinear plants (Sutton and Elliott, 1993, 1995; Hermanski et al., 1995). We assume that a periodic input to the nonlinear plant produces a periodic output with the same period and so even if the plant has memory and a dynamic response, its steady state behaviour at a given excitation frequency is completely described by the transformation of the input harmonic amplitudes to the harmonic amplitudes of the output. The general arrangement is shown in Fig. 8.18, in which the periodic input to the plant is synthesised from the weighted sum of a series of harmonic components, so that in the steady state

$$u(t) = \sum_{n=0}^{N} \left(w_n \cos(n\omega_0 t) + v_n \sin(n\omega_0 t) \right) , \tag{8.4.3}$$

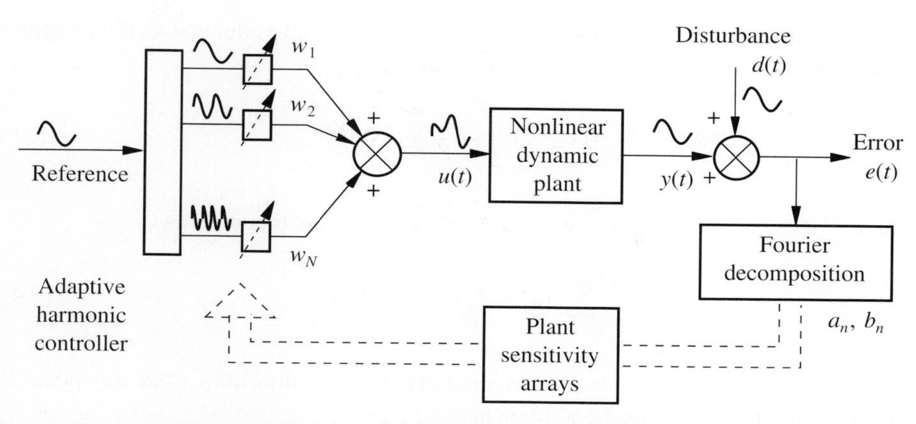

Figure 8.18 Block diagram of a frequency-domain feedforward system for the adaptive control of periodic disturbances through a nonlinear dynamic plant.

where w_n and v_n are the controller coefficients in this case, ω_0 is the fundamental frequency of the disturbance and the zeroth-order term describes the dc input, which will generally affect the operating point of the nonlinear system. The steady state error signal, measured at the output of the plant and which includes the disturbance, is also expanded as a Fourier series,

$$e(t) = \sum_{n=0}^{L} \left(a_n \cos(n\omega_0 t) + b_n \sin(n\omega_0 t) \right) . \tag{8.4.4}$$

The plant response can be represented for a given excitation frequency, ω_0, and disturbance by the multidimensional surface that represents the response of the output amplitudes $a_1, b_1, \ldots, a_L, b_L$ to the input amplitudes $w_1, v_1, \ldots, w_N, v_N$. The local behaviour of this surface can be represented by the matrix of local derivatives, or *sensitivity matrix*, given by

$$\mathbf{\Phi}_{aw}(\mathbf{u}) = \begin{bmatrix} \frac{\partial a_0}{\partial w_0} & \frac{\partial a_1}{\partial w_0} & \cdots & \frac{\partial a_L}{\partial w_0} \\ \frac{\partial a_1}{\partial w_1} & \frac{\partial a_1}{\partial w_1} & & \\ \vdots & & & \\ \frac{\partial a_0}{\partial w_N} & \cdots & \frac{\partial a_L}{\partial w_N} & \end{bmatrix} , \tag{8.4.5}$$

and the correspondingly defined matrices $\mathbf{\Phi}_{bw}(\mathbf{u})$, $\mathbf{\Phi}_{av}(\mathbf{u})$ and $\mathbf{\Phi}_{bv}(\mathbf{u})$, about the operating point $\mathbf{u} = \left[v_0 \cdots v_N \, , \; w_0 \cdots w_N \right]^{\mathrm{T}}$.

Strictly speaking, these matrices of local derivations are, in general, functions of the disturbance as well as the control inputs, but by assuming that the error signal is the superposition of the plant output and disturbance, as in Fig. 8.15, this complication is avoided. The diagonal terms in the sensitivity matrices play a similar role here to the *describing function* used to analyse feedback control systems with nonlinear elements, as described by Banks (1986), for example.

The cost function to be minimised by the control algorithm is defined as the integral of the squared error over a period of the disturbance:

$$J = \frac{1}{T_0} \int_0^{T_0} e^2(t)\, \mathrm{d}t \ , \qquad (8.4.6)$$

where $T_0 = 2\pi/\omega_0$. This can also be written, using Parseval's theorem, as

$$J = \frac{1}{2} \sum_{n=0}^{L} \left(a_n^2 + b_n^2\right) = \frac{1}{2}\left(\mathbf{y}^{\mathrm{T}}\mathbf{y}\right) \ , \qquad (8.4.7)$$

where $\mathbf{y} = \left[a_0 \ \cdots \ a_L \ , \ b_0 \ \cdots \ b_L\right]^{\mathrm{T}}$. The gradient descent algorithm used to update the controller amplitudes can thus be written as

$$\mathbf{u}(\text{new}) = \mathbf{u}(\text{old}) - \mu \frac{\partial J}{\partial \mathbf{u}} \ , \qquad (8.4.8)$$

where in this case, following the general principle described by equations (8.4.1) and (8.4.2),

$$\frac{\partial J}{\partial \mathbf{u}} = \mathbf{\Phi}(\mathbf{u})\, \mathbf{u} \ , \qquad (8.4.9)$$

where

$$\mathbf{\Phi}(\mathbf{u}) = \begin{bmatrix} \mathbf{\Phi}_{aw}(\mathbf{u}) \, \mathbf{\Phi}_{bw}(\mathbf{u}) \\ \mathbf{\Phi}_{av}(\mathbf{u}) \, \mathbf{\Phi}_{bv}(\mathbf{u}) \end{bmatrix} . \qquad (8.4.10)$$

If the plant were entirely linear, the harmonic sensitivity matrices, $\mathbf{\Phi}_{aw}$ etc., would be diagonal. Also there would be some redundancy, since for a linear system, $\partial a_p / \partial w_p = \partial b_p / \partial v_p$ and $\partial a_p / \partial v_p = -\partial b_p / \partial w_p$. For a nonlinear system, however, these matrices are generally fully populated and the relationships above do not hold in general. Also, for a nonlinear system each element of these matrices is a function of the current operating point, so that, in principle, their values should be re-identified after each iteration of the control algorithm. In practice, only estimates of the plant's differential harmonic response, obtained prior to control, are generally used in the implementation of the gradient descent algorithm, but the convergence of gradient descent algorithms is known to be robust to errors in these estimates.

Sutton and Elliott (1995) describe the result of an experiment in which a nonlinear harmonic controller was used to ensure that the displacement output of a magnetostrictive actuator was sinusoidal, even though it exhibited significant hysteretic behaviour. The physical arrangement for this experiment is shown in Fig. 8.19, in which the bottom of the magnetostrictive actuator was driven with a sinusoidal displacement from the shaker. A real-time harmonic controller implemented on a PC was used to adjust the input to the magnetostrictive actuator, $u(t)$, to minimise the mean-square value of the error signal $e(t)$ by implementing the control algorithm given by equations (8.4.8) and (8.4.9). The error signal was derived from an accelerometer on the top of the actuator. The results of these experiments are shown in Fig. 8.20 in terms of the input signal to the actuator, $u(t)$, together

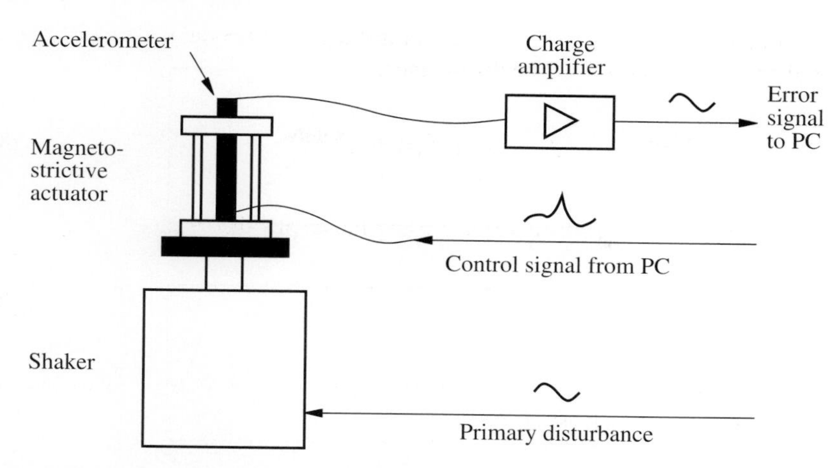

Figure 8.19 Experiment in which the magnetostrictive actuator was controlled to produce a sinusoidal displacement so that when driven by the primary displacement generated by the shaker under the actuator, the error signal measured at the accelerometer tended to zero.

with the residual error signal after control, $e(t)$. Figure 8.20(a) shows the results obtained with a linear controller, in which only the fundamental component of the control signal $u(t)$ was adjusted. The error signal clearly has residual harmonic distortion as a result of the nonlinear behaviour of the magnetostrictive actuator. Figure 8.20(b) shows the results obtained using the nonlinear harmonic control system to generate an input signal, $u(t)$, with 7 harmonics. The control signal is non-sinusoidal in this case, to compensate for the actuator's hysteresis, but the error signal is now much reduced, which indicates that the magnetostrictive actuator is now producing an almost sinusoidal displacement, which is out of phase with that produced by the shaker underneath it.

In this experiment only a single set of the harmonic plant responses, given by the matrix $\mathbf{\Phi}$, was identified for the magnetostrictive actuator, which consisted of the average responses over many different input signals about the expected operating point. The fact that the control system was always stable illustrates the robustness of gradient descent methods to errors in the assumed plant response, even in this nonlinear case. It was also found that if the model of the plant's response was simplified even further, so that only the diagonal elements of the matrices that would be present in a linear system were identified, the harmonic controller was still able to converge under many conditions. This 'linear' harmonic controller was, however, unstable when controlling some nonlinear systems, for example those with a saturating nonlinearity, in which case a fully coupled harmonic plant model was required for stable operation.

A similar harmonic approach has also been taken by Hermanski et al. (1995) for the control of vibrations in nonlinear printing machines, and by Blondel and Elliott (1999) in the adaptive compensation of a compressed-air loudspeaker, as also described by Elliott (1999). Such a loudspeaker is formed by modulating the area of an opening from a plenum chamber in which compressed air is stored. If the pressure in the plenum chamber is much greater than atmospheric pressure, the volume velocity of the air flow is linearly related to the area of the opening, but the overall efficiency of the source is very low, particularly when used in active control applications. As the plenum pressure is lowered, the efficiency

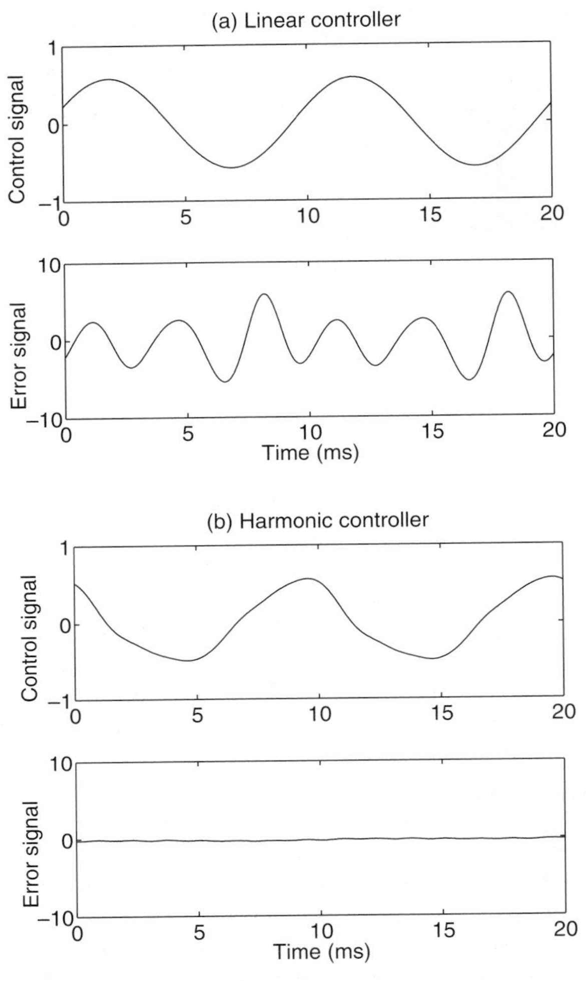

Figure 8.20 Results from the experimental arrangement shown in Fig. 8.19 in which (a) the magnetostrictive actuator is first driven by a sinusoidal input signal to minimise the mean-square error and the resulting error signal, and (b) when the magnetostrictive actuator is driven by the output of the harmonic controller and the resulting error signal in that case.

of the source becomes much greater but the volume velocity of the air flow then becomes a nonlinear function of the instantaneous pressure outside the plenum chamber and the area of the opening. An interesting feature of the compressed-air loudspeaker when used as a secondary source for active control is that if the alternating acoustic pressure at the opening of the plenum chamber can be cancelled, its volume velocity again becomes a linear function of the opening area. In order to approach this condition, however, the control system must cope with the nonlinear nature of the acoustic source during adaptation.

Blondel and Elliott (1999) describe a two-stage control strategy for using this secondary source to control a tonal disturbance. First, only the fundamental component of the opening area is adjusted, which reduces the mean-square pressure outside the plenum pressure, but still leaves significant harmonic distortion in the external pressure waveform. This is analogous to the situation shown in Fig. 8.20(a) for the magnetostrictive actuator. The alternating external pressure is much less than the plenum pressure after this first stage of adaptation, however, and so the source then becomes almost linear. The first five harmonics of the opening area can then be adjusted to minimise the external pressure, using only a linear model for the sensitivity matrix in equation (8.4.10). This two-stage control process results in a robust and stable convergence of the opening area waveform, which gives more than 30 dB attenuation of the external disturbance at the fundamental frequency, with the amplitudes of all the harmonics below this level, which is analogous to the waveforms shown in Fig. 8.20(b) for the magnetostrictive actuator. The control of periodic signals using time-domain neural networks has also been described by, for example, Fuller et al. (1991) and Snyder and Tanaka (1992).

8.4.3. Neural Controller for Random Disturbances

Finally in this section, we will consider an extension of the backpropagation algorithm that can be used for the adaptive feedforward control of nonlinear dynamic plants using time-domain signals. The formulation presented here is based on that put forward by Narendra and Parthasarathy (1990), although a direct input–output model for the plant is assumed below, whereas Narendra and Parthasarathy used a state-space model. Referring back to the block diagram shown in Fig. 8.15, we assume that the plant response can be described by the general NARMA model

$$y(n) = F\left[u(n), \cdots u(n-J), y(n-1) \cdots y(n-K)\right] , \tag{8.4.11}$$

where $F[\ ]$ is a nonlinear function, which we will assume to be differentiable. Equation (8.4.11) can be thought of as defining a surface with $J + K + 1$ dimensions.

The nonlinear controller is again assumed to have a general structure with coefficients $w_0, \ldots, w_I$, which we are seeking to adjust using the method of steepest descents to minimise the *instantaneous* value of the squared error, as in the LMS algorithm. We thus need to calculate the derivative

$$\frac{\partial e^2(n)}{\partial w_i} = 2\,e(n)\frac{\partial e(n)}{\partial w_i} = 2\,e(n)\frac{\partial y(n)}{\partial w_i} , \tag{8.4.12}$$

where w_i is one of the variable coefficients, or weights, in the nonlinear controller.

The current plant output $y(n)$ is assumed to be a differentiable function of a finite number of current and past inputs and past outputs, equation (8.4.11), and so the derivative of the current plant output with respect to the controller coefficient can be written as

$$\frac{\partial y(n)}{\partial w_i} = \sum_{j=0}^{J} \frac{\partial y(n)}{\partial u(n-j)} \frac{\partial u(n-j)}{\partial w_i} + \sum_{k=1}^{K} \frac{\partial y(n)}{\partial y(n-k)} \frac{\partial y(n-k)}{\partial w_i} , \tag{8.4.13}$$

in which we assume that the controller coefficients do not change too quickly, so that the rate of change of w_i with time does not have to be taken into account. We now define the local derivatives of the surface defined by equation (8.4.11), to be

$$\frac{\partial y(n)}{\partial y(n-k)} = a_k(u,y) , \tag{8.4.14}$$

and

$$\frac{\partial y(n)}{\partial u(n-j)} = b_j(u,y) . \tag{8.4.15}$$

The dependence of these coefficients on the state of the nonlinear plant is written in the abbreviated form (u, y), even though it is recognised that they depend on all the signals $u(n), \ldots, u(n-J)$ and $y(n-1), \ldots, y(n-K)$. If a model of the nonlinear plant described by equation (8.4.11) can be identified, using for example a neural network, then estimates of the coefficients $a_k(u,y)$ and $b_j(u,y)$ could be calculated by linearising this model about the current operating point.

Equation (8.4.13) represents a time-varying signal that, for convenience, we will write as

$$\frac{\partial y(n)}{\partial w_i} = r_i(n) , \tag{8.4.16}$$

and assuming w_i does not change too rapidly, i.e. it is quasi-static, it is approximately true that

$$\frac{\partial y(n-k)}{\partial w_i} = r_i(n-k) . \tag{8.4.17}$$

We also define the signal $t_i(n)$ to be equal to the derivative of the plant input with respect to the controller coefficient w_i, so that

$$\frac{\partial u(n)}{\partial w_i} = t_i(n) , \tag{8.4.18}$$

and we can again make the approximation, assuming that the controller coefficients change only slowly, that

$$\frac{\partial u(n-j)}{\partial w_i} = t_i(n-j) . \tag{8.4.19}$$

The signal $t_i(n)$ can be obtained by backpropagation through the potentially nonlinear controller and is thus known. If the controller were a linear FIR filter driven by delayed versions of the reference signal $x(n)$, for example, $t_i(n)$ would be equal to $x(n-i)$. In general, for a purely feedforward controller, $t_i(n)$ is equal to the signal that acts as the input to the weight w_i, filtered by the linearised response from the output of this weight to the output of the controller.

Using these definitions, equation (8.4.13) is recognised as a filtered reference signal that can be written as

$$r_i(n) = \sum_{k=1}^{K} a_k(u,y) r_i(n-k) + \sum_{j=0}^{J} b_j(u,y) t_i(n-j) , \tag{8.4.20}$$

which represents the reference signal $t_i(n)$, recursively filtered by the linearised response of the plant. The final adaptation algorithm for the controller coefficient w_i using the instantaneous steepest-descent method is thus

$$w_i(n+1) = w_i(n) - \alpha e(n) r_i(n) , \qquad (8.4.21)$$

which is a form of filtered-reference LMS algorithm. It has been shown by Marcos et al. (1992) and Beaufays and Wan (1994) that this algorithm is directly analogous to the real-time backpropagation algorithm (Williams and Zipser, 1989).

The adaptation algorithm of equation (8.4.21) can be readily extended to the case of multiple errors by direct analogy with the linear case. Instead of filtering the reference signal with a linearised model of the forward dynamics of the controller and plant, it is also possible to filter the error signal using the reverse plant dynamics, to derive a form of filtered-error algorithm (Marcos et al., 1992; Beaufays and Wan, 1994) which is analogous to the backpropagation-through-time algorithm (Rumelhart and McClelland, 1986; Nguyen and Widrow, 1990). Algorithms that can give faster training rates than those based on gradient-descent methods, as described above, are discussed by Bouchard et al. (1999).

8.4.4. Control of a Backlash Function

An example of the use of this time-domain control strategy is provided by Sutton and Elliott (1993), who describe the feedforward control of a backlash function for a random disturbance. To invert a backlash it is necessary to add or subtract a 'jump' to or from the input signal, whenever the rate of change of the input signal changes sign. This behaviour can, in principle, be synthesised by a nonlinear controller that operates on the current and immediately past values of the reference signal, so there is no need for the controller to be recursive. In these simulations, a multilayer perceptron was used as a controller, which had a single hidden layer with sigmoid nonlinearities and a linear output layer, and was driven by the current and delayed reference signals. The derivatives of the plant response, $a_k(u,y)$ and $b_j(u,y)$, were obtained from a recursive neural model of the backlash function, as described in Section 8.3, which was identified prior to control. The block diagram for the controller is shown in Fig. 8.21 and the results of the simulation are shown in Fig. 8.22.

The disturbance signal was equal to the reference signal in these simulations, which was a band-limited random signal. The inverted disturbance signal is shown in Fig. 8.22, together with the output of the backlash function after convergence of the adaptive neural controller. The tracking of the disturbance by the backlash is not perfect, but is very much better than that obtained if a linear controller is used. The input to the plant that is required to produce this output is also illustrated in Fig. 8.22. This has a strong 'differential action', as required to overcome the lost motion in the plant's backlash, and so has much stronger high-frequency components than are present in the disturbance.

8.5. CHAOTIC SYSTEMS

In this section we will briefly introduce the main properties of chaotic systems and the terms used to describe the behaviour of such systems. This is in preparation for

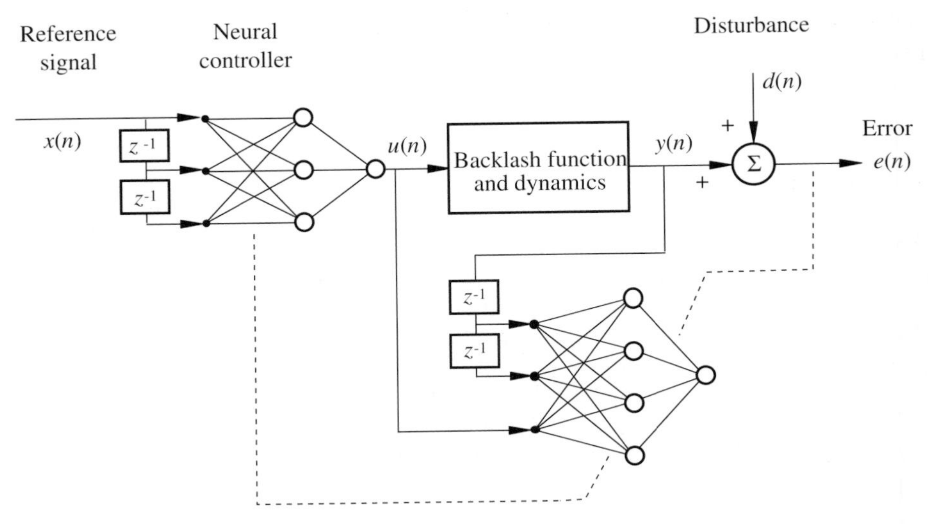

Figure 8.21 Block diagram of a time-domain neural controller used for feedforward control of a plant with backlash and second-order dynamic behaviour.

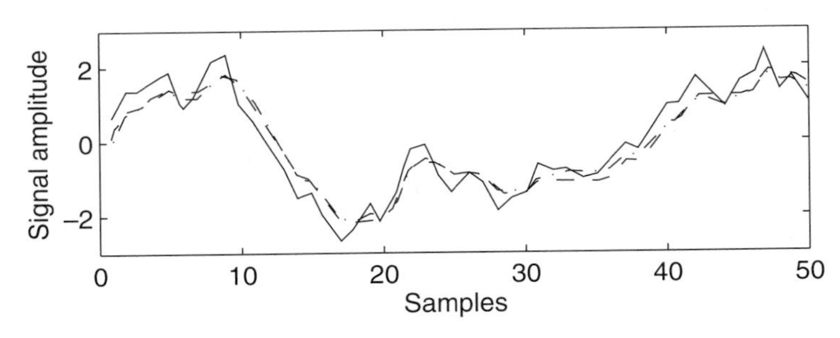

Figure 8.22 Results of the simulation shown in Fig. 8.21, in which a feedforward neural controller was used to generate a drive signal, $u(n)$ (solid line) for the nonlinear plant with backlash so that the output of the plant $y(n)$ (dashed line) best matched the inverted narrowband random disturbance $-d(n)$ (dot-dashed line).

Section 8.6 in which some of the methods that could be used to control such systems will be reviewed. We will describe how chaotic systems have an exponential sensitivity to perturbations. This exponential sensitivity to perturbations in a chaotic weather system can give rise to the so-called 'butterfly effect', whereby the beating of a butterfly's wing on one side of the world could give rise to severe storms on the other side of the world a week later (as described, for example, in Ott, 1993). It is this exponential sensitivity which makes the prospect of controlling chaotic systems so intriguing, since it suggests that very significant changes in the behaviour of a chaotic system can be achieved with only very small control signals, if only we knew how to apply such control signals. Chaos has been observed in a variety of nonlinear mechanical and fluid systems (Moon, 1992), in neural networks that have nonlinear elements (van der Maas et al., 1990), feedback controllers applied to nonlinear systems (Golnaraghi and Moon, 1991), and adaptive control systems in which the adaptation equations are nonlinear (Mareels and Bitmead, 1986).

8.5.1. Attractors

Consider a continuous-time dynamic system governed by a set of N state equations which can be written as

$$\dot{\mathbf{s}}(t) = \mathbf{F}[\mathbf{s}(t)] \ , \tag{8.5.1}$$

where $\mathbf{s}(t)$ is a vector of the waveforms of the N state variables, $\dot{\mathbf{s}}(t)$ denotes the vector of time derivatives of these waveforms, and $\mathbf{F}$ is a nonlinear function. A simple example of the state space description can be derived for an undriven *linear* second-order system, whose differential equation is given by

$$\ddot{y}(t) + c\dot{y}(t) + y(t) = 0 \ , \tag{8.5.2}$$

where $y(t)$ could represent the displacement of a mechanical system, whose damping depends on c and whose mass and stiffness have been normalised to unity. Defining the state variables to be

$$s_1(t) = y(t) \qquad \text{and} \qquad s_2(t) = \dot{y}(t) \ , \tag{8.5.3a,b}$$

they are related by the two first-order equations

$$\dot{s}_1(t) = s_2(t) \ , \tag{8.5.4}$$

and

$$\dot{s}_2(t) = -s_1(t) - c\,s_2(t) \ . \tag{8.5.5}$$

Equation (8.5.1) is a linear matrix equation in this case, which can be written as

$$\begin{bmatrix} \dot{s}_1(t) \\ \dot{s}_2(t) \end{bmatrix} = \begin{bmatrix} 0 & 1 \\ -1 & -c \end{bmatrix} \begin{bmatrix} s_1(t) \\ s_2(t) \end{bmatrix} . \tag{8.5.6}$$

The dynamic behaviour of such a system can be represented as the locus of the variation of one state variable against another in a *phase diagram*. A typical phase diagram for the undriven linear second-order system forms a spiral path into the origin as the underdamped transient due to the initial conditions decays away. If the linear second-order system above were driven by a sinusoidal force, the phase trajectory would be an ellipse in the steady state, indicating that the displacement and velocity were both sinusoidal, but in quadrature. Any initial conditions which generated a transient would again give rise to a spiralling phase trajectory, but this would always end up on the ellipse representing the steady state response, which is said to act as an *attractor* for this system.

The Duffing oscillator introduced in Section 8.1 can be written as

$$\ddot{y}(t) + c\,\dot{y}(t) - \tfrac{1}{2}\,y(t) + \tfrac{1}{2}\,y^3(t) = A\cos(\omega_d\,t) \ . \tag{8.5.7}$$

Defining the state variables in this case to be

$$s_1(t) = y(t), \qquad s_2(t) = \dot{y}(t), \qquad s_3(t) = \omega_d t \,, \tag{8.5.8a,b,c}$$

equation (8.5.7) can be written in terms of the three first-order differential equations given by (Moon, 1992)

$$\dot{s}_1(t) = s_2(t) \tag{8.5.9a}$$

$$\dot{s}_2(t) = \tfrac{1}{2} s_1(t) - \tfrac{1}{2} s_1^3(t) - c\, s_2(t) + A \cos\big(s_3(t)\big) \tag{8.5.9b}$$

$$\dot{s}_3(t) = \omega_d \,. \tag{8.5.9c}$$

The driven second-order system thus has three state variables and the phase trajectory obtained by plotting velocity, $s_2(t)$, against displacement, $s_1(t)$, can be thought of as a projection of this three-dimensional phase trajectory onto two dimensions.

The phase trajectories for the steady-state outputs of the Duffing oscillator when driven at four different amplitudes are shown in Fig. 8.23, and these correspond to the four waveforms shown in Fig. 8.5. Figure 8.23(a) shows an almost elliptical trajectory, since for small amplitude excitations the Duffing oscillator behaves almost like a linear system. Figure 8.23(b) shows the distortion introduced with a higher-amplitude driving signal, which increases the effect of the nonlinearity, but the phase trajectory still goes once round the attractor per cycle, since it has the same fundamental frequency as the driving signal. At higher driving amplitudes the period of the response is twice that of the driving signal and in Figure 8.23(c), for example, it takes two cycles of the driving frequency for the phase trajectory to travel round a closed orbit. The phase trajectory corresponding to chaotic behaviour of the Duffing oscillator is shown in Fig. 8.23(d) and this is considerably more complicated than the previous cases, because the waveform never repeats itself.

Although the phase trajectory for the chaotic motion shown in Fig. 8.23(d) is complicated, it too acts as an attractor, because trajectories close to it, caused by initialising equation (8.5.7) with slightly different conditions, will also end up following the same path in the steady state. The geometry of such chaotic attractors can be very intricate however, and can have a similar structure on arbitrarily small scales, i.e. they are *fractal* (Moon, 1992). These systems are said to have *strange attractors*. Although it is possible for systems with chaotic dynamics not to have strange attractors, and for a system with a strange attractor not to be chaotic, for most cases involving differential equations, strange attractors and chaos occur together (Ott, 1993). Although the behaviour of a chaotic system will never exactly repeat itself, and for most of the time such behaviour appears to be almost random, occasionally the behaviour appears to be periodic for a short time, before diverging again into a more complex form. These periodic orbits are said to be *embedded* within the chaotic attractor, but are unstable, so that although their presence can be observed for a short time, their amplitude grows until the more unpredictable chaotic behaviour returns.

For nonlinear systems with many state variables, the continuous-time phase trajectories can get very complicated. One way of representing these continuous-time curves is to take a 'slice' through the phase plane. Every time the phase trajectory passes through this plane, a dot is plotted. This is called a *Poincaré section* and because it represents discrete-time events, it is a special case of a *map*. If a slice of the three-dimensional phase diagram of the Duffing oscillator were taken for values of the phase of the driving signal, $s_3(t)$, which

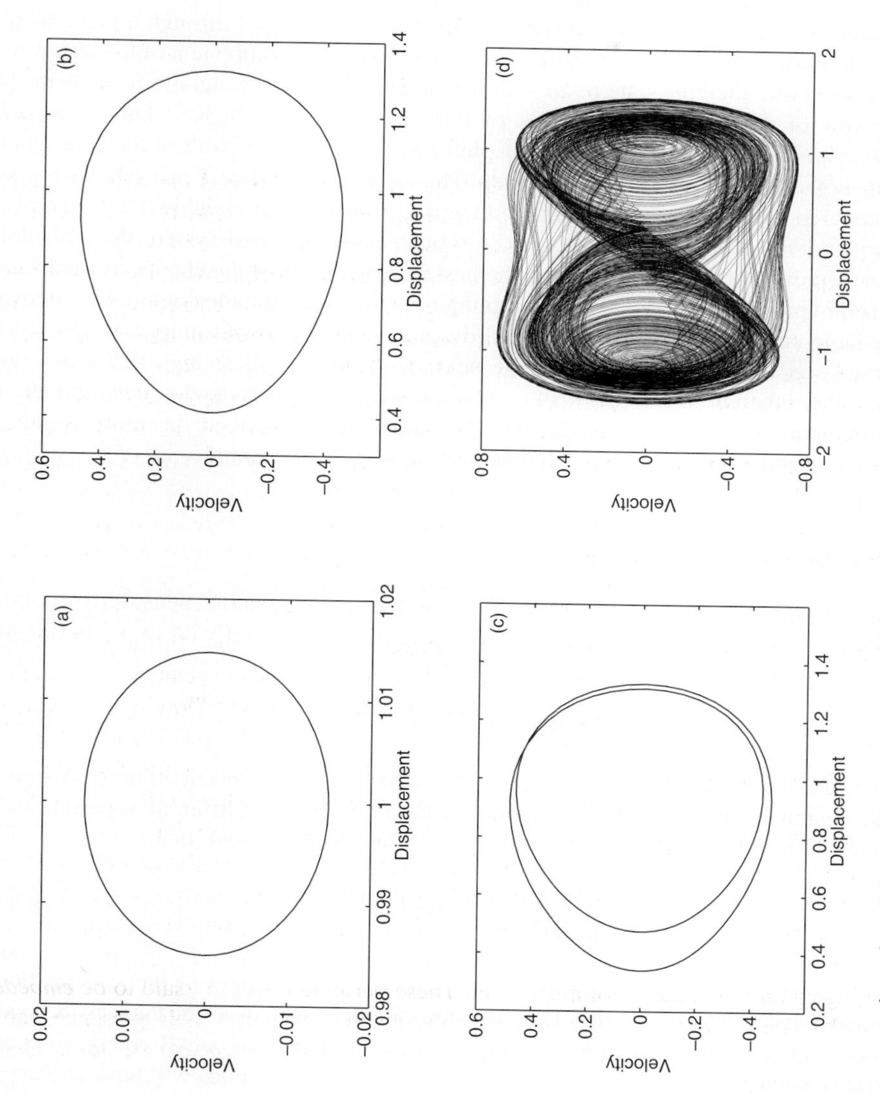

Figure 8.23 Phase-space trajectories of the steady state response of the Duffing oscillator for different amplitudes of the driving signal, A. The four cases are (a) $A = 0.00025$, (b) $A = 0.174$, (c) $A = 0.194$, (d) $A = 0.2348$, and correspond to the time-histories shown in Fig. 8.5.

were integer multiples of 2π, then this would sample the two-dimensional projections shown in Fig. 8.23 every cycle of the driving frequency. This would produce a single point for the period-1 oscillations shown in Figs. 8.23(a) and (b), two points for the period-2 oscillation shown in Fig. 8.23(c), and a dense grid of points for the chaotic motion shown in Fig. 8.23(d). It is the displacement component of such a map which was plotted for different driving amplitudes in the bifurcation diagram shown in Fig. 8.6.

Instead of characterising all the state variables when they pass through a plane in state space, they could be sampled at regular intervals. If only one component of the state vector can be observed, then the state of an N-dimensional system can generally be described by the vector of $D > 2N + 1$ past values of this one variable, which is known as *delay coordinate vector*. The map generated by plotting the observed variable at one time against that at another time is called the *return map*. The fact that the full state-space behaviour can be recovered from this vector of sampled signals from a single variable is an example of the *embedding theorem* (Takens, 1980). Although in a noise-free system the embedding theorem applies for any sampling rate, the best representation of the chaotic system can be obtained in practice by choosing the sampling time to give a suitable compromise between being large enough so that all the system dynamics have an opportunity to influence the observed signal from one sample to the next, and being small enough that when noise excites the inherent instabilities of a chaotic system this does not contaminate the measurements over the timescale of observation, as described in more detail by Broomhead and King (1986) and Abarbanel et al. (1998) for example.

8.5.2. Lyapunov Exponents

The occurrence of chaotic motion is defined by an exponential sensitivity to initial conditions. We assume that equation (8.5.1) is initialised at $t = 0$ with two sets of states, $\mathbf{s}_A(0)$ and $\mathbf{s}_B(0)$, which are close together, so that

$$\|\mathbf{\Delta}(t)\| = \|\mathbf{s}_A(t) - \mathbf{s}_B(t)\| , \tag{8.5.10}$$

is very small if $t = 0$, where $\| \ \|$ denotes the Euclidean norm, as described in the Appendix. If the system is chaotic then the separation between the two different solutions to the dynamic equations will diverge exponentially as time progresses, so that

$$\frac{\|\mathbf{\Delta}(t)\|}{\|\mathbf{\Delta}(0)\|} = \alpha \exp(\lambda t) . \tag{8.5.11}$$

For a one-dimensional system, λ is the single *Lyapunov exponent* of the system. For an N-dimensional system there are N Lyapunov exponents, which describe the way the system dynamics amplify small changes in initial conditions when the small changes occur in different directions in phase space. A system with chaotic dynamics will have at least one Lyapunov exponent whose magnitude is greater than zero. Although their experimental measurement is fraught with difficulty, as described by Abarbanel et al. (1998) for example, the Lyapunov exponents are perhaps the most important indicator of chaotic behaviour.

The suppression of chaotic behaviour is discussed below, but it may also be possible to use the underlying chaotic dynamics of some types of disturbance to predict the future value of the disturbance waveform and thus achieve better control than would be possible with a linear controller (Strauch and Mulgrew, 1997). Such a nonlinear prediction system using neural networks has been described for the active control of fan noise by Matsuura et al. (1995), which showed a clear improvement in performance over a purely linear controller.

8.6. CONTROL OF CHAOTIC BEHAVIOUR

For the Duffing oscillator described by equation (8.5.7), the system could be moved in and out of chaos by slowly adjusting the driving amplitude A. The reliable control of chaos using this method requires considerable knowledge of the underlying dynamics of the system, and the ability to make potentially large changes in A. In this section we will briefly introduce some other methods for the control of chaos that require only small variations in a parameter of the system, although typically they do require that these changes can be made on the timescale of the chaotic motion.

Although these methods are able to control the chaotic behaviour in the system, they do not generally suppress the system's dynamic response. It would therefore be more correct to say that the chaos is being suppressed rather than being controlled. After suppression, the system will generally still display oscillations, and whether this behaviour is any more beneficial than the original chaotic motion will depend on the application and the form of the oscillation. A chaotic attractor will typically have a number of periodic orbits embedded within it, although these orbits are unstable in the sense that the smallest perturbation will push them off into another orbit, which is part of the normal chaotic behaviour. One can potentially select which of these orbits to force the system to follow, and at an initial design stage the behaviour of each of these orbits would be examined to determine which would be most beneficial in the particular application.

A number of different methods of suppressing chaotic motion have been suggested and their relative advantages and disadvantages will depend very much on the particular chaotic system being considered (Chen and Dong, 1993; Shinbrot, 1995; Vincent, 1997). For periodically excited systems, chaotic behaviour can be suppressed by introducing small perturbations to the periodic excitation, which are typically at a subharmonic of the driving frequency (Braiman and Goldhirsch, 1991; Meucci et al., 1994). The amplitude and/or phase of such a perturbation can also be modified using closed-loop controllers to maintain control, as described for example by Aguirre and Billings (1995b). Other methods of feedback use direct feedback control and the most popular feedback technique is known as the OGY method, as described below.

8.6.1. The OGY Method

Following Shinbrot et al. (1993) we assume that the chaotic system under control obeys the discrete-time N-dimensional map

$$s(n+1) = \mathbf{F}[s(n), p] \ , \qquad (8.6.1)$$

where $s(n)$ is an N-dimensional state vector, p is some system parameter with nominal value p_0, and $\mathbf{F}$ is the nonlinear system function. We also assume that we have decided at the design stage to stabilise the system about a fixed-point attractor of the nominal system, which has a constant state given by the vector s_F, so that

$$s_F = \mathbf{F}[s_F, p_0] \ . \tag{8.6.2}$$

Stabilisation into higher-order orbits can also be considered if the state variable s includes sufficient past values for equation (8.6.2) to describe this orbit (Shinbrot et al., 1993).

For state vectors close to s_F and the system parameter close to p_0, we can approximate the dynamic behaviour of the system by the *linear* map

$$[s(n+1) - s_F] = \mathbf{A}[s(n) - s_F] + \mathbf{b}[p - p_0] \ , \tag{8.6.3}$$

where $\mathbf{A}$ is an $N \times N$-dimensional Jacobian matrix, $\mathbf{A} = \partial\mathbf{F}/\partial s$, $\mathbf{b}$ is an N-dimensional column vector, $\mathbf{b} = \partial\mathbf{F}/\partial p$, and the partial derivatives are evaluated about s_F and p_0. The eigenvalues of the matrix $\mathbf{A}$ can be associated with the Lyapunov exponents of the system, so that, by definition, at least one of the eigenvalues of $\mathbf{A}$ will be greater than unity and hence be associated with an unstable mode in a chaotic system.

To force the system from the chaotic orbit into the fixed one, we wait for the state vector to fall close to s_F, so that equation (8.6.3) is reasonably valid, and then apply a change to the system parameter p at each iteration of the map that is proportional to the distance of $s(n)$ from s_F, so that

$$[p(n) - p_0] = -\mathbf{k}^T[s(n) - s_F] \ , \tag{8.6.4}$$

where $\mathbf{k}$ is vector of feedback gains. By substituting equation (8.6.4) into (8.6.3), we can see that the dynamics of the system near the fixed point, with feedback to the parameter p, is now described by

$$[s(n+1) - s_F] = [\mathbf{A} - \mathbf{b}\mathbf{k}^T]\ [s(n) - s_F] \ . \tag{8.6.5}$$

If a feedback gain matrix can be found that ensures that the modulus of all the eigenvalues of $[\mathbf{A} - \mathbf{b}\mathbf{k}^T]$ is less than unity, then equation (8.65) will describe a stable system, whose transients will decay away until the system continuously follows the orbit described by s_F. The problem of choosing an appropriate gain vector $\mathbf{k}$ is the same as that encountered in linear state-space control theory, and if $\mathbf{A}$ and $\mathbf{b}$ are *controllable* (Franklin et al., 1994) then the eigenvalues of $[\mathbf{A} - \mathbf{b}\mathbf{k}^T]$ can, in principle, be arbitrarily chosen using pole placement methods. This method of control was originally described by Ott, Grebogi and Yorke (1990) and is generally referred to as the OGY method. For low-dimensional systems, a trial and error procedure has also been successfully used to design $\mathbf{k}$.

8.6.2. Targeting

Only small perturbations in the system parameter p are used to control the system using the OGY method, and so no perturbation is applied until the states of the system are observed to be close to the desired orbit. Specifically, if Δp_{max} is the maximum perturbation that can be applied, then the system is allowed to evolve until the following condition is satisfied:

$$\left|\mathbf{k}^T\left[s(n)-s_{\mathrm{F}}\right]\right|<\Delta p_{\max}\ , \tag{8.6.6}$$

at which point the control law given by equation (8.6.4) is switched on.

One of the practical problems with this kind of control in a high-dimensional system is that it may take a long time for the states of the system to come close to the desired orbit. If a sufficiently good model of the overall dynamics of the system is available, of the form given in equation (8.6.1), it may be possible to make changes in p when $s(n)$ is far from s_F, such that the system evolves to be closer than s_F than it would otherwise be. This technique is known as *targeting* (Shinbrot et al., 1990), and similar techniques have been used to control the chaotic motion of a spacecraft using only very small rocket bursts (as described by Shinbrot et al., 1993).

Whereas targeting requires a global model of the system under control, as given by equation (8.6.1), the design of the feedback controller for the OGY method only requires knowledge of the local dynamics of the system, as given by equation (8.6.3). It is possible to identify these local dynamics experimentally with only very limited prior knowledge. The local dynamics are best determined from an examination of the state vector during the chaotic motion of the uncontrolled system, and although the state vector can rarely be observed directly, it can be reconstructed from time histories of only a single observed variable using the embedding theorem. If two successive vectors are observed that are both close to the region of interest, s_F, these are recorded and the observation continues. When a large number of such pairs of state vectors have been measured, then the matrix $\mathbf{A}$ in equation (8.6.3) can be estimated using least-squares techniques. If the nonlinear parameter p is then perturbed and the resultant change in the state vector is measured a number of times, the vector $\mathbf{b}$ can be similarly identified.

This method of identification and control has been used to stabilise lasers (Roy et al., 1992) and similar methods have been used to control the arrhythmias artificially introduced into an animal heart, which are found to be chaotic in form (Garfinkel et al 1992). The heart was stabilised into its normal regular beating pattern by applying an electrical stimulus at a time determined from the observed state of the heart. It has also been argued that chaotic behaviour is more desirable in some biological systems than periodic behaviour. In experiments on hippocampal slices from rats brains, for example, it was found that when the potassium level in the cerebrospinal fluid was too high, the slice exhibited bursts of synchronised neural behaviour similar to those that characterise brain seizures (Schiff et al, 1994). This behaviour was prevented in a laboratory experiment by applying a destabilising control signal so that the behaviour of the system returned to a more chaotic form.

8.6.3. Application to a Vibrating Beam

Figure 8.24 shows the results of applying the OGY method of control to a simulation of the Duffing oscillator described by equation (8.5.7), as discussed by Sifakis and Elliott (2000). Delayed samples of the output waveform were used to construct the state vector in this case and Fig. 8.24 shows the results when the OGY method was used to control the system about the unstable period-2 orbit embedded in the chaotic attractor. Although the OGY control method begins to work at the zeroth sample in Fig. 8.24, it takes until the 360th

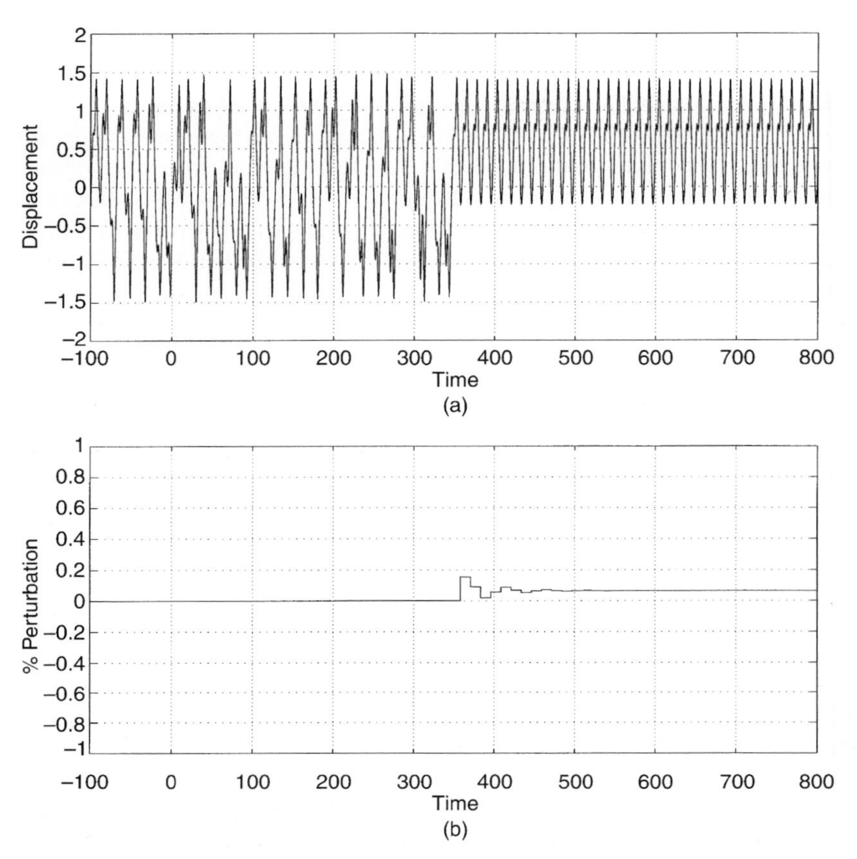

Figure 8.24 Time history of (a) the normalised displacement and (b) the perturbation in the driving signal when the OGY method was used to suppress the chaotic motion in a simulation of the Duffing oscillator, and drive the response into a period-2 orbit.

sample before the trajectory is close enough to the fixed point of the period-2 orbit for the linear approximation given by equation (8.6.3) to be valid. At this time the feedback controller is activated and the system is stabilised into the periodic orbit. After an initial transient during which control is established, the control perturbations, which were made to the driving amplitude in this case and are shown in the lower part of Fig. 8.24, become very small. Perturbations of less than 0.1% of the driving amplitude are required to suppress the chaos in this case.

The Duffing oscillator was motivated in Section 8.1 by showing how it describes the behaviour of a beam with a nonlinear restoring force. An investigation into an experimental version of such a beam has been reported by Ditto et al. (1990) and Spano et al. (1991), who used a thin magnetoelastic beam, whose stiffness could be controlled with an external magnetic field. The beam was mounted vertically in a magnetic field that had a dc component, which was modulated for control, and an ac component, which drove the system with a sinusoidal excitation of frequency 0.85 Hz. The position of the ribbon was measured with a sensor near its base, and the return map was experimentally observed by sampling the sensor signal at the excitation rate. The local dynamics about the fixed points

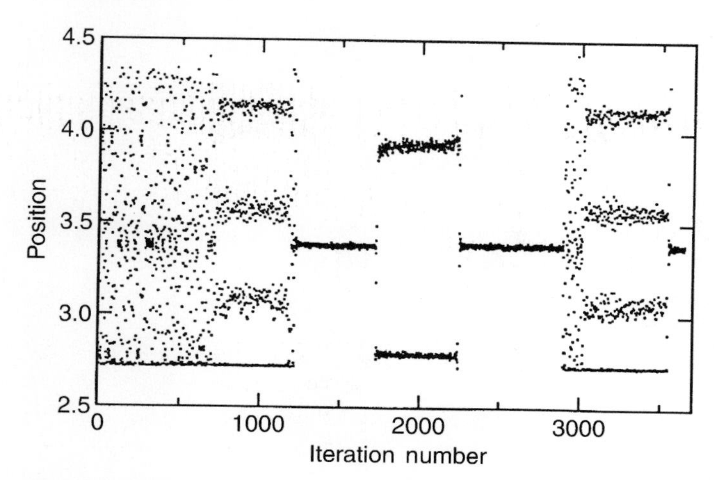

Figure 8.25 Time series of the sensor output of an experimental beam, sampled at the excitation frequency of 0.85 Hz, as the control system was switched between no control, control about the period-1 orbit, period-2 orbit and the period-4 orbit. (After Spano et al., 1991).

in the return map associated with the period-1, period-2 and period-4 orbits were then identified, and feedback controllers were designed which could stabilise the system about each of these points.

Figure 8.25 shows the time series of the measured position, sampled at the excitation frequency, as the control system is switched between no control, control about the period-1 orbit, the period-2 orbit and the period-4 orbit. Control about each of these orbits was achieved by perturbing the magnetic field by less than 0.4% of its mean value. Spano et al. (1991) went on to investigate the robustness of the control algorithm to external random disturbances to the magnetic field and showed that although the control system could occasionally be destabilised by levels of noise that were greater than about 1% of the alternating component of the magnetic field, it was robust for disturbance levels below this level.

Thus, we can see that even systems with very severe nonlinear behaviour can be actively controlled, and in some cases the extreme sensitivity of some nonlinear systems to initial conditions can be used to our advantage in being able to control the system with very little control effort.

9

Optimisation of Transducer Location

9.1. THE OPTIMISATION PROBLEM

Transducers are used for three main purposes in most active control systems:

(1) as reference sensors, which measure the primary disturbance before it reaches the secondary sources, and provide the reference signals;
(2) as secondary actuators, which generate the sound or vibration field that reduces the original disturbance; and
(3) as error sensors, which measure the residual field at selected locations due to both the primary and secondary sources, and whose outputs are used to adjust the characteristics of the controller in adaptive systems.

The typical positions of these transducers in a feedforward system for the active control of random road noise in a car, for example, are illustrated in Fig. 9.1. The design of the

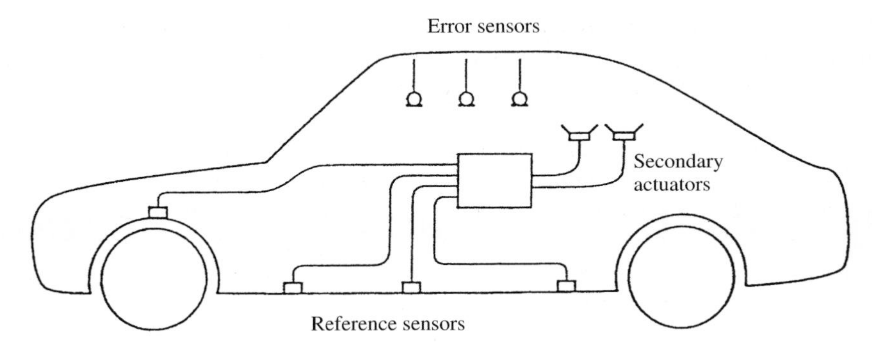

Figure 9.1 An example of a feedforward system for the active control of road noise in cars, showing the position of the transducers used for reference sensors, secondary actuators and error sensors.

electronic controller for such an active control system has been discussed in previous chapters under the assumption that the positions of the sensors and sources are known. The problem addressed in this chapter is where to locate the sensors and secondary sources for the best control performance.

Some guidance as to where it would be sensible to position the transducers can be gained from the physics of the active control problem. If, for example, only a small number of lightly-damped acoustic modes need to be controlled in an enclosure, the loudspeakers used as secondary actuators should be distributed such that they are close to a pressure maximum (antinode) of at least one of these modes, so that the combination of actuators can effectively couple into each of these modes and control them independently. To control the low-order acoustic modes in a rectangular enclosure with rigid walls, the secondary loudspeakers could be placed near the corners of the enclosure, for example. Similarly, the error microphones should be placed so that all the acoustic modes have an influence on their outputs, so that the modes are observed.

In practice, however, the shape of the modes being controlled is rarely known with any accuracy and it is not obvious on physical grounds which combination of actuators would give the best overall control across a range of modes. The practical problem of transducer location thus often comes down to a numerical search, to select the subset of locations which gives the best performance from a much larger set of possible locations.

The location of the reference sensors for the feedforward control of road noise in cars presents something more of a dilemma. It is clear that the primary disturbance in this case originates at the wheels of the vehicle. In order to obtain the greatest time advance in the reference signals, the reference sensors, which are typically accelerometers, would have to be positioned on the wheel hubs. Unfortunately the transmission path for the road vibrations from the wheel hubs to the vehicle body through the suspension system is not very linear. A linear feedforward controller excited by reference signals on the wheel hubs and driving loudspeakers in the cabin will thus have a limited performance due to this nonlinearity. Alternatively, the reference sensors can be positioned on the vehicle body, on the other side of the suspension system. In this case the path of the primary disturbance from the body vibration to the interior sound is almost entirely linear, but there is less delay between the body vibration and the sound inside, which limits the performance of a practical controller with a finite processing delay because of the requirement that the controller must be causal.

The positioning of the reference sensors in this case is thus a compromise between putting them too far upstream, which allows relatively long delays in the controller, but suffers from a transmission path which is nonlinear, and putting them too close to the error sensors, in which case the transmission path is nearly linear but the controller delays must be extremely small (Sutton et al., 1994). Apart from these basic physical compromises, there is generally a far larger number of possible locations at which reference sensors could be positioned than the number of reference sensors which could be used by a practical feedforward controller. Also, the cost of such a control system is partly determined by that of the transducers used as reference sensors, and so an economically viable system must use the minimum possible number of such transducers.

9.1.1. The Combinatorial Explosion

As with the locations of the secondary actuators and error sensors, the problem of the optimisation of the location of the reference sensors is thus partly guided by physical considerations, but in practice generally comes down to a combinatorial problem of how to select the K best reference sensor positions from N possible locations. The number of possibilities of selecting K positions from a possible N locations, where the order of the selection of the positions is obviously not important, is given by

$$ {}_N C_K = \frac{N!}{K!(N-K)!}\,, \tag{9.1.1}$$

where $N!$ denotes N factorial, i.e., $N(N-1)(N-2)\cdots 1$. The number of combinations of reference signals can become very large, even if choosing only a moderate number of actual sensor positions from a somewhat larger number of possibilities. This *combinatorial explosion* can be illustrated by using equation (9.1.1) to calculate the number of possible combinations of choosing 10 sensor locations from various numbers of possibilities, as illustrated in Fig. 9.2. There are about 8.5×10^8 ways of choosing 10 sensors from 40

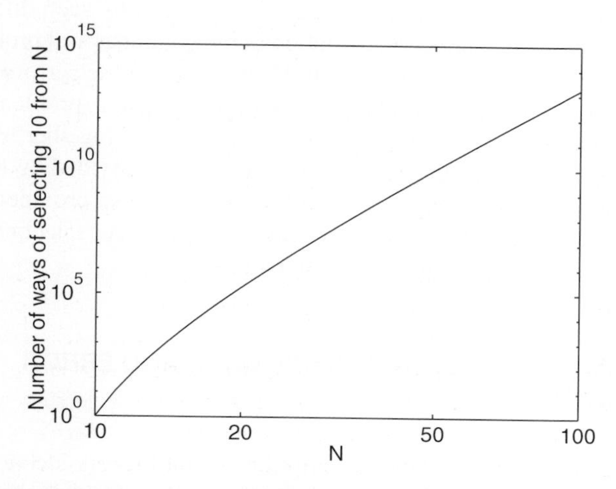

Figure 9.2 The number of possible ways of selecting 10 practical transducer locations for use in an active control system from N possible locations.

possible combinations, and over 1.6×10^{12} ways of choosing 10 sensors from 80 possible combinations.

Given a set of recordings of the signals from reference sensors in all practical locations on a car, and a simultaneous set of recordings of the pressure waveforms in the car, it would be possible to calculate the potential performance of a feedforward active control system using any subset of these reference signals, using the methods outlined in Chapter 5. If there were 80 possible locations, but only 10 of these were to be used by the control system, for example, and the calculation of the performance of each control system took 10 seconds, it would still take over 500,000 years to calculate the performance of all possible combinations of reference signals.

9.1.2. Chapter Outline

The main purpose of this chapter is to introduce some of the methods by which practical solutions to this combinatorial problem can be found. The formulation of the combinatorial problem for selecting both actuators and sensors in a feedforward control system is detailed in Section 9.2. The methods described in this chapter are examples of *guided random search* methods. In other words, they are to some extent stochastic in operation, and repeated application of the same search method on the same data set would not necessarily produce the same result. Unguided random searches could be used to give solutions to such problems, but it has been found that incorporating an element of guidance into the random search can enormously reduce the number of possibilities that have to be searched before a 'good' solution is found. Generally a 'good' solution in this context is not necessarily the very best one, which could, in principle, be found by an exhaustive search, but one which has a performance that is close enough to the best for engineering purposes.

Two families of such guided random search methods are described here: *genetic algorithms*, which are based on biological selection, in Section 9.3, and *simulated annealing* methods, which are inspired by the physical organisation of crystals during cooling, in Section 9.4. These methods are also called *natural algorithms*, because they are both inspired by nature. It is striking how methods motivated by such different fields can both provide a very efficient engineering solution to the combinatorial problems described here and, for the example described in Section 9.5, how similar the performance of the two methods can be. A comparison of these two algorithms for a problem involving the optimisation of a structure to minimise vibration transmission, together with a discussion of several other types of guided random search algorithm is provided by Keane (1994). A recent survey of actuator and sensor placement problems is also provided by Padula and Kincaid (1999). Finally, in Section 9.5 the need for the transducer selection to be robust to realistic changes in operating conditions is also emphasised.

9.2. OPTIMISATION OF SECONDARY SOURCE AND ERROR SENSOR LOCATION

Before embarking on a more detailed description of transducer selection using natural algorithms, some of the physical aspects particularly associated with secondary source and error sensor location will be addressed in this section.

9.2.1. The Performance Surface

In order to understand the various methods that have been used to solve the optimisation problem, the distinction between two classes of problem must be made clear. The first class is that in which the optimisation problem is a choice between a limited number of fixed possible locations, as described in Section 9.1. This may be called *combinatorial optimisation*, and is the one on which we concentrate in the remaining sections of this chapter. The second class of optimisation problem is when each of the secondary actuators, for example, can be moved to *any* position in a one-, two- or three-dimensional space. This may be called *continuous-domain optimisation*. An example is illustrated in Fig. 9.3(a) for the case of a single acoustic secondary source positioned in a two-dimensional plane in a three-dimensional enclosure, which may for example represent an aircraft passenger cabin,

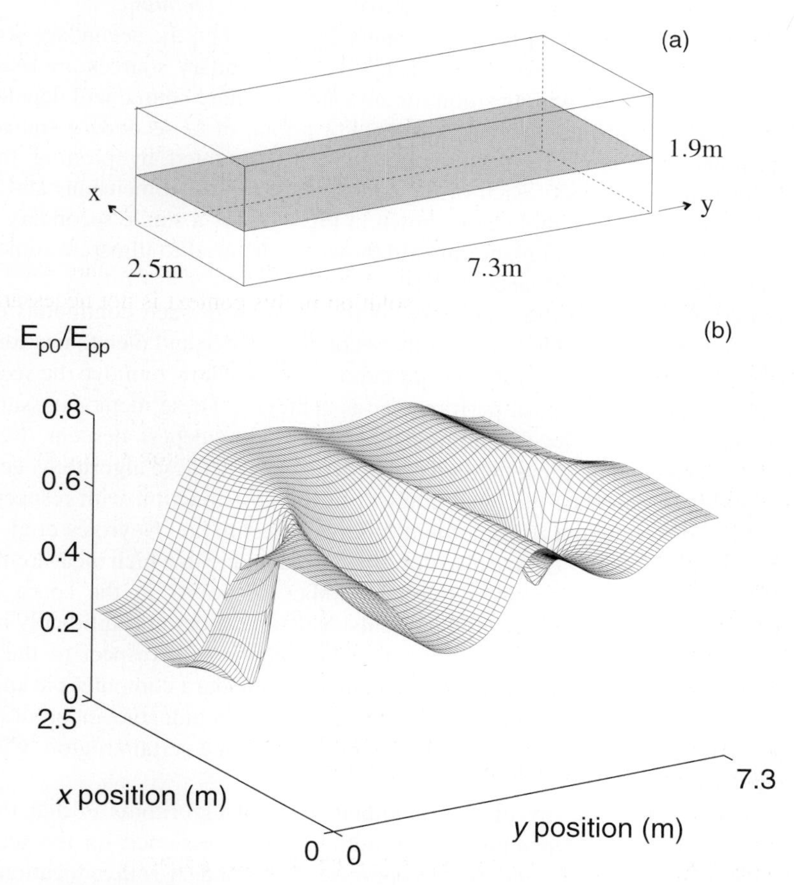

Figure 9.3 A rectangular enclosure (a) in which the pure tone sound field is driven by a combination of primary sources and controlled by a single secondary source positioned on the plane parallel to the *x–y* axes shown shaded, and (b) the fractional change in the total acoustic potential energy in the enclosure, E_{p0}/E_{pp}, as the secondary source is moved to each position on this plane and adjusted to minimise the total acoustic potential energy in the enclosure.

in which several pure-tone primary sources are located. At each location of the secondary source in this two-dimensional plane, the total acoustic potential energy within the enclosure, the cost function in this case, is minimised by adjusting the amplitude and phase of the secondary source. Figure 9.3(b) shows a contour map of the resultant minimised potential energy in the whole enclosure, E_{p0}, divided by the total acoustic potential energy in the enclosure before control, E_{pp}, as a function of the location of the single secondary source in the two-dimensional plane. Clearly, there are three regions in which positioning the secondary source gives significant reductions in acoustic energy within the room. With the secondary source close to the positions in the xy plane in Fig. 9.3 given by either (0.0, 2.4) or (1.2, 5.9), the acoustic energy in the enclosure can be reduced to about 30% or 40% of its initial value. If the secondary source is positioned close to the coordinates (1.4, 0.9), however, the total acoustic potential energy in the enclosure can be reduced to about 15% of its initial value This region thus contains the *global minimum* for this optimisation problem, while the other regions referred to above contain *local minima*.

It is more difficult to represent the performance surface when the secondary source is allowed to move in three dimensions or when several secondary sources are used. The variation of the cost function with position for any one secondary source will depend upon the positions of all the other secondary sources. For a total of M secondary sources in a three-dimensional enclosure, the performance surface generated by plotting the cost function against the positions of each of the sources has $3M + 1$ dimensions and is thus impossible to visualise. The simple case shown in Fig. 9.3, for a single secondary source positioned in a two-dimensional plane can still, however, be used to illustrate some of the important features of the performance surface.

One method of optimising the secondary source location in such continuous-domain problems is to choose an initial location for the secondary source and then to use some sort of *gradient descent* or *iterative improvement* method to decide how to move the secondary source so that better active control performance is achieved. These methods assume that the performance surface is locally unimodal, and include steepest descent, Newton's method and the conjugate gradient method (Press et al., 1987). These algorithms generally rely on being able to calculate the local gradient of the cost function with respect to the spatial variables, or estimate it using a finite-difference technique. Nayroles et al. (1994) and Martin and Benzaria (1994) have also described a technique in which measurements of the secondary source responses at a limited number of points in the space can be interpolated to estimate their values at other points. Newton's method additionally requires that the second derivatives of the cost function be known with respect to the spatial variables, which are the elements of the Hessian matrix. Standard computer packages can be used to implement these algorithms, which can also include the incorporation of constraints, such that the secondary source can only be within a certain region (Clark and Fuller, 1992; Yang et al., 1994).

One major problem with any of these gradient descent algorithms is that they can converge to local minima. Depending on the initial position assumed for the secondary source, a gradient descent algorithm used to optimise the secondary source location for the problem illustrated in Fig. 9.3 may find either the global minimum or either of the two local minima. This is an inherent problem with any deterministic search routine that alters the position of the secondary source using purely local information about the shape of the performance surface. The only way to ensure that the deepest of several minima is reliably selected is to use a method that surveys the whole of the performance surface in some way.

To calculate the cost function for every location on a fine grid would generally take a prohibitively long time, however, and some more efficient search method must be used. Some search methods, such as dynamic hill climbing (Yuret and de la Maza, 1993), achieve this by initialising a gradient-based algorithm from a number of different starting points. There is, however, a danger with any gradient-based method that the solution found will correspond to a very sharp minimum in the performance surface, which makes the performance very sensitive to small changes in the position of the actuator. By their nature, natural algorithms tend not to find these sharp minima and so retain a degree of robustness to uncertainties in the position, as will be further discussed in Section 9.5.

If natural algorithms were applied to a continuous-domain problem such as that shown in Fig. 9.3, the position of the secondary source would generally be coded as a finite-size word. The continuous-domain problem then reduces to a combinatorial problem, albeit an enormous one. Natural algorithms are stochastic, in that they are based on making random changes to the secondary source position, but then use rules to decide whether to retain such changes. The random search is thus guided towards the correct solution. An important property of natural algorithms, in view of the discussion above, is that the average size of the changes they induce in the source position tends to decrease with time. At the start of the optimisation process the selected changes in position are large, and so the algorithm obtains some information about the shape of the entire performance surface. As the optimisation proceeds, the selected changes in position tend to become smaller in magnitude and, hopefully, the algorithm homes in on the global minimum. The word 'hopefully' is significant in this context. These algorithms are driven by a random process and so it is possible that a single application of the algorithm may, by chance, completely miss the global minimum. By careful choice of the parameters used in the algorithm, however, the chance of this happening can be reduced. In order to check its performance, a natural algorithm can be run several times for the same optimisation problem, and the reliability of the result assessed by the spread of final results from the different runs. Benzaria and Martin (1994) have discussed using a combination of a such a guided random search technique, for coarse positioning of the actuators in an active noise control problem, and gradient descent methods, for fine positioning, and have shown good results in a model problem in which it was possible to move the actuators to any position within the enclosure. In most practical problems there are only a finite number of possible locations in which the secondary actuators or the error sensors can be positioned, even though this number may be relatively large, and so we will concentrate in this chapter on the use of a guided random search algorithm for the solution of the combinatorial problem outlined above. The connection between the combinatorial problem of choosing between discrete actuator locations and the variable reduction and subset selection techniques used in the statistics literature have been discussed by Ruckman and Fuller (1995).

9.2.2. Feedforward Control Formulation with Reduced Sets of Secondary Actuators

We now consider the selection of a set of secondary sources and error sensors from a larger but finite set of possible locations for these transducers for a feedforward control system at a single frequency. This formulation demonstrates how the cost function required to choose the best subset of all of these possible secondary sources and error sensor locations can be calculated from experimental estimates of transfer responses, measured from a large number of secondary sources to a large number of error sensors, and the measured primary disturbances at these error sensors. The vector of complex signals at a single frequency measured at the total number of error sensor locations can be written as

$$\mathbf{e}_{\mathrm{T}} = \mathbf{d}_{\mathrm{T}} + \mathbf{G}_{\mathrm{TT}}\,\mathbf{u}_{\mathrm{T}} \ , \tag{9.2.1}$$

where $\mathbf{d}_{\mathrm{T}}$ is the vector of complex disturbances, measured at the total number of error sensors, L_{T}, in the absence of control, $\mathbf{u}_{\mathrm{T}}$ is the vector of complex input signals which drive the total number of possible secondary sources, M_{T}, and $\mathbf{G}_{\mathrm{TT}}$ denotes the $L_{\mathrm{T}} \times M_{\mathrm{T}}$ matrix of transfer responses from each of the M_{T} secondary sources to each of the L_{T} possible error sensors. A method of manipulating the full matrix of transfer functions from all possible actuators to all possible sensors to estimate optimum transducer locations is described by Heck (1995) and Heck et al. (1998). These authors use a technique for estimating the number of significant singular values in $\mathbf{G}_{\mathrm{TT}}$ to determine the required number of transducers, and a pivoting of the columns of its QR factorisation to sequentially estimate their positions. For an active sound control system in an enclosure, the number of non-zero singular values in such a matrix should, in principle, be equal to the number of acoustic modes that can be excited, which can be controlled and observed by an equal number of loudspeakers and microphones. In practice, there are an infinite number of modes excited in such systems and there may not be a clear division between the singular values which correspond to the significant modes and those corresponding to the modes which can safely be ignored, as was seen in Fig. 4.5 of Section 4.3. In the active control of sound radiation, however, the singular values of $\mathbf{G}_{\mathrm{TT}}$ can fall more clearly into a group which is significant for control and a group which is not. It is also observed that the control performance is far more dependent on actuator placement when only controlling the components of the primary field corresponding to the significant singular values than when controlling a larger number, and in this case the deterministic methods outlined above can be very successful (Heck, 1999).

A conceptually similar method of selecting the number and location of the secondary actuators is discussed by Asano et al. (1999). These authors emphasise that if one column of the transfer matrix, $\mathbf{G}_{\mathrm{TT}}$, is linearly dependent on the others, then the corresponding actuator has the same effect as the other actuators. The actuators are thus selected sequentially using Gram–Schmidt orthogonalisation so that the column vector of $\mathbf{G}_{\mathrm{TT}}$ corresponding to the currently selected actuator has the maximum independence from the column vectors of the previously selected actuators. They also emphasise that this technique ensures that the final system is not ill-conditioned.

Returning to the formulation of the cost function for the selection procedure, the measurements taken to define $\mathbf{d}_{\mathrm{T}}$ and $\mathbf{G}_{\mathrm{TT}}$ are initially used to determine the best selection of the possible secondary sources to minimise some practical cost function, for example

$$J_{\mathrm{T}} \ = \ \mathbf{e}_{\mathrm{T}}^{\mathrm{H}}\mathbf{e}_{\mathrm{T}} + \beta\mathbf{u}_{\mathrm{T}}^{\mathrm{H}}\mathbf{u}_{\mathrm{T}} \ , \tag{9.2.2}$$

where β is the effort weighting, which can be used to discriminate against control solutions requiring very large actuator drive voltages, as described in Chapter 4. If a reduced subset of, say, M_{R} secondary sources is selected, so that the inputs to all the others are assumed to be zero, the total vector of error signals can be written as

$$\mathbf{e}_{\mathrm{T}} = \mathbf{d}_{\mathrm{T}} + \mathbf{G}_{\mathrm{TR}}\mathbf{u}_{\mathrm{R}} \ , \tag{9.2.3}$$

where $\mathbf{e}_T$ and $\mathbf{d}_T$ are defined as in equation (9.2.1), $\mathbf{u}_R$ is the vector of inputs to the selected subset of M_R secondary sources, and $\mathbf{G}_{TR}$ is an $L_T \times M_R$ matrix of transfer responses made up from the columns of the matrix $\mathbf{G}_{TT}$ that correspond to the selected subset of secondary sources.

Given a subset of M_R secondary sources, then the cost function J_T, given by equation (9.2.2), is minimised if the vector of secondary source inputs is given by

$$\mathbf{u}_{R,opt} = -\left[\mathbf{G}_{TR}^H \mathbf{G}_{TR} + \beta \mathbf{I}\right]^{-1} \mathbf{G}_{TR}^H \mathbf{d}_T , \tag{9.2.4}$$

in which case the minimum value of the cost function is given by

$$J_{T,min} = \mathbf{d}_T^H \left(\mathbf{I} - \mathbf{G}_{TR}\left[\mathbf{G}_{TR}^H \mathbf{G}_{TR} + \beta \mathbf{I}\right]^{-1} \mathbf{G}_{TR}^H\right)\mathbf{d}_T , \tag{9.2.5}$$

and the control effort required to achieve this minimum is given by

$$\mathbf{u}_{R,opt}^H \mathbf{u}_{R,opt} = \mathbf{d}_T^H \mathbf{G}_{TR}\left[\mathbf{G}_{TR}^H \mathbf{G}_{TR} + \beta \mathbf{I}\right]^{-2} \mathbf{G}_{TR}^H \mathbf{d}_T . \tag{9.2.6}$$

Having obtained an equation for the minimum value of the cost function for a certain selection of M_R secondary sources, equation (9.2.5), we can now address the combinatorial problem of choosing the best such selection of M_R secondary source locations from the M_T possible locations. For relatively small problems this can be performed using an exhaustive search, and the set of secondary sources that produces the best reduction in the total cost function can be found.

Figure 9.4 shows the result of such a calculation, in which the best single secondary source location, the best pair of locations, etc., are selected from a total of 16 possible loudspeaker locations in a room, using the measured transfer response matrix from these 16 loudspeaker positions to 32 microphone locations at an excitation frequency of 88 Hz (Baek

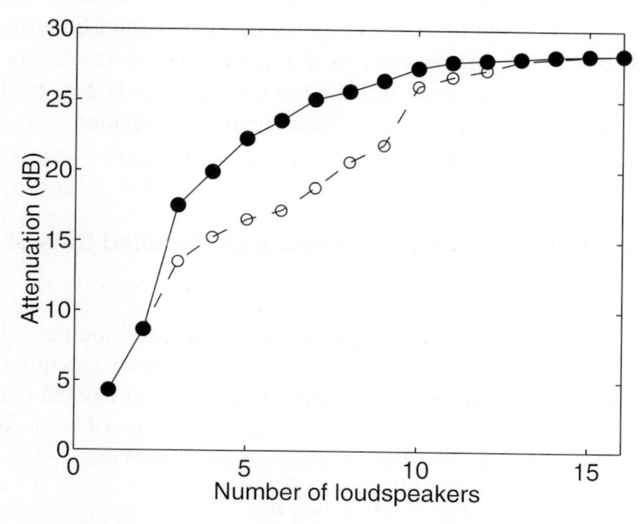

Figure 9.4 Maximum attenuations in the sum of squared outputs of 32 microphones using various numbers of secondary loudspeakers obtained from a sequential (○) and an exhaustive (●) search of 16 possible locations.

and Elliott, 1995). The cost function is equal to the sum of modulus-squared microphone outputs in this case. If these optimum locations are found using an exhaustive search, the distribution of attenuation values for the various combinations of a fixed number of actuators can also be calculated. There are 12,870 different combinations of 8 loudspeakers when chosen from a set of 16, for example, and although the best combination gives an attenuation of 29 dB, as shown in Fig. 9.4, two combinations give an attenuation of only 4 dB and most combinations give between 11 dB and 20 dB of attenuation (Baek and Elliott, 1995), so that significantly better performance than the average can be obtained from a carefully selected combination of actuators. The other curve shown in Fig. 9.4 shows the result of not performing the optimisation for each set of secondary sources independently, but as the number of secondary sources in operation is increased from M_R to M_{R+1}, the M_R secondary source locations previously identified are retained, and only the position of the extra source is sought. This 'sequential' algorithm is clearly more computationally efficient than an independent exhaustive search for each value of M_R, but as can be seen from Fig. 9.4, it can lead to sets of secondary source locations that do not perform as well as those identified with an independent exhaustive search. For a total of eight secondary sources ($M_R = 8$), for example, the sequential search gives a combination of locations that achieves an attenuation in the sum of the squares of the outputs of the 32 microphones of about 20 dB, whereas an independent search reveals a combination of 8 sources which is able to achieve attenuations of about 28 dB. The difference between the complete exhaustive search and the combination of individually good candidates is sometimes expressed by saying that 'the *N* best are not as good as the best *N*'.

A plot of the maximum attenuation that can be achieved, against the number of secondary sources used, is a very important tool in the initial design of an active control system in order to estimate the likely size of the system. For the relatively small enclosure and the 88 Hz pure tone excitation used to produce the data for Fig. 9.4, for example, it is clear that little extra attenuation is obtained if more than 8 to 10 well-positioned secondary sources are used. In this application, the choice of 8 to 10 secondary sources will thus provide a reasonable compromise between good attenuation and controller complexity. Although an exhaustive search has been used to calculate the results for the relatively small-scale problem described above, the optimal locations for larger-sized problems can be estimated using natural algorithms, and the maximum attenuation can again be plotted against a number of secondary sources.

9.2.3. Feedforward Control Formulation with Reduced Sets of Error Sensors

Having determined the number and location of the secondary sources which achieve a good attenuation at all the original error microphone locations, we can turn our attention to the question of whether all these microphones are required in a practical control system. If the number of error microphones is reduced to L_R, the vector of error signals at these sensors due to the M_R selected secondary sources can be written as

$$\mathbf{e}_R = \mathbf{d}_R + \mathbf{G}_{RR}\mathbf{u}_R , \tag{9.2.7}$$

in which $\mathbf{G}_{RR}$ is an $L_R \times M_R$ matrix of transfer responses made up of the rows of the matrix $\mathbf{G}_{TR}$ that correspond to the selected subset of error microphones.

The control system would now minimise a cost function which could include the sum of these error signals and the control effort, given by

$$J_{\mathrm{R}} = \mathbf{e}_{\mathrm{R}}^{\mathrm{H}} \mathbf{e}_{\mathrm{R}} + \beta \mathbf{u}_{\mathrm{R}}^{\mathrm{H}} \mathbf{u}_{\mathrm{R}} \; . \tag{9.2.8}$$

The set of secondary source signals that minimise this cost function is given by

$$\mathbf{u}_{\mathrm{R,opt\,R}} = -\left[\mathbf{G}_{\mathrm{RR}}^{\mathrm{H}} \mathbf{G}_{\mathrm{RR}} + \beta \mathbf{I}\right]^{-1} \mathbf{G}_{\mathrm{RR}}^{\mathrm{H}} \mathbf{d}_{\mathrm{R}} \; , \tag{9.2.9}$$

where the subscript opt R denotes the values found by the practical control system with a reduced number of error sensors.

At the design stage, however, the full matrix of transfer responses, $\mathbf{G}_{\mathrm{TT}}$, is still known, and so we can calculate the effect of the secondary sources, adjusted to minimise the output of the microphones in the practical control system according to equation (9.2.9), on the cost function that includes all the error sensors, J_{T} in equation (9.2.2). In this way, the global performance of practical control systems with different numbers and locations of error sensors can be calculated and compared. In particular, the locations of a given number of error sensors which give the greatest global reduction, J_{T}, can be determined either by an exhaustive search for small-scale problems, or by using natural algorithms for those of a larger size. Alternatively, the set of practical error microphones can be selected that, when their sum of squared outputs is minimised, give the maximum reduction at another subset of error microphones, whose response is measured at the design stage but which would be impractical to use in a production system. Some care must be exercised with such a selection, however, because of the assumption that the primary field will be the same during the design phase and in final operation, as discussed in more detail in Section 9.5. Figure 9.5 shows the maximum reductions in J_{T}, calculated

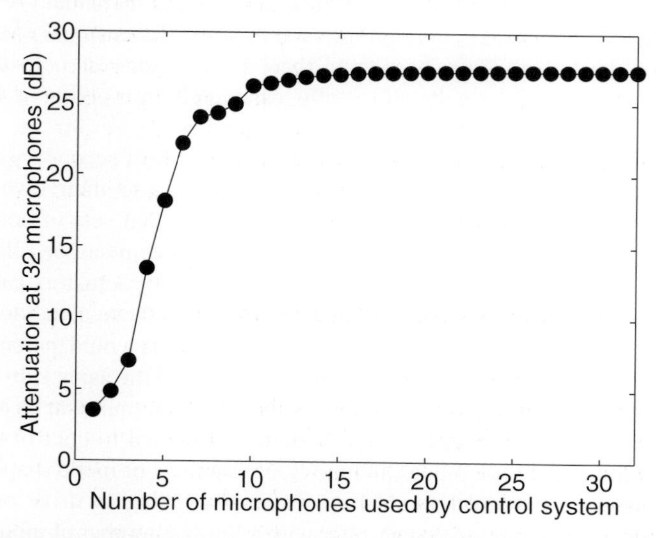

Figure 9.5 Maximum attenuations in the sum of squared outputs of all 32 microphones obtained with a practical control system in which the best 10 secondary sources are adjusted to minimise the sum of squared outputs from a reduced number of microphones, the positions of which are optimised for each number of microphones in the practical control system.

using all 32 microphone outputs, which are obtained using the 10 best secondary sources, as selected above, which are driven to minimise the sum of squared outputs from the best combination of various numbers of error sensors, as selected by another exhaustive search (Baek, 1996).

In this simulation the attenuation at all 32 microphones has been calculated even when the control system minimises the output at fewer than 10 microphones, so that the control problem is underdetermined and the additional constraint of minimum control effort has to be used to obtain a unique solution, as described in Section 4.2. In this simulation, reductions of about 38 dB at all 32 microphones are possible with 10 secondary sources even if the control system only minimises the sum of squared outputs from about 20 microphones. In fact, the number of error microphones can be reduced even further with only a small degradation in global performance, provided that the positions of these microphones are very carefully selected. In order to increase the robustness of the control system, and to reduce the degradation in performance if any of these error sensors fail, then it is common to use a larger number of error sensors than is strictly necessary, and many active sound control systems use about twice as many error sensors as secondary actuators.

It should be noted that mean-square error at the error sensors is not the only criterion that could be used as the cost function in the search for optimal transducer locations. In the feedback control of vibration on a truss structure, Lammering et al. (1994), for example, suggest that in controlling a fixed number of structural modes, a combination of the control effort and the control energy that spills over into the uncontrolled modes should be taken into account in the optimisation procedure. Hać and Liu (1993) suggest that the degree of controllability and observability should be taken into account in defining a measure of performance, and these considerations are reviewed, for example, by Baruh (1992) and Hansen and Snyder (1997). Sergent and Duhamel (1997) have also shown that if the maximum pressure is minimised in a one-dimensional enclosure, rather than the sum of the mean-square pressures, then the locations of both the secondary actuators and error sensors can be efficiently calculated by solving a unique linear programming problem.

There are more sophisticated methods of using a fixed array of secondary actuators than controlling them all independently, and these methods lead to their own optimisation problems. Carneal and Fuller (1995), for example, suggest that sets of actuators can be driven by the same controller channel, so that the control system can be relatively simple but some of the benefits of having large numbers of secondary actuators can be retained. The optimisation problem then involves selecting subsets of those actuators that are best driven with each independent control channel. The actuators could potentially also be driven in or out of phase compared with the other actuators in the same subset by a simple wiring change and the subsets may overlap if the two control signals can be added together before being applied to the actuators. If the system is designed to control only a limited number of structural or acoustic modes and the eigenfunctions or mode shapes of these are known, the actuators and sensors could be connected together to drive or measure the modal amplitudes of these modes. A controller with a limited number of inputs and outputs could then be used to control many actuators and sensors, but the transducer optimisation problem would now be simplified because considerable *a priori* knowledge of the system under control could be incorporated into their placement (as described, for example, by Baruh, 1992).

9.3. APPLICATION OF GENETIC ALGORITHMS

In this section, we first describe how genetic algorithms work, and then discuss their application to the combinatorial problem of transducer selection. As their name implies, genetic algorithms are broadly based on the biological mechanisms of Darwinian evolution, and the terminology used to describe their action reflects these biological roots. They form a family of guided random-search methods, with numerous variations in the detail of their implementation, but all have certain features in common. For a more detailed general description of genetic algorithms, the reader is referred to the textbooks by Goldberg (1989) or Mitchell (1996) or the proceedings of the International Conferences on Genetic Algorithms (1987 onwards). Genetic algorithms are examples of the techniques of evolutionary computing, as reviewed by Fogel (2000), for example. The operation of genetic algorithms and their application to various signal processing problems, including the design of IIR filters, is also reviewed by Tang et al. (1996). All genetic algorithms are fundamentally *parallel* search schemes, in which a number of prototype solutions are considered at any one iteration, or *generation*. The prototype solutions are coded as finite-length chromosomes or *strings*, made up of characters, or *genes*, chosen from a well-defined alphabet. The generation of the basic genetic algorithm will be explained below using the most simple alphabet of binary numbers for the genes. An example of a string having 16 binary genes is shown in Fig. 9.6.

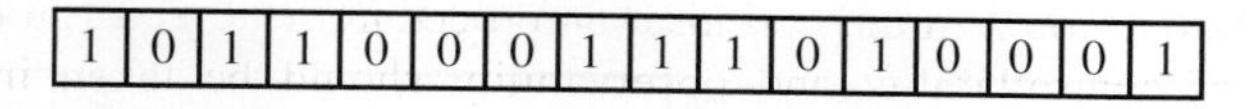

Figure 9.6 An example of a prototype solution to an optimisation problem coded as a string with 16 binary elements or 'genes'.

9.3.1. A Brief Introduction to Genetic Algorithms

It is conventionally assumed that for each string that represents a prototype solution, a measure of *fitness* can be deduced, which the algorithm is attempting to *maximise*. This definition of the performance goal can be contrasted to that generally used above, in which some 'cost function' is generally being *minimised*. The mapping of the cost function into a fitness value is a process that will generally involve a procedure known as scaling, but in the simplest case we can assume that the fitness is just the negation of the cost function, for example, and that a suitable cost function and hence fitness value can be calculated for each string. More complicated scaling of the fitness values is necessary to prevent a few strings, which may have a particularly high fitness, from dominating the selection process and reducing the diversity of subsequent populations. In later generations, scaling is also used to increase the range of fitness values within a population and encourage convergence. In the most common form of scaling (Goldberg, 1989, p. 77), the new fitness values are linearly scaled from the old ones so that strings with an average fitness are selected once, and those with the maximum fitness are selected twice, on average. In practice the calculation of the fitness of the selected strings generally takes much longer than the subsequent scaling and the implementation of the search algorithm itself, and any search algorithm can be run much faster using an efficient method of calculating the fitness. This is one of the reasons why the efficient calculation of the exact least-squares solution was emphasised in previous chapters.

At each iteration of the genetic algorithm, the strings in one generation, the current *population*, are used to create another set of strings in a new generation. The strings in the first generation are generally chosen at random from all the possible string combinations. Pairs of strings in one generation are chosen by a process of *selection*, and are then combined together, using processes described below, to provide a pair of strings for the next generation. This procedure is illustrated in Fig. 9.7, for a small population size, in which, for example, Strings 1 and 3 in the first generation happen to be used as parents for both Strings 1 and 2 in the second generation. The simplest method of selecting the parent strings is at random, but with a probability proportional to the fitness of the string. This process can be thought of as using a roulette wheel, with the number of the slots that correspond to the selection of each individual string being proportional to their fitness. A modification to this completely random selection method is to choose each string as many times as the integer part of the number formed by multiplying the probability of selection by the population size, and then making the remaining selections using the biased roulette wheel described above. This semi-deterministic selection method is called *stochastic remainder sampling without replacement* by Goldberg (1989, p. 121). It has become widely used in a variety of applications and is the method of selection used in all the genetic algorithm simulations discussed below.

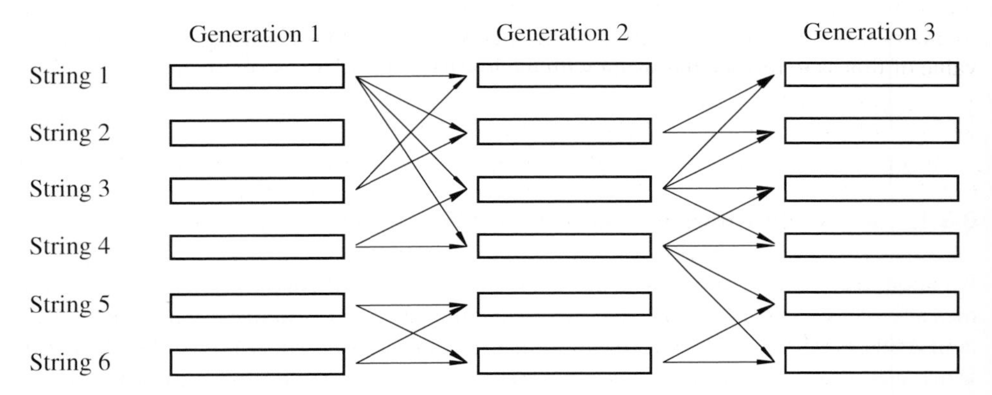

Figure 9.7 Three generations of a genetic algorithm in which pairs of strings from one generation are used to create pairs of strings in the next generation.

Once a pair of parent strings have been selected, they are combined together using *crossover* and *mutation*. The conventional process of crossover is illustrated in Fig. 9.8 for a pair of 6-gene strings. The crossover site, shown by the dashed line in Fig. 9.8, is selected at random along the length of the strings and two new strings are created by swapping all the genes to the right of the crossover site (called the 'tail' of the string) between the two strings in each generation. This process has the effect of mixing together the attributes of the two parent strings. It should be noted, however, that the way in which the mixing occurs in the conventional crossover process does depend on the way in which the attributes of a string are ordered in the string's genes. The 'coding' of the string then affects the

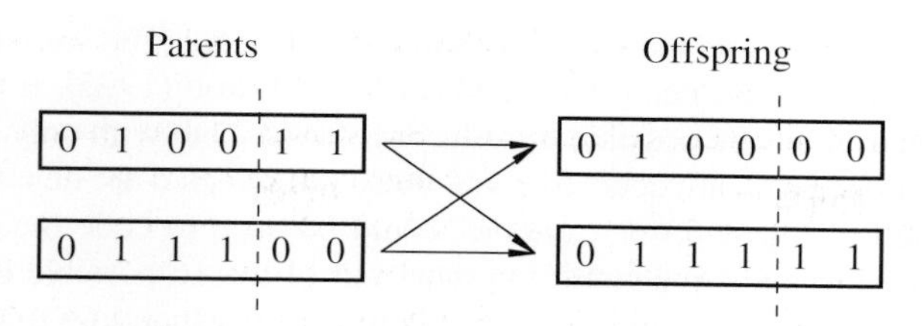

Figure 9.8 An example of the conventional process of crossover, in which the crossover site is shown as a dashed line and the two strings in the offspring generation are generated by swapping the genes in the 'tails' to the right of the crossover site in the parent generation.

performance of the genetic algorithm. Some guidance on good coding methods is provided by Goldberg (1989, p. 80).

The conventional process of mutation involves the random changing of one gene in a string in moving from one generation to another. A small probability of mutation (typically from about 1 in 100 to 1 in 1000 for each gene) helps prevent the genetic algorithm from converging to local rather than global minima in the search space by introducing potentially significant random variations in the population. It is possible to vary the probability of mutation so that it is higher in the early generations of the genetic algorithm, for example, by making it proportional to the ratio of the average fitness to the maximum fitness in each generation. Another refinement to the basic genetic algorithm which was found to be beneficial in the work reported below was to allow the string with the greatest value of fitness to replace the string with the lowest value of fitness in the next generation. This is similar to the *elitist model* described by Goldberg (1989, p. 115) and ensures that the best solution found so far is always passed to the next generation.

At this point it may be worthwhile repeating the comment made by Harp and Samad (1991), in their review of Goldberg's book, that 'The sceptical reader .... , may be wondering how such a seemingly ad hoc procedure could have practical value'. At present there appears to be no strong theoretical justification for many of the procedures adopted in practical implementations of genetic algorithms. The numerous refinements to the basic genetic algorithm are generally not, however, crucial to its success, but only improve the convergence rate for certain classes of problem. In the absence of a theoretical justification, we can only reiterate that genetic algorithms have been found to provide an efficient solution to a wide range of combinatorial problems (Goldberg, 1989), and present the results of applying the algorithm to the problem in hand.

9.3.2. Application to Transducer Selection

The first issue which arises in applying genetic algorithms to the transducer selection problem is how to code the selection into a string. There are at least two options. One method, used for example by Onoda and Hanawa (1993) and Tsahalis et al. (1993) uses a string with as many genes as the allowed number of transducers, with each gene representing one of all the possible transducer locations. The alphabet used to code each gene is thus potentially very large, but the string can be quite short. The order of the genes in the string is not significant using this coding, so that many strings can represent the same selected set of transducers, and the searching space can be large. The second coding option, used for example by Rao and Pan (1991) and Baek and Elliott (1995), is to have as many genes in the string as there are possible transducer locations, but with only binary genes, so that the selected transducers are denoted 1 and those not selected are denoted 0. The string that was illustrated in Fig. 9.6, for example, could be used to code the positions of the 8 loudspeakers used in an active control system, out of the 16 possible locations. In this case the selected transducers would be in the positions corresponding to the 1st, 3rd, 4th, 8th, 9th, 10th, 12th and 16th genes in the string. In order to maintain the same number of selected transducers from one generation to another when using this method of coding, a constraint must be introduced into the conventional crossover and mutation methods. Otherwise, for the conventional crossover process illustrated in Fig. 9.8 for example, the two parent strings would represent two combinations of 3 transducers in 6 possible locations, but after crossover, the number of transducers selected in the two offspring would be 1 or 5. Without any constraint, the genetic algorithm would tend to select as many transducers as possible to improve its performance. One method of introducing a constraint on the total number of transducers is by significantly reducing the fitness of any string with more than the required number of transducers in it. This is described as the penalty method by Goldberg (1969, p. 85), but it has the effect of making the searching process extremely redundant, since a large number of strings can be generated in the new generation with very low values of fitness. The constraint method adopted by Baek and Elliott (1995) was to identify the total number of pairs of genes that were dissimilar in the two parent strings, and then to swap a random but even number of these pairs, which were also selected at random, such that the total number of selected transducers is constant. The process is illustrated in Fig. 9.9, in which the last 4 genes in the two parent strings have dissimilar values. It was randomly selected that two of these genes would be swapped during crossover. The 3rd gene was first selected at random from these pairs for crossover, and so in order to preserve the total number of selected transducers, the second random choice must be made between crossing over either the 4th, 5th or the 6th gene. In Fig. 9.9, it is the 5th gene that is also selected for crossover and so both of the strings generated for the next generation still have three selected transducer locations out of the possible six. This procedure is an example of a 'repair' algorithm. It is interesting to note that adopting this crossover procedure means that the process is not dependent on the ordering of the possible transducer locations as genes within the strings. A similar constraint must be introduced into the mutation operation, with each random mutation of a gene in a single string being accompanied by a mutation of another gene of a dissimilar value. The crossover and mutation operations are thus completely robust to the coding used to determine which possible location is represented by which gene.

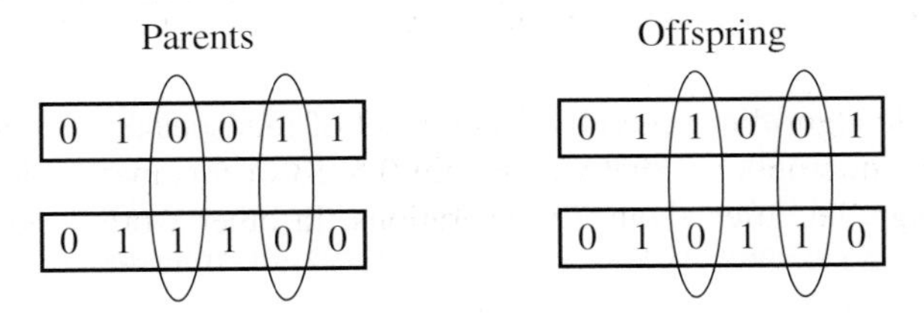

Figure 9.9 An example of a modified crossover process suggested by Baek and Elliott (1995), which retains the same number of genes with equal value from generation to generation.

As an example of the application of genetic algorithms, Fig. 9.10 shows some results presented by Baek and Elliott (1995) for the selection of 8 secondary loudspeaker positions from a possible 32 locations. The maximum attenuation in the sum of squared pressures at 32 microphones using 8 loudspeakers is shown, calculated using subsets of the measured matrix of transfer responses at 88 Hz from all 32 possible loudspeaker locations to all 32 microphone locations in a small room. This is plotted as a function of the total number of strings searched in each successive generation of the genetic algorithm. In this calculation, 100 strings per generation were used with a final mutation rate of 1 in 1000 per gene, and the results were averaged over 20 independent applications of the genetic algorithm to reduce the statistical variation. The fitness of each string was calculated as the attenuation at the microphones in this case.

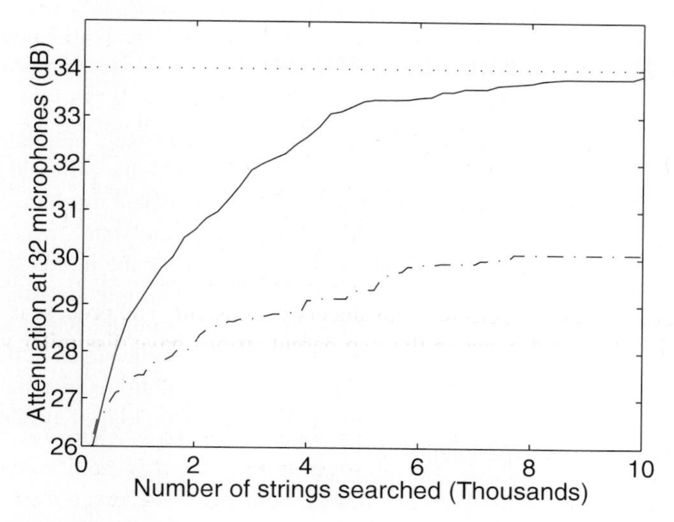

Figure 9.10 The maximum attenuation at the 32 microphone locations using 8 secondary loudspeakers chosen from 32 possible locations using a genetic algorithm (solid line) and using an unguided random search (dot-dashed line). The dotted line is the best possible attenuation determined by searching all 10,518,300 possible solutions to this problem.

There are 10,518,300 possible ways of choosing 8 loudspeaker positions from the 32 possible locations, and so an exhaustive search would normally be prohibitively expensive. In order to assess the performance of the genetic algorithm in this test case, however, an exhaustive search was undertaken, which required the equivalent of a month's running time on a PC! The very best attenuation that could be achieved was found to be about 34 dB. The genetic algorithm generally found a set of loudspeaker locations which gave attenuations at the microphones that was within 0.5 dB of this maximum after searching about 7,000 strings, i.e. after about 70 generations. In other words, the guided random search provided by the genetic algorithm found a solution that would be acceptable, in practical engineering terms, after searching only about 0.07% of the total number of strings. For comparison, the results of an unguided random search are also shown in Fig. 9.10, in which each new string was selected independently of the previous one. In this case the attenuation achieved by the best set of selected sources was, on average, about 4 dB below the best possible even after 10,000 strings had been selected. Although advances in computer technologies will inevitably reduce the time taken for an exhaustive search, it is also likely that active control problems with larger and larger numbers of transducers will be designed. The combinatorial explosion described in Section 9.1 will probably ensure that an exhaustive search will continue to be computationally infeasible and the efficiency of guided random search algorithms such as the genetic algorithm will still be required to provide practical engineering solutions. Genetic algorithms have also been applied to transducer location problems in active noise control by Hamada et al. (1995), Concilio et al. (1995) and Simpson and Hansen (1996); and in active vibration control by Furuya and Haftka (1993) and Zimmerman (1993), for example.

9.4. APPLICATION OF SIMULATED ANNEALING

Whereas genetic algorithms are inspired by biological processes, simulated annealing methods are based on the *statistical mechanics* used to describe the physical behaviour of condensed matter (Kirkpatrick et al., 1983; Cerny, 1985). Statistical mechanics can be used to describe the probability of a physical system being in a given 'state', which can include the arrangement of the atoms in a crystal, or the excitation of the atoms within a given arrangement. Each state has an associated energy level and simple statistical arguments lead to the concepts of temperature and entropy (see, for example, Kittel and Kroemer, 1980). Thermal motion always generates a distribution of possible states of the system at a given temperature. There are, however, such a large number of atoms in a practically sized sample (about 10^{29} atoms/m^3) that only the most probable behaviour of systems in thermal equilibrium is observed in experiments.

9.4.1. A Brief Introduction to Simulated Annealing

The probability of a physical system in thermal equilibrium at a temperature T being in a particular state $\mathbf{x}$, with associated energy $E(\mathbf{x})$, is proportional to the *Boltzmann probability factor*

$$P(\mathbf{x}) = \exp\left(-E(\mathbf{x})/k_{\mathrm{B}}T\right) , \tag{9.4.1}$$

where k_{B} is Boltzmann's constant. This probability factor is plotted in Fig. 9.11 against the energy associated with a particular state, $E(\mathbf{x})$, for two values of the temperature. At high temperatures, there is a reasonable chance of finding the system in a state with a relatively high energy, but at low temperatures, it is very unlikely that the system will be in any state other than those whose energy is very low. There are typically very few states with the lowest possible energy levels associated with them, and these are called *ground states*. In many materials, the arrangement of atoms in a perfect crystal would correspond to the ground state. It is very unlikely, however, that if a molten material was rapidly reduced in temperature through its freezing point, the state of the system would happen to be the ground state when the material solidified, because it is so unusual. The chances of

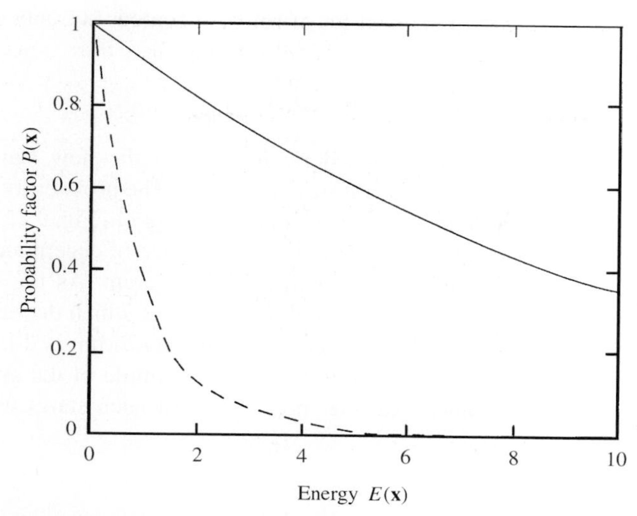

Figure 9.11 The Boltzmann probability factor, which is proportional to the probability of finding a system in a state $\mathbf{x}$, which has an energy $E(\mathbf{x})$, for two different values of temperature corresponding to $k_B T = 1$ (dashed) and $k_B T = 10$ (solid).

achieving a state close to this minimum-energy configuration are considerably increased, however, if the temperature is reduced very slowly. In this case the distribution of states will always remain close to that at equilibrium, and so as the temperature is gradually reduced, the lowest energy states will have a greater and greater probability of occurrence, as illustrated in Fig. 9.11.

The process of gradually reducing the temperature to obtain a well-ordered, low-energy, atomic arrangement is known as *annealing*. It can be contrasted with a very rapid reduction in temperature, quenching, in which case the atomic arrangement tends to get frozen into a randomly-ordered, relatively high-energy state; a glass, for example. The physical processes involved in annealing can be simulated, in numerical optimisation problems of the type we are interested in here, by using the cost function as the analogue of the energy. The objective is to find the extremely rare combinatorial arrangement, or state, of the system that has a particularly low value of the cost function. This can be achieved by simulating the process that generates equilibrium at a particular temperature for long enough that the probability of a particular state is governed by the Boltzmann distribution. This is initially implemented with a relatively high temperature, so that all states have a chance of selection. The temperature is then gradually reduced so that the distribution of states is gradually squeezed into the region where their energy is particularly low.

The most commonly used simulation of the process by which such an equilibrium is generated is due to Metropolis et al. (1953). They showed that a system would evolve into one obeying the Boltzmann distribution if small random perturbations were introduced into the state of the system, and the acceptance of this change in the system was made to depend in a particular way on its associated change of energy. Specifically, if the energy of the system is decreased by the change, it is accepted unconditionally. If the change in the state of the system causes the energy to rise, it may also be accepted, but only at random and with a probability of

$$P(\Delta E) = \exp\left(-\Delta E/k_B T\right) , \tag{9.4.2}$$

where ΔE is the change of energy caused by moving to the new state, and $k_B T$ is proportional to the temperature of the resulting distribution. The probability of accepting a change is plotted as a function of its associated change in energy in Fig. 9.12 for two values of the temperature. At high temperatures there is a fair chance of accepting the change of state even if it significantly increases the energy of the system. As the temperature is lowered, this becomes increasingly unlikely and only changes which decrease the energy, or increase it only slightly, are likely to be accepted. This algorithm is different from the genetic algorithm described above in that only a single example of the system's state is considered at any one time, rather than the 'population' of such states used by genetic algorithms.

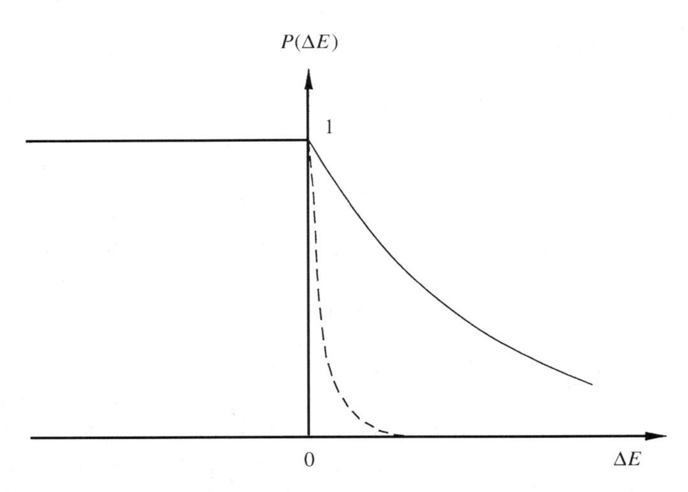

Figure 9.12 The probability of accepting a random change in state with an associated change in energy ΔE according to the Metropolis algorithm, for temperatures corresponding to $k_B T = 10$ (solid) and $k_B T = 1$ (dashed).

The process by which the Metropolis algorithm can be used to implement simulated annealing on a combinatorial selection problem can now be described. An initial *temperature* is chosen to begin the annealing. This is equivalent to the numerical value of $k_B T$ in equation (9.4.1) above, and if the energy of a particular arrangement, $E(\mathbf{x})$, is replaced by its calculated cost function, J, then the temperature variable, $k_B T$, has the same units as those of the cost function. The choice of the initial temperature will thus depend on the range of values of the cost function for the problem of interest, but typically would be set equal to the average value of the cost function over all arrangements. The Metropolis algorithm is then implemented for a fixed number of iterations at this temperature, or until a convergence criterion is satisfied, for example that a certain number of changes have been accepted. The temperature is then reduced by a multiplicative factor less than one, called the *cooling coefficient*, and the Metropolis algorithm is implemented until equilibrium is again achieved at the new temperature. The temperature is thus gradually reduced so that the distribution of combinations is narrowed down to only those with the lowest cost function.

In the early stages of the algorithm, when the temperature is high, the simulated annealing algorithm will accept most of the random changes of configuration, and so will tend to 'explore' the whole of the parameter space. As the temperature is reduced, the algorithm will generally 'home in' on the configurations which have cost functions close to the global minimum, and then refine the solution to get the cost function even closer to the global minimum as the temperature is further reduced. It should be emphasised, however, that even at these later stages of convergence, the configuration of the system is still subject to random changes, so that it could still move to a remote region of the parameter space with a lower cost function, if there was one, or even to one with a slightly higher value of cost function. The convergence behaviour of the simulated annealing algorithm is determined by the size of the random perturbations and the value of the cooling rate, which is controlled by the cooling coefficient. If the cooling coefficient is too high, the probability of converging to a local optimum is increased, but too low a value needlessly increases the computation time. Cooling rates of 0.9 to 0.99 are commonly used.

9.4.2. Application to Transducer Selection

In applying the simulated annealing algorithm to the problem of selecting 8 secondary loudspeakers from 32 possibilities, we can use the coding and random perturbation methods discussed in Section 9.5. The arrangement of loudspeakers can thus again be represented as eight binary 1s in a string of 32 binary numbers, each of which represents the presence (1) or absence (0) of a loudspeaker at a particular location. The random perturbation to the configuration required by the Metropolis algorithm is provided by changing the value of the bit at a random location on the string, but then also changing the value of another bit, again randomly chosen, but having a different value from the initially selected bit. This ensures that the same number of 1s are retained in the perturbed string. The method is the same as that used for mutation in the genetic algorithm. Another method of random perturbation is widely used in the literature, which consists of randomly selecting a section of the string and reversing the order of the bits within it. This method of rearrangement also preserves the total number of 1s in the string, and has been found to work well for a commonly studied paradigm called the 'travelling salesman problem', in which the string represents the order in which a salesman visits a number of cities, and the cost function is the cost of the trip (Lin, 1965; Kirkpatrick et al., 1983). It is interesting to note that this 'reverse and exchange' method of rearrangement is also used in genetic algorithms, where it is known as *inversion*, and where it mimics the way in which the genetic code is reordered in nature (Goldberg, 1989, p. 166). The reverse and exchange method is the most widely used method of rearrangement in the simulated annealing algorithms presented in the 'numerical recipes' books (Press et al., 1987). For the combinatorial problem of transducer selection, however, it was found that the reverse and exchange method of random rearrangement did not work as well as the constrained mutation operation described above. This is illustrated in Fig. 9.13, which shows the attenuation at the 32 microphone locations obtained by using different selections of 8

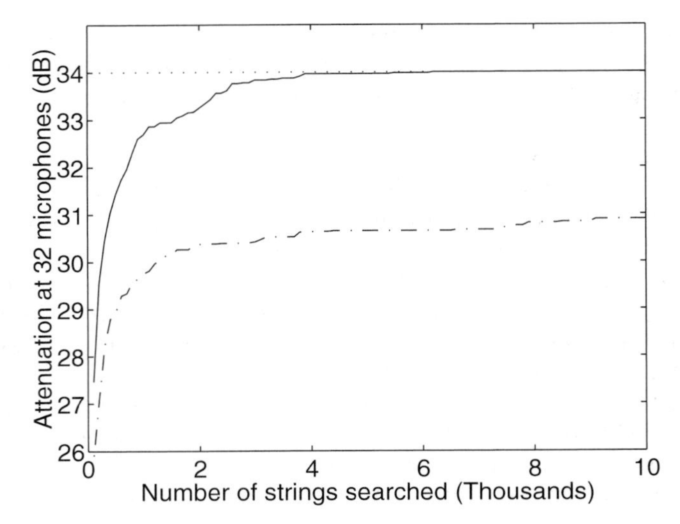

Figure 9.13 The maximum attenuation at the 32 microphone locations using 8 secondary loudspeakers selected from 32 possible locations using a simulated annealing program, with random rearrangement using constrained bit changing (solid line) or reversing and exchanging (dashed line). The dotted line corresponds to the best possible attenuation, as in Fig. 9.10.

loudspeakers in the same arrangement as that described at the end of the previous section (Baek and Elliott, 1995). The cooling coefficient for these simulations was 0.98, but the result was not critically dependent on this, or any of the other parameters which must be selected in the simulated annealing programme. Simulated annealing has also been applied to the problem of actuator location in active vibration control by Chen et al. (1991), for example.

It is clear from Fig. 9.13 that the convergence of the simulated annealing algorithm with the constrained bit changing method of rearrangement is much better for this problem than using the reverse and exchange method. The simulated annealing method with constrained bit changing converges to within 0.5 dB of the best possible solution after searching only about 2,500 strings. When genetic algorithms were used for the same problem, Fig. 9.10, a similar convergence was not reached until about three times as many strings had been evaluated. One should not be too hasty in dismissing genetic algorithms for this problem, however, since the difference in performance is far less when some realistic constraints are introduced into the optimisation problem, as noted by Onoda and Hanawa (1993), for example, and further discussed below.

9.5. PRACTICAL OPTIMISATION OF SOURCE LOCATION

For a feedforward control problem at a single excitation frequency, it was noted in Section 9.2 how the maximum attenuation at all the error sensors could be determined with a reduced set of sources and sensors by partitioning the transfer matrix from a large number of secondary sources to a large number of error sensors and the vector of primary disturbances at the error sensors. The residual level at all the error sensors could then be used as a measure of the cost function to be minimised when selecting the elements of this reduced set of sources and sensors. The definition of the fitness function used by a natural algorithm can also include other indicators of performance or practical constraints that are introduced in real control systems.

If, for example, the full transfer matrix and vector of disturbances at a *number* of frequencies is available, the cost function could include the sum of the residual levels at the error sensors after least-squares minimisation at *all* the excitation frequencies. In this way, the source and sensor locations can be optimised using natural algorithms over all the frequency components that are important. In an example of such an optimisation, Baek and Elliott (1995) calculated the best 8 loudspeaker positions, from 32 possible locations, which minimised the sum of the residual pressures at three harmonic frequencies: 88 Hz, 176 Hz and 264 Hz. An experiment was then performed in which this set of 8 loudspeakers was used by a real-time control system to minimise the sum of the squared pressures at the 32 error microphones for each of these excitation frequencies in turn. The measured reductions were within 1 dB of those predicted from the exact least-squares prediction, with the measured reductions being slightly less than those predicted at 88 Hz, but slightly more at 264 Hz. These differences were thought to be due mainly to an inconsistency in the measurement of the primary disturbance vector. The full 32×32 transfer matrix used in this experiment was, in fact, measured as two 32×16 transfer matrices, with 16 loudspeakers moved from one set of positions to another between the measurements. Because of the slightly different physical geometry within the enclosure with these two loudspeaker arrangements, the disturbance due to the primary source was observed to change at individual microphone locations by up to 1 dB. The cost function for the genetic algorithm used in this case was calculated from the average of the two disturbance vectors measured with the loudspeakers in the two sets of locations, and the disturbance vector was thus not entirely consistent with the primary field used by the real-time controller. The differences in the resulting attenuations were small, however, which demonstrates that the optimisation is not overly sensitive to small changes in the primary sound field. This raises the important issue of the *robustness* of the optimisation of source location.

9.5.1. Performance Robustness of the Selected Locations

If the transducer locations in a practical active control problem are chosen only on the basis of a single set of plant transfer response and disturbance data, there is a danger that the control system will not perform well if small changes occur in the system under control.

Transducer locations that do not generalise well to different conditions are the result of over-optimisation with the original data set and have been referred to as being 'highly-strung' (Langley, 1997). What is required in practice is that the attenuation provided by the selected sources be robust to the kinds of changes that are likely to occur in the plant and the disturbance under realistic operating conditions. Errors in the estimation of the transfer matrix and disturbance vector will inevitably cause changes in the predicted attenuations. These effects were discussed in Section 4.4, where it was shown that a small effort-weighting term in the cost function being minimised improves the robustness of the least-squares solution. Another change in operating conditions that can cause a change in the transfer responses and disturbance signals is the small variations in excitation frequency which typically occur in practice. Baek and Elliott (1995) verified the robustness of the 8 loudspeaker locations found in the optimisation described above to this effect, by measuring the attenuations achieved with the real-time control system as the fundamental frequency was altered by ±1 Hz. They found that the measured attenuation changed by 0.2 dB at 88 Hz, 0.6 dB at 176 Hz and up to 1.8 dB at 274 Hz. The larger variations at higher frequencies are consistent with the much more complicated nature of the sound field at these frequencies, as discussed, for example, by Nelson and Elliott (1992).

A test which could be incorporated into the evaluation of the cost function used by the natural algorithm is whether the least-squares solution required to calculate the minimum value of the cost function is ill-conditioned. Such ill-conditioning can lead to large control efforts being required and also to long convergence times for steepest-descent controllers, as discussed in Chapter 4. A common measure of the extent of the ill-conditioning is the *condition number*, often defined to be the ratio of the maximum to minimum singular values in the matrix being inverted to obtain the least-squares solution. The size of this matrix is $2M \times 2M$ for a frequency-domain controller, where M is the number of secondary sources and $MKI \times MKI$ for a time-domain controller, where K is the number of reference signals and I is the number of coefficients in each of the FIR control filters (see Section 5.2). For a reasonable-size control system, it becomes much more computationally expensive to evaluate the singular values and so determine the condition number than to calculate the least-squares solution itself. If the condition number had to be determined for every set of source locations chosen by the natural algorithm, the source selection programme would be significantly slower. Heatwole and Bernhard (1994) suggest that this computational problem can be reduced, for time-domain controllers, by estimating the condition number from the spectral data.

Rather than explicitly calculating the condition number for each new secondary source arrangement, the cost function could alternatively be modified to ensure that ill-conditioned solutions are not selected. The simplest method of achieving this is to incorporate a degree of effort weighting into the cost function being minimised, as provided by the term proportional to β in equation (9.2.2). The least-squares solution is then modified by adding the constant β to the diagonal elements of the matrix to be inverted, as in equation (9.2.4). If β is small compared with the largest eigenvalue of this matrix, the magnitude of these eigenvalues will not be significantly changed by the incorporation of this degree of effort weighting. The smallest eigenvalue of this new matrix is now guaranteed to be greater than β, however, and since a rise in the condition number is generally caused by particularly low values of the smallest eigenvalues, ill-conditioning can be avoided. A small effort weighting also reduces the sensitivity of an adaptive feedforward control system to measurement errors, as noted in Chapter 4. The most obvious physical effect of incorporating an effort weighting term into the cost function is to prevent the selection of solutions that give good attenuations but require excessive driving signals to be supplied to the secondary sources. An example of this behaviour is provided by Baek and Elliott (1995), who calculated both the attenuation at the error sensors, and the sum of squared secondary drive signals, the effort, for the 100 combinations of 8 secondary loudspeakers that give the greatest attenuation. These values are plotted in Fig. 9.14. For these 100 sets of secondary source positions, the attenuations vary from about 29 dB to 27 dB, but the control effort varies from about 0.2 to 2.5 in the linear units used in Fig. 9.14, which could be proportional to the actuator power requirement in watts, for example. It can thus be seen that very good attenuations can be

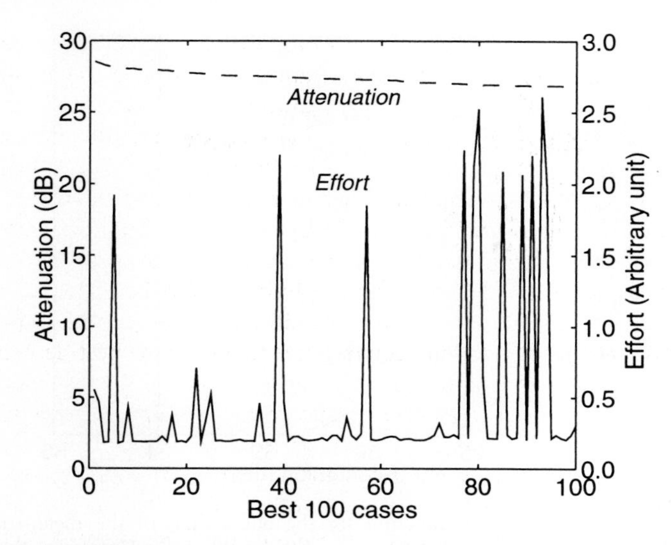

Figure 9.14 The attenuation achieved at the error sensors for the 100 best combinations of 8 secondary sources from 16 locations, and the corresponding control effort required to achieve these attenuations.

achieved with much lower control effort for some loudspeaker combinations than others. It is interesting to note, however, that many loudspeaker combinations require a control effort of about 0.2 units, and that this appears to be the minimum control effort required to get significant attenuation in this case. If the search algorithm for the loudspeaker positions was run using a cost function which incorporated a value of β chosen to weight the residual mean-square error and minimum effort approximately equally, only the loudspeaker combinations shown in Fig. 9.14 with low values of effort would be selected.

9.5.2. Robust Design

Another method of avoiding actuator locations whose performance is not robust to operating conditions is to search for combinations that minimise the average of a set of cost functions each calculated for a different operating condition. In the simplest case, in which the changes in the transfer matrix and primary field between operating conditions are independent random numbers, caused by identification errors for example, Baek and Elliott (1997, 2000) have shown that the selected actuators are similar to those selected with an appropriate amount of effort weighting in a single cost function calculated under nominal, noise free, conditions. The fact that the selection of the secondary sources can be very sensitive to such errors is illustrated in Fig. 9.15, which shows the probability distribution functions of the attenuation achieved at 32 error microphones with two different sets of 8 secondary loudspeakers, denoted Set A and Set B (Baek and Elliott, 2000). Two distribution functions are shown for each set of secondary loudspeakers. The first is with no errors assumed in the plant matrix, so that the probability distributions are delta functions, which indicates that the attenuation obtained with loudspeaker Set A is about 1 dB greater than that with loudspeaker Set B under nominal conditions. If independent

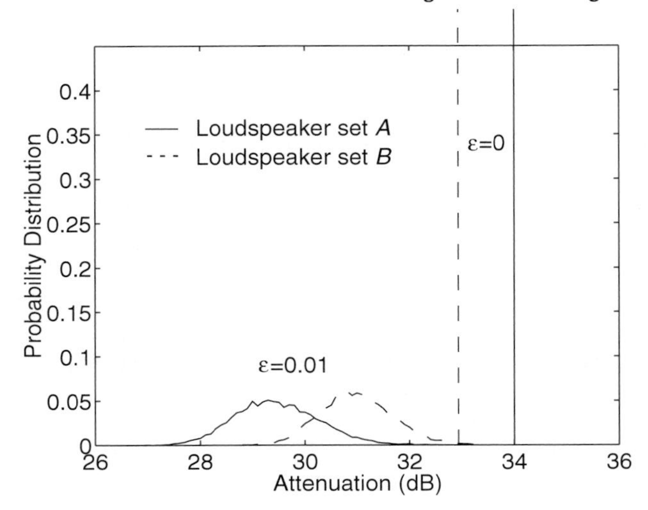

Figure 9.15 Probability distribution function for the attenuation of the mean-square levels at 32 error microphones using two sets of 8 secondary loudspeakers, *A* (solid line) and *B* (dashed line), with various percentages of rms random error, ε, in the elements of the plant matrix. Under nominal conditions, with no error, $\varepsilon = 0$, loudspeaker Set *A* gives the best performance, but with an rms error of only 1%, $\varepsilon = 0.01$, a significantly better average attenuation can be obtained with loudspeaker Set *B*, which is thus the more robust choice of secondary loudspeaker position in practice.

random numbers are now added to the real and imaginary parts of the elements of the disturbance vector and plant matrix to generate a population of perturbed operating conditions, a more interesting distribution of attenuations is observed. Figure 9.15 shows the probability distribution functions for the attenuation with a set of 1% rms random plant and disturbance errors, which represents an error level that could easily be generated by a practical identification method. It can be seen that now the average attenuation achieved with loudspeaker Set *B* is about 2 dB better than that of loudspeaker Set *A*, and loudspeaker Set *B* is thus the more robust choice of loudspeakers in practice. Further analysis shows that if there are only changes in the disturbance then this change in the selection of the best actuator positions does not occur (Martin and Gronier, 1998; Baek and Elliott, 1997, 2000); it is the changes in the plant response that cause these effects.

Bravo and Elliott (1999) have considered the effect of optimising both loudspeaker positions and microphone positions using a cost function given by the averaged attenuation achieved over a set of six measured plant response matrices and disturbance vectors. This set of measurements was taken in the enclosure described above at 88 Hz, but with different numbers of people standing in different positions within the enclosure. The best single loudspeaker position, pair of loudspeaker positions, etc., was then found which minimised this cost function, and a graph was generated of the maximum attenuation against the number of loudspeakers used, as in Fig. 9.4. It was found that for this, averaged, cost function the increase in performance with the number of loudspeakers was more linear than that shown in Fig. 9.4, which was calculated using a cost function employing only a single plant response matrix and disturbance vector. In this case 12 loudspeakers were required to give an averaged attenuation within 1 dB of the attenuation available with all 16 loudspeakers, whereas only 9 loudspeakers were required to give this level of attenuation in Fig. 9.4. Bravo and Elliott (1999) also calculated the attenuation at all 32 microphones, averaged across this set of plant response matrices and disturbance vectors, when the 10 best loudspeakers were used to minimise the sum of squared pressures across different numbers of microphones. Interestingly, the results in this case were almost the same as those shown in Fig. 9.5, which was calculated for a single plant matrix and disturbance vector, and good attenuations were again achieved with only 10 well-placed microphones.

For structured variations in the operating conditions, where it is possible to characterise these variations using the matrices of perturbed singular values described in Section 4.4, an ensemble of perturbed transfer response matrices can be generated from the nominal response matrix and the known structure of the operating condition variations. The average cost function calculated over this ensemble can then be used in the search algorithm to find a set of transducer locations whose performance is robust to these variations (Baek and Elliott, 2000). When optimising over a continuous domain, solutions that were not robust to small parameter variations would show up as sharp peaks in the error surface and are said to be *brittle*, whereas robust solutions would show up as broader peaks. Gradient descent methods may easily converge to one of these brittle solutions, whereas the operation of a genetic algorithm makes it inherently unlikely that such a solution will be selected. Various methods of modifying a genetic algorithm to further discriminate against such sharp peaks have been proposed, for example by adding random perturbations to physical variables represented by the strings when they are used to calculate the fitness function, but not changing the strings themselves (Tsutsui and Ghosh, 1997). In the context of loudspeaker placement, an analogous modification to the genetic algorithm could be made by randomly perturbing the plant matrix every time the attenuation of each loudspeaker set was evaluated.

Finally, transducer positions could be considered that make the performance of the control system robust to transducer failures. This optimisation is very problem-specific since it depends on the control algorithm being used, the types of actuator and sensor and the criterion for performance. One method of achieving such robustness is to use back-up transducers to achieve redundancy (Baruh, 1992). Provided the control algorithm can cope with such failures, however, the performance can be made robust to transducer failures by using a larger number of transducers than would otherwise be required, which are reasonably uniformly distributed in the space being controlled.

9.5.3. Final Comparison of Search Algorithms

By modifying the cost function used in the natural algorithm to include an effort term, we can discriminate against those source combinations that have an excessive control effort and may also be ill-conditioned. The convergence behaviour of both the genetic algorithm and simulated annealing program when using this modified cost function to select 8 source locations from a possible 32 is plotted in Fig. 9.16 (Baek and Elliott, 1995). It is interesting to observe that for this optimisation problem the convergence behaviour of the two algorithms is much more similar than that observed above (Figs. 9.10 and 9.13, in which the attenuation only was accounted for). The size of the searching problem is the same in this case as it was above, but now the simulated annealing program can find a source combination that gives attenuations within 0.5 dB of the best obtained in this simulation,

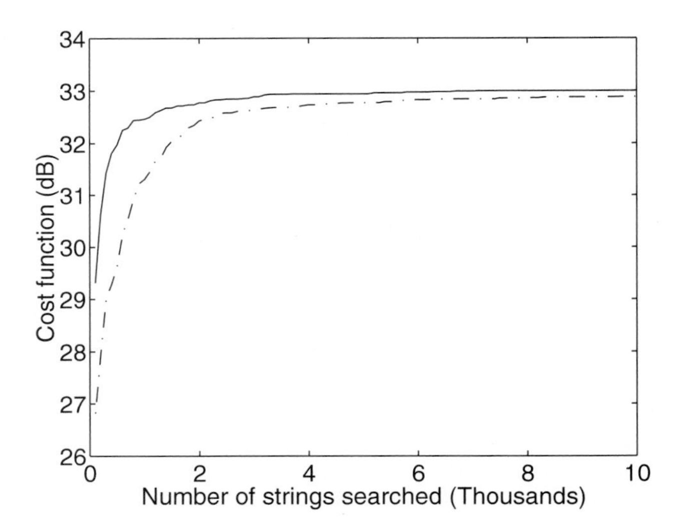

Figure 9.16 The maximum attenuations in the modified cost function, which includes effort weighting, when using either the genetic algorithm (dashed) or the simulated annealing method (solid) to choose 8 secondary loudspeaker positions from a possible 32 locations.

after searching only about 1,000 strings, instead of the 2,500 above, and the genetic algorithm now only needs to search about 2,000 strings instead of the 7,500 above. The search has thus also become more efficient by reducing the number of possible combinations that could be selected by excluding those that are ill-conditioned.

One final comment should be made about the results presented in Fig. 9.16. The motivations behind the genetic algorithm and the simulated annealing method are very different: one is inspired by biological processes and one by physical processes. The ways in which the algorithms work would also appear to be completely different, with the genetic algorithm maintaining a pool of possible solutions, but the simulated annealing algorithm only ever dealing with a single solution. Despite these apparently profound differences, the performance of the two algorithms in the well-conditioned problem shown in Fig. 9.16 is remarkably similar. Similar observations have been made in studies of structural actuator placement (Hakim and Fuchs, 1996). This suggests that there is some common element in both of these algorithms which gives rise to their efficiency in solving this optimisation problem. We have seen in Fig. 9.13 that the genetic algorithm was considerably more efficient than a simple, unguided, random search, which suggests that it is the combination of an intelligent, deterministic guidance mechanism working in cooperation with the random search that makes both the genetic algorithms and the simulated annealing method so effective.

10

Hardware for Active Control

10.1. INTRODUCTION

The theoretical basis for the algorithms used in active control systems has been discussed in previous chapters. In this chapter we concentrate on some of the more practical aspects involved in implementing such systems.

10.1.1. Chapter Outline

After a brief discussion in this section of the reasons for implementing controllers digitally, the next three sections concentrate on the characterisation of the analogue filters generally required to implement a digital active control system. Section 10.2 is concerned with the specification of the anti-aliasing filters, which act on the error and reference signals, and Section 10.3 is concerned with the specification of the reconstruction filters, which are generally necessary to eliminate the higher harmonics of the sampling frequency in the output of a digital control system. The inherent delay in the control system caused by the phase shift of these analogue filters can limit the performance of a practical active control system, and this important topic is addressed in Section 10.4.

After a very brief introduction to the technology of analogue-to-digital and digital-to-analogue converters in Section 10.5, the influence on a control system of quantisation effects in these converters is discussed in Section 10.6. The main effect of this quantisation, which is caused by the finite number of bits in the converters, is to generate measurement noise, and it is shown how this measurement noise may be much more important in the reference signal than in the error signal of an adaptive feedforward control system. Section 10.7 then provides a brief introduction to the properties of the digital processors required for active control applications. Although some general remarks can be made on this subject, the performance of commercial digital signal processing devices is improving very rapidly, and the detailed implementation of an algorithm on a particular processor can be very device-specific. Finally, some comments are made in Section 10.8 about the effects of the finite-precision arithmetic on control systems implemented using these devices. These effects can be troublesome if one is not aware of their consequences, particularly for adaptive filters with very small convergence coefficients.

10.1.2. Advantages of Digital Controllers

At this point it is worth reiterating the reasons for concentrating so much on the design of digital, i.e. sampled, control systems in this book, since most of the practical aspects discussed in this chapter have to do with the implementation of such digital systems. Perhaps the most important aspect of a digital system is its *adaptability*. It is relatively easy to change the coefficients of a digital filter, since they can be stored as numbers held in a memory device. We have seen that it is important in many active control applications to be able to adjust the response of the controller to maintain acceptable performance by compensating for changes in the disturbance spectrum or the plant response. The use of a digital system makes such adjustment possible. It is also often possible to run the program that implements the adaptation algorithm on the same hardware as is used to implement the digital filters.

In addition to adaptability, a modern digital system, in which the filter coefficients are typically specified with a precision of at least 16 bits, can also implement a digital filter very *accurately*. This is important in active control where high levels of attenuation require precise adjustment of the amplitude and phase of the signals driving the secondary actuators. Finally, the *cost* of digital control systems has dropped dramatically over the past decade or so, with the continuing development of processing devices especially designed for digital signal processing, commonly called DSP chips. This is partly reflected in the unit cost of the devices themselves, but also in the cost of developing the software that must run on these devices. There has been a significant shift from the majority of this software being written in a low-level language, such as assembly code, to the majority being written in a high-level language, such as C. This can be partly attributed to the development of floating-point devices, rather than the earlier fixed-point devices whose code often had to be 'hand crafted' to ensure that no overflow or other numerical problems occurred. The effects of finite precision arithmetic will be further discussed in Section 10.8.

10.1.3. Relationship to Digital Feedback Control

Many of the issues discussed in this chapter are the same as those that need to be addressed in more conventional digital control systems, as discussed for example by Franklin et al. (1990) or Åström and Wittenmark (1997). Most of the systems described by these authors are mechanical servo-control systems. Franklin et al. (1990), for example, provides a complete case study for the design of a servo-control system for a computer disc drive. In such systems, the response of the mechanical system under control has a natural low-pass characteristic and the level of high-frequency disturbances is relatively low, so that the system can be designed without reconstruction or anti-aliasing filters. This enables the servo loop to operate more quickly, because the system under control does not contain the delays associated with these filters. The sampling rate is also made much higher than the bandwidth required for control, partly to reduce the delays associated with the processing and data converters. In the case study quoted by Franklin et al. (1990), for example, the aim was to design a control system with a closed-loop bandwidth of 1 kHz, and the sample rate was selected to be 15 kHz.

In the active control of sound and vibration, the residual error can often be experienced as acoustic noise. It is thus important to ensure that in controlling one tone, for example, the other harmonic tones that are generated by the zero-order hold of the digital-to-analogue converters are filtered out. The tones at higher harmonics can be particularly audible because the A-weighting curve, which gives a reasonable approximation to the frequency response involved in human hearing, falls off rapidly at low frequencies, as shown in Fig. 10.1 (IEC Standard 651, 1979, as discussed for example by Kinsler et al., 1982, and Bies and Hansen, 1996). If the tone being actively controlled has a frequency of 100 Hz, for example, a harmonic at 500 Hz will be perceived as being equally loud if its level is some 16 dB lower than that at 100 Hz, because of the A-weighting characteristic. It is thus important to use effective *reconstruction filters* in most active control systems.

Also, it cannot generally be guaranteed that the disturbance in an active sound or vibration control system can be neglected at any frequency below several tens of kHz. It is important that these high-frequency disturbances do not alias down and appear in the bandwidth of the control system, and so *anti-aliasing filters* are also generally required in such systems. In contrast to mechanical servo systems, it is thus generally necessary in active noise and vibration systems to include both anti-aliasing and reconstruction filters in the design, and the specification of these filters is discussed in the following two sections.

The selection of the sampling rate is a critical decision in the design of any digital control system and often this decision comes down to a trade-off between performance and expense. A faster sampling rate will allow smaller delays in the system under control, and hence better steady state performance if the disturbance is broadband, but will require a faster and hence more expensive processor. Even for tonal disturbances, delays in the

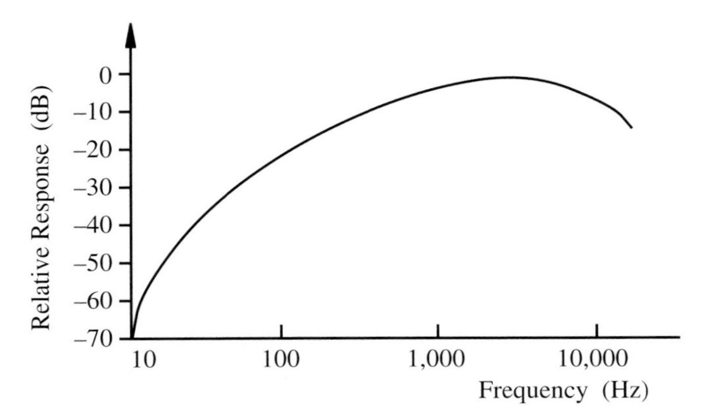

Figure 10.1 Frequency characteristics of the A-weighting curve, which approximates the response of the human ear at low sound levels and is widely used to estimate the subjective effect of sounds.

adaptation loop must be minimised if the controller is required to adapt very quickly. Sample rates of 3 to 10 times the fundamental frequency are often used in active control systems for tonal disturbances, and sample rates of up to 100 times the maximum frequency to be controlled are sometimes used in active control systems for the control of random disturbances. It is, however, important to ensure that the conditioning of the solution for the optimal controller does not become too poor as the sample rate is increased, since otherwise the convergence rate of an adaptive algorithm can be reduced and the system is more susceptible to numerical errors (Snyder, 1999). In digital servo systems, the controllers used may involve several elements, such as state estimators and various feedback loops, but these elements are typically of relatively low order (the state vector had 13 elements in the case study described by Franklin et al., 1990). In contrast, systems for the active control of broadband sound and vibration are typically implemented using FIR digital filters with hundreds of filter coefficients. The computational cost of increasing the sampling rate in such a system is two-fold. First, the number of coefficients required to implement an FIR filter with an impulse response of a given duration will rise in proportion to the sampling rate. Second, the time available to perform the calculations involved in implementing this filter will decrease in proportion to the sampling rate. For such controllers, the required processing power thus increases as the square of the sampling rate.

10.2. ANTI-ALIASING FILTERS

In the following two sections we discuss the specification and design of the filters required to prevent aliasing of the error and reference signals, and to smooth the signals driving the secondary actuators. These filters are usually analogue devices, and although we will see in Section 10.5 that for one class of data converter digital filters may also be used, we will concentrate here on the design compromises involved in analogue filters.

To begin with the design of the anti-aliasing filters, Fig. 10.2 shows the idealised response of a low-pass filter used for this purpose, together with that of the 'aliased'

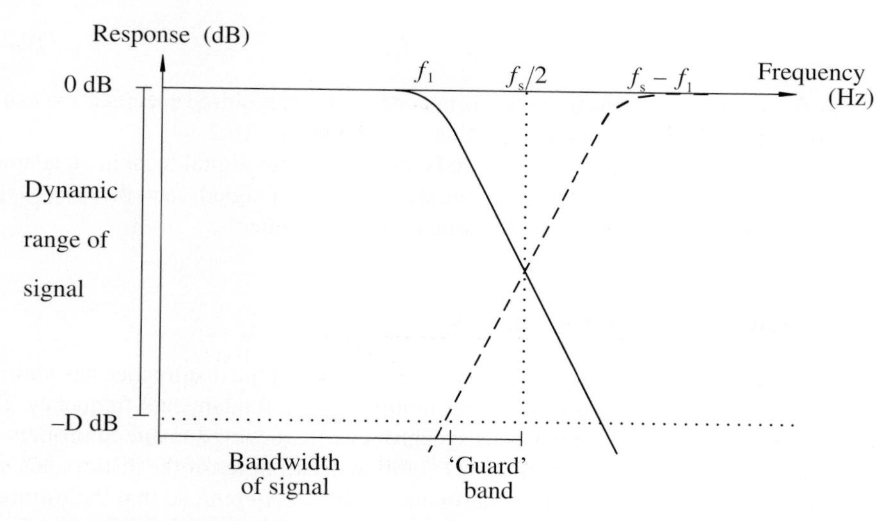

Figure 10.2 Magnitude of the frequency response of an anti-aliasing filter (solid line) and the 'aliased' characteristic (dashed line) used to calculate the required cut-off rate.

characteristic, i.e. the frequency response reflected about the vertical axis and shifted by the sample rate, f_s. If the measured signal has a reasonably flat spectrum up to about half the sample rate, $f_s/2$, then the spectrum of the signal passed to the analogue-to-digital converter is given by the magnitude of the filter response itself up to $f_s/2$. Above half the sample frequency, the signal is aliased, as described for example by Rabiner and Gold (1975), and Åström and Wittenmark (1997), with continuous-time signals at $f_s/2 + \Delta f$ appearing in the sampled signal at $f_s/2 - \Delta f$, so that a signal at $f_s - f_1$ will alias down to a frequency of f_1. The frequency components of the continuous-time signal above $f_s/2$ will thus appear in the sampled signal below $f_s/2$, with an amplitude proportional to the dashed line in Fig. 10.2 which we are calling the 'aliased' characteristic.

We initially assume that the bandwidth of the signal that must be converted into digital form is equal to f_1 and that the dynamic range of this signal is D dB. Figure 10.2 can be used to calculate the relationship between the fall-off rate of the anti-aliasing filter and the sample rate so that the aliased components fall below the dynamic range of the required signal. It can be seen from Fig. 10.2 that the anti-aliasing filter must have a cut-off frequency of f_1 and have attenuated the signal at $f_s - f_1$ by D dB.

Assuming that the anti-aliasing filter falls off at a constant rate of R dB/octave, the filter response will have fallen by D dB in D/R octaves, and the ratio of $f_s - f_1$, to f_1 must thus be given by

$$\frac{f_s - f_1}{f_1} = 2^{D/R} . \tag{10.2.1}$$

For a given filter fall-off rate, R, the sampling frequency must thus be

$$f_s = \left(1 + 2^{D/R}\right) f_1 , \tag{10.2.2}$$

or for a given sampling frequency the fall-off rate of the filter must be

$$R = D/\log_2\left[(f_s - f_1)/f_1\right] \,. \tag{10.2.3}$$

If the fall-off rate of the anti-aliasing filter is not constant, the required specification can be obtained by graphical construction using a chart similar to Fig. 10.2.

Two sets of signals will potentially have to be converted into digital form in an adaptive feedforward control system, the reference signals and the error signals, and these two types of signal will generally have different dynamic range requirements.

10.2.1. Applications to Error Signals

To begin with the error signals, the worst case will occur when the disturbance has multiple harmonics and the sampling rate is an integer multiple of the fundamental frequency. This may be the case if waveform synthesis or synchronous frequency-domain controllers are used, as described in Chapter 3. Under these conditions, harmonics of the disturbance may occur at exactly f_1 and $f_s - f_1$, and these harmonics will be *coherent*, so that the estimated amplitude of the harmonic at f_1 is directly affected by the aliased version of that at $f_s - f_1$. The appropriate value for the dynamic range in the expressions above is given in this case by the expected attenuation in the harmonic at f_1. If, for example, an attenuation of 30 dB is expected and anti-aliasing filters with a fall-off rate of 24 dB/octave are being used, then using equation (10.2.2), the sample rate must be greater than $3.4\,f_1$.

If the disturbance were completely random, however, the signal at $f_s - f_1$ would be incoherent with that at f_1. Aliasing would then have no effect on the cross-correlation function between the sampled reference signal, which is assumed not to be aliased for now, and the sampled error signal. There would, however, be a drop in the coherence between the error signal and the reference signal. The fact that there is no change in the cross-correlation function between the reference signal and error signal, which is driven to zero by the adaptive algorithm in approaching the optimal controller, means that aliasing in the error signal does not necessarily affect the attenuation of the analogue error signal, since the aliased sampled error signal is only used to adapt the controller. The potential steady-state performance of a digital feedforward control system is thus unaffected by aliasing of the error signal if the disturbance is entirely random. Aliasing will, however, have the effect of adding random noise to the sampled error signals, and this may affect the behaviour of the algorithm used to adapt the control filter. If an instantaneous steepest-descent algorithm, such as a variant of the LMS algorithm, is used for example, then the filter coefficients will be subject to increased levels of random perturbation because of this aliasing, causing an increase in the mean-square error due to 'misadjustment', as discussed in Chapter 2. Unless the controller is being adapted very quickly, however, the degradation in performance due to this effect is likely to be small.

10.2.2. Application to Reference Signals

Returning to a consideration of aliasing in the reference signals, these signals are filtered by the control filter and used to drive the secondary sources. It is thus not only necessary to maintain the dynamic range of the reference signals within the working bandwidth of the control system but also to prevent aliased components being broadcast from the secondary actuators. The dynamic range of the reference signals must be maintained, since otherwise the aliased components will appear as 'sensor noise' in the reference signal and reduce the achievable attenuation, as discussed in Section 3.3. The limiting value of attenuation is given by equation (3.3.13), and if the required attenuation is A dB, then the dynamic range, D in equation (10.2.1), can be put equal to A to calculate the characteristics of the anti-aliasing filter required to prevent this loss of performance. The result is, however, a less stringent filter than that required to prevent audibility of the aliased components. To calculate the characteristics of this latter filter, however, we must make some assumptions about the spectrum of the reference signal, and the response of the controller and plant.

If the active control system attenuates the disturbance, by A dB, then the aliased components must clearly be below this level in order not to be audible. If the continuous-time reference signal had a flat A-weighted spectrum, for example, and both the controller and plant have reasonably uniform frequency responses, then we would require that the anti-aliasing filter fall off by A dB at $f_s/2$ in order for aliased signals above this frequency to probably be inaudible, so that in this case

$$\frac{f_s/2}{f_1} = 2^{A/R} . \tag{10.2.4}$$

For a given filter fall-off rate of R dB/octave, the sample rate must thus be

$$f_s = \left(2^{(A/R+1)}\right) f_1 , \tag{10.2.5}$$

or, for a given sampling frequency, the fall-off rate must be

$$R = A \Big/ \log_2\left(\frac{f_s}{2f_1}\right) . \tag{10.2.6}$$

If the attenuation provided by the control system is 30 dB, for example, and the anti-aliasing filter has a fall-off rate of 24 dB/octave, then the sample rate must be greater then 4.8 times the control bandwidth for the aliased components to be inaudible. A similar argument will apply in a feedback system, in which, from the discussion of internal model control in Section 6.4, it can be seen that the outputs from the sensors are effectively used to provide reference signals in an equivalent feedforward control system.

10.3. RECONSTRUCTION FILTERS

We now turn our attention to the specification required for the reconstruction filter. The aim of this filter is to 'smooth' the output of the digital-to-analogue converter, which generally has an inbuilt zero-order hold device and thus has a continuous-time output which, under fine magnification, would look like a staircase. It is convenient, however, to consider the performance of the reconstruction filter in the frequency domain. The original spectrum of the sampled signal used to drive the digital-to-analogue converter, $U\left(e^{j\omega T}\right)$, is periodic, with a period of $\omega T = 2\pi$, as shown in Fig. 10.3(a). In passing through the zero-order hold device the sampled signal is effectively convolved with an impulse response that is of

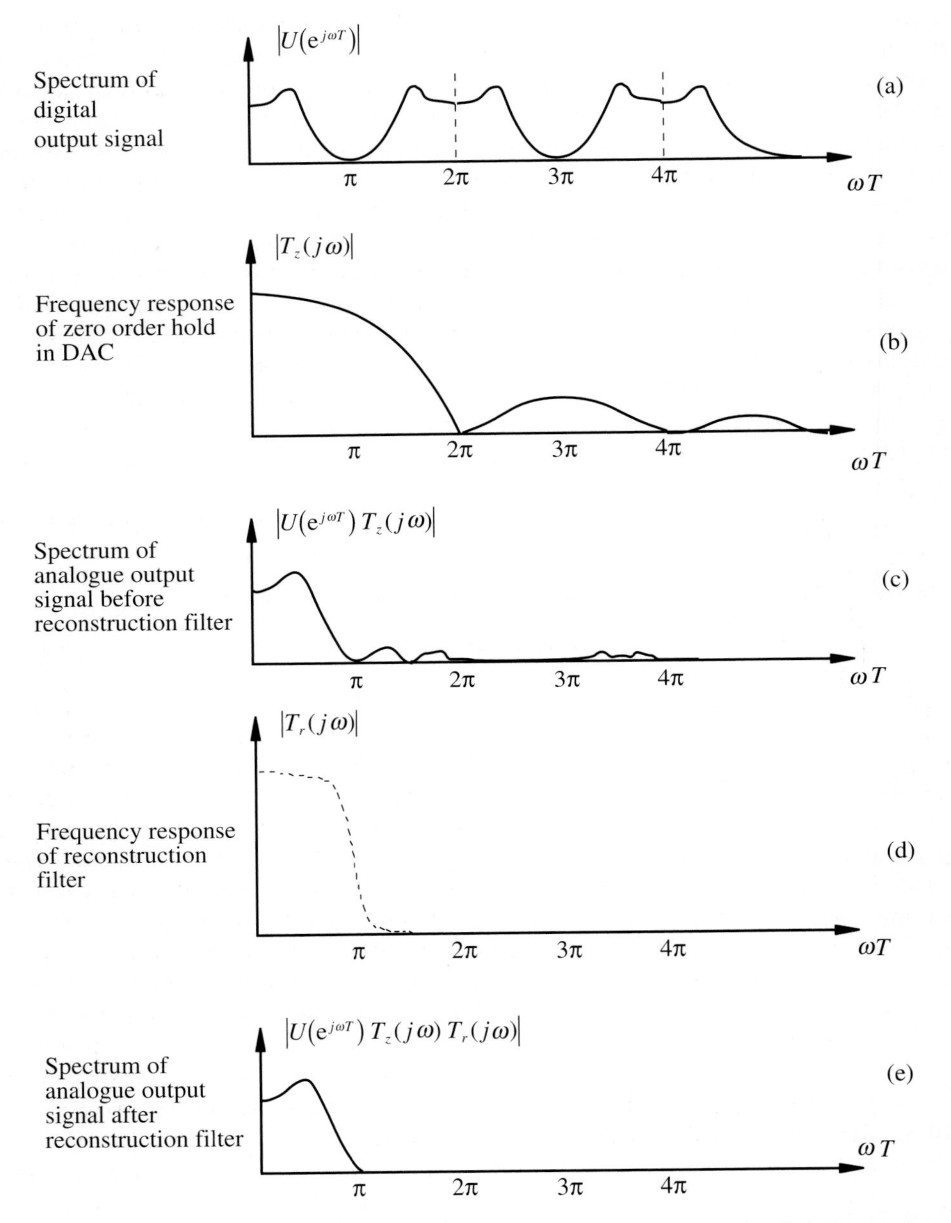

Figure 10.3 Changes that occur in the spectrum of the signal sent to the digital-to-analogue converter due to the zero-order hold and reconstruction filters.

unit amplitude and has a duration equal to one sample period T. In the frequency domain, the effect of the zero-order hold is thus the same as a linear filter whose frequency response is given by the Fourier transform of this impulse response. The modulus of this frequency response is proportional to the sinc function shown in Fig. 10.3(b). It should not be forgotten, however, that the frequency response also has a linear phase component, which corresponds to a delay of half a sample period, $T/2$, and this delay will contribute to the total delay in the sampled system under control. The complete continuous-time frequency response of the zero-order hold can be obtained by manipulating equation (3.1.1) to give

$$T_z(j\omega) \;=\; T\,\mathrm{sinc}\big(\omega T/2\big)\,\mathrm{e}^{-j\omega T/2}\;. \tag{10.3.1}$$

The main effect of the zero-order hold in the frequency domain is to suppress the frequency components in the original sampled signal at frequencies around $\omega T \approx 2\pi$, i.e. $f \approx f_s$. The spectrum of a sampled signal is periodic, and so components of the original spectrum at frequencies of about $f_s/4$, $\omega T \approx \pi/2$, also appear at $3f_s/4$, $\omega T \approx 3\pi/2$. Whereas the former are only attenuated by about 1 dB in passing through the zero-order hold, the latter are attenuated by over 10 dB. All frequency components above $f_s/2$, $\omega T \approx \pi$, are images of the required spectrum and an ideal reconstruction filter should perfectly attenuate these images. Reconstruction filters are therefore sometimes called *anti-imaging filters*.

In practice, the specification of the reconstruction filter will depend on the spectrum of the input signal to the digital-to-analogue converter, $u(n)$, and the degree of attenuation required in the frequency images. If, for example, we assume that the spectrum of $u(n)$ is reasonably uniform up to $f_s/4$ and decays naturally between $f_s/4$ and $f_s/2$, then it will be the imaged components at about $3f_s/4$ that will need to be most heavily attenuated. If these components need to be suppressed by A dB, for example, then the reconstruction filter must have decayed by $(A - 10)$ dB at $3f_s/4$, since the zero-order hold has already attenuated this component by about 10 dB.

If the reconstruction filter has a uniform fall-off rate of R dB/octave and a cut-off frequency of $f_s/4$, then the ratio of the frequency that must be attenuated to the cut-off frequency in this example must be equal to

$$\frac{3f_s/4}{f_s/4} \;=\; 3 \;=\; 2^{(A-10)/R}\,, \tag{10.3.2}$$

so that in this example

$$R \;\approx\; \frac{A-10}{1.6}\;. \tag{10.3.3}$$

If an attenuation of 40 dB is required, for example, then the reconstruction filter must fall off at about 18 dB/octave.

Table 10.1 provides a summary of the required filter attenuation and the frequency at which this attenuation is required for the different cases discussed in the previous two sections.

10.4. FILTER DELAY

As well as the amplitude response discussed above, both the anti-aliasing and reconstruction filters will have a certain phase response, $\phi(j\omega)$. This is important because it contributes to the delays in the system under control. This contribution is most conveniently quantified using the *group delay* which is defined as

Table 10.1 The specification of the various anti-aliasing and reconstruction filters required in an active control system assuming that the cut-off frequency is equal to the control bandwidth, f_1, the required controller attenuation is A dB and the sampling frequency is f_s

Application	Criterion	Required filter attenuation (dB)	Frequency of required attenuation
Anti-aliasing for harmonic error signals	No degradation in control performance	A	$f_s - f_1$
Anti-aliasing for random reference signals	Inaudibility	A	$f_s/2$
Reconstruction filters	Inaudibility	$A - 10$	$3f_s/4$

$$\tau_g(j\omega) = \frac{d\phi(j\omega)}{d\omega} \tag{10.4.1}$$

Different types of analogue filter have different amplitude and phase responses, and hence different group delay characteristics. The amplitude and group delay responses of three different types of analogue filter, the Butterworth, Chebyshev and elliptic filters, are shown in Fig. 10.4. The characteristics of these filters are described more fully by Horowitz and Hill (1989), for example. All filters are fourth order, i.e. they have 4 poles, and hence the Butterworth and Chebyshev filters have an asymptotic fall-off rate of 24 dB/octave. The Chebyshev and elliptic filters have been designed to have 1 dB ripple in the pass band and the elliptic filter to have at least 40 dB attenuation in the stop band. The group delay characteristics of each of the filter responses shown in Fig. 10.4 are reasonably uniform at low frequencies, and the magnitude of the group delay in this region is approximately $0.5/f_c$ seconds where f_c is the cut-off frequency of the filter. This is consistent with the approximation that each pole contributes about 45° of phase shift or 1/8 cycle of delay at the cut-off frequency (Ffowcs-Williams et al., 1985). If we assume that for a complete digital controller there are a total of n poles in both the analogue anti-aliasing and reconstruction filters, which each have a cut-off frequency of f_c, these filters will have a low-frequency group delay of about $n/8f_c$ seconds. Further assuming that the cut-off frequency is set to be one-third the sampling rate, and that an additional one-sample delay is present in the digital controller, together with the half-sample delay in the zero-order hold, then the total group delay through the complete controller, τ_A, when the digital filter is set to directly feed out the input signal, can be written as

$$\tau_A \approx \left(1.5 + \frac{3n}{8}\right)T , \tag{10.4.2}$$

where T is the sampling time. More precise calculations of the minimum controller delay can be made for different cut-off frequencies compared with the sampling rate, or for various filter types, but equation (10.4.2) has been found to be a useful rule of thumb in the initial design of an active control system. Additional care may need to be taken at frequencies close to the cut-off frequency, however, particularly when using Chebyshev or elliptic filters, because of the significant peak in the group delay characteristic at this frequency, as seen in Fig. 10.4.

It is good practice to select the filter with the lowest order which can give the required attenuation above the specified frequency, f_2, listed in Table 10.1. Table 10.2 shows the

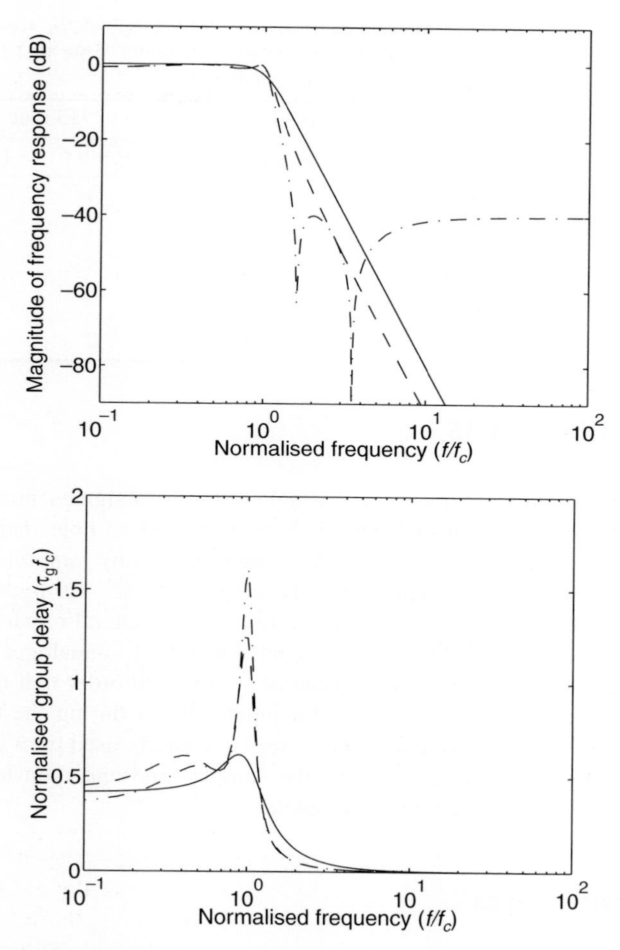

Figure 10.4 Magnitude and group delay responses of various fourth-order analogue filters: Butterworth (solid line), Chebyshev I (dashed line) and elliptic filters (dot-dashed line), all with the same cut-off frequency, f_c. The Chebyshev and elliptic filters are designed to have a pass band ripple of 1 dB and the elliptic filter is designed to have a stop band attenuation of 40 dB.

required orders of either Butterworth or elliptic filters (with 1 dB of passband ripple) which would meet these specifications, assuming the sampling rate was three times the control bandwidth ($f_s = 3f_1$), that the cut-off frequency of the filters is equal to the control bandwidth and the maximum attenuation of the control system (A) is 30 dB. The low-frequency group delay through these filters, in samples, is also shown in Table 10.2. Clearly, a lower order can be used for the elliptic than for the Butterworth filters to meet these specifications, but this advantage is somewhat offset by the greater complexity involved in implementing an elliptic filter, which has zeros as well as poles, and the much larger delay close to the cut-off frequency. It should also be noted that the high-frequency response of an elliptic filter does not give increasing attenuation as the frequency rises, and this may give rise to problems if very high-frequency disturbances are present.

Table 10.2 Required filter order and associated low-frequency group delay for Butterworth and elliptic filters designed to meet the specifications listed in Table 10.1 for an example in which it is assumed that $f_s = 3 f_1$ and $A = 30$ dB

	Butterworth Filter		Elliptic Filter	
Application	Order	Delay (samples)	Order	Delay (samples)
Anti-aliasing for harmonic error signals	5	1.6	3	1.1
Anti-aliasing for random reference signals	9	2.8	5	1.6
Reconstruction filters	3	1.0	2	0.5

10.5. DATA CONVERTERS

In order to convert continuous-time reference signals and error signals into digital form, an analogue-to-digital converter must be used. It is important to note that such a device divides up the continuous-time input both along the time axis by *sampling*, whose effects were discussed in Section 10.2, and along the amplitude axis by *quantisation*, whose effects are discussed in Section 10.6. In this section, a few general remarks will be made about different technologies for analogue-to-digital and digital-to-analogue converters, and the way they are typically used in active control systems. In order that the analogue-to-digital converter is presented with a constant input voltage during the time it takes to perform the conversion, a *sample-and-hold device* is generally used prior to the converter, whose output tracks the input signal until the sample time and then holds the output constant at this level until conversion is complete.

10.5.1. Converter Types

The analogue-to-digital converters with the speed and accuracy required for active control typically use one of two different principles: successive approximation or sigma delta conversion. These techniques, and several other data conversion methods that are not typically used in active control, are described, for example, by Horowitz and Hill (1989), Hauser (1991) and Dougherty (1995). Both methods use an internal digital-to-analogue converter, but in the successive approximation technique this is a multi-bit device operating at the sampling rate multiplied by the number of bits being converted, whereas in the sigma delta or 'over-sampling' device the digital-to-analogue converter is a one-bit device, typically operating at 64 times the sample rate. The output of the sigma delta device is effectively the addition, Σ, of many small increments, Δ, which may be denoted $\Sigma\Delta$, and this is how the converter derives its name. The requirements for the converters used in active control are similar to those encountered in digital audio, and a comprehensive treatment of such converters is provided by Watkinson (1994).

The difference between the successive approximation and sigma delta converters from the user's point of view is that the successive approximation converter requires an external anti-aliasing filter whereas sigma delta converters generally have an in-built anti-aliasing action, achieved with integrated digital filters operating at the over-sampled frequency.

This approach has the advantage that the cut-off frequency of the anti-aliasing filter automatically tracks changes in the sampling frequency. Unfortunately, the filter responses typically programmed into sigma delta converters are optimised for audio use, with very low ripple and nearly linear phase response, and these are not necessarily the properties which are best for control work. In particular, the anti-aliasing filter in a typical sigma delta converter is of very high order and has a latency, which includes delays in data transmission, of about 40 times the reciprocal of the cut-off frequency (Agnello, 1990), which compares with delay of about 0.5 times the reciprocal of the cut-off frequency seen for the fourth-order analogue filters discussed in Section 10.4. This large delay is due to the extremely high attenuation and linear phase of the in-built digital anti-aliasing filter which is required for 16-bit audio work but not for active control work. It is sometimes possible to access the output of such converters before this filter, giving lower precision but also smaller delays (Agnello, 1990).

The digital-to-analogue converters typically used in active control systems also come in two varieties; the conventional devices which use weighted current networks (Horowitz and Hill, 1989; Watkinson, 1994), and over-sampling devices which use sigma delta techniques similar to those discussed above. Most digital-to-analogue converters have an in-built zero-order hold. Conventional digital-to-analogue converters also require an external reconstruction filter, whereas this is built into the sigma delta devices, even though again its properties may not be ideal for control purposes.

If separate data converters are used on all reference signals $x_1(t), \ldots, x_n(t)$, all error signals $e_1(t), \ldots, e_L(t)$ and all output signals to the secondary actuators, $u_1(t), \ldots, u_M(t)$, then the complete block diagram of the data conversion stage in an active control system may be represented as in Fig. 10.5. With this arrangement, it is fairly straightforward to ensure that all the samples are taken simultaneously by timing all the sample-and-hold devices on the inputs, and digital-to-analogue converters on the outputs, from the same sample rate clock. The clock generator is often implemented in hardware, rather than in software, to avoid sampling jitter due to variable-length loops. It has been noted by Goodman (1993) that in severe cases such jitter can cause a significant loss of performance and may even cause an adaptive controller to become unstable.

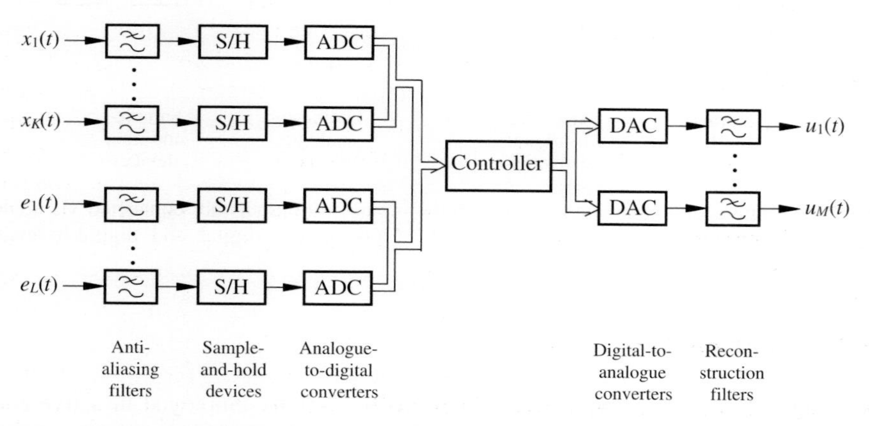

Figure 10.5 Data acquisition system in which separate analogue-to-digital and digital-to-analogue converters are used for each input and output.

An alternative approach to the design of the data conversion system is shown in Fig. 10.6, in which analogue multiplexers are used to reduce the numbers of data converters needed. Separate analogue-to-digital converters have been retained for the reference signals and the error signals because of the difference in the required accuracy, as discussed in the next section, but an even more compact arrangement could be achieved if a single converter was shared between these two sets of signals. Clearly the data converters must work more quickly if they have multiplexed inputs or outputs like this, but since the converters are typically optimised to work at frequencies of tens of kHz, and the sample rates required in active systems are typically a few kHz, then multiplexing can be a cost-effective option. Hardware architectures for active noise control systems have also been discussed by Mangiante (1996).

All the sample-and-hold circuits in Fig. 10.6 can be timed using a single clock to achieve synchronous sampling on all input and output channels. Alternatively, only a single sample-and-hold device can be used before each analogue-to-digital converter and the conversion of each of the input signals performed sequentially, within one sample period. Provided the timing of this 'staggered' sampling arrangement is exactly the same when the plant is being identified and controlled, the extra delay of a fraction of a sample time between the input channels will be accounted for in the plant model, and the behaviour of an adaptive controller will be almost identical to one in which the inputs are sampled synchronously. In control systems with many input channels, the saving in cost, size and weight can be significant when using a single rather than multiple sample-and-hold devices with an input multiplexer.

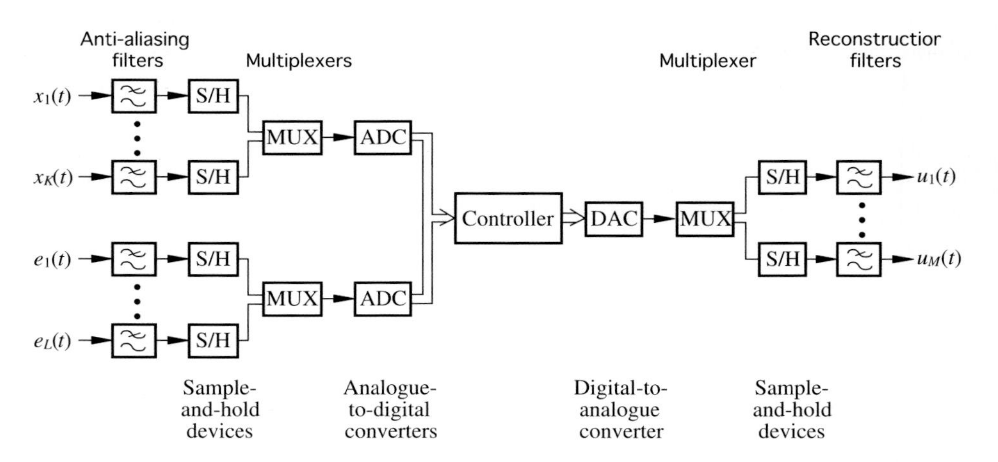

Figure 10.6 Data acquisition system in which the inputs and outputs are connected via analogue multiplexers to a reduced number of analogue-to-digital and digital-to-analogue converters.

10.5.2. Oversampling

Another approach to data acquisition which can be usefully employed in active control systems is to initially sample the reference and error signals at a higher sampling rate than required, but with relatively low-precision converters and low-order anti-aliasing filters, and then to use the processing power of the controller to digitally low-pass filter and decimate these signals. The relatively high precision, lower sample rate signal can then be processed by the control algorithm. The requirements on the analogue-to-digital converters and anti-aliasing filters are considerably relaxed even if the initial sample rate is only four times that used in the control algorithm. This *over-sampling* technique can be seen as being part way between conventional sampling with a high-precision converter at the control sample rate and the sigma delta philosophy of initially sampling at much higher rates with a single-bit converter. Another difference between this over-sampling technique and the use of the sigma delta converter is that, using this method, the digital filtering must be performed by the user, rather than being integrated within the sigma delta device. Although this technique requires extra processing power, it does allow the response of the digital filter to be tailored to the control application, by ensuring that it has a minimum group delay for example. The design of these digital filters broadly parallels that of the analogue anti-aliasing filters described above, except that generally a low-order analogue filter has already been used to limit the level of the input signal at frequencies above half the original sampling rate.

With the general trend towards less expensive digital signal processing hardware, the trade-off between using more expensive analogue hardware with less digital processing, and using cheaper analogue hardware with more digital processing will be pushed further in the direction of the latter as time goes on, and so the over-sampling techniques discussed above will become increasingly attractive.

10.6. DATA QUANTISATION

Modern digital-to-analogue converters behave in a way which is very close to that of an ideal device. An N-bit digital word is accurately converted to an analogue representation of this word, whose value is maintained by the zero-order hold device until the next digital word and the next clock signal are received. Analogue-to-digital converters, on the other hand, must approximate an analogue waveform, whose amplitude can vary continuously, with a set of discrete levels which can be represented by a digital word of finite length. Thus, a practical analogue-to-digital converter, as shown in Fig. 10.7(a), not only samples the analogue signal in the time domain, but also quantises it on an amplitude scale. The quantisation effect can be thought of as the operation of a memoryless nonlinear 'staircase' function, which saturates at either end, as shown in Fig. 10.7(b). The two functions of a practical analogue-to-digital converter are represented separately in Fig. 10.7(b), by a quantiser and an ideal sampler, which is shown as a switch, as commonly used in the digital control literature (see for example Franklin et al., 1990).

10.6.1. Quantisation Noise

The output of the quantiser, $y(t)$, can be defined to be equal to the sum of the input, $x(t)$, and an 'error' signal, $v(t)$, so that

$$y(t) = x(t) + v(t) , \tag{10.6.1}$$

Practical analogue-to-digital converter

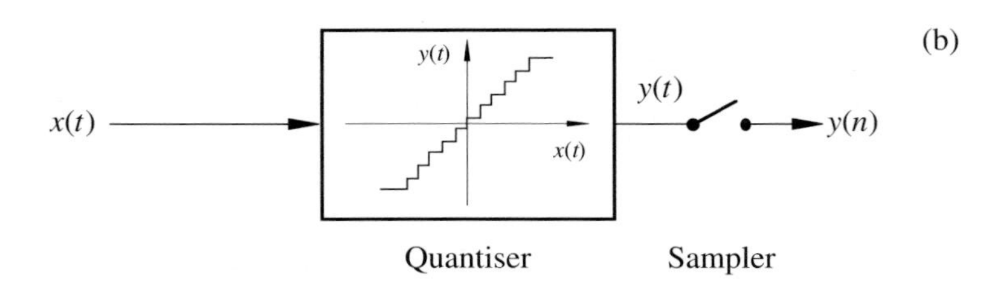

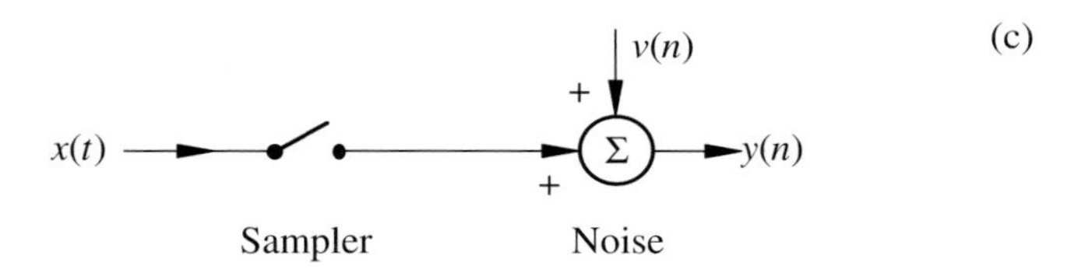

Figure 10.7 Representation of a practical analogue-to-digital converter (a) in terms of its two functions of quantisation and sampling (b) and an equivalent circuit (c) in terms of an ideal sampler and an additional noise signal, $v(n)$, which is almost white if the input signal changes by many quantisation levels between samples.

and if this signal is passed through an ideal sampler, with sample period T, the output sequence is

$$y(nT) = x(nT) + v(nT) \ , \tag{10.6.2}$$

where the explicit dependence of this sequence on T can be dropped, as in all the previous chapters, in order for the output sequence to be represented as

$$y(n) = x(n) + v(n) \ , \tag{10.6.3}$$

as shown in Fig. 10.7(c).

The statistical properties of the error sequence $v(n)$ will depend on those of the input, $x(n)$. If the input were entirely constant, for example, $v(n)$ would also be constant, but would be bounded by half a quantisation level Q, so that

$$|v(n)| \leq Q/2 \ , \tag{10.6.4}$$

where the quantisation level is the range of the converter divided by the number of steps in the staircase function in Fig. 10.7(b), which is equal to

$$Q = \frac{2V}{\left(2^N - 1\right)} \approx 2^{1-N} \, V \ , \tag{10.6.5}$$

where the converter is assumed to have a range of ± V volts and a resolution of N bits. For a 12-bit converter with a range of ± 1 V, the quantisation level is thus about 0.5 mV.

In most applications, the input signal changes by much more than the quantisation level between individual samples. In this case, the position of each sample on an individual step in the staircase function is almost completely random and becomes uncorrelated from one sample to the next. The error signal then becomes very similar to a white noise sequence, which is uniformly distributed from $-Q/2$ to $Q/2$ and thus has a mean value of zero. Specifically (Widrow, 1956), if

$$\left[x(n) - x(n-1)\right] \gg Q \ , \tag{10.6.6}$$

then

$$E\left[v(n)\right] \approx 0 \ , \tag{10.6.7}$$

where E denotes the expectation operator and the signals are assumed to be ergodic. The error sequence is thus zero mean, and

$$E\left[v(n)\, v(n-m)\right] \approx 0 \qquad \text{if} \quad m \neq n \ , \tag{10.6.8}$$

so that the autocorrelation function of error sequence is a delta function and hence the error sequence is white noise. Also, under these conditions,

$$E\left[v(n)\, x(n-m)\right] \approx 0 \ , \tag{10.6.9}$$

so that the error sequence is uncorrelated with the input sequence, and

$$e\left[v^2(n)\right] \approx Q^2/12 \ , \tag{10.6.10}$$

i.e. the mean-square value of the error sequence is proportional to the square of the quantisation level (as described in more detail by Franklin et al., 1990, for example). The rms level of the *quantisation noise* for a 12-bit converter with a range of ± 1 V under these conditions is thus about 0.14 mV.

10.6.2. Signal-to-Noise Ratio

Very stringent precautions must be taken with high-precision converters to ensure that the electrical noise picked up by the converter does not swamp this quantisation noise. If these precautions have been observed then the *signal-to-noise ratio* of the sampled signal will be determined by the statistical properties of the input signal and the quantisation noise. Specifically, the signal-to-noise ratio, or SNR, in decibels is equal to

$$\text{SNR (dB)} = 10 \ \log_{10} \frac{E\left[x^2(n)\right]}{E\left[v^2(n)\right]} \ . \tag{10.6.11}$$

Considering the most optimistic case first, if the input signal is a perfectly scaled square wave, then $E\left(x^2(n)\right) = V^2$ where V is the range of the converter, and using equations (10.6.5) and (10.6.10), the signal-to-noise ratio can be calculated to be

$$\text{SNR (dB)} \approx 6N + 5 \quad \text{dB}, \tag{10.6.12}$$

which confirms the well-known result that the signal-to-noise ratio is improved by about 6 dB for every extra bit of resolution in the converter. Equation (10.6.12) can only be taken as a best case signal-to-noise ratio, however, since it has been assumed that the input is perfectly scaled and has a peak value that is equal to its rms value. The ratio of the peak to the rms value of a signal is called its *crest factor*, which is unity for a square wave and $\sqrt{2}$ for a sinusoid. The definition of the crest factor can be difficult for random signals whose probability distribution functions are said to be Gaussian, because if the expectation is taken over a large enough ensemble, the peak value is, in principle, unbounded. Clearly, an engineering judgement must be made when converting such signals from analogue-to-digital form, and it is common to assume a crest factor of about 3, in which case the signal to quantisation noise ratio will be

$$\text{SNR (dB)} \approx 6N - 5 \qquad \text{dB}. \tag{10.6.13}$$

The signal-to-noise ratio can be even further degraded if there are significant impulsive components in the input signal, or if the input is poorly scaled or non-stationary.

10.6.3. Application to Active Control Systems

In broadband active control applications, analogue-to-digital converters are generally required for both reference and error signals, and the requirements of the converters for these two signals will be considered separately. In Chapter 3 we saw how noise in the reference signal could degrade the performance of a feedforward control system. It is difficult to perform a detailed calculation without knowing the spectrum of the reference signal, but if we assume for the moment that this is white, and has an assumed crest factor of 3, then the signal-to-noise ratio is given by equation (10.6.13). The maximum attenuation of a feedforward control system is inversely proportional to the linear version of the signal-to-noise ratio, as discussed for the electrical noise cancellation case by Widrow and Stearns (1985), so that the limiting level of attenuation in dB is approximately equal to SNR (dB) in equation (10.6.13). If an attenuation of 30 dB is to be achieved, the analogue-to-digital converter would apparently only need about 6 bits. One problem with this argument is the assumption that the reference signal is white, and so the signal-to-noise ratio is the same at all frequencies. In practice, the power spectral density of the reference signal may have a large range, and attenuation may still be required even in frequency ranges in which the reference signal has little power. In this case, the signal-to-noise ratio, and hence the maximum attenuation, must be calculated at each frequency. If the dynamic range of the reference signal's power spectral density is 30 dB for example, then for an achievable attenuation of 30 dB at all frequencies the signal-to-noise ratio must be 60 dB, which requires an analogue-to-digital converter with at least 11 bits. In some applications the reference signal is also non-stationary, with a mean-square value that varies significantly over time. This occurs, for example, in the case of accelerometer reference signals for road noise control, whose mean-square output changes significantly as the car is driven over different road surfaces. Fortunately, however, good active control performance is not usually required when the road is relatively smooth, in which case the level of the accelerometer signals may not be large compared with the quantisation noise, and the system can be designed for good performance in the high noise level case.

As far as the analogue-to-digital converters for the error signal in a feedforward system are concerned, the requirements are somewhat different to those of the reference signals, and the limitations are similar to those discussed in Section 10.2. Since the error signals are principally used in the adaptation algorithm to adjust the response of the controller, their waveforms never directly appear at the output to the secondary actuators. Any uncorrelated noise present in the error signals will thus not influence the exact least-squares controller and hence has no effect on the potential performance of the system. In practice, however, this potential performance can only be approached with an adaptive controller in a stationary environment. If a stochastic gradient algorithm, such as the LMS, is used to adapt the controller, the convergence coefficient must be small to reduce the misadjustment error, caused by small random perturbations in the coefficients of the control filter, as discussed in Chapter 2. Noise in the error signals would make this misadjustment worse and will thus degrade the performance of a practical controller, although this degradation is typically small if the controller is not being adapted very rapidly.

The requirements on the analogue-to-digital converters are thus considerably less stringent for the error signals than they were for the reference signals. Indeed, some variants of the LMS algorithm only use the sign of the error signal in the adaptation (Sondhi, 1967), which is equivalent to having only a one-bit analogue-to-digital converter. Some care must be exercised in this case to determine the properties of the noise signal, $v(n)$ in equation (10.6.3). The assumption that the input changes by much greater than a quantisation level, for example, is clearly not valid for a one-bit converter in which the quantisation level is equal to the range of the converter. Apart from being used to adjust the controller, the error signals are also used in the identification of the plant response, which is necessary in most adaptive systems. Many identification algorithms are not biased by uncorrelated output noise, but the accuracy of the estimated response would suffer if this noise were excessive.

10.7. PROCESSOR REQUIREMENTS

Several different types of processor could, in principle, be used to implement an active control system, ranging from a general-purpose microprocessor to a dedicated microcontroller. In general, however, general-purpose microprocessors have many more functions than are needed to implement an active control system, and microcontrollers do not have sufficient precision for this application. The type of processor that is most widely used for implementing active control systems is optimised for the calculations involved in digital signal processing, which are generally called DSP devices. Arrays of such devices can be used for very processor-intensive applications. Darbyshire and Kerry (1996), for example, describe a system designed for a levitated machinery raft using 96 actuators and 500 sensors which uses 160 DSP chips and is capable of 9.6×10^9 floating-point operations per second.

It is the rapid development in the processing power of DSP devices over the last two decades that has made many active control systems practical. This bold generalisation is supported by the very schematic graph shown in Fig. 10.8, which shows how the processing power of a 'typical' DSP device has increased over this timescale (Rabaey et al., 1998; EDN, 2000; BDTI, 2000). The maximum number of operations per second has been

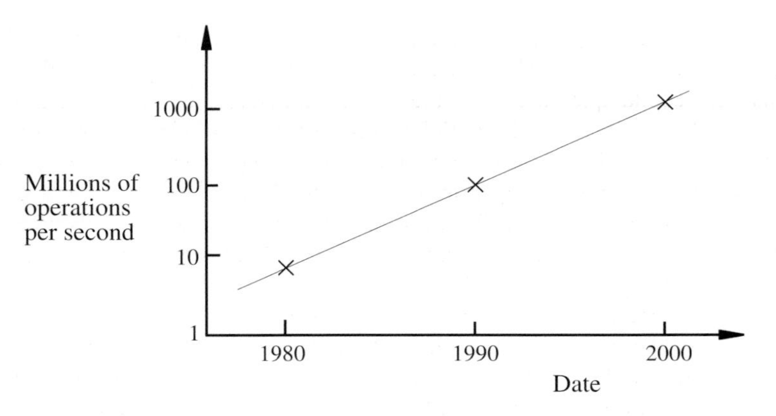

Figure 10.8 The increase in processing power of a typical DSP device over a 20-year period.

doubling about every 3 years. This is not quite so rapid as the doubling every 12 months which Gordon Moore originally predicted for the number of components in an integrated circuit in 1964, but nearly four decades later it is impressive that the processing power of integrated circuits is still developing at an exponential rate.

It would be impossible to discuss all the intricacies involved in matching control algorithms to DSP processors, but some general comments are worthwhile to put the real-time implementation of the algorithms discussed above into context. It is also not possible to describe the many different kinds of DSP devices that are available, but surveys and tutorials are published regularly in several journals (see, for example, Piedra and Fritsch, 1996) and several conferences are dedicated to this topic (see for example Proceedings of ICSPAT, 2000), to which the reader may refer for up-to-date information. The processing requirement of the various algorithms has been broadly quantified in the previous chapters in terms of the number of operations they require per sample. Clearly the number of operations per second that the processor has to perform is this number multiplied by the sampling frequency. Generally speaking, the processing power required to implement algorithms to control broadband random noise is considerably greater than that required to control tones. This is because the number of coefficients required in the controller and in the plant model is considerably larger for broadband noise than it is for tones. We consider two specific examples of control systems which use the time domain filtered-reference LMS algorithm to illustrate this point. This algorithm requires of the order of $(I + J) \times K \times L \times M$ operations per sample where I is the number of coefficients in the control filters, J is the number of coefficients in the plant model, K is the number of reference signals, L is the number of error sensors and M is the number of secondary actuators.

A system to control three harmonics on a propeller aircraft using the filtered-reference LMS algorithm, which has 16 secondary actuators and 32 error sensors, and needs to be sampled at about 1 kHz, requires a processing power of approximately 6×10^6 operations/second, since both the control filter and plant model need have only two coefficients per harmonic. To control broadband road noise at 8 error sensor locations in a vehicle, however, using 4 secondary actuators and 6 reference signals, again with a sample rate of 1 kHz, requires a processing power of nearly 50×10^6 operations/second since the control filters and each plant model typically require 128 coefficients. Although, in principle, the control algorithm only requires a processing power given by the number of operations per second quoted above, there is an inevitable overhead on any practical processor, which is associated with initialisation, data acquisition and formatting, etc., and this can easily multiply the number of operations required per second by a factor of 2–5. In fact it is not unusual to have a separate processor solely dedicated to data preprocessing. The processor in a practical controller must also perform various higher-level tasks, such as monitoring the performance of the transducers and detecting any faults, which further contributes to the computational overhead. The number of operations calculated above thus often turns out to underestimate the required processing power by perhaps a factor of 10. It is possible to use different types of processor for different operations in the overall control architecture, which could then be described as *heterogeneous*, in contrast to systems that use the same type of processor for all tasks, which could be described as having a *homogeneous* architecture, as discussed by Tokhi et al. (1995). These authors also discuss the computational speed and communication time of a number of different processors and the way in which the structure of the control algorithm can be mapped onto parallel architectures using these processors.

The examples considered above have assumed that the control filters are adapted at the same sample rate as the output signal. In order to retain real-time operation of the control system, it is clearly important that the signals feeding the secondary actuators be generated at the full sample rate of the control system by filtering the reference signal, but in some applications it may not be necessary to adapt the control filters at the same rate. One example of such a reduced-rate adaptation is the frequency-domain technique discussed in Chapters 3 and 5. Here blocks of data are accumulated and used to calculate the required changes in the control filter, which are then updated once per block. All the data within this block is used to calculate the required change in the filter coefficients, and although the frequency domain technique is more computationally efficient, the algorithm has almost the same effect as updating the control filter every sample in the time domain. In other applications it may not be necessary to adjust the filter coefficients as quickly as this, in which case although one block of data would be acquired and used to calculate the changes in the control filter, this calculation could be spread over the time it would take to acquire 10 blocks of data, for example. Provided the processor could work on the adaptation as a background task without too much overhead, while generating the real-time control signals as the foreground task, the computational requirements of the control system can be substantially reduced by this 'intermittent updating' technique, particularly for systems with many channels. Since the controller is being updated less frequently, the magnitude of each update can be made larger without risking the instability caused by the delay in the plant, and so the adaptation speed does not deteriorate as quickly with intermittent updating as might be expected (Elliott and Nelson, 1986). The convergence coefficient for a sparsely updated algorithm can thus be greater than it would be otherwise, and so the attenuation limits due to the numerical stalling effects, described in Section 10.8, may be less of a problem for intermittent updating than for sample-by-sample updating, as emphasised by Snyder (1999).

It is also possible to use the same technique to more gradually update the control filters in the time domain using a *sparse adaptation* algorithm, as discussed by Elliott (1992) and illustrated in Fig. 10.9. A number of different approaches to sparse update algorithms have been described in Section 5.4. It should, however, be recognised that for these

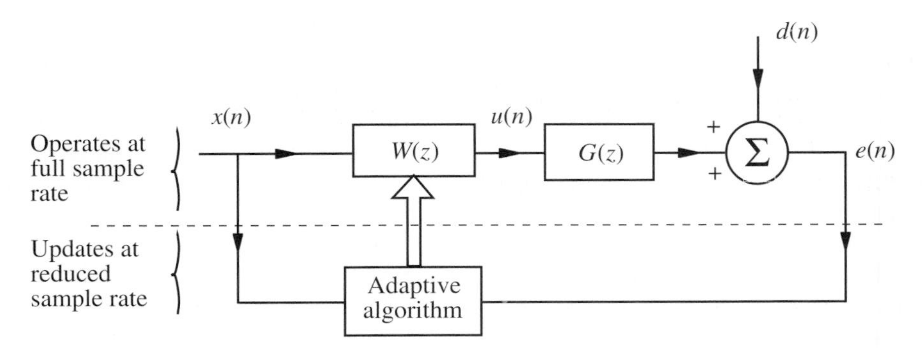

Figure 10.9 Block diagram of an active control system in which the coefficients of the control filter, $W(z)$, are updated at a rate that is slower than the sample rate used to generate the signal, $u(n)$, fed to the secondary actuator.

techniques, the error signal that is used in the adaptation process is effectively sampled at a lower rate than that of the control filter, and so aliasing of these error signals can occur. This is not as much of a problem for random signals since the aliased components may not be coherent with the reference signal, but for synchronously sampled harmonic controllers, the higher harmonics will alias down and corrupt the lower harmonics of the error signals.

10.7.1. Fixed- and Floating-Point Devices

An important distinction between different DSP processor types is whether they operate using fixed-point or floating-point arithmetic. The earlier, fixed-point, devices are considerably more difficult to program than those that use floating-point arithmetic since the programmer must ensure that the result of each arithmetic operation does not overflow, and that reasonable precision is maintained within the fixed word length. An example of a 16-bit fixed-point number representation is shown in Fig. 10.10(a), in which S denotes the sign bit and there are 15 bits that represent the magnitude of the number being represented. Further details of the number represented in DSP chips are provided, for example, by Marvin and Ewers (1993). The programmer must have a very detailed knowledge of the size of the intermediate variables in any fixed-point calculation and keep account of the shifting operations that need to be performed to maintain good precision. Fixed-point processors can be programmed to emulate floating-point calculations but only at the cost of a considerably reduced computation speed. Efficient programs for fixed-point processors thus tend to be written in low-level languages such as assembler. In many cost-sensitive applications it is still desirable to use cheaper, fixed-point DSP devices and, provided the production runs of the final product are large enough, the extra time spent optimising and testing the programs for fixed-point devices will not involve a significant cost compared with that of the individual devices. Fixed-point processors are particularly useful for tonal noise control systems, where the reference signals have a constant level and so fixed scaling of the adaptive coefficients can be used.

Floating-point DSP processors, on the other hand, are considerably easier to program since the magnitude of the mantissa is always normalised by the exponent, and so overflow

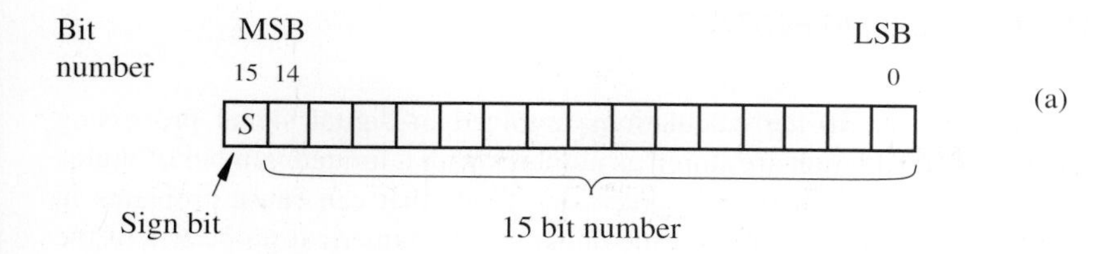

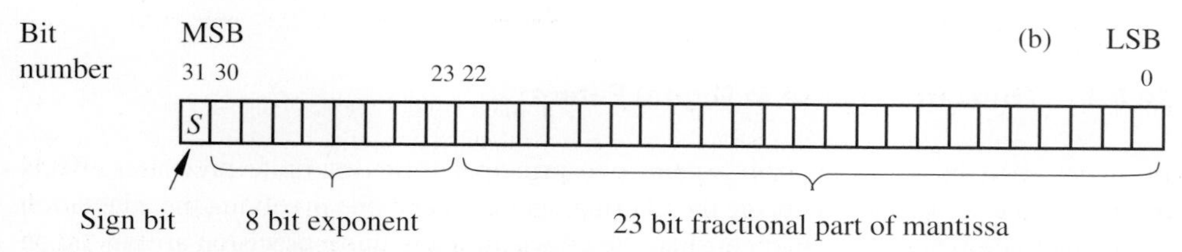

Figure 10.10 An illustration of the format of (a) a 16-bit fixed-point and (b) a 32-bit floating-point number.

in the mantissa just causes an increase in the exponent. An example of the format for a 32-bit floating point number is shown in Fig. 10.10(b). This format is specified by the IEEE 754 Standard (as also discussed for example by Marvin and Ewers, 1993) and has a sign bit, an 8-bit exponent and a 22-bit mantissa. The programmer is thus required to have a far less intimate knowledge of the size of the intermediate results in any floating-point calculation, and higher-level programming languages, such as C, can be used with some confidence. DSP devices have a relatively simple set of commands, and so high-level programs can operate without too great an overhead compared with low-level programs, provided efficient compilers are used. It is sometimes the case, however, that a small part of the program for the control algorithm takes up a large proportion of the computation time. It is then worthwhile to 'hand craft' this part of the program in assembler to minimise the time it takes to run.

The length of the program for the actual control algorithm is often only a relatively small fraction of the total program length. This is because many other functions such as initialisation, data collection from the converters, and error checking have to be performed. In a commercial product there is generally also another level of programming which monitors the functioning and performance of the algorithm to ensure fail-safe operation. It is much easier to write the majority of this program in a higher-level language, especially since this generally constitutes the largest part of the code that has to be written. Franklin et al. (1990) quote a rule of thumb for the size of the program that needs to be written to implement a reliable digital control system, which is

$$\text{Total code size (memory size)} \approx 7 \times \text{control code size,} \tag{10.7.1}$$

and this seems to be a reasonable approximation for production active control systems. The ease of programming and the ability to easily change the program when written in a high-level language for a floating-point processor also makes these processors extremely attractive for experimental and development systems.

10.8. FINITE-PRECISION EFFECTS

Finite-precision effects occur in the calculations involved in digital signal processing because the result of the calculation are stored as numbers with a limited number of digits. In this section, we will focus on the finite-precision effects that can cause problems in active control systems implemented in the time domain. The numerical properties of the FFT algorithm, often used to implement controllers in the frequency domain, are discussed, for example, by Rabiner and Gold (1975).

10.8.1. Truncation Noise in Digital Filters

In a time-domain active control system, two principal forms of finite precision effects occur: one due to truncation during the filtering operation and one involving the adaptation of the filter coefficients. We first consider the effects of using finite-precision arithmetic on digital filtering. If the output of an FIR digital filter is calculated accurately but then truncated or rounded to N-bit precision, its output signal can be written as

$$y(n) = \sum_{i=0}^{I-1} w_i \, x(n-i) + v(n) \,, \qquad (10.8.1)$$

where $v(n)$ is the error in the output due to truncation. Assuming that the fractional part of the output signal is represented with N-bit accuracy, either in fixed-point arithmetic or because this is the length of the mantissa in a floating-point arithmetic, then the smallest number that can be represented is

$$Q \approx 2^{-N} \,, \qquad (10.8.2)$$

which is equal to the number represented by the least-significant bit (LSB). Assuming that the output signal has a magnitude of order 1, then typically the change in the output signal every sample will be a great deal larger than the number represented in equation (10.8.2), i.e.

$$\left[y(n) - y(n-1)\right] \gg Q \,. \qquad (10.8.3)$$

In this case, the statistical properties of $v(n)$ are the same as those deduced for converter quantisation noise in Section 10.6, i.e. $v(n)$ is white, uncorrelated with $y(n)$, and has a mean-square value of

$$E\left[v^2(n)\right] = \frac{Q^2}{12} \,. \qquad (10.8.4)$$

The effect of finite-precision arithmetic on the operation of a well-scaled FIR filter is thus to add white noise to its output. The level of this additional white noise signal is generally very small, however. If 16-bit arithmetic were being used, for example, the level of the noise would typically be about 90 dB below that of the filter output. Notice that we have only considered a single truncation error, even though the calculation involved in equation (10.8.1) involves I separate multiplications. The reason for this is that in most DSP devices, the intermediate results of such a convolution are held in a high precision *accumulator*, which preserves the accuracy of intermediate results and the output of this accumulator is then truncated once at the end of the calculation to produce the output in normal precision. If the output of the filter is rounded, the mean value of $v(n)$ is zero, but if it is truncated it has a mean value of $Q/2$, although such small dc levels feeding the secondary actuators rarely give rise to problems in practice.

The effects of finite-precision arithmetic in IIR digital filters are more complicated, since the error due to truncation is circulated round the filter, due to its recursive nature. This can lead to amplification and coloration of the additive noise and, in extreme cases, limit cycle behaviour, as discussed at length by Rabiner and Gold (1975), for example. Once again, however, with modern processors using typically 16-bit arithmetic these effects do not generally present too great a problem.

10.8.2. Effect on Filter Adaptation

The other important effect of using finite-precision arithmetic in an active control system is on the adaptation of the controller. Assuming that the control filter for a single-channel system is being adapted using the filtered-reference LMS algorithm, then the adaptation equation for the i-th coefficient can be written, from Chapter 3, as

$$w_i(n+1) = w_i(n) - \alpha\, e(n)\, r(n-i) \, , \tag{10.8.5}$$

where α is the convergence coefficient, $e(n)$ is the measured error signal and $r(n)$ is the reference signal, filtered by an estimate of the plant response. This coefficient of the filter is thus updated by a quantity equal to

$$\Delta w_i(n) = -\alpha\, e(n)\, r(n-i) \, , \tag{10.8.6}$$

which must be calculated by multiplying α by $e(n)$ and then by $r(n-i)$, using finite-precision arithmetic. If α is chosen to be a factor of two, the product $\alpha\, e(n)$ can be formed without an explicit multiplication by shifting the number representing $e(n)$. The total update quantity must, however, be truncated before it is added to the filter coefficient. This will generate some level of noise in the update, as described, for example, by Caraiscos and Liu (1984), Haykin (1996) and Kuo and Morgan (1996). Even worse than this, however, is the fact that both the filter coefficient and the update quantity must be represented in the same fractional representation to implement equation (10.8.5). In other words, if $w_i(n)$ is represented as a 16-bit word and the convergence coefficient is small, the update quantity may only be represented in a few of the least-significant bits of this word. This effect will also occur when using floating-point arithmetic unless the size of the mantissa is much larger than the size of the fractional part of the fixed-point number.

If the convergence coefficient is made even smaller, or the error signal becomes small, the update quantity will finally become less than the LSB of the number representing the filter coefficient, and the adaptation of the filter coefficients will stop entirely. The adaptation of the filter is then said to *stall* or *lock-up*, as originally described for adaptive electrical filters by Gitlin et al. (1973). To obtain an approximate estimate of the value of convergence coefficient for which this stalling behaviour occurs in a single-channel controller adapted using equation (10.8.5), we assume that $e(n)$ and $r(n-i)$ are reasonably well correlated and so, on average, $e(n)r(n-i)$ is approximately equal to the product of the rms value of $e(n)$, e_{rms}, and the rms value of $r(n)$, r_{rms}. The adaptive algorithm will stall when the average value of the update term, equation (10.8.6), is equal to the smallest number which can be represented using finite-precision arithmetic, equation (10.8.2), so that

$$|\Delta w_i(n)| \approx \alpha_s e_{\text{rms}} r_{\text{rms}} = Q \;, \tag{10.8.7}$$

where α_s is the value of convergence coefficient below which the algorithm stalls, which, using equation (10.8.2), can be seen to be equal to

$$\alpha_s = \frac{2^{-N}}{e_{\text{rms}} r_{\text{rms}}} \;. \tag{10.8.8}$$

The variation of the residual mean-square error with convergence coefficient for a fully converged LMS control algorithm implemented with finite-precision arithmetic is illustrated in Fig. 10.11. For values of convergence coefficient greater than α_s, the mean-square error grows almost linearly with α due to the increasingly large random perturbations of the filter coefficients which give rise to misadjustment errors, as discussed in Chapter 2. If the adaptation algorithm were implemented with perfect precision arithmetic, then the mean-square error would converge to the exact least-square error, $J_{\min}$, as α tended to zero. As we have seen, however, the algorithm stalls before α can be made this small, as shown in Fig. 10.11. In practice, however, the variation of mean-square error with α due to misadjustment is very shallow for these small values of convergence coefficient, and $J_{\min}$ is very nearly reached. This can be confirmed by an order-of-magnitude calculation. It is assumed that the control filter coefficients are normalised so that there is approximately no change in the rms value of a signal in passing through the control filter. The reversal of transfer functions described in Section 3.3 can then be used to show that, provided a reasonable degree of control is obtained under these circumstances, then

$$E\left[r^2(n)\right] \approx E\left[d^2(n)\right] = J_p \;, \tag{10.8.9}$$

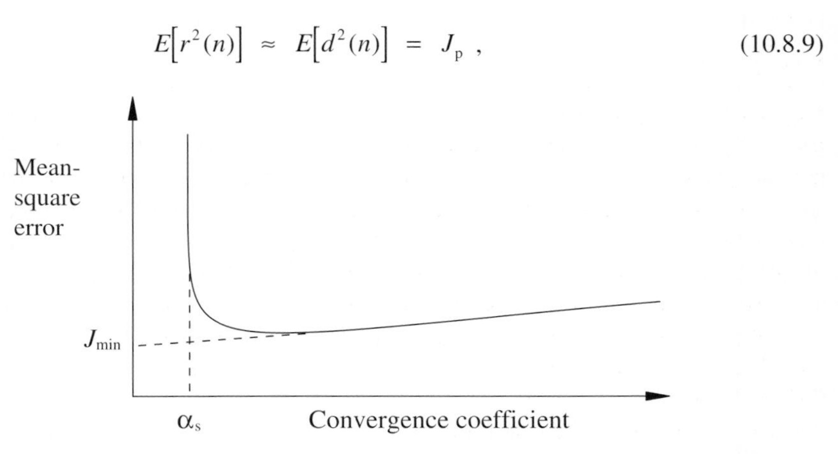

Figure 10.11 Variation of the mean-square error with convergence coefficient for an adaptive controller implemented with finite-precision arithmetic. If the convergence coefficient is below α_s the algorithm 'stalls'. If it is above this value, the mean-square error gradually increases because of misadjustment.

where J_p is the level of mean-square error due to the primary source alone, i.e. with no control. Thus the value of convergence coefficient for which the algorithm stalls is given approximately by

$$\alpha_s \approx \frac{2^{-N}}{\left(J_{\min} J_p\right)^{1/2}} . \qquad (10.8.10)$$

The effect of misadjustment on the steady state error in an adaptive filter was discussed in Section 2.6, from which it can be seen that the mean-square error with misadjustment is given by

$$J_\infty = (1 + M) J_{\min} , \qquad (10.8.11)$$

where M is the misadjustment. If the convergence coefficient is small, the misadjustment is given by equation (3.4.15),

$$M \approx \alpha I E\left[r^2(n)\right]/2 \approx \alpha I J_p/2 , \qquad (10.8.12)$$

where I is the number of the filter coefficients and equation (10.8.9) has again been used. The steady state mean-square error just before the algorithm is stalled is thus approximately equal to

$$J_\infty \approx \left(1 + 2^{-(N+1)} I \left(\frac{J_p}{J_{\min}}\right)^{1/2}\right) J_{\min} . \qquad (10.8.13)$$

If an infinite-precision control system can achieve 30 dB of attenuation, for example, so that $J_p / J_{\min} = 10^3$ and assuming that $I = 128$, and $N = 16$ bits, J_∞ is only about 3% greater than $J_{\min}$, so that the final steady-state attenuation of the finite-precision controller will be limited to about 29.9 dB. The attenuation achieved with a practical controller can thus be very close to the maximum possible attenuation, providing that the signals are properly scaled.

10.8.3. DC Drift

Finally, it is worth commenting on the potential problem of *dc drift*, which may be caused by finite-precision effects (Cioffi, 1987), but can also be caused by offsets in the data converters. Referring back to equation (10.8.5), we note that if the mean value of $e(n)$ and the mean value of $r(n-i)$ are not zero, then the average value of the filter coefficients will slowly drift, causing a dc output to be fed to the secondary actuator. Since most plant responses that are encountered in active control applications have at least one zero at dc, the dc output to the actuators will not affect the error signals, so that, in the worst case, the dc output of the converters will increase until the actuator saturates. This problem can be solved with a simple digital high-pass filter acting on the error or reference signal, or the inclusion of a leakage term in the coefficient adaptation.

Although one must be aware of the potential for finite-precision effects in designing practical active control systems, their effects do not generally cause a significant degradation in performance, provided that elementary precautions are taken. This is particularly true when using modern DSP devices for which the numbers are represented with at least 16-bit precision.

Appendix

Linear Algebra and the Description of Multichannel Systems

In this appendix we briefly review the definition of vectors and matrices and some of the important properties of linear algebra. We have seen that the use of linear algebra considerably simplifies the formulation of multiple-channel control problems. Our purpose here is to provide a convenient reference for the important properties of matrix manipulation and for the definitions of the terms used in the main text. We will concentrate on the case where all variables are complex, unless otherwise stated. Most of the results are taken from Noble and Daniel (1977), Golub and Van Loan (1996), Datta (1995) and Skogestad and Postlethwaite (1996), in which rigorous developments of the various properties can be found.

A.1. VECTORS

A *vector* is an ordered series of numbers, x_1, x_2, …, x_N, which are generally complex. A vector is denoted here by a lower case, bold variable and, unless otherwise stated, corresponds to the numbers being listed as a single *column*:

$$\mathbf{x} = \begin{bmatrix} x_1 \\ x_2 \\ \cdot \\ \cdot \\ \cdot \\ x_N \end{bmatrix} \tag{A1.1}$$

The *transpose* of a vector, denoted by the superscript T, is the corresponding *row* of ordered numbers:

$$\mathbf{x}^{\mathrm{T}} = \begin{bmatrix} x_1 & x_2 & \dots & x_N \end{bmatrix} . \tag{A1.2}$$

The *Hermitian* or conjugate transpose of a vector, denoted by the superscript H is defined to be

$$\mathbf{x}^{\mathrm{H}} = \begin{bmatrix} x_1^* & x_2^* & \dots & x_N^* \end{bmatrix} , \tag{A1.3}$$

where the superscript * denotes complex conjugation. The *inner product* (strictly the Euclidean inner product or scalar product) of two vectors, **x** and **y**, of equal size, is defined to be

$$\mathbf{x}^{\mathrm{H}}\mathbf{y} = x_1^* y_1 + x_2^* y_2 + x_3^* y_3 + \cdots + x_N^* y_N , \tag{A1.4}$$

and is sometimes written as (**x**,**y**). Note that $\mathbf{y}^{\mathrm{H}}\mathbf{x} = (\mathbf{x}^{\mathrm{H}}\mathbf{y})^*$. If $\mathbf{x}^{\mathrm{H}}\mathbf{y} = 0$, then the vectors **x** and **y** are said to be *orthogonal.* The inner produce $\mathbf{x}^{\mathrm{H}}\mathbf{x}$ is equal to the sum of the modulus squared elements of the vector **x**

$$\mathbf{x}^{\mathrm{H}}\mathbf{x} = \sum_{n=1}^{N} |x_n|^2 . \tag{A1.5}$$

A.2. MATRICES

A *matrix* is a set of numbers, generally complex, arranged in a rectangular array. A matrix is denoted here by an upper case, bold, variable. For example:

$$\mathbf{A} = \begin{bmatrix} a_{11} & a_{12} & a_{13} & \dots & a_{1M} \\ a_{21} & a_{22} & a_{23} & \dots & a_{2M} \\ \cdot & \cdot & \cdot & \cdot & \cdot \\ \cdot & \cdot & \cdot & \cdot & \cdot \\ a_{L1} & a_{L2} & a_{L3} & \dots & a_{LM} \end{bmatrix} , \tag{A2.1}$$

is an $L \times M$ matrix (it has L rows and M columns). If the $L \times M$ matrix **A** has elements that are real, this is sometimes written as $\mathbf{A} \in \mathbb{R}^{L\times M}$ and if **A** has complex elements this can be designated by $\mathbf{A} \in \mathbb{C}^{L\times M}$. A vector may be considered to be a matrix with a single column so that $\mathbf{x} \in \mathbb{C}^{N\times 1}$ in equation (A1.1) if **x** has complex elements. The element in the l-th row and m-th column is denoted a_{lm}. If $L = M$, the matrix is said to be *square*. The matrix **0** has all zero elements and is called the *null* or *zero* matrix. If all the off-diagonal elements of a square matrix are zero it is said to be *diagonal*. The diagonal matrix

$$\mathbf{I} = \begin{bmatrix} 1 & 0 & . & . & . \\ 0 & 1 & . & . & . \\ . & . & . & . & . \\ . & . & . & 1 & 0 \\ . & . & . & 0 & 1 \end{bmatrix}, \qquad \text{(A2.2)}$$

with all elements on the leading (or principal) diagonal being unity is called the *identity* or *unit* matrix. The matrices **A** and **B** are said to be *equal* if, and only if, (a) they both have the same number of rows and columns, and (b) all corresponding elements are the same in the two matrices, i.e.

$$a_{lm} = b_{lm} \qquad \text{for all } l \text{ and } m \; . \qquad \text{(A2.3)}$$

The *sum* of two matrices, **A** and **B**, is defined only if they both have the same number of rows and columns, and is itself a matrix, **C**, of the same size whose elements are the sum of the corresponding elements of **A** and **B**:

$$c_{lm} = a_{lm} + b_{lm} \; . \qquad \text{(A2.4)}$$

The following laws apply to matrix addition:

$$(\mathbf{A} + \mathbf{B}) + \mathbf{C} = \mathbf{A} + (\mathbf{B} + \mathbf{C}) \qquad \text{associative law,} \qquad \text{(A2.5)}$$

$$\mathbf{A} + \mathbf{B} = \mathbf{B} + \mathbf{A} \qquad \text{commutative law.} \qquad \text{(A2.6)}$$

The *multiplication* of a matrix **A** by a matrix **B** to form the matrix product **AB** is defined only if the number of columns of **A** is equal to the number of rows of **B**. If **A** is an $L \times N$ matrix with elements a_{ln} and **B** is an $N \times M$ matrix with elements b_{nm}, then $\mathbf{C} = \mathbf{AB}$ is an $L \times M$ matrix whose elements are given by

$$c_{lm} = \sum_{n=1}^{N} a_{ln} \, b_{nm} \; , \qquad \text{(A2.7)}$$

i.e. the elements are the sum of the products of the elements of the l-th row of **A** and the elements of the m-th column of **B**. In the product **AB**, the matrix **A** is said to pre-multiply **B**, or **B** post-multiplies **A**. The order of multiplication is important since in general

$$\mathbf{AB} \neq \mathbf{BA}, \qquad \text{(A2.8)}$$

even if **A** and **B** are square. Although the commutative law is not in general true for matrix multiplication, the following associative and distributive laws do hold, i.e.

$$(\mathbf{AB})\mathbf{C} = \mathbf{A}(\mathbf{BC}) \; , \qquad \text{(A2.9)}$$

$$\mathbf{A}(\mathbf{B} + \mathbf{C}) = \mathbf{AB} + \mathbf{AC} \; . \qquad \text{(A2.10)}$$

Note also that $\mathbf{AB} = \mathbf{0}$ does not imply that either **A** or **B** is zero, so if we have an equation of the form $\mathbf{AB} = \mathbf{AC}$, we cannot, in general, conclude that either $\mathbf{A} = \mathbf{0}$ or $\mathbf{B} = \mathbf{C}$.

The identity matrix has particularly simple properties under multiplication with a square matrix (**A**), such that

$$\mathbf{IA} = \mathbf{AI} = \mathbf{A} . \tag{A2.11}$$

The *transpose* of the $L \times M$ matrix **A** is the $M \times L$ matrix denoted $\mathbf{A}^{\mathrm{T}}$, obtained by interchanging the rows and columns of **A**. If $a_{lm} = a_{ml}$ then $\mathbf{A} = \mathbf{A}^{\mathrm{T}}$ and the square matrix **A** is said to be *symmetric*. The *Hermitian transpose* of a matrix denoted $\mathbf{A}^{\mathrm{H}}$ is the complex conjugate of the transpose. So if

$$\mathbf{A} = \begin{bmatrix} a_{11} & a_{12} & \cdots & a_{1M} \\ a_{21} & a_{22} & \cdots & a_{2M} \\ \cdot & \cdot & \cdot & \cdot \\ \cdot & \cdot & \cdot & \cdot \\ a_{L1} & a_{L2} & \cdots & a_{LM} \end{bmatrix}, \quad \mathbf{A}^{\mathrm{H}} = \begin{bmatrix} a_{11}^* & a_{21}^* & \cdots & a_{L1}^* \\ a_{12}^* & a_{22}^* & & \\ \vdots & & & \\ a_{1M}^* & a_{2M}^* & \cdots & a_{LM}^* \end{bmatrix} . \tag{A2.12}$$

The properties of the Hermitian transpose, from which the properties of the normal transpose are readily deduced, are that,

$$(\mathbf{A} + \mathbf{B})^{\mathrm{H}} = \mathbf{A}^{\mathrm{H}} + \mathbf{B}^{\mathrm{H}} , \tag{A2.13}$$

$$(\mathbf{A}^{\mathrm{H}})^{\mathrm{H}} = \mathbf{A} , \tag{A2.14}$$

$$(\mathbf{A}\,\mathbf{B})^{\mathrm{H}} = \mathbf{B}^{\mathrm{H}}\,\mathbf{A}^{\mathrm{H}} . \tag{A2.15}$$

If a square matrix **A** has the property that

$$\mathbf{A}^{\mathrm{H}}\mathbf{A} = \mathbf{A}\,\mathbf{A}^{\mathrm{H}} = \mathbf{I} ,$$

it is said to be *unitary*, and can be interpreted as having columns that are mutually orthogonal, but normalised such that their inner products with themselves are unity, i.e. they are *orthonormal*. Unitary matrices have the important property that their Hermitian transpose is equal to the inverse, i.e. $\mathbf{A}^{\mathrm{H}} = \mathbf{A}^{-1}$. If **A** is *real*, i.e. it has entirely real elements, and $\mathbf{A}^{\mathrm{T}}\mathbf{A} = \mathbf{I}$, it is said to be *orthogonal*.

If a square matrix **A** is equal to its own Hermitian transpose, so that $a_{lm} = a_{ml}^*$, it is said to be a *Hermitian* matrix, i.e.

$$\text{if } \mathbf{A} = \mathbf{A}^{\mathrm{H}}, \text{ then } \mathbf{A} \text{ is Hermitian} . \tag{A2.16}$$

A simple example of a Hermitian matrix, used in the main text, is $\mathbf{A} = \mathbf{G}^{\mathrm{H}}\mathbf{G}$. Note that a Hermitian matrix must have real elements along its principal diagonal. If **A** is Hermitian, then products of the form $\mathbf{x}^{\mathrm{H}}\mathbf{A}\mathbf{x}$ will be entirely real scalars. **A** is also *positive definite* if

$$\mathbf{x}^{\mathrm{H}}\mathbf{A}\mathbf{x} > 0 \qquad \text{for all } \mathbf{x} \neq \mathbf{0} , \tag{A2.17}$$

or *positive semi-definite* if

$$\mathbf{x}^{\mathrm{H}}\mathbf{A}\mathbf{x} \geq 0 \qquad \text{for all } \mathbf{x} \neq \mathbf{0} . \tag{A2.18}$$

A.3. DETERMINANTS AND THE INVERSE MATRIX

The *determinant* of a 2×2 matrix, $\mathbf{A}$, is a complex scalar given by

$$\det(\mathbf{A}) = \det\begin{bmatrix} a_{11} & a_{12} \\ a_{21} & a_{22} \end{bmatrix} = a_{11}\, a_{22} - a_{12}\, a_{21} \ . \qquad \text{(A3.1)}$$

The *minor*, M_{lm}, of the element a_{lm} of the general square matrix $\mathbf{A}$ is the determinant of the matrix formed by deleting the l-th row and m-th column from $\mathbf{A}$. For example, the minor M_{21} of the 3×3 matrix

$$\mathbf{A} = \begin{bmatrix} a_{11} & a_{12} & a_{13} \\ a_{21} & a_{22} & a_{23} \\ a_{31} & a_{32} & a_{33} \end{bmatrix}, \qquad \text{(A3.2)}$$

is given by striking out the second row and first column to give

$$M_{21} = \begin{bmatrix} a_{12} & a_{13} \\ a_{32} & a_{33} \end{bmatrix} . \qquad \text{(A3.3)}$$

The *cofactor* C_{lm} of the element a_{lm} of a matrix $\mathbf{A}$ is defined to be

$$C_{lm} = (-1)^{l+m} M_{lm} \ . \qquad \text{(A3.4)}$$

The determinant of a square matrix of any size can be expanded as the sum of the products of elements and their cofactors along any row or column. So the determinant of the 3×3 matrix above is, for example, equal to

$$\det(\mathbf{A}) = a_{11}C_{11} + a_{12}C_{12} + a_{13}C_{13} \ . \qquad \text{(A3.5)}$$

The determinant of a square matrix of arbitrary size can thus be broken up into smaller and smaller matrices, and so can be readily calculated. Note that if $\mathbf{A}$ and $\mathbf{B}$ are square and of the same size,

$$\det(\mathbf{A}\,\mathbf{B}) = \det(\mathbf{A})\det(\mathbf{B}) \ . \qquad \text{(A3.6)}$$

If the determinant of a matrix is zero, the matrix is said to be *singular*.

The *inverse* $\mathbf{A}^{-1}$ of a matrix $\mathbf{A}$ is defined by the equation

$$\mathbf{A}\mathbf{A}^{-1} = \mathbf{A}^{-1}\mathbf{A} = \mathbf{I} \ . \qquad \text{(A3.7)}$$

This inverse matrix exists only if (a) the matrix $\mathbf{A}$ is square and (b) the determinant of the matrix is not zero, i.e. it is not singular.

The inverse of a matrix $\mathbf{A}$ can be derived by first defining the *adjugate*, or 'classical adjoint' (Skogestad and Postlethwaite, 1996), of the matrix $\mathbf{A}$ as the transpose matrix of cofactors of $\mathbf{A}$. For example, the adjugate of an $N \times N$ matrix $\mathbf{A}$ is

$$\text{adj}(\mathbf{A}) = \begin{bmatrix} C_{11} & C_{21} & \cdots & C_{N1} \\ C_{12} & C_{22} & & \\ \cdot & & & \\ \cdot & & & \\ C_{1N} & \cdot & \cdot & C_{NN} \end{bmatrix} . \tag{A3.8}$$

The inverse matrix $\mathbf{A}^{-1}$ is now equal to the reciprocal of the determinant of $\mathbf{A}$, multiplied by its adjugate,

$$\mathbf{A}^{-1} = \frac{\text{adj}(\mathbf{A})}{\det(\mathbf{A})} . \tag{A3.9}$$

The inverse of the 2×2 matrix

$$\mathbf{A} = \begin{bmatrix} a_{11} & a_{12} \\ a_{21} & a_{22} \end{bmatrix} , \tag{A3.10}$$

for example, is

$$\mathbf{A}^{-1} = \frac{1}{a_{11}\,a_{22} - a_{12}\,a_{21}} \begin{bmatrix} a_{22} & -a_{12} \\ -a_{21} & a_{11} \end{bmatrix} . \tag{A3.11}$$

It should be emphasised that although equation (A3.9) is a useful theoretical representation of a matrix inverse, it is not very efficient for numerical evaluation. For larger N, it requires of the order of $N!$ operations to compute the inverse, compared to the order of N^3 operations for other general methods (Noble and Daniel, 1977). Matrices with particular structures may have even more efficient algorithms for their inversion. If the matrix is *Toeplitz* for example, i.e. all the elements along any diagonal are equal, the Levinson recursion equation can be used, which only requires of the order of N^2 multiplications (Markel and Grey, 1976).

Note that provided $\mathbf{A}$ and $\mathbf{B}$ are non-singular,

$$[\mathbf{AB}]^{-1} = \mathbf{B}^{-1}\,\mathbf{A}^{-1} , \tag{A3.12}$$

and

$$\left(\mathbf{A}^{\mathrm{H}}\right)^{-1} = \left(\mathbf{A}^{-1}\right)^{\mathrm{H}} , \tag{A3.13}$$

which can also be written as $\mathbf{A}^{-\mathrm{H}}$.

If $\mathbf{A}$ is square but singular, or more generally if $\mathbf{A}$ is an $L \times M$ matrix with $L > M$, the vectors that satisfy $\mathbf{Ax} = \mathbf{0}$ are said to be in the *null space* of $\mathbf{A}$ and vectors which do not satisfy this condition are said to be in the *rank space* of $\mathbf{A}$.

A general formula for the inverse of a modified matrix is given by Noble and Daniel (1977, p. 194) as

$$[\mathbf{A} + \mathbf{BC}]^{-1} = \mathbf{A}^{-1} - \mathbf{A}^{-1}\mathbf{B}\left[\mathbf{I} + \mathbf{CA}^{-1}\,\mathbf{B}\right]^{-1}\mathbf{CA}^{-1} , \tag{A3.14}$$

which is a form of the Woodbury formula (Datta, 1995), and of the matrix inversion lemma

(Haykin, 1996) and assumes that $\left[\mathbf{I} + \mathbf{C}\mathbf{A}^{-1}\mathbf{B}\right]$ is non-singular. Equation (A3.14) can be shown to be correct by multiplying both sides by $\mathbf{A} + \mathbf{BC}$ and a little manipulation.

Consider the special case of the matrix $\left[\mathbf{I} + \boldsymbol{\Delta}\right]$ in the Woodbury formula, in which case $\mathbf{A} = \mathbf{I}$, $\mathbf{B} = \mathbf{I}$ and $\mathbf{C} = \boldsymbol{\Delta}$, and so

$$\left[\mathbf{I} + \boldsymbol{\Delta}\right]^{-1} = \mathbf{I} - \left[\mathbf{I} + \boldsymbol{\Delta}\right]^{-1} \boldsymbol{\Delta} . \tag{A3.15}$$

Repeated application of the equation (A3.15) leads to

$$\left[\mathbf{I} + \boldsymbol{\Delta}\right]^{-1} = \mathbf{I} - \boldsymbol{\Delta} + \boldsymbol{\Delta}^2 - \boldsymbol{\Delta}^3 + \cdots , \tag{A3.16}$$

which could be considered as the matrix form of a binomial expansion.

A.4. TRACE OF A MATRIX AND ITS DERIVATIVES

The *trace* of a square matrix is a scalar quantity, equal to the sum of its diagonal elements and for the $N \times N$ matrix $\mathbf{A}$ is written as

$$\operatorname{trace}(\mathbf{A}) = \sum_{n=1}^{N} a_{nn} . \tag{A4.1}$$

The trace has the properties that

$$\operatorname{trace}(\mathbf{A} + \mathbf{B}) = \operatorname{trace}(\mathbf{A}) + \operatorname{trace}(\mathbf{B}) , \tag{A4.2}$$

where $\mathbf{A}$ and $\mathbf{B}$ must be square. Even if $\mathbf{A}$ and $\mathbf{B}$ are not square but are of the same dimensions, then

$$\operatorname{trace}\left(\mathbf{A}\mathbf{B}^{\mathrm{T}}\right) = \operatorname{trace}\left(\mathbf{B}^{\mathrm{T}}\mathbf{A}\right) = \operatorname{trace}\left(\mathbf{A}^{\mathrm{T}}\mathbf{B}\right) = \operatorname{trace}\left(\mathbf{B}\mathbf{A}^{\mathrm{T}}\right) . \tag{A4.3,4,5}$$

It is useful to be able to differentiate the trace of a matrix function with respect to the elements of one of the matrices in this function. For the $L \times M$ matrices $\mathbf{A}$ and $\mathbf{B}$, for example, we can write

$$\operatorname{trace}\left(\mathbf{A}\mathbf{B}^{\mathrm{T}}\right) = \sum_{l=1}^{L} \sum_{m=1}^{M} a_{lm}\, b_{lm} , \tag{A4.6}$$

so that

$$\frac{\partial \operatorname{trace}\left(\mathbf{A}\mathbf{B}^{\mathrm{T}}\right)}{\partial a_{lm}} = b_{lm} . \tag{A4.7}$$

If the elements given by equation (A4.7) for all l and m are arranged into an $L \times M$ matrix, then this can be written as

$$\frac{\partial \operatorname{trace}\left(\mathbf{A}\mathbf{B}^{\mathrm{T}}\right)}{\partial \mathbf{A}} = \mathbf{B} . \tag{A4.8}$$

Using the properties of the trace of a matrix in equations (A4.3) to (A4.5), it can be shown that

$$\frac{\partial \operatorname{trace}\left(\mathbf{B}^{\mathrm{T}} \mathbf{A}\right)}{\partial \mathbf{A}} = \mathbf{B} , \tag{A4.9}$$

$$\frac{\partial \operatorname{trace}\left(\mathbf{A}^{\mathrm{T}} \mathbf{B}\right)}{\partial \mathbf{A}} = \mathbf{B} , \tag{A4.10}$$

$$\frac{\partial \operatorname{trace}\left(\mathbf{B} \mathbf{A}^{\mathrm{T}}\right)}{\partial \mathbf{A}} = \mathbf{B} , \tag{A4.11}$$

$$\frac{\partial \operatorname{trace}\left(\mathbf{B} \mathbf{A} \mathbf{C}^{\mathrm{T}}\right)}{\partial \mathbf{A}} = \mathbf{B}^{\mathrm{T}} \mathbf{C} , \tag{A4.12}$$

and

$$\frac{\partial \operatorname{trace}\left(\mathbf{C} \mathbf{A}^{\mathrm{T}} \mathbf{B}^{\mathrm{T}}\right)}{\partial \mathbf{A}} = \mathbf{B}^{\mathrm{T}} \mathbf{C} . \tag{A4.13}$$

It can also be shown (Skelton, 1988) that

$$\frac{\partial \operatorname{trace}\left(\mathbf{A} \mathbf{B} \mathbf{A}^{\mathrm{T}}\right)}{\partial \mathbf{A}} = \mathbf{A}\left(\mathbf{B} + \mathbf{B}^{\mathrm{T}}\right) , \tag{A4.14}$$

$$\frac{\partial \operatorname{trace}\left(\mathbf{A}^{\mathrm{T}} \mathbf{B} \mathbf{A}\right)}{\partial \mathbf{A}} = \left(\mathbf{B} + \mathbf{B}^{\mathrm{T}}\right)\mathbf{A} , \tag{A4.15}$$

and that

$$\frac{\partial \operatorname{trace}\left(\mathbf{C} \mathbf{A} \mathbf{B} \mathbf{A}^{\mathrm{T}} \mathbf{C}^{\mathrm{T}}\right)}{\partial \mathbf{A}} = \mathbf{C}^{\mathrm{T}} \mathbf{C} \mathbf{A}\left(\mathbf{B} + \mathbf{B}^{\mathrm{T}}\right) . \tag{A4.16}$$

If the elements of the matrices are complex we can also differentiate the trace of a matrix function with respect to the real and imaginary parts of one of the matrices. We will restrict ourselves to the differentiation of matrix functions which are entirely real, so that we do not get involved in the complications described, for example, in Jolly (1995) or Appendix B of Haykin (1996). As an example of a real matrix function we can write

$$J = \operatorname{trace}\left(\mathbf{A} \mathbf{B}^{\mathrm{H}} + \mathbf{B} \mathbf{A}^{\mathrm{H}}\right) = 2 \sum_{l=1}^{L} \sum_{m=1}^{M} \left(a_{\mathrm{R}lm}\, b_{\mathrm{R}lm} + a_{\mathrm{I}lm}\, b_{\mathrm{I}lm}\right) , \tag{A4.17}$$

where the complex elements of the $L \times M$ matrices $\mathbf{A}$ and $\mathbf{B}$ have been written in terms of their real and imaginary parts

$$a_{lm} = a_{\mathrm{R}lm} + j\, a_{\mathrm{I}lm} \quad \text{and} \quad b_{lm} = b_{\mathrm{R}lm} + j\, b_{\mathrm{I}lm} , \tag{A4.18, 19}$$

so that

$$\frac{\partial J}{\partial a_{\mathrm{R}lm}} + j\frac{\partial J}{\partial a_{\mathrm{I}lm}} = 2\left(b_{\mathrm{R}lm} + j\,b_{\mathrm{I}lm}\right) = 2\,b_{lm} \ . \qquad \text{(A4.20)}$$

The complex elements given by equation (A4.20) for all l and m can be arranged into an $L \times M$ matrix of derivatives as

$$\frac{\partial J}{\partial \mathbf{A}_{\mathrm{R}}} + j\frac{\partial J}{\partial \mathbf{A}_{\mathrm{I}}} = 2\mathbf{B} \ . \qquad \text{(A4.21)}$$

Using the rules for determinants above and assuming that the matrices have compatible dimensions, we can also see that if the real scalar J is equal to

$$J = \mathrm{trace}\left(\mathbf{B}\,\mathbf{A}\,\mathbf{C}^{\mathrm{H}} + \mathbf{C}\,\mathbf{A}^{\mathrm{H}}\,\mathbf{B}^{\mathrm{H}}\right) = \mathrm{trace}\left(\mathbf{A}\,\mathbf{C}^{\mathrm{H}}\,\mathbf{B} + \mathbf{B}^{\mathrm{H}}\,\mathbf{C}\,\mathbf{A}^{\mathrm{H}}\right) , \qquad \text{(A4.22)}$$

then

$$\frac{\partial J}{\partial \mathbf{A}_{\mathrm{R}}} + j\frac{\partial J}{\partial \mathbf{A}_{\mathrm{I}}} = 2\,\mathbf{B}^{\mathrm{H}}\,\mathbf{C} \ , \qquad \text{(A4.23)}$$

and if the real scalar J is equal to

$$J = \mathrm{trace}\left(\mathbf{C}\,\mathbf{A}\,\mathbf{B}\,\mathbf{A}^{\mathrm{H}}\,\mathbf{C}^{\mathrm{H}}\right) , \qquad \text{(A4.24)}$$

where **B** is Hermitian, then

$$\frac{\partial J}{\partial \mathbf{A}_{R}} + j\frac{\partial J}{\partial \mathbf{A}_{I}} = 2\,\mathbf{C}^{\mathrm{H}}\,\mathbf{C}\,\mathbf{A}\,\mathbf{B} \ . \qquad \text{(A4.25)}$$

A.5. OUTER PRODUCTS AND SPECTRAL DENSITY MATRICES

The *outer product* of the $N \times 1$ vector of complex elements $\mathbf{x}$ and the $M \times 1$ vector of complex elements $\mathbf{y}$ is the $N \times M$ matrix of complex elements $\mathbf{x}\mathbf{y}^{\mathrm{H}}$. If $M = N$ then the outer product is square and its trace is equal to the inner product, so that if $\mathbf{y} = \mathbf{x}$, for example, then

$$\mathrm{trace}\left(\mathbf{x}\,\mathbf{x}^{\mathrm{H}}\right) = \mathbf{x}^{\mathrm{H}}\,\mathbf{x} \ . \qquad \text{(A5.1)}$$

Real scalar quantities of the form defined by equation (A5.1) are very useful as cost functions when analysing problems involving multichannel random data. We now define

$$\mathbf{S}_{xx}\left(\mathrm{e}^{j\omega T}\right) = \lim_{L\to\infty} \frac{1}{L} E\left[\mathbf{x}_m\left(\mathrm{e}^{j\omega T}\right)\mathbf{x}_m^{\mathrm{H}}\left(\mathrm{e}^{j\omega T}\right)\right] , \qquad \text{(A5.2)}$$

to be the matrix of power spectral densities and cross spectral densities between the elements of a vector of stationary and ergodic discrete-time random sequences of duration L samples, where

$$\mathbf{x}_m\left(\mathrm{e}^{j\omega T}\right) = \left[X_1\left(\mathrm{e}^{j\omega T}\right) X_2\left(\mathrm{e}^{j\omega T}\right) \cdots X_N\left(\mathrm{e}^{j\omega T}\right)\right]^{\mathrm{T}} \tag{A5.3}$$

is the vector of their spectra for the m-th record and E denotes the expectation (Grimble and Johnson, 1988). It is notationally convenient to abbreviate equation (A5.2) as

$$\mathbf{S}_{xx}\left(\mathrm{e}^{j\omega T}\right) = E\left[\mathbf{x}\left(\mathrm{e}^{j\omega T}\right)\mathbf{x}^{\mathrm{H}}\left(\mathrm{e}^{j\omega T}\right)\right], \tag{A5.4}$$

and this use of the expectation operator in the frequency domain, which is also used for single-channel signals in Section 2.4, will be followed below. This definition of the *spectral density matrix* is consistent with that of Youla (1976) and Grimble and Johnson (1988), but not with the usage by Bongiorno (1969) or Bendat and Piersol (1986), for example, who would define the matrix of power and cross spectral densities to be

$$\mathbf{\Phi}_{xx}\left(\mathrm{e}^{j\omega T}\right) = E\left[\mathbf{x}^{*}\left(\mathrm{e}^{j\omega T}\right)\mathbf{x}^{\mathrm{T}}\left(\mathrm{e}^{j\omega T}\right)\right], \tag{A5.5}$$

in the present notation, which is equal to the transpose of $\mathbf{S}_{xx}\left(\mathrm{e}^{j\omega T}\right)$ in equation (A5.4) and to its conjugate. Both versions reduce to the standard definition in the single channel case. If

$$\mathbf{y}\left(\mathrm{e}^{j\omega T}\right) = \mathbf{H}\left(\mathrm{e}^{j\omega T}\right)\mathbf{x}\left(\mathrm{e}^{j\omega T}\right), \tag{A5.6}$$

then

$$\mathbf{S}_{yy}\left(\mathrm{e}^{j\omega T}\right) = \mathbf{H}\left(\mathrm{e}^{j\omega T}\right)\mathbf{S}_{xx}\left(\mathrm{e}^{j\omega T}\right)\mathbf{H}^{\mathrm{H}}\left(\mathrm{e}^{j\omega T}\right), \tag{A5.7}$$

as used by Youla (1976) and Grimble and Johnson (1988). We extend this notation here to include the matrix of cross spectral densities between the elements of $\mathbf{x}\left(\mathrm{e}^{j\omega T}\right)$ and $\mathbf{y}\left(\mathrm{e}^{j\omega T}\right)$, which is similarly defined to be

$$\mathbf{S}_{xy}\left(\mathrm{e}^{j\omega T}\right) = E\left[\mathbf{y}\left(\mathrm{e}^{j\omega T}\right)\mathbf{x}^{\mathrm{H}}\left(\mathrm{e}^{j\omega T}\right)\right], \tag{A5.8}$$

so that if $\mathbf{x}\left(\mathrm{e}^{j\omega T}\right)$ is related to $\mathbf{y}\left(\mathrm{e}^{j\omega T}\right)$ by equation (A5.6), then

$$\mathbf{S}_{xy}\left(\mathrm{e}^{j\omega T}\right) = \mathbf{H}\left(\mathrm{e}^{j\omega T}\right)\mathbf{S}_{xx}\left(\mathrm{e}^{j\omega T}\right), \tag{A5.9}$$

and assuming that $\mathbf{S}_{xx}\left(\mathrm{e}^{j\omega T}\right)$ is non-singular for all ωT then

$$\mathbf{H}\left(\mathrm{e}^{j\omega T}\right) = \mathbf{S}_{xy}\left(\mathrm{e}^{j\omega T}\right)\mathbf{S}_{xx}^{-1}\left(\mathrm{e}^{j\omega T}\right). \tag{A5.10}$$

If we were to follow the convention of Bendat and Piersol (1986) and define

$$\mathbf{\Phi}_{xy} = E\left[\mathbf{x}^{*}\left(\mathrm{e}^{j\omega T}\right)\mathbf{y}^{\mathrm{T}}\left(\mathrm{e}^{j\omega T}\right)\right], \tag{A5.11}$$

and if $\mathbf{x}\left(\mathrm{e}^{j\omega T}\right)$ and $\mathbf{y}\left(\mathrm{e}^{j\omega T}\right)$ are related by equation (A5.6), then we obtain the rather more ungainly form for $\mathbf{H}\left(\mathrm{e}^{j\omega T}\right)$ of

$$\mathbf{H}\left(e^{j\omega T}\right) = \mathbf{\Phi}_{xy}^{\mathrm{T}}\left(e^{j\omega T}\right) \mathbf{\Phi}_{xx}^{-\mathrm{T}}\left(e^{j\omega T}\right) . \tag{A5.12}$$

Algebraic manipulation using the definition of the cross spectral matrix given by equation (A5.8) is thus much neater than that using (A5.11), because transposition and complex conjugation always occur in the same terms of an equation, but the price one has to pay is the need to remember that $\mathbf{S}_{xy}$ is defined to be the outer product of $\mathbf{y}$ with $\mathbf{x}$. This clumsiness could have been avoided if we had defined the single-channel cross spectral density between x and y to be $E\left[X\left(e^{j\omega T}\right) Y^{*}\left(e^{j\omega T}\right)\right]$, in which case the multichannel generalisation for the spectral density matrix between $\mathbf{x}$ and $\mathbf{y}$ would be $E\left[\mathbf{x}\left(e^{j\omega T}\right) \mathbf{y}^{\mathrm{H}}\left(e^{j\omega T}\right)\right]$ (Morgan, 1999). This Hermitian transposed form of the cross spectral density is consistent with defining the single-channel cross-correlation function between $x(n)$ and $y(n)$ to be $E\left[x(n)\, y(n-m)\right]$, as in Papoulis (1977) and much of the statistics literature, instead of $E\left[x(n)\, y(n+m)\right]$, as we have done here to be consistent with the acoustics and most of the signal processing literature. All of the equations here could be simply converted to this other notation by replacing $\mathbf{S}_{xy}$ by $\mathbf{S}_{yx}$, for example.

In order to be consistent with equation (A5.4), the matrix of correlation functions between the zero-mean sequences of real signals that are elements of the vector

$$\boldsymbol{x}(n) = \left[x_1(n)\ x_2(n)\ \cdots\ x_N(n)\right]^{\mathrm{T}} , \tag{A5.13}$$

is defined to be

$$\boldsymbol{R}_{xx}(m) = E\left[\boldsymbol{x}(n+m)\boldsymbol{x}^{\mathrm{T}}(n)\right] , \tag{A5.14}$$

whose Fourier transform is then equal to $\mathbf{S}_{xx}(e^{j\omega T})$. Note that vectors and matrices of transform-domain quantities are denoted throughout the book by **bold roman** variables, whereas the vectors and matrices of time-domain quantities are denoted by ***bold italic*** variables. The matrix of cross correlation functions between the elements of $\boldsymbol{x}(n)$ and $\boldsymbol{y}(n)$ is similarly defined to be

$$\boldsymbol{R}_{xy}(m) = E\left[\boldsymbol{y}(n+m)\boldsymbol{x}^{\mathrm{T}}(n)\right] , \tag{A5.15}$$

whose Fourier transform is equal to $\mathbf{S}_{xy}\left(e^{j\omega T}\right)$.

The z-transforms of these matrices of auto and cross correlation functions can also be written in terms of the expectation over the z-transforms of individual data samples as their length tends to infinity, so that again using the expectation in the sense explained above,

$$\mathbf{S}_{xx}(z) = E\left[\mathbf{x}(z)\mathbf{x}^{\mathrm{T}}(z^{-1})\right] , \tag{A5.16}$$

(Grimble and Johnson, 1988), and

$$\mathbf{S}_{xy}(z) = E\left[\mathbf{y}(z)\mathbf{x}^{\mathrm{T}}(z^{-1})\right] . \tag{A5.17}$$

A.6. MATRIX AND VECTOR QUADRATIC EQUATIONS

The properties of the derivatives of the trace of a complex matrix can be used to calculate the optimum matrix of control filters, $\mathbf{W}_{\text{opt}}$, which minimise the sum of squared errors in the multichannel feedforward control problem, as used in Chapter 5. This involves the minimisation of the trace of a matrix of the form of equation (A5.4) for the vector of error signals that can be written as

$$J = \text{trace}\left[\mathbf{S}_{ee}\left(\mathrm{e}^{j\omega T}\right)\right] , \tag{A6.1}$$

where

$$\mathbf{e}\left(\mathrm{e}^{j\omega T}\right) = \mathbf{d}\left(\mathrm{e}^{j\omega T}\right) + \mathbf{G}\left(\mathrm{e}^{j\omega T}\right)\mathbf{W}\left(\mathrm{e}^{j\omega T}\right)\mathbf{x}\left(\mathrm{e}^{j\omega T}\right) , \tag{A6.2}$$

and $\mathbf{d}\left(\mathrm{e}^{j\omega T}\right)$ is the vector of disturbance spectra, $\mathbf{G}\left(\mathrm{e}^{j\omega T}\right)$ the matrix of plant frequency responses, $\mathbf{W}\left(\mathrm{e}^{j\omega T}\right)$ the matrix of controller frequency responses and $\mathbf{x}\left(\mathrm{e}^{j\omega T}\right)$ is the vector of reference signal spectra. Dropping the explicit dependence on $\left(\mathrm{e}^{j\omega T}\right)$ for clarity, equation (A6.1) can be written, using equation (A6.2) and the definitions of the spectral density matrices above, as a *matrix quadratic equation*, which has the form

$$J = \text{trace}\left[E\left(\mathbf{e}\,\mathbf{e}^{\mathrm{H}}\right)\right] = \text{trace}\left[\mathbf{G}\,\mathbf{W}\,\mathbf{S}_{xx}\,\mathbf{W}^{\mathrm{H}}\,\mathbf{G}^{\mathrm{H}} + \mathbf{G}\,\mathbf{W}\,\mathbf{S}_{xd}^{\mathrm{H}} + \mathbf{S}_{xd}\,\mathbf{W}^{\mathrm{H}}\,\mathbf{G}^{\mathrm{H}} + \mathbf{S}_{dd}\right] . \tag{A6.3}$$

The derivative of equation (A6.1) with respect to the real and imaginary parts of $\mathbf{W}$ can be obtained using equations (A4.23) and (A4.25) as

$$\frac{\partial J}{\partial \mathbf{W}_{\mathrm{R}}} + j\frac{\partial J}{\partial \mathbf{W}_{\mathrm{I}}} = 2\mathbf{G}^{\mathrm{H}}\,\mathbf{G}\,\mathbf{W}\,\mathbf{S}_{xx} + 2\mathbf{G}^{\mathrm{H}}\,\mathbf{S}_{xd} . \tag{A6.4}$$

Provided that both $\mathbf{G}^{\mathrm{H}}\mathbf{G}$ and $\mathbf{S}_{xx}$ are non-singular matrices, equation (A6.4) can be set to zero and solved to give the matrix of optimum controller responses, which is equal to

$$\mathbf{W}_{\text{opt}} = -\left[\mathbf{G}^{\mathrm{H}}\,\mathbf{G}\right]^{-1}\,\mathbf{G}^{\mathrm{H}}\,\mathbf{S}_{xd}\,\mathbf{S}_{xx}^{-1} . \tag{A6.5}$$

A simplified form of these derivatives of a trace function can also be used to calculate the derivatives of a vector *Hermitian quadratic form*. In Chapter 2 the vector of complex error signals for the multichannel tonal control problem is written in terms of the vector of complex control signals, $\mathbf{u}$, as

$$\mathbf{e} = \mathbf{d} + \mathbf{G}\mathbf{u} , \tag{A6.6}$$

and the objective is to minimise a cost function that is the inner product of $\mathbf{e}$,

$$J = \mathbf{e}^{\mathrm{H}}\mathbf{e} = \mathbf{u}^{\mathrm{H}}\mathbf{G}^{\mathrm{H}}\mathbf{G}\,\mathbf{u} + \mathbf{u}^{\mathrm{H}}\mathbf{G}^{\mathrm{H}}\mathbf{d} + \mathbf{d}^{\mathrm{H}}\mathbf{G}\,\mathbf{u} + \mathbf{d}^{\mathrm{H}}\mathbf{d} , \tag{A6.7}$$

with respect to the real and imaginary parts of the elements of $\mathbf{u}$.

Using the properties of the trace of an outer product, this cost function can also be written as

$$J = \mathrm{trace}\left(\mathbf{e}\mathbf{e}^{\mathrm{H}}\right) = \mathrm{trace}\left(\mathbf{G}\mathbf{u}\mathbf{u}^{\mathrm{H}}\mathbf{G}^{\mathrm{H}} + \mathbf{G}\mathbf{u}\mathbf{d}^{\mathrm{H}} + \mathbf{d}\mathbf{u}^{\mathrm{H}}\mathbf{G}^{\mathrm{H}} + \mathbf{d}\mathbf{d}^{\mathrm{H}}\right) \tag{A6.8}$$

and again using equations (A4.23) and (A4.25), the vector of derivatives of J with respect to the real and imaginary parts of $\mathbf{u}$ can be written as

$$\frac{\partial J}{\partial \mathbf{u}_{\mathrm{R}}} + j\,\frac{\partial J}{\partial \mathbf{u}_{\mathrm{I}}} = 2\,\mathbf{G}^{\mathrm{H}}\mathbf{G}\,\mathbf{u} + 2\,\mathbf{G}^{\mathrm{H}}\mathbf{d}\ , \tag{A6.9}$$

so that provided $\mathbf{G}^{\mathrm{H}}\mathbf{G}$ is non-singular, the optimum vector of control signals, $\mathbf{u}_{\mathrm{opt}}$, can be derived by setting this equation to zero to give

$$\mathbf{u}_{\mathrm{opt}} = -\left[\mathbf{G}^{\mathrm{H}}\mathbf{G}\right]^{-1}\mathbf{G}^{\mathrm{H}}\mathbf{d}\ . \tag{A6.10}$$

Although equation (A6.10) provides a convenient solution to this least-squares problem for analytic purposes, a more efficient and robust numerical solution is provided by using the QR decomposition of $\mathbf{G}$, as discussed for example by Noble and Daniel (1977).

The vector Hermitian quadratic form can be written more compactly as

$$J = \mathbf{x}^{\mathrm{H}}\mathbf{A}\,\mathbf{x} + \mathbf{x}^{\mathrm{H}}\mathbf{b} + \mathbf{b}^{\mathrm{H}}\mathbf{x} + c\ , \tag{A6.11}$$

where $\mathbf{A}$ is a Hermitian matrix. The derivatives of J with respect to the real and imaginary parts of $\mathbf{x}$ can then be shown to be (Nelson and Elliott, 1992)

$$\frac{\partial J}{\partial \mathbf{x}_{\mathrm{R}}} + j\,\frac{\partial J}{\partial \mathbf{x}_{\mathrm{I}}} = 2\,\mathbf{A}\,\mathbf{x} + 2\,\mathbf{b}\ , \tag{A6.12}$$

which is referred to as the *complex gradient vector*, $\mathbf{g}$, in the main text.

The minimum value of J is obtained by setting this gradient vector to zero, and assuming that $\mathbf{A}$ is non-singular, this gives

$$\mathbf{x}_{\mathrm{opt}} = -\,\mathbf{A}^{-1}\,\mathbf{b}\ . \tag{A6.13}$$

The vector of optimal control signals for the multichannel tonal control equation (A6.10) is clearly a special case of equation (A6.13). The minimum value of the vector Hermitian quadratic form can be obtained by substituting equation (A6.13) into equation (A6.11) to give

$$J_{\mathrm{min}} = c - \mathbf{b}^{\mathrm{H}}\mathbf{A}^{-1}\,\mathbf{b}\ . \tag{A6.14}$$

A.7. EIGENVALUE/EIGENVECTOR DECOMPOSITION

The *eigenvalues* λ_i and corresponding *eigenvectors* $\mathbf{q}_i$ (strictly the right eigenvectors) of an $N \times N$ square matrix $\mathbf{A}$ are given by solutions of the equation

$$\mathbf{A}\mathbf{q}_i = \lambda_i\mathbf{q}_i\ , \tag{A7.1}$$

which will have non-trivial solutions only if the determinant in equation (A7.2) is zero,

$$\det\left(\mathbf{A} - \lambda_i\mathbf{I}\right) = 0\ . \tag{A7.2}$$

The resulting equation for λ_i is called the *characteristic equation*. If the eigenvalues are distinct, i.e. all different from each other, then the eigenvectors are linearly independent. A matrix without distinct eigenvalues is called degenerate or *defective*. The eigenvectors can be expressed as the columns of the matrix **Q**, so that

$$\mathbf{Q} = \begin{bmatrix} \mathbf{q}_1 & \mathbf{q}_2 & \cdots & \mathbf{q}_N \end{bmatrix}, \tag{A7.3}$$

and the eigenvectors can be written as the diagonal elements of the matrix **Λ**, so that

$$\mathbf{\Lambda} = \begin{bmatrix} \lambda_1 & 0 & \cdots & 0 \\ 0 & \lambda_2 & & \\ \vdots & & & \\ 0 & & & \lambda_N \end{bmatrix}, \tag{A7.4}$$

in which case equation (A7.1) can be written as

$$\mathbf{A}\,\mathbf{Q} = \mathbf{Q}\,\mathbf{\Lambda}\,. \tag{A7.5}$$

If the eigenvalues are distinct, so that the eigenvectors are linearly independent, then the inverse of **Q** also exists, and we can write

$$\mathbf{A} = \mathbf{Q}\,\mathbf{\Lambda}\,\mathbf{Q}^{-1}\,, \tag{A7.6}$$

and

$$\mathbf{\Lambda} = \mathbf{Q}^{-1}\,\mathbf{A}\,\mathbf{Q}\,. \tag{A7.7}$$

The eigenvalues of the matrix **A** multiplied by itself P times, $\mathbf{A}^P$, are thus λ_i^P. This property also holds for the special case in which $P = -1$ in which the eigenvalues of $\mathbf{A}^{-1}$ are equal to λ_i^{-1}. The inverse exists provided none of the eigenvalues is zero, in which case the square matrix is also said to be *full rank*. If a square matrix is not full rank it is also singular. Even if none of the eigenvalues of **A** is zero, $\mathbf{A}^{-1}$ can have some very large terms if any of the eigenvalues of **A** are very close to zero, in which case the matrix is said to be *ill-conditioned*.

The sum and product of all the eigenvalues of **A** are equal to its trace and determinant respectively, so that

$$\text{trace}\,(\mathbf{A}) = \sum_{n=1}^{N} \lambda_n\,, \tag{A7.8}$$

$$\det\,(\mathbf{A}) = \lambda_1 \lambda_2 \cdots \lambda_N\,. \tag{A7.9}$$

The i-th eigenvalue of a matrix **A** is sometimes denoted as $\lambda_i(\mathbf{A})$, and using this notation another property of eigenvalues may be written as

$$\lambda_i(\mathbf{I} + \mathbf{A}) = 1 + \lambda_i(\mathbf{A})\,, \tag{A7.10}$$

so that

$$\det\,(\mathbf{I} + \mathbf{A}) = (1 + \lambda_1)(1 + \lambda_2)\cdots(1 + \lambda_N)\,. \tag{A7.11}$$

The eigenvalue that has the largest modulus is called the *spectral radius* of the associated matrix and is denoted

$$\rho(\mathbf{A}) \;=\; \max_i \left|\lambda_i(\mathbf{A})\right| \,. \tag{A7.12}$$

If $\mathbf{A}$ is an $M \times N$ matrix and $\mathbf{B}$ is an $N \times M$ matrix then the $M \times M$ matrix $\mathbf{AB}$ has the same non-zero eigenvalues as the $N \times N$ matrix $\mathbf{BA}$. If $M > N$ then $\mathbf{AB}$ has the same N eigenvalues as $\mathbf{BA}$, which is assumed to be full rank, and $M - N$ eigenvalues that are zero. If $\mathbf{A}$ and $\mathbf{B}$ correspond to matrices of frequency responses in a feedback loop, then the eigenvalues of $\mathbf{AB}$ are sometimes said to be the *characteristic gains* of the loop transfer function $\mathbf{AB}$. The characteristic gains quantify the gain in the loop for the special case in which the vectors of both input and output signals are proportional to a single eigenvector of $\mathbf{AB}$. The eigenvalues are thus a rather poor measure of generalised 'gain', and hence the performance, of a multichannel system but are very useful in analysing the system's stability.

The eigenvalues of a Hermitian matrix are entirely real, and if the matrix is positive definite, the eigenvalues are all greater than zero (Noble and Daniel, 1977) so that a positive definite matrix is also non-singular. The eigenvectors corresponding to the distinct eigenvalues of a Hermitian or symmetric matrix can be normalised so that

$$\mathbf{q}_i^{\mathrm{H}}\, \mathbf{q}_j = 1 \qquad \text{if } i = j\,, \qquad \text{else} = 0\,, \tag{A7.13}$$

and the eigenvectors form an orthonormal set. The matrix of eigenvectors, $\mathbf{Q}$ in equation (A7.3), is then unitary, i.e.

$$\mathbf{Q}^{\mathrm{H}}\,\mathbf{Q} = \mathbf{Q}\,\mathbf{Q}^{\mathrm{H}} = \mathbf{I}\,, \tag{A7.14}$$

so that

$$\mathbf{A} = \mathbf{Q}\,\boldsymbol{\Lambda}\,\mathbf{Q}^{\mathrm{H}}\,, \tag{A7.15}$$

$$\boldsymbol{\Lambda} = \mathbf{Q}^{\mathrm{H}}\,\mathbf{A}\,\mathbf{Q}\,. \tag{A7.16}$$

If $\mathbf{A}$ is real and symmetric, equations (A7.13) to (A7.16) still hold, but with the normal transpose replacing the Hermitian transpose.

Finally, we return to the matrix version of the binomial expansion, equation (A3.16), which is

$$\left[\mathbf{I} + \boldsymbol{\Delta}\right]^{-1} \;=\; \mathbf{I} - \boldsymbol{\Delta} + \boldsymbol{\Delta}^2 \;-\boldsymbol{\Delta}^3 + \cdots\,, \tag{A7.17}$$

and note that if the magnitude of all the eigenvalues of $\boldsymbol{\Delta}$ are much smaller than unity, i.e. $\rho(\boldsymbol{\Delta}) \ll 1$ then the magnitude of $\boldsymbol{\Delta}^2$ will be much less than that of $\boldsymbol{\Delta}$ and so

$$\left[\mathbf{I} + \boldsymbol{\Delta}\right]^{-1} \;\approx\; \mathbf{I} - \boldsymbol{\Delta}\,. \tag{A7.18}$$

A.8. SINGULAR VALUE DECOMPOSITION

Any complex $L \times M$ matrix $\mathbf{A}$ may be factorised using the *singular value decomposition*

$$\mathbf{A} = \mathbf{U}\,\boldsymbol{\Sigma}\,\mathbf{V}^{\mathrm{H}}\,, \tag{A8.1}$$

where the $L \times L$ matrix $\mathbf{U}$ and the $M \times M$ matrix $\mathbf{V}$ are generally complex, but unitary so that $\mathbf{U}^{\mathrm{H}}\mathbf{U} = \mathbf{U}\mathbf{U}^{\mathrm{H}} = \mathbf{I}$ and $\mathbf{V}^{\mathrm{H}}\mathbf{V} = \mathbf{V}\mathbf{V}^{\mathrm{H}} = \mathbf{I}$, and if $L > M$, then the $L \times M$ matrix $\boldsymbol{\Sigma}$ can be written as

$$\boldsymbol{\Sigma} = \begin{bmatrix} \sigma_1 & 0 & \cdots & 0 \\ 0 & \sigma_2 & & \vdots \\ \vdots & & \ddots & \vdots \\ 0 & \cdots & \cdots & \sigma_M \\ 0 & \cdots & \cdots & 0 \\ \vdots & & & \vdots \\ 0 & \cdots & \cdots & 0 \end{bmatrix}, \tag{A8.2}$$

where σ_i are the non-negative *singular values* of $\mathbf{A}$, which are arranged so that

$$\sigma_1 \geq \sigma_2 \cdots \geq \sigma_M \ . \tag{A8.3}$$

The i-th singular value of a matrix $\mathbf{A}$ is denoted $\sigma_i(\mathbf{A})$. The largest singular value, σ_1, is sometimes denoted $\overline{\sigma}$, and the smallest, σ_M, as $\underline{\sigma}$.

The *rank* of a matrix $\mathbf{A}$ is equal to the number of its non-zero singular values in equation (A8.2). If $\mathbf{A}$ is square then this is the same as the number of its non-zero eigenvalues. In the example above where $\mathbf{A}$ is an $L \times M$ matrix and $L > M$, then if $\mathbf{A}$ has N non-zero singular values, so that $\sigma_{N+1}, \ldots, \sigma_M$ are zero, its rank is N. If the rank of $\mathbf{A}$ were less than M in this case it would be *rank deficient*. Assuming that $\mathbf{A}$ is not rank deficient, so that $N = M$, then $\mathbf{A}$ and $\mathbf{A}^{\mathrm{H}}\mathbf{A}$ are *full rank* and $(\mathbf{A}^{\mathrm{H}}\mathbf{A})^{-1}$ exists, but $\mathbf{A}\mathbf{A}^{\mathrm{H}}$ is not full rank and so it is singular.

If $L > M$ then the singular values of $\mathbf{A}$, which is assumed to be full rank, are the positive square roots of the eigenvalues of the $M \times M$ matrix $\mathbf{A}^{\mathrm{H}}\mathbf{A}$, so that

$$\sigma_i(\mathbf{A}) = \left[\lambda_i\left(\mathbf{A}^{\mathrm{H}}\mathbf{A}\right)\right]^{1/2} . \tag{A8.4}$$

We can also write

$$\mathbf{A}^{\mathrm{H}}\mathbf{A} = \mathbf{V}\,\boldsymbol{\Sigma}^{\mathrm{T}}\boldsymbol{\Sigma}\,\mathbf{V}^{\mathrm{H}} , \tag{A8.5}$$

where $\mathbf{V}$ is equal to the unitary vector of eigenvectors of the Hermitian matrix $\mathbf{A}^{\mathrm{H}}\mathbf{A}$ and $\boldsymbol{\Sigma}^{\mathrm{T}}\boldsymbol{\Sigma}$ is the $M \times M$ diagonal matrix of its M eigenvalues. Similarly the $L \times L$ matrix $\mathbf{A}\mathbf{A}^{\mathrm{H}}$ can be written as

$$\mathbf{A}\mathbf{A}^{\mathrm{H}} = \mathbf{U}\,\boldsymbol{\Sigma}\boldsymbol{\Sigma}^{\mathrm{T}}\,\mathbf{U}^{\mathrm{H}} , \tag{A8.6}$$

where $\mathbf{U}$ is the matrix of eigenvectors of $\mathbf{A}\mathbf{A}^{\mathrm{H}}$ and the $L \times L$ diagonal matrix $\boldsymbol{\Sigma}\,\boldsymbol{\Sigma}^{\mathrm{T}}$ has M non-zero elements equal to the M eigenvalues of $\mathbf{A}^{\mathrm{H}}\mathbf{A}$ and $L - M$ diagonal elements which are zero.

Some further properties of singular values are that for a square matrix of full rank

$$\overline{\sigma}(\mathbf{A}^{-1}) = 1/\underline{\sigma}(\mathbf{A}) , \tag{A8.7}$$

and more generally

$$\overline{\sigma}(\mathbf{A}\,\mathbf{B}) \geq \overline{\sigma}(\mathbf{A})\,\overline{\sigma}(\mathbf{B}) , \tag{A8.8}$$

$$\overline{\sigma}(\mathbf{A}) \geq |\lambda_i(\mathbf{A})| \geq \underline{\sigma}(\mathbf{A}) \ , \tag{A8.9}$$

and also

$$\rho(\mathbf{AB}) \geq \overline{\sigma}(\mathbf{A})\,\overline{\sigma}(\mathbf{B}) \ . \tag{A8.10}$$

If the rank of the $L \times M$ matrix $\mathbf{A}$ is N, then the singular value decomposition can be written as

$$\mathbf{A} \;=\; \mathbf{U}\,\boldsymbol{\Sigma}\,\mathbf{V}^{\mathrm{H}} \;=\; \sum_{n=1}^{N} \sigma_n\,\mathbf{u}_n\,\mathbf{v}_n^{\mathrm{H}} \tag{A8.11}$$

where $\mathbf{u}_n$ and $\mathbf{v}_n$ are the n-th column vectors of $\mathbf{U}$ and $\mathbf{V}$, respectively, which are orthonormal and are called the output and input singular vectors. Notice that the output and input singular vectors for $n > N$ do not appear in equation (A8.11). Thus $\mathbf{A}$ can also be expressed as the *reduced* or *economy-size singular value decomposition* (Skogestad and Postlethwaite, 1996), which has the form

$$\mathbf{A} = \mathbf{U}_{\mathrm{r}}\,\boldsymbol{\Sigma}_{\mathrm{r}}\,\mathbf{V}_{\mathrm{r}}^{\mathrm{H}} \ , \tag{A8.12}$$

where $\mathbf{A}$ may be $L \times M$, but if it only has rank N, then $\mathbf{U}_{\mathrm{r}}$ is $L \times N$, $\boldsymbol{\Sigma}_r$ is $N \times N$ and $\mathbf{V}_{\mathrm{r}}^{\mathrm{H}}$ is $N \times M$. Notice that if $L > M$ and the matrix is not rank deficient, $\mathbf{U}_{\mathrm{r}}$ is $L \times M$, $\boldsymbol{\Sigma}_{\mathrm{r}}$ is $M \times M$ and $\mathbf{V}_{\mathrm{r}}$ is $M \times M$ in equation (A8.12), which is the form of singular value decomposition used by Maciejowski (1989) for example.

The Moore–Penrose *generalised inverse* or *pseudo-inverse* of the $L \times M$ matrix $\mathbf{A}$ is defined to be the $M \times L$ matrix $\mathbf{A}^{\dagger}$ given by

$$\mathbf{A}^{\dagger} \;=\; \mathbf{V}\,\boldsymbol{\Sigma}^{\dagger}\,\mathbf{U}^{\mathrm{H}} \tag{A8.13}$$

where

$$\boldsymbol{\Sigma}^{\dagger} \;=\; \begin{bmatrix} 1/\sigma_1 & 0 & \cdots & \cdots & \cdots & \cdots & 0 \\ 0 & 1/\sigma_2 & & & & & \vdots \\ \vdots & & \ddots & & & & \vdots \\ \vdots & & & 1/\sigma_N & & & \vdots \\ \vdots & & & & 0 & & \vdots \\ \vdots & & & & & \ddots & \vdots \\ 0 & & & & & 0 \;\cdots & 0 \end{bmatrix} . \tag{A8.14}$$

Note that the N non-zero singular values in $\boldsymbol{\Sigma}$ have been inverted in $\boldsymbol{\Sigma}^{\dagger}$, but the singular values of $\boldsymbol{\Sigma}$ that are zero remain zero in $\boldsymbol{\Sigma}^{\dagger}$. In practice, the low-order singular values of a matrix derived from measured data will rarely be exactly zero, even if the underlying matrix is rank deficient. A measure of the severity of this problem can be obtained from the *condition number* of the matrix $\mathbf{A}$, which is defined here to be the ratio of its largest to its smallest singular values,

$$\kappa(\mathbf{A}) \;=\; \frac{\sigma_1}{\sigma_M} \;=\; \frac{\overline{\sigma}(\mathbf{A})}{\underline{\sigma}(\mathbf{A})} \ . \tag{A8.15}$$

If the condition number is very large, the matrix $\mathbf{A}$ is close to being rank deficient.

In the numerical calculation of the pseudo-inverse it is common to define a threshold or tolerance value for the singular values of **A**, such that if the singular values are above this value they are taken to be data and inverted in equation (A8.14), and if they are below this value they are assumed to be due to noise and set to zero. If the uncertainty in the data is due to numerical precision effects then the threshold can be calculated from a knowledge of the arithmetic precision, and the number of singular values above this threshold is known as the *numerical rank* of the matrix (Golub and Van Loan, 1996). Alternatively, a small positive number could be added to each singular value before inversion to 'regularise' the solution. The threshold level for the singular values or the level of the regularisation parameter is related to the signal-to-noise ratio of the measured data, and by choosing an appropriate value for this threshold level, the calculation of the pseudo-inverse can be made robust to measurement and computational noise.

The pseudo-inverse plays an important role in the multichannel least-squares problems discussed in Chapter 4, and if the plant matrix is full rank the pseudo-inverse can be expressed explicitly as a function of the plant matrix itself, as listed in Table 4.1 for the overdetermined, fully-determined and underdetermined cases. If the plant matrix is not full rank, however, the more general form of the pseudo-inverse given by equation (A8.13) can still be used to calculate the least-squares solution. Although they are useful for analysing the solution to least-squares problems, the SVD and pseudo-inverse can be computationally expensive and alternative methods of computing least-squares solutions, such as using QR decomposition, are described for example by Golub and Van Loan (1996).

The singular values give a good measurement of the 'gain' in a multichannel system. Suppose that **A** were the frequency response matrix of such a system at a single frequency, for example, and write equation (A8.1) as

$$\mathbf{A}\,\mathbf{V} = \mathbf{U}\,\boldsymbol{\Sigma}\,. \tag{A8.16}$$

Using the individual singular values and input and output singular vectors, we can also write

$$\mathbf{A}\,\mathbf{v}_n = \sigma_n\,\mathbf{u}_n\,, \tag{A8.17}$$

so that an input $\mathbf{v}_n$ produces an output $\mathbf{u}_n$ with a 'gain' of σ_n. In contrast to the eigenvalues, the 'gains' deduced from the singular values do not assume that the input and output vectors are equal, and thus give a more precise indication of the performance of the system which has the frequency response matrix **A**.

A.9. VECTOR AND MATRIX NORMS

The *norm* of a matrix **A** is a real scalar which is generally denoted $\|\mathbf{A}\|$, and is a measure of the overall size or magnitude of **A**. There are several types of norms for both vectors and matrices but they must all satisfy the following conditions:

(i) $\|\mathbf{A}\| \geq 0$, so the norm is non-negative.

(ii) $\|\mathbf{A}\| = 0$ if and only if all the elements of **A** are zero, so the norm is positive.

(iii) $\|\alpha \mathbf{A}\| = |\alpha| \|\mathbf{A}\|$ for all complex α, so the norm is homogeneous.

(iv) $\|\mathbf{A} + \mathbf{B}\| \leq \|\mathbf{A}\| + \|\mathbf{B}\|$, so the norm satisfies the triangle inequality.

A further property, which does not apply to vector norms but is considered an additional property of the norm of a matrix by some authors (e.g. Golub and Van Loan, 1996) and as the definition of a 'matrix norm' by others (e.g. Skogestad and Postlethwaite, 1996), is that

(v) $\|\mathbf{A}\,\mathbf{B}\| \leq \|\mathbf{A}\| \|\mathbf{B}\|$, which is called the multiplicative property or consistency condition.

All the matrix norms defined below satisfy this condition, but norms are also used to quantify the response of systems, and these operator-induced or *system norms* do not necessarily satisfy condition (v). Unfortunately, the notation used for matrix norms is not entirely consistent and can be confusing, since the same symbols tend to be used for the matrix norm and the induced or system norms, for example.

We begin with the norm of a constant, complex vector, $\mathbf{x}$, for which we define the *p-norm* to be

$$\|\mathbf{x}\|_p = \left(\sum_{n=1}^{N} |x_n|^p \right)^{1/p} . \tag{A9.1}$$

An important special case is obtained when $p = 2$ so that the *2-norm*, which is also called the *Euclidean norm*, has the squared value

$$\|\mathbf{x}\|_2^2 = \sum_{n=1}^{N} |x_n|^2 , \tag{A9.2}$$

which can also be written as the inner product $\mathbf{x}^{\mathrm{H}}\mathbf{x}$. The vector *infinity-norm*, or *maximum norm* could be written as $\|\mathbf{x}\|_\infty$, but will be expressed here as

$$\|\mathbf{x}\|_{\max} = \max_n |x_n| . \tag{A9.3}$$

The most widely used norm of a constant matrix is the *Frobenius* or Euclidean matrix norm, which is equal to the square root of the sum of its modulus squared elements, so for the $L \times M$ matrix in equation (A2.1),

$$\|\mathbf{A}\|_{\mathrm{F}} = \sqrt{\sum_{l=1}^{L} \sum_{m=1}^{M} |a_{lm}|^2} . \tag{A9.4}$$

This is also equal to the square root of the trace of $\mathbf{A}^{\mathrm{H}}\mathbf{A}$, and thus the square root of the sum of squared singular values of $\mathbf{A}$, so that for a matrix of rank N,

$$\|\mathbf{A}\|_{\mathrm{F}} = \sqrt{\sum_{n=1}^{N} \sigma_n^2(\mathbf{A})} . \tag{A9.5}$$

A number of other matrix norms may be defined that are called *induced norms*; these are connected with the vector norms defined above and are related to signal amplification in multichannel systems. The vector of complex outputs of such a system at a single frequency could be written as

$$\mathbf{y} = \mathbf{A}\,\mathbf{x}\,, \tag{A9.6}$$

where $\mathbf{A}$ is the matrix of frequency responses at this frequency and $\mathbf{x}$ is the vector of complex input signals at this frequency. It is particularly important to define the maximum 'amplification' through the system at this single frequency for all inputs of equal 'magnitude' but this requires a definition of the magnitude of $\mathbf{x}$ and $\mathbf{y}$. If the vector p-norm is used to define the magnitude of $\mathbf{x}$ and $\mathbf{y}$, then the maximum amplification is the maximum value of $\|\mathbf{y}\|_p / \|\mathbf{x}\|_p$ for all $\mathbf{x}$ which are not equal to $\mathbf{0}$. This quantity is called the *induced p-norm* of the matrix $\mathbf{A}$ in equation (A9.6) and can be written as

$$\|\mathbf{A}\|_{ip} = \max_{\mathbf{x}\neq\mathbf{0}} \frac{\|\mathbf{A}\,\mathbf{x}\|_p}{\|\mathbf{x}\|_p}\,. \tag{A9.7}$$

These induced norms satisfy the multiplicative property defined at the beginning of this section since equation (A9.7) implies that

$$\|\mathbf{y}\|_p \le \|\mathbf{A}\|_{ip}\,\|\mathbf{x}\|_p\,. \tag{A9.8}$$

If the vector 2-norm for the input and output signals are used in equation (A9.6), the induced 2-norm of the system is equal to

$$\|\mathbf{A}\|_{i2} = \|\mathbf{A}\|_s = \overline{\sigma}(\mathbf{A})\,, \tag{A9.9}$$

which is also called the singular value, Hilbert or *spectral norm*. More generally, we can show from the fact that the input and output singular vectors are orthonormal that

$$\sigma_n(\mathbf{A}) = \frac{\|\mathbf{A}\,\mathbf{v}_n\|_2}{\|\mathbf{v}_n\|_2}\,, \tag{A9.10}$$

where $\mathbf{v}_n$ is the n-th input singular vector. Thus, in addition to equation (A9.9), which satisfies equation (A9.7) when $\mathbf{x} = \mathbf{v}_1$, we also have that when $\mathbf{x} = \mathbf{v}_N$,

$$\underline{\sigma}(\mathbf{A}) = \min_{\mathbf{x}\neq\mathbf{0}} \frac{\|\mathbf{A}\,\mathbf{x}\|_2}{\|\mathbf{x}\|_2}\,, \tag{A9.11}$$

and so for all $\mathbf{x}$

$$\overline{\sigma}(\mathbf{A}) \ge \frac{\|\mathbf{A}\,\mathbf{x}\|_2}{\|\mathbf{x}\|_2} \ge \underline{\sigma}(\mathbf{A})\,. \tag{A9.12}$$

If the vector ∞-norm is used for the input and output signals, the induced ∞-norm of the system is

$$\|\mathbf{A}\|_{i\infty} = \max_l \left(\sum_{m=1}^{M} |a_{lm}| \right), \tag{A9.13}$$

or the maximum row sum.

When the signal amplification, or 'gain', of a system subject to excitation over a range of frequencies is considered, yet another type of norm must be introduced, which is called the *system norm* or *operator norm*. We return for a moment to a single-channel continuous-time system with impulse response $a(t)$, whose output is given by

$$y(t) = \int_{-\infty}^{\infty} a(\tau)\, x(t-\tau)\, \mathrm{d}\tau \,, \tag{A9.14}$$

where $x(t)$ is the input. Assuming that it converges, we define the *temporal p-norm* or l_p norm of $y(t)$ as

$$\|y(t)\|_p = \left(\int_{-\infty}^{\infty} |y(\tau)|^p \, \mathrm{d}\tau \right)^{1/p} , \tag{A9.15}$$

so that the l_2 norm or integrated squared value of a real signal with a finite energy is

$$\|y(t)\|_2 = \sqrt{\int_{-\infty}^{\infty} y(\tau)^2 \, \mathrm{d}\tau} \,. \tag{A9.16}$$

By Parseval's theorem, the temporal 2-norm of the output signal can also be written as

$$\|y(t)\|_2 = \left(\frac{1}{2\pi} \int_{-\infty}^{\infty} |Y(j\omega)|^2 \, \mathrm{d}\omega \right)^{1/2} , \tag{A9.17}$$

where $Y(j\omega)$ is the Fourier transform of $y(t)$. From equation (A9.14), $Y(j\omega)$ is also equal to

$$Y(j\omega) = A(j\omega)\, X(j\omega) \,, \tag{A9.18}$$

where $A(j\omega)$ is the frequency response of the system and $X(j\omega)$ is the spectrum of its input, so that

$$\|y(t)\|_2^2 = \frac{1}{2\pi} \int_{-\infty}^{\infty} |A(j\omega)|^2 \, |X(j\omega)|^2 \, \mathrm{d}\omega \,. \tag{A9.19}$$

If we assume that $|X(j\omega)|^2$ is constant with frequency, which would be the case if the input were a delta function, then the mean square output is proportional to

$$\|A(s)\|_2 = \left(\frac{1}{2\pi} \int_{-\infty}^{\infty} |A(j\omega)|^2 \, \mathrm{d}\omega \right)^{1/2} , \tag{A9.20}$$

where $A(s)$ is the Laplace-domain transfer function of the continuous-time system and $\|A(s)\|_2$ is called the H_2 *norm* of the system transfer function, which is a measure of the 'average' gain of the system.

If, on the other hand, we know nothing about the spectrum of the input signal and wish to calculate the 'worst case' temporal 2-norm of the output given the temporal 2-norm of the input, then this is equal to the H_∞ norm of the system transfer function given by

$$\|A(s)\|_\infty = \sup_{\omega} |A(j\omega)| , \tag{A9.21}$$

where sup indicates the supremum or least upper bound. The H_∞ norm has the property that

$$\|A(s)\|_\infty = \max_{x(t)\neq 0} \frac{\|y(t)\|_2}{\|x(t)\|_2} . \tag{A9.22}$$

The terms H_2 and H_∞ come from the use of the symbol H to represent a Hardy space (Skogestad and Postlethwaite, 1996). The H_2 norm is finite provided that $A(s)$ is stable and strictly proper, i.e. $\lim_{\omega\to\infty} A(j\omega) = 0$, and the H_∞ norm is finite provided that $A(s)$ is stable and proper, i.e. $\lim_{\omega\to\infty} A(j\omega) < \infty$.

For multichannel systems, the operator H_2 norm of a matrix of transfer functions is equal to (Skogestad and Postlethwaite, 1996)

$$\|\mathbf{A}(s)\|_2 = \left(\frac{1}{2\pi} \int_{-\infty}^{\infty} \text{trace} \left(\mathbf{A}^{\mathrm{H}}(j\omega)\, \mathbf{A}(j\omega) \right) \mathrm{d}\omega \right)^{1/2} . \tag{A9.23}$$

Using the properties of the trace of a matrix and those of its singular values, this may be written as

$$\|\mathbf{A}(s)\|_2 = \left(\frac{1}{2\pi} \int_{-\infty}^{\infty} \sum_{n=0}^{N} \sigma_n^{\,2} \left(\mathbf{A}(j\omega) \right) \mathrm{d}\omega \right)^{1/2} , \tag{A9.24}$$

or

$$\|\mathbf{A}(s)\|_2 = \left(\frac{1}{2\pi} \int_{-\infty}^{\infty} \sum_{l=1}^{L} \sum_{m=1}^{M} |a_{lm}(j\omega)|^2 \, \mathrm{d}\omega \right)^{1/2} . \tag{A9.25}$$

It should be noted, however, that this norm does not satisfy the multiplicative property and is thus not a true matrix norm. The operator H_∞ norm of a multichannel analogue system is equal to

$$\|\mathbf{A}(s)\|_\infty = \sup_{\omega} \overline{\sigma} \left(\mathbf{A}(j\omega) \right) . \tag{A9.26}$$

For sampled-time signals the relevant operator norms can be defined as

$$\|\mathbf{A}(z)\|_2 = \left(\frac{1}{2\pi} \int_{-\pi}^{\pi} \sum_{n=0}^{N} \sigma_n^2 \left(\mathbf{A}(\mathrm{e}^{j\omega T}) \right) \mathrm{d}\omega T \right)^{1/2} , \tag{A9.27}$$

and

$$\|\mathbf{A}(z)\|_\infty = \sup_{\omega T} \overline{\sigma} \left(\mathbf{A}(\mathrm{e}^{j\omega T}) \right) , \tag{A9.28}$$

as mentioned in Chapter 6.

References

Abarbanel H.D.I., Frison E.W. and Tsimring L.S. (1998) Obtaining order in a world of chaos—time-domain analysis of nonlinear and chaotic signals. *IEEE Signal Processing Magazine*, **15**(3), 49–65.

Agnello A. (1990) 16-bit conversion paves the way to high-quality audio for PCs. *Electric Design*, July 26.

Aguirre L.A. and Billings S.A. (1995a) Nonlinear chaotic systems: approaches and implications for science and engineering—a survey. *Applied Signal Processing*, **2**, 224–248.

Aguirre L.A. and Billings S.A. (1995b) Closed-loop suppression of chaos in nonlinear driven oscillators. *Journal of Nonlinear Science*, **3**, 189–206.

Akiho M., Haseyama M. and Kitajima H. (1998) Virtual reference signals for active noise cancellation system. *Journal of the Acoustical Society of Japan*, **19**(2), 95–103.

Anderson B.D.O. and Moore J.B. (1989) *Optimal Control, Linear Quadratic Methods*. Prentice Hall.

Arelhi R., Wilkie J. and Johnson M.A. (1996) On stable LQG controllers and cost function values. *UKACC International Conference on Control*, CONTROL'96, 270–275.

Arzamasov S.N. and Mal'tsev A.A. (1985) Adaptive algorithm for active attenuation of broadband random wave fields. *Isvestiga Vyssikh Uchebuyka Zavedenn,* **28**(8), 1008–1016 (in Russian).

Asano F., Suzuki Y. and Swanson D.C. (1999) Optimisation of control system configuration in active control systems using Gram–Schmidt orthogonalisation. *IEEE Transactions on Speech and Audio Processing*, **7**, 213–220.

Åström K.J. (1987) Adaptive feedback control. *Proc. IEEE*, **75**, 185–217.

Åström K.J. and Wittenmark B. (1995) *Adaptive Control*. Addison-Wesley.

Åström K.J. and Wittenmark B. (1997) *Computer Controlled Systems: Theory and Design*, 3rd edn, Prentice Hall.

Auspitzer T., Guicking D. and Elliott S.J. (1995) Using a fast-recursive-least squared algorithm in a feedback controller. *Proc. IEEE Workshop on Applications of Signal Processing to Audio and Acoustics*, New Paltz, NY.

Baek K-H. (1996) *Non-linear optimisation problems in active control*. PhD thesis, University of Southampton.

Baek K-H. and Elliott S.J. (1995) Natural algorithms for choosing source locations in active control systems. *Journal of Sound and Vibration*, **186**, 245–267.

Baek K-H. and Elliott S.J. (1997) Unstructured uncertainty in transducer selection for multichannel active control systems. *Digest IEE Colloquium on Active Sound and Vibration Control*, London, 97/385, 3/1–3/5.

Baek K-H. and Elliott S.J. (2000) The effects of plant and disturbance uncertainties in active control systems on the placement of transducers. *Journal of Sound and Vibration*, **230**, 261–289.

Bai M. and Lee D. (1997) Implementation of an active headset by using the H_∞ robust control theory. *Journal of the Acoustical Society of America*, **102**(4), 2184–2190.

Banks S.P. (1986) *Control Systems Engineering*. Prentice Hall.

Bao C., Sas P. and van Brussel H. (1992) A novel filtered-x LMS algorithm and its application to active noise control. *Proceedings of Eusipco 92, 6th European Signal Processing Conference,* 1709–1712.

Bardou O., Gardonio P., Elliott S.J. and Pinnington R.J. (1997). Active power minimisation and power absorption in a panel with force and moment excitation. *Journal of Sound and Vibration*, **208**, 111–151.

Baruh H. (1992) Placement of sensors and actuators in structural control. *Control and Dynamic Systems*, **52** (ed. C.T. Leondes). Academic Press.

Baumann W.T., Saunders W.R. and Robertshaw H.H. (1991) Active suppression of acoustic radiation from impulsively excited structures. *Journal of the Acoustical Society of America*, **88**, 3202–3208.

BDTI (1999) Berkley Design Technology Inc. www.bdti.com.

Beatty L.G. (1964). Acoustic impedance in a rigid-walled cylindrical sound channel terminated at both ends with active transducers. *Journal of the Acoustic Society of America*, **36**, 1081–1089.

Beaufays F. (1995) Transform-domain adaptive filters: an analytical approach. *IEEE Trans on Signal Processing*, **43**, 422–431.

Beaufays F. and Wan E.A. (1994) Relating real-time back propagation and backpropagation-through-time: An application of flow graph interreciprocity. *Neural Computation*, **6**, 296–306.

Bendat J.S. and Piersol A.G. (1986) *Random Data*, 2nd edn. Wiley.

Benzaria E. and Martin V. (1994) Secondary source location in active noise control: selection or optimisation? *Journal of Sound and Vibration*, **173**, 137–144.

Berkman F., Coleman R., Watters B., Preuss R. and Lapidot N. (1992) An example of a fully adaptive SISO feedback controller for integrated narrowband/broadband active vibration isolation of complex structures. *Proceedings of ASME Winter Meeting*.

Bershad N. and Macchi O. (1989) Comparison of RLS and LMS algorithms for tracking a chirped signal. *Proc. Int. Conf. on Acoustics, Speech and Signal Processing ICASSP89*, **2**, 896–899.

Bies D.A. and Hansen C.H. (1996) *Engineering Noise Control*, 2nd edn, Unwin Hyman.

Billet L. (1992) *Active noise control in ducts using adaptive digital filters.* MPhil thesis, University of Southampton.

Billings S.A. (1980) Identification of nonlinear systems—a survey. *IEE Proc.* **D-127**(6).

Billings S.A. and Chen S. (1998) The determination of multivariable nonlinear models for dynamic systems. *Control and Dynamic Systems*, Vol.7 (ed. C.T. Leondes). Academic Press, 231–277.

Billings S.A. and Voon W.S.F. (1984) Least squares parameter estimation algorithms for nonlinear systems. *International Journal of Systems Science*, **15**, 601–615.

Billout G, Galland M.A. and Sunyach M. (1989) The use of time algorithms for the realisation of an active sound attenuator. *Proc. Int. Conf. on Acoustics, Speech and Signal Processing ICASSP89*, **A2.1**, 2025–2028.

Billout G., Galland M.A., Huu C.H. and Candel S. (1991) Adaptive control of instabilities. *Proc. First Conference on Recent Advances in the Active Control of Sound and Vibration*, 95–107.

Billout G., Norris M.A. and Rossetti D.J. (1995) System de controle actif de bruit Lord NVX pour avions d'affaire Beechcraft Kingair, un concept devanu produit. *Proc. Active Control Conference*, Cenlis.

Bjarnason E. (1992) Active noise cancellation using a modified form of the filtered-x LMS algorithm. *Proceedings of Eusipco 92, 6th European Signal Processing Conference,* 1053–1056.

Black M.W. (1934) Stabilised feedback amplifiers. *Bell Systems Technical Journal*, **13**, 1–18.

Blondel L.A. and Elliott S.J. (1999) Electropneumatic transducers as secondary actuators for active noise control. Part I Theoretical analysis; Part II Experimental analysis of the subsonic source; Part III Experimental control in ducts with the subsonic source. *Journal of Sound and Vibration*, **219**, 405–427, 429–449, 451–481.

Bode H.W. and Shannon C.E. (1950) A simplified derivation of linear least square smoothing and prediction theory. *Proc. IRE*, **38**, 417–425.

Bode M.W. (1945) *Network Analysis and Feedback Amplifier Design*. Van Nostrand.

Bodson M. and Douglas S.C. (1997) Adaptive algorithms for the rejection of sinusoidal disturbances with unknown frequency. *Automatica*, **33**(12), 2213–2221.

Bongiorno J.J. (1969a) Minimum sensitivity design of linear multivariable feedback control systems by matrix spectral factorisation. *IEEE Transactions on Automatic Control*, **AC-14**, 665–673.

Bongiorno J.J. (1969b) Minimum sensitivity design of linear multivariable feedback control systems by matrix spectral factorisation. *IEEE Transactions on Automatic Control*. **AC-14**, 665–673.

Borchers I.U. et al. (1994) Advanced study of active noise control in aircraft (ASANCA), in *Advances in Acoustics Technology* (ed. J.M.M. Hernandez). Wiley.

Borgiotti G.V. (1990). The power radiated by a vibrating body in an acoustic fluid and its determination from boundary measurements. *Journal of the Acoustical Society of America*, **88**, 1884–1893.

Bouchard M. and Paillard B. (1996) A transform domain optimisation to increase the convergence time of the multichannel filtered-x least-mean-square algorithm. *Journal of the Acoustical Society of America*, **100**, 3203–3214.

Bouchard M. and Paillard B. (1998) An alternative feedback structure for the adaptive active control of periodic and time-varying periodic disturbances. *Journal of Sound and Vibration*, **210**, 517–527.

Bouchard M., Paillard B. and Le Dinh C.T. (1999) Improved training of neural networks for the nonlinear active control of sound and vibration. *IEEE Transactions on Neural Networks*, **10**, 391–401.

Boucher C.C. (1992) *The behaviour of multichannel active control systems for the control of periodic sound.* PhD thesis, University of Southampton.

Boucher C.C., Elliott S.J. and Nelson P.A. (1991) The effect of errors in the plant model on the performance of algorithms for adaptive feedforward control. *Proceedings IEE-F*, **138**, 313–319.

Boucher C.C., Elliott S.J. and Baek K-H. (1996) Active control of helicopter rotor tones. *Proceedings of InterNoise 96*, 1179–1182.

Boyd S.P. and Barratt C.H. (1991) *Linear Controller Design, Limits of Performance*. Prentice Hall.

Boyd S.P. and Vandenberghe L. (1995) *Introduction to convex optimisation with engineering applications*. Lecture notes for EE392X, Electrical Engineering Department, Stanford University.

Boyd S.P., Balakrishnan V., Barratt C.H., Khraishi N.M., Li X., Meyer G., and Norman S.A. (1988) A new CAD method and associated architectures for linear controllers. *IEEE Transactions on Automatic Control*, **33**, 268–283.

Braiman Y. and Goldhirsch I. (1991) Taming chaotic dynamics with weak periodic perturbations. *Physical Review Letters*, **66**(20), 2545–2548.

Brammer A.J., Pan G.J. and Crabtree R.B. (1997) Adaptive feedforward active noise reduction headset for low frequency noise. *Proc. ACTIVE97*, Budapest, 365–372.

Bravo T. and Elliott S.J. (1999) The selection of robust and efficient transducer locations for active control. *ISVR Technical Memorandum No. 843.*

Brennan M.J., Pinnington R.J. and Elliott S.J. (1994) Mechanisms of noise transmission through gearbox support struts. *Transactions of the American Society of Mechanical Engineering Journal of Vibration and Acoustics*, **116**, 548–554.

Brennan M.J., Elliott S.J. and Pinnington R.J. (1995) Strategies for the active control of flexural vibration on a beam. *Journal of Sound and Vibration*, **186**, 657–688.

Brennan M.J., Elliott S.J. and Pinnington R.J. (1996) A non-intrusive fluid-wave actuator and sensor pair for the active control of fluid-borne vibrations in a pipe. *Smart Materials and Structures*, **5**, 281–296.

Bronzel M. (1993) *Aktive Beeinflussung nicht-stationärer schallfelder mid adaptiven Digitalfiltern.* PhD thesis, University of Göttingen, Germany.

Broomhead D.S. and King G.P. (1986) Extracting qualitative dynamics from experimental data. *Physica D*, **20**, 217–236.

Broomhead D.S. and Lowe D. (1988) Multivariable function interpolation and adaptive networks. *Complex Systems* **2**, 321–355.

Brown M. and Harris C.J. (1995) *Neuro-Fuzzy Adaptive Modelling and Control.* Prentice Hall.

Buckingham M.J. (1983) *Noise in Electronic Devices and Systems.* Ellis Horwood.

Bullmore A.J., Nelson P.A., Elliott S.J., Evers J.F. and Chidley B. (1987) Models for evaluating the performance of propeller aircraft active noise control systems. *Proceedings of the AIAA 11th Aeroacoustics Conference*, Palo Alto, CA, Paper AIAA-87-2704.

Burdisso R.A., Thomas R.H., Fuller C.R. and O'Brien W.F. (1993) Active control of radiated inlet noise from turbofan engines. *Proc. Second Conference on Recent Advances in the Active Control of Sound and Vibration*, 848–860.

Burgess J.C. (1981) Active adaptive sound control in a duct: a computer simulation. *Journal of the Acoustical Society of America*, **70**, 715–726.

Cabell R.H. (1998) *A principle component algorithm for feedforward active noise and vibration control*. PhD thesis, Virginia Tech, Blacksburg.

Caraiscos C. and Liu B. (1984) A roundoff error analysis of the LMS adaptive algorithms, *IEEE Transactions on Acoustics Speech and Signal Processing*, **ASSP-32**, 34–41.

Carme C. (1987) *Absorption acoustique active dans les cavites*. These presentée pour obtenir le titre of Docteur de l'Université d'Aix-Marzeille II, Faculte des Sciences de Luminy.

Carneal J.P. and Fuller C.R. (1995) A biologically inspired controller. *Journal of the Acoustical Society of America*, **98**, 386–396.

Casali D.G. and Robinson G.S. (1994) Narrow-band digital active noise reduction in a siren-cancelling headset: real-ear and acoustic manikin insertion loss. *Noise Control Engineering Journal*, **42**, 101–115.

Casavola A. and Mosca E. (1996) LQG multivariable regulation and tracking problems for general system configurations. *Polynomial Methods for Control Systems Design*, (ed. M.J. Grimble and V. Kučera). Springer.

Cerny V. (1985) Thermodynamical approach to the travelling salesman problem: an efficient simulation algorithm. *Journal of Optimisation Theory and Applications*, **45**, 41–51.

Chaplin G.B.B. (1983) Anti-sound—the Essex breakthrough. *Chartered Mechanical Engineer*, **30**, 41–47.

Chen G. and Dong X. (1993) From chaos to order—perspectives and methodologies in controlling chaotic nonlinear dynamical systems. *International Journal of Bifurcation and Chaos*, **3**(6), 1363–1409.

Chen G-S, Bruno R.J. and Salama M. (1991) Optimal placement of active/passing members in truss structures using simulated annealing. *AIAA Journal*, **29**, 1327–1334.

Chen S. and Billings S.A. (1989) Representations of nonlinear systems: the NARMAX model. *International Journal of Control* **49**, 1013–1032.

Chen S., Billings S.A. and Grant P.M. (1992) Recursive hybrid algorithms for non-linear system identification using radial basis function networks. *International Journal of Control*, **55**(5), 1051–1070.

Cioffi J.M. (1987) Limited-precision effects in adaptive filtering. *IEEE Transactions on Circuits and Systems*, **CAS-72**, 821–833.

Clark G.A., Mitra S.K. and Parker S.R. (1980) Block adaptive filtering. *Proc. IEEE Int. Symp. Circuits and Systems*, 384–387.

Clark G.A., Mitra S.K. and Parker S.R. (1981) Block implementation of adaptive digital filters. *IEEE Transactions on Circuits and Systems*, **CAS-28**, 584–592.

Clark R.L. (1995) Adaptive feedforward modal space control. *Journal of the Acoustic Society of America*, **98**, 2639–2650.

Clark R.L. and Bernstein D.S. 1998. Hybrid control: separation in design. *Journal of Sound and Vibration*, **214**, 784–791.

Clark R.L. and Fuller C.R. (1992) Optimal placement of piezoelectric actuators and polyvinylidiene floride sensors in active structural acoustic control approaches. *Journal of the Acoustical Society of America*, **92**, 1521–1533.

Clark R.L., Saunders W.R. and Gibbs G.P. (1998) *Adaptive Structures, Dynamics and Control*. Wiley.

Clarkson P.M. (1993) *Optimal and Adaptive Signal Processing*. CRC Press.

Coleman R.B. and Berkman E.F. (1995) Probe shaping for on-line plant identification. *Proc. ACTIVE 95*, 1161–1170.

Concilio A., Lecce L. and Ovallesco A. (1995) Position and number optimisation of actuators and sensors in an active noise control system by genetic algorithms. *Proc. CEAS/AIAA Aeroacoustics Conference 95–084*, 633–642.

Conover W.B. (1956) Fighting noise with noise. *Noise Control*. **2**, 78–82.

Cook J.G. and Elliott S.J. (1999) Connection between multichannel prediction error filter and spectral factorisation. *Electronics Letters*, **35**, 1218–1220.

Crabtree R.B. and Rylands J.M. (1992) Benefits of active noise reduction to noise exposure in high-risk environments. *Proc. Inter-Noise '92*, 295–298.

Crawford D.H. and Stewart R.W. (1997) Adaptive IIR filtered-*v* algorithms for active noise control. *Journal of the Acoustical Society of America*, **101**, 2097–2103.

Crawford D.H., Stewart R.W. and Toma E. (1996) A novel adaptive IIR filter for active noise control. *Proc. Int. Conf. on Acoustics, Speech and Signal Processing ICASSP96*, **3**, 1629–1632.

Cremer L. and Heckl M. (1988). *Structure-Borne Sound*, 2nd edn. (trans. E.E. Ungar). Springer-Verlag.

Cunefare K.A. (1991). The minimum multimodal radiation efficiency of baffled finite beams. *Journal of the Acoustical Society of America*, **90**, 2521–2529.

Curtis A.R.D., Nelson P.A., Elliott S.J. and Bullmore A.J. (1987). Active suppression of acoustic resonance. *Journal of the Acoustical Society of America*, **81**, 624–631.

Darbyshire E.P. and Kerry C.J. (1996) A real-time computer for active control. *GEC Journal of Research*, **13**, 138–145.

Darlington P. (1987) *Applications of adaptive filters in active noise control*. PhD thesis, University of Southampton.

Darlington P. (1995) Performance surfaces of minimum effort estimators and controllers. *IEEE Transactions on Signal Processing*, **43**, 536–539.

Darlington P. and Nicholson G.C. (1992) Theoretical and practical constraints on the implementation of active acoustic boundary elements. *Proc. 2nd Inc. Congress on Recent Developments in Air and Structure-borne Sound and Vibration*.

Datta A. (1998) *Adaptive Internal Model Control*. Springer.

Datta A. and Ochoa J. (1996) Adaptive internal model control: Design and stability analysis. *Automatica*, **32**, 261–266.

Datta A. and Ochoa J. (1998) Adaptive internal model control: H_2 optimisation for stable plant. *Automatica*, **34**, 75–82.

Datta B.N. (1995) *Numerical Linear Algebra and Applications*. Brooks/Cole Publishing Co.

Davis M.C. (1963) Factoring the spectral matrix. *IEEE Transactions on Automatic Control*, **AC-8**, 296–305.

Dehandschutter W., Herbruggen J.V., Swevers J. and Sas P. (1998) Real-time enhancement of reference signals for feedforward control of random noise due to multiple uncorrelated sources. *IEEE Transactions on Signal Processing*, **46**(1), 59–69.

Ditto W.L., Rauseo S.N. and Spano M.L. (1990) Experimental control of chaos. *Physical Review Letters*, **65**(26), 3211–3214.

Doelman N. (1993) *Design of systems for active sound control*. PhD thesis, Technische Universiteit Delft, The Netherlands.

Doelman N.J. (1991) A unified control strategy for the active reduction of sound and vibration. *Journal of Intelligent Materials, Systems and Structures*, **2**, 558–580.

Dorf R.C. (1990) *Modern Control Systems*. 6th edn. Addison-Wesley.

Dorling C.M., Eatwell G.P., Hutchins S.M., Ross C.F. and Sutcliffe S.G.C. (1989) A demonstration of active noise reduction in an aircraft cabin. *Journal of Sound and Vibration*, **128**, 358–360.

Dougherty K.M. (1995) *Analogue-to-Digital Conversion*. McGraw-Hill.

Douglas S.C. (1997) Adaptive filters employing partial updates. *IEEE Transactions on Circuits and Systems II Analog and Digital Signal Processing*, **44**, 209–216.

Douglas S.C. (1999) Fast implementations of the filtered-x LMS and LMS algorithms for multichannel active noise control. *IEEE Transactions on Speech and Audio Processing*, **7**(4), 454–465.

Douglas S.C. and Olkin J.A. (1993) Multiple-input multiple-error adaptive feedforward control using the filtered-*z* normalised LMS algorithm. *Proc. Second Conference on Recent Advances in the Active Control of Sound and Vibration*, 743–754.

Doyle J.C. (1983) Synthesis of robust controllers and filters. *Proc. IEEE Conference on Decision and Control*, 109–114.

Doyle J.C. and Stein G. (1979) Robustness with observers. *IEEE Transactions on Automatic Control*, **AC-24**, 607–611.

Doyle J.C., Francis B.A. and Tannenbaum A.R. (1992) *Feedback Control Theory*. Maxwell MacMillan International.

Eatwell G.P. (1995) Tonal noise control using harmonic filters. *Proceedings of ACTIVE'95*, Newport Beach, CA, USA, 1087–1096.

EDN (1999) *Electronic Design News*. www.ednmag.com.

Elk A. (1972) quoted in *Monty Python's Flying Circus, Just the Words*, Volume II. Methuen (1989).

Elliott S.J. (1992) DSP in the active control of sound and vibration. *Proc. Audio Engineering Soc.* UK DSP Conference, London, 196–209.

Elliott S.J. (1993) Active control of structure-borne sound. *Proceedings of the Institute of Acoustics*. **15**, 93–120.
Elliott S.J. (1994). Active control of sound and vibration. *Proc. ISMA19*.
Elliott S.J. (1998a) Active noise and vibration control. *Journal of Applied Mathematics and Computer Science*, **8**, 213–251.
Elliott S.J. (1998b) Adaptive methods in active control. *Proc. MOVIC '98*, Zurich, 41–48.
Elliott S.J. (1998c) Filtered reference and filtered error LMS algorithms for adaptive feedforward control. *Mechanical Systems and Signal Processing*, **12**, 769–781.
Elliott S.J. (1999) Active control of nonlinear systems. *Proc. ACTIVE99*, 3–44.
Elliott S.J. (2000) Optimum controllers and adaptive controllers for multichannel active control of stochastic disturbances. *IEEE Trans. Signal Processing*, **48**, 1053–1060.
Elliott S.J. and Baek K-H. (1996) Effort constraints in adaptive feedforward control. *IEEE Signal Processing Letters*, **3**, 7–9.
Elliott S.J. and Billet L. (1993) Adaptive control of flexural waves propagating in a beam. *Journal of Sound and Vibration*, **163**, 295–310.
Elliott S.J. and Boucher C.C. (1994) Interaction between multiple feedforward active control systems. *IEEE Transactions on Speech and Audio Processing*, **2**, 521–530.
Elliott S.J. and Darlington P. (1985) Adaptive cancellation of periodic, synchronously sampled interference. *IEEE Transactions on Acoustics Speech and Signal Processing*, **ASSP-33**, 715–717.
Elliott S.J. and David A. (1992) A virtual microphone arrangement for local active sound control. *Proc. 1st Int. Conf. on Motion and Vibration Control (MOVIC)*, 1027–1031.
Elliott S.J. and Johnson M.E. (1993). Radiation modes and the active control of sound power. *Journal of the Acoustical Society of America*, **94**, 2194–2204.
Elliott S.J. and Nelson P.A. (1984) Models for describing active noise control in ducts. *ISVR Technical Report No. 127*.
Elliott S.J. and Nelson P.A. (1985a) Algorithm for multichannel LMS adaptive filtering. *Electronics Letters*, **21**, 979–981.
Elliott S.J. and Nelson P.A. (1985b) The active minimisation of sound fields. *Proc. InterNoise85*, 583–586.
Elliott S.J. and Nelson P.A. (1986) Algorithms for the active control of periodic sound and vibration. *ISVR Technical Memorandum No. 679*, University of Southampton.
Elliott S.J. and Nelson P.A. (1989) Multiple point equalisation in a room using adaptive digital filters. *Journal of the Audio Engineering Society*. **37**(11), 899–908.
Elliott S.J. and Nelson P.A. (1993) Active noise control. *IEEE Signal Processing Magazine*, October 1993, 12–35.
Elliott S.J. and Rafaely B. (1997) Frequency-domain adaptation of feedforward and feedback controllers. *Proc. ACTIVE97*, 771–788.
Elliott S.J. and Rafaely B. (2000) Frequency domain adaptation of causal digital filters. *IEEE Transactions on Signal Processing*, **48**, 1354–1364.
Elliott S.J. and Rex J. (1992) Adaptive algorithms for underdetermined active control problems. *Proc. Int. Conf. on Acoustics, Speech and Signal Processing ICASSP92*, 237–240.
Elliott S.J. and Sutton T.J. (1996) Performance of feedforward and feedback systems for active control. *IEEE Transactions on Speech and Audio Processing*, **4**, 214–223.
Elliott S.J., Stothers I.M. and Nelson P.A. (1987) A multiple error LMS algorithm and its application to the active control of sound and vibration. *IEEE Transactions on Acoustics, Speech and Signal Processing*, **ASSP-35**, 1423–1434.
Elliott S.J., Joseph P., Bullmore A.J. and Nelson P.A. (1988). Active cancellation at a point in a pure tone diffuse field. *Journal of Sound and Vibration*, **120**, 183–189.
Elliott S.J., Nelson P.A., Stothers I.M. and Boucher C.C. (1989) Preliminary results of in-flight experiments on the active control of propeller-induced cabin noise. *Journal of Sound and Vibration*, **128**, 355–357.
Elliott S.J., Nelson P.A., Stothers I.M. and Boucher C.C. (1990) In-flight experiments on the active control of propeller-induced cabin noise. *Journal of Sound and Vibration*, **140**, 219–238.
Elliott S.J., Joseph P., Nelson P.A. and Johnson M.E. (1991). Power output minimisation and power absorption in the active control of sound. *Journal of the Acoustical Society of America*, **90**, 2501–2512.

Elliott S.J., Boucher C.C. and Nelson P.A. (1992) The behaviour of a multiple channel active control system. *IEEE Transactions on Signal Processing,* **40**, 1041–1052.

Elliott S.J., Sutton T.J., Rafaely B. and Johnson M. (1995) Design of feedback controllers using a feedforward approach. *Proc. Int. Symp. on Active Control of Sound and Vibration, ACTIVE 95*, 561–572.

Elliott S.J., Boucher C.C. and Sutton T.J. (1997) Active control of rotorcraft interior noise. *Proceedings of the Conference on Innovations in Rotorcraft Technology*, Royal Aeronautical Society, London, 15.1–15.6.

Emborg U. and Ross C.F. (1993) Active control in the SAAB 340, *Proceedings of Recent Advances in the Active Control of Sound and Vibration*, 567–573.

Eriksson L.J. (1991a) Development of the filtered-*u* algorithm for active noise control. *Journal of the Acoustical Society of America*, **89**, 257–265.

Eriksson L.J. (1991b) Recursive algorithms for active noise control. *Proc. Int. Symposium on Active Control of Sound and Vibration*, Tokyo, 137–146.

Eriksson L.J. and Allie M.C. (1989) Use of random noise for on-line transducer modelling in an adaptive active attenuation system. *Journal of the Acoustical Society of America*, **85**, 797–802.

Eriksson L.J., Allie M.C. and Greiner R.A. (1987) The selection and application of an IIR adaptive filter for use in active sound attenuation. *IEEE Transactions on Acoustics, Speech and Signal Processing*, **ASSP-35**, 433–437.

Eriksson L.J., Allie M.C., Hoops R.H. and Warner J.V. (1989) Higher order mode cancellation in ducts using active noise control. *Proc. InterNoise 89*, 495–500.

Fedorynk M.V. (1975) The suppression of sound in acoustic waveguides. *Soviet Physics Acoustics*, **21**, 174–176.

Feintuch P.L. (1976) An adaptive recursive LMS filter. *Proc IEEE,* **64**, 1622–1624.

Ferrara E.R. (1980) Fast implementation of LMS adaptive filters. *IEEE Transactions on Acoustics, Speech and Signal Processing*, **ASSP-28**, 474–475.

Ferrara E.R. (1985) Frequency-domain adaptive filtering. *Adaptive Filters* (ed. C.F.N. Cowan and P.M. Grant). Prentice Hall, 145–179.

Feuer A. and Cristi R. (1993) On the steady state performance of frequency domain LMS algorithms. *IEEE Transactions on Signal Processing*, **41**, 419–423.

Ffowcs-Williams J.E. (1984) Review Lecture: Anti-Sound. *Proceedings of the Royal Society of London*, **A395**, 63–88.

Ffowcs-Williams J.E. (1987) The aerodynamic potential of anti-sound. *Journal of Theoretical and Applied Mechanics*. Special Supplement to Volume **6**, 1–21.

Ffowcs-Williams J.E., Roebuck I. and Ross C.F. (1985) Antiphase noise reduction, *Physics in Technology*, **6**, 19–24.

Fletcher R. (1987) *Practical Methods of Optimisation*. Wiley.

Flockton S.J. (1991) Gradient-based adaptive algorithms for systems with external feedback paths. *IEE Proceedings F*, **138**, 308–312.

Flockton S.J. (1993) Fast adaptation algorithms in active noise control. *Proc. Second Conference on Recent Advances in the Active Control of Sound and Vibration,* 802–810.

Fogel D.B. (2000) What is evolutionary computation? *IEEE Spectrum Magazine*, **37**(2), 26–32.

Forsythe S.E., McCollum M.D. and McCleary A.D. (1991) Stabilization of a digitally controlled active-isolation system. *Proc. First Conference on Recent Advances in Active Control of Sound and Vibration*, Virginia, 879–889.

Franklin G.F., Powell J.D. and Workman M.L. (1990) *Digital Control of Dynamic Systems*, 2nd edn. Addison-Wesley.

Franklin G.F., Powell J.D. and Emani-Naeini A. (1994) *Feedback Control of Dynamic Systems,* 3rd edn. Addison-Wesley.

Freudenberg J.S. and Looze D.P. (1985) Right hand plane poles and zeros and design trade-offs in feedback systems. *IEEE Transactions on Automatic Control*, **AC-30**, 555–565.

Friedlander B. (1982) Lattice filters for adaptive processing. *Proc. IEEE*, **70**, 829–867.

Fuller C.R. (1985) Experiments on reduction of aircraft interior noise using active control of fuselage vibration. *Journal of the Acoustical Society of America*, **78** (S1), S88.

Fuller C.R. (1988). Analysis of active control of sound radiation from elastic plates by force inputs. *Proceedings of Inter-Noise '88*, Avignon, **2**, 1061–1064.

Fuller C.R. (1997) Active control of cabin noise—lessons learned? *Proc. Fifth International Congress on Sound and Vibration*, Adelaide.

Fuller C.R., Cabell R.H., Gibbs G.P. and Brown D.E. (1991) A neural network adaptive controller for nonlinear systems. *Proc. Internoise 91*, 169–172.

Fuller C.R., Maillard J.P., Meradal M. and von Flotow A.H. (1995) Control of aircraft interior noise using globally detuned vibration absorbers. *Proceedings of the First Joint CEAS/AIAA Aeroacoustics Conference*, Munich, Germany, Paper CEAS/AIAA-95–082, 615–623.

Fuller C.R., Elliott S.J. and Nelson P.A. (1996) *Active Control of Vibration*. Academic Press.

Furuya H. and Haftka R.T. (1993) Genetic algorithms for placing actuators on space structures. *Proc. 5th Int. Conf. on Genetic Algorithms*, 536–543.

Gabor D., Wilby W.P.L. and Woodcock R. (1961) A universal non-linear filter, predictor and simulator which optimises itself by a learning process. *Proc. IEE(B)*, **108**, 422–439.

Gao F.X. and Snelgrove W.M. (1991) Adaptive linearisation of a loudspeaker. *Proc. Int. Conf. on Acoustics, Speech and Signal Processing* (ICASSP), 3589–3592.

Garcia-Bonito J. and Elliott S.J. (1996). Local active control of vibration in a diffuse bend wave field. *ISVR Technical Memorandum No. 790.*

Garcia-Bonito J., Elliott S.J. and Boucher C.C. (1997) Generation of zones of quiet using a virtual microphone arrangement. *Journal of the Acoustical Society of America*, **101**, 3498–3516.

Garfinkel A., Spano M.L., Ditto W.L. and Weiss J.N. (1992) Controlling cardiac chaos. *Science*, **257**, 1230–1235.

Gitlin R.D., Mazo J.E. and Taylor M.G. (1973) On the design of gradient algorithms for digitally implemented adaptive filters. *IEEE Transactions on Circuit Theory*, **CT-20**, 125–136.

Glentis G-O., Berberidis K. and Theodoridis S. (1999) Efficient least squares adaptive algorithms for FIR transversal filtering. *IEEE Signal Processing Magazine*, July 1999.

Glover J.R. (1977) Adaptive noise cancellation applied to sinusoidal noise interferences. *IEEE Transactions on Acoustics, Speech and Signal Processing*, **ASSP-25**, 484–491.

Goldberg D.E. (1989) *Genetic Algorithms in Search, Optimisation and Machine Learning*. Addison Wesley.

Golnaraghi M.F. and Moon F.C. (1991) Experimental evidence for chaotic response in a feedback system. *Journal of Dynamic Systems, Measurement and Control*, **113**, 183–187.

Golub G.H. and Van Loan C.F. (1996) *Matrix Computations*, 3rd edn. The Johns Hopkins University Press.

Gonzalez A. and Elliott S.J. (1995) Adaptive minimisation of the maximum error signal in an active control system. *IEEE Workshop in Applications of Signal Processing to Audio and Acoustics*, Mohonk, New Paltz, New York.

Gonzalez A, Albiol A. and Elliott S.J. (1998) Minimisation of the maximum error signal in active control. *IEEE Transactions on Speech and Audio Processing*, **6**, 268–281.

Goodman S.D. (1993) Electronic design considerations for active noise and vibration control systems, *Proc. Second Conference on Recent Advances in Active Control of Sound and Vibration*, 519–526.

Goodwin G.C. and Sin K.S. (1984) *Adaptive Filtering Prediction and Control*. Prentice Hall.

Gray R.M. (1972) On the asymptotic eigenvalue distribution of Toeplitz matrices. *IEEE Transactions on Information Theory*, **IT-18**, 725–730.

Grimble M.J. and Johnson M.A. (1988) *Optimal Control and Stochastic Estimation*, Vols 1 and 2. Wiley.

Gudvangen S. and Flockton S.J. (1995) Modelling of acoustic transfer functions for echo cancellers. *IEE Proc. Vision, Image and Signal Processing*, **142**, 47–51.

Guicking D. and Karcher K. (1984). Active impedance control for one-dimensional sound. *American Society of Mechanical Engineers Journal of Vibration, Acoustics, Stress and Reliability in Design*, **106**, 393–396.

Guigou C. and Berry A. (1993) *Design strategy for PVDF sensors in the active control of simply supported plates*. Internal Report, GAUS, Dept. of Mechanical Engineering, Sherbrooke University.

Gustavsson I, Ljung L. and Söderström T. (1977). Survey paper: Identification of processes in closed loop—identifiability and accuracy aspects. *Automatica*, **13**, 59–75.

Hać́ A. and Liu L. (1993) Sensor and actuator locations in motion control of flexible structures. *Journal of Sound and Vibration*, **167**, 239–261.

Hakim S. and Fuchs M.B. (1996) Quasistatic optimal actuator placement with minimum worst case distortion criterion. *Journal of the American Institute of Aeronautics and Astronautics*, **34**(7), 1505–1511.

Hamada H. (1991) Signal processing for active control—Adaptive signal processing. *Proc. Int. Symp. on Active Control of Sound and Vibration*, Tokyo, 33–44.

Hamada H. Takashima N. and Nelson P.A. (1995) Genetic algorithms used for active control of sound—search and identification of noise sources. *Proc. ACTIVE95, the 1995 Int. Symp. on Active Control of Sound and Vibration*, 33–37.

Hansen C.H. and Snyder S.D. (1997) *Active Control of Noise & Vibration*. E&FN Spon.

Harp S.A. and Samad T. (1991) Review of Goldberg (1989). *IEEE Transactions on Neural Networks*, **2**, 542–543.

Harteneck M. and Stewart R.W. (1996) A fast converging algorithm for Landau's output-error method. *Proc. MMAR*, Poland, 625–630.

Hauser M.W. (1991) Principles of oversampling A/D conversion. *Journal of the Audio Engineering Society*, **39**, 3–26.

Haykin S. (1996) *Adaptive Filter Theory*, 3rd edn. Prentice Hall.

Haykin S. (1999) *Neural Networks, A Comprehensive Foundation.* 2nd edn. Macmillan.

Heatwole C.M. and Bernhard R.J. (1994) The selection of active noise control reference transducers based on the convergence speed of the LMS algorithm. *Proc. Inter-Noise '94*, 1377–1382.

Heck L.P. (1995) Broadband sensor and actuator selection for active control of smart structures. *Proc. SPIE Conference on Mathematics and Control in Smart Structures*, Vol. 2442, 292–303.

Heck L.P. (1999) Personal communication.

Heck L.P., Olkin J.A. and Naghshineh K. (1998) Transducer placement for broadband active vibration control using a novel multidimensional QR factorization. *ASME Journal of Vibration and Acoustics*, **120**, 663–670.

Hermanski M., Kohn K-U and Ostholt H. (1995) An adaptive spectral compensation algorithm for avoiding flexural vibrations of printing cylinders. *Journal of Sound and Vibration*, **187**, 185–193.

Hon J. and Bernstein D.S. (1998) Bode integral constraints, collocation and spillover in active noise and vibration control. *IEEE Transactions on Control Systems Technology*, **6**(1), 111–120.

Horowitz P. and Hill W. (1989) *The Art of Electronics*, 2nd edn. Cambridge University Press.

Hunt K.J. and Sbarbaro D. (1991) Neural networks for nonlinear internal model controller. *IEE Proceedings-D*, **138**, 431–438.

Hush D.R. and Horne B.G. (1993) Progress in supervised neural networks. *IEEE Signal Processing Magazine*, January 1993, 8–38.

IEC Standard 651 (1979) Sound level meters.

IEEE (1987) *Proceedings of the First International Conference on Neural Networks.*

Imai H. and Hamada H. (1995) Active noise control system based on the hybrid design of feedback and feedforward control. *Proc. ACTIVE95*, 875–880.

Jessel M.J.M. (1968) Sur les absorbeurs actifs. *Proc. 6th International Conference on Acoustics*, Tokyo. Paper F-5–6, 82.

Johnson C.R. Jr. (1988) *Lectures on Adaptive Parameter Estimation*. Prentice Hall.

Johnson C.R. Jr. (1995) On the interaction of adaptive filtering, identification and control. *IEEE Signal Processing Magazine*, 12(2), 22–37.

Johnson C.R. Jr. and Larimore M.G. (1997) Comments on and additions to 'An adaptive recursive LMS filter'. *Proc. IEEE*, **65**, 1399–1401.

Johnson M.E. and Elliott S.J. (1993). Volume velocity sensors for active control. *Proceedings of the Institute of Acoustics*, **15**(3), 411–420.

Johnson M.E. and Elliott S.J. (1995a) Experiments on the active control of sound radiation using volume and velocity sensor. *Proc. SPIE, North American Conference on Smart Structures and Materials*, **2443**, 658–669.

Johnson M.E. and Elliott S.J. (1995b) Active control of sound radiation using volume velocity cancellation. *Journal of the Acoustical Society of America*, **98**, 2174–2186.

Jolly M.R. (1995) On the calculus of complex matrices. *Int. Journal of Control*, **61**, 749–755.

Jolly M.R. and Rossetti D.J. (1995) The effects of model error on model-based adaptive control systems. *Proceedings of ACTIVE'95*, Newport Beach, CA, 1107–1116.

Joseph P., Nelson P.A. and Fisher M.J. (1996) An in-duct sensor array for the active control of sound radiated by circular flow ducts. *Proc. InterNoise 96*, 1035–1040.

Joseph P., Nelson P.A. and Fisher M.J. (1999) Active control of fan tones radiated from turbofan engines. I: External error sensors, and, II: In-duct error sensors. *Journal of the Acoustical Society of America*, **106**, 766–778 and 779–786.

Kabal P. (1983) The stability of adaptive minimum mean-square error equalisers using delayed adjustment. *IEEE Transactions on Communications*, **COM-31**, 430–432.

Kadanoff L.P. (1983) Roads to chaos. *Physics Today*, December 1983, 46–53.

Kailath T. (1974) A view of three decades of linear filter theory. *IEEE Transactions on Information Theory*, **IT-20**, 146–181.

Kailath T. (1981) *Lectures on Wiener and Kalman Filtering*. Springer.

Kailath T., Sayed A.M. and Hassibi B. (2000) Linear estimation, Prentice Hall.

Kawato M., Uno Y., Isobe M. and Suzuki R. (1988) Hierarchical neural network model for voluntary movements with application to robotics. *IEEE Control Systems Magazine*, April 1988, 8–16.

Keane A.J. (1994) Experiences with optimisers in structural design. *Proc. Conference on Adaptive Computing in Engineering Design and Control.*

Kewley D.L., Clark R.L. and Southwood S.C. (1995) Feedforward control using the higher-harmonic, time-averaged gradient descent algorithm. *Journal of the Acoustical Society of America*, **97**, 2892–2905.

Kim I-S, Na H-S, Kim K-J and Park Y. (1994) Constraint filtered-x and filtered-u least-mean-square algorithms for the active control of noise in ducts. *Journal of the Acoustical Society of America* **95**, 3379–3389.

Kinsler L.E., Frey A.R., Coppens A.B. and Sanders J.V. (1982) *Fundamentals of Acoustics*, 3rd edn. Wiley.

Kirkeby O., Nelson P.A., Hamada H. and Orduna-Bustamante F. (1998) Fast deconvolution of multichannel systems using regularisation. *IEEE Transactions on Speech and Audio Processing*, **6**, 189–194.

Kirkpatrick S., Gelatt C.D. and Vecchi M.P. (1983) Optimisation by simulated annealing. *Science*, **220**, 671–680.

Kittel C. and Kroemer H. (1980) *Thermal Physics,* 2nd edn. W.H. Freeman and Co.

Klippel W.J. (1995) Active attenuation of nonlinear sound. *Proc. ACTIVE95*, Newport Beach, 413–422.

Klippel W.J. (1998) Adaptive nonlinear control of loudspeaker systems. *Journal of the Audio Engineering Society*, **46**, 939–954.

Knyazev A.S. and Tartakovskii B.D. (1967). Abatement of radiation from flexurally vibrating plates by means of active local vibration dampers. *Soviet Physics Acoustics*, **13**, 115–116.

Konaev S.I., Lebedev V.I. and Fedorynk M.V. (1977) Discrete approximations of a spherical Huggens surface. *Soviet Physics Acoustics*, **23**, 373–374.

Kučera V. (1993a) Diophantine equations in control—a survey. *Automatica*, **29**, 1361–1375.

Kučera V. (1993b) The algebraic approach to control system design. *Polynomial Methods in Optimal Control and Filtering*. (ed. K.J.Hunt). Peter Peregrinus Ltd.

Kuo B.C. (1980) *Digital Control Systems*. Holt, Rinehart and Wilson.

Kuo S.M. and Morgan D.R. (1996) *Active Noise Control Systems, Algorithms and DSP Implementations*. Wiley.

Kwakernaak H. and Sivan R. (1972) *Linear Optimal Control Systems*. Wiley.

Lammering R., Jianhu J. and Rogers C.A. (1994) Optimal placement of piezoelectric actuators in adaptive truss structures. *Journal of Sound and Vibration*, **171**, 67–85.

Landau I.D. (1976) Unbiased recursive identification using model reference adaptive techniques. *IEEE Transactions on Automatic Control*, **AC-21**, 194–202.

Langley A. (1997) Personal communication.

Laugesen S. (1996) A study of online plant modelling methods for active control of sound and vibration. *Inter-Noise 96*, 1109–1114.

Lee C.K. and Moon F.C. (1990). Modal sensors/actuators. *American Society of Mechanical Engineers Journal of Applied Mechanics*, **57**, 434–441.

Levinson N. (1947) The Wiener rms (root mean-square) error criterion in filter design and prediction. *Journal of Mathematical Physics*, **25**, 261–278.

Liavas A.P. and Regalia P.A. (1998) Acoustic echo cancellation: do IIR models offer better modelling capabilities than their FIR counterparts? *IEEE Transactions on Signal Processing*, **46**, 2499–2504.

Lin S. (1965) Computer solutions to the travelling salesman problem. *Bell System Technical Journal*, **44**, 2245–2269.

Lippman R.P. (1987) An introduction to computing with neural networks. *IEEE Signal Processing Magazine*, April 1987, 4–22.

Ljung L. (1999) *System Identification: Theory for the User*, 2nd edn. Prentice Hall.

Ljung L. and Sjöberg J. (1992) A system identification perspective on neural nets. *Proc. IEEE Conf. on Neural Networks for Signal Processing II*, Copenhagen, 423–435.

Long G., Ling F. and Proakis J.G. (1989) The LMS algorithm with delayed coefficient adaptation. *IEEE Transactions on Acoustics, Speech and Signal Processing*, **37**, 1397–1405. See also *IEEE Transactions on Signal Processing*, **SP-40**, 230–232.

Lueg P. (1936) Process of silencing sound oscillations. *U.S. Patent*, No. 2,043,416.

Luenberger O.G. (1973) *Introduction to Linear and Nonlinear Programming*. Addison-Wesley.

Macchi O. (1995) *Adaptive Processing, the Least Mean-Squares Approach with Applications in Transmission*. Wiley.

Maciejowski J.M. (1989) *Multivariable Feedback Design*. Addison-Wesley.

Maeda Y. and Yoshida T. (1999) An active noise control without estimation of secondary-path-ANC using simultaneous perturbation. *Proc ACTIVE99*, Fort Lauderdale, FL., 985–994.

Mangiante G. (1994). The JMC method for 3D active absorption: A numerical simulation. *Noise Control Engineering*, **41**, 1293–1298.

Mangiante G. (1996) An introduction to active noise control, *International Workshop on Active Noise and Vibration Control in Industrial Applications*, CETIM, Senlis, France, Paper A.

Marcos S., Macchi O., Vignat C., Dreyfus G., Personnaz L. and Roussel-Ragot P. (1992) A united framework for gradient algorithms used for filter adaptation and neural network training. *International Journal of Circuit Theory and Applications*, **20**, 159–200.

Mareels I.M.Y. and Bitmead R.R. (1986) Non-linear dynamics in adaptive control: chaos and periodic stabilization. *Automatica*, **22**, 641–655.

Markel J.D. and Gray A.H. Jr. (1976) *Linear Prediction of Speech*. Springer.

Martin V. and Benzaria E. (1994) Active noise control: optimisation of secondary source locations (harmonic linear range). *Journal of Low Frequency Noise and Vibration*, **13**, 133–138.

Martin V. and Gronier C. (1998) Minimum attenuation guaranteed by an active noise control system in the presence of errors in the spatial distribution of the primary field. *Journal of Sound and Vibration*, **215**, 827–852.

Marvin C. and Ewers G. (1993) *A Simple Approach to Digital Signal Processing*. Texas Instruments.

Matsuura T., Hiei T., Itoh H. and Torikoshi K. (1995) Active noise control by using prediction of time series data with a neural network. *Proc. IEEE Conf. on Systems, Man and Cybernetics*, 2070–2075.

Maurer M. (1996) *An orthogonal algorithm for the active control of sound and the effects of realistic changes in the plant*. MSc thesis, University of Southampton.

Meeker W.F. (1958) Active ear defender systems: Component considerations and theory. *WADC Technical Report*, **57–368** (I).

Meeker W.F. (1959) Active ear defender systems: Development of a laboratory model. *WADC Technical Report*, **57–368** (II).

Meirovitch L. (1990). *Dynamics and Control of Structures*. Wiley.

Metrolopolis N., Rosenbluth A.W., Rosenbluth M.N., Teller A.H. and Teller E. (1953) Equation of state calculations by fast computing machines. *Journal of Chemical Physics,* **21**, 1087–1092.

Metzer F.B. (1981) Strategies for reducing propeller aircraft cabin noise. *Automotive Engineering*, **89**, 107–113.

Meucci R., Gadomski W., Giofini M. and Arecchi F.T. (1994) Experimental control of chaos by means of weak parametric perturbations. *Physical Review E*, **49**, 2528–2531.

Meurers A.T. and Veres S.M. (1999) Stability analysis of adaptive FSF-based controller tuning. *Proc. ACTIVE99*, 995–1004.

Minkoff J. (1997) The operation of multichannel feedforward adaptive systems. *IEEE Transactions on Signal Processing*, **45**(12), 2993–3005.

Mitchell M. (1996) *An Introduction to Genetic Algorithms*. Cambridge University Press.

Mørkholt J., Elliott S.J. and Sors T.C. (1997) A comparison of state-space LQG, Wiener IMC and polynomial LQG discrete-time feedback control for active vibration control purposes. *ISVR Technical Memorandum No. 823*, University of Southampton.

Moon F.C. (1992) *Chaotic and Fractal Dynamics*. Wiley.

Moon F.C. and Holmes P.J. (1979) A magnetoelastic strange attractor. *Journal of Sound and Vibration*, **65**, 275–296.

Moore D.F. (1975) *The Friction of Pneumatic Tyres*. Elsevier.

Morari M. and Zafiriou E. (1989) *Robust Process Control*. Prentice Hall.

Morgan D.R. (1980) An analysis of multiple correlation cancellation loops with a filter in the auxiliary path. *IEEE Transactions on Acoustics Speech and Signal Processing*, **ASSP-28**, 454–467.

Morgan D.R. (1991a) A hierarchy of performance analysis techniques for adaptive active control of sound and vibration. *Journal of the Acoustical Society of America*, **89**(5), 2362–2369.

Morgan D.R. (1991b) An adaptive modal-based active control system. *Journal of the Acoustical Society of America*, **89**, 248–256.

Morgan D.R. (1995) Slow asymptotic convergence of LMS acoustic echo cancellers, *IEEE Transactions on Speech and Audio Processing*, **SAP-3**, 126–136.

Morgan D.R. (1999) Personal communication.

Morgan D.R. and Sanford C. (1992) A control theory approach to the stability and transient response of the filtered-x LMS adaptive notch filter. *IEEE Transactions on Signal Processing*, **40**, 2341–2346.

Morgan D.R. and Thi J.C. (1995) A delayless subband adaptive filter architecture. *IEEE Trans. Signal Processing*, **43**, 1819–1830.

Morse P.M. (1948) *Vibration and Sound*, 2nd edn. McGraw-Hill (reprinted 1981 by the Acoustical Society of America).

Narayan S.S., Petersen A.M. and Narasimha M.J. (1983) Transform domain LMS algorithm. *IEEE Transactions on Acoustics Speech and Signal Processing*, **ASSP-31**, 609–615.

Narendra K.S. and Parthasarthy K. (1990) Identification and control of dynamic systems using neural networks. *IEEE Transactions on Neural Networks*, **1**(1), 4–27.

Nayroles B., Touzot G. and Villon P. (1994) Using the diffuse approximation for optimising the locations of anti-sound sources. *Journal of Sound and Vibration*, **171**, 1–21.

NCT (1995) 'Noise-buster' active headset literature.

Nelson P.A. (1996) Acoustical prediction. *Proceedings Internoise 96*, 11–50.

Nelson P.A. and Elliott S.J. (1992) *Active Control of Sound*. Academic Press.

Nelson P.A., Curtis A.R.D., Elliott S.J. and Bullmore A.J. (1987). The minimum power output of freefield point sources and the active control of sound. *Journal of Sound and Vibration*, **116**, 397–414.

Nelson P.A., Sutton T.J. and Elliott S.J. (1990) Performance limits for the active control of random sound fields from multiple primary sources. *Proceedings of the Institute of Acoustics*, **12**, 677–687.

Netto S.L., Dines P.S.R. and Agathoklis P. (1995) Adaptive IIR filtering algorithms for system identification: a general framework. *IEEE Transactions on Education*, **E38**, 54–66.

Newland D.E. (1993) *An Introduction to Random Vibrations, Spectral and Wavelet Analysis*, 3rd edn. Longman Scientific and Technical.

Newton G.C., Gould L.A. . and Kaiser J.F. (1957) *Analytical Design of Linear Feedback Controls*. Wiley.

Nguyen D.H. and Widrow B. (1990) Neural networks for self learning control systems. *IEEE Control Systems Magazine*, **10**(3), 18–33.

Noble B. and Daniel J.W. (1977) *Applied Linear Algebra*, 2nd edn, Prentice Hall.

Norton J.P. (1986) *An Introduction to Identification*. Academic Press.

Nowlin W.C., Guthart G.S. and Toth G.K. (2000) Noninvasive system identification for the multi-channel broadband active noise control. *Journal of the Acoustical Society of America*, **107**, 2049–2060.

Nyquist M. (1932) Regeneration theory. *Bell Systems Technical Journal*, **11**, 126–147.

Olson H.F. (1956) Electronic control of noise, reverberation, and vibration. *Journal of the Acoustical Society of America*, **28**, 966–972.

Olsen H.F. and May E.G. (1953). Electronic sound absorber. *Journal of the Acoustical Society of America*, **25**, 1130–1136.

Omoto A. and Elliott S.J. (1996) The effect of structured uncertainty in multichannel feedforward control systems. *Proc. Int. Conf. on Acoustics, Speech and Signal Processing ICASSP96*, 965–968.

Omoto A. and Elliott S.J. (1999) The effect of structured uncertainty in the acoustic plant on multichannel feedforward control systems. *IEEE Transactions on Speech and Audio Processing*, **SAP-7**, 204–213.

Onoda J. and Hanawa Y. (1993) Actuator placement optimisation by genetic and improved simulated annealing algorithms. *AIAA Journal*, **31**, 1167–1169.

Oppenheim A.V. and Shafer R.W. (1975) *Digital Signal Processing*. Prentice Hall.

Oppenheim A.V., Weinstein E., Zangi K.C. Feder M. and Gauger D. (1994) Single-sensor active noise cancellation. *IEEE Transactions on Speech and Audio Processing*, **2**(2), 285–290.

Orduña-Bustamante, F. (1995) *Digital signal processing for multi-channel sound reproduction*. PhD thesis, University of Southampton.

Orduña-Bustamante F. and Nelson P.A. (1991). An adaptive controller for the active absorption of sound. *Journal of the Acoustical Society of America*, **91**, 2740–2747.

Ott E. (1993) *Chaos in Dynamical Systems* Cambridge University Press.

Ott E., Grebogi C. and Yorke A. (1990) Controlling chaos. *Physical Review Letters*, **64** (11), 1196–1199.

Padula S.L. and Kincaid R.K. (1999) *Optimisation strategies for sensor and actuator placement*. NASA/TM-1999–209126.

Pan J. and Hansen C.H. (1991). Active control of total vibratory flow in a beam, I: Physical system analysis. *Journal of the Acoustical Society of America*, **89**, 200–209.

Papoulis A. (1977) *Signal Analysis*. McGraw Hill.

Park Y.C. and Sommerfeldt S.D. (1996) A fast adaptive noise control algorithm based on the lattice structure. *Applied Acoustics*, **47**, 1–25.

Petitjean B. and Legrain I. (1994) Feedback controllers for broadband active noise reduction. *Proc. Second European Conference on Smart Structures and Materials*, Glasgow.

Photiadis D.M. (1990) The relationship of singular value decomposition to wave-vector filtering in sound radiation problems. *Journal of the Acoustical Society of America*, **88**, 1152–1159.

Piedra R.M. and Fritsch A. (1996) Digital signal processing comes of age. *IEEE Spectrum*, May 1996, 70–74.

Popovich (1994) A simplified parameter update for identification of multiple input multiple output systems. *Proceedings of Inter-Noise 94*, 1229–1232.

Popovich S.R. (1996) An efficient adaptation structure for high speed tracking in tonal cancellation systems. *Proc. Inter-noise 96*, 2825–2828.

Press W.H., Flannery B.P., Tenkolsky S.A. and Vetterling W.T. (1987) *Numerical Recipes in C*. Cambridge University Press. See also www.ulib.org/WebRoot/Books/Numerical_Recipes

Preumont A. (1997) *Vibration Control of Active Structures, An Introduction*. Kluwer Academic Publishers.

Priestley, M.B. (1988) *Nonlinear and Nonstationary Time Series Analysis*. Academic Press.

Proceedings of ICSPAT 2000, International Conference on Signal Processing Applications and Technology.

Proceedings of the International Conference on Genetic Algorithms, 1987 onwards.

Psaltis D., Sideris A. and Yamamura A.A. (1988) A multilayered neural network controller. *IEEE Control Systems Magazine*, April 1988, 17–21.

Quershi S.K.M. and Newhall E.E. (1973) An adaptive receiver for data transmission of time dispersive channels. *IEEE Transactions on Information Theory*, **IT-19**, 448–459.

Rabaey J.M., Gass W., Brodersen R. and Nishitani T. (1998) VLSI design and implementation fuels the signal processing revolution. *IEEE Signal Processing Magazine*, January 1998, **15**, 22–37.

Rabiner L.R. and Gold B. (1975) *Theory and Application of Digital Signal Processing*. Prentice Hall.

Rafaely B. (1997) *Feedback control of sound*. PhD thesis, University of Southampton.

Rafaely B. and Elliott S.J. (1996a). Adaptive plant modelling in an internal model controller for active control of sound and vibration *Proc. Identification in Engineering Systems Conference*.

Rafaely B. and Elliott S.J. (1996b). Adaptive internal model controller—stability analysis. *Proc. Inter-Noise 96*, 983–988.

Rafaely B. and Elliott S.J. (1999) H_2/H_∞ active control of sound in a headrest: design and implementation. *IEEE Transactions on Control System Technology*, **7**, 79–84.

Rafaely B., Garcia-Bonito J. and Elliott S.J. (1997) Feedback control of sound in headrests. *Proc. ACTIVE97*, 445–456.

Rafaely B., Elliott S.J. and Garcia-Bonito J. (1999) Broadband performance of an active headrest. *Journal of the Acoustical Society of America*, **102**, 787–793.

Rao S.S. and Pan T-S (1991) Optimal placement of actuators in actively controlled structures using genetic algorithms. *AIAA Journal*, **29**, 942–943.

Rayner P.J.W. and Lynch M.R. (1989) A new connectionist model based on a nonlinear adaptive filter. *Proc. Int. Conf. on Acoustics, Speech and Signal Processing ICASSP89* Paper D7.10, 1191–1194.

Reichard K.M. and Swanson D.C. (1993) Frequency-domain implementation of the filtered-*x* algorithm with on-line system identification. *Proc. Second Conference on Recent Advances in Active Control of Sound and Vibration*, 562–573.

Ren W. and Kumar P.R. (1989) Adaptive active noise control structures, algorithms and convergence analysis. *Proc. InterNoise 89*, 435–440.

Ren W. and Kumar P.R. (1992) Stochastic parallel model adaptation: Theory and applications to active noise cancelling, feedforward control, IIR filtering and identification. *IEEE Transactions on Automatic Control*, **37**, 566–578.

Rex J. and Elliott S.J. (1992). The QWSIS—a new sensor for structural radiation control. *Proceedings of the International Conference on Motion and Vibration Control (MOVIC)*, Yokohama, 339–343.

Robinson E.A. (1978) *Multichannel Time Series Analysis with Digital Computer Programs* (Revised edition). Holden-Day.

Ross C.F. (1982) An algorithm for designing a broadband active sound control system. *Journal of Sound and Vibration*, **80**, 373–380.

Ross C.F. and Purver M.R.J. (1997) Active cabin noise control. *Proc. ACTIVE 97*, xxxix–xlvi.

Rossetti D.J., Jolly M.R. and Southward S.C. (1996) Control effort weighting in adaptive feedforward control. *Journal of the Acoustical Society of America*, **99**, 2955–2964.

Roure A. (1985) Self adaptive broadband active noise control system. *Journal of Sound and Vibration*, **101**, 429–441.

Roy R., Murphy T.W., Maier T.D., Gills Z. and Hunt E.R. (1992) Dynamic control of a chaotic laser: Experimental stabilisation of a globally coupled system. *Physical Review Letters*, **68**, 1259–1262.

Rubenstein S.P., Saunders W.R., Ellis G.K., Robertshaw H.H. and Baumann W.T. (1991) Demonstration of a LQG vibration controller for a simply-supported plate. *Proc. First Conference on Recent Advances in Active Control of Sound and Vibration*, 618–630.

Ruckman C.E. and Fuller C.R. (1995) Optimising actuator locations in active noise control systems using subset selection. *Journal of Sound and Vibration*, **186**, 395–406.

Rumelhart D.E. and McClelland J.L. (1986) *Parallel Distributed Processing*. MIT Press.

Rupp M. and Sayed A.H. (1998) Robust FxLMS algorithms with improved convergence performance. *IEEE Transactions on Speech and Audio Processing*, **6**, 78–85.

Safonov M.G. and Sideris A. (1985) Unification of Wiener Hopf and state space approaches to quadratic optimal control. *Digital Techniques in Simulation Communication and Control* (ed. S.G. Tzafestas). Elsever Science Publishers.

Saito N. and Sone T. (1996) Influence of modeling error on noise reduction performance of active noise control systems using filtered-X LMS algorithm. *Journal of the Acoustical Society of Japan*, **17**, 195–202.

Saunders T.J., Sutton T.J. and Stothers I.M. (1992) Active control of random sound in enclosures. *Proceedings of the 2nd International Conference on Vehicle Comfort*, Bologna, Italy, 749–753.

Saunders W.R., Robertshaw H.W. and Burdisso R.A. (1996) A hybrid structural control approach for narrow-band and impulsive disturbance rejection. *Noise Control Engineering Journal*, **44**, 11–21.

Schiff S.J., Jerger K., Duong D.H., Chang T., Spano M.L. and Ditto W.L. (1994) Controlling chaos in the brain. *Nature*, **370**, 615–620.

Schroeder M. (1990) *Fractals, Chaos, Power Laws*. W.H. Freeman.

Sergent P. and Duhamel D. (1997) Optimum placement of sources and sensors with the minimax criterion for active control of a one-dimensional sound field. *Journal of Sound and Vibration*, **207**, 537–566.

Serrand M. and Elliott S.J. (2000) Multichannel feedback control of base-excited vibration. *Journal of Sound and Vibration*.

Shaw E.A.G. and Thiessen G.J. (1962) Acoustics of circumaural ear phones. *Journal of the Acoustical Society of America*, **34**, 1233–1246.

Shen Q. and Spanias A. (1992) Time and frequency domain *x*-block LMS algorithms for single channel active noise control. *Proc. 2nd International Congress on Recent Developments in Air and Structure Borne Sound and Vibration,* 353–360.

Shinbrot T. (1995) Progress in the control of chaos. *Advances in Physics*, **44**, 73–111.

Shinbrot T., Ott E., Grebogi, C. and Yorke J.A. (1990) Using chaos to direct trajectories to targets. *Physical Review Letters*, **65**, 3215–3218.

Shinbrot T., Grebogi C., Ott, E. and Yorke J.A. (1993) Using small perturbations to control chaos. *Nature*, **363**, 411–417.

Shynk J.J. (1989) Adaptive IIR Filtering. *IEEE Signal Processing Magazine,* April, 4–21.

Shynk J.J. (1992) Frequency domain and multirate adaptive filtering. *IEEE Signal Processing Magazine*, January, 14–37.

Sievers L.A. and von Flotow A.H. (1992) Comparison and extension of control methods for narrowband disturbance rejection. *IEEE Transactions on Signal Processing*, **40**, 2377–2391.

Sifakis M. and Elliott S.J. (2000) Strategies for the control of chaos in a Duffing-Holmes oscillator. *Mechanical Systems and Signal Processing* (in press).

Silcox R.J. and Elliott S.J. (1990) *Active control of multi-dimensional random sound in ducts*. NASA Technical Memorandum 102653.

Simpson M.T. and Hansen C.H. (1996) Use of genetic algorithms to optimise vibration actuator placement for active control of harmonic interior noise in a cylinder with floor structure. *Noise Control Engineering Journal*, **44**, 169–184.

Simshauser E.D. and Hawley M.E. (1955) The noise cancelling headset—an active ear defender. *Journal of the Acoustical Society of America*, **27**, 207 (Abstract).

Sjöberg J. and Ljung L. (1992) Overtraining, regularization and searching for minimum in neural networks. *Proc. Symp. on Adaptive Systems in Control & Signal Processing*, 73–78.

Skelton R.E. (1988) *Dynamic Systems Control*. Wiley.

Skogestad S. and Postlethwaite I. (1996) *Multivariable Feedback Control, Analysis and Design*. Wiley.

Smith R.A. and Chaplin G.B.B. (1983) A comparison of some Essex algorithms for major industrial applications. *Proceeding Internoise 83*, **1**, 407–410.

Snyder S.D. (1999) Microprocessors for active control: bigger is not always enough. *Proc. ACTIVE99*, 45–62.

Snyder S.D. and Hansen C.H. (1990a) Using multiple regression to optimize active noise control design. *Journal of Sound and Vibration*, **148**, 537–542.

Snyder S.D. and Hansen C.M. (1990b) The influence of transducer transfer functions and acoustic time delays on the LMS algorithm in active noise control systems. *Journal of Sound and Vibration*, **140**, 409–424.

Snyder S.D. and Hansen C.M. (1994) The effect of transfer function estimation errors on the filtered-*x* LMS algorithm. *IEEE Transactions on Signal Processing*, **42**, 950–953.

Snyder S.D. and Tanaka N. (1992) Active vibration control using a neural network. *Proc. 1st International Conf. on Motion and Vibration Control* (*MOVIC*), 86–73.

Snyder S.D. and Tanaka N. (1997) Algorithm adaptation rate in active control: is faster necessarily better? *IEEE Transactions on Speech and Audio Processing*, **5**, 378–381.

Sommerfeldt S.D. and Nasif P.J. (1994) An adaptive filtered-*x* algorithm for energy-based active control. *Journal of the Acoustical Society of America*, **96**, 300–306.

Sommerfeldt S.D. and Tichy J. (1990) Adaptive control of two-stage vibration isolation mount. *Journal of the Acoustical Society of America*, **88**, 938–944.

Sondhi M.M. (1967) An adaptive echo canceller. *Bell Systems Technical Journal,* **46**, 497–511.

Spano M.L., Ditto W.L. and Ranseo S.N. (1991) Exploitation of chaos for active control: an experiment. *Proc. First Conference on Recent Advances in Active Control of Sound and vibration*, 348–359.

Stearns S.D. (1981) Error surfaces for adaptive recursive filters. *IEEE Transactions on Acoustics Speech and Signal Processing*, **ASSP-28**, 763–766.

Stell J.D. and Bernhard R.J. (1991) Active control of higher-order acoustic modes in semi-infinite waveguides. *Transactions of the American Society of Mechanical Engineering Journal of Vibration and Acoustics*, **113**, 523–531.

Stonick V.L. (1995) Time-varying performance surfaces for adaptive IIR filters: geometric properties and implications for filter instability. *IEEE Transactions on Signal Processing*. **43**, 29–42.

Stothers I.M., Saunders T.J., McDonald A.M. and Elliott S.J. (1993). Adaptive feedback control of sunroof flow oscillations. *Proceedings of the Institute of Acoustics*, **15**(3), 383–394.

Stothers I.M., Quinn D.C. and Saunders T.J. (1995) Computationally efficient LMS based hybrid algorithm applied to the cancellation of road noise. *Proceedings of Active 95*, 727–734.

Strauch P. and Mulgrew B. (1997) Nonlinear active noise control in a linear duct. *Proc. Int. Conf. on Acoustics, Speech and Signal Processing ICASSP97*, **1**, 395–398.

Sutton T.J. and Elliott S.J. (1993) Frequency and time-domain controllers for the attenuation of vibration in nonlinear structural systems. *Proceedings of the Institute of Acoustics* **15**, 775–784.

Sutton T.J. and Elliott S.J. (1995) Active attenuation of periodic vibration in nonlinear systems using an adaptive harmonic controller. *ASME Journal of Vibration and Acoustics* **117**, 355–362.

Sutton T.J., Elliott S.J. and McDonald A.M. (1994) Active control of road noise inside vehicles. *Noise Control Engineering Journal*, **42**, 137–147.

Sutton T.J., Elliott S.J., Brennan M.J., Heron K.H. and Jessop D.A. (1997) Active isolation of multiple structural waves on a helicopter gearbox support strut. *Journal of Sound and Vibration*, **205**, 81–101.

Swinbanks M.A. (1973). The active control of sound propagating in long ducts. *Journal of Sound and Vibration*, **27**, 411–436.

Takens F. (1980) Detecting strange attractors in turbulence. *Dynamic Theory and Turbulence* (ed. D.A. Rand and L.S.Young). Springer Lecture Notes in Mathematics, **898**, 366–381.

Tang K.S., Man K.F., Kwon S. and He Q. (1996) Genetic algorithms and their application. *IEEE Signal Processing Magazine*, **13**(6), 22–37.

Tapia J. and Kuo S.M. (1990) New adaptive on-line modelling techniques for active noise control systems. *Proceedings IEEE International Conference on Systems Engineering*. 280–283.

Tay T.T. and Moore J.B. (1991) Enhancement of fixed controllers via adaptive-Q disturbance estimate feedback. *Automatica*, **27**, 39–53.

Therrien C.W. (1992) *Discrete Random Signals and Statistical Signal Processing*. Prentice Hall.

Tichy J. (1988) Active systems for sound attenuation in ducts. *Proc. Int. Conf. on Acoustics, Speech and Signal Processing ICASSP88*, 2602–2605.

Titterton P.J. and Olkin J.A. (1995) A practical method for constrained optimisation controller design: H_2 or H_∞ optimisation with multiple H_2 and/or H_∞ constraints. *Proc. 29th IEEE Asilomer Conf. on Signals Systems and Computing*, 1265–1269.

Tohyama M. and Koike T. (1998) *Fundamentals of Acoustic Signal Processing*. Academic Press.

Tokhi M.O., Hossain M.A., Baxter M.J. and Fleming P.J. (1995) Heterogeneous and homogeneous parallel architectures for real-time active vibration control. *IEE Proceedings on Control Theory Applications*, **142**, 625–632.

Treichler J.R. (1985) Adaptive algorithms for infinite impulse response filters. *Adaptive Filters* (ed. C.F.N. Cowan and P.M. Grant). Prentice Hall, 60–90.

Treichler J.R., Johnson C.R. Jr. and Larimore M.G. (1987) *Theory and Design of Adaptive Filters*. Wiley.

Tsahalis D.T., Katsikas S.K. and Manolas D.A. (1993) A genetic algorithm for optimal positioning of actuators in active noise control: results from the ASANCA project. *Proc. Inter-Noise '93*, 83–88.

Tseng W-K, Rafaely B. and Elliott S.J. (1998) Combined feedback-feedforward active control of sound in a room. *Journal of the Acoustical Society of America,* **104**(6), 3417–3425.

Tsutsui S. and Ghosh A. (1997) Genetic algorithms with a robust solution searching scheme. *IEEE Transactions on Evolutionary Computation*, **1**(3), 201–208.

Twiney R.C., Holden A.J. and Salloway A.J. (1985) Some transducer design considerations for earphone active noise reduction systems. *Proceedings of the Institute of Acoustics*, **7**, 95–102.

van der Maas H.L.J., Vershure P.F.M.J. and Molenaer P.C.M. (1990) A note on the chaotic behaviour in simple neural networks. *Neural Networks* **3**, 119–122.

Veight I. (1988) A lightweight headset with active noise compensation. *Proc. Inter-Noise '88*, 1087–1090.

Vincent T.L. (1997) Control using chaos. *IEEE Control Systems Magazine*, December 1997, 65–76.

Walach E. and Widrow B. (1983) Adaptive signal processing for adaptive control. *Proc. IFAC Workshop on Adaptive Systems in Control and Signal Processing*.

Wallace C.E. (1972). Radiation resistance of a rectangular panel, *Journal of the Acoustical Society of America*, **51**, 946–952.

Wan E.A. (1993) *Finite impulse response neural network with applications in time series prediction*. PhD thesis, Stanford University.

Wan E.A. (1996) Adjoint LMS: an efficient alternative to the filtered-x LMS and multiple error LMS algorithms. *Proc. Int. Conf. on Acoustics, Speech and Signal Processing ICASSP96*, 1842–1845.

Wang A.K. and Ren W. (1999a) Convergence analysis of the filtered-*u* algorithm for active noise control. *Signal Processing*, **73**, 255–266.

Wang A.K. and Ren W. (1999b) Convergence analysis of the multi-variable filtered-*x* LMS algorithm with application to active noise control. *IEEE Transactions on Signal Processing*, **47**, 1166–1169.

Watkinson J. (1994) *The Art of Digital Audio*, 2nd edn. Focus Press.

Wellstead P.E. and Zarrop M.B. (1991) *Self-tuning Systems Control and Signal Processing*. Wiley.

Wheeler P.D. (1986) *Voice communications in the cockpit noise environment—the role of active noise reduction*. PhD thesis, University of Southampton.

Wheeler P.D. (1987) The role of noise cancellation techniques in aircrew voice communications systems. *Proc. Royal Aero. Soc. Symposium on Helmets and Helmet-mounted Displays*.

White S.A. (1975) An adaptive recursive digital filter. *Proc. 9th Asilomar Conference on Circuits, Systems and Computing*, 21–25.

Whittle P. (1963) On the fitting of multivariate autoregressions and the approximate canonical factorisation of a spectral density matrix. *Biometrika*, **50**, 129–134.

Widrow B. (1956) A study of rough amplitude quantisation by means of the Nyquist sampling theory. *IRE Transactions on Circuit Theory*, **CT-3**, 266–276.

Widrow B. (1986) Adaptive inverse control. *Second IFAC Workshop on Adaptive Systems in Control and Signal Processing*.

Widrow B. and Hoff M. (1960) Adaptive switching circuits. *Proc. IRE WESCON Convention Record*, Part 4, Session 16, 96–104.

Widrow B. and Lehr M.A. (1990) 30 years of adaptive neural networks: Perceptron, Madaline, and Backpropagation. *Proceedings IEEE*, **78** (9), 1415–1441.

Widrow B. and McCool J.M. (1977) Comments on an adaptive recursive LMS filter. *Proc. IEEE*, **65**, 1402–1404.

Widrow B. and Stearns S.D. (1985) *Adaptive Signal Processing*. Prentice Hall.

Widrow B. and Walach E. (1996) *Adaptive Inverse Control*. Prentice Hall.

Widrow B., Shur D. and Shaffer S. (1981) On adaptive inverse control. *Proc. 15th ASILOMAR Conference on Circuits, Systems and Computers*, 185–195.

Widrow B., Winter R.G. and Baxter R.A. (1988) Layered neural nets for pattern recognition. *IEEE Transactions on Acoustics Speech and Signal Processing*, **36**(7), 1109–1118.

Wiener N. (1949) *Extrapolation, Interpolation and Smoothing of Stationary Time Series*. Wiley.

Wiggins R.A. and Robinson E.A. (1965) Recursive solution to the multichannel filtering problem. *Journal of Geophysical Research*, **70**, 1885–1991.

Wilby J.F., Rennison D.C., Wilby E.G. and Marsh A.H. (1980) Noise control prediction for high speed propeller-driven aircraft. *Proceedings of American Institute of Aeronautics and Astronautics 6th Aeroacoustic Conference*, Paper AIAA-80-0999.

Williams R.J. and Zipser D. (1989) A learning algorithm for continually running fully recurrent neural networks. *Neural Computation*, **1**(2), 270–280.

Wilson G.T. (1972) The factorisation of matricial spectral densities. *SIAM Journal of Applied Mathematics*, **32**, 420–426.

Winkler J. and Elliott S.J. (1995). Adaptive control of broadband sound in ducts using a pair of loudspeakers. *Acoustica*, **81**, 475–488.

Wright M.C.M. (1989) *Active control in changing environments*. MEng Project Report, ISVR, University of Southampton.

Yang T.C., Tseng C.H. and Ling S.F. (1994) Constrained optimisation of active noise control systems in enclosures. *Journal of the Acoustical Society of America*, **95**, 3390–3399.

Youla D.C. (1961) On the factorisation of rational matrices. *IRE Transactions on Information Theory*, **IT-7**, 172–189.

Youla D.C., Bongiorno J.J. and Jabr M.A. (1976a) Modern Wiener-Hopf design of optimal controllers. Part I—The single input-output case. *IEEE Transactions on Automatic Control*, **AC-21**, 3–13.

Youla D.C., Jabr M.A.and Bangiorno J.J. (1976b) Modern Wiener–Hofp design of optimal controllers. Part II—The multivariable case. *IEEE Transactions on Automatic Control*, **AC-21**, 319–338.

Yuret D. and de la Maza M. (1993) Dynamic hill climbing: overcoming the limitations of optimisation techniques. *Proc. 2nd Turkish Symposium of Artificial Intelligence and Artificial Neural Networks*, 254–260; see also Yuret D. (1994) *From genetic algorithms to efficient optimisation*. MSc Thesis MIT.

Zadek L.A. and Ragazzini J.R. (1950) An extension of Wiener's theory of prediction. *Journal of Applied Physics*, **21**, 645–655.

Zames G. (1981) Feedback and optimum sensitivity: model reference transformations, multiplicative seminorms and approximate inverses. *IEEE Transactions on Automatic Control*, **AC-26**, 301–320.

Zander A.C. and Hansen C.H. (1992) Active control of higher-order acoustic modes in ducts. *Journal of the Acoustical Society of America*, **92**, 244–257.

Zavadskaya M.P., Popov A.V. and Egelskii B.L. (1976). Approximations of wave potentials in the active suppression of sound fields by the Malyuzhinets method. *Soviet Physics Acoustics*, **33**, 622–625.

Zhou K, Doyle J.C. and Glover K. (1996) *Robust and Optimal Control*. Prentice Hall.

Zimmerman D.C. (1993) A Darwinian approach to the actuator number and placement problem with non-negligible actuator mass. *Mechanical Systems and Signal Processing*, **74**, 363–374.

Index